MASS BALANCE OF THE CRYOSPHERE

The cryosphere can be loosely defined as comprising all the frozen water and soil on the surface of the Earth. This book focusses on two key components of this environment: land ice (in the form of ice sheets, caps and glaciers) and sea ice. These components have been identified as important indicators of climate change on timescales ranging from years to millennia.

Early chapters cover the theory behind field-based and satellite observations, and modelling of mass balance, providing the reader with a thorough grounding in all the concepts and issues presented later in the book. The rest of the book reviews our current understanding, from modelling and observational perspectives, of the present and predicted future mass balance of the cryosphere.

This book is an important reference for all scientists working in the fields of climate change, environmental sciences and glaciology, and provides a valuable supplementary text for senior undergraduate and graduate courses in glaciology. It has been written by leading authors in the field, and is fully integrated to provide a coherent, cross-referenced and consistent exposition on the subject.

JONATHAN BAMBER did a degree in Physics at the University of Bristol before undertaking a Ph.D. in glaciology at the University of Cambridge. Since then he has been a lecturer at University College London in the Department of Space and Climate Physics, and is currently Reader in physical geography in Bristol. With some 20 years experience in airborne and satellite remote sensing of the polar regions, he is the author of over 70 articles on this topic, in addition to a number of articles relating to more general applications of Earth observation. He has undertaken fieldwork in Antarctica, the Arctic and Karakoram, in collaboration with the German Polar Research Centre, AWI and NASA. He is currently President of the Cryospheric Sciences section of the European Geosciences Union.

TONY PAYNE is a Reader in the School of Geographical Sciences, University of Bristol. He was previously a Lecturer (and then Reader) at the Department of Geography, University of Southampton, and a postdoctoral researcher at the Universität Bremen, Germany, and the Grant Institute of Geology, University of Edinburgh. He has published widely on the application of numerical models in glaciology. He was a contributing author on ice-sheet modelling for the IPCC 3rd Assessment Report 2001, and contributed to the Department of the Environment, Transport and the Region's Risk Estimation of the collapse of the West Antarctic ice sheets (2000). In addition to modelling, he has undertaken fieldwork on the Langjökull ice cap, Iceland (1999 and 2000), where he was involved with hot-water drilling and subglacial instrumentation.

MASS BALANCE OF THE CRYOSPHERE
Observations and Modelling of Contemporary and Future Changes

edited by

JONATHAN L. BAMBER AND ANTONY J. PAYNE

University of Bristol

CAMBRIDGE
UNIVERSITY PRESS

CAMBRIDGE
UNIVERSITY PRESS

University Printing House, Cambridge CB2 8BS, United Kingdom

One Liberty Plaza, 20th Floor, New York, NY 10006, USA

477 Williamstown Road, Port Melbourne, VIC 3207, Australia

314-321, 3rd Floor, Plot 3, Splendor Forum, Jasola District Centre, New Delhi-110025, India

79 Anson Road, #06-04/06, Singapore 079906

Cambridge University Press is part of the University of Cambridge.

It furthers the University's mission by disseminating knowledge in the pursuit of education, learning and research at the highest international levels of excellence.

www.cambridge.org
Information on this title: www.cambridge.org/9781108457217

© Cambridge University Press 2004

First published 2004
First paperback edition 2018

A catalogue record for this publication is available from the British Library

Library of Congress Cataloging in Publication data
The mass balance of the cryosphere: observations and modelling of contemporary and future changes/edited by Jonathan Bamber and Tony Payne.
p.   cm.
Includes bibliographical references and index.
ISBN 0 521 80895 2
1. Ice sheets – Observations.   2. Ice sheets – Mathematical models.   3. Mass budget (Geophysics).   I. Bamber, Jonathan, 1962–   II. Payne, Tony, 1963–
QC981.8.123M37 2003
551.31 – dc21   2003043472

ISBN 978-1-108-45721-7 Paperback

Contents

Contributors

JONATHAN L. BAMBER
Bristol Glaciology Centre, School of Geographical Sciences,
University of Bristol, University Road, Bristol BS8 1SS, UK
j.l.bamber@bristol.ac.uk

CHARLES R. BENTLEY
Department of Geology and Geophysics, University of
Wisconsin-Madison, 1215 W. Dayton St, Madison, WI 53706,
USA
bentley@geology.wisc.edu

JULIAN A. DOWDESWELL
Scott Polar Research Institute, University of Cambridge,
Lensfield Road, Cambridge CB2 1ER, UK
jd16@cam.ac.uk

MARK B. DYURGEROV
Institute of Arctic and Alpine Research (INSTAAR), University of
Colorado, Campus Box 450, Boulder, CO 80309-0450, USA
Mark.Dyurgerov@Colorado.EDU

GREGORY M. FLATO
Canadian Centre for Climate Modelling and Analysis,
Meteorological Service of Canada, University of Victoria,
PO Box 1700, Victoria, BC V8W 2Y2, Canada
Greg.Flato@ec.gc.ca

CHRISTOPHE GENTHON
Laboratoire de Glaciologie et Géophysique de L'Environnement,
CNRS, 54 Rue Molière, BP96, 38402 Saint Martin d'Hères,
France
genthon@lgge.obs.ujf-grenoble.fr

WOUTER GREUELL
Institute for Marine and Atmospheric Research Utrecht (IMAU),
Utrecht University, Princetonplein 5, 3584 CC Utrecht,
The Netherlands
W.Greuell@phys.uu.nl

WILFRIED HAEBERLI
World Glacier Monitoring Service (WGMS), Glaciology and
Geomorphodynamics Group, Department of Geography,
University of Zurich–Irchel, Winterthurerstraße 190, CH-8057
Zurich, Switzerland
haeberli@geo.unizh.ch

JON OVE HAGEN
Department of Physical Geography, Faculty of Mathematics and
Natural Sciences, University of Oslo, PO Box 1042, Blindern,
0316 Oslo, Norway
j.o.m.hagen@geografi.uio.no

WILLIAM D. HIBLER, III
International Arctic Research Center, University of Alaska at Fairbanks,
930 Koyukuk Drive, PO Box 757335, Fairbanks, Alaska, USA
billh@iarc.uaf.edu

PHILIPPE HUYBRECHTS
Alfred Wegener Institute for Polar and Marine Research,
Columbusstraße, Postfach 120161, D-27515 Bremerhaven,
Germany
phuybrechts@awi-bremerhaven.de

OLA M. JOHANNESSEN
Nansen Environmental and Remote Sensing Center,
Edvard Griegs vei 3a, N-5059 Bergen, Norway
ola.johannessen@nersc.no

RON KWOK

Polar Remote Sensing Group, Jet Propulsion Laboratory, California
Institute of Technology, MS 300-235 4800 Oak Grove Drive, 91109
Pasadena, USA
ron@Radar-Sci.Jpl.Nasa.Gov

SEYMOUR W. LAXON

Centre for Polar Observation and Modelling, Department of Space &
Climate Physics, Pearson Building, University College London,
Gower Street, London WC1E 6BT, UK
swl@mssl.ucl.ac.uk

MARK F. MEIER

Institute of Arctic and Alpine Research (INSTAAR), University of
Colorado, Campus Box 450, Boulder, CO 80309-0450, USA
Mark.Meier@Colorado.EDU

MARTIN MILES

Nansen Environmental and Remote Sensing Center,
Edvard Griegs vei 3a, N-5059 Bergen, Norway
martin.miles@nersc.no

ANTONY J. PAYNE

Bristol Glaciology Centre, School of Geographical Sciences,
University of Bristol, University Road, Bristol BS8 1SS, UK
a.j.payne@bristol.ac.uk

NIELS REEH

Center for Arctic Technology, Byg-DTU, Technical University of
Denmark, Building 204, Kemitorvet, DK-2800 Kgs. Lyngby,
Denmark
nr@emi.dtu.dk

ROBERT H. THOMAS

EG&G Inc., Building N-159, Wallops Flight Facility,
Wallops Island, VA 23337, USA
robert_thomas@hotmail.com

RODERICK S. W. VAN DE WAL

Institute for Marine and Atmospheric Research Utrecht (IMAU),
Utrecht University, Princetonplein 5, 3584 CC, Utrecht,
The Netherlands
R.vandeWal@phys.uu.nl

CORNELIS J. VAN DER VEEN
Byrd Polar Research Center, The Ohio State University,
Columbus, OH 43210, USA
vanderveen.1@osu.edu

PETER WADHAMS
Department of Applied Mathematics and Theoretical Physics,
University of Cambridge, Wilberforce Road, Cambridge CB3
0WA, UK
p.wadhams@damtp.cam.ac.uk

JOHN E. WALSH
Department of Atmospheric Sciences, University of Illinois,
105 South Gregory Avenue, Urbana, IL 61801, USA
walsh@atmos.uiuc.edu

Foreword

The regions of the great ice caps in the Arctic and Antarctic are places of stunning beauty. Also, being tantalizingly remote and largely unspoilt by human interference, they hold compelling fascination and interest. However, these are not the only reasons for their study. Compared with the rest of the Earth's surface, they are of importance far beyond what might be expected from their comparative size. The changing balance in the cryosphere between the accumulation and ablation of ice has dominated the Earth's climatic history through the quasi-regular ice ages of the last million years – extending also to earlier epochs about which rather less is known. The world's coastal regions have been enormously affected as this changing balance has led to large excursions of sea level. For instance, at the end of the last ice age, 20 000 years or more ago, the sea level was lower than today by about 120 metres.

The long-term driving influence on the mass of ice in the polar regions, either in the form of sea ice or locked in the ice caps, has been the regular oscillations in key features of the Earth's orbit around the Sun, namely its eccentricity, the tilt of the Earth's axis and the time of year when the Earth is closest to the Sun. These features change with periods varying from about 20 000 years to about 100 000 years, and combine to cause substantial variations in the amount of solar energy that reaches the polar regions at different times of year, most particularly in the northern summer. It is these variations, as recognized first by James Croll in 1867 and later studied extensively during the 1920s by the Serbian scientist Milutin Milankovitch, that have triggered the growth and decay of the ice sheets. The influence of these ice sheets has extended far beyond the polar regions. For instance, as the ice sheets have grown, large areas of land in the northern hemisphere have been covered over many millennia, and as they have receded the fresh water released has strongly affected the ocean circulation.

Turning to more recent times, during the last two decades we have all become increasingly aware of the way in which the climate is being affected by human activities. It is the burning of fossil fuels that results in emissions of large quantities of the greenhouse gas carbon dioxide into the atmosphere that is leading to substantial warming of the Earth's climate and therefore to climate change at a rate that has probably not been experienced on Earth for at least 10 000 years. It is imperative to find out in as much detail as possible how the sea ice and the ice sheets are reacting to this anthropogenic warming and how they are, in turn, influencing its impact. Are the ice sheets growing because of increased snow fall or are they beginning to melt down because of the increased temperature? Is the stability of the great ice sheets of Antarctica at risk because of the changes that are taking place? How is the deep ocean circulation affected by the changes in fresh-water input? And so on.

This book is written by international experts in the scientific disciplines involved – especially those of physics and dynamics applied particularly to ice but also to the atmosphere and the ocean that surround the ice. The primary tools of observations and modelling feature large in its chapters as a state of the art description is provided about many aspects of the cryosphere, its behaviour and evolution. Answers to some of the important questions are beginning to emerge, and this volume provides an important synthesis of current knowledge.

John Houghton

Preface

In 2000, Tony Payne and I organized a session at the annual congress of the European Geophysical Society on the mass balance of the cryosphere. It was clear from the impressive scientific breakthroughs presented at this meeting and also in the recent literature that major progress has been achieved in this subject over the last decade. This is a result of advances in both observational technology through new satellite and airborne hardware, and our modelling capability, and through improvements in computational power and physical understanding. As a consequence, it was timely and fitting to embark on producing a comprehensive review of what we know about the theory behind measuring and modelling mass balance and the actual results from the latest observations and model simulations. In this respect, the book is unique, in that it combines both the theory and the results in a single text. Twenty-three expert authors have contributed to seventeen chapters covering sea ice, glaciers and ice sheets in five thematic sections. Although this is an edited volume, each chapter is extensively cross-referenced and forms part of a fully integrated text. In addition, the chapters were externally peer-reviewed to ensure the highest scientific standards. Part I of this book is designed to offer a comprehensive, yet compact, reference text on the theory and practice of measuring mass balance. Part II is a parallel section on modelling, and Parts III–V comprise detailed and comprehensive reviews of what we know, from both measurements and modelling, about the current and predicted behaviour of sea ice, ice sheets, glaciers and ice caps.

A book of this kind requires the effort of a group of dedicated and committed people with wide ranging expertise. Many thanks are due to all the authors, some of whom have had to wait a long time to see their material in print. I would also like to thank the editors and staff at CUP, the external reviewers and, in particular, our copy editor Irene Pizzie, for their perseverance and professionalism. Finally, I would like to acknowledge the generous sponsorship of the European Space Agency and National Aeronautics and Space Administration for their contribution to the cost of colour reproduction.

Jonathan Bamber

1

Introduction and background

JONATHAN L. BAMBER AND ANTONY J. PAYNE
School of Geographical Sciences, University of Bristol

1.1 Aims and objectives of the book

The cryosphere can loosely be defined as all frozen water and soil on the surface of the Earth. This definition encompasses a diverse range of ice masses with a vast spectrum of spatial and temporal characteristics. It ranges from ephemeral river and lake ice to the quasi-permanent (on a millennial timescale) ice sheets of Antarctica and Greenland. Included in the definition is seasonal snow cover and permafrost. In compiling this book it was neither possible nor desirable to include all these different components. This is because the processes and interactions at play are as diverse as the components and, in some cases, unrelated. We have focussed here on two key components, which interact with each other and with the rest of the climate system in an inter-related way. They are land ice, in the form of ice sheets, caps and glaciers, and sea ice. Combined, these represent, at any one time, by far the largest component of ice on the planet, both by volume and area, yet respond to climate change over timescales ranging from seasons to millennia. Sea ice has been identified by the Intergovernmental Panel for Climate Change (IPCC) as a key indicator of short-term climate change, while land-based ice masses may have contributed as much as 50% of attributable sea-level rise during the twentieth century, and represent a large uncertainty in our predictions of a future rise (Houghton *et al.*, 2001). In the rest of this book the word cryosphere refers to these two components only.

The goal of this book is to provide, in a single volume, a comprehensive, up to date and timely review of our state of knowledge about the present-day mass balance of the cryosphere from observations and how it might change over the next millennium based on the latest modelling studies. The book is designed as a reference text covering all aspects of both the theory and practice of measuring and modelling the mass balance of land and sea ice. There are several excellent texts that cover the general physical principles of glaciology but none that deal, specifically, with the determination of the mass balance of either land or sea ice. In this respect, therefore, this is a unique contribution. Parts I and II cover the theoretical principles underpinning the methods used to observe and model mass balance, respectively. The subsequent parts present the state of knowledge of the present-day and predicted future

Mass Balance of the Cryosphere: Observations and Modelling of Contemporary and Future Changes, eds. Jonathan L. Bamber and Antony J. Payne. Published by Cambridge University Press. © Cambridge University Press 2003.

mass balance of the cryosphere. These chapters are detailed reviews authored by leading scientists in their field. In all, 23 authors have contributed, but each chapter represents an integral part of the whole book rather than being a stand-alone contribution. This is not, therefore, a set of separate, unconnected contributions but an integrated, coherent treatise presenting (i) background material on the subject, (ii) our best guess as to the present and future state of health of land and sea ice, and (iii) how this information has been derived from both measurement and modelling.

In the 1990s a number of major advances have taken place in (i) our ability to monitor and estimate mass balance, and (ii) the sophistication and accuracy of numerical models of the various components of the cryosphere and more general Earth system models incorporating the cryosphere. Satellite, airborne and terrestrial programmes supported by the European Space Agency (ESA) and the National Aeronautics and Space Administration (NASA)[1] during the 1990s have, in particular, resulted in a quantum improvement in our knowledge of the mass balance of land and sea ice. New satellite programmes, dedicated to studies of the cryosphere, have been announced by both agencies, reflecting the recognition by governments and non-governmental organizations alike of the key role that the cryosphere plays in the Earth system and its vulnerability to changing climate.

1.2 Importance of the cryosphere in the Earth system

1.2.1 Sea level

The mass balance of land ice has a direct impact on sea level and is most likely contributing to sea-level rise, although the error bars on estimates of this contribution are as large as the signal (Church *et al.*, 2001). The Antarctic and Greenland ice sheets contain enough ice to raise global sea level by around 65 m and 6 m, respectively. Even a relatively small imbalance in these ice masses will have a significant effect on sea-level rise, which is currently believed to be between 1.5 and 2 mm per year. The uncertainty in the mass balance of Antarctica represents 1 mm, i.e. one-half, of the total signal (cf. Chapter 12). The level of uncertainty associated with the Antarctic ice sheet's future behaviour is such that we are not even certain of its sign (Houghton *et al.*, 2001). A better understanding of the dynamics of both ice sheets is therefore crucial to our ability to reduce the uncertainty in our predictions of future sea level. Smaller ice masses are, without question, presently contributing to sea-level rise at an ever increasing rate (currently at 0.41 mm per year, cf. Chapters 15 and 16) and represent one of the more sensitive elements of the global climate system, for reasons explained below.

1.2.2 Ice–ocean–atmosphere feedbacks

Land-based ice interacts with the global climate system in a number ways and over a range of different timescales. The principal interaction (shared with sea ice) is due to very high

---

[1] The crucial contributions by ESA and NASA to the study of the cryosphere is reflected in their joint sponsorship of this book.

reflectance characteristics. The albedo of clean snow ranges from 80 to 97%, while that of clean ice varies between 34 and 51% (Paterson, 1994). This is many times greater than the albedos of other naturally occurring surfaces, such as water ($\sim$1%), forest (10 to 25%) and bare soil (5 to 20%). This contrast gives rise to many important feedbacks between the cryosphere, the atmosphere and the underlying ocean or land surface, which can locally exacerbate the effects of global climate change on very short timescales. Increasing snow cover reduces the amount of solar radiation absorbed, producing a cooling effect that leads to increased snow cover. The converse is also true: increased temperatures reduce snow cover, which leads to more solar radiation being absorbed and a further increase in temperature. This is usually called the ice-albedo feedback mechanism and is one of the most important interactions that the cryosphere has with the rest of the Earth system. It is a particularly important factor for sea ice as this is such a dynamic component of the cryosphere, fluctuating in extent by about a factor of 5 between summer and winter in the Southern Ocean (Chapter 8). The albedo of sea-water is about 1% and the effect of sea-ice cover is, therefore, to reduce the amount of solar radiation absorbed at the surface by as much as 95%, which could amount to around $100 \, W/m^2$. By comparison, the estimated radiative forcing effect of the increase in atmospheric CO_2 from 1750 to 2000 is $1.46 \, W/m^2$ (Houghton *et al.*, 2001).

In addition to the albedo effect, land-based ice masses often have a dramatic effect on regional climates. At synoptic scales, the topographic blocking of the large ice sheets has an important influence on atmospheric circulation. At intermediate scales, ice masses often have their own distinctive regional climates, for instance ice masses are normally associated with strong katabatic winds. At local scales, the presence of a glacier in a valley will alter the micro-climate dramatically. In fact, valley glaciers are extremely sensitive indicators of climate change because they cannot buffer the effects of an atmospheric warming by increasing long-wave emission once their surfaces have reached melting point.

It should also be noted that, in addition to being active components of the climate system, ice sheets and glaciers are efficient recorders of past climate change. Information obtained from ice cores has played an important role in characterizing past changes on timescales varying from decades to ice ages (for example, the Vostok ice core from East Antarctica extends back 420 000 years; Petit *et al.*, 1999). Sub-polar and alpine glaciers are often an important source of water for both hydroelectric schemes (in Norway and Iceland, for example) and irrigation/human consumption (such as in the Himalayas and Karakoram). In many areas they are a key resource for the biggest growth industry on the planet: tourism.

Both the ice sheets and sea ice play an important role in ocean circulation, and, in particular, in the formation of deep water that forms part of the ocean conveyor belt (Stossel, Yang and Kim, 2002). Rapid and dramatic changes in climate have been identified from ice-core data during the last glacial in the northern hemisphere (Stocker, 2000). These fluctuations (in particular Heinrich events and Dansgaard–Oeschger oscillations) involve major changes in a few decades, and the former are probably associated with the shutdown of the thermohaline circulation by massive iceberg discharge events from the Laurentide ice sheet (which covered North America). Ice is clearly implicated in rapid climate change during glacial times, and it has been suggested that the Greenland ice sheet could, under conditions of global warming, influence the thermohaline circulation of the North Atlantic

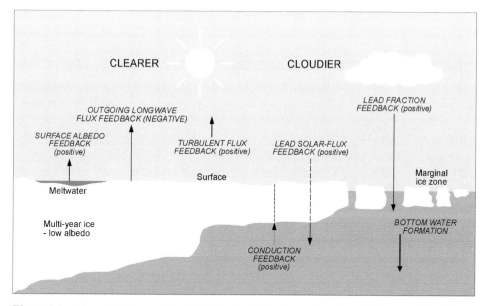

Figure 1.1. Schematic diagram illustrating the key interactions of sea ice with the rest of the climate system. (Courtesy E. Hanna, University of Plymouth.)

in the future. During the present-day, iceberg discharge and bottom melting from floating ice shelves fringing most of the Antarctic ice sheet contribute to the fresh-water budget and mixing in the Southern Ocean.

Sea ice is the most dynamic and variable component of the cryosphere examined in this book and, partly as a consequence, it is strongly coupled to the rest of the climate system. Figure 1.1 is a schematic diagram illustrating some of the key feedbacks that take place, in particular close to the marginal ice zone. Sea ice acts as a thermal blanket, as well as greatly reducing the exchange of moisture and CO_2 between the ocean and the atmosphere. Leads within the pack ice are areas of intense energy and moisture exchange, due to the large contrast between surface oceanic and atmospheric temperatures. Intense cooling of the surface waters take place in leads and polynyas, producing dense, cold water that forms North Atlantic deep water in the northern hemisphere and Antarctic bottom water (ABW) in the south. The Weddell Sea Polynya, for example, is believed to play a key role in ABW formation (Goosse and Fichefet, 2001). A statistically significant reduction in Arctic sea-ice extent and thickness has been observed since about 1970 (see Chapter 8), and this could, if it continues, result in substantial, but largely uncharted, changes to the climate of the North Atlantic through the interactions and feedbacks described above.

1.3 Timescales of variability

This book discusses the mass balance of components of the cryosphere ranging in size from the Antarctic ice sheet, covering an area of 1.3×10^7 km^2, to valley glaciers a few kilometres in length. We also discuss trends in sea-ice cover, which, as mentioned, varies markedly on a seasonal basis. It is important, therefore, to discuss briefly the issue of timescales of variability and to place current fluctuations in context with respect to past variations.

The present-day ice sheets of Greenland and Antarctica cover a total area of approximately 1.36×10^7 km^2. This is about half of the land area covered by ice sheets during their maximum extent some 21 000 years ago (at the last glacial maximum, LGM). Both present-day ice sheets were significantly larger and extended to the edge of the continental shelf. In addition, ice sheets covered much of the northern hemisphere continents north of 45° (the Laurentide, Innuitian and Cordilleran ice sheets in North America, and the Scandinavian and Barents Sea ice sheets in Eurasia). A concomitant increase in sea-ice extent also existed during this period. Any assessment of the present-day cryosphere should be made, therefore, in the context of the global deglaciation that took place at the end of the last Ice Age. This deglaciation started at approximately 14 500 years before present (BP) and was largely completed by 11 500 years BP, although it should be noted that these dates vary significantly on a regional basis (for instance, the Cordilleran ice sheet did not reach its maximum until 4000 years after the LGM; Clague and James, 2002). The transition from the glacial world to the present-day, inter-glacial one was not a gradual process. In particular, it was punctuated by a major re-advance of ice masses during a cool period known as the Younger Dryas (12 000 years BP). More recently, during the Holocene epoch (the last 10 000 years), the world has seen several periods during which global temperatures have changed dramatically and ice masses have temporarily re-advanced or retreated. Examples of the former are the climatic optimum or hypsothermal between 5000 and 6000 years BP (when global temperatures were roughly 1 °C warmer than present) and a secondary optimum at 1000 AD, which saw Viking expeditions to Iceland, Labrador and Greenland, and the establishment of farming in Greenland. A period of relative ice advance separated these two optima, and the second one was terminated by the Little Ice Age, centred around 1700 AD and lasting for approximately 400 years. Our assessment of the present-day mass balance of the cryosphere must therefore been seen against a background of natural variability on a range of timescales varying from the major glacial–inter-glacial cycles to sporadic, centennial events such as the Little Ice Age or even shorter climate fluctuations such as the quasi-decadal North Atlantic oscillation.

Antarctica has response times of the order of 10 000 years or more. It is, therefore, amongst the slowest components of the climate system (along with the deep ocean, see Table 1.1). This timescale is principally associated with the very slow flow of ice within an ice sheet, where velocities are typically of the order of 1 to 10 m per year away from the faster flowing zones such as ice streams (where velocities are of the order of 1 km per year but which only occupy a small fraction of the present-day ice sheets). This has the important implication that Antarctica (and possibly Greenland) is still responding to climate changes associated with the Earth's emergence from the last Ice Age some 8000 years ago (see Chapter 13 for more details). Since the ice sheet was significantly larger at the LGM, this implies that some of the observed recent sea-level rise may be attributable to this long timescale, non-anthropogenic cause. Furthermore, a component of the present-day mass balance of the ice sheets may not be related to climate change during the Holocene.

Table 1.1. *Estimated timescales from various components of the climate system to reach equilibrium (i.e. their response times).*

Component	Response time
Free atmosphere	days
Atmospheric boundary layer	hours
Oceanic mixed layer	months to years
Deep ocean	centuries
Sea ice	days to centuries
Snow and surface ice	hours
Lakes and rivers	days
Soil and vegetation	days to centuries
Glaciers	decades to centuries
Ice sheets	millennia
Mantle's isostatic response	millennia

Source: McGuffie and Henderson-Sellers (1997).

For land ice, the dynamic response time, t, is proportional to the size of the ice mass (Paterson, 1994) and can be approximated by the relationship

$$t \approx H/a_0, \qquad (1.1)$$

where H is the maximum ice thickness and a_0 is the ablation rate. If typical values for Greenland are used in equation (1.1), we find that it has a present-day response time of about 3000 years, while for a valley glacier t can be of the order of few hundred years. Land-ice masses are integrators of the climate. In contrast, sea-ice extent is directly related to the immediate climate and, as a consequence, may be one of the earliest indicators of recent, and possibly anthropogenic, climate change. In the chapters that follow, discussions of mass balance are implicitly linked to the time constants for a response, and it is important, therefore, that the reader keeps this in mind and is aware of the implications for the interpretation of a short-term record from satellite or *in situ* observations. Modelling studies of variability also implicitly incorporate the timescale for various types of response. The emphasis in this book is on variability in mass balance from decadal to millennial timescales. We do not deal with glacial–inter-glacial variations, except where they may be influencing the present-day state, as discussed above.

1.4 Geographical context

The largest ice mass on the planet (by a factor of 10) is the Antarctic ice sheet (see Chapter 12). It contains around 80% of the world's fresh-water supply and covers an area in excess of 1.3×10^7 km^2. As mentioned, the dynamic response time of an ice mass is proportional to its size, and as a consequence it has, in general, the longest response time of any of the

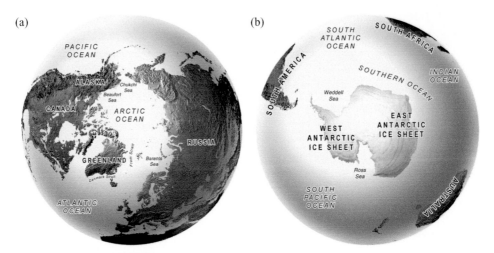

Figure 1.2. Polar stereographic map projections of (a) the Arctic Ocean and surrounding continents and (b) Antarctica and the Southern Ocean.

ice masses discussed in this book. In contrast, Antarctic sea ice covers as much as $1.9 \times 10^7 \, \text{km}^2$ of the Southern Ocean during the austral winter, but has a strong seasonal cycle of growth and decay with a minimum extent of less than $4 \times 10^6 \, \text{km}^2$. Geographically, the Arctic has a very different setting (Figure 1.2).

The Arctic Ocean is almost completely land-locked with relatively narrow channels such as the Bering and Fram Straits providing for ice or water transport into or out of the area. By contrast, the Southern Ocean surrounds a continent and is responsible for the largest current on the planet: the Antarctic Circumpolar Current. These differences are fundamental to the climate of the regions and the behaviour of both the land and sea ice. The continental character of East Antarctica results in it being one of the driest deserts on the planet with annual precipitation as low as 7 cm per year in parts of the interior. Precipitation over Greenland is around 20–100 cm per year. The Greenland ice sheet is around a factor of 10 smaller in area and volume compared with its southern hemisphere counterpart, but it has an extensive ablation area (unlike its southern hemisphere counterpart) which is extremely sensitive to changing climate. The other islands and continental areas surrounding the Arctic Ocean are also glaciated to a larger or greater extent. Almost 60% of the Svalbard archipelago is, for example, glaciated. These smaller ice masses of the Arctic represent about half the total global volume of land ice not contained in the ice sheets. The other half is distributed amongst glaciers, mainly throughout the northern hemisphere, including the European Alps, the Himalayas, Karakoram and other areas in central Asia. There is relatively little land ice in the southern hemisphere outside of Antarctica. The exception to this is the Patagonian ice field that stretches for some 350 km along the spine of South America from a latitude of around 48–51.5°S and is the largest ice cap outside of the polar regions. There are in excess of 160 000 glaciers on the planet, ranging in length from kilometres to hundreds of kilometres. The cryosphere is clearly extremely heterogeneous in its distribution, size, response time and interaction with the rest of the planet. To observe and model such a heterogeneous constituent of the Earth system presents a daunting prospect, and in the following chapters we present the 'state of the art' and our best attempt at tackling this challenge.

References

Church, J. A. *et al.* 2001. Changes in sea-level. In Houghton, J. T. and Yihui, D., eds., *IPCC Third Scientific Assessment of Climate Change*. Cambridge University Press, pp. 640–93.

Clague, J. J. and James, T. S. 2002. History and isostatic effects of the last ice sheet in southern British Columbia. *Quat. Sci. Rev.* **21** (1–3), 71–88.

Goosse, H. and Fichefet. T. 2001. Open-ocean convection and polynya formation in a large-scale ice-ocean model. *Tellus Series a – Dyn. Meteorol. & Oceanography* **53** (1), 94–111.

Houghton, J. T. *et al.* 2001. *Climate Change 2001: The Scientific Basis*. Cambridge University Press.

McGuffie, K. and Henderson-Sellers, A. 1997. *A Climate Modelling Primer*. Chichester, Wiley.

Paterson, W. S. B. 1994. *The Physics of Glaciers*, 3rd edn. Oxford, Pergamon.

Petit, J. R. *et al.* 1999. Climate and atmospheric history of the past 420 000 years from the Vostok ice core, Antarctica. *Nature* **399** (6735), 429–36.

Stocker, T. F. 2000. Past and future reorganizations in the climate system. *Quat. Sci. Rev.* **19** (1–5), 301–19.

Stossel, A., Yang, K. and Kim, S. J. 2002. On the role of sea ice and convection in a global ocean model. *J. Phys. Oceanography* **32** (4), 1194–208.

Part I

Observational techniques and methods

2

In situ measurement techniques: land ice

JON OVE HAGEN

Department of Physical Geography, Faculty of Mathematics and Natural
Sciences, University of Oslo

NIELS REEH

Ørsted-DTU, Electromagnetic Systems, Technical University of Denmark

2.1 Introduction

Measurement of the mass balance of larger glaciers, ice sheets, ice caps and ice fields
requires different field techniques than for the smaller valley glaciers. These larger glaciers
are an integral part of the Earth's interactive ice–ocean–land–atmosphere system, and may
also provide valuable insight into the cause of changes of the Earth's climate system (Meier,
1998).

In this chapter, we deal with *in situ* measurement techniques. However, we include
measurements based on aerial photography, since such measurements for more than half a
century have been used in combination with field studies. Modern, mainly satellite-based,
remote-sensing techniques for measuring glacier mass balance are presented in Chapter 4.

2.2 Mass balance equations

In glacier context, the term 'mass balance' is traditionally used in two ways with different
meanings. At a specific point of the glacier, the local mass balance designates the sum
of accumulation (supply of mass mainly by snow deposition) and ablation (loss of mass
mainly by melting of snow/ice). The local (specific) mass balance may be positive or
negative depending on whether accumulation or ablation dominates. However, the sign of
the specific mass balance does not say anything about the local change of ice thickness
or the local change of mass in a vertical column through the glacier. This is because the
specific mass balance may be compensated for, or even be overruled by, mass input/loss
due to a gradient of the horizontal ice flux.

Also, the local change of thickness or change of mass in a vertical column is often
referred to as the local mass balance. To avoid confusion, we shall use the term 'specific
mass balance' for the sum of local accumulation and ablation, whereas the local mass
change of the column will be termed 'local mass balance'.

Mass Balance of the Cryosphere: Observations and Modelling of Contemporary and Future Changes, eds. Jonathan L. Bamber and
Antony J. Payne. Published by Cambridge University Press. © Cambridge University Press 2003.

The local mass balance at a point of a glacier is expressed by the following equation (Reeh, 1999; Reeh and Gundestrup, 1985):

$$\partial H/\partial t = b_s + b_b - F\,[H(\partial u_s/\partial x + \partial v_s/\partial y) + u_s \partial H/\partial x], \qquad (2.1)$$

where H = ice thickness, t = time, b_s and b_b are specific mass balances at the surface and bottom, respectively, u_s and v_s = horizontal components of the surface velocity (u_s is in the direction of flow) and $F = \bar{u}/u_s$ ($\bar{u}$ = depth-averaged velocity). The ice flow is assumed to be in the x-direction. All quantities in equation (2.1) are expressed in ice equivalents. For land-based glaciers outside geothermal (volcanic) active regions, we can assume that b_b is zero or negligible.

The change of surface elevation with time at a fixed position on the glacier gives the local net mass balance (and a minimum value of the specific net balance). It is determined by the kinematic boundary condition, relating the rate of change of surface elevation $\partial h/\partial t$ to the vertical ice-particle velocity w_s, the horizontal velocity vector $\vec{u}_s$, the surface gradient $\vec{grad}h$, and the specific mass balance b_h measured in metres of ice per year (see Figure 2.1):

$$\frac{\partial h}{\partial t} = b_s + w_s - \vec{u}_s \cdot \vec{grad}h. \qquad (2.2a)$$

The term $w_s - \vec{u}_s \cdot \vec{grad}h$ is called the emergence velocity. It represents the vertical flow of ice relative to the glacier surface; see e.g. Paterson (1994, p. 258). The net balance can therefore be estimated if one knows the emergence velocity and assumes that the density does not change with depth during the period. This assumption should at least be valid in the ablation area. The true cumulative net balance will be more negative than that estimated from

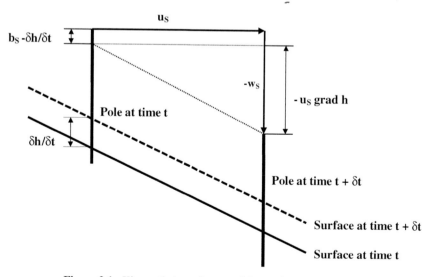

Figure 2.1. Kinematic boundary condition at the glacier surface.

the geometry changes alone in the ablation area and more positive than in the accumulation area due to the emergence/submergence velocity. Since we have $h = H + B$, where B is the bed elevation, the local mass balance is found from equation (2.2a) to be

$$\frac{\partial H}{\partial t} = b_s - \frac{\partial B}{\partial t} + w_s - \vec{u}_s \cdot \vec{grad} h. \tag{2.2b}$$

Equations (2.1) and (2.2b) can both be used to estimate the local mass balance. Use of equation (2.2b) requires that $\partial B/\partial t$ is known or can be neglected, and that both the horizontal and vertical velocity components are measured. This is not always the case. Velocity measurement by photogrammetry and synthetic aperture radar interferometry (InSAR), for example, only provides the horizontal velocity component. In such cases, the more complicated equation (2.1) can be applied, assuming that the ice thickness is also measured.

The term 'mass balance' is also used globally to mean the total mass change of the glacier or a region of a glacier. The total mass balance is found by integrating either the local or the specific mass balance over the total area of the glacier and subtracting the loss through possible vertical boundary surfaces such as calving fronts.

The total mass balance is expressed by the equation

$$\partial V/\partial a = M_a - M_m - M_c \pm M_b, \tag{2.3}$$

where $V = $ ice volume, $a = $ one year, $M_a = $ the annual surface accumulation, $M_m = $ the annual loss by glacial surface runoff, $M_c = $ the annual loss by calving of icebergs, and $M_b = $ the annual balance at the bottom (melting or freeze-on of ice). All volumes are expressed in terms of ice equivalents.

Equation (2.3) suggests that the total mass balance can be determined by two entirely different methods: (i) by direct measurement of the change in volume by monitoring surface elevation change, and (ii) by the budget method, by which each term on the right hand side of the mass balance equation is determined separately.

In order to obtain the local mass balance by means of equation (2.1), measurement of specific mass balance, horizontal velocity and ice thickness is required, whereas determination of the specific mass balance from equation (2.2b) requires measurement of both the vertical and horizontal velocity components as well as the surface gradient.

2.3 Direct measurement of surface elevation change

The average total net mass balance of a glacier over a given time period can be calculated from the geometry changes of the glacier, i.e. the change in altitude integrated over the whole area of the glacier. The best way to obtain reliable volume change results is by subtracting two digital elevation models of the glacier at the beginning and end of the time period. This can be obtained with different levels of accuracy depending on available resources: surface geodetic methods by point measurements and profiles, or by scanning laser altimetry by aircraft or aerial photography.

2.3.1 Traditional surveying methods

Ordinary surveying techniques by theodolite and electronic distance meter have a long tradition in surface mapping and ice-flow measurements on valley glaciers. Marked points on the glacier are surveyed from fix-points outside the glacier. Angles and distances are measured and thus point positions (x,y,z) can be calculated. The method requires that the operator follows careful procedures to minimize errors. The method is time-consuming, but can give very precise data. It has been a commonly used method of studying valley glaciers where distances between fix-points and the points on the glacier are less than 10 km. The primary objective of such measurements has, in most cases, been to derive glacier motion, and few studies of surface elevation change by these methods have been published. For larger ice masses, such as ice caps and ice sheets that cannot be surveyed from fixed points on bedrock, the measurement must be done by repeated levelling along profile lines marked with poles (e.g. Koerner, 1986; Seckel, 1977). If horizontal positions are also measured along the profile line, then surface elevation change is simply calculated from the elevation differences of the repeated levelling. If surface elevations are determined at the pole sites only, the surface elevation change must be corrected for horizontal ice motion, which requires measurement of horizontal ice velocity and local surface slope; see Figure 2.1.

On the big ice sheets, few accurate surface elevation profiles have been measured with traditional surveying methods because of the high manpower and logistics costs involved. An example is the EGIG line across the central Greenland ice sheet (Figure 2.2), along which surface elevation was measured by trigonometric levelling in 1959 and 1968. In the 9-year period from 1959 to 1968, the centre part of the profile experienced a mean increase in surface elevation of 9 cm per year (Seckel, 1977). A re-measurement of the profile in 1992 by using GPS observations revealed that, in the subsequent 24-year period from 1968 to 1992, the centre part of the profile experienced a mean surface lowering of 15 cm per year (Kock, 1993; Möller, 1996). Repeated airborne laser altimeter profiling in 1993/94 and 1998/99

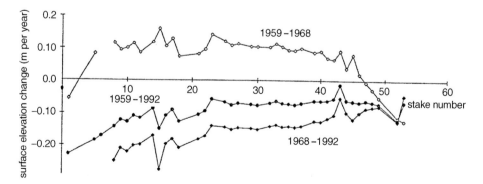

Figure 2.2. Surface elevation change along the EGIG line as determined by repeated trigonometric levelling in 1959, 1968 and 1992. Stake number refers to the original T-poles marking the EGIG line at 10 km distances. (After Kock (1993).)

confirm that quite rapid short-term thinning or thickening occur locally in the higher elevation regions of the Greenland ice sheet (see Chapter 11). These observations, as well as those from the EGIG line, illustrate that fluctuations in surface elevation on a decadal timescale, probably caused by variations of snow accumulation and/or snow/firn temperature (Arthern and Wingham, 1998; Van der Veen 1993), are so large that they may prevent long-term trends from being detected even by observational programmes running over more than three decades.

2.3.2 Cartographic method: comparison of topographic maps from different years

In general, point or profile measurements provide limited information about the spatial distribution of glacier surface elevation change. By comparisons of topographic maps from different years, the spatial distribution and total volume change can be calculated, and thus the average annual net balance can be estimated. Digital elevation models (DEMs) must be created from digitized contour maps. Length, area and volume change can then be found by subtracting the DEMs. This is standard procedure in different geographical information system software. The result is given in a grid with altitude change in snow/firn and blue ice areas. These altitude changes must be converted to volume change given in water equivalents, and therefore the density and thickness of snow/firn in the accumulation areas should be known. Other significant errors are the map accuracy derived from the geodetic network and the quality of aerial photos, especially in the accumulation area where bad contrasts on the white snow surface make it difficult to construct contour lines. The time of the year when aerial photography is taken should be as late in the ablation season as possible. Andreassen (1999) analysed maps in the scale 1:10 000 from five different years over the period 1941 to 1997 for Storbreen (5.4 km^2) in southern Norway. An assessment of the different sources for altitude errors gave an estimated accuracy of the maps to be in the range of ±1.0–2.0 m. The results could be compared with annual mass balance measurements carried out over the period 1948 to 1997. The map analysis gave less negative mass balance over the whole period 1941 to 1997 than the direct measurements: −10.9 m water equivalent (−0.2 m per year) versus −16.8 m (−0.3 m per year). Considering the sources of errors in both methods, Andreassen concluded that one cannot use the cartographic method to verify the other, or vice versa. Krimmel (1999) conducted a similar study from South Cascade glacier (2 km^2), USA, where five DEMs from the period 1970 to 1997 were compared with direct mass balance measurements. In this study, the geodetic method gave *more* negative mass balance than the direct method, and largest uncertainties were found in the direct method. The study indicated that the geodetic or cartographic method was valid and could even be used with confidence on annual basis if care is used to assure that the control datum for photogrammetric stereo models is consistent. The cartographic method is definitively useful for monitoring long-term changes and spatial distribution of changes.

Maps can also be used in a different way to assess changes of glacier surface elevation, i.e. by comparing the height of the trim lines along a glacier to the height of the adjacent glacier margin. The elevation difference is a measure of the glacier thinning in the period since the glacier extended to the trim line. The method was applied by Weidick (1968) who measured heights of the trim lines and ice margins along glacier lobes from the ice sheet in south-west Greenland. The maximum extent in historical time of the ice-sheet margin in south-west Greenland was attained in the late 19th century (Weidick, 1995), i.e. approximately 100 years before the mapping. From the height differences between trim lines and ice margins, Weidick (1968) calculated the mean annual thinning of the ice-sheet margin to be approximately 0.7 m per year. If extrapolated to the entire Greenland ice-sheet margin, this thinning would mean an annual loss of *c.* 200 km³ of ice equivalent per year, corresponding to an average sea-level rise of *c.* 0.5 mm per year. This is about one-quarter of the average annual sea-level rise over the last 100 years as estimated from tide gauge data (Warrick *et al.*, 1996).

A cost-efficient way to create a DEM is by digital photogrammetry based on repeated aerial photography. Alternative methods are based on remote sensing techniques such as airborne laser altimetry (see e.g. Echelmeyer *et al.*, 1996; Favey *et al.*, 1999; Krabill *et al.*, 2000) or satellite-borne radar altimetry (Davis, Kluever and Haines, 1998); see Chapter 4. Although processing techniques for interferometric synthetic aperture radar (InSAR) measurement of surface elevation (see Chapter 4) have improved greatly in recent years, the precision of this method ($\pm$10 m) is still too low to make it a generally useful method for mapping glacier surface elevation change.

2.3.3 Repeated altitude profiles by GPS

Point positions and longitudinal profiles of elevation changes along selected profiles can be obtained from global positioning system (GPS) surveying. The GPS survey can be conducted using two methods (Eiken, Hagen and Melvold, 1997):

(1) static GPS survey of poles drilled into the ice for flow velocity measurements;
(2) kinematic GPS survey to measure longitudinal profiles of surface elevation.

Comparisons with traditional surveying by theodolite and electronic distance meter show that the GPS survey is applicable and gives sufficient accuracy to replace traditional methods, and is thus especially useful on large glaciers where traditional surveying is impossible. An example of elevation changes measured by GPS from 1991 to 1995 on Kongsvegen in Svalbard is shown in Figure 2.3. GPS profiling can be used on large glaciers in remote areas to monitor geometry changes, ice flow and net mass balance changes. However, it requires that the centre-line profile changes are representative for the area/altitude intervals, i.e. that the accumulation and ablation pattern is evenly distributed (Hagen *et al.*, 1999). In addition, rough surface topography with sastrugi can make kinematic measurements difficult. Sampling intervals are recommended to be as short as three seconds, and measurements

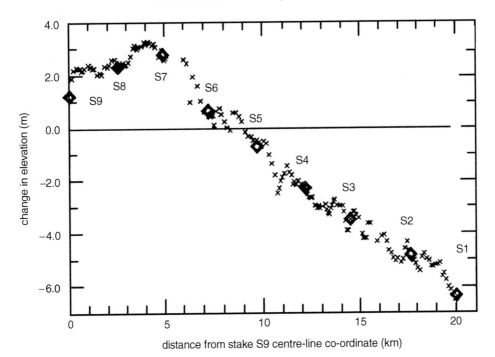

Figure 2.3. Observed changes in elevation from GPS profiles in 1991 and 1995 on the Kongsvegen glacier, Svalbard. The emergence velocity is very small, almost negligible, and thus the profile mirrors the net mass balance distribution. (From Eiken, Hagen and Melvold (1997).)

must be carried out in periods with more than five available satellites and good geometry between satellites and receivers. When this is fulfilled the kinematic profile can be established with an accuracy better than ±5 cm in the horizontal position and ±10 cm in the vertical (altitude) (Eiken *et al.*, 1997).

GPS re-measurements of profiles have shown that it is very important to follow the same route on the glacier to avoid errors due to local (lateral) surface undulations on the glacier. For re-measurements of profiles, the use of real-time-kinematic mode is recommended, where data are transmitted by a radio link from the reference station to the rover, where the differential position can be computed in near real time. Thus the same track can more easily be followed on the glacier.

More extensive kinematic differential GPS profiling with a dense net of profiles can be used as input to grid-based DTMs to replace repeated aerial photography (Jacobsen and Theakstone, 1997).

2.3.4 Coffee-can method

The local mass balance as outlined in equation (2.2b) can be measured directly on a spot on an ice sheet by a combined measurement of vertical velocity and snow accumulation (Hulbe and Whillans, 1994). By this method, called the 'submergence velocity' or 'coffee-can' method, a firn anchor is drilled down 10–20 m. The anchor is connected to a surveyable point above the snow surface by a non-stretchable steel cable. (This procedure ensures that the survey point is not affected by the short-term fluctuations in snow fall or snow densification processes in the upper firn.) A precise GPS survey will then give submergence velocity. Disregarding a generally insignificant vertical velocity contribution from isostatic adjustment, the thickness change is the difference between submergence velocity and long-term rate of snow accumulation b_s (measured in metres per year of water equivalent), which is obtained from dated horizons in firn cores and the depth-integrated density above that horizon. The local rate of thickness change is then obtained from equation (2.2a) as

$$\frac{\partial H}{\partial t} = b_h/\rho + w + \delta w - \vec{u}_s \cdot \vec{grad}h, \tag{2.4}$$

in which ρ is the density at the marker depth, and δw is the relative vertical velocity of the marker with respect to a reference pole anchored in the near-surface layer. Other symbols are explained in connection with equations (2.1) and (2.2). The terms δw and $\vec{u}_s \cdot \vec{grad}h$ are usually small.

The method has been tested in Antarctica and in Greenland, and has shown promising results, giving a net uncertainty of the thickness change of about 0.02 m per year over a 5-year interval (Hamilton, Whillans and Morgan, 1998).

2.4 Measurement of mass balance components

In general it is easier to measure volume changes over a time period, and thus estimate the mean annual net balance, than to obtain annual data and especially seasonal data. A change in the glacier mass balance can be a result of melting or accumulation changes. Observation of individual components, surface accumulation and ablation is therefore necessary to understand how glaciers respond to climate variability. In addition, mass exchange can occur as freezing/melting under floating glaciers and ice shelves, and mass is lost by calving of tide-water glaciers.

A database of the global network of glaciers where direct measurements of specific mass balance components are or have been measured is operated by the World Glacier Monitoring Service (WGMS) (www.wgms.ch). Monitoring of glacier mass balance has been conducted both for climate change purposes and for hydrological purposes connected to hydro power plants in partly glacier covered basins. Several measurement methods have been developed and used in the mass balance monitoring programmes. Attempts have been made to agree on a uniform methodology, and Østrem and Stanley (1969) and Østrem and Brugman (1991) have produced manuals for field work and office work for mass balance measurements. These methods are, however, mainly suitable for fairly small valley glaciers. The bulk of long time series glacier monitoring have been on these glaciers. For logistical and economic reasons (accessibility, crevasses, remote areas) a full surface mass balance programme can be hard to accomplish on many glaciers. On larger glaciers, ice caps and ice sheets, different methods are required. In the following we will outline some of the most commonly used methods in specific mass balance component studies.

2.4.1 Accumulation and ablation rate

Stake readings

The traditional method in mass balance monitoring, often called the direct glaciological method, is based on a dense array of point measurements on the surface, usually comprising stakes drilled into the ice. The stake nets can be used both for winter snow accumulation and for summer ablation measurements. The field methods are thoroughly described in the field manual by Østrem and Brugman (1991). In addition to stake readings, density measurements are carried out in snow pits or on shallow cores. Snow accumulation pattern is often obtained by snow probing in addition to stake readings. In this way snow accumulation maps for the entire glacier can be made as well as the ablation pattern during summer. Contour lines of equal balances are then drawn by hand to calculate the specific mass balance in the areas between contour lines, and these are then integrated for the whole glacier. The results are usually reported as mean specific mass balances, summer, winter and net values, for the entire glacier based on a 'stratigraphic' system. The reference layer is the previous summer surface. The maximum winter balance is measured at the end of the winter accumulation season giving the snow accumulated on top of the previous summer surface. The minimum balance, the summer balance, is measured at the end of the melting season. When a clear summer surface is formed all over the glacier it is, in principle, possible to measure both winter and summer balances during the spring fieldwork on the glacier, providing that the stakes from the previous summer are still visible (have not melted out or snowed down).

When point numbers $b(x, y)$ are found, the mean specific balances b_{winter}, b_{summer} and b_{net} for the entire glacier area S can be calculated using

$$\bar{b} = \frac{1}{S} \int b(x, y) \, dx \, dy. \tag{2.5}$$

An ideal glacier for mass balance monitoring has a suitable shape, is easily accessible, is clean of debris and has a clear summer and winter season. A number of small alpine glaciers are measured by this method. Alternatively a 'fixed-date' system is used. Then the summer and winter balances are measured on selected dates every autumn and spring.

World-wide, these ideal glaciers are only a subset of glaciers that mainly occur in mid and high latitudes. Most glaciers are more complicated; some exist in regions where snow accumulation may occur the year round, heavy snow fall may accumulate in the upper part of the glaciers even during summer, and melting may occur in the lower part of some glaciers during winter (e.g. Himalayan glaciers). The stratigraphic method will then collapse. In low latitude (tropical and subtropical) areas ablation takes place throughout the whole year, while accumulation is limited to periods of precipitation and to higher areas only (Kaser, Hastenrath and Ames, 1996). The mass balance regime and mass balance profiles are thus complicated and different from higher latitude glaciers. The shape and location of the glaciers may also make it very difficult to conduct direct measurements. In some regions, especially in polar and sub-polar areas, the formation of superimposed ice and internal ice layers is very important. Some glaciers are partly debris covered, some are steep and heavily crevassed and some have a calving front. However, for climate-change monitoring purposes it is still of great value to select a glacier as ideal as possible because detections of long-term trends need reliable information that can be followed by simple methods over a long time.

There has been much discussion regarding the number of stakes that are required to measure the mass balance of a glacier. The optimal number of stakes depends on local conditions. The ablation is easier to measure than the accumulation since it is mainly dependent on the altitude, while the accumulation pattern varies more due to wind re-distribution and side effects such as avalanches. However, five to ten stakes are usually sufficient for small valley glaciers ($S < 20\,km^2$) and 10 to 20 for larger glaciers ($S > 20$–$500\,km^2$). A number of scientists have discussed these problems; see for example Fountain and Vecchia (1999).

Index methods

It can be very difficult to cover large areas on ice sheets and ice caps by traditional stake measurements. In order to monitor long-term trends in the accumulation and ablation, it could therefore be better to use spot measurements. Koerner (1986) suggests two different methods. The first method is to use a balance/elevation integration in which a single line of poles is placed from the lower to the highest part of the glacier. Accumulation and ablation are then monitored in each altitude represented by a stake following the traditional stratigraphic method. The total volume change is then not calculated and area per altitude is not required. By the second method a group of measuring points (stakes) are drilled into the ice over a small area (few hundred square metres) of a glacier or an ice cap. Then time series of either accumulation or ablation could be established, depending on the location of the stake farm. The number of stakes in such a stake farm must be large enough to avoid small, local variability and statistical noise in the data, usually four or five stakes. Both methods could provide an efficient and reliable way of monitoring the climatic effect on glaciers as both melting rate trends and accumulation trends can be detected.

Pit studies, firn and ice cores

In all mass balance calculations, information about the depth/density is required. Density is needed to transform snow accumulation depths into comparable water equivalent units. The densification process is quite different for glaciers in variable climate regions. In cold, polar ice sheets, where no surface melting occurs during the summer, the metamorphism is slow, and firn can be found at 100 m depth, while on temperate glaciers the melting and re-freezing processes quickly change the density. In Figure 2.4 two depth–density profiles are shown. The previous year's density profiles can be measured in a snow pit where a known volume sample is weighed. For deeper density profiles snow/firn cores are taken.

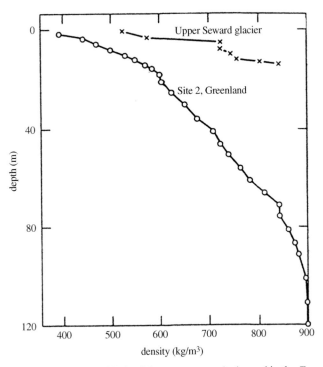

Figure 2.4. Variation of firn density with depth in a temperate glacier and in the Greenland ice sheet. (From Paterson (1994, fig. 2.2).)

Detailed studies of structure, internal refreezing in ice layers, dust layers, free water content etc. can be studied in the pits. At glaciers where detailed mass balance studies are carried out, the density must be known at different altitudes at the end of the accumulation season, and again the remaining snow density at the end of the ablation season must be known.

Annual cycles – oxygen isotopes, dust, chemistry

In cold regions on glaciers where little melting occurs during the summer, the annual cycles can be recorded in deep firn cores by oxygen isotope studies. There is a well known relationship between the oxygen isotopes ^{16}O and ^{18}O in sea-water and thus also in water vapour. Since ^{18}O is heavier than ^{16}O, it will start to condense from vapour to water earlier than ^{16}O. Snow formed from condensation of water vapour at $-30\,°C$ will therefore contain more ^{18}O than snow formed at $-20\,°C$. The fraction $^{18}O/^{16}O$ will therefore vary by the temperature when the snow was formed. Thus, snow formed during cold winter periods can be distinguished from snow formed at less cold summer temperatures. In cold firn, where little melting and percolation of melt water occurs, the thickness of winter and summer snow can be found by analysing the ratio $^{18}O/^{16}O$ in thin layers. This has been done in a number of firn and ice-core studies on the polar ice sheets to detect the accumulation history at the drill site (Hammer *et al.*, 1978). At favourable locations, more than 10 000-year-long

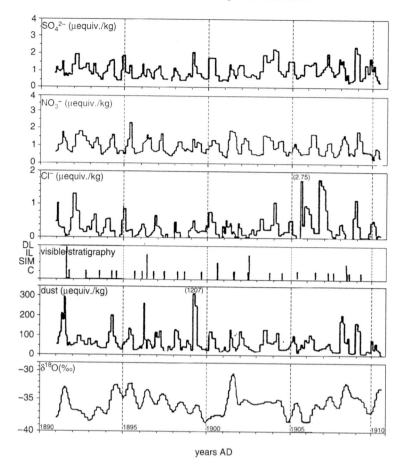

Figure 2.5. Various ice-core parameters, which are often used for dating. The records are from a site in central Greenland with moderately high accumulation rate that conserves the seasonal $^{18}O/^{16}O$ variation. However, the record is somewhat smoothed by diffusion in the firn. Nitrate seems to be the parameter with the clearest seasonal variation. (From Steffensen (1988).)

annual layer records can be established by stratigraphic methods (e.g. Hammer, Clausen and Tauber, 1986), for example by using seasonal variations of quantities such as the $^{18}O/^{16}O$ ratio, concentration of micro-particles, conductivity, or chemical composition; see Figure 2.5. For the near-surface layers, a reliable accumulation record can be reconstructed if the detection of annual layers is combined with density measurements and preferably also with dating of reference horizons. With increasing depth, corrections must be made both for flow-induced thinning of the layers and for possible spatial and temporal variations of the accumulation rate upstream of the drill site (Reeh, 1989). This is done by means of ice-flow model calculations. The precision depends on how well the relevant glaciological parameters are known, and decreases the further back in time the correction procedure is taken.

The trend in the general level of the $^{18}O/^{16}O$ ratio can also be used as an indicator of general temperature trends in the area (Isaksson *et al.*, 1996). However, also in this case, corrections must be made for possible variations of the $^{18}O/^{16}O$ ratio upstream of the drill site.

Reference layers

Nuclear weapons tests in the 1950s, 60s and 70s and the Chernobyl accident in 1986 emitted large amounts of man-made radioisotopes into the atmosphere. Some of these depositions have shown a global signal that can be detected both in the northern and southern hemispheres (Chernobyl fallout only in the northern) providing permanent records of deposition. These radioactive signatures can be used to determine snow accumulation rates. Shallow snow and firn cores can be drilled and analysed to detect the radioactive reference layers. The different events usually have a characteristic fallout composition so that the source event can be identified. The layers can be detected by analysis of caesium (^{134}Cs and ^{137}Cs) and tritium (Figure 2.6). These signals can only be used in the accumulation area of the glaciers, and will give information about the mean specific net balance at a specific point. Due to percolation of melt water, the fallout elements can be spread downwards in the snow and firn. The uppermost layer must therefore be used in the calculations even if the maximum level could be seen deeper down (Pinglot *et al.*, 1999).

The shallow cores only give data from the accumulation area, and only mean specific net accumulation values over the time period. When cores are drilled in different altitudes of a glacier, the specific mass balance gradient for the accumulation area can be found (see Figure 14.10, this volume). The average equilibrium-line altitude (ELA) can thus also be detected.

Automatic registrations

Detailed spot measurements of accumulation and ablation can be obtained by different automatic techniques for detailed studies.

An acoustic sounder measures the distance from the sensor to the snow surface by measuring the time for an emitted sound beam to be reflected from the snow surface. This sensor must be mounted on a construction that is drilled down into the ice so that it does not sink into the ice. It is often connected to an energy balance station on which all meteorological parameters are measured. The energy balance at the spot can then be calculated and directly compared to snow/ice melting recorded by the acoustic sensor. This is most easily achieved in the ablation area of a glacier where direct ice melt is recorded by the acoustic sensor. In the accumulation area the method is less reliable due to the variable snow density.

A snow pillow is a small, flat rubber tank (*c.* 2 m^2) filled with antifreeze. A pressure sensor in the tank measures the weight of the snow accumulating on the tank. The system is commonly used in snow mapping in high remote mountain areas. Reliable results can be obtained during single snow-fall events, but when hard layers are formed in thick snow packs this may unload the weight, and give errors. The advantage is that the weight, and thus the water equivalent, are given directly.

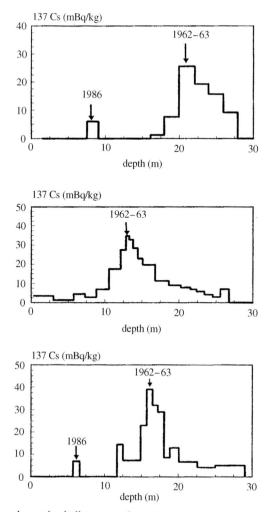

Figure 2.6. Reference layers in shallow cores from Svalbard glaciers used to detect mean net accumulation in the accumulation area of the glaciers. On these glaciers caesium fallout from the nuclear tests in 1962–63 and the Chernobyl accident in 1986 have been used. (From Pinglot *et al.* (1999).)

Thermistor strings (a number of thermistors mounted on a pole drilled into the ice) have been used to measure accumulation. The temperature in the air above the snow surface can easily be distinguished from thermistor temperatures down in the snow. Thus both the accumulation rate and the melt rate can be recorded during a year. The density is not recorded and must be measured or estimated in addition.

Ground-penetrating radar (GPR)

The accumulation measurements obtained by snow-probe sounding to the previous summer surface is time-consuming work that only gives point values, and in the accumulation area

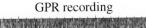

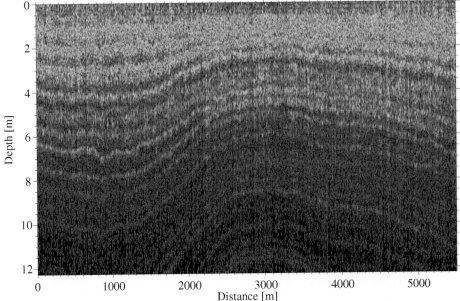

Figure 2.7. Snow radar registration from Dronning Maud Land, East Antarctica, recorded 150 km from the coast. Several snow layers are clearly seen in the registration. (From Richardson-Näslund (2001).)

it is also an unreliable method for inhomogeneous surfaces. GPR surveying can give the snow distribution variability over long distances.

A GPR system with a high frequency (usually 200 MHz or higher) can be used to image previous summer surfaces (Richardson *et al.*, 1997; Kohler *et al.*, 1997). GPR provides detailed pictures of the snow pack, both with depth and laterally, and thus allows continuous tracking of horizons (Figure 2.7). In the ablation and superimposed ice areas, the echo from the contact between the winter snow and the underlying ice surface is sufficiently strong to reduce the problem of calibration to one of determining radar wave velocity as a function of snow density. In the accumulation areas the interpretation is more complex. The radar receives signals from different reflectors in the snow pack. In order to obtain quantitative snow accumulation measurements from the GPR images, one needs to convert time-dependent radar return signals to a depth. The previous winter snow–ice or snow–firn interface must be identified and separated from the other reflectors. A time-varying gain function used to amplify the signal at depth must be performed. The wave speed, v, in the snow and firn is determined by the dielectric constant, ε, so that

$$v = c_0/ \left(|\varepsilon| \right)^{1/2}, \tag{2.6}$$

where c_0 is the wave speed in a vacuum. The dielectric constant varies with the density and temperature in the snow. In dry polar firn it is given by the empirical equation (Kovacs, Gow and Morey, 1995)

$$\varepsilon = (1 + 0.845\rho)^2. \tag{2.7}$$

Conversion of the GPR image's timescale to a depth-scale is thus not straightforward, and independent depth measurements to previous summer surfaces are needed to calibrate the radar and obtain quantitative mass balance data (Kohler *et al.*, 1997). GPR can be used as an excellent complement to traditional surveying. On large glaciers and ice sheets it is a useful tool, as the reflective horizons can be traced over long distances and over short times. Richardson *et al.* (1997) and Richardson and Holmlund (1999) showed that significant regional spatial variations in accumulation rate, with high local variability, could be measured by radar in two long transects (1040 km and 500 km) from the ice shelf up to the polar plateau in Antarctica. They used a radar with frequencies of 800–2300 MHz. However, the method cannot yet replace completely traditional methods as some calibration points are needed from shallow cores or snow probing.

Usually, two dominant signals can be seen: (1) the reflection returned from the surface of the snow cover, which serves as the time reference; (2) the echo from the snow–firn or snow–ice interface. In the ablation area, the snow–ice reflector can easily be identified by use of layer reflection intensity, since the winter snow cover lies on top of uniform cold ice. In the accumulation area the snow–firn interface is less distinct, due to a less homogeneous and less developed summer surface. The snow cover lies on top of heterogeneous firn (firn, ice layers and superimposed ice) as can be seen in pits and ice-core stratigraphy. The previous winter snow layer reflection must be correlated to the core stratigraphy and manual probing.

2.4.2 Superimposed ice and internal accumulation

The formation of superimposed ice at the surface of High Arctic glaciers and high altitude mountain glaciers in lower latitudes has an important impact on the surface mass balance. The response of Arctic glaciers to climate warming is also complicated by the formation of superimposed ice. Melt water from the snow percolates into the cold snow pack and refreezes, either as ice lenses in the snow and firn or as a layer of superimposed ice on top of cold impermeable ice. In the firn area the melt water may also penetrate below the previous year's summer surface and freeze as internal accumulation. Superimposed ice formation is important in many Arctic glaciers, and in some it is even the dominant form of accumulation (Koerner, 1970). Superimposed ice that survives the summer season becomes a part of the glacier net accumulation. By artificial reference horizons or poles drilled into the ice it is possible to measure directly the thickness of the superimposed ice layer. The amount of ice layers is almost impossible to measure directly. The formation of superimposed ice is closely linked to meteorological parameters, mainly air temperature and near-surface ice temperature (related to mean annual air temperature). The amount of superimposed ice is therefore often modelled, and as early as the beginning of the 1950s Ward and Orvig (1953) developed a model for superimposed ice formation on the Barnes ice cap in northern Canada. Woodward, Sharp and Arendt (1997) developed this further and came up with a simple empirical relationship between the mean annual air temperature (or the ice temperature at 14 m depth), θ, and the maximum thickness X (in centimetres) of the superimposed ice layer:

$$X = -0.69\theta + 0.0096. \tag{2.8}$$

More studies of the formation of superimposed ice and its impact on glacier mass balance during climate change should be recommended. It is also very difficult to distinguish between the superimposed ice zone and glacier ice by remote-sensing techniques (Engeset *et al.*, 2002).

2.4.3 Error analysis

An assessment of errors is critical in mass balance programmes, but it has been given little attention. Some authors have discussed the optimal number of measuring points (Cogley, 1999; Fountain and Vecchia, 1999). Funk, Morelli and Stahel (1997) concluded that for a small valley glacier a stake density of the order of one stake per square kilometre is sufficient. However, they also concluded that this low density network required that a measuring point density of ten per square kilometre is necessary over a short period to obtain reliable interpolation functions that could be used with the less dense network. Cogley (1999) and Fountain and Vecchia (1999) tested a method using one-dimensional regression of mass balance with altitude based on a few measurement points along the central flow line and compared that with the traditional contouring method. They found that their method yielded mass balance values equivalent to contouring methods. An advantage of the regression method is that the input is a sparse data set and can provide an objective error of the resulting estimate. The contouring requires large data sets, and error estimates are often ambiguous. The error estimate can be small (less than 10%) with an experienced observer who knows the glacier. This level is obtained at Storglaciären (3 km^2) in Sweden where Jansson (1999) estimated an accuracy of ± 0.1 m water equivalent. The statistical analysis of the Swiss glacier Griesgletscher (6 km^2) mass balance data and comparisons with geodetic data indicate a net balance accuracy of ± 0.06 m water equivalent. (Funk *et al.*, 1997). However, both these glaciers are small with a well developed and performed mass balance programme. This level of accuracy can only be obtained on such glaciers. The most important point for long-trend analysis is that the glacier is measured by the same method year by year. Simpler mass balance methods, such as the index method by a 'stake farm' recommended by Koerner (1986), can be better for reliable trend studies, but do not give the overall mass balance for the glacier. All analyses point to the dominant effect of the gradient of mass balance with altitude of alpine glaciers compared with transverse variations. The number of mass balance points required to determine the glacier balance appears to be invariant for small glaciers (*c.* 5–10 km²) and five to ten stakes are sufficient. Larger glaciers need a smaller stake density than smaller glaciers. Small valley and cirque glaciers are more sensitive to edge effects such as avalanches and wind distribution. On larger glaciers (∼100 km²) the altitude variations are more dominant, and each stake measurement is more representative for its altitude interval; therefore a stake density of one per 10 km² can be sufficient. Annual mass balance measurements should always be compared with volume-change estimates based on geodetic methods over a number of years. As the mass balance changes, the area and dynamics will also change, and the glaciers will always tend toward a zero mass balance as they adjust to climate.

2.4.4 Balance velocity

This method relates the mass flux through the cross-section under the equilibrium line to the net mass balance of the accumulation area. Assuming steady-state conditions ($\partial h/\partial t = 0$), the balance velocity ($\bar{u}$) can be defined by the conditions that the volume flux through the cross-sectional area A under the equilibrium line must equal the volume of ice deposited in the accumulation area (S_c):

$$(\bar{u}A = b_c S_c \rho_{\text{water}}/\rho_{\text{ice}}), \tag{2.9}$$

where b_c is given in metres of water equivalent per year (Kuhn *et al.*, 1999). The method can be used to evaluate if the glacier dynamics is in balance with the climate. If the actual velocity is higher than the calculated balance velocity the glacier is under-nourished, and vice versa. Melvold and Hagen (1998) used the method to show that the glacier Kongsvegen in Svalbard is building up toward a surge. The surface mass balance was close to zero, but the actual velocity was much lower than the balance velocity, and thus the gradient is gradually increasing and a surge is likely to be the result.

2.4.5 Calving

Calving of icebergs constitutes roughly 45% and 90% of the mass loss from the Greenland and Antarctic ice sheets, respectively. Iceberg calving is a significant ablation mechanism for other glacierized areas such as Svalbard and coastal Alaska. For all the world's glaciers and ice sheets, calving of icebergs constitutes *c.* 2.36×10^{15} kg per year i.e. *c.* 70% of the total accumulation of 3.4×10^{15} kg per year (Warrick *et al.*, 1996). Hence, iceberg calving accounts for the major mass loss from the land-ice masses of the world, but, unfortunately, is still the term of the mass balance equation with the largest uncertainty.

Calving of icebergs is a discrete process. Consequently, it makes sense only to define the calving flux M_c (the rate of production of icebergs) as a quantity averaged over a period of time T that is large compared with the average interval between major calving events.

As illustrated in Figure 2.8, the change with time t of the position of the calving front (or the length of the glacier) L consists of a smooth forward motion with a speed equal to

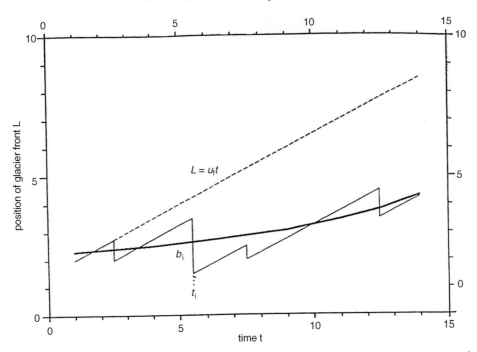

Figure 2.8. Position of glacier front as a function of time. The thin full line is the actual change of position consisting of a smooth forward motion with a speed equal to the ice velocity at the calving front u_f interrupted at intervals by abrupt backward steps of magnitude b_i due to the break-off of icebergs. The dashed line shows the change in front position if there were no calving. The thick full line represents the smoothed long-term position of the calving front.

the ice velocity at the calving front u_f interrupted at intervals by abrupt backward steps of magnitude b_i due to the break-off of icebergs. In mathematical terms, this behaviour can be expressed as

$$L = L_0 + \int_0^t u_f \, dt - \sum_i Hv(t - t_i)b_i, \qquad (2.10)$$

where $Hv(\tau) = 0$ for $\tau < 0$ and $Hv(\tau) = 1$ for $\tau \geq 0$, and u_f and b_i are average values over the width of the calving front.

A direct way of determining the calving flux is by observing the b_i during a period T and calculating the calving flux as $M_c = 2WH\Sigma b_i/T$, where H is ice thickness averaged over the width of the calving front. In practice, however, this is seldom feasible, and instead the ice flux at the calving front is often determined by observing ice thickness H, ice velocity u_f and glacier width $2W$, and calculating the flux as $M_f = 2\,WHu_f$. In most cases, u_f is determined by observations during a relatively short period, of the order of a few days or weeks.

In general, however, the calving flux M_c is different from the flux of ice at the calving front M_f. The two fluxes are related by the equation (Meier, 1994)

$$M_c = M_f - 2WH\partial L/\partial t. \tag{2.11}$$

A similar relationship is valid between calving speed u_c and ice velocity at the calving front u_f:

$$u_c = u_f - \partial L/\partial t. \tag{2.12}$$

All quantities are averaged over the width $2W$.

Obviously, the $\partial L/\partial t$ term in equations (2.11) and (2.12) must be understood as the time derivative of the front position history smoothed over a period which is large compared with the average interval between major calving events; see Figure 2.8. Only in the case that the long-term average position of the calving front is constant ($\partial L/\partial t = 0$) are the calving and front terms equal. If this is not the case, the calculated ice flux at the calving front must be corrected for the influence from long-term changes in front position in order to obtain a correct value for the calving flux. Front position histories of Greenland calving glaciers have been studied by Weidick (1994) and Weidick *et al.* (1996). These studies show that, in some periods, the $\partial L/\partial t$ term was significant, and consequently that the term cannot in general be neglected. However, in other periods, the glacier fronts have remained stationary so that the usual procedure of neglecting the $\partial L/\partial t$ term is justified.

The traditional way of measuring calving flux can be illustrated by the study of Olesen and Reeh (1969), who studied the glaciers calving into the northern part of the Scoresby Sund fjord system. Ice velocities in the frontal region of the glaciers were determined over a 6-week period in the summer of 1968 by means of repeated theodolite intersections from fixed points on firm ground beside the glaciers. The thickness of the glaciers was estimated from observations of overturned icebergs and measurements of the surface altitude in transverse profiles in near-front regions where the glaciers are afloat or close to being afloat (Reeh and Olesen, 1986). In 1972, a similar campaign was carried out on the glaciers in the southern part of the Scoresby Sund fjord system (Henriksen, 1973).

Measurement of glacier velocity by repeated theodolite surveying is an extremely time-consuming process, and terrestrial or aerial photogrammetry has therefore often been used to measure glacier velocity. Pioneering work applying terrestrial photogrammetry was made during the German Spitzbergen expeditions in the 1930s (Pillewizer, 1939).

For Greenland, extensive photogrammetric measurements of calving front thickness and velocity have been published by Bauer *et al.* (1968), Carbonnell and Bauer (1968) and Higgins (1991). Bauer *et al.* (1968) determined velocities in the frontal region of glaciers calving into Disko Bugt and Uummannaq Bugt in central West Greenland, using repeated vertical aerial photographs taken in 1957 at 4- to 5-day intervals. The mean thickness near the glacier fronts was estimated from overturned icebergs displayed on the aerial photographs or from bathymetry of the fjords near the glacier fronts.

In 1964 a similar campaign was undertaken for the same glaciers (Carbonnell and Bauer, 1968). This time, the interval between the two photographs was about two weeks. Frontal heights were determined from the aerial photographs, but this new information did not result in revised ice thicknesses as compared with the 1957 survey.

The results of the two studies vary by about 10% (82 km^3 of water equivalent per year in 1957 and 93 km^3 of water equivalent per year in 1964). It is uncertain whether this difference reflects an increase of the calving flux from 1957 to 1964, or whether it is simply due to measurement uncertainty.

Repeated aerial photographs were also used by Higgins (1991) to derive the loss by calving from all North Greenland fjord glaciers. Long-term average velocities of the floating segments of the glaciers were determined by means of preserved distinctive patterns of meandering streams and melt-water pools recognizable on vertical aerial photographs taken decades apart, i.e. in the period 1959 to 1963 and again in 1978. The glacier thickness was derived from photogrammetric measurement of the glacier-surface altitude in transverse profiles across each of the major floating ice tongues close to their calving fronts. The derived calving flux (3.4 km^3 of water equivalent per year) is about eight times less than the estimated calving flux derived from applying equation (2.3) with $M_b = 0$ to the northern sector of the Greenland ice sheet (Reeh, 1994). The derived large imbalance of the North Greenland ice sheet initiated speculations on bottom melting as a significant ablation mechanism for Greenland floating glacier tongues (Reeh, 1994), an idea that was soon confirmed by detailed mass budget studies of North Greenland glacier drainage basins (Rignot *et al.*, 1997).

Observations of thickness, width and ice velocity at the front of Greenland calving glaciers and derived calving fluxes are listed by Reeh (1994) and Weidick (1995).

2.4.6 Bottom mass balance (floating glaciers and ice shelves)

For a glacier on bedrock, the amount of basal melting is limited by the heat supplied by geothermal heat flux and ice deformation. In general, this limits basal melting to at most a few centimetres per year, which in most cases is negligible compared with the specific mass balance at the glacier surface. An exception is those glaciers in volcanically active areas, for example the Grimsvötn area of Vatnajökull, Iceland (Björnsson, Björnsson and Sigurgeirsson, 1982). For floating glaciers, bottom mass balance can be orders of magnitude larger than for grounded glaciers, and thus is the dominant mass balance term.

The mass balance at the bottom of Antarctic floating glaciers and ice shelves has been studied for decades. Theoretical work by Robin (1979) based on oceanographic soundings in front of the Filchner ice shelf, Antarctica (Foldvik and Kvinge, 1977), showed that thermohaline circulation beneath floating glaciers and ice shelves is likely to occur. This circulation would very likely result in basal melting near the grounding line, and possibly freeze-on of sea ice at shallower depths nearer to the ice-shelf front. The thermohaline circulation model could explain the layer of frozen sea-water found in an ice core drilled at a location on the Amery ice shelf and the absence in the core of ice deposited far inland (Morgan, 1972). Later studies have confirmed large annual melt rates of several metres per year near the grounding line of Antarctic floating glaciers (e.g. Jenkins *et al.*, 1997), and

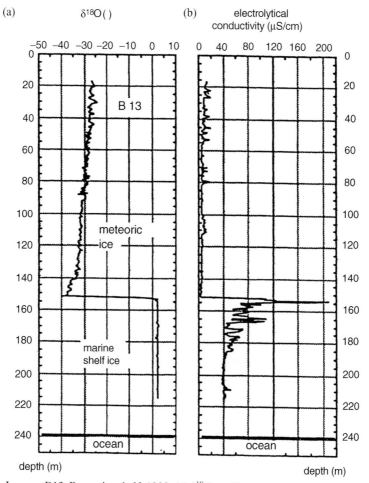

Figure 2.9. Ice core B13, Ronne ice shelf, 1990. (a) $\delta^{18}O$ profile measured at drill chipping samples. (b) Profile of the electrolytic conductivity, measured on melted drill chipping samples. The boundary between meteoric ice and marine ice is displayed with the sharp increase of $\delta^{18}O$ and electrolytic conductivity. The bottom of the ice shelf was at 239 m. (From Oerter et al. (1992).)

also that freeze-on of sea ice is a widespread phenomenon beneath Antarctic ice shelves (e.g. Oerter et al., 1992); see Figure 2.9.

As mentioned in the previous section, balance calculations for drainage basins feeding floating outlet glaciers from the North Greenland ice sheet have shown that these glaciers are also subject to substantial bottom melting that, at the grounding line, may reach values of 30 per year (Rignot et al., 1997; Thomsen et al., 1997).

Bottom freeze-on and melting rates are difficult to measure directly. The layer of frozen-on sea ice found in the bottom part of some ice cores retrieved from floating ice shelves represents ice that is accumulated along the ice-shelf section upstream of the drill site.

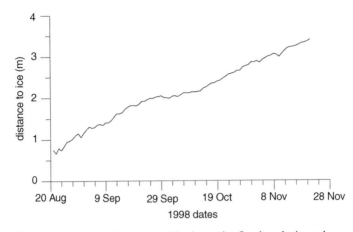

Figure 2.10. Direct measurement of bottom melting beneath a floating glacier: echo-sounding record (24 h means) from beneath Nioghalvfjerdsfjorden glacier. Sounder directed upward toward the bottom of the floating glacier. (From Olesen (1999).)

Derivation of annual freezing rates from such data therefore requires knowledge about conditions upstream of the drilling site.

Upward-pointing echo sounder

Attempts have been made to measure directly basal melt rates beneath Antarctic ice shelves by installing upward-pointing echo sounders under the ice (Lambrecht, Nixdorf and Zürn, 1995; Nixdorf *et al.*, 1995). The same method was successfully applied to the *c.* 80 km long and 20 km wide, floating section of Nioghalvfjerdsfjorden glacier, north-east Greenland (Olesen, 1999); see Figure 2.10. The echo sounder was mounted via a *c.* 130 m long, 12 cm diameter, hot-water drilled hole through the glacier located 31 km behind the glacier front, about 5 km from the northern glacier margin (Figure 2.12). The sounder operated from August 20 to November 28, 1998. During this period, *c.* 2.5 m of ice was melted from the bottom of the glacier. If we assume that there is no seasonal variation of the basal melt rate, this would correspond to an annual melt rate of the order of 10 m per year.

Thickness change in bore holes, combined with strain-rate and surface balance measurements

Figure 2.11 illustrates the principle behind another direct method of measuring the mass balance at the bottom of a floating glacier. The mass balance at the bottom can be determined from the measured change in ice thickness from one year to another, provided that strain rate and surface mass balance are also measured.

This method, which is based on the application of equation (2.1), was used at a location (79°32.5′N, 20°00.0′W) on Nioghalvfjerdsfjorden glacier (Figure 2.12). In 1996, five holes about 200 m apart were drilled along a line parallel with the ice movement. The ice thickness was measured in the holes by using a kevlar cable attached to a folding anchor, which had been constructed for this purpose. By using a spring balance, the tension of the kevlar

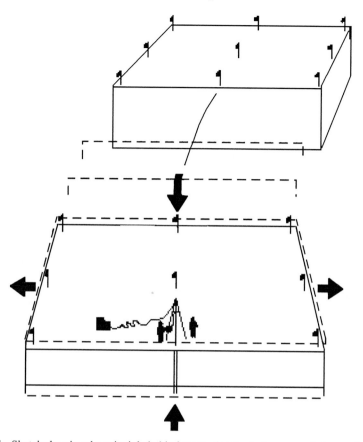

Figure 2.11. Sketch showing the principle behind measuring the bottom mass balance. A slab of ice delimited by a local strain net moves downstream as part of the glacier tongue. During this movement, strain effects (marked by black arrows) and surface mass balance effects influencing the ice thickness are measured. Mass balance conditions at the bottom of the ice can be expressed as changes in ice thickness after correction for these factors. Ice thickness is repeatedly measured by hot-water drilling. (From Thomsen *et al.* (1997).)

line was maintained at 10 kPa during all measurements. In all cases the water level in the drill holes dropped immediately to sea level after the drill had penetrated through the ice. The distance from the ice surface to the sea level was measured at all drill locations. The holes were re-drilled in the 1997 summer season and the ice thickness measurements were repeated. The measurements show a small change in the ice thickness of −0.6 m per year ±0.6 m per year (Olesen *et al.*, 1998). Surface strain rates were measured by repeated GPS observations in an approximately 800 × 800 m square centred around the drilling sites. The derived ice-thickness change due to ice deformation was 0.35 m per year. The surface ablation from 1996 to 1997 was measured as 0.8 m per year. Based on these data, the mass balance at the bottom is calculated to be −0.2 ± 0.6 m per year, suggesting a bottom balance close to zero at this location, which is in accordance with the flux divergence calculations described below.

The method for measuring bottom mass balance described above is highly dependent on the holes being straight and vertical, which is not so easy to achieve and check in practice.

Mass flux divergence calculations

Detailed measurements of ice thickness and velocity distributions allow derivation of the specific mass balance distribution by flux divergence calculations, if the glacier is assumed to be in steady-state; see equation (2.1). At Nioghalvfjerdsfjorden glacier, a detailed 15-year-average velocity field derived from repeated aerial photography (A. K. Higgins, personal communication) combined with an ice thickness distribution derived from photogrammetric mapping of glacier surface elevations has been used to calculate the distribution over the glacier area of the quantity $\partial H/\partial t + b_S + b_B$; see Figure 2.12. The surface balance, b_S, on the floating section of the glacier is of the order of -0.8 m ice per year. Moreover, the temporal thickness change $\partial H/\partial t$ is believed to be modest; therefore Figure 2.12 primarily displays the distribution of the bottom melt rate b_B. It appears that bottom melt rates are highly variable, with high values in the main glacier channel and decreasing values toward the ice front, in agreement with the directly measured bottom melt rates described above.

The cavity beneath the glacier

Knowledge of the shape of the cavity beneath a floating glacier is important for establishing basal melt rate models. Whereas glacier ice thickness is most feasibly measured by ice radar, seismic soundings are required for measuring water depths beneath floating glacier tongues, because the ice–water interface is a perfect reflector of radar signals. Seismic soundings carried out from rough glacier surfaces require a considerable logistic effort in terms of moving heavy equipment from point to point on the surface. Unmanned automated submarines may in the future be used for investigation of the cavities beneath floating ice shelves and glaciers with the potential of providing profile measurements of water depths instead of point measurements, and at the same time performing other oceanographic measurements in the water column beneath the floating glacier.

2.5 Local mass balance equation

The information required for calculating the local mass balance from equation (2.1) is specific mass balance, ice thickness and its gradient in the direction of flow, horizontal surface velocity and total surface strain rate (the sum of the longitudinal and transverse strain rates). Moreover, the ratio of the depth-averaged horizontal velocity to the horizontal surface velocity must be known. At the Dye 3 deep drilling site on the South Greenland ice sheet, horizontal velocity and strain rates were measured by using both traditional surveying methods and Doppler satellite positioning, ice-thickness profiles were measured by surface ice radar, and specific mass balance and horizontal velocity depth profile from ice-core

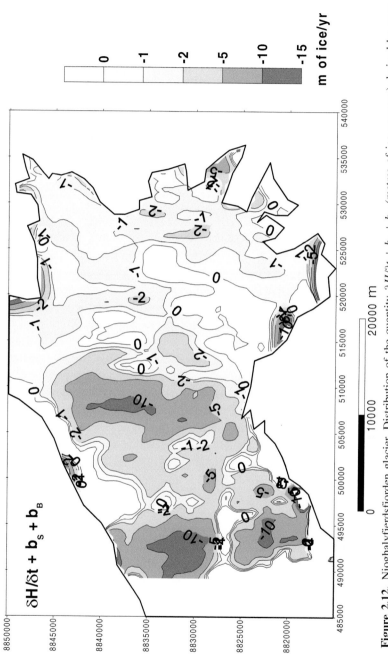

Figure 2.12. Nioghalvfjerdsfjorden glacier. Distribution of the quantity $\partial H/\partial t + b_S + b_B$ (metres of ice per year) derived by mass flux divergence modelling.

stratigraphy and bore-hole inclinometry. By means of equation (2.1), the local balance was calculated as 0.03 ± 0.03 m per year (Reeh and Gundestrup, 1985).

Kääb and Funk (1999) used equation (2.1) to derive the mass balance distributions on the tongue of Griesgletscher in the Swiss Alps. Horizontal velocities and surface-elevation change were measured by photogrammetric methods, and glacier thickness was measured by surface ice radar. The error estimates and comparison with traditional stake measurements showed an average accuracy of ± 0.3 m per year for the calculated vertical velocity and ± 0.7 m per year for the calculated mass balance. The ratio between the column-averaged vertical strain rate and the strain rate at the surface (corresponding to the F-factor in equation (2.1)) was estimated at 0.75.

Application of the local mass balance equation to estimate the distribution of bottom melt rate of the floating Nioghalvfjerdsfjorden glacier is mentioned in section 2.4.6.

2.6 Conclusion

With the increasing capability and precision of remote-sensing methods to measure glacier and ice-sheet volume change and the different mass balance components, the role of *in situ* measurements will probably in the future primarily be to provide ground control for remote-sensing measurements. As described in section 2.4, ice-sheet surface elevation, horizontal surface motion and ice thickness can all be measured from airborne and space-borne plat-forms with high accuracy. However, other quantities that are important for mass balance assessments, for example specific accumulation and ablation rate, cannot yet be determined with sufficient accuracy by remote-sensing measurements, and *in situ* measurements are still needed.

It is also important to notice that the optimal technique used for mass balance mea-surements is different for ice sheets, ice caps and glaciers. Whereas air- and space-borne measurements are definitely needed for studies of the big ice sheets, the use of surface vehicles for kinematic GPS measurement of surface elevation, snow-radar measurement of accumulation rate, and ice-radar measurement of ice thickness may still be an attractive alternative for smaller glaciers and ice caps. It is also worth stressing that, in order not to break the continuity of mass balance records of glaciers that have been studied for a long time with traditional methods, such studies should be continued at least for a period long enough to provide the necessary overlap to a recording period using modern techniques.

In situ studies are also needed to retrieve the long-term mass balance history at specific locations of cold ice caps and ice sheets by ice coring. The ice cores provide seasonal and annual resolution for more than 40 000 years in Greenland and with coarser resolution up to 200 000 years in Antarctica.

References

Andreassen, L. M. 1999. Comparing traditional mass balance measurements with long-term volume change extracted from topographical maps. A case study of

Storbreen glacier in Jotunheimen, Norway, for the period 1940–1977. *Geograf. Ann.* **81 A** (4), 467–76.

Arthern, R. A. and Wingham, D. J. 1998. The natural fluctuations of firn densification and their effect on the geodetic determination of ice sheet mass balance. *Climate Change* **40**, 605–24.

Bauer, A., Baussart, M., Carbonnell, M., Kasser, P., Perraud, P. and Renaud, A. 1968. Missions aériennes de reconnaissance au Groenland 1957–1958. *Meddr. Grønland* **173** (3), 116 pp.

Björnsson, H., Björnsson, S. and Sigurgeirsson, Th. 1982. Penetration of water into hot rock boundaries of magma at Grímsvötn. *Nature* **295**, 580–881.

Carbonnell, M. and Bauer, A. 1968. Exploitation des couvertures photographiques aériennes répétées du front des glacier vêlant dans Disko Bugt et Umanak Fjord, Juin–Juillet 1964. *Meddr. Grønland* **173** (5), 77 pp.

Cogley, J. G. 1999. Effective sample size for glacier mass balance. *Geograf. Ann.* **81 A** (4), 497–507.

Davis, C. H., Kluever, C. A. and Haines, B. J. 1998. Elevation change of the Southern Greenland Ice Sheet. *Science* **279**, 2086–8.

Echelmeyer, K. A. *et al.* 1996. Airborne surface profiling of glaciers: a case-study in Alaska. *J. Glaciol.* **142**, 538–47.

Eiken, T., Hagen, J. O. and Melvold, K. 1997. Kinematic GPS-survey of geometry changes on Svalbard glaciers. *Ann. Glaciol.* **24**, 157–63.

Engeset, R. V., Kohler, J., Melvold, K. and Lundén, B. 2002. Change detection and monitoring of glacier mass balance and facies using ERS SAR winter images over Svalbard. *Int. J. Remote Sensing* **23** (10), 2023–50.

Favey, E., Geiger, A., Gudmundsson, G. H. and Wehr, A. 1999. Evaluating the potential of an airborne laser-scanning system for measuring volume changes of glaciers. *Geograf. Ann.* **81 A** (4), 555–61.

Foldvik, A. and Kvinge, T. 1977. Thermohaline convection in the vicinity of an ice shelf. In Dunbar, M. J., ed., *Polar Oceans*. Proceedings of the Polar Oceans Conference, McGill University, Montreal, May 1974. Calgary, Alberta, Arctic Institute of North America, pp. 247–55.

Fountain, A. G. and Vecchia, A. 1999. How many stakes are required to measure the mass balance of a glacier? *Geograf. Ann.* **81 A** (4), 563–72.

Funk, M., Morelli, R. and Stahel, W. 1997. *Z. Gletscherkd. Glazialg.* **33**, 41–56.

Hagen, J. O., Melvold, K., Eiken, T., Isaksson, E. and Lefauconnier, B. 1999. Mass balance methods on Kongsvegen, Svalbard. *Geograf. Ann.* **81 A** (4), 593–601.

Hamilton, G., Whillans, I. M. and Morgan, P. J. 1998. First point measurements of ice-sheet thickness change in Antarctica. *Ann. Glaciol.* **27**, 125–9.

Hammer, C. U., Clausen, H. B., Dansgaard, W., Gundestrup, N., Johnsen, S. J. and Reeh, N. 1978. Dating of Greenland ice cores by flow models, isotopes, volcanic debris, and continental dust. *J. Glaciol.* **20**, (82), 326.

Hammer, C. U., Clausen, H. B. and Tauber, H. 1986. Ice-core dating of the Pleistocene/Holocene boundary applied to a calibration of the ^{14}C time scale. *Radiocarbon* **28**, (2A), 284–91.

Henriksen, N. 1973. Regional mapping and palaeomagnetic and glaciological investigations in the Scoresby Sund region, Central East Greenland. *Report Grønlands Geol. Unders.* **55**, 42–7.

Higgins, A. K. 1991. North Greenland glacier velocities and calf ice production. *Polarforschung* **60**, 1–23.

Hulbe, C. I. and Whillans, I. M. 1994. A method for determining ice-thickness change at remote locations using GPS. *Ann. Glaciol.* **20**, 263–8.

Isaksson, E., Karlén, W., Gundestrup, N., Mayewski, P., Whitlow, S. and Twickler, M. 1996. A century of accumulation and temperature changes in Dronning Maud Land, Antarctica. *J. Geophys. Res.* **101** (D3), 7085–94.

Jacobsen, F. M. and Theakstone, W. 1997. Monitoring glacier changes using a global positioning system in differential mode. *Ann. Glaciol.* **24**, 314–19.

Jansson, P. 1999. Effect of uncertainties in measured variables on the calculated mass balance of Storglaciären. *Geograf. Ann.* **81 A** (4), 633–42.

Jenkins, A., Vaughan, D. G., Jacobs, S. S., Helmer, H. H. and Keys, J. R. 1997. Glaciological and oceanographic evidence of high melt rates beneath Pine Island Glacier, West Antarctica. *J. Glaciol.* **43** (143), 114–21.

Kääb, A. and Funk, M. 1999. Modelling mass balance using photogrammetric and geophysical data: a pilot study at Griesgletscher Swiss alps. *J. Glaciol.* **45** (151), 575–83.

Kaser, G., Hastenrath, S. and Ames, A. 1996. Mass balance profiles on tropical glaciers. *Z. Gletscherk. Glazialg.* **32**, 91–9.

Kock, H. 1993. Height determinations along the EGIG line and in the GRIP area. In Reeh, N. and Oerter, H., eds., *Mass Balance and Related Topics of the Greenland Ice Sheet.* Open File Series Grønlands Geologiske Undersøgelse, 93/5, pp. 68–70.

Koerner, R. M. 1970. Some observations on superimposition of ice on the Devon Island ice cap, N. W. T. Canada. *Geograf. Ann.* **52A** (1), 57–67.

1986. A new method for using glaciers as monitors of climate. *Mater. Glyatsiol. Issled.* **57**, 175–9.

Kohler, J., Moore, J., Kennett, M., Engeset, R. and Elvehøy, H. 1997. Using ground penetrating radar to image previous years' summer surfaces for mass balance measurements. *Ann. Glaciol.* **24**, 355–60.

Kovacs, A., Gow, A. J. and Morey, R. M. 1995. The in-situ dielectric constant of polar firn revisited. *Cold Regions Sci. & Technol.* **23** (3), 245–56.

Krabill, W. *et al.* 2000: Greenland Ice Sheet: high-elevation balance and peripheral thinning. *Science* **289**, 428–30.

Krimmel, R. M. 1999. Analysis of differences between direct and geodetic mass balance measurements at South Cascade glacier, Washington. *Geograf. Ann.* **81 A** (4), 653–8.

Kuhn, M., Dreiseitl, E., Hofinger, S., Markl, G., Span, N. and Kaser, G. 1999. Measurements and models of the mass balance of Hintereisferner. *Geograf. Ann.* **81 A** (4), 659–70.

Lambrecht, A., Nixdorf, U. and Zürn, W. 1995. Ablation rates under the Ekström Ice Shelf deduced from different methods. *Filchner-Ronne Ice Shelf Programme, report no. 9*, pp. 50–6. Bremerhaven, Alfred-Wegener Institut für Polar- und Meeresforschung.

Meier, M. 1994. Columbia glacier during rapid retreat: interaction between glacier flow and iceberg calving dynamics. In Reeh, N., ed., *Report on the Workshop on the Calving Rate of West Greenland Glaciers in Response to Climate Change.* Copenhagen, Danish Polar Center, pp. 63–84.

1998. Monitoring ice sheets, ice caps and large glaciers. In Haeberli, W., Hoelzle, M. and Suter, S., eds., *Into the Second Century of Worldwide Glacier Monitoring – Prospects and Strategies.* UNESCO, Studies and reports in hydrology no. 56, pp. 209–14.

Melvold, K. and Hagen, J. O. 1998. Evolution of a surge-type glacier in its quiescent phase: Kongsvegen, Spitsbergen, 1964–1995. *J. Glaciol.* **44** (147), 394–404.

Möller, D. 1996. Die Höhen und Höhenänderungen des Inlandeises. Die Weiterfürung der geodätischen Arbeiten der Internationalen Glaziologischen Grönland-Expedition (EGIG) durch das Institut für Vermessungskunde der TU Braunschweig 1987–1993. Deutsche Geodätische Kommission bei der Bayrischen Akademie der Wissenschaften, Reihe B, Angewandte Geodäsie, Heft Nr. 303. Verlag der Bayrischen Akademie der Wissenschaften, pp. 49–58.

Morgan, V. I. 1972. Oxygen isotope evidence for bottom freezing on the Amery Ice Shelf. *Nature* **238** (5364), 321–5.

Nixdorf, U., Rohardt, G., Lambrecht, A. and Oerter, H. 1995. Deployment of oceanographic-glaciological strings under the Filchner-Ronne Schelfeis in 1995. *Filchner-Ronne Ice Shelf Programme, report no. 9.* Bremerhaven, Alfred-Wegener Institut für Polar- und Meeresforschung, pp. 87–90.

Oerter, H. *et al.* 1992. Evidence for basal marine ice in the Filcher-Ronne ice shelf. *Nature* **358**, 399–401.

Olesen, O. B. 1999. The GEUS/AWI programs during Polarstern Cruise XV/2.

Olesen, O. B. and Reeh, N. 1969. Preliminary report on observations in Nordvestfjord, East Greenland. *Report Grønlands Geol. Unders.* **115**, 107–11.

Olesen, O. B., Thomsen, H. H., Reeh, N. and Bøggild, C. E. 1998. Attempts at measuring bottom melting at Nioghalvfjerdsfjorden glacier. In Dowdeswell, J. A., Dowdeswell, E. K. and Hagen, J. O., eds., *International Arctic Science Committee (IASC), Arctic Glaciers Working Group Meeting and Workshop on Arctic Glaciers Mass Balance held at Gregynog, Wales 29–30 January 1998.* Centre for Glaciology Report 98–01. Prifysgol Cymru Aberystwyth, The University of Wales.

Østrem, G. and Brugman, M. 1991. *Glacier and Mass Balance Measurements – A Manual for Field and Office Work.* Canadian National Hydrology Research Institute (NHRI) Science report no. 4.

Østrem, G. and Stanley, A. D. 1969. *Glacier and Mass Balance Measurements – A Manual for Field and Office Work.* Canadian Inland Water Branch, reprint series No. 66.

Paterson, W. S. B. 1994. *The Physics of Glaciers*, 3rd edn. Oxford, Pergamon.

Pillewizer, W. 1939. Die kartographischen und gletscherkundlichen Ergebnisse der Deutschen Spitzbergenexpedition 1938. *Peterm. Geogr. Mitteilungen Erg. H.* **238**, 36–8.

Pinglot, J. F. *et al.* 1999. Accumulation in Svalbard glaciers deduced from ice cores with nuclear tests and Chernobyl reference layers. *Polar Res.* **18** (2), 315–21.

Reeh, N. 1989. Dating by ice flow modelling: a useful tool or an exercise in applied mathematics?. In Oeschger, H. and Langway Jr., C. C., eds., *The Environmental Record in Glaciers and Ice Sheets.* John Wiley & Sons Limited, pp. 141–59.

    1994. Calving from Greenland glaciers: observations, balance estimates of calving rates, calving laws. In Reeh, N., ed., *Report on the Workshop on the Calving Rate of West Greenland Glaciers in Response to Climate Change.* Copenhagen, Danish Polar Center, pp. 85–102.

    1999. Mass balance of the Greenland ice sheet: can modern observation methods reduce the uncertainty? *Geograf. Ann.* **81 A** (4), 735–42.

Reeh, N. and Gundestrup, N. 1985. Mass balance of the Greenland ice sheet at Dye 3. *J. Glaciol.* **31** (108), 198–200.

Reeh, N. and Olesen, O. B. 1986. Velocity measurements on Daugaard-Jensen Gletscher, Scoresby Sund, East Greenland. *Ann. Glaciol.* **8**, 146–50.

Richardson, C. E. and Holmlund, P. 1999. Spatial variability at shallow snow depths in central Dronning Maud Land, East Antarctica. *Ann. Glaciol.* **29**, 10–16.

Richardson, C., Aarholt, E., Hamran, S. E., Holmlund, P. and Isaksson, E. 1997. Spatial distribution of snow in western Dronning Maud Land, east Antarctica, mapped by a ground-based snow radar. *J. Geophys. Res.* **102** (B9), 20 343–53.

Richardson-Näslund, C. 2001. Spatial distribution of snow in Antarctica and other glaciers using ground-penetrating radar. Doctoral dissertation, Department of Physical Geography and Quaternary Geology, Stockholm University: Dissertation series, no. 18.

Rignot, E. J., Gogineni, S. P., Krabill, W. B. and Ekholm, S. 1997. North and Northeast Greenland ice discharge from satellite radar inferometry. *Science* **276**, 934–7.

Robin, G. de Q. 1979. Formation, flow, and disintegration of ice shelves. *J. Glaciol.* **24** (90), 259–71.

Seckel, H. 1977. Das geometrische Nivellement über das Grönländische Inlandeis der Gruppe Nivellement an der internationalen glaziologischen Grönland-Expedition 1967–68. *Meddr. Grønland* **187** (3).

Steffensen, J. P. 1988. Analysis of the seasonal variation in dust, Cl^-, NO_3^-, and SO_4^{2-} in two Central Greenland firn cores. *Ann. Glaciol.* **10**, 171–7.

Thomsen, H. H. *et al.* 1997. The Nioghalvfjerdsfjorden glacier project, North-East Greenland: a study of ice sheet response to climatic change. *Geol. Greenland Surv. Bull.* **176**, 95–103.

Van der Veen, C. J. 1993. Interpretation of short-term ice sheet elevation changes inferred from satellite altimetry. *Climate Change* **23**, 383–405.

Ward, W. H. and Orvig, S. 1953. The glaciological studies of the Baffin Island Expedition, 1950. Part IV: The heat exchange at the surface of the Barnes Ice Cap during the ablation period. *J. Glaciol.* **2** (13), 158–68.

Warrick, R. A., Provost, C. le, Meier, M. F., Oerlemans, J. and Woodworth, P. L. 1996. Changes in sea level. In Houghton, J. T. *et al.*, eds., *Climate Change 1995*. Cambridge University Press, pp. 358–405.

Weidick, A. 1968. Observation on some Holocene glacier fluctuations in West Greenland. *Meddr. Grønland* **165** (6).

1994. Fluctuations of West Greenland calving glaciers. In Reeh, N., ed., *Report on the Workshop on the Calving Rate of West Greenland Glaciers in Response to Climate Change*. Copenhagen, Danish Polar Center, pp. 141–68.

1995. *Satellite Image Atlas of Glaciers of the World, Greenland*. US Geological Survey Professional Paper 1386-C. Washington, United States Government Printing Office.

Weidick, A., Andreasen, C., Oerter, H. and Reeh, N. (1996). Neoglacial glacier changes around Storstrømmen, North-East Greenland. *Polarforschung* **64** (3), 95–108.

Woodward, J., Sharp, M. and Arendt, A. 1997. The influence of superimposed-ice formation on the sensitivity of glacier mass balance to climate change. *Ann. Glaciol.* **24**, 186–90.

3

In situ measurement techniques: sea ice

PETER WADHAMS

Department of Applied Mathematics and Theoretical Physics,
University of Cambridge

3.1 Current techniques

We can learn a great deal about sea ice from satellite and aircraft surveys – its extent, its type, its surface features. But its thickness is hard to measure by remote sensing, because the brine cells in the ice give it a high electrical conductivity such that electromagnetic waves do not easily penetrate. The radio-echo sounding methods which have been used to measure the thickness of terrestrial ice sheets and glaciers cannot therefore be used for sea ice.

So far, five direct techniques have been commonly employed for measuring ice-thickness distribution. In decreasing order of total data quantity, they are:

(1) submarine sonar profiling;
(2) moored upward sonars;
(3) airborne laser profilometry;
(4) airborne electromagnetic techniques;
(5) drilling.

3.1.1 Submarine sonar profiling

Most synoptic data to be published so far have been obtained by upward sonar profiling from submarines. Beginning with the 1958 voyage of *Nautilus* (Lyon, 1961; McLaren, 1988), which was the first submarine to the North Pole, many tens of thousands of kilometres of profile have been obtained in the Arctic by US and British submarines, and our present knowledge of Arctic ice-thickness distributions derives largely from the analysis and publication of data from these cruises. Problems include the necessity of removing the effect of beamwidth where a wide-beam sonar has been employed (Wadhams, 1981), and the fact that the data are sometimes obtained during military operations, which necessitates restrictions on the publication of exact track lines. For the same reason, the data set is not systematic in time or space.

Mass Balance of the Cryosphere: Observations and Modelling of Contemporary and Future Changes, eds. Jonathan L. Bamber and Antony J. Payne. Published by Cambridge University Press. © Cambridge University Press 2003.

The chief advantage of upward sonar (mobile or moored) as a way of generating ice-thickness distributions is that it is still the only direct and accurate means of measuring the draft of sea ice, from which the thickness distribution can be inferred with very little error, while submarine-mounted sonar allows basin-scale surveys to be carried out on a single cruise, giving the geographical variation in ice-thickness characteristics. Submarine-mounted sonar also permits the shape of the ice bottom to be determined accurately, including pressure ridges and the roughness of undeformed multi-year ice, allowing spectral and fractal studies to be undertaken and an understanding of the mechanics of the ridge-building process to be gained. By the use of additional sensors and concurrent airborne studies, a submarine can be used as a powerful vehicle for validating remote-sensing techniques, including laser, passive and active microwave. The chief drawbacks of submarines are that they cannot generate a systematic time series of ice thickness at a point in space, and they cannot carry out surveys safely in very shallow water, so that many interesting aspects of ice deformation near shore cannot be studied. For instance, the data set obtained by USS *Gurnard* north of Alaska in 1976 and analysed by Wadhams and Horne (1980) began at the 100 m isobath and so did not cover the whole of the Alaskan shear zone. Finally, it is unlikely that a military submarine would be available in the Antarctic, both because of remoteness and because the Antarctic Treaty requires that military vessels used in the Antarctic must be available for international inspection. The Antarctic is, however, a very suitable region for the use of sonar on an autonomous underwater vehicle (AUV).

3.1.2 Moored upward sonars

The second technique is the use of upward sonar mounted on moorings, so as to obtain a time series of $g(h)$ at a fixed location. Experiments have involved bottom-mounted systems in shallow water in the Beaufort Sea (Hudson, 1990; Melling and Riedel, 1995; Pilkington and Wright, 1991) and Chukchi Sea (Moritz, 1991), and systems in deeper water in the Fram Strait and the southern Greenland Sea (Vinje, 1989). Work of this kind has been taking place since 1991 using lines of sonars which span the East Greenland current in the Fram Strait, at 75 °N and in the Denmark Strait. In conjunction with the use of advanced very high resolution radiometer (AVHRR) or ERS-1 synthetic aperture radar (SAR) imagery to yield ice velocity vectors, this technique permits the time dependence of ice mass flux to be measured, and hence the fresh-water input to the Greenland Sea at different latitudes (Kwok and Rothrock, 1999; Vinje, Nordlund and Kvambekk, 1998). The observations involve mounting the sonars at 50 m depth (together with a current meter) in water up to 2500 m deep on a taut wire mooring. Similar work has been done in the Weddell Sea, Antarctica, using a greater mooring depth and an armoured cable to protect against iceberg damage (Strass and Fahrbach, 1998).

3.1.3 Airborne laser profilometry

Laser profiling of sea ice in the Arctic Ocean has been carried out extensively during three decades (e.g. Ketchum, 1971; Krabill, Swift and Tucker, 1990; Tucker, Weeks and

Frank, 1979; Wadhams, 1976; Weeks, Kovacs and Hibler, 1971), while limited studies have also been carried out in the Antarctic (Weeks, Ackley and Govoni, 1989). The aim has been to delineate the frequency and height distributions of pressure ridge sails and the spatial distribution of surface roughness. On only two occasions has it been posssible to match a laser profile against a coincident profile of ice draft over substantial lengths of joint track. The first was a joint aircraft–submarine experiment (Lowry and Wadhams, 1979; Wadhams, 1980, 1981; Wadhams and Lowry, 1977), and the second was a similar experiment in May 1987 which involved a NASA P-3A aircraft equipped with an airborne oceanographic lidar (AOL) and a British submarine equipped with narrow-beam, upward looking sonar (Comiso *et al.*, 1991; Wadhams, 1990; Wadhams *et al.*, 1991). Only the more recent experiment permitted a direct comparison of the probability density functions (pdfs) of draft and freeboard to be made, since the AOL has a superior capability over earlier lasers in the removal of the sea-level datum from the record. Comiso *et al.* (1991) found from this experiment that over a 60 km sample of track the overall pdfs of ice freeboard and draft could be brought to a close match across the entire range of data by a simple co-ordinate transform of the AOL data based on the ratio of mean densities of ice and water. Specifically, they showed that if R is the ratio of mean draft to mean freeboard, then matching of the freeboard pdf with the draft pdf is achieved by expanding the elevation scale of the freeboard pdf by a factor of R, and diminishing the magnitude of the pdf per metre by the same factor. This is equivalent to saying that if a fraction $F(h)$ of the ice cover has an elevation in the range h to $(h + dh)$, then the same fraction $F(h)$ will have a draft in the range Rh to $R(h + dh)$. R is related to mean material density (ice plus snow) ρ_m and near-surface water density ρ_w by

$$R = \rho_m/(\rho_w - \rho_m). \tag{3.1}$$

The success of this correlation prompted an analysis of the entire 300 km of coincident track (Wadhams *et al.*, 1992) divided into six 50 km sections, all from the north of Greenland, within the zone 80.5–85°N, 2–35°W. The results of the analysis were as follows.

(i) Despite variations in mean draft from 3.6 to 6 m, the six values of R all lay within a narrow range, of mean 8.04 ± 0.19. This corresponds to a mean material density of $910.7 \pm 2.3 \, \text{kg/m}^3$.

(ii) When each section was subjected to a co-ordinate transform based on its own value of R, the pdfs matched the sonar pdfs extremely well when plotted on a semi-logarithmic scale (Figure 3.1).

(iii) When plotted on a linear scale, the agreement was less good, in that mid-range depth probabilities are enhanced by the transformation, while very thin and very thick ice probabilities are reduced (Figure 3.2). This is comprehensible on the basis of considering a uniform snow cover, which would give a low value of R for thin ice and a high value for thick ice. The use of a single average value for the transformation causes thin ice to be moved into thicker categories, and thick ice into thinner, thus making the converted distribution more narrow and peaked than the real ice draft distribution. In principle, a freeboard-dependent R could be used for the transform, but this would involve a sacrifice of simplicity.

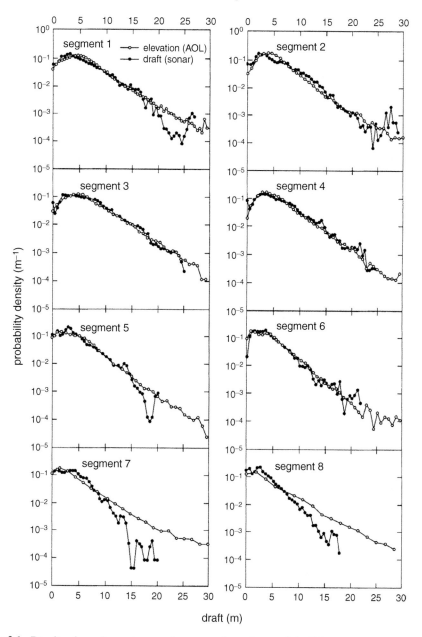

Figure 3.1. Results of carrying out a co-ordinate transformation on eight 50 km sections of coincident laser and sonar track, using the mean draft–elevation ratio. Segments 7 and 8, with a poor fit, did not comprise coincident data. (After Wadhams *et al.* (1992).)

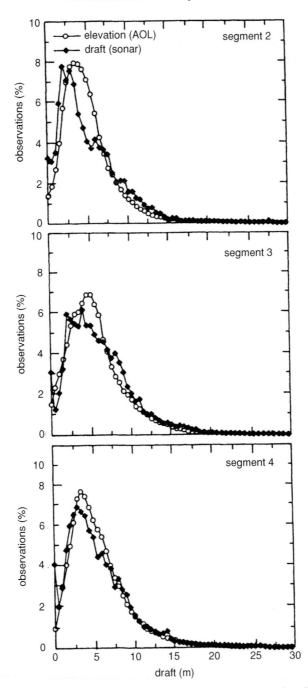

Figure 3.2. Distributions of observed drafts and drafts predicted from laser elevations using a mean draft–elevation ratio of 7.89. (After Wadhams *et al.* (1992).)

The question arises of how R might vary with time of year and location. It is difficult to develop even a simple model, since snow thickness on Arctic and Antarctic sea ice is poorly known, and there are few systematic measurements of ice density, especially of any fundamental difference between the densities of first- and multi-year ice. Near-surface water density must be known to high accuracy, since R depends on the difference between water and ice density, and we know that it diminishes during summer because of dilution by melt water. One set of results, applicable to the Arctic Basin, is shown in Figure 3.3 based on the seasonal snow depth assumptions that were used in the Maykut–Untersteiner model. We see that there is a large seasonal variation in R, mainly due to snow load. R is at its highest value on August 20, at the start of the snow season, with bare ice. R falls rapidly during the autumn snow falls of September and October, then diminishes only very slowly between November and the end of April, when little snow falls. A further onset of spring snow brings R to its lowest level at the beginning of June. As soon as surface snow melt begins, R rises rapidly until, by the end of June, it has risen again almost to its August value. The final slow drift is due to surface water dilution. In fact there will be one or more higher peaks for R during the summer period, as melt-water pools form on the ice surface (increasing R), then drain (decreasing R), but no useful data exist on mean melt-water pool coverage, mean depth, or draining dates. In any case the results show that snow load is a critical parameter, and that the best time to carry out surveys is when dR/dt is at its lowest value, i.e. between November and the end of April.

We can conclude that it is indeed worthwhile considering the use of an airborne laser as a way of surveying both the mean ice thickness and the thickness distribution over the Arctic. The model suggests that a survey during early spring would be most useful, with annual repetition in order to examine inter-annual fluctuations and trends (possibly climate-induced) in the mean ice thickness or the form of the distribution. However, better data are required on the seasonal and spatial variability of snow load, ice density and surface water density before full confidence can be attached to the results. Even in the absence of improved background data, the present results suggest that in regions with mean ice thickness in the range 4–6 m, the ratio R will lie in the vicinity of 8.0 in spring, and can be estimated to an accuracy of ±2.4% in 300 km of track. This yields an accuracy of about ±12 cm in mean thickness over 300 km of track (30 cm in 50 km), neglecting other sources of error. This is an acceptable accuracy for detecting and mapping variability. In the Antarctic, much more validation would be needed before a laser could be employed in this way, since snow has a much stronger influence on mean density than in the Arctic, and there is a major difference between snow load on first-year and multi-year ice (since the snow cover does not all melt during summer). However, if such validation were done, this technique could be a highly effective mapping tool for the vast expanse of Antarctic sea-ice cover in winter. Most valuable of all for synoptic purposes would be a laser mounted in a satellite. Such a system (ICESat) has been launched by NASA.

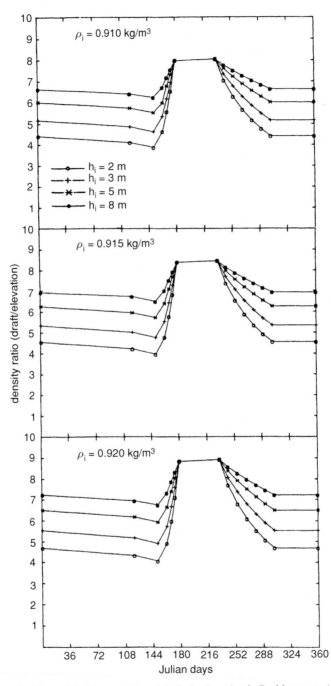

Figure 3.3. Results of a model of the variation of draft–freeboard ratio R with season. (After Wadhams *et al.* (1992).)

3.1.4 Airborne electromagnetic techniques

The first electromagnetic technique to be applied to sea ice was impulse radar (Kovacs and Morey, 1986), in which a nanosecond pulse (centre frequency about 100 MHz) is applied to the ice via a paraboloidal antenna on a low and slow flying helicopter. The technique was found to have many limitations besides the slowness of data gathering. Superimposition of surface and bottom echoes means that it does not function well over ice less than 1 m thick, while absorption and scattering by brine cells means that better results are obtained over multi-year than first-year ice, with, in any case, a fading of the return signal at depths beyond 10–12 m. Thus the full profile of deeper pressure ridges is not obtained. Such instruments can therefore be regarded as of restricted application to local surveys in ice regimes of a favourable kind.

A more recent development is the use of electromagnetic induction. The technique was devised by Aerodat Ltd of Toronto, and has subsequently been developed by CRREL (Kovacs and Holladay, 1990) and by Canpolar, Toronto (Holladay, Rossiter and Kovacs, 1990). The technique involves towing a 'bird' behind a helicopter flying at normal speed. The bird contains a coil which emits an electromagnetic field in the frequency range 900 Hz–33 kHz, inducing eddy currents in the water under the ice, which in turn generate secondary electromagnetic fields. The secondary fields are detected by a receiver in the bird; their strength depends on the depth of the ice–water interface below the bird. The bird also carries a laser profilometer to measure the depth of the ice–air interface below the bird, and the difference gives the absolute thickness of the ice. The method appears promising, but requires further validation. Figure 3.4 illustrates the principle and shows some results obtained in the Labrador Sea during the LIMEX experiment (Holladay *et al.*, 1990). A similar system has been fitted into a Twin Otter aircraft of the Finnish Geological Survey and used in the Baltic (Multala *et al.*, 1995).

3.1.5 Drilling

Drilling is the traditional method of measuring ice thickness. The first systematic measurements of ice thickness in the Arctic were made by Nansen (1897), who drilled through undeformed ice during the drift of 'Fram'. Many methods have been used: manual drilling is the most painful, since in very thick ice a large number of extensions are needed to the drill, and if the bit sticks there is little that the driller can do except spend many hours chipping downwards with a chisel to free it. The use of a gasoline-powered head is an improvement, while the most rapid technique is the hot-water drill, where water is heated in a boiler and pumped through a hose and out as a jet through a heavy bronze probe, which thus melts its way quickly down through the ice. A tape with a self-opening set of scissors at the bottom is sent down the hole, and the draft, the ice thickness and the snow thickness are read off as quickly as possible before the hole refreezes.

Drilling as a technique was considered statistically by Rothrock (1986), who estimated that 62 independent random holes would give a mean thickness with a standard deviation

(a)

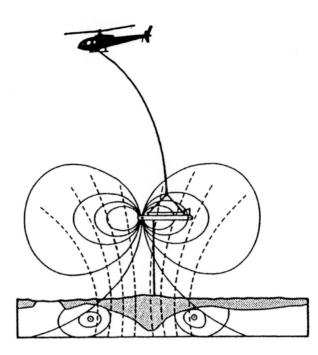

(b)

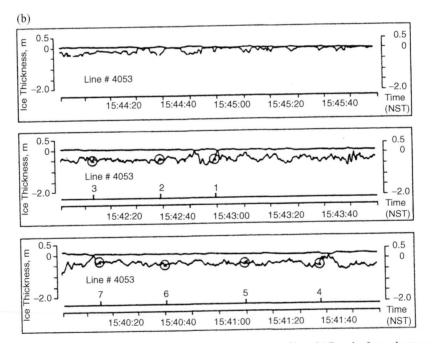

Figure 3.4. (a) Principle of electromagnetic induction sounding of ice. (b) Results from electromagnetic induction sounding over first-year ice in the Labrador Sea. (After Holladay *et al.* (1990).)

of 30 cm, while 560 holes would give a 10 cm error. Eicken and Lange (1989) followed this approach, and obtained a reasonable approximation to an Arctic ice-thickness distribution. In general, however, we must conclude that the technique is reasonable as a means of estimating mean ice thickness, but is poor as a way of giving the shape of $g(h)$. It is an essential validation for any other technique.

3.2 Possible future techniques

There are several new techniques which show some promise for measuring $g(h)$, or parts of it, under certain conditions. We may mention the following:

(1) sonar mounted on AUVs or neutrally buoyant floats;
(2) acoustic tomography or thermometry;
(3) inference from a combination of microwave sensors.

3.2.1 Sonar on AUVs and floats

The purpose of mounting upward sonar systems on mobile platforms other than military submarines is to obtain systematic data sets along a repeatable grid of survey tracks, enabling inter-seasonal and inter-annual comparisons to be carried out for identical geographical locations. Possible platforms with synoptic potential include autonomous underwater vehicles (AUVs), of which several are under development, including long-range systems with basin-wide capability (e.g. the Danish Maridan vehicles and the UK Autosub); long-range civilian manned submersibles, of which the Canadian–French *Saga I* was an example (Grandvaux and Drogou, 1989); and neutrally buoyant floats, on which an upward sonar could be mounted to generate and store a pdf which can then be transmitted acoustically to an Argos readout station on a floe.

3.2.2 Acoustic tomography

Acoustic tomography is a technique for monitoring the structure of the ocean within an area of order $10^6 \, km^2$ by measuring acoustic travel times between the elements of a transducer array enclosing that area (Munk and Wunsch, 1979). Guoliang and Wadhams (1989) showed in a theoretical study that the presence of a sea-ice cover should measurably decrease travel times by an amount which is dependent on the modal ice thickness. The way that this works is that in polar seas sound is always refracted upward toward the surface, since sound velocity increases monotonically from the surface downward. The sound rays are reflected downward, then arch back upward and undergo a number of such 'bounces' before reaching the receiver. If the sea surface is covered by ice, each 'bounce' occurs at a depth equal to the ice draft, reducing the travel path for the sound ray. There are complications caused by the phase change suffered by the ray at reflection, but on the whole the travel time should decrease as the ice thickness increases.

Data were obtained during the 1988–89 Scripps-WHOI tomography experiment in the Greenland Sea (Jin *et al.*, 1993), and this technique may have a more general application within the Arctic Ocean itself, although the additional scattering due to under-ice roughness has been found to have an important limiting effect on resolution. A development of tomography is called ATOC (acoustic thermometry of ocean climate), which uses much lower frequencies (e.g. 57 Hz) to achieve greater transmission lengths between a single transmitter and receiver (Johannessen *et al.*, 1993), and this may allow the technique to be used to estimate a modal ice draft for the Arctic Basin. The first experiments on transmitting sound across the Arctic Basin from north of Svalbard to an ice camp in the Beaufort Sea were conducted successfully in April, 1994 (Michalevsky, Gavrilov and Baggeroer, 1999).

3.2.3 *The use of microwave sensors*

Possible ice sounding techniques involving microwave sensors have been reviewed by Wadhams and Comiso (1992). In principle, after sufficient validation, we might hope to find empirical relationships between the distribution of passive microwave brightness temperatures, of SAR backscatter levels, and of aspects of $g(h)$ such as the mean thickness. To date, however, the only quantitative validation of a microwave sensor against ice thickness has been a 1987 joint survey between a submarine, an aircraft equipped with the STAR-2 X-band SAR system, and a second aircraft equipped with a laser profilometer and passive microwave radiometers. The SAR system operates at HH polarization and 9.6 GHz, with a swath width of 63 km and resolution of 16.8 m. This provided an opportunity to examine correlations between ice draft, as measured by the sonar, and backscatter level along the same track measured by the SAR. The question which can be addressed is: To what extent can SAR brightness variability be used to infer the shape of the ice thickness distribution? The results were discussed in Comiso *et al.* (1991) and Wadhams *et al.* (1991).

In these studies an initial qualitative examination of the profile of SAR backscatter along the tracks of the submarine and aircraft showed that there was a clear positive correlation with both the draft and elevation profiles. This is to be expected since pressure ridges in particular give strong returns on account of their geometry (Burns *et al.*, 1987; Livingstone, 1989; Onstott *et al.*, 1987). Next, 125 km of matched SAR and laser data were examined. With each set of data averaged over a window length of 1 km, the correlation coefficient between elevation and SAR backscatter reached 0.51. Figure 3.5 shows the mean elevation and mean backscatter using this window. Clearly there are regions of both good and bad correlation, the best correlation often appearing to occur in areas of low elevation. This is probably because these correspond to open water, young ice and first-year ice, which on X-band SAR offer a high contrast with multi-year and ridged ice. Evidence indicates that SAR backscatter for ridges is dependent on look-angle and angle of incidence (Leppäranta

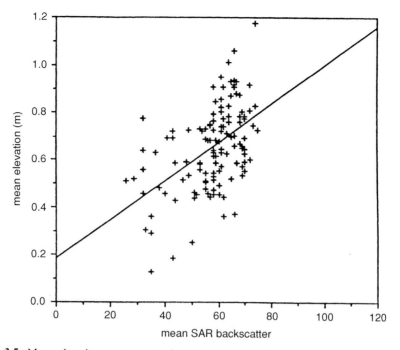

Figure 3.5. Mean elevation versus mean SAR backscatter with best-fit regression line for 1 km averaging window.

and Thompson, 1989), while other results (Holt, Crawford and Carsey, 1990) show that lower frequency SAR (0.44 GHz) may well be superior in its resolution of ridges to high frequency SAR such as X-band.

The correlation between SAR backscatter and sonar ice draft was carried out over a more restricted 22 km section of track (Wadhams *et al.*, 1991) where there was excellent matching between tracks. Once again it was found that the correlation coefficient depends on the window length, but in fact the highest correlation was obtained with a lower window length. At an averaging length of 252 m (15 SAR pixels), the correlation between draft and backscatter reached 0.68, which is better than the best correlation with elevation, and which implies that 46% of the backscatter variance can be explained by draft differences. Figure 3.6 shows the scatter plot of draft versus SAR backscatter.

It is clear that SAR brightness alone cannot be used to infer the complete shape of the ice pdf. Firstly, less than half of the variance of the SAR brightness can be explained by draft variations, so it cannot be used alone as a predictor. Secondly, this correlation is developed over averaging lengths of 252 m, indicating that to some extent we are relating a mean ice draft to a mean SAR backscatter, rather than obtaining an algorithm which can generate a fine-resolution pdf as in the case of the laser technique. Thirdly, the experiment described above was specific to X-band SAR, whereas the types of satellite SAR which produce routine synoptic data from the polar oceans are of different frequencies (e.g. C-band, 5.3 GHz, on

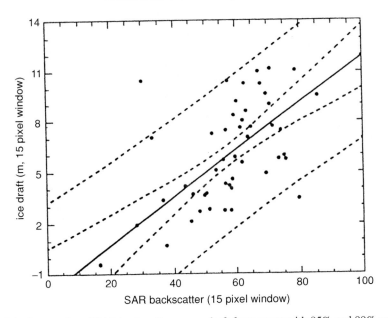

Figure 3.6. Scatter plot of SAR backscatter versus draft from sonar with 95% and 99% confidence limits.

ERS-1). Nevertheless, the results show some promise for the use of SAR, in conjunction with other sensors, as an element in an empirical scheme for determining ice thickness using microwave backscatter or emission properties as a proxy. A recent approach toward such a scheme, by Kwok *et al.* (1999), uses SAR to track ice velocity and define ice types, and an ice model to predict growth and melt rates, in this way obtaining an ice thickness distribution and following its evolution.

It is possible that with further validation we may be able to use microwave signatures as a way of defining ice roughness. This does not necessarily mean establishing a more perfect correlation between ice thickness and SAR brightness – it is likely that the level of correlation obtained so far is about the limit of the natural correlation which really exists. It is more likely that progress will come from improving our ability to discriminate between multi-year and first-year ice in SAR images or in passive microwave. It is recognized that multi-year ice is characterized on SAR by a greater average brightness and a more speckled texture, due to the greater surface roughness and greater variability in volume scattering. Kerman (1998) suggested that by examining the distribution of brightness *differences* between neighbouring pixels a distribution might be obtained which bears a quantitative relationship to the ice thickness distribution. Some success was achieved with X-band airborne SAR, but tests of such an approach with ERS-2 data proved inconclusive (Doble and Wadhams, 1999), possibly because of the intrinsic speckle level in ERS-2 SAR. In the case of passive microwave, there was developed an approach to a finer scale ice classification by using a cluster analysis of data obtained from all seven frequency–polarization combinations in the special sensor microwave images (SSM/I), allowing the Arctic or Antarctic to be divided into radiometrically distinct 'ice regimes' (see Chapter 4). The possibility exists that each regime may correspond to a distinct mean ice-thickness value, allowing this technique to be used as a proxy for mean thickness or roughness.

References

Burns, B. A. *et al.* 1987. Multisensor comparison of ice concentration estimates in the marginal ice zone. *J. Geophys. Res.* **92** (C7), 6843–56.

Comiso, J. C., Wadhams, P., Krabill, W.B., Swift, R.N., Crawford, J.P. and Tucker, W.B. III. 1991. Top/bottom multisensor remote sensing of Arctic sea ice. *J. Geophys. Res.* **96** (C2), 2693–709.

Doble, M. and Wadhams, P. 1999. Analysis of concurrent SAR images and submarine ice draft profiles in the Arctic Ocean. *Proceedings of POAC'89, Port & Ocean Engineering under Arctic Conditions.* Helsinki University of Technology.

Eicken, H. and Lange, M. A. 1989. Sea ice thickness: the many vs the few. *Geophys. Res. Lett.* **16** (6), 495–8.

Grandvaux, B. and Drogou, J.-F. 1989. Saga 1, une première etape vers les sous-marins autonomes d'intervention. International Colloquium: *Arctic Technology and Economy* – Present Situation, Problems and Future Issues, Paris, Feb. 15–17, 1989. Paris, Banque Nationale de Paris. (In French with English summary.)

Guoliang, J. and Wadhams, P. 1989. Travel time changes in a tomography array caused by a sea ice cover. *Prog. Oceanography* **22** (3), 249–75.

Holladay, J. S., Rossiter, J. R. and Kovacs, A. 1990. Airborne measurement of sea ice thickness using electromagnetic induction sounding. In Ayorinde, O. A. *et al.*, eds., *Proceedings of the 9th International Conference on Offshore Mechanical and Arctic Engineering.* American Society of Mechanical Engineers, pp. 309–15.

Holt, B., Crawford, J. and Carsey, F. 1990. Characteristics of sea ice during the Arctic winter using multifrequency aircraft radar imagery. In Ackley, S. F. and Weeks, W. F., eds., *Sea Ice Properties and Processes.* Monograph 90–1, Hanover, NH, US Army Cold Regions Research and Engineering. Laboratory, p. 224 (abstract).

Hudson, R. 1990. Annual measurement of sea-ice thickness using an upward-looking sonar. *Nature* **344**, 135–7.

Jin, Guoliang, Lynch, J. F., Pawlowicz, R., Wadhams, P. and Worcester, P. 1993. Effects of sea ice cover on acoustic ray travel times, with application to the Greenland Sea Tomography Experiment. *J. Acoust. Soc. Am.* **94** (12), 1044–56.

Johannessen, O. M. *et al.* 1993. *ATOC – Arctic: Acoustic Thermometry of the Ocean Climate in the Arctic Ocean.* Report by the Nansen Environmental and Remote Sensing Centre, Bergen.

Kerman, B. R. 1998. On the relationship of pack ice thickness to the length of connectivity trees in SAR imagery. In Shen, H.T., ed., *Ice in Surface Waters*, vol. 2. Potsdam, NY, Clarkson University.

Ketchum, R. D. 1971. Airborne laser profiling of the Arctic pack ice. *Remote Sensing Environ.* **2**, 41–52.

Kovacs, A. and Holladay, J. S. 1990. Airborne sea ice thickness sounding. In Ackley, S. F. and Weeks, W. F., eds., *Sea Ice Properties and Processes.* Monograph 90–1, Hanover, NH, US Army Cold Regions Research and Engineering Laboratory, pp. 225–9.

Kovacs, A. and Morey, R. M. 1986. Electromagnetic measurements of multiyear sea ice using impulse radar. *Cold Regions Sci. Technol.* **12**, 67–93.

Krabill, W. B., Swift, R. N. and Tucker, W. B. III 1990. Recent measurements of sea ice topography in the Eastern Arctic. In Ackley, S. F. and Weeks, W. F., eds., *Sea Ice Properties and Processes.* Monograph 90–1, Hanover, NH, US Army Cold Regions Research and Engineering Laboratory, pp. 132–6.

Kwok, R. and Rothrock, D. A. 1999. Variability of Fram Strait ice flux and North Atlantic Oscillation. *J. Geophys. Res.* **104** (C3), 5177–89.

Kwok, R., Cunningham, G. F., LaBelle-Hamer, N., Holt, B. and Rothrock, D. 1999. Ice thickness derived from high-resolution radar imagery. *EOS, Trans. Am. Geophys. U.* **80** (42), pp. 495, 497.

Leppäranta, M. and Thompson, T. 1989. BEPERS-88 sea ice remote sensing with synthetic aperture radar in the Baltic Sea. *EOS, Trans. Am. Geophys. U.* **70** (28), 698–9, 708–9.

Livingstone, C. E. 1989. Combined active/passive microwave classification of sea ice. *Proc. IGARSS-89* **1**, 376–80.

Lowry, R. T. and Wadhams, P. 1979. On the statistical distribution of pressure ridges in sea ice. *J. Geophys. Res.* **84** (C5), 2487–94.

Lyon, W. K. 1961. Ocean and sea-ice research in the Arctic Ocean via submarine. *Trans. N.Y. Acad. Sci.*, series 2, **23**, 662–74.

McLaren, A. S. 1988. Analysis of the under-ice topography in the Arctic Basin as recorded by the USS *Nautilus* during August 1958. *Arctic* **41** (2), 117–26.

Melling, H. and Riedel, D. A. 1995. The underside topography of sea ice over the continental shelf of the Beaufort Sea in the winter of 1990. *J. Geophys. Res.* **100** (C7), 13 641–53.

Mikhalevsky, P. N., Gavrilov, A. N. and Baggeroer, A. B. 1999. The trans-arctic acoustic propagation experiment and climate monitoring in the Arctic. *IEEE J. Oceanic Engng.* **24** (2), 183–201.

Moritz, R. E. 1991. Sampling the temporal variability of sea ice draft distribution. *EOS supplement*, fall AGU Meeting, pp. 237–8 (abstract).

Multala, J. *et al.* 1995. Airborne electromagnetic surveying of Baltic sea ice. University of Helsinki, Department of Geophysics, Report Series in Geophysics no. 31, 58pp.

Munk, W. H. and Wunsch, C. 1979. Ocean acoustic tomography: a scheme for large scale monitoring. *Deep-Sea Res.* **26**, 123–61.

Nansen, F. 1897. *Farthest North.*, vol. 1. London, Constable, p. 299.

Onstott, R. G., Grenfell, T. C., Maetzler, C., Luther, C. A. and Svendsen, E. A. 1987. Evolution of microwave sea ice signatures during early and mid summer in the marginal ice zone. *J. Geophys. Res.* **92** (C7), 6825–37.

Pilkington, G. R. and Wright, B. D. 1991. Beaufort Sea ice thickness measurements from an acoustic, under ice, upward looking ice keel profiler. *Proceedings of the 1st International Offshore & Polar Engineering Conference.* Edinburgh, 11–16 Aug. 1991.

Rothrock, D. A. 1986. Ice thickness distribution – measurement and theory. In Untersteiner, N., ed., *The Geophysics of Sea Ice.* New York, Plenum, pp. 551–75.

Strass, V. H. and Fahrbach, E. 1998. Temporal and regional variation of sea ice draft and coverage in the Weddell Sea obtained from upward looking sonars. In Jeffries, M. O., ed., *Antarctic Sea Ice. Physical Processes, Interactions and Variability.* Antarctic Research Series 74. Washington, American Geophysical Union, pp. 123–40.

Tucker, W. B. III, Weeks, W. F. and Frank, M. 1979. Sea ice ridging over the Alaskan continental shelf. *J. Geophys. Res.* **84** (C8), 4885–97.

Vinje, T. E. 1989. An upward looking sonar ice draft series. In Axelsson, K. B. E. and Fransson, L. A., eds., *Proceedings of the 10th International Conference on Port & Ocean Engineering under Arctic Conditions*, vol.1. Luleå University of Technology, pp. 178–87.

Vinje, T., Nordlund, N. and Kvambekk, A. 1998. Monitoring ice thickness in Fram Strait. *J. Geophys. Res. Oceans.* **103** (C5), 10 437–49.

Wadhams, P. 1976. Sea ice topography in the Beaufort Sea and its effect on oil containment. *AIDJEX Bull.* **33**, 1–52. Division of Marine Resources, University of Washington, Seattle.

1980. A comparison of sonar and laser profiles along corresponding tracks in the Arctic Ocean. In Pritchard, R. S., ed., *Sea Ice Processes and Models*. Seattle, University of Washington Press, pp. 283–99.

1981. Sea ice topography of the Arctic Ocean in the region 70°W to 25°E. *Phil. Trans. Roy. Soc., Lond.*, **A302** (1464), 45–85.

1990. Evidence for thinning of the Arctic ice cover north of Greenland. *Nature* **345**, 795–7.

Wadhams, P. and Comiso, J. C. 1992. The ice thickness distribution inferred using remote sensing techniques. In Carsey, F., ed., *Microwave Remote Sensing of Sea Ice*. Geophysical Monograph 68, Washington, American Geophysical Union, ch. 21, pp. 375–83.

Wadhams, P. and Horne, R. J. 1980. An analysis of ice profiles obtained by submarine sonar in the Beaufort Sea. *J. Glaciol.* **25** (93), 401–24.

Wadhams, P. and Lowry, R. T. 1977. A joint topside-bottomside remote sensing experiment on Arctic sea ice. *Proceedings of the 4th Canadian Symposium on Remote Sensing*. Quebec, 16–18 May, 1977. Canadian Remote Sensing Society, pp. 407–23.

Wadhams, P. *et al.* 1991. Concurrent remote sensing of Arctic sea ice from submarine and aircraft. *Int. J. Remote Sensing* **12** (9), 1829–40.

Wadhams, P., Tucker, W. B. III, Krabill, W. B., Swift, R. N., Comiso, J. C. and Davis, N. R. 1992. Relationship between sea ice freeboard and draft in the Arctic Basin, and implications for ice thickness monitoring. *J. Geophys. Res.* **97** (C12), 20 325–34.

Weeks, W. F., Kovacs, A. and Hibler, W. D. III 1971. Pressure ridge characteristics in the Arctic coastal environment. In Wetteland, S. S. and Bruun, P., eds., *Proceedings of the 1st International Conference on Port and Ocean Engineering under Arctic conditions*. Trondheim, Technical University of Norway, pp. 152–83.

Weeks, W. F., Ackley, S. F. and Govoni, J. 1989. Sea ice ridging in the Ross Sea, Antarctica, as compared with sites in the Arctic. *J. Geophys. Res.* **94** (C4), 4984–8.

4

Remote-sensing techniques

J O N A T H A N L. B A M B E R

School of Geographical Sciences, University of Bristol

R O N K W O K

Jet Propulsion Laboratory, California Institute of Technology

4.1 Introduction

The cryosphere covers a vast expanse of the polar oceans and land surfaces. The area of the Southern Ocean covered by sea ice fluctuates between about 3.4 and 19.1×10^6 km^2 over the period of one year. The Antarctic ice sheet covers an area of some 13×10^6 km^2, greater than the conterminous USA. The number of glaciers on the planet is not well known, but certainly exceeds 160 000. Monitoring such large areas, often in remote and hostile environments, is ideally suited to satellite-based observations, which provide the only practical means of obtaining synoptic, timely coverage. Due to the importance of satellite remote sensing to observations of the cryosphere, we provide here a brief introduction to the subject, covering the satellites and sensors most commonly employed and describing how they can be used to derive information on mass balance. These instruments and techniques are referred to extensively in the subsequent chapters covering observational data on mass balance. This section is a primer in the subject. For comprehensive coverage of the general principles of remote sensing of the environment, the reader is referred to a number of excellent textbooks on remote sensing, referenced in this chapter. In section 4.2 we provide an overview of the general principles of satellite remote sensing, which are common to both land- and sea-ice measurements. In section 4.3 the characteristics and pertinent operating principles of the satellites and sensors relevant to cryospheric studies are reviewed. Their use for a range of mass balance related studies is presented in the subsequent sections on measurement methodologies. Section 4.4 deals specifically with satellite measurements of land ice, while section 4.5 does the same for sea ice.

4.2 Electromagnetic theory and basic principles

Observations of the Earth from space utilize a relatively small number of wavebands where modulation of the electromagnetic wave by the atmosphere is low. These wavebands are known as atmospheric windows. Three main windows exist in the visible, infra-red and microwave parts of the spectrum (Figure 4.1). For observing the cryosphere, perhaps the

Mass Balance of the Cryosphere: Observations and Modelling of Contemporary and Future Changes, eds. Jonathan L. Bamber and Antony J. Payne. Published by Cambridge University Press. © Cambridge University Press 2003.

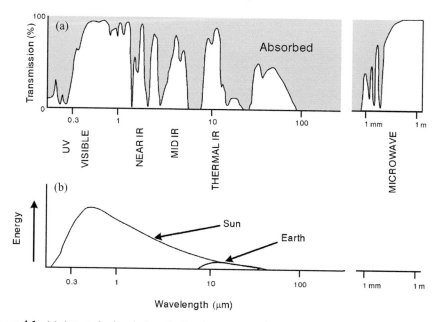

Figure 4.1. (a) Atmospheric windows in the electromagnetic spectrum. (b) This shows the strength of the thermally emitted radiation from the Sun and the Earth as a function of wavelength.

most useful waveband is the latter, as clouds are transparent in this part of the spectrum and microwave sensors can be used day or night. Nonetheless, all three of these regions are used in remote-sensing studies of snow and ice, and here we present the principal satellites, sensors and techniques that are useful in determining the mass balance of land and sea ice.

Imaging sensors operating in any part of the electromagnetic spectrum are defined by their spatial, radiometric, spectral and temporal resolutions. These four characteristics, primarily, determine the use and relevance of a particular sensor. Spatial resolution is usually associated with the size of picture elements or pixels recorded by the instrument, which are, in general, a function of the instantaneous field of view (IFOV) of the sensor (the area from which radiation is received at any instant). With some instruments, such as passive microwave radiometers, however, the IFOV can be substantially larger than the pixel size in the recorded image, and one footprint, or IFOV, has a large overlap with the next. This is not the case for most visible and infra-red imaging instruments. Radiometric resolution is determined by the sensitivity of the sensor and the number of bits used in the digitization of the analogue signal at the receiver. Eight-bit digitization, for example, provides a radiometric range of 0–255. Many sensors employ an automatic gain control in the receiver subsystem to ensure that the digitized signal does not saturate or is not too weak. Spectral resolution is determined both by the number of channels or wavelengths at which measurements are made, and also by the bandwidth of each channel. Temporal resolution is defined, primarily, by the satellite orbit, which is another key mission characteristic, discussed next.

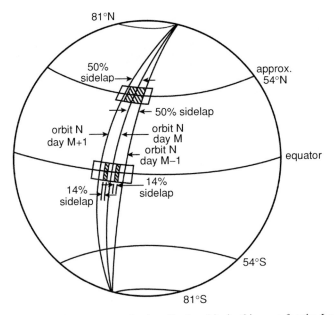

Figure 4.2. Example of the geometry of a low Earth orbit, in this case for the Landsat series of satellites.

Most of the satellites discussed below are in what are known as low Earth orbits, at altitudes of typically 750–1000 km. They are also typically in exact repeat cycles. This means that after a certain number of days the pattern of orbit tracks on the ground is repeated exactly. An example of this type of orbit is shown in Figure 4.2, which illustrates the characteristics of the Landsat orbit (see section 4.3). The inclination of the orbit (its angle with respect to the equator) is 98.2°, which means that the latitudinal limit of the satellite is ±81.8°. The exact repeat period for Landsat 7 is 16 days, and the orbital period is about 100 minutes (giving ∼14.4 orbits/day). As a consequence, the satellite makes 230 orbits every repeat period. This gives an across-track spacing at the equator of about 174 km (40 000/230). The swath width (the width of the image in the across-track direction) of the Landsat sensor is 185 km, resulting in a ∼5 km overlap in images at the equator. At higher latitudes the orbits converge, producing a greater overlap (Figure 4.2), coalescing at ∼82° where very dense track spacing of less than 30 km results in almost daily temporal sampling. It should be remembered, however, that visible sensors rely on cloud-free, daylight conditions to view the ground. The surface of the Earth is, at any time, covered by about 50% cloud, and the polar regions suffer extended periods of darkness during their winters. Consequently, the re-visit time of a satellite sensor, operating in the visible or infra-red, does not necessarily indicate its temporal sampling rate. Microwave instruments, however, can function day or night and in cloudy conditions. Their temporal resolution is, therefore, directly related to the re-visit interval. For this reason alone, microwave sensors have proved particularly valuable for cryospheric remote sensing.

Imaging, passive sensors (i.e. ones which do not emit their own energy source) record one of two types of radiation: reflected solar radiation and/or thermally emitted radiation. The latter is emitted by every object above 0 K, and the amount of radiation emitted (the radiance, L) is defined by Planck's law:

$$L_\lambda = \frac{2hc^2}{\lambda^5}(e^{hc/\lambda kT} - 1)^{-1}, \qquad (4.1)$$

where h is Planck's constant, k is the Boltzmann constant, T is the temperature of the body, λ is its wavelength and c is the speed of electromagnetic radiation. Equation (4.1) defines the radiance for a perfect emitter or black body. The strength of this emission, as a function of λ, is indicated by the curves in Figure 4.1(b). A black body absorbs all radiation falling on it (hence its name) and re-emits the radiation according to equation (4.1). If we differentiate Planck's law, we can find the peak of the curves in Figure 4.1(b):

$$\lambda_{max} = C_w/T, \qquad (4.2)$$

where $C_w \sim 2.9 \times 10^{-3}$ K m is a constant. This is known as Wien's law, and indicates that the maximum radiance emitted by the surface of the Earth (which is at a temperature of around 300 K) takes place at a wavelength of $\sim 10\,\mu m$. It can be seen from Figure 4.1 that this lies in an atmospheric window, in a region known as the thermal infra-red, where most satellite infra-red radiometers make measurements of surface temperature.

The total outgoing radiance of a black body is simply

$$L = \int_0^\infty L_\lambda \, d\lambda, \qquad (4.3)$$

and the total radiant exitance, M, is

$$M = L \int_0^{2\pi} \cos\theta \, d\Omega = \pi L = \sigma T^4 \,(\text{W/m}^2). \qquad (4.4)$$

This is Stefan's law, and σ is the Stefan–Boltzmann constant. Natural surfaces are not black bodies and do not absorb all the electromagnetic energy falling on them. This deviation from equation (4.1) can be described by the emissivity, $\varepsilon(\lambda)$, of an object, which ranges between 0 and 1. For a body to remain in equilibrium it must emit as much energy as it absorbs. Hence the emissivity is equivalent to the absorptance and is the converse of the reflectivity or albedo, α, of a surface such that $\alpha = 1-\varepsilon$. Thus, the emissivity of a black body is 1 and that of a perfect reflector is 0. A grey body is an object where emissivity is less than unity but is independent of wavelength. A selective radiator is one where $\varepsilon = \varepsilon(\lambda)$. All natural surfaces are selective radiators.

In the microwave part of the electromagnetic spectrum, $hc/\lambda kT \ll 1$, in which case

$$L_\lambda \approx 2kT/\lambda^2. \qquad (4.5)$$

This is known as the Rayleigh–Jeans approximation, and it means that the signal measured by the sensor, often termed the brightness temperature, T_b, is a linear function of the surface temperature, T_s:

$$T_b = \varepsilon T_s, \tag{4.6}$$

where ε is the microwave emissivity of the surface.

Visible sensors receive solar radiation that has been reflected by the surface but also scattered by the atmosphere back to the sensor. This latter component of the signal, known as sky noise or skylight, is an unwanted signal due to scattering by air molecules and more significantly particulate matter, which can produce 'haze' in an image. In general, however, sky noise does not seriously affect discrimination of surface types over glacierized terrain, but it can influence, for example, calculation of the albedo of a surface (Stroeve, Nolin and Steffen, 1997). Infra-red sensors measure a combination of reflected solar radiation (below a wavelength of about 3 μm) and thermal radiation emitted by the surface of the Earth and the atmosphere.

4.3 Satellites and sensors

There are a number of books on the general principles of remote sensing that cover, in detail, the concepts and the characteristics of the more common and ubiquitous sensors/satellites (e.g. Lillesand and Kiefer, 2000; Rees, 2001). These textbooks, however, do not necessarily carry details of the sensors relevant to cryospheric applications. Thus, here we provide only brief details of established sensors and technology with appropriate references where necessary. Greater detail is provided on recently launched and upcoming missions, especially those that have a particular emphasis on polar applications, and on those instruments that are particularly relevant to mass balance studies of land and sea ice that are not well represented in the existing literature.

4.3.1 Visible and infra-red sensors

The relevance of visible and infra-red imaging instruments is primarily for mapping the extent and, to a lesser degree, the characteristics of ice on the planet. Because of the relatively long time series provided by some of these instruments, they are particularly valuable for determining variations in areal extent of glaciers, ice caps and sea ice. Surface velocities of land ice have also been obtained using 'feature tracking' methods (section 4.4). Given below are some brief details pertaining to the instruments that have been used most extensively for cryospheric monitoring applications.

Landsat

The Landsat series of satellites has been providing visible and near infra-red imagery of the Earth's surface (up to a latitudinal limit of $\sim \pm 82.5°$) since 1972. The original instrument

Table 4.1. *Landsat satellite characteristics and history.*

Satellite	Launch (end) (month/year)	Instrument	Resolution (m)	Altitude (km)	Repeat period (days)
Landsat 1	7\72 (1\78)	RBV\MSS	80\80	917	18
Landsat 2	1\75 (2\82)	RBV\MSS	80\80	917	18
Landsat 3	3\78 (3\83)	RBV\MSS	80\80	917	18
Landsat 4	7\82 (8\93)	MSS\TM	80\80	705	16
Landsat 5	3\84	MSS\TM	80\80	705	16
Landsat 6	failed	ETM	15\30	705	16
Landsat 7	4\99	ETM+	15\30	705	16

Table 4.2. *Characteristics of the ETM + channels.*

Band number	Spectral range (microns)	Resolution (m)
1	0.45−0.515	30
2	0.525−0.605	30
3	0.63−0.690	30
4	0.75−0.90	30
5	1.55−1.75	30
6	10.40−12.5	60
7	2.09−2.35	30
Pan	0.52−0.90	15

was known as the multi-spectral scanner (MSS) and had a resolution of 79 m and a repeat cycle of 18 days. MSS was superseded by the thematic mapper (TM), which has a resolution of 30 m, improved dynamic range in digitization (0–255) and seven channels in the visible, near and thermal infra-red (Table 4.2).

Landsat 7 was launched in April, 1999, with the enhanced thematic mapper plus (ETM+) on board. The ETM+ includes a number of new features that make it a more versatile and effective instrument compared with its predecessors. The primary new features on Landsat 7 are: a panchromatic band with 15 m spatial resolution; onboard, full aperture, 5% absolute radiometric calibration; and a thermal infra-red (TIR) channel with 60 m spatial resolution (Lillesand and Kiefer, 2000). A summary of the principal characteristics of the Landsat series of satellites is given in Tables 4.1 and 4.2.

Because of their relatively high spatial resolution, Landsat data have proved useful in mapping and monitoring changes in the areal extent of glacier cover, although the data are of relatively limited value for providing quantitative estimates of mass balance as it can be difficult to convert a change in area into a mass change. A preliminary assessment of the glaciological utility of the ETM+ has been undertaken for both sea- and land-ice applications (Bindschadler *et al.*, 2001).

SPOT

Another visible sensor system known as the Systeme Probatoire de l'Observation de la Terre (SPOT) has been flown since 1986, and provides the additional benefits of 10 m resolution in panchromatic mode and a stereo-imaging capability, which has been used to derive digital elevation models (DEMs) of glaciers through photogrammetric techniques. SPOT has three channels in the visible and near infra-red (NIR) and a spatial resolution of 20 m in multi-spectral mode (Lillesand and Kiefer, 2000).

ASTER

ASTER (advanced spaceborne thermal emission and reflection radiometer), an imaging radiometer flying on board *Terra*, was launched in December, 1999, as part of NASA's Earth observing system (EOS). The instrument consists of three different subsystems: the visible and near infra-red (VNIR), the short-wave infra-red (SWIR), and the thermal infra-red (TIR). The characteristics of each of these are detailed in Table 4.3. ASTER is the prime instrument that will be used to produce an inventory of glaciers around the world as part of the GLIMS (Global Land Ice Measurements from Space) project (see Chapter 15) due to the high resolution of the VNIR and the capability of producing stereo pairs allowing the derivation of DEMs. The VNIR consists of two telescopes, one that is at nadir and another that looks backward, which records an image in band 3 (Table 4.3) for use in stereo photogrammetry.

An example of a nadir, VNIR image of part of the southern Patagonian ice field is shown in Figure 4.3 as a false colour composite, with red indicating vegetation. Areas of debris-covered ice can be clearly seen and, although the sensor has saturated over the higher snow-covered areas, some variations in surface snow properties are visible. This nadir view image has been combined with the backward view to produce a DEM of much of the glacier for elevation change studies (see section 4.4).

AVHRR

The advanced very high resolution radiometer (AVHRR) is a broad-band, four- or five-channel (depending on the model) scanner, sensing in the visible, NIR and TIR portions of the electromagnetic spectrum over a swath width of 2400 km. This sensor is carried on the National Oceanic and Atmospheric Association (NOAA) polar orbiting environmental satellites (POES), beginning with TIROS-N in 1978. Table 4.4 outlines the AVHRR bands and general applications of each. Channels 4 and 5 are usually used together to determine surface temperatures and to correct for the effects of atmospheric attenuation, using what is termed the split-window technique (Qin *et al.*, 2001). The average resolution of the channels at nadir is 1.1 km. There has been continuous coverage by at least one satellite since 1978. At the time of writing, four satellites are operational.

Table 4.3. *Characteristics of the three ASTER instruments.*

Characteristic	VNIR	SWIR	TIR
Spectral range	band 1: 0.52 − 0.60 µm nadir looking	band 4: 1.600− 1.700 µm	band 10: 8.125− 8.475 µm
	band 2: 0.63− 0.69 µm nadir looking	band 5: 2.145− 2.185 µm	band 11: 8.475− 8.825 µm
	band 3: 0.76− 0.86 µm nadir looking	band 6: 2.185− 2.225 µm	band 12: 8.925− 9.275 µm
	band 3: 0.76− 0.86 µm backward looking	band 7: 2.235− 2.285 µm	band 13: 10.25− 10.95 µm
		band 8: 2.295− 2.365 µm	band 14: 10.95− 11.65 µm
		band 9: 2.360− 2.430 µm	
Ground resolution (m)	15	30	90
Data rate (Mbits/s)	62	23	4.2
Cross-track pointing (km)	±318	±116	±116
Swath width (km)	60	60	60
Quantization (bits)	8	8	12

Table 4.4. *Channels of the advanced very high resolution radiometer (AVHRR).*

Band	Wavelength (µm)	Application
1	0.58−0.68	daytime cloud/surface mapping
2	0.725−1.0	surface water delineation, ice and snow melt
3A	1.58−1.64	snow/ice discrimination (NOAA satellites: K, L, M)
3	3.55−3.93	sea surface temperature, night cloud detection
4	10.3−11.3	sea surface temperature, day/night cloud detection
5	11.5−12.5	sea surface temperature, day/night cloud detection

AVHRR provides a useful complement to the higher resolution visible sensors such as the Landsat ETM as it has a wide swath width and moderate resolution, providing daily (or better, particularly at high latitudes) coverage of cloud-free sea- and land-ice areas. The instrument has been used to examine sea-ice extent, although cloud discrimination is a serious problem for this application, and, as a consequence, passive microwave radiometers have often been preferred for this application, despite their much poorer resolution (section 4.3.4). AVHRR has also been used to map the margins, surface characteristics and morphology (such as albedo, flow stripes and surface temperature) of the ice sheets (Fujii *et al.*, 1987;

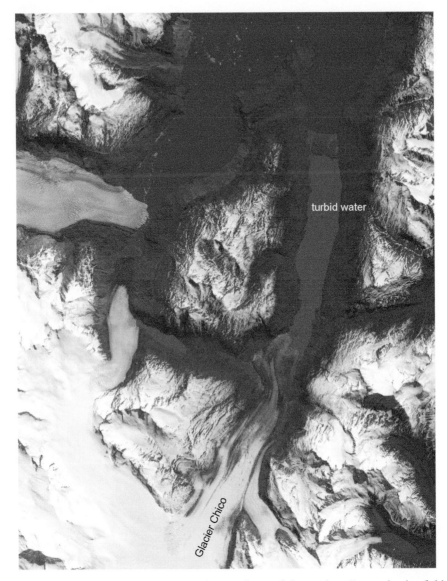

Figure 4.3. A false colour, VNIR, ASTER image of part of the southern Patagonian ice field. A terminal moraine can be seen to be blocking the fjord, which Glacier Chico drains into, marking the limit of the glacier at an earlier date. The water between the moraine and glacier is a different colour from the lake water due to the high sediment concentration of the former. This is a nadir view and has been combined with the backward view data to produce a digital elevation model of the glaciers. (Image courtesy of EOS Data Center.)

Scambos and Bindschadler, 1991; Steffen *et al.*, 1993; Stroeve *et al.*, 1997). Some of these applications, where relevant to mass balance studies, are discussed later. Landsat TM and ETM offer a similar capability for smaller ice masses such as valley glaciers. AVHRR can also be used to determine the size distribution of icebergs calving from the floating margins of Antarctica, but this is limited by the difficulty in discriminating cloud and snow in the wavebands available. The instrument does, however, offer a relatively long and continuous time series of observations that are, in general, freely available to the science community.

4.3.2 Synthetic aperture radars and scatterometers

As mentioned earlier, clouds are transparent in the microwave part of the spectrum, which means that radar imaging systems have the dual advantage over visible or infra-red imaging sensors (such as Landsat TM, SPOT, AVHRR and ASTER) that they can offer day/night coverage, even in cloudy conditions. In the polar regions, where cloud cover is often extensive and permanent darkness occurs for many months of the year, microwave sensors have the potential to provide the only continuous, synoptic satellite data sets. A problem, however, with most imaging microwave sensors is their relatively poor resolution (see, for example, Table 4.6). This is due to the fact that resolution is, usually, proportional to wavelength, which is about 10^4 times longer in the microwave compared to the visible. Here, we explain how this limitation has been overcome for one type of microwave sensor. The instrument in question is known as a synthetic aperture radar (SAR) and has, since the launch of the European Space Agency's (ESA) first European remote sensing satellite, ERS-1, in 1991, become a key tool for monitoring both land and sea ice.

Unlike visible and infra-red sensors, a radar measures the range to a target as well as the radiation scattered by that target in the direction of its antenna (backscatter). The intensity of the backscatter is dependent on the polarization and frequency of the transmitted wave and the complex interactions between the radar wave and the target. There is a wide range of scattering mechanisms that contribute to the observed intensity, which include the electromagnetic and physical properties of the target. If the target is not a Lambertian scatterer (one which scatters equally in all directions), then the orientation of the radar relative to the target is also important. For remote sensing of the Earth surface, this orientation is described by an incidence angle and an azimuth angle (Figure 4.4).

The spatial resolution in azimuth of a real-aperture radar is dictated by the beamwidth of the antenna, and is therefore linearly proportional to the distance between the sensor and the surface. Larger antennas provide narrower beamwidths and better spatial resolution. In the range direction, the resolution is controlled by the bandwidth of the transmitted pulse. Therefore, very high resolution in the range can be achieved, compared to that in azimuth. One way to obtain comparable resolutions in azimuth, without using a large physical antenna, is by aperture synthesis. In SARs, large apertures in azimuth are formed using the along-track motion of the radar platform. As a platform flies over the surface of interest, the range and Doppler history of each target is sampled by a sequence of radar

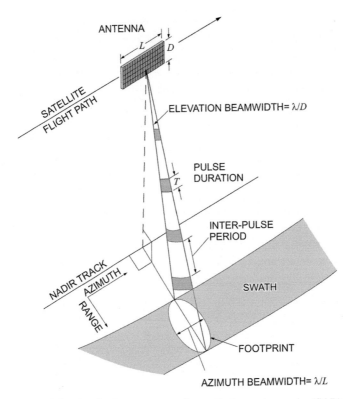

Figure 4.4. The viewing geometry of a synthetic aperture radar (SAR).

pulses. The equivalent size of the synthesized antenna is equal to the product of the platform velocity and the time the target remains within the physical footprint of the antenna on the ground (Figure 4.4). Post-processing of the data, on the ground, then separates the radar signal in range and azimuth to produce a high resolution image. To achieve the aperture synthesis, the instrument needs to record both the phase and amplitude of the returned signal. The typical resolution of current space-borne SARs is ∼25 m by 25 m. Table 4.5 lists the SARs and scatterometers relevant here.

Scatterometers, like SARs, are active microwave sensors. They transmit a microwave pulse to the surface, which is then measured by the sensor at several different angles. The backscatter coefficient (σ^0) of the ocean surface, in the microwave, is sensitive to azimuthal angle due to the presence of capilliary waves caused by winds. This sensitivity is used to determine wind-speed direction as well as its magnitude. For the ERS-1 and 2 satellites the wind scatterometer and SAR were combined within a single active microwave instrument (AMI), utilizing the same radiofrequency subsystem and power supply, etc. The scatterometer had three antennas that viewed the surface in the forward, nadir and aft directions at incidence angles ranging between 18° and 59° with a spatial resolution of ∼45 km along and across-track. The shallow incident angles allow for significant penetration

Table 4.5. *Details of SAR and scatterometer missions suitable for remote sensing of the cryosphere.*

	Time period	Frequency (GHz)	Incidence angle (deg.)	Polarization
SAR				
ERS-1	1991–1995	5.3	23	VV
ERS-2	1995–2002	5.3	23	VV
ENVISAT	2002–	5.3	23	VV, HH, VH
JERS-1	1992–1998	1.2	35	HH
RADARSAT-1	1995–2003	5.3	20–50	HH
RADARSAT-2	2003 launch	5.3	20–50	HH
Scatterometers				
ERS-1	1991–1995	5.3	18–59	VV
ERS-2	1995–2002	5.3	18–59	VV
ENVISAT	2002–	5.3	18–59	VV
NSCAT	1996 (6 months)	13.9	18–65	VV, HH
SeaWINDs	1999–	13.4	45, 54	VV, HH

into the surface layers of the dry snow zone of the Antarctic and Greenland ice sheets. The σ^0 of these zones is dependent on the microwave properties of the snow pack and, in particular, grain size and stratification of snow layers. The σ^0 of the surface can potentially, therefore, provide information about relative patterns and variations in accumulation rate. Data from several scatterometer missions have, for example, been used in empirical studies of accumulation rate over Greenland (Drinkwater, Long and Bingham, 2001).

4.3.3 Satellite altimetry

The use of an active ranging system for mapping elevation of ice sheets was first postulated as far back as the early 1960s. (Robin, 1966). It was not, however, until the launch of Seasat in 1978 that the concept was tested and its value for cryospheric applications more fully explored. Further details about relevant missions are given in section 4.4.1. Here, we describe the basic measurement principles and properties of the current fleet of altimeters, which differ significantly from other satellite systems described so far, as an altimeter is not an imaging instrument. For an extended discussion of these and related issues, the reader is referred to Chelton *et al.* (2001).

Satellite radar altimeters (SRAs) are active instruments that transmit a microwave pulse (typically at Ku band, 13.5 GHz) to the surface and measure the two-way time delay of the pulse as it travels from the satellite to the ground and back (Figure 4.5). With sufficiently good orbit determination, it is possible to measure the elevation of the sea surface with

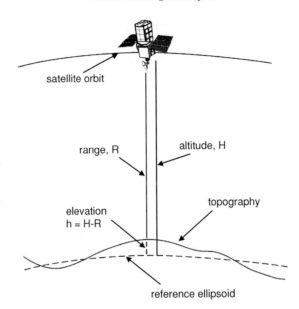

Figure 4.5. The measurement principle of a satellite radar altimeter (SRA).

an accuracy of about 3 cm (Chelton *et al.*, 2001). SRAs were designed, primarily, for oceanographic applications, and details of the specific problems related to their use over ice sheets are discussed later. It should be noted here, however, that, due to their large footprint size on the ground (several kilometres) and their relatively narrow antenna beamwidth (typically 1.3°), the current fleet of instruments can only provide useful data over the ice sheets of Greenland and Antarctica and, even then, not for some of the steeper marginal areas.

The measurement concept is illustrated in Figure 4.5, where R is the range to the surface, as measured by the SRA, H is the altitude of the instrument above the ellipsoid, and h is the elevation of the surface with respect to the ellipsoid. The SRA measures a time delay

$$t = 2r/c, \tag{4.7}$$

where c is the speed of the electromagnetic radiation through the atmosphere, and the elevation $h = H - R$.

Atmospheric corrections

The speed of the electromagnetic radiation is not a constant and is affected by three factors that need to be corrected for: (i) the dry tropospheric correction, ΔR_{dry}; (ii) the wet tropospheric correction, ΔR_{wet}, and (iii) the ionospheric correction, ΔR_{iono}. The method for calculating each of these corrections depends on the type of surface that the altimeter is ranging to. For example, over the ocean, a separate instrument (a microwave radiometer for example) is used to calculate ΔR_{wet}. The TOPEX and ENVISAT SRAs (known as the RA-2) are dual-frequency instruments allowing calculation of ΔR_{iono} over the ocean. Due to variations in the microwave properties of ice-sheet surfaces, a different approach must be used, however, for estimating these two corrections.

The dry tropospheric correction is a function of the total mass of dry air between the satellite and the ground, and this is, in turn, a function of the surface air pressure, and can therefore be estimated with a high degree of accuracy ($\sim$1 cm) using meteorological or forecast data on surface pressure. ΔR_{dry} is the largest atmospheric correction, but has the smallest variation (225–235 cm) over the ocean. Over the ice sheets, surface pressure is rarely known with any accuracy, and it is necessary to extrapolate from the coast, applying a correction for elevation and change in temperature with altitude (Bamber and Huybrechts, 1996).

ΔR_{wet} is a function of the total columnar water vapour content in the atmosphere (Chelton *et al.*, 2001). In the polar regions this is relatively small, and an analysis of 17 000 radiosonde launches from coastal sites in Antarctica found that it had a mean value of 2 ± 3 cm (Bamber and Huybrechts, 1996), while ΔR_{dry} was found to be 225 ± 2 cm. The combined root-mean-squared (RMS) atmospheric variability in the wet and dry corrections around the coast of Antarctica is, therefore, only 3.6 cm. For Greenland, the wet correction is likely to be larger and to have a higher variance, as moisture levels are, in general, higher than in Antarctica.

ΔR_{iono} is a function of the electron content of the ionosphere and the frequency of the microwave signal. At the frequency of the ERS SRAs (13.5 GHz), the correction is typically in the range 2–40 cm, and can be calculated with reasonable accuracy using look-up tables for the electron content at a particular time of day, season and location (Cudlip *et al.*, 1994).

Orbits

It is clear from Figure 4.5 that the accuracy of h is affected both by errors in R and in H. Accurate determination of the orbit is, therefore, crucial for accurate height measurements. Two components are involved: a satellite tracking system and an orbit model. Improvements in both these components over the 1980s and 90s have led to steadily increasing accuracies. Seasat relied solely on a sparse network of ground-based, laser ranging stations and a fairly crude Earth gravity model.[1] The ERS, ENVISAT and TOPEX/POSEIDON missions have onboard satellite tracking systems and greatly improved gravity field models. The best orbits, to date, have been achieved for TOPEX/POSEIDON, which uses a satellite-based tracking system known as DORIS (Doppler orbitography and radiopositioning integrated by satellite) combined with a relatively high altitude of 1336 km compared with $\sim$800 km for ERS, Seasat and Geosat. The higher altitude reduces the influence of variations in atmospheric drag and has resulted in orbit accuracies of 2–3 cm (Chelton *et al.*, 2001)

---

[1] The gravity field is a crucial component of the orbit model as it defines the variations in the gravitational acceleration experienced by the satellite about its orbit. Data from SRA missions during the 1990s have, to a large degree, been responsible for the improvements to the gravity models.

compared with 20–30 cm for Seasat and 10–20 cm for Geosat. Orbit accuracies for ERS-1 and 2 are about 5 cm (Scharroo and Visser, 1998). Further details of these missions are provided in section 4.4.1.

CryoSat

SRAs have proved a useful tool for determining volume changes of the ice sheets and, as a consequence, inferring mass balance. Their use is, however, not without problems, particularly those associated with coverage and footprint size (section 4.4.1). For example, in a study of the mass balance of the Antarctic ice sheet using ERS-1 data, reliable observations were achieved for only 63% of the ice sheet (Wingham *et al.*, 1998) due to (i) the latitudinal limit of the satellite and (ii) poor accuracy and coverage in the steeper marginal parts of the ice sheet (see Chapter 12 for a fuller discussion). Similar difficulties were encountered in the marginal zone of the Greenland ice sheet, where surface melting produces an additional complication through changes to the microwave properties of the snow pack (Davis, Kluever and Haines, 1998).

A new type of SRA has been proposed that could address some of these problems by combining the principles of operation of an altimeter with synthetic aperture processing in the across-track direction (Raney, 1998). The mission is part of ESA's Earth explorer programme; it is called CryoSat and has an anticipated launch date of late 2004. Among the key objectives of this mission is to derive sea-ice freeboard and improved estimates of elevation changes over the ice sheets. Unlike the current fleet of SRAs, CryoSat will not only provide reliable elevation estimates for the marginal portions of the ice sheets but also for smaller ice masses such as ice caps in the Arctic with an area of $\sim 10\,000\,\mathrm{km}^2$ or more.

The ice, clouds and elevation satellite, ICESat

ICESat, launched in January 2003, is the first ever mission to carry an autonomous satellite laser altimeter: the geosciences laser altimeter system (GLAS). This instrument is designed to measure ice-sheet topography and associated temporal changes, as well as cloud and atmospheric properties. For ice-sheet applications, the laser altimeter will measure height from the spacecraft to the surface, with an intrinsic precision of better than 10 cm and a 70 m surface spot size. Accurate orbit determination will be achieved using onboard global positioning system (GPS) receivers. Characteristics of the return pulse will be used to determine surface roughness. In addition, it is hoped that the GLAS will provide estimates of sea-ice freeboard from the relative difference in elevation between leads and floes.

During the commissioning phase of the mission, ICESat will be placed in an eight-day repeat orbit. Following this it will be placed in a 183-day repeat for the remainder of its three-year mission. The orbit inclination will be 94°, providing coverage to ±86° latitude. This will give complete coverage of Greenland and Arctic ice masses and much of Antarctica. A number of airborne laser altimeter missions, and in particular the airborne topographic mapper (ATM), have demonstrated the capability of this type of technology for measuring the mass balance of glaciers and ice sheets. Results from the ATM are discussed in detail in Chapter 10.

Table 4.6. *Frequency and approximate resolution of the channels on the SMMR, SSM/I and AMSR instruments.*

Each frequency has channels for both vertical and horizontally polarized radiation unless otherwise stated.

Frequency (GHz)	Cell size (km)
SMMR (1978–1987).	swath width: 780 km
6.6	136 × 89
10.7	87 × 58
18	54 × 35
21	44 × 29
32	28 × 18
SSM/I (1987–2003)	swath width: 1400 km
19.4	69 × 43
22.3 (V only)	60 × 37
37	36 × 22
85.5	15 × 13
AMSR (2002–2003)	swath width: 1445 km
6.9	76 × 44
10.7	49 × 28
18.7	28 × 16
23.8	22 × 13
36.5	14 × 8
89	6 × 4

4.3.4 Passive microwave radiometers (PMRs)

These instruments measure the thermally emitted microwave radiation from a surface or the atmosphere. The first instrument of use to glaciology was the electrically scanned microwave radiometer (ESMR), launched in 1972 and used successfully to determine a time series of sea-ice extent for both the Southern and Arctic Oceans (see Chapter 8). EMSR was superseded by the scanning multi-spectral microwave radiometer (SMMR), flown on Seasat and the NIMBUS satellite series from 1978 to 1987. It operated at five frequencies, each with dual polarization. Table 4.6 lists the frequencies and resolutions of SMMR and the radiometers that followed it.

The first special sensor microwave/imager (SSM/I) on board the Defense Meteorological Satellite Program (DMSP) series of satellites was launched in 1987, providing a valuable inter-calibration period between it and SMMR. The SSM/I is a seven-channel, four-frequency, linearly polarized, passive microwave radiometer. DMSP satellites are in a sun-synchronous polar orbit at a nominal altitude of 830 km. The swath width of SSM/I is 1400 km.

A PMR with an improved resolution (by a factor of 2, Table 4.6), utilizing a greater range and number of frequencies, was launched on Aqua in May, 2002. The instrument is known as AMSR-E and is similar to the AMSR that will fly on board the ADEOS II satellite, launched in 2002.

The principal use of PMR data, relevant here, has been in monitoring sea-ice extent and concentration, providing a continuous 30-year record extending back to 1972 (Chapter 8). These instruments have also been used to monitor melt area extent over the Greenland ice sheet (Chapter 10). Furthermore, the lowest frequency channel on board SMMR (and now on AMSR-E) has been shown to be related to accumulation rates in the dry snow zones of Greenland (Winebrenner, Arthern and Shuman, 2001) and Antarctica (Zwally, 1977) and has been used as a 'background field' for interpolating the sparse *in situ* accumulation data in Antarctica (Vaughan *et al.*, 1999). The course resolution of these instruments renders them of limited value for smaller ice masses.

4.4 Land-ice mass balance

There are two distinct approaches to solving the local mass balance (equation 2.1) or the total mass balance (equation 2.3), as discussed in Chapter 2, and repeated here:

$$\partial V/\partial a = M_a - M_m - M_c + M_b, \qquad (2.3)$$

where V = ice volume, a = one year, M_a = the annual surface accumulation, M_m = the annual loss by glacial surface runoff, M_c = the annual loss by calving of icebergs and M_b = the annual balance at the bottom (melting or freeze-on of ice). The first method is to determine the left hand side of equation (2.3) by direct measurement of volume or height change. In the next section we present the satellite and airborne methods that have been used in this approach.

4.4.1 Direct measurement of volume changes

It is possible, by accurately determining changes in the elevation of the surface of an ice mass over time, to infer changes in volume which, assuming a constant density, can be used to determine a mass change. This assumption is reasonable for sub-polar ice masses where the depth of the firn layer (the layer of compacted snow overlying solid ice) is only a few metres. For Antarctica and central Greenland, however, this layer can be more than 100 m deep, and variations in densification rate within the firn, due, primarily, to fluctuations in temperature, can produce as large a variation in surface height as the expected signal (Arthern and Wingham, 1998). Nonetheless, satellite radar altimeters (SRAs) have been used to examine volume changes in Greenland and Antarctica, and the principles of this approach are discussed below. Conventional SRAs cannot be used on smaller ice masses, as mentioned earlier, due to their large footprint and their inability to range reliably to slopes greater than about $1°$.

Radar altimetry

There are four satellite missions that have satisfied the dual requirement of having accurate enough orbit determination and an orbital inclination that provided useful coverage of the ice sheets. The first of these was Seasat, launched in 1978. This satellite had a latitudinal limit of 72° providing coverage of the southern half of Greenland and about one-fifth of Antarctica (Figure 4.6). Geosat, which flew from 1985 to 1989, extended the temporal record but not the spatial coverage. In 1991, ERS-1 was placed in an orbit that provided coverage to 81.5°, and in 1995 this satellite was superseded by ERS-2, which had similar characteristics and an identical SRA. There is, thus, a continuous record of elevation change from 1991, covering the whole of the Greenland ice sheet and four-fifths of Antarctica. These data, along with the discontinuous and spatially less extensive Seasat and Geosat measurements, have been used to investigate variations in the surface elevation of the ice sheets in an attempt to infer their mass balance (Davis, 1995; Davis *et al.*, 1998; Wingham *et al.*, 1998; Zwally, 1989; Zwally *et al.*, 1989). Results of these analyses are discussed in Chapters 10 and 12. Table 4.7 provides details of past, present and future altimeter missions.

Over non-ocean surfaces a number of corrections need to be applied due to the undulating, non-uniform nature of the surface. The first of these is a range-estimate refinement procedure known as waveform retracking.

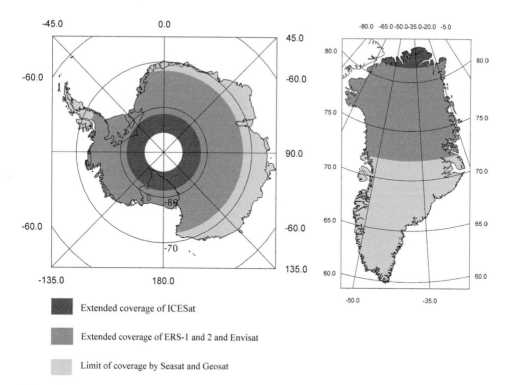

Figure 4.6. Coverage of the Antarctic and Greenland ice sheets by past and present satellite altimeter missions.

Table 4.7. *Satellite altimeter missions relevant to studies of the cryosphere.*

Satellite	Operation	Properties	Coverage	Repeat period (days)
Seasat	June–Aug. 1978	Ku band SRA	±72°	24
Geosat		Ku band SRA	±72°	17, ~540
ERS-1	1991–1995	Ku band SRA with ice mode tracking	±81.5°	3, 35, 336
ERS-2	1995–2002	Ku band SRA with ice mode tracking	±81.5°	35
ENVISAT	2002–	dual-frequency SRA with adaptive tracking	±81.5°	35
ICESat	Jan. 2003	dual-frequency steerable laser	±86°	8, 183
Cryosat	2004	imaging Ku band SRA	±82°	2, 30, 354

Waveform retracking An SRA sends a series of pulses down to the ground, which are reflected by the surface and received back at the satellite. For ERS-1 the pulse repetition frequency was about 1 kHz. Onboard software averaged 50 consecutive pulses to reduce the noise in the waveforms, resulting in a recording rate of 20 Hz. With a ground track speed of about 6.6 km/s, this produced range measurements at about 330 m intervals. To reduce the data volume, each pulse or waveform is sampled within a narrow band of time delay known as the range window (Figure 4.7), which contains the waveform samples or bins (typically 64). The waveform is made up of a leading edge, where the power rises with time, as the pulse illuminates an increasing area of the surface (Figure 4.7). When the pulse has fully illuminated the surface, the waveform reaches its maximum amplitude. The pulse then expands over the surface, forming an annulus, and the power in the waveform gradually decreases due to the influence of the receiver antenna pattern. This part of the waveform is known as the trailing edge. For a Gaussian distribution of surface slopes, the half power point (i.e. the point at which the power is one-half of the maximum amplitude) on the leading edge represents the location of the mean surface height within the antenna footprint. This point is often assumed to be where the surface lies within the waveform (Bamber, 1994), and is known as the retrack point.

The range recorded by the SRA is referenced to the mid-point of the range window. If the half power point (or retrack point) of the leading edge is not at this mid-point (as it always is over the ocean), then there is an offset between the recorded and actual range to the surface. This offset is known as the retrack correction and is determined through post-processing of the waveform data on the ground. Figure 4.8(a), is a sequence of ERS-1 waveforms from the south-east margin of the Greenland ice sheet (location shown in Figure 4.8b) and demonstrates that waveforms can have complex shapes due to the breakdown of the assumption of

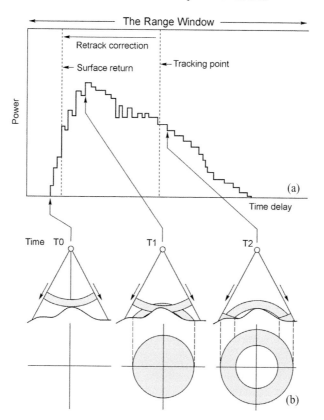

Figure 4.7. (a) Radar altimeter waveform return from a non-ocean surface. (b) This shows how the pulse illuminates the surface at the three points in time represented in part (a).

a Gaussian distribution of slopes and other effects. In areas of rough terrain, the radar pulse can be illuminating several different patches or facets of ground simultaneously. This can lead to multiple maxima within the waveform making interpretation ambiguous. The key step in the retracking procedure is the method used for determining where the mean surface lies within the leading edge. Various different approaches have been proposed, ranging from fitting a mathematical model, based on ocean-like returns, to the shape of the waveform (Zwally *et al.*, 1983), to a model-independent approach that makes no assumptions about the properties of the return (Bamber, 1994) known as a threshold retracker. In this approach, the retrack position is the first point on the leading edge that exceeds some threshold of the amplitude of the waveform (such as 50%, for example). Threshold retrackers require a robust method for determining the waveform amplitude in the presence of substantial noise (Figure 4.8).

To ensure that the waveform is captured within the timespan represented by the range window, the SRA needs to have an onboard tracker that predicts the range of the next return. Over the ocean, this tracker works well and is able to centre the leading edge of

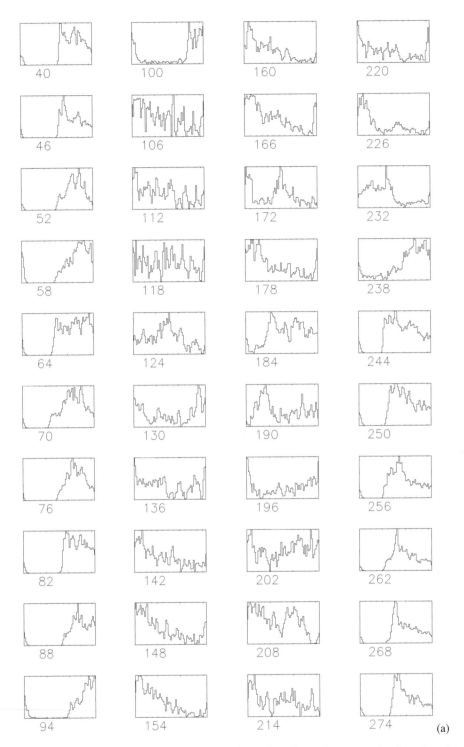

Figure 4.8. (a) Sequence of ERS-1 altimeter waveforms from the south-east margin of the Greenland ice sheet. The spacing between the plotted waveforms is about 2 km (every sixth waveform of the full 20 Hz data are plotted). (b) The location of the track.

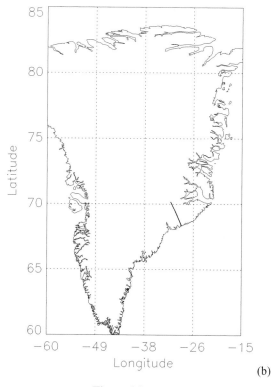

(b)

Figure 4.8. (*cont.*)

the waveform within the range window (Figure 4.7). Over undulating surfaces, however, this is not necessarily the case, and the leading edge can wander back and forward within the window (Figure 4.8) or disappear altogether. This is known as 'loss of lock', and when this happens the SRA goes into 'acquisition mode', where it searches, in time delay, for a signal. It can take several seconds for the SRA to recover the surface return, during which time no valid data are recorded. Loss of lock takes place most commonly over rugged terrain where the range is changing rapidly; this is illustrated in Figure 4.8(a). Every sixth waveform from the 20 Hz samples has been plotted so that the distance separating each one shown is about 2 km. Near the coast the waveforms are fairly simple in shape with a distinct leading edge (numbers 40–88), although there are some multi-peaked echoes (52 and 70 for example). Moving inland, the terrain becomes more rugged and the leading edge disappears from the range window, during which time no range measurement can be obtained (echoes 106–232). The surface is lost for some 40 km before being acquired again in waveform 238, although this has a long and noisy leading edge. It should also be noted that, although the subsequent waveforms have a fairly unambiguous leading edge, some of them display a complex, irregular shape (e.g. 262, 268 and 274). These waveforms have a distinct peak in them, perhaps due to a strong scatterer within the footprint such as a supraglacial lake, and will not conform to any retracking model that is based on a mathematical description of the waveform shape. The noisy nature of the returns is also evident, and Figure 4.8(a) illustrates the difficulties associated with accurately determining a consistent and meaningful retrack point in the presence of this noise.

ERS-1 and 2 were unique in having a range window that could be increased by a factor of 4, from 28.8 to 115.2 m width. The two modes were known as ocean and ice mode, the latter having the larger window and a subsequently decreased range resolution (the number of samples remained the same in both modes, but the sample width increased from 0.45 to 1.82 m between ocean and ice mode). A geographic mask was used to switch automatically between the two modes. The concept behind the ice mode was that it would provide better coverage over the undulating terrain of non-ocean surfaces by employing a more robust, model-independent, onboard tracker and a larger window for the waveform to lie in. The cost was a factor of 4 reduction in range resolution.

For the purposes of elevation change detection (often referred to as dh/dt), it is not the absolute elevation that is required but the difference between two range estimates. dh/dt estimates are obtained from comparing range estimates at crossing points of descending and ascending tracks obtained at different epochs. In this case we do not necessarily need to find the mean surface within each waveform but some unambiguous point that is relatively insensitive to the effects of noise. The point that best fits these criteria is the first return, where the pulse first intercepts the surface. As the two waveforms being compared are from almost identical locations (within 100–200 m), and assuming that the surface properties have not changed in the intervening time, the first return should be a reliable estimator for determining dh/dt, and a number of studies have used this approach (Davis *et al.*, 1998). A review of the use of different retrackers for the determination of height changes has been undertaken (Davis, 1996a), and a robust scheme developed based on identifying a point on the leading edge that is 10% of the amplitude of the waveform (Davis, 1997).

The assumption of invariant surface properties is questionable, however. A theoretical study of the interaction of a radar pulse with snow pack demonstrated that there is both a surface and subsurface (or volume) component to the backscattered signal, and that the ratio of the power in these two components affects the shape of the leading edge (Ridley and Partington, 1988). As a consequence, a retracker has been developed that attempts to determine the proportion of volume to surface scattering (Davis and Moore, 1994). If this ratio changes, it will affect both the shape of the leading edge and the amplitude of the waveform. Both these factors influence the location of the retrack point, even when attempting to retrack the first return. As a consequence, a correction was applied for variations in backscatter coefficient of the surface (a measure of the waveform amplitude) for ERS-1 data analysed over Antarctica (Wingham *et al.*, 1998). Variations in backscatter coefficient (known as σ^0) can be caused by changes in the density and/or grain size of the surface and subsurface snow and, in Greenland, surface melting can have a dramatic effect on σ^0. Backscatter properties in Greenland have been shown to vary seasonally and inter-annually, with the implication that this is likely to affect dh/dt estimates derived from radar altimetry over the ice sheet (Davis, 1996b).

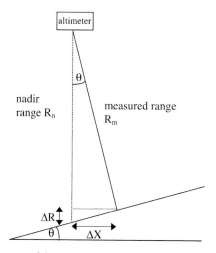

Figure 4.9. The geometry of the measurement made by an SRA over a sloping surface.

Slope-induced error SRAs range to the nearest point on the ground rather than the nadir point (Figure 4.9). For a slope of 1° the difference in range between the nadir point and the nearest point is about 120 m. The magnitude of the range difference is

$$\Delta R = R_{\mathrm{n}}(1 - \cos\theta), \tag{4.8}$$

where R_{n} is the range to the nadir point and θ is the regional slope. The relevant length-scale for estimating θ is the approximate diameter of the beam-limited footprint of the SRA, which over ice sheets is typically of the order of 4–5 km. The slope-induced error can be large, but for a cross-over analysis it is, in general, not relevant as the slope at the cross-over is identical for ascending and descending tracks. The value of ΔR, in equation (4.8), can be seen, however, to be a function of the range, and hence altitude, of the satellite. Thus, if data from different missions, such as Geosat and Seasat, are being compared, the slope-induced error may be different due to the difference in altitudes of the two satellite overpasses. For a slope of 0.5° (which is relatively high for an ice sheet) a variation of 10 km in the altitude of ERS-1 affects the slope-induced error by less than one millimetre.

Laser altimetry

Airborne laser altimetry has now been used in a number of studies to examine height changes of the surface of an ice mass (Krabill *et al.*, 1995a, 1999). The concept is similar to that of an SRA except that instead of using cross-overs, flight-lines are re-flown at a later date and elevation changes determined along the flight-line. This approach was employed in Greenland using a system known as the airborne topographic mapper (ATM; see section 4.3.3) that collected dense 200 m wide swaths of points. The laser spot diameter on the surface was about 1 m, in contrast to the footprint of an SRA, which is several kilometres, and the laser data accuracy was generally of the order of 10 cm (Krabill *et al.*, 1995b). The ATM results for Greenland are discussed in Chapter 10.

The first satellite laser mission, the ice, cloud and elevation satellite (ICESat), was launched in January 2003, and carried the geoscience laser altimeter system (GLAS). One of the primary objectives of this mission is to measure dh/dt over ice-covered areas. The GLAS also has the capability to measure sea-ice freeboard from the height difference between leads and adjacent ice. If the mean ice density and mean sea surface height are known, ice thickness can then be determined.

Other methods of determining volume change

It is also possible to infer a volume change by deriving digital elevation models (DEMs) of the surface from two different epochs. DEMs derived from satellite stereo photogrammetry (e.g. from SPOT) have an accuracy of about 10 m, which, in general, limits their usefulness for mass balance determination, and DEMs derived from terrestrial or airborne photographic or cartographic data are more commonly used (Reynolds and Young, 1997; Theakstone and Jacobsen, 1997). In a novel approach, repeated airborne photography has been used to derive both a surface DEM and velocity field simultaneously (Kaab, 2000; Kaab and Funk, 1999) for a glacier in the Swiss Alps. The surface DEMs were obtained with an accuracy of about 0.2 m and a resolution of 25 m, and were combined with velocity and ice-thickness data to derive the flux divergence and hence mass balance over the glacier. Terrestrial photogrammetric techniques are discussed further in Chapter 2.

Changes in areal extent can be used to infer a volume change if information about the ice-thickness distribution is available (Bayr, Hall and Kovalick, 1994). Visible imagery (aerial photography, Landsat, SPOT and, most recently, ASTER) have been used to map the areal extent of glaciers over a period of time and so determine retreat rates. The length of the time series available for examining changes has recently been extended back to the early 1960s with the aid of declassified satellite photography (Bindschadler and Vornberger, 1998).

GLIMS (Global Land Ice Measurements from Space) is a project designed to monitor the world's glaciers primarily using data from the ASTER (advanced spaceborne thermal emission and reflection radiometer) instrument aboard NASA's Earth Observing System (EOS) *Terra* spacecraft, launched in December, 1999. GLIMS objectives are to establish a global inventory of land ice, including surface topography, to measure the changes in extent of glaciers and, where possible, their surface velocities. This work will also establish a digital baseline inventory of ice extent for comparison with inventories at later times. Data from Landsat 7 will also be used as part of the programme, where necessary, and the GLIMS project is discussed further in Chapter 15.

4.4.2 Measurement of mass balance components: budget approach

As mentioned at the beginning of section 4.4, there are two possible approaches to deter-mining the total mass balance. The first, discussed above, determines the left hand side of equation (2.3). The second is based on determining the right hand side of the equation, which requires estimating each component of the equation and summing them. In the following sections we discuss methods for directly (or more often indirectly) measuring these components.

Accumulation rates

As mentioned, the brightness temperature measured by a passive microwave radiometer is proportional to the physical temperature of the surface and its emissivity (equation (4.6)). The latter quantity is related to the grain size of snow crystals. This in turn is related to the age of the firn in areas where there is no summer melting (i.e. in the dry snow zone of an ice sheet). Thus, the emissivity is, in part, a function of the accumulation rate, A. If the mean annual air temperature, T_m, can be determined, it is possible, in principle, to obtain an estimate of accumulation. This has been done for central Greenland (Winebrenner *et al.*, 2001) semi-analytically, and for Antarctica by deriving a relationship between A, T_b and T_m based on a simplified radiative transfer model and grain-growth function (Zwally, 1977). The constants in the resulting hyperbolic function were found empirically using measured, *in situ*, estimates of A (Zwally and Giovinetto, 1995) and data from SMMR. This approach relies on a number of assumptions, the most limiting of which is that the microwave characteristics of the snow pack are spatially and temporally invariant. As a consequence, the fit to the *in situ* data, although showing a reasonable correlation (of 0.82) with T_b, had a relatively large residual. In the most recent compilation of accumulation rates for Antarctica, the SMMR data were used as a background field for interpolating between sparse *in situ* data rather than as an independent estimate of A (Vaughan *et al.*, 1999). The use of PMR data for this application relies on the presence of a relatively low frequency channel (around 6 GHz). The last PMR with a channel at this frequency was SMMR, which ceased operation in 1987. However, the recently launched AMSR instrument has redressed this omission and also improved the spatial resolution of the data (Table 4.6).

Ablation

Ablation can either be modelled using a surface energy balance model (EBM) or a positive degree-day (PDD) approach (see Chapter 5) or it can be measured directly or inferred indirectly. Satellite observations of surface temperature may be useful if a PDD method is adopted but, if based on thermal infra-red data (from e.g. AVHRR), are limited to cloud-free times only. This makes the data, in practice, difficult to use as it introduces a bias to any estimate of temporally averaged surface temperature that is difficult to correct for. Infra-red radiometers can also provide information on some of the parameters required to drive an EBM such as long-and short-wave radiation fluxes. In particular, sensors such as Landsat, AVHRR and ATSR-2, which possess several channels in the visible, have been used to determine broad-band albedos of snow-covered terrain (Haefliger, Steffen and Fowler, 1993).

Attempts have also been made to use Landsat and, more recently, SAR data to identify the equilibrium-line altitude (ELA) of glaciers, which is a measure of the relative mass balance of the glacier for that year. It is difficult, however, to distinguish the snow line from the equilibrium line and, although they are related, this adds some uncertainty to the use of this approach for relative mass balance studies.

Satellite sensors cannot, however, provide much useful information related to the sensible and latent heat fluxes at the surface, which are dependent on parameters such as wind speed, humidity and near-surface air temperature. As a consequence, remote-sensing data are of relatively limited value for determining ablation rates using a modelling approach and require extensive supplementation with *in situ* data.

There has been no success in inferring ablation rates directly from space, although PMR data from SSM/I has been used to infer the length and extent of melting over the Greenland ice sheet (Abdalati and Steffen, 1995), which is, to some extent, related to ablation rates. Over larger ice masses (such as ice caps in the Arctic, Patagonia and Greenland), it is hoped that the accuracy and relatively narrow footprint size of GLAS may enable direct measurement of ablation rates under cloud-free conditions.

Iceberg calving

Visible and radar imagery has been used to determine the size and frequency distribution of icebergs calving from the Greenland and Antarctic coastlines (Frezzotti, 1997; Higgins, 1990; Young *et al.*, 1998). Figure 4.10 is part of the RADARSAT mosaic (RAMP; Jezek, Sohn and Noltimier, 1998) of Antarctica showing a large tabular iceberg that calved from the Filchner ice shelf, West Antarctica. A number of smaller icebergs can also be identified as bright patches within the darker fast-ice. Translating these data into ice flux, however, is difficult due to (i) the temporal variability in calving rates, especially for large, tabular icebergs that are produced on an irregular, decadal timescale, and (ii) limited information on keel depths/ice thickness.

Bottom mass balance of floating ice

Until recently, the importance of basal melting under floating ice tongues had been largely overlooked. Recent estimates of melt rates, derived indirectly from satellite measurements (see p. 88) indicate, however, that melt rates can be significantly higher than previously believed and may reach values of as much as 60 m per year near the grounding line (Rignot, 1998; Rignot *et al.*, 1997a). Due to the difficulty in measuring sub-ice melt rates, this part of the mass balance can introduce large errors into an estimate of the total mass balance from a sum of the individual components. For northern Greenland the measured grounding line flux exceeded the estimate of iceberg flux by a factor of 4.5, the difference being largely attributed to bottom melting (Rignot *et al.*, 1997a). For the Amery ice shelf, it has been estimated that as much as 60% of the meteoric ice is lost by bottom melting, while there are substantial areas of marine ice accretion near the ice front (Fricker, Warner and Allison, 2000; Fricker *et al.*, 2001). Combined with the difficulties in determining iceberg volumes,

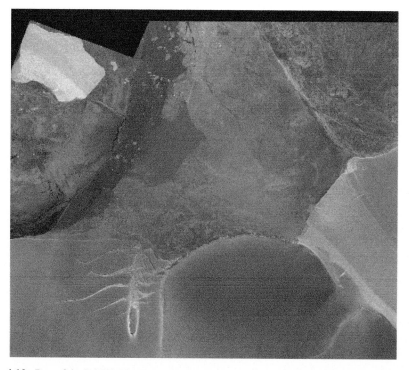

Figure 4.10. Part of the RADARSAT mosaic of Antarctica covering the northern limit of the Filchner ice shelf, showing the presence of a large tabular iceberg and a number of small icebergs.

errors in estimating the state of balance of the ice sheets using the component approach, as described above, are now felt to be too large to provide a useful result.

Grounding-line fluxes

The difficulties in determining calving flux and bottom mass balance can be circumvented if a flux divergence or grounding-line (GL) flux estimate approach is used. In the latter case, discharge is calculated at the GL from observations and compared with the net mass balance for the drainage basin upstream of the GL. Figure 4.11 illustrates schematically the measurement concept and the parameters that need to be determined. This approach is most suitable for outlet glaciers and ice streams in Antarctica and Greenland, where a well defined floating tongue exists. It is not appropriate for a glacier tongue that is grounded on bedrock. The key parameters that are required are (i) the position of the grounding line, (ii) ice velocity, (iii) ice thickness and (iv) balance flux at the GL. This latter quantity is simply the integral of the net accumulation for the basin upstream of the GL. The area of the basin is usually determined from a high accuracy DEM (Hardy, Bamber and Orford, 2000; Vaughan *et al.*, 1999) and net mass balance from *in situ* observations, as discussed earlier. Methods for determining the ice thickness and velocity are discussed next. A number of significant mass balance studies have adopted this GL flux approach using satellite radar

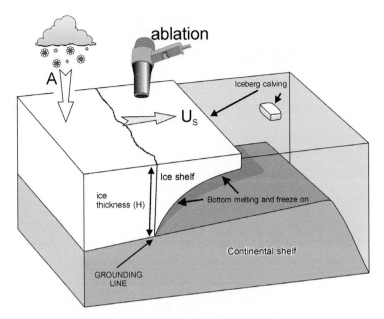

Figure 4.11. Idealized grounding line, indicating the parameters that need to be estimated to determine the mass balance, namely: accumulation, *A*, ice thickness across the grounding line, *Z*, the location of the grounding line, and the velocity of the ice across it, *U*.

interferometry to derive both the velocity field and GL location for a swathe of outlet glaciers and ice streams draining the Greenland and Antarctic ice sheets (Joughin and Tulaczyk, 2002; Rignot, 1998, 2002; Rignot *et al.*, 1997a, 2000).

Determination of ice thickness

As mentioned above, to determine the GL ice flux it is necessary to know the ice thickness. For floating tongues or ice shelves that have a width several times greater than the thickness it is a reasonable assumption that, at some distance (of the order of a few ice thicknesses) downstream from the GL, the ice is in hydrostatic equilibrium. If the ice and sea-water densities and mean sea level are known it is then possible to determine the thickness from the surface elevation (Bamber and Bentley, 1994; Rignot *et al.*, 1997a). Due to high basal melt rates near the GL and the effect of shear stresses acting around the edge of the floating ice, this may not be an accurate estimate of the ice-thickness profile across the GL (Reeh *et al.*, 1997; Rignot *et al.*, 1997b). The other difficulty with this approach is that the density of the ice must be corrected for the overlying, lower density, firn layer. The thickness of this layer can vary between a few metres and about 120 m (Paterson, 1994), introducing a potentially large uncertainty for the averaged, column ice density.

If it is not possible to invert surface elevation, then the most commonly used method of determining thickness is by airborne or ground-based radio-echo sounding (RES) (Bogorodskiy and Bentley, 1985). Ice is semi-transparent to radio waves between about

50 and 400 MHz. It is possible to send radio wave pulses through the ice and obtain a reflection from the ice/bed interface (Bogorodskiy and Bentley, 1985). If the speed of the radio wave in ice is known, it is possible to estimate thickness from a measurement of the two-way travel time of the radio pulse. The accuracy of this technique is primarily dependent on knowledge of the density of the column of firn/ice being sounded. The accuracy of airborne measurements over the Greenland ice sheet are about ± 10 m, for example (Gogineni *et al.*, 1998). Typical grounding-line thicknesses are about 500–1000 m, so the measurement error is likely to be only 1–2%. There are, however, rarely RES data along the GL, as they more often follow the flow direction (i.e. perpendicular to the GL), providing a thickness at only one or possibly two points across the GL.

Velocity and grounding-line estimation

There are two principal methods for determining ice motion from space: feature tracking and interferometric SAR (InSAR). Radar interferometry is a technique that combines coherent radar images recorded by antennas at different locations and/or at different times to form interferograms that permit the detection of small differences in range to a target from two points of observation. The resolution in the range differences is much better than the wavelength (λ) of the transmitted signal. At microwave frequencies, λ is of the order of centimetres. InSAR measurements are extremely useful for construction of high resolution topography and surface change or motion fields (Kwok and Fahnestock, 1996). For a given set of repeat-pass observations, from the *i*th and *j*th epochs, with baseline B_{ij} (Figure 4.12) and look-angle θ, the interferometric phase difference at each sample is given by

$$\Delta\phi_{ij} = (4\pi/\lambda)B_{ij}\sin(\theta - \alpha_{ij}) + (4\pi/\lambda)\Delta\rho_{ij} = \phi_{\text{top}} + \phi_{\text{disp}}, \qquad (4.9)$$

where B is the distance separating the two points and α is the tilt of the baseline with respect to the horizontal. The difference in the range path length $|\mathbf{r}_i - \mathbf{r}_j|$ is approximated by $\mathbf{B} \cdot \mathbf{r}$. Clearly, to obtain information on surface motion the *i*th and *j*th observations must be made at different epochs, while the baseline must be less than a few hundred metres, to prevent what is known as phase ambiguity problems. This is termed repeat-pass interferometry, and the most useful mission(s) to date for this type of InSAR were ERS-1 and 2. This is because, for reasons discussed below, a relatively short time interval (of a few days) is desirable and ERS-1 was, for part of its life, in a 3-day repeat cycle (Table 4.7); additionally, the two satellites were placed in orbits that followed each other by one day (known as the tandem phase) for nine months during 1995.

   The term ϕ_{top} in equation (4.9) contains phase contributions from the topography of the Earth surface relative to the interferometric baseline. The sensitivity of the measurements to surface relief is directly proportional to the length of the baseline. If the targets are displaced by $\Delta\rho_{ij}$ in the range direction between two observations, then the observed phase will include a second contribution of $4\pi/\lambda\Delta\rho_{ij}$ due to this displacement. This additional term, ϕ_{disp}, is independent of the spatial baseline. When the *i*th and *j*th observations are acquired at the same time (single-pass interferometry), only the first term is relevant. If the *i*th and *j*th observations are separated by a time interval, and the surface has moved in this interval,

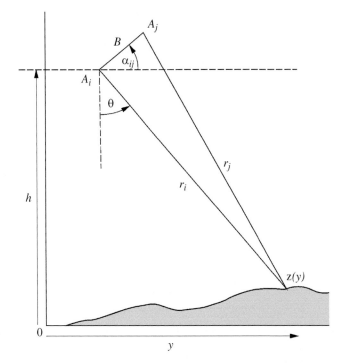

Figure 4.12. The geometry of repeat-pass interferometric SAR observations.

then both terms on the right hand side of equation (4.9) contribute to the interferogram. The reader is referred to the following articles for more detailed descriptions of the general principles of imaging radar interferometry: Goldstein, Zebker and Werner (1988), Li and Goldstein (1990), Rodriguez and Martin (1992), Zebker and Goldstein (1986). Although the technique has provided a wealth of new information about ice dynamics and mass balance (some of which is discussed further in Chapters 10 and 12) it has certain limitations. Firstly, a single pair of repeat-pass images provides the displacement in the look-direction of the SAR only. If this direction happens to be more than about 60° away from the flow direction, then the data are of little use (Bamber, Hardy and Joughin, 2000a). In addition, if the displacement between consecutive pixels on the ground translates to more than 2π change in phase (equivalent to 7.6 cm horizontal displacement for ERS-1), then a phase ambiguity exists, making determination of velocity more difficult. Finally, there is the problem of temporal de-correlation of the surface over time. This is caused by changes to the scattering characteristics of the surface between image acquisitions due, for example, to a snow-fall event, melting or wind-blown snow, and results in loss of fringes in the interferogram. This particular problem severely limits the usability of repeat-pass interferometry where melting is taking place (i.e. summer-time data for most glaciers outside of Antarctica).

The second technique for determining ice velocities is known as feature tracking and, as the name suggests, involves measuring the displacement of uniquely identifiable 'features'

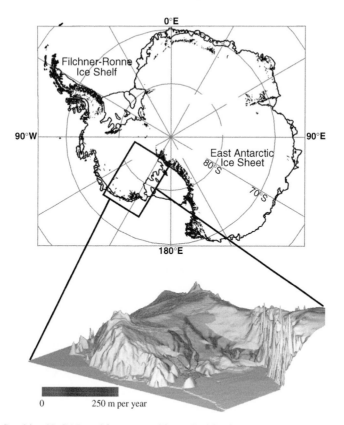

Figure 4.13. Combined InSAR and feature-tracking velocities for the Siple Coast ice streams in West Antartica. The velocities are superimposed on a digital elevation model to indicate the pattern of flow with respect to the topography.

on the surface of the ice. Using visible imagery from Landsat or SPOT, for example, limits this technique to areas where there is sufficient contrast/features present, which is not the case for much of the interior of Antarctica and Greenland. However, the method has been successfully used to derive velocities for ice streams and outlet glaciers around the margins of the ice sheets (Bindschadler and Scambos, 1991; Bindschadler *et al.*, 1996; Lucchitta and Ferguson, 1986; Scambos *et al.*, 1992). Recently, an extension of this approach has been developed that uses as 'features' the pattern of speckle found in SAR data over snow (Gray *et al.*, 1998). An example of velocities derived from both speckle tracking and InSAR for part of the Siple Coast ice streams in West Antarctica is shown in Figure 4.13.

As mentioned earlier, to estimate grounding line flux, it is not only necessary to determine the velocity of the ice but also the position of the grounding line (Figure 4.11). The gradient in ice thickness close to the GL is usually large, which means that a small error in GL location can result in a relatively large error in ice flux. Attempts to locate the GL have been made by identifying the break in slope in visible imagery or radar altimeter-derived topography. This is, however, an indirect estimate of its location and possesses an associated uncertainty of at least several times the ice thickness. However, in addition to determining surface velocity, InSAR measurements can also be used to obtain the location of the hinge line (Rignot, 1996) if multiple interferograms are available. The method is based on the assumption that the horizontal motion is the same in each interferogram and differences in fringe density are due to vertical motion caused by tidal flexure. The hinge line thus defines the limit of tidal flexure, which is closely related to the position of the GL and can be located with an accuracy of about 100 m (Rignot, 1996).

4.4.3 Balance velocities and fluxes

Over an ice sheet, the gravitational driving force that makes the ice flow is a function of the surface slope and ice thickness. If the slope is estimated over an appropriate distance (typically 20 times the ice thickness) it is reasonable to assume that the ice flows downhill (Paterson, 1994). It is therefore possible to trace particle paths down-slope, from an ice divide to the coast. If the net mass balance (accumulation−ablation) is integrated along these flow lines, then the ice flux at any point can be estimated. The depth-averaged velocity at a point is simply this flux divided by the ice thickness. This estimate of velocity is what is required to keep the ice mass in steady-state. Hence the name: balance velocity. To determine the net mass balance of a glacier or ice stream at the GL, an InSAR-derived flux estimate must be subtracted from the balance flux described above (see also Chapter 2). This can be determined at the GL by integrating the net mass balance over the glacier catchment area. If the integral is calculated at every point within the drainage basin (rather than just at the GL), using the flow directions obtained from the surface slope, it is possible to determine the spatial pattern and vector values of balance flux (Budd and Warner, 1996). This has been done for the whole of the grounded portions of the Antarctic and Greenland ice sheets using satellite-derived topography, combined with terrestrial measurements of accumulation and ice thickness to derive both balance fluxes and velocities (Bamber *et al.*, 2000a; Bamber, Vaughan and Joughin, 2000b). A finite-difference model was used to integrate ice flux along particle paths (Budd and Warner, 1996). The result for Antarctica is shown in Figure 4.14, and provides the most detailed and comprehensive picture available of flow over the whole ice sheet. Floating ice shelves were not included as the assumptions concerning the influence of longitudinal stresses break down in these areas. Although these data cannot be used on their own to determine mass balance, when combined with observations of the present-day velocity field from InSAR data, for example, they can highlight important changes in flow regime and, hence, mass balance (Bamber and Rignot, 2002; Bamber *et al.*, 2000b).

4.5 Sea-ice mass balance: introduction

This section reviews current and promising remote-sensing techniques that address the questions of sea-ice mass balance. Sea-ice mass balance is defined as the difference between

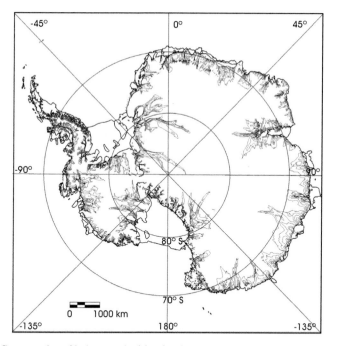

Figure 4.14. Contour plot of balance velocities for the grounded portion of the Antarctic ice sheet. The contour intervals plotted are 25, 50, 100, 200, 300 and 500 m per year.

ice production and the sum of ablation and ice export. Over an annual cycle, there is no net accumulation of sea-ice mass if ablation and export removes an equivalent volume of ice produced in the winter. Since the ice is a moving medium, the regional mass balance and seasonal distributions of sea-ice growth and melt are strongly dependent on its motion. At any given time, the net ice volume is simply the integral of the spatially varying sea-ice thickness over the ice-covered ocean. Thus, the three essential measurements that are crucial for monitoring the mass balance are the areal ice coverage, ice thickness and ice motion. As we shall see, we have varying success in making these measurements due to limitations in sensor technology and to our understanding of available remote-sensing data sets.

The polar sea-ice covers have always been data-poor areas compared with the more accessible lower latitudes of the Earth. Climatological studies, however, require a long data record. In the past, the processes of the Arctic Ocean have typically received more attention than the ice-cover of the Southern Ocean due to a variety of reasons. However, the situation is rapidly changing with advances in remote-sensing capabilities. For remote sensing of sea ice, active and passive microwave sensors are the current sensors of choice because of their day/night viewing capabilities, and these measurements are relatively uncontaminated by clouds, water vapour and other forms of atmospheric moisture. The longest continuous satellite record of sea ice, over 20 years, comes from passive microwave radiometers and contributes to its importance for multi-decadal climate studies (section 4.3.4 and Table 4.6).

Even though routine maps of the sea-ice cover from scatterometers and SARs have provided a much shorter record (since 1996), their contributions to sea-ice remote sensing is growing as these data sets become more readily available. Furthermore, SAR imagery has the added attraction of offering high resolution maps capable of resolving leads and floes.

In this section, our ability to measure sea-ice coverage, sea-ice motion and deformation, and sea-ice thickness is discussed. Of the three, ice motion and deformation in terms of displacements in sequential imagery is the only direct observable in remote-sensing data. Sea-ice coverage and sea-ice thickness estimates depend on inferences from radiometry, estimates of ice draft and deformation. The present goal is to provide an overview of the observational issues in the transformation of remote-sensing measurements into parameters of geophysical utility. The focus of this section is mainly on microwave measurement techniques and the quality of these estimates. The treatment of active and passive microwave remote sensing here might seem unbalanced, but this is a reflection of the relative maturity of the applications using passive microwave data compared with the newer developments in active microwave measurements. Procedures for retrieval of sea-ice parameters vary between the Arctic and Southern Oceans, the seasonal and perennial ice zones, and summer and winter. Some are better developed than others. This is by no means an exhaustive review; we consider here only those techniques that are relevant to mass balance calculations. Basic information on the spatial and spectral characteristics of the various satellite sensors discussed here can be found in section 4.3.

4.5.1 Sea-ice coverage – extent, concentration and type

Retrieval of ice concentration and extent

The commonly used measures of sea-ice coverage are ice concentration and ice extent. The former is the fractional coverage of an enclosed area, while the latter is the area covered by sea ice above a certain ice concentration threshold. This threshold is selected to remove noisy estimates of ice concentration near the ice edge.

Analysed fields of SMMR (scanning multi-channel microwave radiometer) and SSM/I (special sensor microwave/imager) ice concentration of the Arctic and Southern Oceans (Figure 4.15) on a 25 km grid have provided a continuous record for more than 20 years (Gloersen et al., 1992; Parkinson et al., 1987; Zwally et al., 1987). Due to the size of the radiometer footprints of tens of kilometres, a mixture of open water and ice is typically contained within each resolution element. Retrieval procedures generally attempt to solve for their relative coverage based on assumptions of the behaviour of their radiometric signature. The widely available ice concentration data sets are produced using two different algorithms. One approach, the NASA team algorithm (Cavalieri, Gloersen and Campbell, 1984; Cavalieri, et al., 1999), takes advantage of the high polarized emission of open water (OW) compared with that of ice, and the decreasing contrast between multi-year (MY) and first-year (FY) ice with increasing wavelength. These characteristics are captured by the polarization

Average Seasonal Ice Concentrations, 1996

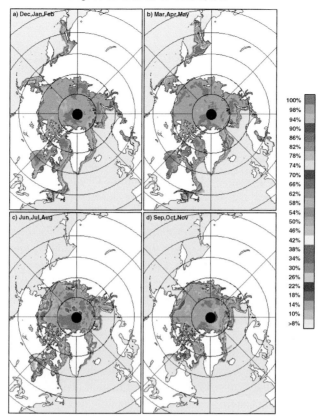

Average Seasonal Ice Concentrations, 1996

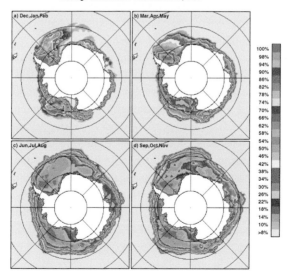

Figure 4.15. Seasonal sea-ice concentration fields of the Arctic (top panel) and Southern (bottom panel) Oceans derived from the SSM/I sensor. (Courtesy of J. Comiso, NASA Goddard Space Flight Center.)

ratio (*PR*) and gradient ratio (*GR*), viz.

$$PR = \frac{T_b^V(19) - T_b^H(19)}{T_b^V(19) + T_b^H(19)}; \qquad GR = \frac{T_b^V(37) - T_b^V(19)}{T_b^V(37) + T_b^V(19)}. \tag{4.10}$$

With the additional constraints that observed brightness temperature $T_b = C_{OW}T_{OW} + C_{FY}T_{CY} + C_{MY}T_{MY}$ and $C_{OW} + C_{FY} + C_{MY} = 1$, the fractional coverage of three surface types (C_{MY}, C_{FY}, and C_{OW}) is resolved in the solution. To determine the constants in the retrieval equations, this technique depends on fixed reference microwave signatures (tie-points) of OW, FY and MY ice at the two frequencies (37 GHz, 19 GHz) and polarizations (V, H). These signatures are estimated from pure samples taken from regions known to have high concentrations of the three surface types. These signatures are assumed to be constant over all seasons and locations.

In another approach, the bootstrap algorithm (Comiso *et al.*, 1997), the ice concentration is given as the ratio of the difference between the observed brightness temperature (T_b) and T_b of open water and the difference between T_b of ice and open water in the multi-channel data. The name 'bootstrap' stems from the fact that the reference T_b of pure open water and ice are determined from the data and therefore more adaptive to the spatially and seasonally varying signatures. This algorithm does not estimate ice type coverage. Comparison between the two algorithms (Comiso *et al.*, 1997) shows only small differences in the central Arctic in the winter, while larger disagreements can be found in the seasonal ice zones and during the summer. Recently, Hanna and Bamber (2001) reported on the advantages of a new passive microwave ice concentration algorithm that systematically accounts for the effect of changing environmental conditions on the tie-points used above.

Uncertainties in the retrievals are affected by varying surface and atmospheric conditions. As satellite passive microwave radiometers measure the emitted radiation of the surface at the top of the atmosphere, the observed brightness temperature is contaminated by atmospheric emissions (clouds, rain and water vapour). These weather effects have been reported (Maslanik, 1992; Oelke, 1997) and modifications have been made to the algorithms (Cavalieri, St Germain and Swift, 1995, Comiso *et al.*, 1997) to mitigate these effects. Due to the difficulty in obtaining ice concentration measurements, the estimates have been validated against only a limited number of measurements from aircraft radars, aircraft radiometers and Landsat imagery. A summary of these validation results can be found in Cavalieri (1992). The uncertainties in the ice concentration vary. In the Arctic and Antarctic winter, the uncertainties are approximately 6% with possible biasses of similar size. In the summer, snow wetness, surface melt and melt ponds are additional complications in the retrieval process (Comiso and Kwok, 1996). Melt ponds have been shown to reduce ice concentration estimates because of the presence of open water over thicker ice. In the summer, the estimated uncertainty is much higher (over 10%), and the ice concentration data are of less value. Using the 15% concentration isopleth as the ice edge gives an edge location uncertainty of approximately 12 km. In the future, the advanced microwave scanning radiometer (AMSR) onboard NASA's Aqua and the Japanese ADEOS II satellites will provide higher resolution and sensor performance (Table 4.6). It is anticipated that planned validation efforts will provide a more complete characterization of the ice concentration retrieval approaches.

Ice types

Since satellites sense only the surface and near-surface properties of the ice, one can relate the 'sensed' surface properties (e.g. salinity, roughness, etc.) to ice thickness, through the intermediary of 'ice type'. Thus, it is sometimes useful to describe the ice cover in terms of its fractional coverage by multi-year (MY) and first-year (FY) ice. MY ice is sea ice that has survived at least one summer's melt, whereas FY ice is seasonal and is removed during the summer. Simple ice types, FY and MY ice, as inferred from active (Kwok and Cunningham, 1992) or passive microwave (Cavalieri *et al.*, 1984) data sets can be used as a proxy, albeit crude, indicator of ice thickness.

An important climate property that distinguishes MY ice from FY ice is thickness. MY ice is generally thicker because its greater age corresponds to a larger cumulative energy deficit at the surface and therefore more growth by freezing. The climatic significance of MY ice coverage of the Arctic Ocean can be attributed to its strong relation to the summer ice concentration (Thomas and Rothrock, 1993). If changes in the climate cause persistent decreases in the summer ice concentration, it would be reflected in decreases in the winter MY ice coverage. This reduction would be due to increased melt during the previous summer or ice export through the Fram Strait. At the same time, the outflow of thick MY ice into the Fram Strait also represents a major source of surface fresh water for the Greenland–Iceland–Norwegian Seas, which are source regions of much of the deep water in the world's oceans (Aagaard and Carmack, 1989). An accurate record of the MY ice coverage and its variability is therefore crucial in understanding the relationship between climate and MY ice balance. An adequate description of the sea-ice cover requires the relative proportions of FY and MY ice to be known as a function of time. Even though this distinction between the two ice types is relatively simple, accurate estimates of the relative coverage of the two ice types in the Arctic Ocean have been difficult to obtain.

Ice types from passive microwave data

Ice type retrieval from satellite passive microwave data (Cavalieri *et al.*, 1984), using the NASA team algorithm discussed earlier, has been shown to be unreliable (Carsey, 1982; Comiso, 1986; Kwok and Comiso, 1998; Thomas, 1993) due to the following shortcomings. The estimated MY coverage in the winter is much lower than the minimum ice area coverage near the end of summer. If ice that survives the summer is classified as MY ice, then the MY coverage during the winter should be nearly equivalent to the ice coverage during the previous summer's minima, differing by an area due to melt, ridging, new/young ice formation and Arctic ice export. This mismatch between the winter and summer estimates creates a seasonal inconsistency and points to possible shortcomings in the analysis. In the passive microwave ice type retrievals, the variability in ice type fractions appears to be caused by spatial and temporal variations in the assumed signatures (Thomas, 1993). Based on models and observations (Kwok and Comiso, 1998; Thomas and Rothrock, 1993), the uncertainty in the MY ice coverage from passive microwave data could have uncertainties of up to 20%, with biasses of similar magnitude.

Ice types from active microwave data

SAR and scatterometer data have the advantage that they are relatively free of weather effects that sometimes plague sea-ice type retrievals from passive microwave data. Under cold dry winter conditions, there is significant contrast and stability in the radar signatures of MY ice and FY ice from C band (Kwok and Cunningham, 1992) to Ku band (Yueh and Kwok, 1998) frequencies. The importance of particular sea-ice characteristics and their effects on observed microwave signatures can be found in Winebrenner *et al.* (1989). Overall, the higher contribution of scattering from the MY ice volume, due to its lower salinity, accounts for its higher backscatter compared with that of the more lossy (saline) FY ice. The superimposed dry snow layer does not significantly affect the observed backscatter at C band and probably has a minor effect at Ku band. The signatures of younger and thinner ice types, however, are more complex and less well defined from a uniqueness point of view. Typically, only the principal ice types, FY and MY, are identified in classification procedures. These properties are exploited in the design of ice type classifiers.

At this time, ice type data sets are being produced in large volume from RADARSAT data (Kwok, 1998). Because of the higher resolution samples, the assumption of a mixture of ice types within a pixel is not necessary. The implementation uses a winter ice type algorithm consisting of a maximum likelihood classifier and a look-up table of expected backscatter signature to guide the labelling of each image pixel as one of two ice types (MY or FY). Wind-roughened open water and the presence of frost flowers on new ice are problematic. Both of these surface types exhibit highly variable radar backscatter, sometimes causing these pixels to be mislabelled when their signatures overlap (Fetterer, Gineris and Kwok, 1994; Steffen and Heinrichs, 1994.) To minimize the errors caused by these signature issues, the technique also includes a time series analysis of MY ice area from the same region. The assumption is that the area of MY ice should remain constant (no MY ice is created) over a winter, and any anomaly which shows up as a transient spike or hump could be filtered out. The preliminary results look promising. Based on two winters of RADARSAT analysis the MY ice coverage over the Arctic Ocean, as expected, remains quite stable throughout both winters. But more extensive validation is required; the next step would be to demonstrate the seasonal consistency of these estimates.

At moderate resolutions (~kilometres), scatterometers measure the radar backscatter from the ice cover at a range of incidence and azimuth angles. Scatterometer data are available at 5.3 GHz (European Earth remote sensing scatterometer – EScat) and 14 GHz (NSCAT and QuikSCAT). The Seawinds scatterometer on QuikSCAT, launched in May, 1999, is an NSCAT replacement (Table 4.5). The MY and FY ice signatures have also been shown to have good contrast and seasonally stable signatures that can be used for identification of sea ice types (Gohin and Cavanie, 1994; Long and Drinkwater, 1999;

Yueh and Kwok, 1998; Yueh *et al.*, 1997). Similar to passive microwave data, the size of the scatterometer footprints (>10 km) will necessitate a mixing formulation to resolve the ice types within a pixel sample. This approach has yet to be examined. Some investigators have moved toward blending the active and passive microwave data sets to eliminate some of the shortcomings of the passive microwave retrieval procedures discussed above (Grandell, Johannessen and Hallikainen, 1999). However, no operational procedure has yet been devised. Instead of ice type classification, Kwok, Cunningham and Yueh (1999a) used a simple backscatter-based procedure to delineate the winter perennial ice zone (PIZ – area with higher than 90% MY fraction) in NSCAT and QuikSCAT data. They used a simple area balance of PIZ coverage within the Arctic Ocean to examine the validity of the results. Over the winter, the PIZ coverage within the Arctic Ocean is a function of the ice deformation and export. After accounting for export, they showed that the PIZ area could be balanced to within 1% of the area of the Arctic Ocean. From the two years of NSCAT and QuikSCAT analyses, the Arctic PIZ coverage at fall freeze-up is comparable to the minimum passive microwave ice extent during the summer, satisfying the seasonal consistency constraint. There is considerable work to be done to develop and validate these data sets.

The MY ice coverage differs between the Arctic and Antarctic. Because Antarctic ice is so efficiently advected from the coastal regions north to the marginal areas where it melts, MY ice appears to be scarce around most of Antarctica. Large areas of Antarctic sea ice remain virtually unstudied (Comiso *et al.*, 1992). Region-scale ice typing using active microwave data has not been attempted to the degree of that of the Arctic Ocean (Drinkwater, 1998). In the Arctic, as in the Antarctic, sea-ice typing cannot be carried out reliably in the summer due to surface melt. The presence of liquid water in the snow layer prevents any penetration of the radio waves into the ice layer and therefore eliminates the contribution of any scattering from the ice layer which serves to assist in ice type discrimination.

4.5.2 Sea-ice motion and deformation

The large-scale circulation of sea ice determines the advective part of the ice balance and its spatial gradients determine the deformation of the ice cover. At small scale, divergent motion controls the abundance of thin ice and the many surface processes dependent on thin ice, such as turbulent heat flux to the atmosphere and salt flux into the ocean. Convergent motion increases the mass of the local ice cover by rafting and ridging. These openings and closings in the ice cover are typically located along linear fractures up to hundreds of kilometres in length. Because of these fractures, the field of ice motion is spatially discontinuous as a result of these features.

Retrieval of sea-ice motion

Sea-ice motion can be readily obtained from satellite data by tracking the displacement of common features in time sequential imagery. The only requirement is that these features can be identified over a sampling interval. The quality of these measurements depends more on the geometric fidelity and resolution of the imagery than on a thorough physical understanding of the ice signatures themselves (Holt, Rothrock and Kwok, 1992). The spatial scale of the ice motion that can be resolved in remote-sensing data sets is dependent on the resolution of the available imagery, and images that can resolve the floes and leads are attractive because they allow dense sampling of the motion of the ice cover. Timescale is an important factor in the sampling of ice motion. On short timescales (one day), ice motion responds largely to wind forcing. Tidal and inertial effects on ice motion occupy an even shorter timescale. Repeat coverage of the ice-covered seas, crucial to the adequate sampling of ice motion, however, is usually constrained by the orbit characteristics and swath width of the sensor. At this point, it is difficult to have better than daily sampling frequency on a large scale. Clearly, temporal sampling and spatial resolution are important considerations in the quality of ice motion measurements.

Ice motion fields have been produced from a variety of sensors. It has been demonstrated (Agnew, Le and Hirose, 1997; Emery, Fowler and Maslanik, 1997; Kwok *et al.*, 1998; Liu and Cavalieri, 1998) that despite antenna footprints of ten or more kilometres, data from low resolution passive microwave radiometers and scatterometers can provide rather coarse measurements of ice motion (Figure 4.16). The combination of daily ice motion from the 85 GHz channel of SSM/I and 2 day ice motion from the 37 GHz channel of SSMR has provided an ice motion data record dating back to 1978. The quality of the scatterometer motion fields ($\sim$5 km uncertainty in displacement), obtained from NSCAT and QuikSCAT is comparable with that derived from the 85 GHz SSM/I data. The daily ice motion measurements from QuikSCAT seems complementary to the passive microwave observations. The relative merits of the scatterometer versus the passive microwave ice motion remain to be examined. Summer ice motion from scatterometers and radiometers are unreliable due to surface melt and, in the case of passive microwave data, the added contamination by increased atmospheric water content with increasing temperature. With the level of uncertainty from these low resolution data, these measurements are more suited to the study of large-scale circulation patterns (Emery *et al.*, 1997; Kwok, 2000) and ice export (Kwok and Rothrock, 1999; Martin and Augstein, 2000), rather than the small-scale processes associated with openings and closings of the ice cover.

Sequential AVHRR imagery provides moderate resolution ice motion from its visible, near infra-red, and thermal infra-red bands with uncertainties of 1 km and 5 km, depending on the data products. The thermal bands can be used in winter darkness. The only drawback is that clouds obscure the surface, moderately during the winter and quite substantially during the summer. An ice motion data set dating back to 1982 is being compiled.

High resolution ice motion from SAR

In the early 1990s, the availability of small volumes of high resolution ice motion data from SAR data (ERS-1, ERS-2 and JERS-1) allowed a more detailed look at the small-scale deformation of the ice cover. The uncertainty in SAR ice motion, approximately 100–300 m, is at least an order of magnitude better than that from other satellite sensors and

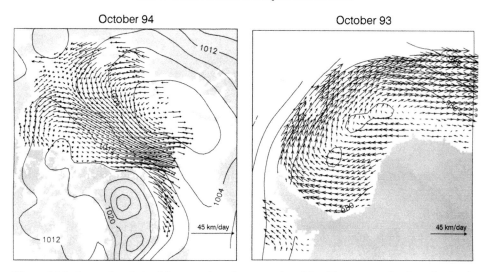

Figure 4.16. Ice motion derived from passive microwave data. Monthly mean motion from the Arctic Ocean and the Weddell Sea. Sea-level pressure contours are overlaid (contour interval: 4 hPa).

is comparable to that obtained from buoy drift. These data sets have been put to a variety of uses. Fily and Rothrock (1990) examined digital methods to determine the linear openings and closings in the ice cover. Stern, Rothrock and Kwok (1995) studied the parameterization of open water production based on area-averaged deformation. They showed that the small-scale (several kilometre) lead activity can be parameterized fairly well in terms of the large-scale (several hundred kilometre) strain invariants. Cunningham, Kwok and Banfield (1994) measured the orientation of newly opened leads and their relation to large-scale forcing in a small number of SAR images. Overland *et al.* (1998) used SAR ice motion to examine the granular plastic properties of sea ice.

At the time of writing, the largest source of SAR imagery of the Arctic Ocean is from Canada's RADARSAT, launched in November, 1995. The orbit configuration and wide swath mode (460 km) of the C band radar provide near repeat coverage of the high latitudes at three- and seven-day intervals. Beginning in November, 1996, and continuing today, 3-day SAR maps of the Arctic Ocean are being acquired. These RADARSAT images are routinely fed into the RADARSAT geophysical processor system (RGPS), a data system that converts the sequential SAR maps into basin-wide fields of ice motion and deformation, and estimates of ice age and thickness (Kwok, 1998; Kwok *et al.*, 1995). The uniqueness of this system is that it tracks the ice motion on a grid that moves and deforms with the ice cover, beginning with a uniformly spaced 10 km grid at some initial time. Thus, a trajectory identical to that from buoy drift can be formed from a sequence of RGPS motion observations of any grid point. Moreover, this approach allows a detailed description of the deformation history of Lagrangian elements enclosed by points on the grid. The use of

RGPS deformation measurements to estimate ice thickness is described in the following section. These near three-day repeat RADARSAT mappings allow the routine observation of small-scale ice motion and deformation over a large part of the Arctic Ocean. Tracking ice motion in SAR imagery during the summer, though sometimes difficult due to atmospheric and ice conditions, is possible because features in the high resolution data remain distinct and are not entirely masked by snow melt.

Small-scale ice motion and deformation

As the RGPS measurements are relatively new, we digress here to discuss the relative significance of small-scale deformation in sea-ice studies. For the first time, quasi-linear fracture patterns of the scale of kilometres to hundreds of kilometres are revealed in the high resolution deformation fields of the sea-ice cover produced by the RGPS (Figure 4.17). They appear as discontinuities separating regions of uniform ice motion. These features are expressions of one of five processes: opening, closing, shear, shear with closing and shear with opening. Open water is created during an opening event and ridges are formed during a closing event. Shear, however, does not always result in convergence or divergence which modify the sea-ice thickness distribution. The patterns observed in RGPS products are clearly related to past ice deformation, and their patterns and orientations are related to the constitutive material behaviour of the ice pack and the way the pack interacts with the coast (Kwok, 2001). At this point, there is certainly nothing that we could hold out as a climatology of fracture patterns properties – abundance, spatial distribution, length, orientation and development.

The current generation of models assumes the sea ice to be an isotropic viscous–plastic material. Over large length-scales, say several hundred kilometres, it is often assumed that as more leads of random orientation appear in each neighbourhood, the behaviour can be approximated as isotropic. However, it is evident from the RGPS observations (Figure 4.17) that at the length-scales of kilometres, fracture patterns are spatially oriented, organized and sometimes persistent. When a fracture is present, it has weaker strength across it, yet it retains its strength along its length. The dynamical behaviour of this strongly oriented, or anisotropic, material must differ in these directions (Coon *et al.*, 1998; Hibler and Schulson, 2000). Thus, pack ice is an anisotropic material for most of the year, with the ice dynamics affected by fractures that appear as velocity discontinuities or slip lines in RGPS ice motion products.

With the advent of high resolution coupled ice–ocean models (e.g. Zhang, Maslowski and Semtner, 1999; Zhang *et al.*, 1998) that approaches the widths of leads, the distinction between isotropy and anisotropy has become even more significant. The consequence of an anisotropic ice cover is not well understood in terms of the modelled surface heat and mass balance. The small-scale ice motion available in the RGPS data set is a crucial component in the testing of new models that account for the spatial and temporal characteristics of the linear kinematic features observed.

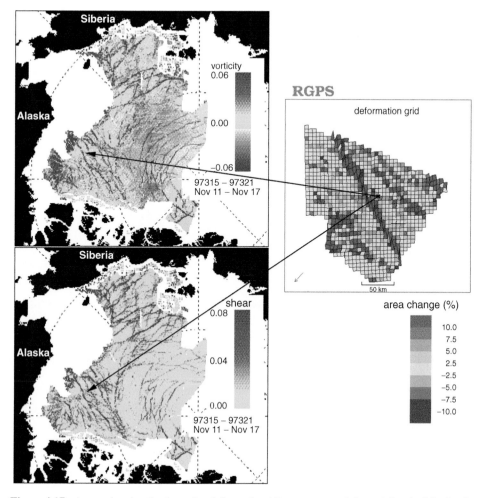

Figure 4.17. A map showing the three-day deformation (divergence, vorticity and shear) of the Arctic Ocean ice cover from the RADARSAT geophysical processor system (RGPS). Each sample represents the deformation of a grid element approximately 10 km on a side.

4.5.3 Sea-ice thickness

Two phenomena act to alter the thickness characteristics of floating ice: thermodynamic processes are responsible for mass changes at the upper and lower boundaries of the ice; and mechanical processes, resulting from non-uniform motion of the ice, cause the formation of leads and pressure ridges. Thus, understanding the ice-thickness distribution of sea ice is essential for quantifying its mass balance.

Because of the variability of ice thickness on all scales from metres to kilometres, it is difficult to construct useful averages using limited data from moored and submarine upward looking sonars. Remote determination of ice thickness at almost any spatial scale

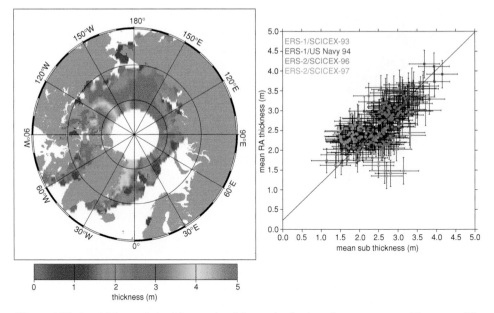

Figure 4.18. Ice thickness derived from radar altimeter ice freeboard measurements. (Courtesy of S. Laxon, University College London.)

has long been desired. Satellites, however, do not see the lower surface of the ice. Current sensors can see only radiation emitted or scattered from the top surface or the volume of the top few tens of centimetres of the ice. Still, three promising approaches that utilize only surface measurements, discussed below, have emerged in recent years that allow us to address different ranges of the thickness distribution. These techniques use measurements from radar altimeters, small-scale ice motion from SAR and infra-red imagery.

Radar altimetry

Sea-ice freeboard can be estimated by careful analysis of altimeter waveforms from the sea-ice cover. Laxon (1994) pointed out that height deflections associated with diffuse echoes over ice-covered seas can be used to estimate ice freeboard. The first example of ice freeboard measurements is given in Peacock *et al.* (1998). Since only 11% of the floating ice is above the ocean surface, freeboard measurement errors are magnified when applied to estimating ice thickness. Comparisons of altimetric ice-thickness estimates with observations of ice draft from moored and submarine mounted upward looking sonars are under way (Figure 4.18). Presently, the best estimate of the uncertainty in the retrieved ice thickness is approximately 0.5 m. Figure 4.18 shows an example of a gridded (100 km) field of altimeter ice thickness derived from the ERS-2 radar altimeter. The radar altimeter measurements address the mean ice thickness over length-scales of perhaps 100 km.

There are still a number of difficulties associated with understanding the achievable accuracy of the freeboard measurement. Some examples include the possible height biases introduced by the snow layer and the dependence of the height measurement on the location of the scattering centre (snow–ice boundary or within the snow layer). Another issue is the variability in the estimates due to ice advection and deformation, since height estimates within a grid cell are taken from measurements obtained over one month. Nevertheless, this technique represents a significant advancement in the measurement of sea-ice thickness. The impact of this method for climate and sea-ice studies would be enormous if continual long-term direct observations of ice freeboard and thence ice thickness can be realized. In the near term, the NASA ICESat/GLAS laser altimeter and the planned ESA Cryosat radar altimeter missions are designed to address these observational need (see section 4.3.3).

Seasonal ice-thickness estimates from kinematics

This approach utilizes the small-scale ice motion measurements from the RGPS (Kwok, 1998; Kwok *et al.*, 1995). Ice age and thickness are estimated from repeated observations of Lagrangian elements or cells of sea ice in sequential SAR imagery. Cells are ice areas enclosed by line segments connecting points on the deforming grid. The drift and deformation of a cell over time are obtained by the displacement of its vertices obtained from the ice motion. A time sequence of area changes is used to estimate the ice thickness within each cell. An increase in area is interpreted as the creation of an area of open water where new ice is assumed to grow. A decrease in area is taken up by the ridging or rafting of the youngest ice in the cell. The assumption is that once ridging starts, the deformation tends to be localized in the recently formed thinner and weaker ice in leads. In conjunction with an ice growth model and a ridging/rafting model, it is possible to maintain a seasonal ice-thickness distribution within each cell. The growth rate is approximated as a function of the number of freezing-degree days associated with each thickness category using Lebedev's parameterization (discussed in Maykut, 1986) with $h = 1.33F^{0.58}$, where h is thickness and F is the accumulated freezing-degree days derived from 2 m air temperature. This relationship is based on 24 station years of observations from various locations in the Soviet Arctic, and describes ice growth under 'average' snow conditions. Volume is conserved when ice is ridged or rafted (thinner ice rafts, while thicker ice ridges). Rafted ice is twice the original thickness and half the area, whereas ridged ice is five times its original thickness and occupies one-quarter of the area (Parmerter and Coon, 1972).

The process is illustrated in Figure 4.19. The nominal repeat sampling of the cell over the six-month period shown here is ~3 days. The decrease in cell area between days 312 and 335 created ice ridges in the cell. On day 339, the opening of two new leads introduced a large area of open water and thus a category of thin ice. This is followed by more openings between days 339 and 359. The cell is now covered with five different categories of sea ice in the 10–60 cm thickness range. A series of closings after day 359 ridged most of the thin ice in the cell. Only small changes in the cell area are evident after day 30. At the end of the 171-day period, the one largest category of undeformed ice is 1.5 m thick and occupies ~40% of the initial area.

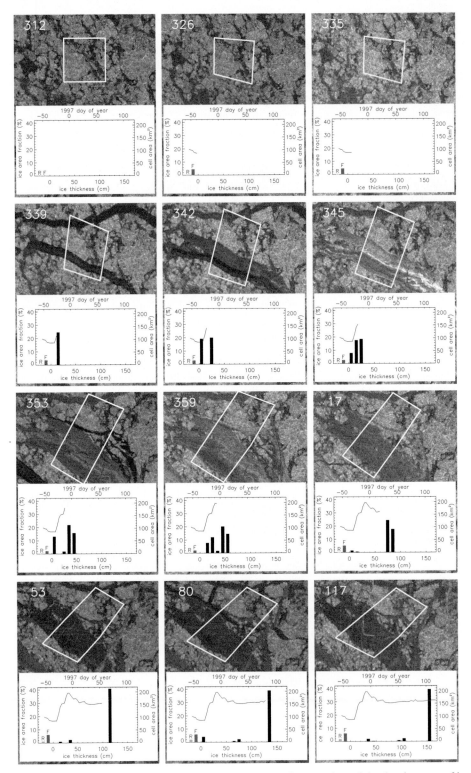

Figure 4.19. The deformation of a Lagrangian element over five months and the development of its seasonal thickness distribution derived from its deformation history. (After Kwok and Cunningham (2002).)

At start up, the ice-thickness distribution within each cell is not known. In this process, the ice volume accumulated over the length of the RGPS record represents the seasonal ice volume produced by kinematics and thermodynamics since initialization. The uncertainty in the thickness estimates is dependent on the length of the sampling interval. There are a number of sources of uncertainty in the ice-volume and ice-thickness calculations (Kwok and Cunningham, 2002). Several investigators are moving toward a comparison of the estimated ice thickness with AVHRR retrievals (discussed below) and submarine ice draft measurements.

The thickness distribution estimated by the RGPS analysis includes only that of the seasonal undeformed and deformed ice, since the procedures work only during the growth season. This provides only one component of the entire thickness distribution. Still, this is the crucial thickness range that produces the most ice growth, the most turbulent heat flux to the atmosphere and the most salt flux to the ocean. This is a unique approach for estimating basin-scale ice production from observations of ice motion in SAR images of the Arctic Ocean.

Ice surface temperature and ice thickness

The basis of this approach is that thin ice is thermally warm and dark, and that thick ice is cold and bright. The surface temperature of thin ice is closer to the freezing point at the bottom boundary-while the surface temperature of thick ice is close to that of the air temperature. During the winter, this large contrast in the ice surface temperature between thick and thin ice can be used to estimate ice thickness of up to one metre. Yu and Rothrock (1996) demonstrate that this thin ice thickness can be estimated using surface measurements from AVHRR and a surface energy balance model. The surface measurements include surface temperature, the near-surface air temperature, both from the thermal channels, and the surface albedo from the visible channels. With these measurements, the ice thickness can be solved with appropriate parameterizations of the remainder of the terms in the surface energy balance equation. These retrieved AVHRR thickness estimates have been shown to be comparable to the thickness measurements from moored upward looking sonars. The uncertainty in the cumulative thickness distribution is several per cent for thinner ice and higher for ice closer to one metre thick. Issues associated with this retrieval procedure include the effects of clouds, snow depth and the relatively low resolution of the AVHRR sensor (~1 km) for resolving leads. This method has not been used to produce large data sets.

4.6 Summary

A number of remote-sensing techniques germane to the subject of land- and sea-ice mass balance have been reviewed here. The importance of validation and understanding of the uncertainties of retrieved parameters should be emphasized. Remote sensing provides synoptic-scale observations and presents a challenge for extensive validation of such observations. This is especially true of remote-sensing data of the Arctic and Southern Ocean sea-ice covers with their large seasonal and inter-annual variability and for the huge and remote expanse of the Antarctic ice sheet. Two avenues are available. One of these is the comparison of various types of remote-sensing data with each other using arguments of internal consistency. Another approach has been that of comparing aircraft, satellite and *in situ* data for given seasons and locales. The former is by far the more popular approach due to the cost of field programmes associated with collecting *in situ* and aircraft data sets. Therefore, it takes a number of years and considerable work to develop these data sets and to quantify and understand the uncertainties of the retrieved parameters.

It is clear from this review that the quality of the measurements are far from adequate for addressing all the questions of land- and sea-ice mass balance. Remote-sensing techniques for determining accumulation and ablation rates of land ice are in their infancy, and there are, currently, no satisfactory methods for measuring sub-polar and alpine glacier mass balance directly from space.

Remote sensing of summer sea-ice cover is a problem, and progress has been slow. Development of the ice coverage and ice motion data sets have progressed in recent years. Ice thickness remains an issue. The several ice-thickness determination techniques discussed here are producing promising results. They will no doubt have important roles and will contribute to the subject of sea-ice mass balance in the coming years.

References

Aagaard, K. and Carmack, E. 1989. The role of sea ice and other fresh water in the Arctic circulation. *J. Geophys. Res.* **94** (C10), 14 485–98.

Abdalati, W. and Steffen, K. 1995. Passive microwave-derived snow melt regions on the Greenland ice-sheet. *Geophys. Res. Lett.* **22** (7), 787–90.

Agnew, L., Le, H. and Hirose, T. 1997. Estimation of large scale sea ice motion from SSM/I 85.5 GHz imagery. *Ann. Glaciol.* **25**, 305–11.

Arthern, R. J. and Wingham, D. J. 1998. The natural fluctuations of firn densification and their effect on the geodetic determination of ice sheet mass balance. *Climatic Change*, **40** (3–4), 605–24.

Bamber, J. L. 1994. Ice sheet altimeter processing scheme. *Int. J. Remote Sensing* **14** (4), 925–38.

Bamber, J. L. and Bentley, C. R. 1994. A comparison of satellite altimetry and ice thickness measurements of the Ross ice shelf, Antarctica. *Ann. Glaciol.* **20**, 357–64.

Bamber, J. L. and Huybrechts, P. 1996. Geometric boundary conditions for modelling the velocity field of the Antarctic ice sheet. *Ann. Glaciol.* **23**, 364–73.

Bamber, J. L. and Rignot, E. 2002. Unsteady flow inferred for Thwaites Glacier and comparison with Pine Island Glacier, West Antarctica. *J. Glaciol.* **48** (161), 237–46.

Bamber, J. L., Hardy, R. J. and Joughin, I. 2000a. An analysis of balance velocities over the Greenland ice sheet and comparison with synthetic aperture radar interferometry. *J. Glaciol.* **46** (152), 67–72.

Bamber, J. L., Vaughan, D. G. and Joughin, I. 2000b. Widespread complex flow in the interior of the Antarctic ice sheet. *Science* **287** (5456), 1248–50.

Bayr, K. J., Hall, D. K. and Kovalick, W. M. 1994. Observations on glaciers in the eastern Austrian Alps using satellite data. *Int. J. Remote Sensing* **15** (9), 1733–42.

Bindschadler, R. A. and Scambos, T. A. 1991. Satellite-image-derived velocity-field of an Antarctic ice stream. *Science* **252** (5003), 242–6.

Bindschadler, R. and Vornberger, P. 1998. Changes in the west Antarctic ice sheet since 1963 from declassified satellite photography. *Science* **279** (5351), 689–92.

Bindschadler, R. *et al.* 1996. Surface velocity and mass balance of ice streams D and E, West Antarctica. *Science* **42** (142), 461–75.

Bindschadler, R. *et al.* 2001. Glaciological applications with Landsat-7 imagery: early assessments. *Remote Sensing Environ.* **78** (1–2), 163–79.

Bogorodskiy, V. and Bentley, C. 1985. *Radioglaciology*. Kluwer Academic Publishers.

Budd, W. F. and Warner, R. C. 1996. A computer scheme for rapid calculations of balance-flux distributions. *Ann. Glaciol.* **23**, 21–7.

Carsey, F. D. 1982. Arctic sea ice distribution at end of summer 1973–1976 from satellite microwave data. *J. Geophys. Res.* **87** (C8), 5809–35.

Cavalieri, D. J. 1992. The validation of geophysical products using multisensor data. In Carsey, F. D., ed., *Microwave Remote Sensing of Sea Ice*. Washington, D.C., American Geophysical Union, pp. 233–42.

Cavalieri, D. J., Gloersen, P. and Campbell, W. J. 1984. Determination of sea ice parameters from Nimbus 7 SMMR. *J. Geophys. Res.* **89** (D4), 5355–69.

Cavalieri, D. J., St. Germain, K. M. and Swift, C. T. 1995. Reduction of weather effects in the calculation of sea ice concentration in DMSP SSM/I. *J. Glaciol.* **41** (139), 455–64.

Cavalieri, D. J., Parkinson, C. L., Gloersen, P., Comiso, J. C. and Zwally, H. J. 1999. Deriving long-term time series of sea ice cover from satellite passive-microwave multisensor data sets. *J. Geophys. Res.* **104** (C7), 15 803–14.

Chelton, D. B. *et al.* 2001. Satellite altimetry. In Fu, L.-L. and Cazenave, A., eds., *Satellite Altimetry and the Earth Sciences*. New York, Academic Press, pp. 1–132.

Comiso, J. C. 1986. Characteristics of Arctic winter sea ice from satellite multispectral microwave observations. *J. Geophys. Res.* **91** (C1), 975–94.

Comiso, J. C. and Kwok, R. 1996. Surface and radiative characteristics of the summer Arctic sea cover from multisensor satellite observations. *J. Geophys. Res.* **101** (C12), 28 397–416.

Comiso, J. C., Grenfell, T. C., Lange, M. M., Lohanick, M., Moore, R. K. and Wadhams, P. 1992. Microwave remote sensing of the Southern Ocean ice cover. In Carsey, F. D., ed., *Microwave Remote Sensing of Sea Ice*. Washington, D. C., American Geophysical Union, pp. 243–59.

Comiso, J. C., Cavalieri, D. J., Parkinson, C. L. and Gloersen, P. 1997. Passive microwave algorithms for sea ice concentration: a comparison of two techniques. *Remote Sensing Environ.* **60**, 357–84.

Coon, M. D., Knoke, G. S., Echert, D. C. and Pritchard, R. S. 1998. The architecture of an anisotropic elastic-plastic sea ice mechanics constitutive law. *J. Geophys. Res.* **103** (C10), 21 915–25.

Cudlip, W. *et al.* 1994. Corrections for altimeter low-level processing at the Earth observation data center. *Int. J. Remote Sensing.* **15** (4), 889–914.

Cunningham, G. F., Kwok, R. and Banfield, J. 1994. Ice lead orientation characteristics in the winter Beaufort Sea. *Proceedings of IGARSS*. Pasadena, CA, 1994.

Davis, C. H. 1995. Growth of the Greenland ice-sheet – a performance assessment of altimeter retracking algorithms. *IEEE Trans. Geosci. & Remote Sensing* **33** (5), 1108–16.

1996a. Comparison of ice-sheet satellite altimeter retracking algorithms. *IEEE Trans. Geosci. & Remote Sensing* **34** (1), 229–36.

1996b. Temporal change in the extinction coefficient of snow on the Greenland ice sheet from an analysis of Seasat and Geosat altimeter data. *IEEE Trans. Geosci. & Remote Sensing* **34** (5), 1066–73.

1997. A robust threshold retracking algorithm for measuring ice-sheet surface elevation change from satellite radar altimeters. *IEEE Trans. Geosci. & Remote Sensing* **35** (4), 974–9.

Davis, C. H. and Moore, R. K. 1994. A combined surface and volume scattering model for ice sheet radar altimetry. *J. Glaciol.* **39** (133), 675–86.

Davis, C. H., Kluever, C. A. and Haines, B. J. 1998. Elevation change of the southern Greenland ice sheet. *Science* **279** (5359), 2086–8.

Drinkwater, M. R. 1998. Satellite microwave observations of Antarctic sea ice. In Tsatsoulis, C. and Kwok, R., eds., *Analysis of SAR data of the Polar Oceans: Recent Advances*. Berlin, Springer Verlag, pp. 145–87.

Drinkwater, M. R., Long, D. G. and Bingham, A. W. 2001. Greenland snow accumulation estimates from satellite radar scatterometer data. *J. Geophys. Res.* **106** (D24), 33 935–50.

Emery, W. J., Fowler, C. W. and Maslanik, J. A. 1997. Satellite-derived maps of Arctic and Antarctic sea ice motion, 1988–1994. *Geophys. Res. Lett.* **24** (8), 897–900.

Fetterer, F., Gineris, D. and Kwok, R. 1994. Sea ice type maps from Alaska synthetic aperture radar facility imagery: an assessment. *J. Geophys. Res.* **99** (C11), 22 443–58.

Fily, M. and Rothrock, D. A. 1990. Opening and closing of sea ice leads: digital measurements from synthetic aperture radar. *J. Geophys. Res.* 95 (C1), 789–96.

Frezzotti, M. 1997. Ice front fluctuation, iceberg calving flux and mass balance of Victoria Land glaciers. *Antarct. Sci.* **9** (1), 61–73.

Fricker, H. A., Warner, R. C. and Allison, I. 2000. Mass balance of the Lambert glacier–Amery ice shelf system, East Antarctica: a comparison of computed balance fluxes and measured fluxes. *J. Glaciol.* **46** (155), 561–70.

Fricker, H. A. *et al.* 2001. Distribution of marine ice beneath the Amery ice shelf. *Geophys. Res. Lett.* **28** (11), 2241–4.

Fujii, Y. *et al.* 1987. Comparison of surface conditions of the inland ice sheet, Dronning Maud Land, Antarctica, derived from NOAA AVHRR data with ground observations. *Ann. Glaciol.* **9**, 72–5.

Gloersen, P., Campbell, W. J., Cavalieri, D. J., Comiso, J. C., Parkinson, C. L. and Zwally, H. J. 1992. *Arctic and Antarctic Sea Ice, 1978–1987: Satellite Passive-Microwave Observation and Analysis*, NASA SP-511. Washington, D.C., National Aeronautics and Space Administration.

Gogineni, S. *et al.* 1998. An improved coherent radar depth sounder. *J. Glaciol.* **44** (148), 659–69.

Gohin, F. and Cavanie, A. 1994. A first try at identification of sea ice using the three beam scatterometer of ERS-1. *Int. J. Remote Sensing* **15** (6), 1221–8.

Goldstein, R. M., Zebker, H. A. and Werner, C. 1988. Satellite radar interferometry: two-dimensional phase unwrapping. *Radio Sci.* **23** (4), 713–20.

Grandell, J., Johannessen, J. A. and Hallikainen, M. T. 1999. Development of a synergistic sea ice retrieval method for the ERS-1 AMI wind scatterometer and SSM/I radiometer. *IEEE Trans. Geosci. Remote Sensing* **37** (2), 668–79.

Gray, A. L. *et al.* 1998. InSAR results from the RADARSAT Antarctic mapping mission data: estimation of glacier motion using a simple registration procedure. In Stein, T., ed., *1998 International Geoscience and Remote Sensing Symposium (IGARSS 98) on Sensing and Managing the Environment.* Seattle, WA, 6–10 July 1998. Piscataway, NJ, IEEE Service Center, pp. 1638–40.

Haefliger, M., Steffen, K. and Fowler, C. 1993. AVHRR surface temperature and narrow-band albedo comparison with ground measurements for the Greenland ice sheet. *Ann. Glaciol.* **17**, 49–54.

Hanna, E. and Bamber, J. 2001. Derivation and optimization of a new Antarctic sea-ice record. *Int. J. Remote Sensing* **22** (1), 113–39.

Hardy, R. J., Bamber, J. L. and Orford, S. 2000. The delineation of major drainage basins on the Greenland ice sheet using a combined numerical modelling and GIS approach. *Hydrol. Process* **14** (11–12), 1931–41.

Hibler, W. D. III and Schulson, E. M. 2000. On modeling the anisotropic failure and flow of flawed sea ice. *J. Geophys. Res.* **105**, 17 105–20.

Higgins, A. K. 1990. North Greenland glacier velocities and calf ice production. *Polarforschung* **60** (1), 1–23.

Holt, B. D., Rothrock, A. and Kwok, R. 1992. Determination of sea ice motion from satellite images. In Carsey, F. D., ed., *Microwave Remote Sensing of Sea Ice.* Washington, D.C., American Geophysical Union, pp. 343–54.

Jezek, K. C., Sohn, H. G. and Noltimier, K. F. 1998. The radarsat Antarctic mapping project. In Stein, T., ed., *1998 International Geoscience and Remote Sensing Symposium (IGARSS 98) on Sensing and Managing the Environment.* Seattle, WA, 6–10, July 1998. Piscataway, NJ, IEEE Service Center, pp. 2462–4.

Joughin, I. and Tulaczyk, S. 2002. Positive mass balance of the Ross Ice Streams, West Antarctica. *Science* **295** (5554), 476–80.

Kaab, A. 2000. Photogrammetric reconstruction of glacier mass balance using a kinematic ice-flow model: a 20 year time series on Grubengletscher, Swiss Alps. *Ann. Glaciol.* **31**, 45–52.

Kaab, A. and Funk, M. 1999. Modelling mass balance using photogrammetric and geophysical data: a pilot study at Griesgletscher, Swiss Alps. *J. Glaciol.* **45** (151), 575–83.

Krabill, W. *et al.* 1995a. Greenland ice-sheet thickness changes measured by laser altimetry. *Geophys. Res. Lett.* **22** (17), 2341–4.

Krabill, W. B. *et al.* 1995b. Accuracy of airborne laser altimetry over the Greenland ice sheet. *Int. J. Remote Sensing* **16** (7), 1211–22.

Krabill, W. *et al.* 1999. Rapid thinning of parts of the southern Greenland ice sheet. *Science* **283** (5407), 1522–4.

Kwok, R. 1998. The RADARSAT geophysical processor system, In Tsatsoulis, C. and Kwok, R., eds., *Analysis of SAR data of the Polar Oceans: Recent Advances.* Berlin, Springer Verlag, pp. 235–57.

  2000. Recent changes of the Arctic Ocean sea ice motion associated with the north Atlantic oscillation. *Geophys. Res. Lett.* **27** (6), 775–8.

  2001. Deformation of the Arctic Ocean sea ice cover: November 1996 through April 1997. In Dempsey, J. and Shen, H. H., eds., *Scaling Laws in Ice Mechanics and Dynamics.* Dordrecht, Kluwer Academic, pp. 315–23.

Kwok, R. and Comiso, J. C. 1998. The perennial ice cover of the Beaufort Sea from active and passive microwave observations. *Ann. Glaciol.* **25**, 376–81.

Kwok, R. and Cunningham, G. F. 1992. Backscatter characteristics of the winter sea ice cover in the Beaufort Sea. *J. Geophys. Res.* **99** (C4), 7787–803.

2002. Seasonal ice area and volume production of the Arctic Ocean: November 1996 through April 1997. *J. Geophys. Res.* **107** (C10), 8038.

Kwok, R. and Fahnestock, M. A. 1996. Ice sheet motion and topography from radar interferometry. *IEEE Trans. Geosci. Remote Sensing* **34** (1), 189–200.

Kwok, R. and Rothrock, D. A. 1999. Variability of Fram Strait ice flux and North Atlantic oscillation. *J. Geophy. Res.* **104**, 5177–89.

Kwok, R., Rothrock, D. A., Stern, H. L. and Cunningham, G. F. 1995. Determination of Ice Age using Lagrangian observations of ice motion. *IEEE Trans. Geosci. Remote Sensing* **33** (2), 392–400.

Kwok, R., Schweiger, A., Rothrock, D. A., Pang, S. and Kottmeier, C. 1998. Sea ice motion from satellite passive microwave data assessed with ERS SAR and buoy data. *J. Geophys. Res.* **103** (C4), 8191–214.

Kwok, R., Cunningham, G. F. and Yueh, S. 1999. Area balance of the Arctic Ocean perennial ice zone: Oct 1996–April 1997. *J. Geophys. Res.* **104** (C11), 25 747–9.

Laxon, S. W. 1994. Sea ice altimeter processing scheme at the EODC. *Int. J. Remote Sensing* **15** (4), 915–24.

Li, F. K. and Goldstein, R. M. 1990. Studies of multi-baseline interferometric synthetic aperature radars. *IEEE Trans. Geosci. & Remote Sensing* **28**, 88–97.

Lillesand, T. M. and Kiefer, R. W. 2000. *Remote Sensing and Image Interpretation*, 4th edn. New York, John Wiley.

Liu, A. and Cavalieri, D. J. 1998. On sea ice drift from the wavelet analysis of DMSP SSM/I data. *Int. J. Remote Sensing* **19** (7), 1415–23.

Long, D. G. and Drinkwater, M. 1999. Cryosphere applications of NSCAT data. *IEEE Trans. Geosci. Remote Sensing* **37** (3), 1671–84.

Lucchitta, B. and Ferguson, H. 1986. Antarctica-measuring glacier velocity from satellite images. *Science* **234** (4780), 1105–8.

Martin, T. and Augstein, E. 2000. Large-scale drift of Arctic Sea ice retrieved from passive microwave satellite data. *J. Geophys. Res.* **105** (C4), 8775–88.

Maslanik, J. A. 1992. Effects of weather on the retrieval of sea ice concentration from passive microwave data. *Int. J. Remote Sensing* **13**, 37–54.

Maykut, G. A. 1986. The surface heat and mass balance. In Untersteiner, N., ed., *Geophysics of Sea Ice*. Series B: Physics Vol. 146. Plenum Press, pp. 395–463.

Oelke, C. 1997. Atmospheric signatures in sea ice concentration estimates from passive microwave data: modeled and observed. *Int. J. Remote Sensing* **18**, 1113–36.

Overland, J. E., McNutt, S. L., Salo, S., Groves, J. and Li, S. S. 1998. Arctic sea ice as a granular plastic. *J. Geophys. Res.* **103** (C10), 21 845–67.

Parkinson, C., Comiso, J. C., Zwally, H. J., Cavalieri, D. J., Gloersen, P. and Campbell, W. J. 1987. *Arctic Sea Ice, 1973–1976: Satellite Passive-Microwave Observations*, NASA SP-489. Washington, D.C., National Aeronautics and Space Administration.

Parmerter, R. R. and Coon, M. 1972. Model of pressure ridge formation in sea ice. *J. Geophys. Res.* **77**, 6565–75.

Paterson, W. S. B. 1994. *The Physics of Glaciers*, 3rd edn. Oxford, Pergamon Press.

Peacock, N. R., Laxon, S. W., Maslowski, W., Winebrenner, D. P. and Arthern, R. J. 1998. Geophysical signatures from precise altimetric height measurements in the Arctic Ocean. In Stein, T., ed., *1998 International Geoscience and Remote Sensing*

Symposium (IGARSS 98) on Sensing and Managing the Environment. Seattle, WA, July 6–10, 1998. Piscataway, NJ, IEEE Service Center, pp. 1964–6.

Qin, Z. H. *et al.* 2001. Derivation of split window algorithm and its sensitivity analysis for retrieving land surface temperature from NOAA- advanced very high resolution radiometer data. *J. Geophys. Res.* **106** (D19), 22 655–70.

Raney, R. K. 1998. The delay/Doppler radar altimeter. *IEEE Trans. Geosci. Remote Sensing* **36** (5), 1578–88.

Reeh, N. *et al.* 1997. Mass balance of North Greenland. *Science* **278** (5336), 207–9.

Rees, W. G. 2001. *Physical Principles of Remote Sensing.* Cambridge University Press.

Reynolds, J. R. and Young, G. J. 1997. Changes in areal extent, elevation and volume of Athabasca Glacier, Alberta, Canada, as estimated from a series of maps produced between 1919 and 1979. *Ann. Glaciol.* **24**, 60–5.

Ridley, J. K. and Partington, K. C. 1988. A model of satellite altimeter return from ice sheets. *Int. J. Remote Sensing* **9** (4), 601–24.

Rignot, E. 1996. Tidal motion, ice velocity and melt rate of Petermann Gletscher, Greenland, measured from radar interferometry. *J. Glaciol.* **42** (142), 476–85.

   1998. Fast recession of a West Antarctic glacier. *Science* **281** (5376), 549–51.

   2002. Mass balance of East Antarctic glaciers and ice shelves from satellite data. *Ann. Glaciol.* **34**, 217–27.

Rignot, E. J. *et al.* 1997a. North and northeast Greenland ice discharge from satellite radar interferometry. *Science* **276** (5314), 934–7.

   1997b. Mass balance of North Greenland – response. *Science* **278** (5336), 209.

   2000. Mass balance of the northeast sector of the Greenland ice sheet: a remote-sensing perspective. *J. Glaciol.* **46** (153), 265–73.

Robin, G. de Q. 1966. Mapping the Antarctic ice sheet by satellite altimetry. *Can. J. Earth Sci.,* **3** (6), 893–901.

Rodriguez, E. and Martin, J. 1992. Theory and design of interferometric SARs. *Proc. IEE Radar & Signal Processing* **139** (2), 147–59.

Scambos, T. A. and Bindschadler, R. A. 1991. Feature maps of ice streams C, D and E, West Antarctica. *Antarc. J. US* **26** (5), 312–13.

Scambos, T. A. *et al.* 1992. Application of image cross-correlation to the measurement of glacier velocity using satellite image data. *Remote Sensing of Environment* **42** (3), 177–86.

Scharroo, R. and Visser, P. 1998. Precise orbit determination and gravity field improvement for the ERS satellites. *J. Geophys. Res. Oceans* **103** (C4), 8113–27.

Steffen, K. and Heinrichs, J. 1994. Feasibility of sea ice typing with SAR: merging of Landsat TM and ERS-1 SAR satellite imagery. *J. Geophys. Res.* **99** (C11), 22 413–24.

Steffen, K. *et al.* 1993. Snow and ice application of AVHRR in polar regions. *Ann. Glaciol.* 1–16.

Stern, H. L., Rothrock, D. A. and Kwok, R. 1995. Open water production in Arctic sea ice: satellite measurements and model parameterizations. *J. Geophys. Res.* **100** (C10), 20 601–12.

Stroeve, J., Nolin, A. and Steffen, K. 1997. Comparison of AVHRR-derived and in situ surface albedo over the Greenland Ice Sheet. *Remote Sensing Environ.* **62** (3), 262–76.

Theakstone, W. H. and Jacobsen, F. M. 1997. Digital terrain modelling of the surface and bed topography of the glacier Austre Okstindbreen, Okstindan, Norway. *Geogr. Ann. Ser. A-Phys. Geogr.* **79A** (4), 201–14.

Thomas, D. R. 1993. Arctic sea ice signatures for passive microwave algorithms. *J. Geophys. Res.* **98** (C6), 10 037–52.

Thomas, D. R. and Rothrock, D. A. 1993. The Arctic Ocean ice balance: a Kalman filter smoother estimate. *J. Geophys. Res.* **98** (C6), 10 053–67.

Vaughan, D. G. *et al.* 1999. Reassessment of net surface mass balance in Antarctica. *J. Climate* **12** (4), 933–46.

Winebrenner, D. P., Arthern, R. J. and Shuman, C. A. 2001. Mapping Greenland accumulation rates using observations of thermal emission at 4.5-cm wavelength. *J. Geophys. Res.* **106** (D24), 33 919–34.

Winebrenner, D. P., Tsang, L. Wen, B. and West, R. 1989. Sea-ice characterization measurements needed for testing of microwave remote sensing models. *IEEE J. Oceanic. Engng.* **14** (2), 149–57.

Wingham, D. J. *et al.* 1998. Antarctic elevation change from 1992 to 1996. *Science* **282** (5388), 456–8.

Young, N. W. *et al.* 1998. Near-coastal iceberg distributions in East Antarctica, 50–145 degrees E. *Ann. Glaciol.* **27**, 68–74.

Yu, Y. and Rothrock, D. A. 1996. Thin ice thickness from satellite thermal imagery. *J. Geophys. Res.* **101** (C10), 25 753–66.

Yueh, S. and Kwok, R. 1998. Arctic sea ice extent and melt onset from NSCAT observations. *Geophys. Res. Lett.* **25** (23), 4369–72.

Yueh, S., Kwok, R., Lou, S. H. and Tsai, W. Y. 1997. Sea ice identification using dual-polarized Ku-band scatterometer data. *IEEE Trans. Geosci. Remote Sensing* **35** (3), 560–9.

Zebker, H. A. and Goldstein, R. 1986. Topographic mapping from interferometric SAR observations. *J. Geophys. Res.* **91** (B5), 4993–9.

Zhang, J., Hibler, W., Steele, M. and Rothrock, D. 1998. Arctic ice-ocean modeling with and without climate restoring. *J. Phys. Oceanography* **28** 191–217.

Zhang, Y., Maslowski, W. and Semtner, A. J. 1999. Impact of mesoscale ocean currents on sea ice in high-resolution Arctic ice and ocean simulations. *J. Geophys. Res.* **104** (C8), 18 409–30.

Zwally, H. J. 1977. Microwave emissivity and accumulation rate of polar firn. *J. Glaciol.* **18**, 195–215.

1989. Growth of Greenland ice sheet: interpretation. *Science* **246**, 1589–91.

Zwally, H. J. and Giovinetto, M. B. 1995. Accumulation in Antarctica and Greenland derived from passive-microwave data: a comparison with contoured compilations. *Ann. Glaciol.* **21**, 123–30.

Zwally, H. J. *et al.* 1983. Surface elevation contours of Greenland and Antarctic ice sheets. *J. Geophys. Res.* **88** (C3), 1589–96.

Zwally, H. J., Comiso, J. C., Parkinson, C., Cavalieri, D. J., Gloersen, P. and Campbell, W. J. 1987. *Antarctic Sea Ice, 1973–1976: Satellite Passive-Microwave Observations*, NASA SP-459. Washington, D. C., National Aeronautics and Space Administration.

Zwally, H. J. *et al.* 1989. Growth of Greenland ice sheet: measurement. *Science* **246**, 1587–9.

Part II

Modelling techniques and methods

5

Modelling land-ice surface mass balance

WOUTER GREUELL
Institute for Marine and Atmospheric Research Utrecht
CHRISTOPHE GENTHON
Laboratoire de Glaciologie et Géophysique de l'Environnement,
CNRS/Université de Grenoble

5.1 Introduction

The topic of this chapter is the surface mass balance, often called the specific balance, specific mass balance (as in Chapter 2) or simply the mass balance. The surface mass balance is defined as the total change in mass in a vertical column of glacier material during an undefined amount of time (in this chapter, the term 'glacier' refers to small glaciers, ice caps and ice sheets). Glacier material may include snow, ice and water. Changes in mass due to divergence or convergence of the ice velocity field are excluded, as well as mass changes due to processes occurring at the bedrock. Positive contributions to the mass balance are called accumulation, and negative contributions are called ablation. Snow fall is the dominant process causing accumulation on glaciers. In the lower parts of glaciers, ice caps and the Greenland ice sheet, melt followed by runoff of the melt water is the dominant ablation process. Other processes that contribute to the mass balance are evaporation (change from liquid to gas), condensation (change from gas to liquid) and sublimation (change from gas to solid and vice versa), removal and deposition of snow by avalanches and wind, and rain water that does not run off. In Antarctica and the higher parts of the Greenland ice sheet, sublimation forms the dominant contribution to ablation. Mass balance is often expressed in kg/m^2, but the most widely employed unit is mm water equivalent or m water equivalent. In many cases, the interesting quantity is the change in mass over a year, called the annual mass balance, which may also be expressed as the mass balance rate, using the unit mm water equivalent per year.

Obviously the local climatic conditions during the period considered determine the mass balance at a specific location. As a consequence, spatial variations in mass balance can be explained by spatial variations in climate conditions. An example is the typical increase with elevation of the mass balance for mid-latitude glaciers. It also follows that the surface mass balance is sensitive to climate change. Therefore, to understand both spatial and temporal variations in the surface mass balance, we need to understand the relationship between climate and mass balance and, if quantification is desired, we must use models.

Mass Balance of the Cryosphere: Observations and Modelling of Contemporary and Future Changes, eds. Jonathan L. Bamber and Antony J. Payne. Published by Cambridge University Press. © Cambridge University Press 2003.

In this chapter we will discuss how the surface mass balance of glaciers can be modelled as a function of climatic variables. Hock (1998) recently reviewed the topic. We start with the treatment of surface energy balance models (section 5.2). These models compute each of the relevant energy fluxes between the atmosphere and the surface of the glacier. They are forced by meteorological measurements and observations, such as near-surface wind speed, near-surface temperature, near-surface humidity, radiative fluxes, cloud amount and precipitation. Using energy-budget and mass-budget equations, energy balance models will produce the mass balance (section 5.4), provided they are linked to a subsurface module, or assumptions are made about the state of the surface or subsurface.

An alternative approach is the degree-day method (section 5.3), which relates ablation to atmospheric temperature only. This method will be discussed as a simplification of the energy balance approach. Both in energy balance and in degree-day models, precipitation is prescribed and not computed. Therefore, in this chapter we will employ 'ablation models' as a term covering these two types of models.

In sections 5.2–5.4 it is assumed that the forcing at the location of the calculations is known. This is indeed the case when a glacio-meteorological experiment has been carried out during one or a few summer seasons at one or some locations on a glacier (Figure 5.1). Simulations of the mass balance with such 'local forcing' may be helpful in the development stage of a model and in order to obtain insight into relevant processes. However, if we want to calculate the mass balance on a grid covering an entire glacier and for many years, local forcing data are not available at most or all grid points. In that case, several strategies may be employed, which will be discussed. Firstly, parameterizations can be used to convert data from the free atmosphere (defined as the part of the atmosphere that is not influenced by the glacier itself; this definition includes the atmosphere at climate stations that are not located on a glacier) into the forcing needed by ablation models (section 5.6). Alternatively, the mass balance can be computed directly by means of a global or regional atmospheric model (section 5.7). Note at this point that these atmospheric models do not only incorporate the equations describing the surface energy balance, but also simulate processes higher up in the atmosphere, clouds and the atmospheric circulation. Therefore, they calculate the amount of precipitation and accumulation, given certain boundary conditions, whereas precipitation is prescribed in ablation models. Finally, the mass balance of an entire glacier can be computed with regression models, which are statistically derived relations between data from a climate station and the mass balance (section 5.8).

This chapter will focus on energy balance and degree-day models, on modelling with atmospheric models and on regression models. We will discuss advantages and drawbacks of the different types of models in section 5.9. We intend to give an overview of the state of the art of these models. We will give more attention to how the models can be constructed and to the advantages and disadvantages of the different possibilities, rather than elaborate on the physics underlying the equations. Therefore, this chapter may be valuable for those readers who are interested in modelling glacier mass balance and have some basic knowledge of glaciology and meteorology.

Figure 5.1. Automatic weather station on the Morteratschgletscher (Switzerland). The meteorological mast on the left carries sensors which measure the four radiative fluxes separately and further sensors for the measurement of temperature, humidity, wind speed and wind direction at a single level. Making some assumptions, this suffices to calculate the surface energy balance and ablation. The latter can be compared with the change in height measured with the sonic ranger (shown on the right) or along stakes (seen in the middle of the picture).

Because of space restrictions, several aspects of mass balance modelling will be mentioned but not treated, although they are certainly important, or even crucial, in large parts of the cryosphere. For example, we will not pay specific attention to snow outside glaciers, to sea ice and to permafrost. Nevertheless, this chapter may also be useful for readers interested in modelling snow and sea ice, because most aspects of energy exchange between the atmosphere and surface are basically the same for glaciers, snow and sea ice. Processes occurring at the bottom constitute the largest difference between glaciers on the one hand and snow and sea ice on the other hand. For the surface mass balance of glaciers, energy exchange with the bed can be ignored. For snow and sea ice, energy exchange with the underlying medium must be considered. Brun *et al.* (1992) and Male and Granger (1981) form valuable contributions on ablation modelling of snow, and Maykut and Untersteiner (1971) consider ablation modelling of sea ice.

Aspects of glacier mass balance ignored here are avalanches (Hutter, 1995; Issler, 1998) and the effect of debris cover on the mass balance (Bozhinskiy, Krass and Popovnin, 1986; Mattson and Gardner, 1989). We will also disregard drifting and blowing snow, though they may be important in terms of mass redistribution (Pomeroy, Gray and Landine, 1993), mass loss through sublimation (Bintanja, 1998; Schmidt, 1982) and influence on the turbulent fluxes of heat and water vapour (Bintanja, 2000; Déry, Taylor and Xiao, 1998).

5.2 The surface energy balance

5.2.1 Introduction

We will now discuss how the surface mass balance can be computed by means of the surface energy balance approach. This means that the focus is on energy fluxes between atmosphere and glacier, but while reading the sections about these fluxes, one should not forget that the mass balance is the main aim of the calculations. In section 5.2.8 we will discuss how runoff can be evaluated from the computations, and in section 5.4 we see how the other components of the mass balance can be computed.

Surface energy balance models consist of two parts, which feed back on each other. The first computes the energy fluxes between atmosphere and glacier as a function of meteorological variables and the state of the surface. The second part deals with the state (temperature, density, water content) of the subsurface and is forced by the energy exchange with the atmosphere.

The total energy flux from the atmosphere toward the glacier surface (Q_0) consists of the following components:

$$Q_0 = S{\downarrow}(1 - \alpha) + L{\downarrow} - L{\uparrow} + Q_H + Q_L + Q_R, \tag{5.1}$$

where $S{\downarrow}$ is the short-wave incoming radiative flux, α is the albedo of the surface, $L{\downarrow}$ is the long-wave incoming radiative flux, $L{\uparrow}$ is the long-wave outgoing radiative flux, Q_H is the sensible heat flux, Q_L is the latent heat flux and Q_R is the heat flux supplied by rain.

Measurements (Ambach, 1979; Greuell and Smeets, 2001; Hock and Holmgren, 1996; Van den Broeke, 1996) show that over melting glaciers net short-wave radiation ($S{\downarrow}(1 - \alpha)$) is generally the dominant term, that the latent heat flux is relatively small, and that net long-wave radiation ($L{\downarrow} - L{\uparrow}$) and the sensible heat flux are of intermediate magnitude (see Figure 5.2). In most cases, the heat flux supplied by rain is negligible. We will now discuss how these variables can be computed from the available input data.

5.2.2 The incoming short-wave radiative flux

When measurements of $S{\downarrow}$ in a plane parallel to the surface are available, the measurements can be used as input without having to make any correction. However, measurements are often performed in the horizontal plane (when $S{\downarrow}$ is called global radiation), whereas the surface is inclined. In this case, as in the more general case where surface and sensor are not parallel to each other, a correction must be made. Equations designed to make this correction can be found in Garnier and Ohmura (1968). The correction depends on the ratio of diffuse to total short-wave incoming radiation, which, according to measurements made by Ohmura (1981), and Konzelmann and Ohmura (1995), may vary between 0.13 and 0.21

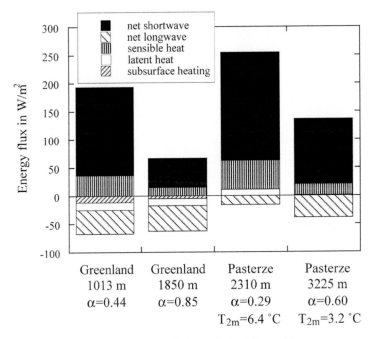

Figure 5.2. Components of the surface energy balance derived from mid-summer measurements on the western part of the Greenland ice sheet and on the Pasterze (Austria). The figure depicts averages over 38 days at Camp IV (just below the equilibrium-line altitude, Greenland, 1013 m above sea level, a.s.l.; Ambach, 1979), over 70 days at Carrefour (accumulation area of the Greenland ice sheet, 1850 m a.s.l.; Ambach, 1979), and over 46 days at the Pasterze, where the equilibrium line is located between the two sites (Greuell and Smeets, 2001). The mean albedo and 2 m temperature (not available for the sites on the Greenland ice sheet) over the entire period are given below the graph. The sum of the fluxes depicted in the graph is used to melt snow and ice. At Carrefour melt is negligible.

for clear-sky conditions. Oerlemans (1992) suggested a linear increase of the ratio from 0.15 for clear skies to 0.8 for overcast skies.

When measurements are not available, $S\!\downarrow$ can be estimated by means of the following parameterization:

$$S\!\downarrow = I_0 \cos(\theta_s) T_R T_{wv} T_g T_{as} F_{ms} F_{ho} F_{rs} T_c, \tag{5.2}$$

where I_0 is the flux at the top of the atmosphere on a surface normal to the incident radiation and θ_s is the solar zenith angle. If the surface is inclined, θ_s must be replaced by the incidence angle, which is the angle between the normal to the surface and the direct solar ray. Equations can be found in Garnier and Ohmura (1968). This replacement of θ_s applies only to the direct part of the incoming radiation, which can be estimated from the cloud amount (see above). The factor T_c represents the reduction of $S\!\downarrow$ due to clouds. All the other factors on the right hand side of the equation represent gains or losses due to different processes in a cloud-free atmosphere. The factors T_R, T_{wv} and T_g are transmission coefficients for

Rayleigh scattering, absorption by water vapour and absorption by other gases, respectively. Formulae for the first five terms on the right hand side of equation (5.2) can be found in, e.g., Kasten (1983) and Meyers and Dale (1983). Input variables for these formulae are the Julian day, the time of the day, latitude, longitude, surface pressure, atmospheric humidity near the surface and total ozone column. The uncertainty in each of these five terms is generally smaller than in the other terms.

The factor T_{as} accounts for aerosol extinction, and was given by Meyers and Dale (1983) as

$$T_{as} = k_a^m,$$

with k_a an empirical constant depending on aerosol type and amount and m the optical air mass. Gueymard (1993) gave a value of 0.94 for k_a for Weissfluhjoch (Switzerland, 2667 m above sea level, a.s.l.), and Greuell, Knap and Smeets (1997) found values between 0.94 and 0.965 for five locations on the Pasterze (Austria, 2205–3225 m a.s.l.).

The three factors F_{ms}, F_{ho} and F_{rs} may be important on glaciers having a high surface albedo and/or those surrounded by steep slopes. The factor F_{ms} accounts for multiple reflections between the surface and the atmosphere. Its clear-sky value can be estimated by means of the following equation (Konzelmann *et al.*, 1994):

$$F_{ms} = \frac{1}{1 - \alpha \alpha_{cs}},$$

where α_{cs} ($= 0.075$) is the albedo of the clear sky. So, for a location in the dry-snow zone of an ice sheet ($\alpha = 0.85$), $F_{ms} = 1.07$. The effect of terrain occupying the upper hemisphere, as seen from the location of interest, is two-fold. The terrain may cast shadows onto the surface, thereby reducing $S\!\downarrow$ as represented by F_{ho}, but will also reflect short-wave radiation, in turn increasing $S\!\downarrow$ as represented by F_{rs}. Greuell *et al.* (1997) estimated F_{ms}, F_{ho} and F_{rs} for five locations on the Pasterze (Austria) and found values in the ranges 1.008–1.030 (F_{ms}), 0.909–0.995 (F_{ho}) and 1.026–1.073 (F_{rs}).

The term causing the largest uncertainty in the estimate of $S\!\downarrow$ is the cloud factor T_c. It is equal to unity for clear skies and decreases with increasing cloud amount and cloud optical thickness. However, in much of the literature T_c is parameterized as a (polynomial) function of cloud amount only (Bintanja and Van den Broeke, 1996; Greuell *et al.*, 1997; Konzelmann *et al.*, 1994), because it is easier to estimate cloud amount than cloud optical thickness. If cloud optical thickness is not taken into account, two problems arise. Firstly, a large uncertainty is introduced into the calculation of instantaneous values with the parameterizations. The uncertainty should decrease when averages over longer periods of time are considered. The second problem introduced by the neglect of cloud optical thickness in the parameterizations is the limited applicability in space. This is because mean cloud optical thickness (for a given cloud amount) varies in space. Examples are the tendencies of clouds to become thinner with latitude and elevation. The effect of elevation is accounted for in the parameterizations of Greuell and Oerlemans (1986), based on data from Austria, and of Konzelmann *et al.* (1994), based on data from West Greenland. However, the latter parameterization is not valid for locations on and near Antarctica, as demonstrated by Bintanja and Van den Broeke (1996).

It should be realized that T_c represents the net effect of clouds on $S\!\downarrow$. Clouds have an important effect on some of the other terms in equation (5.2). For example, clouds considerably enhance multiple reflection between the surface and the atmosphere (Rouse, 1987; Wendler, Eaton and Ohtake, 1981). This effect is considered in the parameterization of F_{ms} proposed by Kayastha, Ohata and Ageta (1999). However, in most parameterizations F_{ms} does not depend on cloud amount, and the enhancement of multiple scattering by clouds is therefore implicit in the formulation of T_c. Clouds also increase the ratio of diffuse to total incoming short-wave radiation. This effect must be considered in the use of the solar zenith angle in equation (5.2), which should not be replaced by the incidence angle if all of the incoming short-wave radiation is diffuse.

5.2.3 Surface albedo

The basic reason why a correct estimate of the surface albedo (α) is crucial is the large sensitivity of melt to variations in α combined with the enormous temporal and spatial variability in α. The α of snow and glacier ice may range from less than 0.1 for very dirty glacier ice (e.g. parts of Vatnajökull, Iceland; see Reijmer, Knap and Oerlemans, 1999) to 0.87 for dry, fine-grained snow in the Antarctic (Grenfell, Warren and Multen, 1994).

Surface albedo can be measured near the ground or inferred from satellite data (Knap and Oerlemans, 1996; Stroeve, Nolin and Steffen, 1997). In the absence of ground measurements and satellite data, α must be modelled. Unfortunately, this is not an easy task. For snow, α depends on surface characteristics such as impurity content, grain size and shape, and water content (Warren, 1982; Wiscombe and Warren, 1980). For ice, α depends on bubbles and cracks, debris and dust, and water. Since short-wave radiation penetrates into snow and ice, it is not only the state of the surface itself that counts but also the subsurface variation of the relevant properties. Furthermore, α depends on characteristics of the incoming radiation, namely its angular and spectral distributions (Warren, 1982). This means that α is a function of the incidence angle and of cloud conditions. Clouds affect both the angular and the spectral distributions.

In α parameterizations two types of independent variables can be distinguished. Physical variables describe surface and atmospheric characteristics that directly control variations in α. Examples are impurity content, grain size and depth of the snow layer. In many models values of these variables are not available. Therefore, they are replaced by proxy variables such as density, the time since the last snow fall or the accumulated amount of daily temperature above 0 °C since the last snow fall. This is justified by the correlation between the physical and proxy variables (Brock, Willis and Sharp, 2000).

In the context of α parameterization, two types of ablation models can be distinguished (see section 5.2.8), namely (a) models with a module that resolves the vertical structure of the snow pack and the ice (multi-layer subsurface module) and (b) models that treat the subsurface as a single layer or as two layers (bulk subsurface module). For the parameterization of α, a multi-layer subsurface module has the advantage that it keeps track of properties like grain size and density of individual layers, which may then be used as independent variables in the parameterization when a specific layer appears at the surface. Albedo parameterizations implemented in multi-layer subsurface modules can be found in Brun *et al.* (1992), who used grain size and snow age in the uppermost grid point as independent variables, and in Greuell and Konzelmann (1994), who used the density of the uppermost grid point and cloud amount as independent variables. We will now discuss some α parameterizations that can be combined with bulk subsurface modules.

(A) Brock *et al.* (2000) validated various α parameterizations with ground measurements obtained on Haut Glacier d'Arolla (Switzerland). The best fit to the data for deep (≥ 0.5 cm water equivalent) snow was found for

$$\alpha_s = c_1 - c_2{}^{10}\log T_{\mathrm{ma}}, \tag{5.3}$$

where T_{ma} is the accumulated amount of daily maximum temperatures above $0\,^{\circ}$C since the last snow fall, $c_1 = 0.71$ and $c_2 = 0.11$. For small T_{ma}, α_s is constrained by a maximum of 0.85. This equation performs better than any of the following.

(i) An exponential function in T_{ma} as proposed by Ranzi and Rossi (1991).

(ii) Logarithmic or exponential functions in the time since the last snow fall (t_s). An exponential function in t_s was proposed by Oerlemans and Knap (1998). Indeed, T_{ma} is expected to be a better predictor than t_s since grains and impurity concentrations grow mainly during melting, whereas little change occurs when the snow is dry. To a good approximation, T_{ma} represents the accumulated amount of melt (M_a) since the last snow fall (see section 5.3).

(iii) Functions using M_a as predictor. This is surprising in view of the argument that T_{ma} only represents M_a, which is the direct cause of the changes in α, but the data from Haut Glacier d'Arolla do not substantiate that M_a is a better predictor than T_{ma}. Note further that using M_a as predictor bears the danger that errors increase owing to the positive feedback between α and M_a.

(B) Simulation of the ice albedo (α_i) is problematic. It is important, though, because α_i may vary considerably from glacier to glacier. Some glaciers at mid-latitudes are largely covered by debris and therefore have a very low α_i. Polar glaciers, on the other hand, generally have much cleaner ice surfaces and hence a higher α_i. Evaluation of satellite data suggests that on some glaciers α_i increases with elevation (Haut Glacier d'Arolla, Switzerland, Knap *et al.*, 1999; Hintereisferner, Austria, Koelemeijer, Oerlemans and Tjemkes, 1993), whereas it is constant with elevation on other glaciers (Pasterze, Austria, Greuell *et al.*, 1997). Owing to medial and lateral moraines, substantial variation in α_i occurs in the direction perpendicular to the flow direction. Furthermore, during the melt season α_i may increase with time, as observed by Oerlemans and Knap (1998) on the Morteratschgletscher (Switzerland), decrease with time, as observed by Reijmer *et al.* (1999) on Vatnajökull (Iceland), or remain constant, as observed by Greuell *et al.* (1997) on the Pasterze, Austria. No explanations have been given for these spatial differences in temporal behaviour, but they must be related to the surface budget of debris and dust. This state of knowledge suggests neglecting temporal variations in α_i in the parameterizations. In most models spatial variations can be described as a function of elevation, but the scheme should be adapted for each individual glacier. Oerlemans (1991/92) proposed to make α_i a function of the difference between the elevation and the equilibrium-line altitude (ELA) on the assumption that on a decadal timescale the surface budget of debris and dust is tied to the ELA.

(C) A zone of relatively low α_i extends from north to south parallel to the ice-sheet margin on the western part of the Greenland ice sheet (see Figure 5.3). The albedo decrease in this zone can be ascribed to melt-water accumulation on the surface having a slope less than 0.02 (Greuell, 2000). Zuo and Oerlemans (1996) successfully simulated the albedo of this 'dark zone' by making α_i a function of the amount of accumulated melt water, which in turn depends on the production of melt water and the surface slope.

(D) Oerlemans and Knap (1998) proposed the following equation in order to simulate the effect of the underlying ice surface on α for shallow snow packs:

$$\alpha = \alpha_s + (\alpha_i - \alpha_s)\exp(-d/d^*), \tag{5.4}$$

where α_s is the snow albedo, α_i is the ice albedo, d is the thickness of the snow pack on top of glacier ice (d may also be zero or infinite) and d^* is a characteristic scale for snow depth (3.2 cm according to Oerlemans and Knap (1998)). The measurements analysed by Brock *et al.* (2000) demonstrate that the effect of the underlying surface fades very rapidly, i.e. it is absent if the snow pack is deeper than 0.5 cm water equivalent. This suggests neglecting the effect. On the other hand, equation (5.4) can be used to simulate α of the mosaic of ice and snow patches occurring around the snow line, which justifies a value for d^* greater than 0.5 cm water equivalent.

The discussed parameterizations (A)–(D) all apply to specific types of surface and sub-surface conditions: Equation (5.3) can be used for 'deep snow packs', α_i can be a function of elevation (B), the effect of accumulation of melt water on ice can be taken into account as suggested under (C), and 'shallow snow packs' can be described by equation (5.4). In a model, these 'partial parameterizations' can be combined where equation (5.3) may be replaced by one of the alternatives mentioned after that equation. Note, though, that none of the parameterizations (A)–(D) considers the effect of the incidence angle and clouds. Based on data from Ohmura (1981), Greuell and Konzelmann (1994) proposed a linear relation between cloud amount (n_c) and α (α increases by 0.05 when n_c increases from 0 to 1).

It should be emphasized that most of the constants in the parameterizations vary from glacier to glacier or from region to region because they depend on the amount of impurities and debris accumulating on the surface. This means that parameterizations have to be tuned per glacier or region by means of ground or satellite data.

The success of most of the α parameterizations is not only determined by the quality of the equations. It is also determined by the accuracy of the independent variables, especially if the model itself generates these. This is, for instance, the case with grain size, density, snow depth, amount of accumulated melt water and whether the snow pack has melted away or not. As already stated, errors in the calculation of α have a significant impact on the mass balance calculations since melt is very sensitive to α variations. The positive feedback between α and melt aggravates this, which, according to calculations for the Greenland ice sheet (Van de Wal, 1996) and for the Pasterze, Austria (Greuell and Böhm, 1998), roughly doubles the sensitivity of the mass balance to temperature variations.

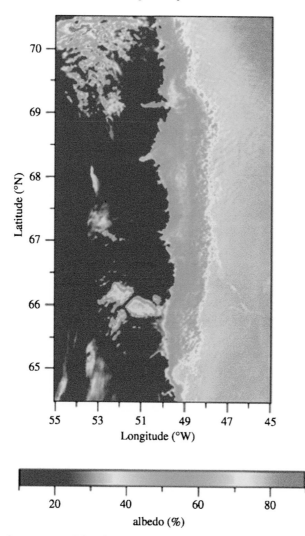

Figure 5.3. Albedo pattern on July 16, 1995, 14:29 GMT, of a section of the western part of the Greenland ice sheet (roughly east of 50 °W), the adjacent tundra area and the sea surface as derived from advanced very high resolution radiometer (AVHRR) data. The eastern part of the ice sheet on the image has albedos higher than 75% and exhibits a relatively small spatial variability in albedo. This area is interpreted as a surface that is uniformly covered by snow. Going westward, the albedo suddenly decreases while the spatial variability increases abruptly. In this part the surface presumably consists of a mosaic of snow, slush, lakes and ice, and, more to the west, of ice only. Within this marginal zone the 'dark zone' is visible. The relatively low albedo in the 'dark zone' is probably due to accumulation of melt water on the surface. (From Greuell and Knap (2000), with permission of the American Geophysical Union.)

5.2.4 The incoming long-wave radiative flux

Unless measurements are available, the long-wave incoming radiative flux ($L\downarrow$) must be estimated by means of a parameterization. For clear skies, $L\downarrow$ is a function of the entire vertical profile of atmospheric temperature and greenhouse gases, of which water vapour is the most effective. However, usually only the 2 m temperature (T_{2m}) and 2 m vapour pressure (e_{2m}), or values of atmospheric temperature (T) and vapour pressure (e) at another height near the surface, are available as input for a parameterization. So, for clear skies, one has to compute $L\downarrow$ as a function of T_{2m} and e_{2m} only. This is justified to some extent by the fact that much of the long-wave radiation received at the surface is emitted by the atmosphere near the surface: $\sim$30% by the lowest 10 m, $\sim$60% by the lowest 100 m and $\sim$90% by the lowest 1000 m (Konzelmann *et al.*, 1994). All the parameterizations of $L\downarrow$ for clear skies read:

$$L\downarrow_{cs} = \varepsilon_{cs}\sigma T_{2m}^4,\tag{5.5}$$

with ε_{cs} the clear-sky emittance, σ the Stefan–Boltzmann constant ($= 5.67 \times 10^{-8}$ W/m² K⁴) and T_{2m} in kelvin. Various expressions for ε_{cs} can be found in the literature (an overview was given by Idso (1981)). Here we give the expression proposed by Konzelmann *et al.* (1994):

$$\varepsilon_{cs} = 0.23 + c_{\mathrm{L}}\left(\frac{e_{2m}}{T_{2m}}\right)^{1/8},\tag{5.6}$$

where c_{L} is a tuning parameter. If temperature and vapour pressure are measured at a height other than 2 m, the measurements must be reduced to values at the 2 m level.

Obviously, $L\downarrow_{cs}$ differs for different profiles of temperature and vapour pressure, even if the profiles have the same values for these variables at the 2 m level. However, the parameterization predicts the same values for $L\downarrow_{cs}$. This is the cause of random error if the equations are applied at the location where the data for a tuning procedure were collected. If the equations are applied at other locations, where atmospheric profiles may systematically deviate from the profiles at the location of tuning, systematic errors occur. This problem can be solved by using different values for c_{L} for different locations and elevations, as discussed by Greuell *et al.* (1997). However, according to Meesters and Van den Broeke (1997) equation (5.5) is not suitable at all over melting surfaces owing to the peculiar shape of the profiles of e and T, but the only alternative these authors offered was coupling of an ablation model to a model with vertical resolution in the atmosphere.

Greuell *et al.* (1997), Kayastha *et al.* (1999) and Plüss and Ohmura (1997) modified equation (5.6) in order to take radiation emitted by surrounding slopes into account. It appeared that for horizontal surfaces surrounding slopes only marginally increased $L\downarrow$, even for steep glacierized or snow-covered valleys in the Alps and the Himalayas. However, Plüss and Ohmura (1997) report that on inclined slopes $L\downarrow$ may increase significantly owing to terrain emissions.

Clouds have a significant impact on $L\!\downarrow$. The impact increases with cloud amount (n_c), cloud emissivity and cloud-base temperature. All these factors are considered in a parameterization proposed by Kimball, Idso and Aase (1982), but, since it is generally difficult to estimate cloud emissivity and cloud-base temperature, many parameterizations of $L\!\downarrow$ only consider n_c. One such parameterization, proposed by Konzelmann *et al.* (1994), reads:

$$L\!\downarrow = \left[\varepsilon_{cs}\left(1 - n_c^a\right) + \varepsilon_{oc}n_c^a\right]\sigma T_{2m}^4,$$

where ε_{oc} (the emittance of the totally overcast sky) and a are tuning parameters. Konzelmann *et al.* (1994) tuned this equation with data from the Greenland ice sheet, and Greuell *et al.* (1997) did the same with data from the Pasterze (Austria). They found different values for ε_{oc} and a for different locations, which primarily reflects spatial differences in mean cloud emissivity and spatial variations in the mean difference between T_{2m} and the cloud-base temperature. Hence, similar to the parameterization of the cloud factor in the parameterization for the incoming short-wave radiation, tuned parameterizations for $L\!\downarrow$ have a limited applicability in space due to spatial variations in the shape of atmospheric profile of temperature and humidity and spatial variations in cloud characteristics. Other work on tuning parameterizations of $L\!\downarrow$ over glaciers has been performed by König-Langlo and Augstein (1994) for Neumayer (Antarctica).

5.2.5 The outgoing long-wave radiative flux

The outgoing long-wave radiative flux ($L\!\uparrow$) can be calculated by means of the following equation:

$$L\!\uparrow = \varepsilon_s\sigma T_s^4 + (1 - \varepsilon_s)L\!\downarrow,$$

where T_s is the surface temperature (in kelvin) and ε_s is the emissivity of the surface. In ablation models T_s is computed with a subsurface module or follows from the 'zero-degree assumption' (see section 5.2.8). Griggs (1968) measured a value of 0.99 for ε_s of snow in the spectral range 8–14 μm, and model calculations by Warren (1982) indicate that ε_s is larger than 0.96 for wide ranges of snow conditions and for the spectral range 4–40 μm. Therefore, the ε_s of snow and glacier ice is probably close to 1.0. Note that small deviations from 1.0 have a negligible effect on the net long-wave radiative flux,

$$L\!\downarrow - L\!\uparrow = \varepsilon_s\left(L\!\downarrow - \sigma T_s^4\right),$$

because $L\!\downarrow - \sigma T_s^4$ hardly ever exceeds 100 W/m². Therefore, it is generally assumed that $\varepsilon_s = 1.0$, so:

$$L\!\uparrow = \sigma T_s^4. \tag{5.7}$$

Sometimes, measurements of $L\uparrow$ are available, but, if a multi-layer subsurface module is available for the calculation of T_s, values of $L\uparrow$ computed with equation (5.7) are more accurate than direct measurements (Greuell and Konzelmann, 1994; Greuell and Smeets, 2001).

5.2.6 The fluxes of sensible and latent heat

The turbulent fluxes of sensible and latent heat can be directly measured by means of eddy-correlation techniques. However, it is impossible to perform such measurements continuously for longer periods of time. The instruments do not work during rain and in fog, and the likelihood of technical failure is large.

However, eddy-correlation measurements are most useful for validation or improvement of other, more practical, methods for the estimation of the turbulent fluxes. One such method is the 'profile method' which uses time-averaged values of wind speed (u), potential temperature (θ) and specific humidity (q) at two or more levels near the surface to estimate the fluxes. The relationships between the profiles and the fluxes are generally established within the framework of the Monin–Obukhov similarity theory (see e.g. Stull, 1988). They read:

$$\tau = \rho_a u_*^2, \tag{5.8}$$

$$u_* = \frac{\kappa z}{\phi_m} \frac{\partial u}{\partial z}, \tag{5.9}$$

$$Q_H = \rho_a C_{pa} u_* \theta_*, \tag{5.10}$$

$$\theta_* = \frac{\kappa z}{\phi_h} \frac{\partial \theta}{\partial z}, \tag{5.11}$$

$$Q_L = \rho_a L_s u_* q_*, \tag{5.12}$$

$$q_* = \frac{\kappa z}{\phi_h} \frac{\partial q}{\partial z}, \tag{5.13}$$

where τ is the momentum flux or shear stress, ρ_a is the density of the air, u_*, θ_* and q_* are velocity, temperature and humidity scales, κ is the Von Karman constant (0.4), z is the height above the surface, C_{pa} is the specific heat capacity of the air (1005 J/(kg K)) and L_s is the latent heat of sublimation (2.84 × 10^6 J/kg). The velocity scale is also called the friction velocity. The stability functions ϕ_m and ϕ_h account for the effect of atmospheric stability on the degree of turbulence. It is assumed that the fluxes are constant with z and hence, in all equations in this section, represent the surface fluxes, which form the aim of the calculations. Equations for ϕ_m and ϕ_h, which differ for unstable and stable conditions, are obtained by relating simultaneous measurements of the fluxes and the gradients of the profile variables. Högström (1988) lists various expressions for ϕ_m and ϕ_h. Stable conditions predominate over melting glaciers. The expressions given by Högström (1988) for ϕ_m and

ϕ_h under stable conditions are all linear in z/L_{ob}, where L_{ob} is the Monin–Obukhov length:

$$\phi_m = 1 + \alpha_m \frac{z}{L_{ob}} \qquad (5.14)$$

$$\phi_h = Pr + \alpha_h \frac{z}{L_{ob}}, \qquad (5.15)$$

$$L_{ob} = \frac{u_*^2}{\kappa \left(\frac{g}{T_0}\right)(\theta_* + 0.62 T_0 q_*)}. \qquad (5.16)$$

In these equations α_m, Pr (Prandtl number) and α_h are empirical constants, which vary from study to study. Further, T_0 is the absolute temperature and g is the acceleration due to gravity ($9.81 \ \mathrm{m/s^2}$). After analysis of measurements over a vegetated area, Högström (1988) concluded that optimal values are $\alpha_m = 6.0$, $Pr = 0.95$ and $\alpha_h = 7.8$. Studies over glaciers are rare. An analysis of data collected over the Greenland ice sheet by Forrer and Rotach (1997) shows that for that location the non-linear (in z/L_{ob}) stability functions introduced by Beljaars and Holtslag (1991) yield better results than linear functions. Measurements at Halley (Antarctica) show that the ϕ-functions (equations (5.14) and (5.15)) have a maximum value of 12 (King *et al.*, 1996).

Once the expressions for ϕ_m and ϕ_h are known, the turbulent fluxes can be determined from the gradients of u, θ and q. This is not straightforward since L_{ob} is a function of u_*, θ_* and q_*. Therefore, the equations can only be solved by iteration, as discussed by Munro (1989).

To determine gradients, profile measurements from at least two levels must be interpolated (Forrer and Rotach, 1997). Alternatively, measurements at two levels may be used directly to calculate the turbulent fluxes. In that case, integral forms of equations (5.9), (5.11) and (5.13) are used, e.g.

$$u_* = \frac{\kappa(u_{L2}) - u_{L1})}{\ln \frac{z_{L2}}{z_{L1}} + \frac{\alpha_m}{L_{ob}}(z_{L2} - z_{L1})},$$

where the subscripts L1 and L2 refer to the two levels. The equations may again be solved by means of Munro's iterative scheme.

Measurements of u, T (for convenience, we neglect the difference between temperature and potential temperature here) and q are often available from one level only. In that case, calculations are carried out with the so-called 'bulk method'. Equations (5.9), (5.11) and (5.13) are integrated from the level in the atmosphere to the surface, resulting in the following equations for the fluxes:

$$Q_H = \rho_a C_{pa} \frac{\kappa^2 u(T - T_s)}{\left(\ln \frac{z}{z_0} + \frac{\alpha_m z}{L_{ob}}\right)\left(\ln \frac{z}{z_T} + \frac{\alpha_h z}{L_{ob}}\right)}, \qquad (5.17)$$

$$Q_L = \rho_a L_s \frac{\kappa^2 u(q - q_s)}{\left(\ln \frac{z}{z_0} + \frac{\alpha_m z}{L_{ob}}\right)\left(\ln \frac{z}{z_q} + \frac{\alpha_h z}{L_{ob}}\right)}, \qquad (5.18)$$

where T_s is the surface temperature, q_s is the specific humidity at the surface, and z_0, z_T and z_q are the roughness lengths for velocity, temperature and water vapour. The roughness lengths are defined as the heights above the surface, where the profiles of u, θ and q defined by equations (5.8)–(5.16) reach their surface values. As with the other methods, iteration is needed to solve the equations.

In ablation models, T_s is computed with a subsurface module or follows from the 'zero-degree assumption' (see section 5.2.8) and q_s follows from T_s because the vapour pressure at the surface is equal to the saturation vapour pressure at T_s $(e_{sat}(T_s))$:

$$q_s = 0.621 \frac{e_{sat}(T_s)}{p},$$

where p is atmospheric surface pressure.

Braithwaite (1995a) and Munro (1989) pointed out that, of all factors, uncertainty in z_0 causes the largest uncertainty in the determination of the turbulent fluxes. The most widely employed method to determine z_0 is by extrapolation of the wind-speed profile to the height where the wind speed vanishes (Denby and Smeets, 2000; Munro, 1989). This is an obvious method, because it corresponds to the definition of z_0. A problem of this method is the uncertainty in the reference height $(z = 0)$, especially on glaciers with a hummocky surface. This may introduce a substantial uncertainty in the determination of z_0 (Munro, 1989), but the problem can be solved by introducing a displacement height into the equations (Smeets, Duynkerke and Vugts, 1999). Hummocks disturb the flow to a height of approximately twice the height of the main roughness elements (Smeets *et al.*, 1999). Only measurements from above the disturbed layer may be used to determine z_0.

Alternatively, z_0 can be estimated from micro-topographical data, a method that does not suffer from the problem of having to determine a zero-reference plane. In principle, the method is feasible because in 'aerodynamically rough flow' (a definition can be found in Garratt, 1992, p. 41) z_0 is only determined by the geometry of the surface. Note, though, that it is not entirely clear which scale of roughness elements determines z_0 (Munro, 1989). Kondo and Yamazawa (1986) demonstrated that z_0 increases with the size of roughness elements with a wavelength smaller than 10 cm. According to the same authors, undulations of a snow surface with a wavelength greater than 10 cm are not important with regard to aerodynamic effects because of their low slope. On the other hand, for two sites in the ablation area of Vatnajökull (Iceland), Smeets *et al.* (1999) reported that z_0 increases in time together with a growth of the largest surface roughness elements to a height of 1.7 m. This suggests that these macroscopic terrain features have an effect on z_0. For snow and glacier ice, Munro (1989) and Smeets *et al.* (1999) compared the micro-topographical method with the profile method. In both papers the authors conclude that the micro-topographical method yields correct results, but the number of different surfaces studied is very limited. Therefore more comparisons of the micro-topographical method with the profile method over various types of surfaces are needed to judge the validity of the micro-topographical method.

If measurements of u, T and q are available from one height only, z_0 must be estimated with the micro-topographical method or a value from the literature must be adopted. Values for z_0 of snow and glacier ice found during various studies were listed by Braithwaite (1995a) and Morris (1989). For glacier ice most z_0-values range between 1 and 10 mm.

These numbers also hold for snow, but values down to 0.1 mm were found on Antarctic ice shelves.

It is frequently assumed that the roughness lengths for temperature and for water vapour are equal to that for momentum. However, theory predicts (Andreas, 1987) that the various roughness lengths are not equal, because, close to the surface, transfer mechanisms for momentum differ from those for scalars (temperature and water vapour). Momentum is transferred by molecular diffusion and by pressure forces against obstacles (form drag), whereas the scalars are transmitted by molecular diffusion only. Andreas (1987) incorporated these differences in transport mechanisms in a 'surface renewal model' in order to derive equations relating z_T and z_q to z_0. For aerodynamically rough surfaces, the equations read as follows:

$$\ln\left(\frac{z_T}{z_0}\right) = 0.317 - 0.565 R_* - 0.183 R_*^2, \tag{5.19}$$

$$\ln\left(\frac{z_q}{z_0}\right) = 0.396 - 0.512 R_* - 0.180 R_*^2, \tag{5.20}$$

$$R_* = \frac{u_* z_0}{v}, \tag{5.21}$$

where R_* is the roughness Reynolds number and v is the kinematic viscosity of air (1.461×10^{-5} m^2/s). Under conditions typical for melting snow and ice ($z_0 = 1$–10 mm and $u_{2m} = 3$–7 m/s), these equations predict z_T and z_q to be one to four orders of magnitude smaller than z_0. Denby and Snellen (2001) analysed field data and concluded that the proposed equations are correct. Other analyses of field data, such as the analysis by Munro (1989), are somewhat inconclusive on this point. Holmgren (1971) found $z_T/z_0 \sim 0.01$, and Smeets, Duynkerke and Vugts (1998) found values of z_T/z_0 ranging between 0.01 and 0.2, thus broadly confirming equations (5.19)–(5.21). On the other hand, measurements by King and Anderson (1994) carried out on an Antarctic ice shelf suggest $z_T/z_0 \sim 1000$, which is clearly in conflict with equations (5.19)–(5.21).

It is noteworthy that the use of equations (5.19)–(5.21) to predict z_T and z_q decreases the sensitivity of the sensible and latent heat flux to uncertainty in z_0, because z_T and z_q decrease with increasing z_0.

Many glacier surfaces, especially those that are or were subjected to melting, are probably aerodynamically rough ($R_* > 55$), so equations (5.19)–(5.21) apply. Over aerodynamically smooth surfaces ($R_* < 5$) form drag is absent, leading to different relations of z_T and z_q to z_0. These equations are given by Andreas (1987).

Strictly speaking, the Monin–Obukhov similarity theory may be applied only in terrain with homogeneous surface conditions and within the so-called constant-flux layer. However, the constant-flux layer may be thin over glaciers, where katabatic winds with a low wind maximum tend to dominate (see Figure 5.4). For instance, during 50% of the time the maximum wind speed over the ablation area of the Pasterze (Austria) was recorded at 4 m or lower (Greuell and Smeets, 2001). For conditions with a low wind-speed maximum, Denby (1999) computed the vertical fluxes of τ and Q_H with a second-order closure model

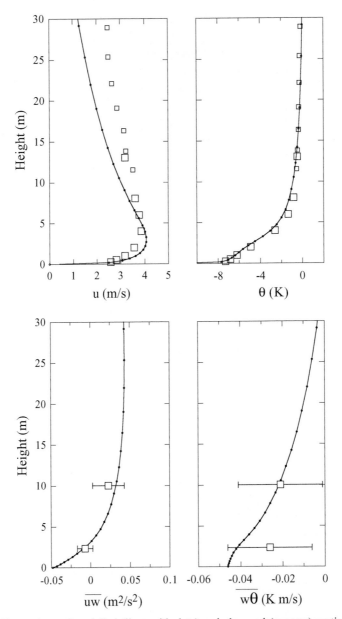

Figure 5.4. Comparison of modelled (lines with dots) and observed (squares) vertical profiles of wind speed (u), potential temperature (θ) and the vertical fluxes of momentum ($\overline{uw}$) and sensible heat ($\overline{u\theta}$) for the Pasterze, Austria. The observations are the mean for a period of two days with melt at the surface and small synoptic forcing. Fluxes were measured by eddy-correlation. The error bars in the measured fluxes give the standard deviation during the observational period. (From Denby (1999), with kind permission of Kluwer Academic Publishers.)

of the katabatic wind layer. The divergence of τ and Q_H within the first few metres from the surface appears to be significant, which implies that the Monin–Obukhov similarity theory is not applicable here. Earlier, Munro and Davies (1978) drew the same conclusion, based on an analysis of profile measurements of u and T. Nevertheless, Denby and Greuell (2000) used the katabatic wind model of Denby (1999) to test the validity of the 'profile method' and the 'bulk method' for conditions with a low wind maximum. They concluded that for cases with a low wind maximum the profile methods severely under-estimate turbulent fluxes, whereas the bulk method yields good results. Thus, the bulk method seems to be applicable even when the wind maximum is within a few metres from the surface. It should be noted that the problem is much less serious for ice shelves and larger glaciers, such as ice sheets, where slopes are generally smaller and the wind maximum occurs higher above the surface. For instance, at the ETH camp, located near the equilibrium line on the Greenland ice sheet, the constant-flux layer has a thickness of 5–10 m (Forrer and Rotach, 1997). For a site on an Antarctic ice shelf, the similarity theory was found to hold up to a height of 5 m (King, 1990) but to fail at greater heights.

It is quite common to simplify equations (5.8)–(5.16) by setting $\phi_m = \phi_h = 1$. Then, the profiles of u, T and q become logarithmic and the turbulent fluxes can be calculated with equations (5.17) and (5.18), setting $\alpha_m = \alpha_h = 0$. Calculations with this simplification were compared with calculations with the full set of equations by Braithwaite (1995a) and Greuell and Oerlemans (1986). Greuell and Oerlemans (1986) showed that the error increases with decreasing elevation owing to an increase in stability.

Some ablation modellers (e.g. Oerlemans, 1992) made an extreme simplification of equation (3.17) (a similar equation is used to simplify equation (5.18)):

$$Q_H = C_H(T - T_s).$$

The heat-transfer coefficient C_H is sometimes used to tune the model. Braithwaite (1995a) provides a list of values. Obviously, the choice of C_H as a constant in time and space corresponds to the neglect of temporal and spatial variations in ρ_a, u, z_0, z_T, ϕ_m and ϕ_h. This may be adequate if all those variations are small or unknown. In any case, it seems useful to compare the value of C_H used in a particular study with a value calculated from estimates of ρ_a, u, z_0, z_T, ϕ_m and ϕ_h.

5.2.7 The heat flux supplied by rain

If rain with a temperature (T_r) that differs from the temperature of the surface falls onto that surface, it supplies the following amount of energy to the glacier:

$$Q_R = \rho_w r C_{pw}(T_r - T_s),$$

where ρ_w is the density of water (1000 kg/m^3), r is the rain-fall rate (in metres per second) and C_{pw} is the specific heat capacity of water (4200 J/(kg K)). This equation implies that the rain is added to the liquid phase at the surface. To a first approximation, T_r can be set equal to T_{2m}. Except for extraordinary circumstances of heavy rain fall at high temperature events (Hay and Fitzharris, 1988), this term can be neglected in the energy balance.

5.2.8 Subsurface processes

Most of the energy exchange between the atmosphere and the glacier affects the surface itself. The only exception is a part of the short-wave radiative flux, which will be absorbed below the surface. In principle, the energy exchange with the atmosphere results in a temperature change of the snow or ice, but, when the temperature is already at the melting point, absorption of additional energy causes the snow or ice to melt. Water formed by melt or deposited on the glacier as rain will penetrate into snow. On its way down, some of the water will be retained in the pores owing to capillary forces. If the water encounters 'cold' snow (i.e. snow with a temperature below the melting point), it will refreeze and locally increase the temperature. The water that refreezes within the snow is called 'internal accumulation'. If the water encounters ice on its way downward, it will generally run off, thus contributing to ablation. However, if the slope is small, part of the water will accumulate on top of the ice. There it may freeze as so-called 'superimposed ice' if the underlying ice is 'cold'. Meanwhile, vertical variations in temperature tend to even out owing to conduction. At the same time, metamorphosis, melting and refreezing cause changes in grain size and shape, and density. These variables describing the structure of the snow and ice affect most of the processes mentioned above.

Some ablation models (Brun *et al.*, 1992, Fujita and Ageta, 2000, Greuell and Konzelmann, 1994, Lehning *et al.*, 1999, Morris, Bader and Weilenmann, 1997) incorporate multi-layer subsurface modules which deal with most or all of these processes. These modules compute temperature, density and water content on a grid extending downward from the surface. The modules that Brun *et al.* (1992) and Lehning *et al.* (1999) developed for avalanche forecasting also simulate grain size and shape. A description of how these modules treat the relevant subsurface processes would be lengthy, and is therefore beyond the scope of this chapter.

Many ablation models treat the subsurface with much simpler modules, which we will call bulk subsurface modules, since they only consider a snow pack of variable depth on top of ice (the snow pack may be infinitely deep or absent). Whether there is snow or ice at the surface in these models matters for two reasons. It affects the surface albedo (see section 5.2.3) and it may affect the calculation of how much melt water refreezes. We will now consider a few simple schemes that represent the influence of subsurface processes on the mass balance.

A widely employed method is the 'zero-degree assumption'. This refers to the surface, which is assumed to be at the melting point. If the zero-degree assumption leads to a positive energy flux from the atmosphere to the glacier ($Q_0 > 0$), the energy is assumed to be entirely consumed in the formation of melt water, which all runs off:

$$\frac{dR_{off}}{dt} = \frac{Q_0}{L_f},$$

where R_{off} is runoff and L_f is the latent heat of fusion (0.334×10^6 J/kg). If the zero-degree assumption leads to a negative value of Q_0, no melt water is formed. The zero-degree assumption is correct if, in reality, the temperature of the entire snow and ice column is at the melting point and, according to the calculations, $Q_0 > 0$. This should be the case at most locations on mid-latitude glaciers during summer. However, when in reality the snow or ice in the uppermost layers is 'cold', energy coming in from the atmosphere will heat the snow or ice directly, through conduction and absorption of short-wave radiation, and indirectly, through refreezing of melt water as internal accumulation or superimposed ice. This is not accounted for by the zero-degree assumption, so under those conditions the calculated mass balance is in error. This applies especially to locations near and above the annual snow line on polar glaciers, where all or almost all of the melt water refreezes (Braithwaite, Laternser and Pfeffer, 1994). The zero-degree assumption also leads to larger errors for spring or very high elevations on mid-latitude glaciers, as quantified by Greuell and Oerlemans (1986). Note further that, if the real surface temperature is below the melting point, the zero-degree assumption results in an over-estimation of the amount of sublimation (Greuell and Oerlemans, 1986). This error can be subdued by setting the amount of sublimation equal to zero if $Q_0 < 0$.

When the zero-degree assumption is expected to lead to substantial errors, one of the multi-layer subsurface modules mentioned above may be incorporated in an ablation model. However, those modules are complex and computer-time-consuming. Oerlemans, Van de Wal and Conrads (1991/92) used a simpler scheme for the simulation of heating of snow and ice and for the simulation of refreezing and runoff. Based on measurements made by Ambach (1963) on the Greenland ice sheet, Oerlemans *et al.* (1991/92) proposed that a layer, equivalent to 2 m of ice, may be heated during the ablation season. At the beginning of the ablation season the temperature of this layer (T_L, in °C) is set equal to the annual mean atmospheric temperature. For the computation of Q_0 the authors assume $T_s = 0$ °C. If $Q_0 > 0$, the following equations apply:

$$L_f \frac{dR_{off}}{dt} = Q_0 \exp(T_L),$$

$$\frac{dT_L}{dt} = \frac{\left(Q_0 - L_f \frac{dR_{off}}{dt}\right)}{(M_L C_{pi})},$$

where M_L is the mass of 2 m^3 of ice (1820 kg) and C_{pi} is the specific heat capacity of ice (2009 J/(kg K)). Therefore, the higher the ice temperature, the more melt water runs off at the expense of refreezing within the snow pack or accumulation on top of the ice. Calculations of the mass balance of the Greenland ice sheet with the zero-degree assumption and with this scheme differ considerably (Oerlemans *et al.*, 1991/92). Note, however, that this scheme does not distinguish between snow and ice, though relevant processes like percolation of melt water and conduction differ substantially between the two media.

Pfeffer, Meier and Illangasekare (1991) accounted for the refreezing effect on mass balance by firstly calculating the runoff limit. The runoff limit is located at the elevation where the annual melt equals the sum of the melt water needed to raise the temperature of the annual accumulation to the melting point and the melt water filling the pore space of the annual accumulation. Computations show that this occurs roughly at the point where the annual melt is 70% of the annual accumulation. Pfeffer *et al.* (1991) then assume that the runoff limit separates the area where no melt water at all runs off from the area where all of the melt water runs off. Janssens and Huybrechts (2000), however, also used the equations of Pfeffer *et al.* (1991) to compute refreezing below the runoff limit. They then applied the adjusted scheme to the entire Greenland ice sheet and compared the amount of refreezing with that from two other simple refreezing schemes. It turns out that, even though the total amount of refreezing for the entire ice sheet varies little among the three models, the spatial variation of refreezing predicted by the three schemes differs considerably. Pfeffer's scheme produced the most realistic spatial variation.

More insight into the merits and disadvantages of the different refreezing parameterizations should be obtained by comparison with more complex multi-layer subsurface modules.

5.3 The degree-day approach

Up to now we have been looking at the energy balance approach to ablation modelling. A simpler approach to ablation modelling are degree-day models, which basically use the following equation to predict ablation (N):

$$N = \beta T_{\text{pdd}}. \tag{5.22}$$

Here, β is the degree-day factor (in millimetres water equivalent per day per kelvin) and T_{pdd} is the sum of all positive daily mean temperatures (in °C) over the period of interest. The degree-day factor is a constant, which must be determined by means of field data of N and T_{pdd}. A list of published degree-day factors can be found in Braithwaite and Zhang (2000).

From a physical point of view, energy balance modelling is the most correct way to treat the relationship between the conditions in the atmosphere and the mass balance. However, the simplicity of the degree-day method has two advantages compared with the energy balance method. Firstly, it is computationally cheaper, and secondly in most energy balance formulations it is necessary to specify values for variables such as extra-terrestrial irradiance ($I_0 \cos(\theta_s)$ in equation (5.2)), wind speed, humidity and cloud amount, and to develop equations that generate values for albedo and the roughness lengths. This problem does not exist in degree-day models; the effect of all these variables is lumped into the degree-day factor.

The degree-day approach may be considered as a simplification of the energy balance approach. If the surface temperature and the subsurface temperature are at the melting point, the ablation rate (dN/dt) can be computed from

$$L_f \frac{dN}{dt} = S\!\!\downarrow(1-\alpha) + \beta_1 T$$

because for $T_s = 0\,^\circ\mathrm{C}$, to a first approximation, the sum of net long-wave radiation and the turbulent fluxes is linear in T (β_1 is a constant). If we assume that $T > 0\,^\circ\mathrm{C}$ and $\beta_2 = \beta_1/L_f$, integration over time yields

$$N = \frac{1}{L_f}\int S\!\!\downarrow(1-\alpha)\,dt + \beta_2 T_{\mathrm{pdd}}. \qquad (5.23)$$

This equation, based on energy balance considerations, has the same form as equation (5.22), except for the appearance of the time integral over net short-wave radiation ($S\!\!\downarrow(1-\alpha)$). However, that term is important because it constitutes the dominant term in the surface energy balance in many cases. Nevertheless, degree-day models, which neglect the short-wave radiation term, work reasonably well, presumably owing to one (or both) of the following reasons:

(a) the mean over the period of interest of $(1-\alpha)$ is positively correlated with T_{pdd} because N increases with T_{pdd} and α decreases with increasing N;
(b) the mean over the period of interest of $S\!\!\downarrow$ is positively correlated with T_{pdd}, or, in other words, sunny summers are warmer.

A drawback of degree-day models is that they can only be used to calculate the sensitivity of the mass balance to variations in temperature. Also, known temporal and spatial variations in variables other than temperature cannot be considered. Seasonal and latitudinal variations in the extra-terrestrial irradiance are important examples of effects not taken into account by degree-day models. Another problem with degree-day models occurs when T is around the melting point. Depending on the radiation balance, the threshold between melt and no melt may be above or below $0\,^\circ\mathrm{C}$. Therefore, some authors (e.g. Braithwaite, 1995b) introduce an intercept (β_0) into equation (5.22):

$$N = \beta_0 + \beta T_{\mathrm{pdd}}.$$

Even with the assumptions that the ablation rate is linear in T and ablation is zero for $T \leq 0\,^\circ\mathrm{C}$, the use of daily means to compute positive degree-day sums will cause errors in the calculated ablation. For example, on a day with a mean T of $0\,^\circ\mathrm{C}$, T probably was above $0\,^\circ\mathrm{C}$ during part of the day. So, according to our assumptions, melt occurred, but no melt is predicted if T_{pdd} is based on daily means of T. These errors can be avoided if the daily mean is computed only over the time intervals with $T > 0\,^\circ\mathrm{C}$ (Hock, 1999).

Van de Wal (1996) compared simulations of the mass balance on the Greenland ice sheet made with an energy balance model with simulations made with degree-day models (see Chapter 11). The results of the two models differed significantly, but it appeared to be hard to tell which model performed best owing to a lack of validation data.

Because degree-day models with constant degree-day factors totally neglect the effect of variations in extra-terrestrial irradiance and albedo on the mass balance, several proposals were made to express the degree-day factor as a function of extra-terrestrial irradiance and albedo. Indeed, it is quite common (e.g. Braithwaite and Zhang, 2000; Hock, 1999; Jóhannesson *et al.*, 1995) to use two different values for the degree-day factor, one for snow (at the surface) and a higher one for ice (at the surface), reflecting differences in albedo.

Hock (1999) proposed the following equation for the degree-day factor:

$$\beta = a_{s/i}I + \beta_2, \tag{5.24}$$

where I is computed clear-sky direct solar radiation (see section 5.2.2), and a_s and a_i are radiation coefficients for snow and ice, respectively. The coefficient β_2 is the same as in equation (5.23). The quantity I is calculated from the extra-terrestrial radiation, the orientation of the surface and the atmospheric transmissivity, which in the calculations by Hock (1999) depends only on elevation and the solar zenith angle. As a result, this variant of the degree-day method is computationally more expensive than the simplest version of the degree-day model (equation (5.22)). Equation (5.24) is appropriate for the study of spatial ablation variations in terrain with much relief, where the orientation of the surface and shadow effects are important. Apart from T, no additional time-dependent meteorological input data are needed. Note that according to equations (5.22) and (5.24) part of the ablation is linearly related to the product of incoming solar radiation and the sum of positive degree days. This seems incorrect in view of energy balance considerations, which suggest that ablation is the sum of a radiative term and a degree-day term (equation (5.23)). The product may be justified by the fact that, as mentioned before, the mean over the period of interest of $(1 - \alpha)$ is positively correlated with T_{pdd}.

If daily mean temperatures are not available, degree-day sums may be estimated from monthly means (Jóhannesson *et al.*, 1995) or from a function describing the annual variation of the temperature by a sinusoidal curve (Reeh, 1989). For the conversion to degree-day sums these authors assume that daily temperatures fluctuate around their monthly means or 'climatological values' according to a Gaussian distribution with a specified standard deviation.

5.4 The mass balance in ablation models

The main aim of simulations with the energy balance and degree-day models is, of course, the surface mass balance. We will now discuss how the different components of the mass balance, namely precipitation, runoff and sublimation follow from the model calculations.

The previous sections about energy balance and degree-day models exclusively dealt with melt, refreezing of water and runoff. Of these processes, only runoff contributes to the mass balance. The way runoff is computed by means of multi-layer subsurface and bulk subsurface modules was briefly discussed in section 5.2.8. This occurs in a rather complicated way in multi-layer subsurface modules. In bulk subsurface modules the mass of melt water that runs off according to the calculations is removed from the upper layer. Once the snow layer is completely removed, the ice appears at the surface (which influences the computation of melt because the albedo of ice differs from that of snow). Thereafter, removal of more mass will be in the form of ice.

In most models, precipitation at a temperature above a threshold (0 °C in Greuell and Böhm (1998) or 2 °C in Oerlemans (1991/92)) is considered as rain, whereas precipitation at a lower temperature is considered as snow. In multi-layer subsurface modules, rain is added to the liquid phase and snow is added to the solid phase of the uppermost grid box. The density of the freshly fallen snow must be specified in these models. The density of the material plays no role in models with a bulk subsurface module, in which snow is added to the snow layer and rain runs off or freezes (see section 5.2.8).

Energy balance models consider the contribution of sublimation to the mass balance through the computation of the latent heat flux. The amount of water vapour deposited on the surface is given by Q_L/L_s (see equation (5.18)) (if Q_L is negative, mass is removed from the surface). Note that, since L_s refers to the transition from the solid to the gas phase, the computed mass is added to or removed from the snow or ice, and not the water. It is impossible to compute sublimation of snow and ice by means of degree-day models.

Ablation models may consider removal and addition of mass by action of the wind (drifting and blowing snow) and avalanches. These topics, however, are beyond the scope of this chapter.

5.5 Introduction to modelling the mass balance at the scale of glaciers

In sections 5.2 and 5.3 it was implicitly assumed that the forcing variables, such as the near-surface temperature and cloud amount, are known. However, in most cases measurements or observations of these variables are not available at most or all of the grid points. In this section we will discuss several strategies that can be employed to solve this problem. These fall into three broad categories.

(1) Calculations are performed with an ablation model (energy balance or degree-day). The forcing at each grid point is derived from measurements at a weather station located on or outside the glacier. Then, functions transferring the data from one location to another are needed. An example is the use of a constant lapse rate to convert the temperature measured at a climate station to the near-surface temperature at the various grid points.
(2) An alternative is the calculation of the mass balance on a grid by means of a global or regional atmospheric model, which incorporates a subsurface component able to compute the mass balance. In fact, this kind of model calculates runoff and sublimation on the basis of the energy balance approach. Accumulation is computed in the model and not prescribed, as is the case in ablation models.
(3) Finally, statistical methods can be used to relate the mass balance to data from a climate station. Such a relation may, for example, be derived from observed time series of the annual mass balance, summer temperature and annual precipitation. Commonly, spatial variations in mass balance are not computed with this method but are derived from actual mass balance measurements.

5.6 Ablation models

5.6.1 Grids and forcing

The computation of the mass balance by means of surface energy balance models or degree-day models is generally done on one of two types of grids. On small glaciers mass balance is primarily a function of elevation. Therefore, in many studies the mass balance is computed on a grid with the grid points at equal distance in elevation. The other type of grid is a two-dimensional grid where the grid points are equidistant in both horizontal directions. This approach is useful for the study of across-glacier variations in mass balance due to across-glacier variations in incoming short-wave radiation related to shadow and surface-orientation effects (Arnold *et al.*, 1996; Hock, 1999). A two-dimensional grid may also be necessary for the calculation of the mass balance of ice caps and ice sheets where different sides may be subject to radically different climatic regimes. An example is the study of the mass balance of the Greenland ice sheet by Van de Wal and Oerlemans (1994).

We will now discuss how the different input variables can be transferred from one point to another or from a climate station outside the glacier to grid points on the glacier. Temperature is generally assumed to vary as a function of elevation only, described by a constant lapse rate (i.e. dT/dz). Jóhannesson *et al.* (1995) used the lapse rate as a tuning parameter, but usually the lapse rate is derived from temperature measurements made at at least two stations or adapted from other studies. Values range roughly between -0.005 and -0.007 K/m. In a study of the mass balance of the Greenland ice sheet, Oerlemans *et al.* (1991/92) prescribed seasonal and regional variations in lapse rate ranging between -0.004 and -0.009 K/m based on the work by Ohmura (1987).

A constant lapse rate is a very simple method for the inter- or extrapolation of the near-surface atmospheric temperature. Reality is, of course, much more complicated, as demonstrated by the mean temperature at five stations on the Pasterze (Austria) during the ablation season (Figure 5.5). Note that the temperature of the atmosphere just above the glacier is significantly lower than the temperature at the same elevation outside the glacier (Sonnblick). This can be ascribed to the cooling effect of the melting glacier surface. This difference has been ignored in many studies! To a large extent, the variation of the temperature with elevation over the glacier itself is determined by the length profile of the glacier, the absolute value of dT/dz increasing with the slope. The measured temperature distribution could well be explained by means of an analytical thermodynamic glacier-wind model (Greuell and Böhm, 1998). However, it must still be established how widely applicable this model is. In any case, data from the ablation zone of the Greenland ice sheet show that the model cannot be applied there.

It is important to note that the choice of the method of computing the temperature distribution is crucial for the sensitivity of the mass balance to temperature variations. Due to the constant surface temperature, temperatures near the surface of a melting glacier vary less than temperatures outside the thermal influence of the glacier. Measurements made over the Pasterze (Austria) yield a ratio of these two temperatures between 0.83 at the crest of the glacier and 0.30 near the terminus, which is well simulated with the model of Greuell

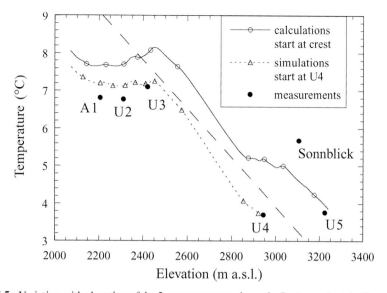

Figure 5.5. Variation with elevation of the 2 m temperature above the Pasterze, Austria (from Greuell and Böhm, 1998). The measurements are the mean over a period of 46 days during mid-summer when the surface was almost permanently melting. Sites A1, U2, U3, U4 and U5 were located at the central flow line of the glacier. Sonnblick is a mountain some 20 km east of the Pasterze and has a permanent climate station. The two curves present simulations with an analytical glacier-wind model that computes the temperature of an air parcel travelling down along the glacier within the glacier-wind layer. The two simulations shown were initiated at the crest (near U5) and at U4. They suggest that the model works quite well below U4, where the glacier wind is an almost permanent feature, but that the model fails in the upper part of the glacier where the glacier wind is less frequent. For comparison, the broken straight line gives a temperature distribution corresponding to a constant lapse of -0.0065 K/m. (Reprinted from the *Journal of Glaciology* with permission of the International Glaciological Society.)

and Böhm (1998); see Figure 5.6. Note that if a constant lapse rate is used, the ratio is equal to unity at all locations.

In most energy balance models wind speed is assumed to be spatially invariant. However, theory predicts (Prandtl, 1942) that the katabatic wind, which generally dominates over glaciers, increases in strength with surface slope and with temperature contrast between the surface and the free atmosphere. The latter dependency implies that the sensitivity of the mass balance to temperature variations is enhanced over melting glacier surfaces (Van den Broeke, 1997).

In many studies, relative humidity and cloud amount are assumed to be constant in time and space. These may seem crude assumptions, but generally the treatment of humidity does not have much effect on calculated ablation because latent heat fluxes are relatively small. Variations in cloud amount also have a limited effect on ablation because clouds have opposite effects on incoming short-wave and incoming long-wave radiation. For high albedo, net radiation may decrease with increasing cloud amount, which, since it is contrary

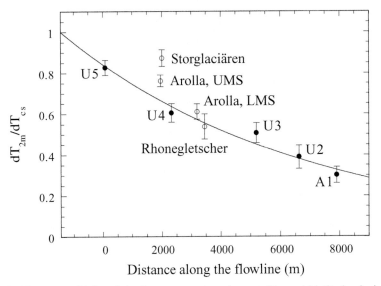

Figure 5.6. Climate sensitivity of the 2 m temperature along melting mid-latitude glaciers (from Greuell and Böhm, 1998). Here the climate sensitivity is defined as the increase of the 2 m temperature above the glacier (T_{2m}) divided by the simultaneous increase in the temperature at a climate station not located on the glacier (T_{cs}). The data indicated by the dots with error bars are based on daily mean values for five locations on the Pasterze (see Figure 5.5), two locations on Haut Glacier d'Arolla (Switzerland), a site on the Rhonegletscher (Switzerland) and a site on Storglaciären (Sweden). The glacier-wind model of Greuell and Böhm (1998) predicts an exponential decay of climate sensitivity along the glacier. The best fit of this model to the data from the Pasterze is given by the curve. It is unclear why the data from Storglaciären, and to a lesser extent those from Arolla UMS, do not concur with those from the other sites. (Reprinted from the *Journal of Glaciology* with permission of the International Glaciological Society.)

to intuition, is called the 'radiation paradox' (Ambach, 1984; Bintanja and Van den Broeke, 1996). In the future, satellite remote-sensing data may provide data sets on clouds describing temporal and spatial variations that could be used for energy balance modelling.

For calculations on a one-dimensional grid, precipitation is mostly prescribed as a linear function of elevation (Greuell and Oerlemans, 1986; Oerlemans and Hoogendoorn, 1989). However, Oerlemans *et al.* (1991/92) proposed other forms of this relationship for various transects on the Greenland ice sheet. Since the spatial (elevation) variation in precipitation for most small glaciers and ice caps is poorly confined by measurements, many authors use it for tuning their models. Ablation models designed to simulate the mass balance of the large ice sheets on a two-dimensional grid generally prescribe precipitation (or accumulation) per grid point. The amounts are derived from maps based on *in situ* measurements (as e.g. in Huybrechts and Oerlemans (1990) for Antarctica and in Van de Wal and Oerlemans (1994) for Greenland). Different authors use different methods to perturb the precipitation (or accumulation) in climate sensitivity experiments. Huybrechts and de Wolde (1999) and Letréguilly, Huybrechts and Reeh, (1991) prescribed a spatially and temporally invariable fractional change in accumulation and precipitation, respectively, with temperature. Van de Wal, Wild and de Wolde (2001) used a general circulation model (GCM) to compute for each grid point the fractional change in accumulation for CO_2-doubling. They then multiplied the observed actual amount of precipitation by the computed fractional change.

If the spatial and temporal variations in cloud amount, temperature and humidity are established, and surface conditions are known, the radiative fluxes can be computed with the parameterizations discussed in sections 5.2.2 and 5.2.4. Two other ways to compute the radiative fluxes should be mentioned. Hock (1999) calculated the incoming short-wave radiation on a two-dimensional grid for Storglaciären (Sweden) with a digital elevation model based on an assumed ratio of direct to diffuse radiation. Measurements of global radiation at one specific location were then used to normalize the calculated radiation field. The other exception is the determination of the albedo from satellite data (Knap and Oerlemans, 1996; Stroeve *et al.*, 1997). In the future, satellite-derived albedos may serve as direct input for ablation models.

5.6.2 *Validation*

Uncertainties in many factors such as albedo, temperature, cloud amount, wind speed and the roughness lengths and their spatial distributions cause uncertainties in mass balance profiles calculated by means of energy balance modelling. However, for many glaciers, precipitation probably causes the largest uncertainty. Mass balance profiles can usually be simulated by adjusting the precipitation and its spatial distribution within the uncertainty margins. This means that surface energy balance models are not validated; they are tuned. In fact, Braithwaite and Zhang (2000), Greuell and Böhm (1998), Oerlemans (1992), Oerlemans and Fortuin (1992) and Oerlemans and Hoogendoorn (1989) all tuned their models by adjusting precipitation only. Other variables and parameters sometimes used for tuning are atmospheric temperature (Oerlemans and Reichert, 2000) and degree-day factors (Hock, 1999; Jóhannesson *et al.*, 1995). Whereas most other authors tuned so that they correctly simulated the annual mass balance profile averaged over many years, Jóhannesson *et al.* (1995) tuned in order to simulate the various mass balance profiles for individual years correctly. The latter procedure gives more confidence in the simulation of inter-annual variations in mass balance and in the simulation of the effect of climatic change.

In some studies (Arnold *et al.*, 1996; Greuell and Böhm, 1998), ice albedo (α_i) and its variation with elevation were derived from field measurements, but in most studies a value for α_i is assumed. This may lead to large errors in the mass balance, because α_i varies considerably along the glacier and from glacier to glacier, and because the mass balance is very sensitive to variations in α_i. This also means that α_i can be used (or could have been used) as the tuning parameter. As long as the measured mass balance profile can be simulated, it may seem irrelevant whether tuning is done by means of precipitation or α_i, but unfortunately the choice affects the sensitivity of the mass balance to temperature change. Higher values of the mass balance can be obtained by increasing the amount of precipitation or the ice albedo. However, the climate sensitivity increases with the amount of precipitation (Oerlemans and Fortuin, 1992) and decreases with increasing α_i, because a smaller contrast between snow and ice albedo decreases the albedo feedback. So, whether a model is tuned by means of precipitation or α_i has opposite effects on the climate sensitivity. Similar arguments apply to tuning procedures with other sets of parameters (Braithwaite and Zhang, 2000).

5.7 Atmospheric models

5.7.1 Introduction

The range of numerical tools that qualify as atmospheric models is wide. It includes zonally averaged heat and moisture diffusion models, one-dimensional radiative–convective models and the much publicized general circulation models (GCMs, also known as global climate models) (e.g. Kiehl, 1992; Thuburn, 2000). Here, we focus on the physically based models that simulate the three-dimensional dynamics and thermodynamics of the atmosphere, radiation transfer and the atmosphere–surface exchanges of momentum, energy and moisture. The meteorological and climate research community widely uses these so-called atmospheric circulation models for both short-term and long-term studies and predictions (hours to decades or even centuries). To meet a high degree of realism and to fulfil the needs of a large community of users, most atmospheric circulation models comprehensively account for interactions of the atmosphere with other components of the climate system, including the cryosphere. The thermal, hydrological, radiative and, in some cases, topographic characteristics of the cryosphere strongly impact on the atmosphere. In turn, the atmosphere determines the surface mass and energy balance of the cryosphere. Spatially distributed surface mass, energy and momentum balance equations are therefore used in atmospheric circulation models. Subsurface modules of varying complexity are also used to close these equations. Thus, atmospheric circulation models basically incorporate surface energy balance models (section 5.2). The energy balance is solved along and consistent with a full three-dimensional meteorological model. This avoids the need for observations and for semi-empirical extrapolations, at least at the scales resolved. This also allows the study of mass balance changes under changing climate conditions. On the other hand, mass balance results are only as good as both the meteorological components of the models and the energy balance. We will now discuss different types of atmospheric circulation models, their suitability and their typical performances in cryospheric mass balance studies. The discussion is extended to the products of meteorological analyses and forecasts which also rely on atmospheric circulation models.

5.7.2 Global and regional atmospheric circulation models

Atmospheric GCMs (AGCMs) make up a large class of atmospheric circulation models. A concise practical definition of an AGCM could be: the most physically based model of the global atmosphere possible, considering existing limitations in available computing resources and in basic knowledge of atmospheric physics. The nature of GCMs has thus evolved in time. A number of textbooks (e.g. Kiehl, 1992; McGuffie and Henderson-Sellers, 1996) amply describe the nature, structure and uses of modern AGCMs. Increasingly, AGCMs are coupled with OGCMs (Ocean GCMs), thus forming AOGCMs, for consistent simulations of the ocean–atmosphere system. The trend toward more comprehensive models of the fully interacting climate system continues, as in particular attempts to model, rather than prescribe, the chemical and biological content of the atmosphere and ocean are made. GCMs already incorporate part of the cryosphere as fully interacting with the atmosphere and the ocean, although in many cases somewhat schematically. In particular, this is the case for snow in AGCMs and sea ice in AOGCMs. The resolved glaciers, ice caps and ice sheets only partially interact with climate in current GCMs. Their extent, volume and altitude do not vary, even though the mass balance at the interface with the atmosphere is simulated. So far, the geometrical evolution of large ice bodies has generally been considered insignificant for climate considerations, except for very large timescales (millennia and more). This view is currently changing. In the future, physically based, thermomechanical glacier and ice-sheet models may become standard parts of GCMs for climate-change studies.

Meanwhile, AGCMs compute atmosphere–surface fluxes for all kinds of surfaces, including snow and ice. These fluxes can be used to evaluate diagnostically the surface mass balance of the cryosphere. The cryosphere is a globally scattered feature of the climate system, since ice can sustain itself only in cold enough environments, that is, mostly at medium and high latitudes and/or altitudes. Therefore, regional climate modelling is an attractive alternative for GCMs, even for the more extensive cryosphere (for instance winter snow cover, polar ice sheets), but more particularly for the smaller one (mountain glaciers, etc.). Regional circulation models (RCMs, also known as regional climate models) of the atmosphere and of the atmosphere ocean system have only relatively recently been developed and successfully applied in cryosphere studies, in particular of meteorology and climate of the main polar ice sheets (e.g. Hines, Bromwich and Liu, 1997; Van Lipzig, 1999). RCMs tend to resemble GCMs, and a similar concise definition could be used: the most physically based regional atmospheric models. Compared with GCMs, RCMs can reach significantly finer spatial resolution at similar computer cost, but they need an additional set of prescribed boundary conditions at the lateral frontiers of the regional domain. Adequately providing the lateral boundary conditions, e.g. damping wave reflections at the frontiers, is a significant additional difficulty (Sashegyi and Madala, 1994). Lateral boundary conditions are generally either extracted from meteorological analysis archives or provided by GCMs. In the latter case, an RCM can be run either in-line with the GCM, thus allowing full ('two-way') interaction between the two models, or off-line, the GCM not being influenced by the RCM. Most RCM climate simulations of large ice bodies have been made with lateral boundary conditions from meteorological analyses. However, meteorological analyses (section 5.7.6) cannot be used for studies of past climate or of climate change beyond simple sensitivity experiments, as in Van Lipzig (1999). Off-line and in-line coupling with GCMs are thus likely to become frequent in the future.

Very high resolution, limited-area atmospheric circulation models have also been devised for process and meteorological studies (timescales of a few hours to a few days) over the cryosphere. At such small timescales, initial conditions may be a bigger issue than lateral boundary conditions. In fact, meso-scale models of such configurations are not climate-oriented, yet they can contribute to the understanding of the processes involved in the cryosphere's mass balance (e.g. Gallée, 1996, 1998). For instance, they may explicitly reproduce mechanisms that must otherwise be parameterized in RCMs and GCMs due to resolution limitations. There is not always a clear boundary between such meso-scale models and RCMs. A particular model, depending on configuration, resolution and region covered, may fall in either category.

Finally, stretchable grid GCMs attempt to combine global coverage and high resolution regional climate modelling in just one numerical tool. The concept may seem recent; however, it is currently used for weather forecasting and also in regional climate-change studies (Déqué, Marquet and Jones, 1998), including pertaining to polar ice sheets (Krinner and Genthon, 1998).

5.7.3 Atmospheric and surface physics in the models

GCMs and RCMs are characterized by an explicit resolution of the hydrodynamic equations of the atmosphere. Currently, all such models use the primitive equations (e.g. Salby, 1992). The highest resolution meso-scale models may even use fully non-hydrostatic dynamical formulations. The hydrodynamics are important for the mass balance of the cryosphere, not least because they determine transport of moisture from the source regions. Meteorological events associated with dynamical features such as low pressure systems significantly influence the mass balance of the cryosphere in many cases. The explicit hydrodynamic part of a circulation model is generally described as 'the dynamics'. Unresolved hydrodynamics, hydrology and radiation transfer make up 'the physics'. Much of the physics is parameterized. Together, the dynamics and physics of an atmospheric circulation model account for atmospheric processes, for the physical exchanges between the atmosphere and the surface, and, to some extent, for processes that are purely relevant to the surface, such as heat and moisture transfer and storage in soils.

GCMs and RCMs traditionally simulate phase changes both in the atmosphere and at the surface, precipitation of the water condensed in the atmosphere and runoff from the surface to the oceans. These are the processes that largely determine the surface mass balance of the cryosphere. Precipitation is a purely atmospheric phenomenon. Atmospheric circulation models simulate condensation and precipitation by large-scale super-saturation, including saturation through mechanical uplift of air by topography. They also formulate convective precipitation through various subgrid parameterizations. However, atmospheric models do not simulate precipitation consistently well. In addition, harsh conditions, snow being blown away by the wind and freezing of devices, make the measurement of precipitation over cryospheric surfaces difficult and prone to errors. Validation of climate model precipitation over ice is thus generally more limited than for other areas. This is unfortunate since precipitation is the major positive term in the mass balance equation for most of the cryosphere (except possibly for lake and sea ice).

Atmospheric circulation models compute full radiation transfer through the atmosphere (e.g. Kiehl, 1992), including the clouds. Background surface albedo is generally prescribed over land and land ice. However, when a pack of snow (or of sea ice for oceanic surfaces in AOGCMs) is simulated, the albedo is affected. The albedo of snow may be modelled as a simple empirical function of age, temperature and depth of the pack (equations (5.3) and (5.4)). Detailed pack models simulate albedo on a stronger physical basis.

Turbulent exchange of water vapour at the surface is often parameterized through bulk aerodynamic formulations (equation (5.18)). This exchange can neither be isolated from turbulent diffusion of vapour through the boundary layer, which is also parameterized (mixing length or higher level of closure, see e.g. McFarlane, 2000), nor from large-scale advection of moisture, which together export or import moisture to or from the surface atmosphere. Heat and momentum are similarly exchanged. Moisture can either sublimate from the ice surface or evaporate from melt water. It may also condense from the atmosphere to the ice if the upward moisture gradient in the surface atmosphere is positive. Because melting and freezing at the surface are essential for the energy balance of the surface atmosphere, they are simulated in atmospheric circulation models. On the other hand, disposal of melt water is not always adequately accounted for. In many models, the cryosphere has no liquid-water-holding capacity, and all melt water over ice runs off into the ocean. As this happens in the model without delay, melt water cannot refreeze. In an increasing number of models, snow holds liquid water, which may refreeze, in a reservoir of limited size. The liquid-water content affects the thermal and radiative properties of snow. In fact, AGCMs account comparatively well for snow as a component of the cryosphere because of its temporal and spatial scales of evolution and impact on the atmosphere. For instance, aspects such as density variations or grain metamorphism are predicted in some models (Royer, 2000) and may determine albedo, heat capacity and conduction. For similar reasons, sea ice receives a relatively favoured treatment in AOGCMs (e.g. Hibler and Flato, 1992; see Chapters 9 and 10 of this book).

Of course, that a model takes a process that influences the mass balance of the cryosphere into account does not guarantee that it is quantitatively, or even qualitatively, correctly represented. In addition, other processes that may influence the mass balance of ice are not simulated in models. This is the case, for instance, for drifting and blowing snow, which may export frozen particles out of a cryospheric area, e.g., from coastal ice sheets to the ocean. Also, the associated suspension of snow particles in the air may increase sublimation by increasing the surface of snow in contact with the atmosphere (e.g. Gallée, 1998).

5.7.4 Scales, resolution and computing cost

The spatial resolution of AGCMs has increased with the expansion of available computing resources. The horizontal resolution on climate timescales is now about 200 to 300 km.

Typical vertical discretization is from 15 to 30 layers across the troposphere and stratosphere. On modern supercomputers, computing time for simulating a year of climate ranges from a couple to several tens of hours. A few AGCMs have already been run at resolutions close to 100 km. However, it has been suggested that, for many large-scale climate studies, additional computing resources should be allocated to developing more complex physical parameterizations, to running longer simulations and to producing ensemble simulations, rather than to simply increasing spatial resolution (HIRETYCS, 1998). On the other hand, regional climate studies can justify higher resolutions. Notably, regions where topography significantly affects climate and hydrology, such as mountain glaciers and polar ice sheets, are very sensitive to resolution (Genthon, Jouzel and Déqué, 1994; HIRETYCS, 1998).

RCMs and stretchable grid AGCMs can be routinely used with horizontal resolutions of ~100 km or finer over regions of particular interest. For instance, a long (10-year) simulation of the climate and mass balance of the Antarctic ice sheet, and several sensitivity experiments for climate change, have been carried out with an RCM with a 55 km resolution (Van Lipzig, 1999). Also, the finest resolution (~100 km) multi-year simulations of an ice-age climate over the polar ice sheets have been carried out with a stretchable grid AGCM (Krinner and Genthon, 1998). Globally fine (~100 km) AGCMs have been used for climate-change experiments (Wild and Ohmura, 2000), but this option is currently restricted to groups with access to considerable computer resources, and further refinement of global resolution is unlikely in the near future.

Many ice bodies, in particular mountain glaciers, are much smaller than the grid steps of even the finest GCMs or RCMs. Even for extensive cryospheric components, subgrid parameterizations are desirable, because fractional coverage and spatial distribution of topography, snow depth, etc., may have subgrid characteristics. For instance, within a 300×300 km grid box, snow may cover only a fraction of the surface, e.g. the forest or plateaux, leaving bare land or valleys free. Another example is formed by the ablation zones of Greenland, which are narrow (a few ten to a few hundred kilometres) bands along the coasts at relatively low elevation.

So far, few attempts have been made to implement the necessary parameterizations in GCMs and RCMs. Subgrid statistics of topography (variance, kurtosis) have been used to improve the simulation of snow in GCMs (Walland and Simmonds, 1996). Modern surface vegetation atmosphere transfer (SVAT) schemes (e.g. Douville, Royer and Mahfouf, 1995) in GCMs increasingly include the impact of vegetation on snow cover and density. Mostly, subgrid surface parameterizations use a mosaic approach in which the surface of each model grid box is divided into a mosaic of unlocated sub-boxes (or, rather, all collocated to the main grid box). Each sub-box has different surface characteristics, e.g. land or ocean (e.g. Hansen *et al.*, 1983), kind of soil or vegetation and/or mean elevation (Leung and Ghan, 1998). Surface exchanges like those of heat, radiation, moisture and momentum, and budgets, like those of soil, water and snow, can then be computed independently over each sub-box, taking into account its specific characteristics. Atmospheric variables may also be corrected over each sub-box. For instance, the mean atmospheric temperature in the first model layer could be adjusted to the surface altitude of each sub-box, with prescribed or observed lapse rates, thus improving the representation of heat and thermal radiation exchange between the surface and the atmosphere. This is expected to improve the way melting in particular is accounted for, e.g. in the ablation zones of Greenland, although this remains to be verified. A climate model usually 'knows' about the mosaic representation of the surface only in the routines that compute surface exchanges and budgets. Most of the model, including the dynamics, works at the nominal resolution.

Because most models are run without such subgrid parameterization, a mosaic approach may be applied off-line to the results of a climate model. This is obviously less accurate than in-line parameterization, mainly because this approach includes no feedback of improved surface modelling to the atmosphere. On the other hand, rather than being unlocated within a model grid box, the sub-boxes can be the grid boxes of a new organized grid having a finer resolution than the model grid (Glover, 1999). This approach is one of the methods for off-line refining over specific regions of moderate resolution model results, a process known as downscaling. It may be related to transferring input variables of ablation models from one point to another, as discussed in Section 5.6. Regression methods as introduced in Section 5.8 may also be applied off-line to atmospheric circulation model results (e.g. Gregory and Oerlemans, 1998).

5.7.5 *Model performances and biasses*

The discussion here is limited to (1) describing the ability of atmospheric circulation models to reproduce the surface mass balance of the two major polar ice sheets, Greenland and Antarctica, and (2) summarizing possible reasons for model failures. Many of the general problems and difficulties affecting simulation of the surface atmosphere above the cryosphere, e.g. albedo, orography, inversions, occur in Greenland and Antarctica. The relatively large extension of the ice sheets limits the scale problems encountered with mountain glaciers, but does not fully remove them. Discussion and visualization of results is easier and more concise for conterminous ice bodies than for scattered glaciers. Being interactive components of modern models, snow and sea ice are increasingly discussed elsewhere in textbooks and review articles on climate modelling.

Figure 5.7 compares the 10-year mean precipitation minus evaporation $(P - E)$ simulated for Greenland by the Arpège (version 0) AGCM at three different resolutions (Genthon *et al.*, 1994). The dynamics of the Arpège model are computed by spectral decomposition. Spectral truncation T21 (on the left of the figure) corresponds to a dynamic smallest wavelength of almost 2000 km but a physical spatial resolution of about 600 km. Truncation T21 is no longer used but was customary in the 1980s and early 1990s. T42 (centre of figure) is twice as fine as T21 and is now standard. T79 (right of figure) is almost twice as fine as T42 (about 160 km resolution for the physics), and is currently considered quite fine. It becomes obvious that the T21 model is useless for studies of the mass balance of Greenland, when we compare Figure 5.7 with the widely referred climatology assembled by Ohmura and Reeh (1991) from snow gauges in coastal regions and glaciological measurements in the

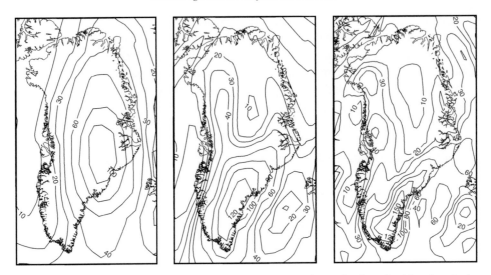

Figure 5.7. Precipitation minus evaporation/sublimation on and near the Greenland ice sheet in the Arpège version 0 AGCM at various spectral truncations: T21 (left), T42 (centre) and T79 (right). In units of centimetres water equivalent per year.

interior, or with a recent revision (Ohmura *et al.*, 1999). A dramatic improvement occurs when resolution is increased to T42: the south-east maximum, the north-east minimum and the relative maximum along the west flank are reproduced. The north-east minimum is too low in latitude, although Friedmann *et al.* (1995) suggested less accumulation than is shown by Ohmura and Reeh (1991) in this particular area. Smaller scale details, such as a double local maximum within the western higher accumulation tongue, or a hooked structure of the high accumulation region around the southern stern of Greenland, appear at T79. By modern standards, the reconstruction of the $(P - E)$ of Greenland by the Arpège model is fair at T42 and quite good at T79.

The atmospheric circulation around Greenland and the elevation-induced condensation of humid air, which determine precipitation and evaporation over the ice sheet, obviously depend on the orography. Figure 5.8 shows the surface orography in and around Greenland in the model. Spectral decomposition induces non-zero surface elevation at the sea in the neighbourhood of a strong topographic feature such as Greenland. A non-spectral model would not be affected by this problem but, at comparable resolutions, would be similarly wrong on the ice sheet itself. In particular, with a horizontal resolution of 600 km, it would also under-estimate the altitude at the summit of Greenland by more than 1 km. Better accounting for orography is clearly a major reason for improving model performances by increasing resolution.

Runoff from the Arpège runs presented above was not archived and is not available. Figure 5.9 shows the difference of simulated runoff in a 10-year simulation by the HADAM2B AGCM at medium ($3.75° \times 2.5°$) and high ($1.1° \times 0.8°$) resolution

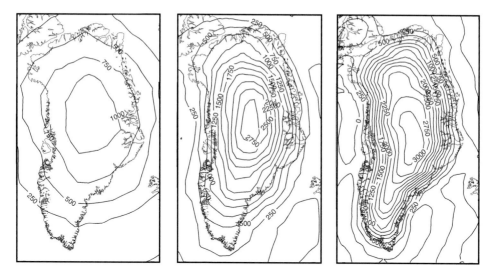

Figure 5.8. The surface elevation on and near the Greenland ice sheet in the Arpège version 0 AGCM at various spectral truncations: T21 (left), T42 (centre) and T79 (right). In units of metres.

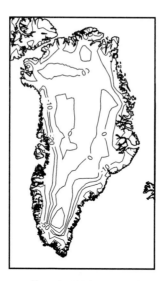

Figure 5.9. Runoff difference in a medium (∼300 km) and a high (∼100 km) resolution simulation by the HADAM2B AGCM. In units of centimetres water equivalent per year.

(HIRETYCS, 1998). This difference can be considerable. It is, again, directly related to the improved representation of topography in the higher resolution model.

Figure 5.10 shows the 'biasses' of the HADAM2B high resolution model in terms of simulated precipitation minus evaporation/sublimation over Antarctica. These biasses are evaluated with respect to the latest compilation of glaciological measurements of the Antarctic

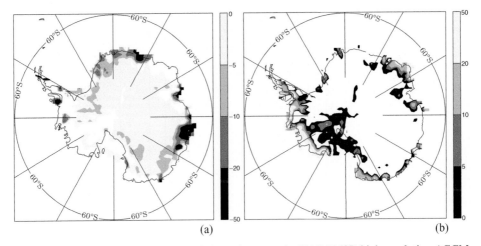

Figure 5.10. Difference in surface mass balance between the HADAM2B high resolution AGCM simulations and the compilation of observations by Vaughan *et al.* (1999). (a) Negative biasses (model under-estimates mass balance). (b) Positive biasses (over-estimates). In units of centimetres water equivalent per year.

surface mass balance, calculated and mapped by Vaughan *et al.* (1999). Except for some of the Antarctic peninsula and some of the ice shelves, and in contrast to Greenland, melting and runoff have a negligible contribution to the surface mass balance of Antarctica. The maps in Figure 5.10 turn out to be typical of biasses in other high resolution models: many of the HADAM2B biasses are systematic biasses in atmospheric circulation models (Genthon and Krinner, 2001). Although some of the biasses may actually reveal problems in the observations rather than in the models, systematic shortcomings in model design should be suspected. Even higher spatial resolution is unlikely to resolve the problem, as a regional circulation model with a 55 km grid step still shows similar biasses (Van Lipzig, 1999). One process missing in all models is drifting and blowing snow. Parameterizations have been proposed (e.g. Gallée, 1998) and remain to be comprehensively tested in atmospheric models. Other processes incorporated in all models, such as the turbulent exchanges in strongly stably stratified boundary layers, may also be inadequately accounted for.

Strong and durable boundary layers are infrequent except over snow or ice. Such boundary layers are not as extensively studied as more common marginally stable or unstable ones, and adequate parameterizations for atmospheric circulation models still have not been comprehensively tested. In addition, surface temperature inversions result in katabatic wind regimes. These also affect turbulent exchanges, and are also not necessarily well validated in models. It appears that models can reproduce the mean katabatic wind regime of Antarctica rather well, as long as the topography is represented in sufficient detail (sufficiently fine resolution), and temperature inversions are well reproduced (e.g. Krinner *et al.*, 1997; Van Lipzig, 1999). This does not guarantee that the surface climate is sufficiently accurately represented in the model to simulate the surface mass balance correctly. The simulation of clouds and radiation in the polar atmosphere, an essential term in the surface energy balance (equation (5.1)), also still needs to be verified comprehensively in climate models.

In fact, atmospheric circulation models are often developed to address a range of questions that do not specifically include climate and mass balance of the cryosphere. Even now, users can expect significant development and validation work before applying existing models to cryosphere studies.

5.7.6 Meteorological analyses and short-term forecasts

Meteorological analyses are the result of the assimilation of observed data from meteorological stations, from satellites and/or from other sources of information into a weather forecasting model. Meteorological analyses are produced by weather forecasting centres to build as complete and accurate snapshots of the atmospheric circulation as possible. These snapshots are then used to initialize weather forecasting runs. Because weather prediction is an operational task, meteorological analyses and weather forecasts are made regularly, on a daily and sometimes subdaily basis. Some weather forecasting centres now have decades of past global meteorological analyses and weather forecasts in store, an attractive source of climate information for cryosphere studies. However, changes in the assimilation tools and in the weather forecasting models are periodically made to improve performances. As a consequence, spurious variability may affect the temporal consistency of operational meteorological analyses and forecasts. This is why major weather forecasting centres such as the European Centre for Medium Range Weather Forecasts (ECMWF, Reading, UK) and the National Centers for Environmental Prediction (NCEP, Washington D.C., USA) have produced re-analyses and re-analyses-based forecasts (e.g., WCRP, 2000). Re-analyses are non-operational meteorological analyses over a given period of the past (e.g. 1979–1993), made with an unmodified numerical (assimilation + meteorological model) package throughout. Re-analyses can also make use of data which, at the time of the operational analyses, were not distributed in real time, were not yet received and properly treated, or for which (in particular, for satellite data) no algorithm had yet been developed to translate efficiently the information in terms of meteorological variables.

Because meteorological models are atmospheric circulation models, either GCMs or RCMs, the previous discussions on the conceptual and physical nature, and on the ability to reproduce the surface mass balance of the cryosphere, apply to weather forecasting models. However, because results are controlled by observations, and because the spatial resolution is generally finer than with climate-oriented GCMs and RCMs, weather forecasting models may be expected to simulate mass balance more accurately. On the other hand, the assimilation of observation will not correct all possible biases of the models, particularly over the cryosphere, where the density of observation is low. In addition, not all meteorological variables of interest are analysed. Winds, surface pressure and atmospheric moisture, for instance, are analysed. Precipitation, evaporation, sublimation or ice melting, for instance, are currently excluded from analysis. However, they are available as forecasts in the meteorological centre databases. If the shorter term (e.g. 6-hour) forecasts may be expected to benefit from observational control at the initialization step, they may also be affected by model adjustment to the initial conditions if data assimilation is not fully consistent with the model physics. This is a problem which is reported by Genthon and Krinner (1998) to affect precipitation over Antarctica in the ECMWF 1979–1993 re-analyses-based forecasts. These authors suggest that subtracting the 6-hour forecast from the 12-hour forecast efficiently reduces the initial condition adjustment problem while maintaining control by the observations.

Genthon and Braun (1995) show that the precipitation minus evaporation/ sublimation $(P - E)$ over Greenland and Antarctica is fairly well reproduced by the ECMWF short-term operational forecasts over the period from May 1985 to April 1991. Over this period, the spectral truncation of the weather prediction model was T106 (about 125 km resolution for the physics). These good results are nonetheless not consistently better than those of some recent high resolution AGCM runs in climate mode, for instance Arpège T79 (Figure 5.7). In addition, some of the biasses affecting the surface mass balance of Antarctica in the operational ECMWF short-term forecasts are similar to those found in several climate models (Genthon and Krinner (2001); see Figure 5.10), a fact which again raises the questions of systematic shortcomings in models, including weather forecasting models, and/or of errors in existing mass balance maps based on the compilation and extrapolation of glaciological observations. Also with a T106 truncation (but with a polar reduction of the zonal resolution in the physics), the ECMWF 1979–1993 re-analyses-based forecasts are found to reproduce the surface mass balance of Antarctica somewhat less well than the abovementioned operational forecasts (Genthon and Krinner, 1998). This result reminds one that one should not take improved performances in all possible respects for granted when a newer and theoretically better product is made available.

In principle, the net moisture budget in a column of atmosphere extending from the surface into space should equal the convergence of moisture across the column walls minus $(P - E)$ at the surface. Thus, on average over time, the convergence should equal $(P - E)$, so that $(P - E)$ may be computed from the analyses of atmospheric winds and moisture, rather than from the forecasts of precipitation and evaporation or sublimation. However, Genthon and Krinner (1998) show that, in the ECMWF re-analyses and re-analyses-based forecasts, the meridional convergence of moisture across the 70 °S parallel, toward the southern polar cap, is significantly larger than forecast $(P - E)$ averaged over the polar cap. Such inconsistency reveals that, because of the assimilation of observations, moisture is not a conserved quantity of time-averaged analyses and forecasts. It can be argued that, because of a stronger observational control and lesser exposition to model deficiencies in the analyses, the calculations based on convergence are more reliable. In addition, convergence methods cannot distinguish between precipitation and evaporation/sublimation, but atmospheric dynamic components of the mass balance, such as the contribution of transient and stationary eddies to the convergence of moisture, can be diagnosed. This is done, for instance, by Cullather, Bromwich and Van Woert (1998), along with a thorough comparison of the mass balance of Antarctica by convergence calculations with other meteorological and glaciological methods.

Presently available re-analyses and re-analyses-based forecasts have satisfied many users (WCRP, 2000). The production (in progress) of newer, longer and hopefully even better re-analysis products attracts much attention, including from those interested in the climate of the cryosphere. Bromwich (2000) provides a concise picture of the perspectives on re-analyses from polar applications, including from polar ice-sheet mass balance applications. Analyses and re-analyses are also commonly used to study and force models of snow and sea-ice cover (e.g., the polar section of WCRP (2000)). However, prospects for other components of the cryosphere remain to be assessed.

5.8 Regression models

The annual mean specific mass balance (M_n), which is the surface mass balance averaged over the entire area of the glacier, may also be estimated by means of the following regression equation:

$$M_n = c_0 + c_1 T_{\text{wcs}} + c_2 P_{\text{wcs}} + \varepsilon_i, \tag{5.25}$$

where T_{wcs} and P_{wcs} are the annual mean temperature and annual precipitation at a climate station located as near as possible to the glacier and ε_i is the error. The coefficients c_0, c_1 and c_2 are found by multiple regression analysis. In calculating T_{wcs}, only the months of the ablation season are considered, with the weights of the individual months varying from study to study. Chen and Funk (1990) calculated T_{wcs} as the mean of the temperatures of the three summer months (June, July, August), whereas Greuell (1992) added May and September with weights of 0.5. Often precipitation is taken over only a part of the year, for instance for the months October to May inclusive (Letréguilly, 1988), but Chen and Funk (1990) used the annual precipitation. Many variants on equation (5.25) have been proposed in the literature. Lefauconnier and Hagen (1990) extended equation (5.25) by a linear term in incoming long-wave radiation. Letréguilly (1988) and Martin (1977) replaced the mean temperature by the means of the daily minimum temperatures and the daily maximum temperatures, respectively. Steinacker (1979) and Hammer (1993) employed rather complicated degree-day methods with corrections for precipitation to compute a temperature and then applied equation (5.25) without the term in P_{wcs}. Finally, Nicolussi (1994) established a linear relationship between the mass balance of the Hintereisferner (Austria) and tree-ring widths. All of these authors used the regression equations for reconstructing the mass balance of a particular glacier for the periods preceding and following the direct glaciological measurements of the mass balance.

A conceptual error in equation (5.25) is the use of M_n as the dependent variable. The problem is that M_n does not depend only on the climate but also on the hypsometry of the glacier, which changes with time (for instance, in steady-state $M_n = 0$ for every climate). This problem can be solved by defining a reference mass balance profile (the mean over many years of measurements) and calculating an annual mass balance anomaly over a fixed elevation interval. The anomaly then replaces M_n in equation (5.25) (Greuell, 1992).

The regression method represents the most essential processes linking climate to mass balance in the simplest possible way. In fact, this includes the transfer of the data measured at the climate station to the glacier. However, several drawbacks of the regression method should be pointed out.

(1) Equation (5.25) is a poor description of reality. In reality, the relation between atmospheric temperature and ablation departs from linearity, and dependency of the mass balance on factors such as cloud amount, albedo and wind speed is not dealt with either (see, e.g., Vincent and Vallon (1997)). Moreover, variations in space (elevation) are not considered.

(2) A regression equation derived for a particular glacier should not be applied to other glaciers. The problem is that many of the factors that determine differences between mass balances of different glaciers, like orientation, hypsometry and ice albedo, are not explicitly taken into account.

(3) It is not obvious what the weights of the individual months for the calculation of T_{wcs} and P_{wcs} must be. In many studies, the weights were chosen by trial and error such that ε_i is minimal. This is a dangerous strategy in view of the short length of the records and the simplicity of equation (5.25). Therefore, Chen and Funk (1990) and Greuell (1992) set the weights *a priori* based on knowledge of the duration of the ablation season. Recently, Oerlemans and Reichert (2000) computed weights by means of energy balance modelling. Large differences between different glaciers appeared. For dry polar glaciers, the temperature of the three summer months (June, July and August) determines T_{wcs} for more than 90%, whereas for the extremely wet Franz-Josef glacier in New Zealand the total contribution of the temperature of the three summer months to T_{wcs} is only 32%. Oerlemans and Reichert (2000) applied the regression method with monthly weights computed through energy balance modelling to Nigardsbreen (Norway). The resulting reconstruction is almost equal (correlation coefficient, 0.97) to a reconstruction directly made with the energy balance model, which justifies some of the assumptions underlying the regression method (linearity in temperature and precipitation, use of monthly means).

5.9 Comparison of the different types of models

Surface energy balance, degree-day, atmospheric and regression models together form a suite of methods for modelling the mass balance of the cryosphere. Table 5.1 summarizes the relative advantages and drawbacks of each method. Depending on available resources and observational data, and on scientific objectives, one will find one or another method more appropriate.

Energy balance, degree-day and regression models can be relatively easily developed from, e.g., the present textbook and references. In addition, they can be run on moderately priced workstations. Whether or not developed by the user, such models can be thoroughly understood, a clear advantage for interpreting results. Conversely, access to atmospheric circulation models and to necessary supercomputing resources may not be easy for all potential users. Atmospheric circulation models are complex, and, although they are more physically based than other models, most users consider at least part of them as black boxes. On the other hand, because of the cost of running an atmospheric model, results

Table 5.1. *Performances of different types of mass balance models in five different aspects.*

A plus sign means that the performance is relatively good, a minus sign that the performance is relatively poor and a '0' that the performance is neither good nor poor.

	Surface energy balance model	Degree-day model	Atmospheric model	Regression model
Surface–atmosphere interaction	+	−	+	−
Interaction between free atmosphere and near-surface atmosphere	−	−	+	0
Atmospheric circulation	−	−	+	−
Spatial resolution	+	+	−	0
Computational cost	0	+	−	+

(e.g. of climate-change experiments) are generally archived and made available to a large community, including scientists who have never worked with atmospheric models.

There is more explicit physics in atmospheric models and in energy balance models than in the other models. This enables the study of how the mass balance is affected by, for instance, the albedo, wind speed, cloud amount and many other factors. Computing costs of atmospheric models limits spatial resolution. Resolution is a lesser problem for the study of most of the Antarctic and Greenland ice sheets, but downscaling or parameterization are inescapable steps for the ablation zones of Greenland and small glaciers and ice caps. Moderate computational cost is a major asset of energy balance, degree-day and regression models when numerous sensitivity studies are requested, for instance to test cryospheric changes in response to a range of scenarios of atmospheric changes.

However, for climate-change studies, the surface mass balance should be related to the free atmosphere. This is physically done with atmospheric models and statistically done with regression models, but in energy balance and degree-day models processes linking the free atmosphere to the near-surface atmosphere are generally ignored. The best way to improve energy balance and degree-day models in this respect is by adding parameterizations describing the relation between the free-atmospheric and the near-surface variables.

The atmospheric models discussed in section 5.7 compute the circulation of the atmosphere as a function of, for instance, the concentration of greenhouse gases in the atmosphere, the Sun–Earth geometry and lateral boundary conditions. Along with the circulation, atmospheric models produce the amount of precipitation as a function of time and location. Therefore, atmospheric models are, in principle, suitable for investigating the effect of climate change on precipitation and accumulation. This is not true for the other types of models, in which precipitation is prescribed.

Potential mass balance modellers need to balance carefully the advantages and limits of the different types of models listed in Table 5.1. Because the models are, in many respects, complementary, a joint use should not be excluded. For instance, atmospheric models may generate the forcing in the free atmosphere and they may be used to develop parameterizations describing the relation between free-atmospheric and near-surface variables. The actual mass balance calculations may then be done on a finer grid with a surface energy balance model.

5.10 List of symbols

a_i	ice radiation coefficient
a_s	snow radiation coefficient
C_H	heat transfer coefficient
C_{pa}	specific heat capacity of the air
C_{pi}	specific heat capacity of ice
C_{pw}	specific heat capacity of water
d	thickness of the snow pack on top of glacier ice
d^*	characteristic scale for snow depth
e	vapour pressure
e_{2m}	2 m vapour pressure
e_{sat}	saturation vapour pressure
F_{ho}	factor describing reduction of the short-wave incoming radiative flux due to shadows cast onto the surface
F_{ms}	factor describing enhancement of the short-wave incoming radiative flux due to multiple reflections between the surface and the atmosphere
F_{rs}	factor describing enhancement of the short-wave incoming radiative flux due to reflections by the surrounding terrain
g	acceleration due to gravity
k_a	empirical constant related to aerosol extinction
I	computed clear-sky direct solar radiation
I_0	flux at the top of the atmosphere on a surface normal to the incident radiation
L_f	latent heat of fusion
L_{ob}	Monin–Obukhov length
L_s	latent heat of sublimation
$L\!\downarrow$	long-wave incoming radiative flux
$L\!\downarrow_{cs}$	long-wave incoming radiative flux for clear skies
$L\!\uparrow$	long-wave outgoing radiative flux
m	optical air mass
M_a	accumulated amount of melt since the last snow fall
M_L	mass of 2 m^3 of ice
M_n	annual mean specific mass balance
n_c	cloud amount
N	ablation
p	atmospheric surface pressure

Pr	Prandtl number
P_{wcs}	weighted annual precipitation at a climate station
q	specific humidity
q_s	specific humidity at the surface
q_*	humidity scale used in flux-profile relationship
Q_0	energy flux from the atmosphere toward the glacier surface
Q_H	sensible heat flux
Q_L	latent heat flux
Q_R	heat flux supplied by rain
r	rain-fall rate
R_{off}	runoff
R_*	roughness Reynolds number
$S\downarrow$	short-wave incoming radiative flux
t	time
t_s	time since the last snow fall
T	atmospheric temperature
T_0	absolute atmospheric temperature
T_{2m}	2 m atmospheric temperature
T_{as}	atmospheric transmission coefficient for aerosol extinction
T_c	factor describing reduction of the short-wave incoming radiative flux due to clouds
T_{cs}	temperature at a climate station
T_g	atmospheric transmission coefficient for absorption by gases except water vapour
T_L	temperature of uppermost 2 m of snow and/or ice
T_{ma}	accumulated amount of daily maximum temperatures above $0\,^{\circ}\mathrm{C}$ since the last snow fall
T_{pdd}	positive degree days
T_r	temperature of the rain
T_R	atmospheric transmission coefficient for Rayleigh scattering
T_s	surface temperature
T_{wcs}	weighted annual mean temperature at a climate station
T_{wv}	atmospheric transmission coefficient for absorption by water vapour
u	wind speed
u_{2m}	2 m wind speed
u_*	friction velocity
w	vertical wind speed
z	height above the surface
z_0	roughness length for velocity
z_q	roughness length for water vapour
z_T	roughness length for temperature
α	surface albedo
α_{cs}	albedo of the clear sky
α_h	empirical function in stability function for sensible heat
α_i	surface albedo for ice
α_m	empirical function in stability function for momentum
α_s	surface albedo for snow

β	degree-day factor
ε_{cs}	clear-sky emittance
ε_{oc}	emittance of the totally overcast sky
ε_s	emissivity of the surface
θ	potential temperature
θ_s	solar zenith angle
θ_*	temperature scale used in flux-profile relationship
κ	Von Karman constant
ν	kinematic viscosity of air
ρ_a	density of the air
ρ_w	density of water
σ	Stefan–Boltzmann constant
τ	momentum flux
ϕ_h	stability function for sensible heat
ϕ_m	stability function for momentum

References

Ambach, W. 1963. Untersuchungen zum Energieumsatz in der Ablationszone des grönländischen Inlandeises. *EGIG 1957–1960*, vol. **4**, (4). Copenhagen, Bianco Lunos Bogtrykkeri.

  1979. Zum Wärmehaushalt des Grönländischen Inlandeises: Vergleichende Studie im Akkumulations- und Ablationsgebiet. *Polarforschung* **49** (1), 44–54.

  1984. The influence of cloudiness on the net radiation balance of a snow surface with high albedo. *J. Glaciol.* **13** (67), 73–84.

Andreas, E. L. 1987. A theory for the scalar roughness and the scalar transfer coefficients over snow and sea ice. *Bound. Layer Meteorol.* **38**, 159–84.

Arnold, N. S., Willis, I. C., Sharp, M. J., Richards, K. S. and Lawson, W. J. 1996. A distributed surface energy-balance model for a small valley glacier. I. Development and testing for Haut Glacier d'Arolla, Valais, Switzerland. *J. Glaciol.* **42** (140), 77–89.

Beljaars, A. and Holtslag, A. 1991. Flux parameterisation over land surfaces for atmospheric models. *J. Appl. Meteorol.* **30**, 327–41.

Bintanja, R. 1998. The contribution of snowdrift sublimation to the surface mass balance of Antarctica. *Ann. Glaciol.* **27**, 251–9.

  2000. Snowdrift suspension and atmospheric turbulence. Part II: Results of model simulations. *Bound. Layer Meteorol.* **95**, 391–417.

Bintanja, R. and Van den Broeke, M. R. 1996. The influence of clouds on the radiation budget of ice and snow surfaces in Antarctica and Greenland in summer. *Int. J. Climatol.* **16**, 1281–96.

Bozhinskiy, A. N., Krass, M. S. and Popovnin, V. V. 1986. Role of debris cover in the thermal physics of glaciers. *J. Glaciol.* **32** (111), 255–66.

Braithwaite, R. J. 1995a. Aerodynamic stability and turbulent sensible-heat flux over a melting ice surface, the Greenland ice sheet. *J. Glaciol.* **41**, 562–71.

  1995b. Positive degree-day factors for ablation on the Greenland ice sheet studied by energy-balance modelling. *J. Glaciol.* **41** (137), 153–60.

Braithwaite, R. J. and Zhang, Y. 2000: Sensitivity of mass balance of five Swiss glaciers to temperature changes assessed by tuning a degree-day model. *J. Glaciol.* **46**, 7–14.

Braithwaite, R. J., Laternser, M. and Pfeffer, W. T. 1994. Variations of near-surface firn density in the lower accumulation area of the Greenland ice sheet, Pâkitsoq, West Greenland. *J. Glaciol.* **40**, 477–85.

Brock, B. W., Willis, I. C. and Sharp, M. J. 2000. Measurement and parameterization of albedo variations at Haut Glacier d'Arolla, Switzerland. *J. Glaciol.* **46** (155), 675–88.

Bromwich, D. H. 2000. Perspectives on reanalyses from polar applications. In *Proceedings of the Second WCRP International Conference on Reanalyses.* WMO/TD-no. 985, pp. 221–4.

Brun, E., David, P., Sudul, M. and Brunot, G. 1992. A numerical model to simulate snow-cover stratigraphy for operational avalanche forecasting. *J. Glaciol.* **38**, 13–22.

Chen, J. and Funk, M. 1990. Mass balance of Rhonegletscher during 1882/83–1986/87. *J. Glaciol.* **36**, 199–209.

Cullather, R. I., Bromwich, D. H. and Van Woert, M. L. 1998. Spatial and temporal variability of Antarctic precipitation from atmospheric methods. *J. Climate* **11**, 334–67.

Denby, B. 1999. Second order modelling of turbulence in katabatic flows. *Bound. Layer Meteorol.* **92** (1), 67–100.

Denby, B. and Greuell, W. 2000. The use of bulk and profile methods for determining surface heat fluxes in the presence of glacier winds. *J. Glaciol.* **46** (154), 445–52.

Denby, B. and Smeets, P. 2000. Derivation of turbulent fluxes profiles and roughness lengths from katabatic flow dynamics. *J. Appl. Meteorol.* **39** (9), 1601–12.

Denby, B. and Snellen, H. 2001. A comparison of surface renewal theory with the observed roughness length for temperature on a melting glacier surface. *Boundary Layer Meteorol.* **103**, 459–68.

Déqué, M., Marquet, P. and Jones, R. G. 1998. Simulation of climate change over Europe using a global variable resolution general circulation model. *Climate Dyn.* **14**, 173–89.

Déry, S. J., Taylor, P. A. and Xiao, J. 1998. The thermodynamic effects of sublimating snow in the atmospheric boundary layer. *Bound. Layer Meteorol.* **89**, 251–83.

Douville, H., Royer, J. F. and Mahfouf, J. F. 1995. A new snow parameterization for the Météo-France climate model. Part 1: Validation in stand-alone experiments. *Climate Dyn.* **12**, 21–5.

Forrer, J. and Rotach, M. W. 1997. On the turbulence structure in the stable boundary layer over the Greenland ice sheet. *Bound. Layer Meteorol.* **85**, 111–36.

Friedmann, A., Moore, J. C., Thorsteinsson, T., Kipfstuhl, J. and Fischer, H. 1995. A 1200 year record of accumulation from northern Greenland. *Ann. Glaciol.* **21**, 19–25.

Fujita, K. and Ageta, Y. 2000. Effect of summer accumulation on glacier mass balance on the Tibetan Plateau revealed by mass-balance model. *J. Glaciol.* **46** (153), 244–52.

Gallée, H. 1996. Mesoscale atmospheric circulations over the southern Ross Sea Sector, Antarctica. *J. Appl. Meteor.* **35**, 1129–41.

1998. Simulation of blowing snow over the Antarctic ice sheet. *Ann. Glaciol.* **26**, 203–6.

Garnier, B. and Ohmura, A. 1968. A method of calculating the direct short-wave radiation income on slopes. *J. Appl. Meteorol.* **7**, 796–800.

Garratt, J. R., 1992. *The Atmospheric Boundary Layer.* Cambridge University Press.

Genthon, C. and Braun, A. 1995. ECMWF analyses and predictions of the surface climate of Greenland and Antarctica. *J. Climate* **8**, 2324–32.

Genthon, C. and Krinner, G. 1998. Convergence and disposal of energy and moisture on the Antarctic polar cap from ECMWF reanalyses and forecasts. *J. Climate* **11**, 1703–16.

　2001. The Antarctic surface mass balance and systematic biases in GCMs. *J. Geophys. Res.* **106**, 20 653–64.

Genthon, C., Jouzel, J. and Déqué, M. 1994. Accumulation at the surface of polar ice sheets: observation and modeling for global climate change. In Desbois, M. and Desalmand, F., eds., *Global Precipitations and Climate Change.* NATO ASI Series I, vol. 26, pp. 53–76.

Glover, R. W. 1999. Influence of spatial resolution and the treatment of orography on GCM estimates of the surface mass balance of the Greenland ice sheet. *J. Climate* **12**, 551–63.

Gregory, J. M. and Oerlemans, J. 1998. Simulated future sea-level rise due to glacier melt based on regionally and seasonally resolved temperature changes. *Nature* **391**, 474–6.

Grenfell, T. C., Warren, S. G. and Mullen, P. C. 1994. Reflection of solar radiation by the Antarctic snow surface at ultraviolet, visible, and near-infrared wavelengths. *J. Geophys. Res.* **99** (D9), 18 669–84.

Greuell, W. 1992. Hintereisferner, Austria: mass-balance reconstruction and numerical modelling of the historical length variations. *J. Glaciol.* **38**, 233–44.

　2000. Melt-water accumulation on the surface of the Greenland ice sheet: effect on albedo and mass balance. *Geograf. Ann.* **82A**, 489–98.

Greuell, W. and Böhm, R. 1998. Two-metre temperatures along melting mid-latitude glaciers and implications for the sensitivity of the mass balance to variations in temperature. *J. Glaciol.* **44** (146), 9–20.

Greuell, W. and Knap, W. H. 2000. Remote sensing of the albedo and detection of the slush line on the Greenland ice sheet. *J. Geophys. Res.* **105** (D12), 15 567–76.

Greuell, W. and Konzelmann, T. 1994. Numerical modelling of the energy balance and the englacial temperature of the Greenland ice sheet. Calculation for the ETH-Camp location (West Greenland, 1155 m a.s.l.). *Global & Planetary Change* **9**, 91–114.

Greuell, W. and Oerlemans, J. 1986. Sensitivity studies with a mass balance model including temperature profile calculations inside the glacier. *Z. Gletscherkd. Glazialgeol.* **22**(2), 101–24.

Greuell, W. and Smeets, C. J. J. P. 2001. Variations with elevation in the surface energy balance on the Pasterze (Austria). *J. Geophys. Res.* **106** (D23), 31 717–27.

Greuell, W., Knap, W. H. and Smeets, P. C. 1997. Elevational changes in meteorological variables along a midlatitude glacier during summer. *J. Geophys. Res.* **102** (D22), 25 941–54.

Griggs, M. 1968. Emissivities of natural surfaces in the 8- to 14-micron spectral region. *J. Geophys. Res.* **73**, 7545–51.

Gueymard, C. 1993. Critical analysis and performance assessment of clear sky solar irradiance models using theoretical and measured data. *Solar Energy* **51**(2), 121–38.

Hammer, N. 1993. Wurtenkees: Rekonstruktion einer 100jährigen Reihe der Gletschermassenbilanz. *Z. Gletscherkd. Glazialgeol.* **29** (1), 15–37.

Hansen, J. *et al.* 1983. Efficient three-dimensional global models for climate studies: models I and II. *Month. Weather Rev.* **111**, 609–62.

Hay, J. E. and Fitzharris, B. B. 1988. A comparison of the energy-balance and bulk-aerodynamic approaches for estimating glacier melt. *J. Glaciol.* **34** (117), 145–53.

Hibler, W. D. and Flato, G. M. 1992. Sea-ice models. In Trenberth, K. E., ed., *Climate System Modeling*. Cambridge University Press, pp. 413–36.

Hines, K. M., Bromwich, D. H. and Liu, Z. 1997. Combined global climate model and mesoscale model simulations of Antarctic climate. *J. Geophys. Res.* **102**, 13 747–60.

HIRETYCS (high-resolution ten-year climate simulations). 1998. Final report, available from Météo-France, Centre National de Recherches Météorologiques, Toulouse, France. http://www.cnrm.meteo.fr/hiretycs/final_report.htm.

Hock, R. 1998. Modelling of glacier melt and discharge. *Zürcher Geographische Schriften*, vol. **70**. ETH Zurich, Geographisches Institut.

    1999. A distributed temperature-index ice- and snowmelt model including potential direct solar radiation. *J. Glaciol.* **45** (149), 101–11.

Hock, R. and Holmgren, B. 1996. Some aspects of energy balance and ablation of Storglaciären, northern Sweden. *Geograf. Ann.* **78A** (2–3), 121–32.

Högström, U. 1988. Non-dimensional wind and temperature profiles in the atmospheric surface layer: a re-evaluation. *Bound. Layer Meteorol.* **42**, 55–78.

Holmgren, B. 1971. Climate and energy exchange on a sub-polar ice cap in summer. Part D: On the vertical turbulent fluxes of water vapour at Ice Cap Station. Meteorologiska Institutionen Uppsala Universitetet, Meddelande no. 110.

Hutter, K. 1995. Avalanche dynamics. In Singh, V. P., ed., *Hydrology of Disasters*. Dordrecht, Kluwer Academic Publishers, pp. 317–94.

Huybrechts, P. and Oerlemans, J. 1990. Response of the Antarctic ice sheet to future greenhouse warming. *Climate Dyn.* **5**, 93–102.

Huybrechts, P. and de Wolde, J. 1999. The dynamic response of the Greenland and Antarctic ice sheets to multiple-century climatic warming. *J. Climate* **12** (8), 2169–88.

Idso, S. B. 1981. A set of equations for full spectrum and 8- to 14-mm and 10.5- to 12.5-mm thermal radiation from cloudless skies. *Water Resources Res.* **17** (2), 295–304.

Issler, D. 1998. Modelling of snow entrainment and deposition in powder-snow avalanches. *Ann. Glaciol.* **26**, 253–8.

Janssens, I. and Huybrechts, Ph. 2000. The treatment of meltwater retention in mass-balance parameterisations of the Greenland ice sheet. *Ann. Glaciol.* **31**, 133–40.

Jóhannesson, T., Sigurdsson, O., Laumann, T. and Kennett, M. 1995. Degree-day glacier mass-balance modelling with applications to glaciers in Iceland, Norway and Greenland. *J. Glaciol.* **41** (138), 345–58.

Kasten, F. 1983. Parametrisierung der Globalstrahlung durch Bedeckungsgrad und Trübungsfaktor. *Ann. Meteorol.* **20**, 49–50.

Kayastha, R. B., Ohata, T. and Ageta, Y. 1999. Application of a mass-balance model to a Himalayan glacier. *J. Glaciol.* **45** (151), 559–67.

Kiehl, J. T. 1992. Atmospheric general circulation modeling. In Trenberth, K. E., ed., *Climate System Modeling*. Cambridge University Press, pp. 319–69.

Kimball, B. A., Idso, S. B. and Aase, J. K. 1982. A model of thermal radiation from partly cloudy and overcast skies. *Water Resources Res.* **18** (4), 931–6.

King, J. C. 1990. Some measurements of turbulence over an Antarctic ice shelf. *Q. J. R. Meteorol. Soc.* **116**, 379–400.

King, J. C. and Anderson, P. S., 1994. Heat and water vapour fluxes and scalar roughness lengths over an Antarctic ice shelf. *Bound. Layer Meteorol.* **69**, 101–21.

King, J. C., Anderson, P. S. Smith, M. C. and Mobbs, S. D. 1996. The surface energy and mass balance at Halley, Antarctica during winter. *J. Geophys. Res.* **101** (D14), 19 119–28.

Knap, W. H. and Oerlemans, J. 1996. The surface albedo of the Greenland ice sheet: satellite-derived and in situ measurements in the Søndre Strømfjord area during the 1991 melt season. *J. Glaciol.* **42**, 364–74.

Knap, W. H., Brock, B. W., Oerlemans, J. and Willis, I. C. 1999. Comparison of Landsat-TM derived and ground-based albedos of Haut Glacier d'Arolla, Switzerland. *Int. J. Remote Sensing* **20** (17), 3293–310.

Koelemeijer, R., Oerlemans, J. and Tjemkes, S. 1993. Surface reflectance of Hintereisferner, Austria, from Landsat 5 TM imagery. *Ann. Glaciol.* **17**, 17–22.

Kondo, J. and Yamazawa, H. 1986. Bulk transfer coefficient over a snow surface. *Bound. Layer Meteorol.* **34**, 123–35.

König-Langlo, G. and Augstein, E. 1994. Parameterisation of the downward long-wave radiation at the Earth's surface in polar regions. *Meteorol. Zeit.* **6**, 343–7.

Konzelmann, T. and Ohmura, A. 1995. Radiative fluxes and their impact on the energy balance of the Greenland ice sheet. *J. Glaciol.* **41**, 490–502.

Konzelmann, T., Van de Wal, R. S. W., Greuell, W., Bintanja, R., Henneken, E. A. C. and Abe-Ouchi, A. 1994. Parameterization of global and longwave incoming radiation for the Greenland ice sheet. *Global & Planetary Change* **9**, 143–64.

Krinner, G. and Genthon, C. 1998. GCM simulations of the last glacial maximum surface climate of Greenland and Antarctica. *Climate Dyn.* **14**, 741–58.

Krinner, G., Genthon, C., Li, Z.-X. and Le Van, P. 1997. Studies of the Antarctic climate with a stretched grid GCM. *J. Geophys. Res.* **102**, 13 731–45.

Lefauconnier, B. and Hagen, J. O. 1990. Glaciers and climate in Svalbard: statistical analysis and reconstruction of the Brøggerbreen mass balance for the last 77 years. *Ann. Glaciol.* **14**, 148–52.

Lehning, M., Bartelt, P., Brown, B., Russi, T., Stöckli, U. and Zimmerli, M. 1999. SNOWPACK model calculations for avalanche warning based upon a new network of weather and snow stations. *Cold Regions Sci. Technol.* **30**, 145–57.

Letréguilly, A. 1988. Relation between the mass balance of western Canadian mountain glaciers and meteorological data. *J. Glaciol.* **34** (116), 1–8.

Letréguilly, A., Huybrechts, P. and Reeh, N. 1991. Steady-state characteristics of the Greenland ice sheet under different climates. *J. Glaciol.* **37** (125), 149–57.

Leung, L. R. and Ghan, S. J. 1998. Parameterizing subgrid orographic precipitation and surface cover in climate models. *Month. Weather Rev.* **126**, 3271–91.

McFarlane, N. 2000. Boundary layer processes. In Mote, P. and O'Neill, A., eds., *Numerical Modeling of the Global Atmosphere in the Climate System.* NATO Science Series vol. 550. Dordrecht, Kluwer Academic Press, pp. 221–38.

McGuffie K. and Henderson-Sellers, A. 1996. *A Climate Modeling Primer.* New York, John Wiley and Sons, inc. CD-ROM.

Male, D. H. and Granger, R. J. 1981. Snow surface and energy exchange. *Water Resources Res.* **17** (3), 609–27.

Martin, S. 1977. Analyse et reconstitution de la série des bilans annuels du Glacier de Sarennes, sa relation avec les fluctuations du niveau de trois glaciers du Massif du Mont-Blanc (Bossons, Argentière, Mer de Glace). *Z. Gletscherkd. Glazialgeol.* **13**, 127–53.

Mattson, L. E. and Gardner, J. S. 1989. Energy exchanges and ablation rates on the debris-covered Rakhiot Glacier, Pakistan. *Z. Gletscherkd. Glazialgeol.* **25** (1), 17–32.

Maykut, G. A. and Untersteiner, N. 1971. Some results from a time-dependent thermodynamic model of sea ice. *J. Geophys. Res.* **76** (6), 1550–75.

Meesters, A. and Van den Broeke, M. 1997. Response of the longwave radiation over melting snow and ice to atmospheric warming. *J. Glaciol.* **43** (143), 66–70.

Meyers, T. P. and Dale, R. F. 1983. Predicting daily insolation with hourly cloud height and coverage. *J. Climate Appl. Meteorol.* **2**, 537–45.

Morris, E. M. 1989. Turbulent transfer over snow and ice. *J. Hydrol.* **105**, 205–23.

Morris, E. M., Bader, H.-P. and Weilenmann, P. 1997. Modelling temperature variations in polar snow using DAISY. *J. Glaciol.* **43**, 180–91.

Munro, D. S. 1989. Surface roughness and bulk heat transfer on a glacier: comparison with eddy correlation. *J. Glaciol.* **35** (121), 343–8.

Munro, D. S. and Davies, J. A. 1978. On fitting the log-linear model to wind speed and temperature profiles over a melting glacier. *Bound. Layer Meteorol.* **15**, 423–37.

Nicolussi, K. 1994. Jahrringe und Massenbilanz. Dendroklimatologische Rekonstruktion der Massenbilanzreihe des Hintereisferners bis zum Jahr 1400 mittels Pinus cembra-Reihen aus den Ötztaler Alpen, Tirol. *Z. Gletscherkd. Glazialgeol.* **30**, 11–52.

Oerlemans, J. 1991/92. A model for the surface balance of ice masses: part I. Alpine glaciers. *Z. Gletscherkd. Glazialgeol.* **27/28**, 63–83.

    1992. Climate sensitivity of glaciers in southern Norway: application of an energy-balance model to Nigardsbreen, Hellstugubreen and Alfotbreen. *J. Glaciol.* **38** (129), 223–32.

Oerlemans, J. and Fortuin, J. P. F. 1992. Sensitivity of glaciers and small ice caps to greenhouse warming. *Science* **258**, 115–17.

Oerlemans, J. and Hoogendoorn, N. C. 1989. Mass-balance gradients and climatic change. *J. Glaciol.* **35** (121), 399–405.

Oerlemans, J. and Knap, W. H. 1998. A 1 year record of global radiation and albedo in the ablation zone of Morteratschgletscher, Switzerland. *J. Glaciol.* **44** (147), 231–8.

Oerlemans, J. and Reichert, B. K. 2000. Relating glacier mass balance to meteorological data by using seasonal sensitivity characteristics. *J. Glaciol.* **46** (152), 1–6.

Oerlemans, J., Van de Wal, R. S. and Conrads, L. A. 1991/92. A model for the surface balance of ice masses: part II. Application to the Greenland ice sheet. *Z. Gletscherkd. Glazialgeol.* **27/28**, 85–96.

Ohmura, A. 1981. Climate and energy balance on Arctic tundra, Axel Heiberg Island, Canadian Arctic Archipelago, spring and summer 1969, 1970 and 1972. *Zürcher Geographische Schriften*, vol. **3**. ETH Zurich, Geographisches Institut.

    1987. New temperature distribution maps for Greenland *Z. Gletscherkd. Glazialgeol.* **23** (1), 1–45.

Ohmura, A. and Reeh, N. 1991. New precipitation and accumulation maps for Greenland. *J. Glaciol.* **37**, 140–8.

Ohmura, A., Calanca, P., Wild, M. and Anklin, M. 1999. Precipitation, accumulation and mass balance of the Greenland ice sheet. *Z. Gletscherkd. Glazialgeol.* **35**, 1–20.

Pfeffer, W. T., Meier, M. F. and Illangasekare, T. H. 1991. Retention of Greenland runoff by refreezing: implications for projected future sea level change. *J. Geophys. Res.* **96** (C12), 22 117–24.

Plüss, C. and Ohmura, A. 1997. Longwave radiation on snow-covered mountainous surfaces. *J. Appl. Meteorol.* **36**, 818–24.

Pomeroy, J. W., Gray, D. M. and Landine, P. G. 1993. The prairie blowing snow model: characteristics, validation, operation. *J. Hydrol.* **144**, 165–92.

Prandtl, L. 1942. *Führer durch die Ströhmungslehre*. Braunschweig, Vieweg u. Sohn.

Ranzi, R. and Rossi, R. 1991. Physically based approach to modelling distributed snowmelt. IAHS Publication 205, pp. 141–52.

Reeh, N. 1989. Parameterisation of melt rate and surface temperature on the Greenland ice sheet. *Polarforschung* **59** (3), 113–28.

Reijmer, C. H., Knap, W. H. and Oerlemans, J. 1999. The surface albedo of the Vatnajökull ice cap, Iceland: a comparison between satellite-derived and ground-based measurements. *Bound. Layer Meteorol.* **92** (1), 125–44.

Rouse, W. R. 1987. Examples of enhanced global solar radiation through multiple reflection from an ice-covered Arctic Sea. *J. Climate Appl. Meteorol.* **26**, 670–4.

Royer, J.-F. 2000. Land surface processes and hydrology. In Mote, P. and O'Neill, A., eds., *Numerical Modeling of the Global Atmosphere in the Climate System*. NATO Science Series vol. 550. Dordrecht, Kluwer Academic Press, pp. 321–53.

Salby, M. L. 1992. The atmosphere. In Trenberth, K. E., ed., *Climate System Modeling*. Cambridge University Press, pp. 53–115.

Sashegyi, K. D. and Madala, R. V. 1994. Initial and boundary conditions. In Pielke, R. A. and Pearce, N. P., eds., *Mesoscale Modeling of the Atmosphere*. Meteorological Monographs vol. 25. American Meteorological Society, pp. 1–12.

Schmidt, R. A. 1982. Vertical profiles of wind speed, snow concentration, and humidity in blowing snow. *Bound. Layer Meteorol.* **23**, 223–46.

Smeets, C. J. P. P., Duynkerke, P. G. and Vugts, H. F. 1998. Turbulence characteristics of the stable boundary layer over a mid-latitude glacier. Part I: A combination of katabatic and large-scale forcing. *Bound. Layer Meteorol.* **87**, 117–45.

1999. Observed wind profiles and turbulence fluxes over an ice surface with changing surface roughness. *Bound. Layer Meteorol.* **92**, 101–23.

Steinacker, R. 1979. Rückrechnung des Massenhaushaltes des Hintereisferners mit Hilfe von Klimadaten. *Z. Gletscherkd. Glazialgeol.* **15**, 101–4.

Stroeve, J., Nolin, A. and Steffen, K. 1997. Comparison of AVHRR-derived and in situ surface albedo over the Greenland ice sheet. *Remote Sensing Environ.* **62**, 262–76.

Stull, R. B. 1988. *An Introduction to Boundary Layer Meteorology*. Dordrecht, Kluwer Academic Publishers.

Thuburn, J., 2000. Use of simplified atmospheric models. In Mote, P. and O'Neill, A., eds., *Numerical Modeling of the Global Atmosphere in the Climate System*. NATO Science Series vol. 550. Dordrecht, Kluwer Academic Press, pp. 105–17.

Van den Broeke, M. R. 1996. Characteristics of the lower ablation zone of the West Greenland ice sheet for energy-balance modelling. *Ann. Glaciol.* **23**, 160–6.

1997. A bulk model of the atmospheric boundary layer for inclusion in mass balance models of the Greenland ice sheet. *Z. Gletscherkd. Glazialgeol.* **33** (1), 73–94.

Van Lipzig, P. M. 1999. The surface mass balance of the Antarctic ice sheet. Ph.D. thesis, University of Utrecht, The Netherlands.

Van de Wal., R. S. W. 1996. Mass-balance modelling of the Greenland ice sheet: a comparison of energy-balance and a degree-day model. *Ann. Glaciol.* **23**, 36–45.

Van de Wal, R. S. W. and Oerlemans, J. 1994. An energy balance model for the Greenland ice sheet. *Global & Planetary Change* **9** (1/2), 115–31.

Van de Wal, R. S. W., Wild, M. and de Wolde, J. R. 2001. Short-term volume changes of the Greenland ice sheet in response to doubled CO_2 conditions. *Tellus* **538**, 94–102.

Vaughan, D. G., Bamber, J. L., Giovinetto, M., Russel, J. and Cooper, A. P. R. 1999. Reassessment of net surface mass balance in Antarctica. *J. Climate* **12**, 933–46.

Vincent, C. and Vallon, M. 1997. Meteorological controls on glacier mass balance: empirical relations suggested by measurements on glacier de Sarennes, France. *J. Glaciol.* **43** (143), 131–7.

Walland. D. J. and Simmonds, I. 1996. Subgrid scale topography and the simulation of northern hemisphere snow cover. *Int. J. Climatol.* **16**, 961–82.

Warren, S. G. 1982. Optical properties of snow. *Rev. Geophys. Space Phys.* **20** (1), 67–89.

WCRP. 2000. *Proceedings of the Second WCRP International Conference on Reanalyses.* WMO/TD-no. 985, 452 pp.

Wendler, G., Eaton, F. D. and Ohtake, T. 1981. Multiple reflection effects on irradiance in the presence of Arctic stratus clouds. *J. Geophys. Res.* **86** (C3), 2049–57.

Wild, M. and Ohmura, A. 2000. Changes in mass balance of the polar ice sheets and sea-level under greenhouse warming as projected in high resolution GCM simulations. *Ann. Glaciol.* **30**, 197–203.

Wiscombe, W. J. and Warren, S. G. 1980. A model for the spectral albedo of snow. I: Pure snow. *J. Atmos. Sci.* **37**, 2712–33.

Zuo, Z. and Oerlemans, J. 1996. Modelling albedo and specific balance of the Greenland ice sheet: calculations for the Søndre Strømfjord transect. *J. Glaciol.* **42**, 305–17.

6

Modelling land-ice dynamics

CORNELIS J. VAN DER VEEN
Byrd Polar Research Center, The Ohio State University
ANTONY J. PAYNE
School of Geographical Sciences, University of Bristol

6.1 Introduction

Glaciers respond dynamically to external forcings, such as climate variations, which cause ice masses to approach a new equilibrium compatible with the new environmental conditions. For example, it has been suggested that greenhouse warming may result in increased snow fall in the interior of Antarctica and increased ablation in the coastal regions of this ice sheet. Model simulations of the ice-sheet response indicate a thickening in the interior and surface lowering near the margins. Thus, the slope of the ice surface becomes steeper, resulting in greater discharge velocities to redistribute excess mass from the interior toward the margins to compensate for mass loss from ablation. Generally, the response of ice sheets to forcings may be complicated because feedback processes become operable that may amplify or mitigate the ice sheet's adjustment to forcing or because of internal instabilities that may cause rapid changes in ice volume due to changes in the dynamical flow regime. To model ice-sheet evolution adequately, it is therefore necessary to identify the important physical controls and processes affecting the flow of glaciers.

In most models, whether numerical time-evolving or analytical, simplifying assumptions are commonly made to allow a solution to be found. Such simplifications are permissible provided the essential physics are retained. A model aimed at simulating the evolution of the Greenland ice sheet over the last few glacial cycles need not explicitly calculate deformation of each individual ice crystal. Instead, such microscopic processes are parameterized through the flow law for glacier ice that captures the bulk behaviour of aggregates of ice crystals. Perhaps the most important step in model development is therefore to decide which processes are to be included in the model. The objective of this chapter is to provide the elementary glaciological background required to model glacier flow without restriction to any particular glacier or ice sheet. Applications pertaining to specific ice masses are discussed elsewhere in this book.

Glacier flow results from internal deformation and from ice sliding over its bed where the basal temperature has reached the melting temperature and a lubricating water-saturated

Antony J. Payne is supported by the Natural Environment Research Council of the UK's Centre for Polar Observations and Modelling. Cornelis J. van der Veen is supported through grants from the National Science Foundation and NASA.

Mass Balance of the Cryosphere: Observations and Modelling of Contemporary and Future Changes, eds. Jonathan L. Bamber and Antony J. Payne. Published by Cambridge University Press. © Cambridge University Press 2003.

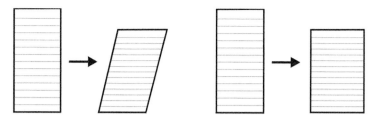

Internal deformation   Basal sliding

Figure 6.1. Illustrating the two modes of glacier movement. Internal deformation (left) results in zero movement at the base of the ice and maximum displacement at the upper surface, while for basal sliding (right) the horizontal displacement is the same throughout an ice column. Generally, divergence of horizontal flow causes thinning of the ice column, as shown.

layer has formed. Basal sliding involves the down-slope movement of a column of ice without changing its vertical orientation. The height and width of the column may change as it is advected down-glacier, but the column remains essentially vertical. Internal deformation, on the other hand, involves vertical shear in ice velocity, with zero speed at the base and maximum speed at the glacier surface, and thus causes an initially vertical column to become slanted (Figure 6.1). Whereas basal sliding depends to a large extent on the properties of the bed under the ice, internal deformation is the inherent manifestation of the deformation of individual ice crystals subjected to stress. This deformation involves the movement of dislocations or irregularities in a crystal and leads to layers of molecules gliding over each other. Referring to Figure 6.2 for notation and definitions, gliding of layers that are parallel to the crystal basal plane, perpendicular to the *c*-axis, is most easily achieved, although non-basal glide may occur at much higher stresses. The amount of deformation experienced by a crystal depends on the component of shear stress parallel to its basal plane, so that crystals that are oriented optimally with respect to an imposed stress deform more rapidly than crystals for which the shear stress resolved on the basal plane is small. The common assumption is that ice is isotropic with the orientations of *c*-axes randomly distributed. In that case, an aggregate of many crystals will deform at a slower rate than a single crystal because many of the crystals are not oriented optimally for basal glide.

One of the consequences of crystals deforming by basal glide is that the orientation of the basal plane changes as deformation progresses, such that the *c*-axis rotates toward the compressive axis and away from the tensile axis (Figure 6.2). Thus, if an aggregate is subjected to one particular stress regime for a sufficiently long time, all crystals would become oriented such that their basal planes are near perpendicular to the principal compressive stress and no further crystal deformation would occur since, in this geometry, the shear stress parallel to the basal plane vanishes. In reality, other processes become important, including displacement between crystals, crystal growth and boundary migration and re-crystallization, that result in more favorably oriented crystals, thereby permitting continued deformation.

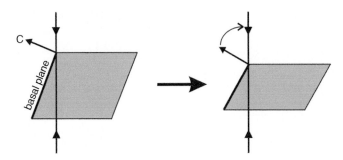

Figure 6.2. Geometrical illustration of the rotation of the crystal c-axis (perpendicular to the basal plane of the crystal) under compression.

For the study and modelling of glacier flow, the mechanisms and processes allowing individual crystals to deform need not be understood in great detail. Instead, the approach is to apply continuum mechanics to model ice flow. The basic premise of continuum mechanics is that the material under consideration is continuous, and that the response to applied stress can be described by a single constitutive relation, thus avoiding the need to model crystals individually. For glacier ice, the response to stress is visco-elastic, that is, the rate of deformation increases non-linearly with applied stress. Provided the aggregate ice remains isotropic, with crystals oriented randomly, the commonly used flow law is that proposed by Glen (1955), which applies to situations in which the ice has undergone sufficient deformation to pass the initial stage of primary creep and reach the stage of secondary creep, during which the rate of deformation remains constant. Primary or transient creep occurs immediately after a stress is applied and is characterized by a rapidly decreasing rate of deformation. Van der Veen (1999a) suggests that this decrease in deformation rate is due to the progressive rotation of crystal *c*-axes toward orientations unfavourable for further deformation by basal glide. Because of the rotation of individual crystals, an aggregate becomes increasingly harder as deformation progresses so that the rate of deformation decreases with increasing strain. If allowed to continue, the rate of deformation would approach zero as all crystals become misaligned. In reality, this does not occur because recrystallization replaces old and strained crystals with new and unstrained ones oriented optimally for deformation by basal glide. Thus, the minimum creep rate characterizing the end of primary creep and the onset of secondary creep may in fact mark the onset of recrystallization. During secondary creep, the rate of deformation remains more or less constant as a result of the balance between hardening of an aggregate due to crystal rotation, and softening resulting from recrystallization.

In some instances, the ice has developed a pronounced fabric, with the majority of *c*-axes aligned in one preferred direction, making the ice 'soft' with respect to some stress and 'hard' with respect to other stresses. For example, the basal ice layers of the Greenland ice sheet are characterized by *c*-axes clustering around the vertical direction, leading to greater rates of vertical shear than would be found in isotropic ice. The correct procedure for including this effect is to apply an anisotropic flow law. In practice, however, a more expedient procedure is to prescribe an enhancement factor to achieve greater shearing rates.

The flow law allows strain rates, and thus velocities, to be calculated from the stresses in the ice. Because glacier flow is sufficiently slow that accelerations can be neglected, Newton's second law of motion reduces to equilibrium of forces. The action force is the gravitational *driving stress* responsible for making the ice flow in the direction of decreasing surface slope (generally there are small-scale exceptions where the direction of flow does not coincide with the downslope direction). This action is opposed by resistive forces acting at the boundaries of the ice mass. These boundaries include the glacier bed (*basal drag*), the lateral margins, where faster moving ice is bounded by rock or near-stagnant ice (*lateral drag*), and the up- and down-glacial ends (*gradients in longitudinal stress*). In general, the distribution and nature of these resistive stresses are not expected to be uniform, and certain sites may play a larger role in governing glacial flow than others. Because of this, the study of a particular glacier requires the identification of the most important sites of resistance and the stress regimes there. The modelled glacier response is, to a large extent, controlled by the various resistances incorporated into the model; thus, careful consideration should be given to identifying the mechanical controls acting on the glacier being modelled. In practice, this requires interpretation of measured surface speeds and geometry to assess the partitioning of flow resistance among the various resistive forces.

The force balance equation and the flow law are diagnostic equations for calculating the ice velocity and discharge for a given geometry. To evaluate the time evolution of glaciers, the prognostic continuity equation is required. In essence, this equation states that no ice may be created or lost and thickness changes at any particular point must be entirely due to ice flow and local snow accumulation or melting. Firn densification in the upper layers is usually neglected in time-evolving glacier models, and the density of glacier ice is considered constant. In that case, conservation of mass as expressed by the continuity equation corresponds to conservation of ice volume. The continuity equation allows changes in glacier thickness to be calculated and is therefore the central equation that is solved in time-marching numerical models.

A complicating matter is that the rate of deformation for a given stress depends on the temperature of the ice. The warmer the ice, the easier deformation becomes. In the flow law, this temperature effect is incorporated by adopting a temperature-dependent rate factor. Consequently, thermodynamics need to be included when modelling glacier flow and, in particular, the depth variation in ice temperature. Generally, deeper ice layers are warmer than layers closer to the surface and, combined with greatest shear stresses near the glacier base (except on floating glaciers and ice streams where basal drag is zero or very small), most vertical shear occurs in the lower ice layers. To account fully for this, many existing numerical models explicitly calculate the ice temperature at discrete depth layers, thus making the model domain two- or three-dimensional, depending on whether flow along a flow line or in both horizontal directions is considered.

The foregoing summary discussion has introduced the basic tools needed to construct models of glacier flow. How these tools are applied depends on the objective of the modelling study.

6.2 Glacier dynamics

6.2.1 Force balance

Balance of forces acting on a section of glacier has been discussed many times, most recently in Hooke (1998, pp. 191–5), Hughes (1998, pp. 51–4) and Van der Veen (1999a, pp. 32–6). In most cases, balance of forces is discussed in terms of stress deviators, defined as the full stress minus the hydrostatic pressure. That is,

$$\sigma'_{ij} = \sigma_{ij} - \frac{1}{3}[\sigma_{xx} + \sigma_{yy} + \sigma_{zz}], \tag{6.1}$$

where the prime denotes the stress deviator and unprimed stresses are full stresses. While these deviatoric stresses are called for in the flow law for glacier ice, their use in discussing balance of forces unnecessarily complicates the interpretation because the longitudinal deviatoric stress in one direction depends on the full normal stresses in all three directions of a cartesian co-ordinate system, as can be seen from the definition in equation (6.1). It is more convenient to consider the stresses in a glacier as the sum of the stress due to the weight of the ice above (*lithostatic stress*) and special stresses due to the flow (*resistive stresses*). This partitioning makes a clear distinction between action and reaction in glacier mechanics. The following development is for force balance of a column of ice extending from the bed to the upper surface, but applies equally to layers that do not extend over the full ice thickness.

Driving stress

Lithostatic stresses are associated with the gravitational force acting on the ice, and horizontal gradients in this gravitational force give rise to the driving stress, the 'action' that makes the glacier flow. In Figure 6.3 the lithostatic components of stress acting on an ice column

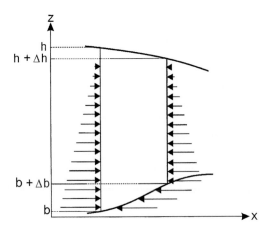

Figure 6.3. Vertical column through a glacier showing lithostatic stresses. Only those acting on the column are shown and summed to give the driving force.

are indicated by the solid arrows. This lithostatic stress is a normal stress that increases linearly with depth (assuming constant ice density, ρ):

$$L(z) = -\rho g(h - z), \tag{6.2}$$

in which h represents the elevation of the ice surface and z is the vertical direction (positive upward). The net right-directed lithostatic force acting on the left hand face of the ice column is obtained by integrating the lithostatic stress over the depth of the column, and is

$$\int_b^h \rho g(h - z)\,dz = \frac{1}{2}\rho g(h - b)^2, \tag{6.3}$$

where $z = b$ denotes the elevation of the glacier bed. On the right hand face of the column, the surface and bed elevation are $(h + \Delta h)$ and $(b + \Delta b)$, respectively, and the total lithostatic force on that face is

$$-\int_{b+\Delta b}^{h+\Delta h} \rho g(h - z)\,dz = -\frac{1}{2}\rho g[(h + \Delta h) - (b + \Delta b)]^2, \tag{6.4}$$

noting that this stress is negative because it is directed in the negative x-direction. The remaining x-component of lithostatic force acts at the base of the column. The average column thickness may be used to evaluate the lithostatic force acting on the sloping glacier sole. This force is equivalent to the lithostatic force acting on a vertical step in the bottom of height Δb, and is

$$-\rho g \left[\left(h + \frac{\Delta h}{2} \right) - \left(b + \frac{\Delta b}{2} \right) \right]. \tag{6.5}$$

Summing all three forces acting on the column provides the gravitational driving force,

$$-\rho g \bar{H} \Delta h, \tag{6.6}$$

in which $\bar{H}$ represents the average thickness of the column. Finally, the driving force per unit map area is

$$-\rho g \bar{H} \frac{\Delta H}{\Delta x}. \tag{6.7}$$

Taking the limit of the length of the column approaching zero, $\Delta x \to 0$, the driving stress is

$$\tau_{dx} = -\rho g H \frac{\partial h}{\partial x}. \tag{6.8}$$

A similar expression can be derived for the driving stress corresponding to the other horizontal direction.

The derivation of the driving stress given above differs from the conventional approach in which stresses in a plane slab are considered (Hooke, 1998, pp. 47–8; Paterson, 1994, pp. 239–41). In this special case of laminar or lamellar flow, the driving stress and basal

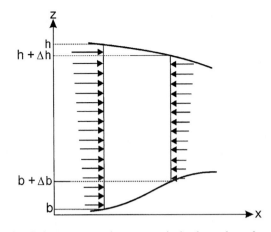

Figure 6.4. Horizontal resistive stresses acting on a vertical column through a glacier. The difference between the net force acting on the right and left faces gives the resistance to flow associated with longitudinal tension or compression.

shear stress are equal, and the driving stress is equated with the vertical shear stress, σ_{xz}. Extension to a more complicated geometry, including ice-shelf flow, in which vertical shear is negligible, is not immediately clear. In the present approach, the term driving stress is reserved for the action due to gravity and is given by the formula above for any glacier geometry. Moreover, there is no theoretical need to average that expression over areas, although, in practice, the driving stress is calculated over horizontal distances of a few ice thicknesses or so to eliminate small-scale flow features that are not important to the large-scale dynamics of glaciers.

Resistive stresses

Glacier flow is driven by gravity and restrained by interaction against the bed, or the sides, or by pressure or tension transmitted along the glacier. The effect of gravity is described by the driving stress derived above, while flow restraint is described by resistive stresses. Because the weight-induced lithostatic stress is accounted for through the driving stress, resistance to flow must originate from the remaining stresses, that is, from the differences between the full stresses and the lithostatic stress. These stress differences are referred to as resistive stresses, defined as

$$R_{ij} = \sigma_{ij} - \delta_{ij} L, \qquad (6.9)$$

in which δ_{ij} represents the Kronecker delta ($\delta_{ij} = 1$ if $i = j$, and zero otherwise).

The net resistive force acting on an ice column is found by summing the forces acting on the faces of the column, as in the derivation above for the driving stress. Referring to Figure 6.4, a column of length Δx and width Δy is considered and the net force in the x-direction evaluated. The first resistive force is associated with longitudinal tension or compression on the left and right faces perpendicular to the x-axis. The longitudinal resistive stress acting on the left face is R_{xx}, and integrating this stress over the thickness and width of the column gives the longitudinal force,

$$\int_b^h R_{xx}(z)\Delta y \, dz = H\bar{R}_{xx}\Delta y, \tag{6.10}$$

where the overbar denotes the depth-averaged resistive stress. The longitudinal resistive stress acting on the right face is negative (oriented in the negative x-direction) and differs from the stress on the left face by an amount ΔR_{xx}. Again, integrating gives the net force acting on the right face of the column,

$$-\int_{b+\Delta b}^{h+\Delta h} (R_{xx}(z) + \Delta R_{xx}(z)) \, \Delta y \, dz$$

$$= -H\bar{R}_{xx}\Delta y - \Delta H\bar{R}_{xx}\Delta y - H\Delta\bar{R}_{xx}\Delta y - \Delta H\Delta\bar{R}_{xx}\Delta y, \tag{6.11}$$

with $\Delta H = \Delta h - \Delta b$ the difference in thickness at the two faces. Summing expressions (6.10) and (6.11), and neglecting the higher order term in equation (6.11), gives the net longitudinal force acting on the column:

$$-\left[\frac{\Delta H}{\Delta x}\bar{R}_{xx} - H\frac{\Delta\bar{R}_{xx}}{\Delta x}\right]\Delta x\Delta y = -\frac{\partial}{\partial x}\left(H\bar{R}_{xx}\right)\Delta x\Delta y, \tag{6.12}$$

in which the limit $\Delta x \to 0$ has been taken.

The second source of flow resistance is lateral drag acting at the two vertical faces parallel to the x-direction. Following a derivation similar to that for the net longitudinal force, net flow resistance from lateral drag is found to be

$$-\frac{\partial}{\partial y}\left(H\bar{R}_{xy}\right)\Delta x\Delta y. \tag{6.13}$$

Equations (6.12) and (6.13) show that resistance to flow is offered by *gradients* in resistive stress, rather than by the *magnitude* of these stresses. This can be readily understood by considering two people pushing on opposite ends of a large object. If both push equally hard, the object will not move because the net pushing force (or longitudinal force) sums to zero. Only if one person pushes harder than the other will the object be displaced. Thus, even on glaciers where the longitudinal stress is large compared with the driving stress, the effect on force balance may be small or negligible when the gradients are much smaller than the driving stress. This distinction between magnitude and gradient appears to have confused some authors. For example, Paterson (1994, p. 312) estimates the longitudinal strain rate on Whillans ice stream[1] and Rutford ice stream to be comparable with those measured on the Ross ice shelf, and concludes that 'it is therefore essential to include the

---

[1] Ice stream B was renamed Whillans ice stream in 2001 in recognition of the late Prof. Ian Whillans' contribution to Antarctic glaciology.

longitudinal stress deviator in analyses of flow in ice streams or other transition zones'. This statement is not supported by the evidence, however. A force balance calculation conducted along the entire length of Whillans ice stream indicates that gradients in longitudinal stress contribute little to the mechanical control of that ice stream and may, therefore, be omitted from any discussion of balance of forces (Whillans and Van der Veen, 1993b).

The third resistance originates from drag at the glacier base as expressed by the basal drag, τ_{bx}. This drag represents force per unit area, so the total resistive force acting at the glacier base is

$$\tau_{bx} \Delta x \Delta y. \tag{6.14}$$

Resistance to flow at the glacier bed may be due to frictional forces (*skin drag*) and resistive stresses acting on a sloping bed (*form drag*) (note that lithostatic stresses acting on a sloping bed are included in the driving stress). If the bed is perfectly lubricated, skin drag is zero and there is no vertical shear in velocity at the glacier base. Form drag is the resistance due to pressure fluctuations generated by flow over an obstacle (see Van der Veen, 1999a, pp. 71–2). The formal definition of basal drag is (Van der Veen, 1999a, p. 35)

$$\tau_{bx} = R_{xz}(b) - R_{xx}(b)\frac{\partial b}{\partial x} - R_{xy}(b)\frac{\partial b}{\partial y}. \tag{6.15}$$

Thus, the analysis of forces presented here does not make any implicit assumptions about the shape of the glacier or the underlying bed, and applies to all geometries.

Force balance in the horizontal direction

The net resistance to flow is given by the sum of equations (6.12)–(6.14) and must balance the action of gravity as described by the driving stress. Equation (6.8) represents the driving force per unit area considered. The resistive force per unit area is obtained from equations (6.12)–(6.14) by dividing by the area of the column ($\Delta x \Delta y$). Equating action and reaction gives the force balance equation

$$\tau_{dx} = \tau_{bx} - \frac{\partial}{\partial x}\left(H\bar{R}_{xx}\right) - \frac{\partial}{\partial y}\left(H\bar{R}_{xy}\right). \tag{6.16}$$

A similar equation can be derived for the balance of forces in the other horizontal direction.

While the balance equation derived here applies to the full ice thickness, the result can also be applied to individual layers within the ice to allow for explicit calculation of the depth variation of resistive stresses (Van der Veen, 1989; Van der Veen and Whillans, 1989). Such a calculation is of particular interest where measurements of surface speed are available. From these, surface strain rates can be calculated and converted to resistive stresses using the flow law for glacier ice. Applying the force balance equations then allows the partitioning of flow resistance to be evaluated to identify the type of flow regime.

Force balance in the vertical direction

So far, balance of forces in the horizontal directions has been considered. Where the weight of the ice is fully supported by the bed underneath or, in the case of floating glaciers, by sea-water, the full vertical stress at any depth equals the lithostatic stress, and the vertical resistive stress, R_{zz}, is zero. For most applications considered in this book, this is a valid approximation. Differences from the lithostatic stress may become important on horizontal scales that are small compared with the ice thickness, for example when considering ice flow in the lee of a subglacial cavity. There, the basal ice may become separated from the bed and, locally, may not be supported from below. Instead, local shear stress gradients transfer the weight to surrounding areas where the ice is in contact with the bed. This transfer of weight is similar to a bridge, where the span is not supported from below and the abutments carry the full weight of the bridge. This so-called bridging effect (Van der Veen and Whillans, 1989) corresponds to the T-term in the commonly used force balance equation (Hooke, 1998, p. 192). This term has been discussed in a number of papers (e.g. Budd, 1970a, b; Kamb and Echelmeyer, 1986), and the general conclusion is that this term is negligible when averaged over longitudinal distances greater than three or four times the ice thickness. This conclusion is based on scale analysis of all terms in the balance equation, but the physical interpretation of the implications of neglecting the T-term are not entirely clear from these analyses. By identifying the T-term with the bridging effect, its meaning becomes more transparent. Neglecting the T-term is equivalent to making the assumption that bridging effects are negligible, that is, the ice is fully supported from directly below.

It may be noted that the horizontal balance equation (6.16) and its counterpart for the other horizontal direction are independent of the balance equation applying to the vertical direction. Thus, whether or not bridging effects are included in the analysis does not affect the horizontal balance equations if formulated in terms of resistive stresses, as above. However, when relating resistive stresses to strain rates, the vertical resistive stress comes into play. In the following discussion, the assumption is made that R_{zz} is zero. How bridging effects can be incorporated into an analysis of glacier flow is discussed in Van der Veen (1999a, section 3.7).

6.2.2 Flow law

According to Alley (1992), 'the flow law relating strain rate to stress, temperature and ice properties is to glaciology what the Holy Grail was to chivalry – an important goal but one that may not be attainable'. Indeed, given the transient nature of processes acting on the scale of individual crystals (dislocation glide, boundary migration, *c*-axis rotation, etc.) it seems impossible, or at least impractical, to formulate a flow law that includes all of these effects. For example, while a relation between strain rate and stress may be formulated for ice with prescribed anisotropy, further fabric development depends on how the ice deforms and thus on the flow law. Consequently, to be fully correct, the flow law should be coupled to a model describing how ice fabric evolves and how this evolution, in turn, affects deformation rates. Apart from the fact that a general form of anisotropic flow law has yet to be proposed, including such a relation plus the required model for fabric development into a numerical model of, say, the Antarctic ice sheet is not a feasible option for modellers. Instead, what is commonly used is *a* flow law – Glen's law – that appears to work reasonably well. That is, results obtained with models based on this flow law agree for the most part with observations on real glaciers. It is important to keep in mind, however, that the form of Glen's law has not been confirmed unambiguously.

Without going into the mathematical details, it can be shown that the general form of the flow law is

$$\dot{\varepsilon}_{ij} = F_2(I_2, I_3)\,\sigma'_{ij} + F_3(I_2, I_3)\left(\sigma'_{ik}\sigma'_{kj} - \frac{2}{3}I_2\delta_{ij}\right), \tag{6.17}$$

where F_2 and F_3 are arbitrary functions of the second and third invariants of the deviatoric stress tensor (Glen, 1958; Van der Veen, 1999a, section 2.6). Deviatoric stresses are used to eliminate the hydrostatic pressure from the flow law. This is based on the experiments of Rigsby (1958), which indicate that the deformation of ice is mostly independent of hydrostatic pressure provided the difference between ice temperature and pressure-melting temperature is kept constant. Note that the first invariant of the stress tensor is zero if ice is considered incompressible and therefore does not enter into the flow law.

The second term on the right hand side of equation (6.17) describes normal stress effects, or dilatancy. Dilatancy manifests itself in a material subjected to simple shear through non-hydrostatic normal stresses perpendicular to the plane of shear and in the plane of flow. While such effects appear to be common in non-Newtonian fluids for which appropriate measurements have been made (Schowalter, 1978), they are commonly neglected when glacier flow is considered. Two studies in which normal stress effects on glaciers are discussed suggest that their manifestation may be small on glaciers (McTigue, Passman and Jones, 1985; Man and Sun, 1987). Van der Veen and Whillans (1990) found that by retaining the second term on the right hand side of equation (6.17) in the flow law, small-scale flow features near Dye 3, Greenland, can be better explained than when these effects are not included. In spite of these results, the role of normal stress effects in large-scale glacier flow appears to be limited and is not included in the flow law used in most modelling studies. That is, F_3 is set equal to zero.

A further commonly made simplification is that the first term on the right hand side of equation (6.17) is a function of the second invariant only. This simplification was suggested by Nye (1953) in an effort to compare experimental creep data with the rate of tunnel closure, and has since gone mostly unchallenged by the glaciological community (an exception is Baker (1987)). For the functional relation, $F_2(I_2)$, an exponential form is used. Defining the effective stress and strain rate as

$$2\tau_e^2 = \sigma'^2_{xx} + \sigma'^2_{yy} + \sigma'^2_{zz} + 2\left(\sigma'^2_{xy} + \sigma'^2_{xz} + \sigma'^2_{yz}\right), \tag{6.18}$$

$$2\dot{\varepsilon}_e^2 = \dot{\varepsilon}^2_{xx} + \dot{\varepsilon}^2_{yy} + \dot{\varepsilon}^2_{zz} + 2\left(\dot{\varepsilon}^2_{xy} + \dot{\varepsilon}^2_{xz} + \dot{\varepsilon}^2_{yz}\right), \tag{6.19}$$

Nye's generalization of Glen's law becomes

$$\dot{\varepsilon}_{ij} = A\tau_e^{n-1}\sigma'_{ij}. \tag{6.20}$$

For stresses commonly found on glaciers, theoretical arguments as well as laboratory and field studies suggest a value $n = 3$ for the exponent (Alley, 1992).

The flow law equation (6.20) relates strain rates to stresses. For some applications, such as inferring stresses from measured surface velocities, the inverse formulation is needed. The inverse flow law is (Van der Veen, 1999a, p. 15)

$$\sigma'_{ij} = B\dot{\varepsilon}_{e}^{1/n-1}\dot{\varepsilon}_{ij}, \tag{6.21}$$

with

$$B = A^{-1/n}. \tag{6.22}$$

The flow parameters, A and B, depend on the temperature of the ice, with the rate of deformation under a given stress increasing in warmer ice. This dependency is discussed in the next section.

The flow law in terms of resistive stresses can be obtained by equating the two schemes for partitioning full stresses (equation (6.1) involving deviatoric stresses and equation (6.9) involving resistive stresses), to yield the following relations between resistive and deviatoric stresses (Van der Veen, 1999a, pp. 37–8):

$$R_{xx} = 2\sigma'_{xx} + \sigma'_{yy} + R_{zz}, \tag{6.23}$$

$$R_{yy} = 2\sigma'_{yy} + \sigma'_{xx} + R_{zz}, \tag{6.24}$$

$$R_{xy} = \sigma'_{xy}, \tag{6.25}$$

$$R_{xz} = \sigma'_{xz}, \tag{6.26}$$

$$R_{yz} = \sigma'_{yz}. \tag{6.27}$$

Applying the flow law equation (6.21), the resistive stresses can be expressed in terms of strain rates as

$$R_{ij} = B\dot{\varepsilon}_{e}^{1/n-1}\dot{\varepsilon}_{ij}, \qquad i \neq j = x, y, z, \tag{6.28}$$

$$R_{ii} = B\dot{\varepsilon}_{e}^{1/n-1}\left(2\dot{\varepsilon}_{ii} + \dot{\varepsilon}_{jj}\right) + R_{zz}, \qquad i \neq j = x, y. \tag{6.29}$$

The second expression for the two horizontal longitudinal resistive stresses contains the vertical resistive stress, R_{zz}. As noted above, for most applications, this stress may be neglected.

6.2.3 Velocities and strain rates

Stresses acting on the ice result in deformation. This deformation is described by the flow law equation (6.20) which expresses strain rates in terms of deviatoric stresses. Strain rates, in essence, characterize how the shape of an ice segment changes as a result of deformation. To illustrate this, consider Figure 6.5, which shows two neighbouring points before and after deformation. The length and orientation of the line segment AB changes to $A'B'$, and

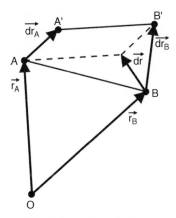

Figure 6.5. Deformation of a line segment.

the actual deformation is given by the vector

$$\vec{dr} = \overrightarrow{A'B'} - \overrightarrow{AB} = (\vec{r_B} + \vec{dr_B}) - (\vec{r_A} + \vec{dr_A}) - (\vec{r_B} - \vec{r_A}) = \vec{dr_B} - \vec{dr_A}. \qquad (6.30)$$

For further discussion, it is convenient to consider deformation in three mutually perpendicular directions. Defining a cartesian co-ordinate system with axes x_j ($j = 1, 2, 3$), a first-order Taylor expansion of each of the three components of the displacement vector gives

$$dr_i = \frac{\partial r_i}{\partial r_j} dx_i, \qquad (6.31)$$

where summation over repeated indices is implied.

As noted, the line segment may be rotated and its length may change. Similarly, if a volume of ice is considered, deformation involves rigid body rotation and distortion, which changes the shape of the volume. From a dynamical point of view, rotation is of no importance and only distortion needs to be considered. To find this distortion, the second-order tensor with components $\partial r_i / \partial r_j$ is split into a symmetric and anti-symmetric part:

$$\frac{\partial r_i}{\partial x_j} = \frac{1}{2} \left(\frac{\partial r_i}{\partial x_j} + \frac{\partial r_j}{\partial x_i} \right) + \frac{1}{2} \left(\frac{\partial r_i}{\partial x_j} - \frac{\partial r_j}{\partial x_i} \right). \qquad (6.32)$$

The anti-symmetric part (the second term on the right hand side) describes the rigid body rotation. The symmetric part describes the actual distortion of the volume and is called the strain tensor, with components ε_{ij}. There are two types of distortion or deformation, namely stretching ($i = j$) and shearing ($i \neq j$).

Rather than considering actual deformation, the rate at which an ice volume changes shape is of interest. This rate is obtained by differentiating the strain tensor with respect to time, t, to give the strain rate tensor

$$\dot{\varepsilon}_{ij} = \frac{1}{2} \left[\frac{\partial}{\partial x_j} \left(\frac{\partial r_i}{\partial t} \right) + \frac{\partial}{\partial x_i} \left(\frac{\partial r_j}{\partial t} \right) \right] = \frac{1}{2} \left(\frac{\partial u_i}{\partial x_j} + \frac{\partial u_j}{\partial x_i} \right).$$ (6.33)

In this expression, the u_i represent the three components of ice velocity in the three orthogonal directions, x_i.

To summarize the above derivation in words, the rate of deformation is described by gradients in ice velocity. Intuitively, this makes sense because a volume of ice encapsulated in a uniform velocity field does not change shape but is simply advected down-flow, and hence is not subject to deformation. Further, the absolute velocity does not matter to strain rates. This point is of importance when measurements of surface speed are used to infer resistive stresses. Often, relative motions are comparatively easy to measure, for example by surveying a grid of poles, or by using interferometry. Obtaining absolute motions may require significant additional effort, either by tying the survey grid to an absolute reference frame, or by identifying stagnant ice in an interferogram. Because the absolute motion is common to all derived velocities, it cancels when spatial gradients in velocity are considered.

6.2.4 Thermodynamics

The rate factor, A, in the flow law varies with ice temperature, T (in kelvin), according to the Arrhenius relation,

$$A = A_0 \exp \left[-\frac{Q}{RT} \right],$$ (6.34)

in which $R = 8.314$ J/(mol K) represents the universal gas constant. Q is the activation energy for creep, and its value is ~60 kJ/mol, for temperatures below 263.15 K, and ~139 kJ/mol at temperatures between 263.15 K and the pressure-melting temperature. The pre-factor A_0 is independent of temperature but may be a function of crystal size, impurity content, orientation of c-axes, etc. (Alley, 1992; Paterson, 1994, pp. 86–95).

A somewhat different temperature dependency was suggested by Hooke (1981), who evaluated available data and found the following best fit:

$$A = A_0 \exp \left[-\frac{Q}{RT} + \frac{3C}{(T_r - T)^k} \right],$$ (6.35)

or

$$B = B_0 \exp \left[\frac{T_0}{T} - \frac{C}{(T_r - T)^k} \right],$$ (6.36)

where $A_0 = 9.302 \times 10^7 / \text{kPa}^3$ per year, $Q = 78.8$ kJ/mol, $C = 0.16612$ K^k, $T_r = 273.39$ K, $k = 1.17$, $B_0 = 2.207 \times 10^{-5}$ kPa yr$^{1/3}$, $T_0 = 3155$ K and the ice temperature, T, is expressed in kelvin. Other values for the rate factor have been recommended (Paterson, 1994, p. 97; Paterson and Budd, 1982), and differences suggest that, for a given stress and temperature, the strain rate may vary by as much as a factor of 5.

The temperature dependency of the rate factor has several consequences for modelling glacier flow. Firstly, in addition to evaluating the change in glacier geometry by solving the continuity equation, the evolution of englacial temperatures must also be calculated using the thermodynamic temperature equation as a prognostic time-stepping equation. Secondly, thermodynamics introduce long timescales into the analysis of glacier flow. This is because transfer of heat throughout a glacier is a slow process. For example, Huybrechts (1994) suggests that current thinning in the central and northern regions of the Greenland ice sheet may be the manifestation of the delayed thermodynamic response following the climate warming after the last glacial maximum, as the warmer temperatures only now are reaching the basal ice layers. Generally, the larger the glacier considered, the longer its 'thermal memory', and consequently the longer the time period over which its evolution must be simulated. For the polar ice sheets, if the most recent glacial cycle is to be modelled, simulations should involve at least a few full glacial cycles to eliminate any effects of initial temperature conditions on the model results.

For a full discussion of glacier thermodynamics, the reader is referred to Paterson (1994, chap. 10), Hooke (1998, chap. 6) or Van der Veen (1999a, chap. 7). In short, conservation of energy can be expressed in terms of the rate of temperature change:

$$
\begin{aligned}
\frac{\partial T}{\partial t} = &-u\frac{\partial T}{\partial x} - v\frac{\partial T}{\partial y} - w\frac{\partial T}{\partial z} \\
&+ K\frac{\partial^2 T}{\partial x^2} + K\frac{\partial^2 T}{\partial y^2} + K\frac{\partial^2 T}{\partial z^2} \\
&+ \frac{1}{\rho C_p}(\dot{\varepsilon}_{xx}\sigma_{xx} + \dot{\varepsilon}_{yy}\sigma_{yy} + \dot{\varepsilon}_{zz}\sigma_{zz}) \\
&+ \frac{2}{\rho C_p}(\dot{\varepsilon}_{xy}\sigma_{xy} + \dot{\varepsilon}_{xz}\sigma_{xz} + \dot{\varepsilon}_{yz}\sigma_{yz}) \\
&+ \frac{L_f M_f}{\rho C_p}.
\end{aligned}
\tag{6.37}
$$

In this equation, $K = k/(\rho C_p)$ represents the thermal diffusivity, k is the thermal conductivity, ρ is the ice density and C_p is the specific heat capacity. This equation states that the temperature of the ice may change as a result of advection by the ice flow (first line), diffusion of heat (second line), heat generated by internal deformation (third and fourth lines). The final term represents heat released through phase changes, where M_f is the amount of melt water that refreezes per unit time and volume, and L_f represents the latent heat of ice.

The coupling between temperature and ice flow can, potentially, lead to creep instability, with a runaway increase in ice velocity. Heat generated by internal deformation increases with the rate of deformation, which in turn increases as the temperature increases, and a positive feedback develops. This mechanism has been studied by a number of authors (Clarke, Nitsan and Paterson, 1977; Schubert and Yuen, 1982; Yuen and Schubert, 1979; Yuen, Saari and Schubert, 1986), but these studies are local and do not take into account increased advection of colder ice from up-glacier as the ice speed increases.

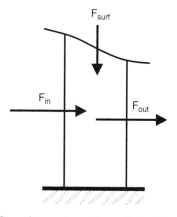

Figure 6.6. Mass fluxes into and out of a vertical column through a glacier.

This process tends to counter the positive feedback, and model simulations suggest that creep instability is not a likely process for leading to the collapse of large ice sheets. A more important feedback is that between basal melting and the onset of sliding, which may lead to periodic short-lived increases in glacier speed and consequent loss of ice volume.

6.2.5 Continuity

Consider an ice column extending from the bed to the surface with length Δx in the direction of flow and width Δy in the other horizontal direction (Figure 6.6). The flux of ice through the left and right vertical faces is equal to the product of ice thickness and depth-averaged ice velocity. Denoting the ice thickness and ice speed at the left face by H and U, respectively, the ice flux into the column is

$$F_{in} = HU\Delta y. \tag{6.38}$$

If ΔH and ΔU represent the change in thickness and speed over the length of the column, the flux through the right face is

$$F_{out} = (H + \Delta H)(U + \Delta U)\,\Delta y. \tag{6.39}$$

The net gain (or loss) of ice due to advection is then

$$F_{adv} = -(H\Delta U + U\Delta H)\,\Delta y, \tag{6.40}$$

if the second-order term is neglected.

Mass gain or loss may occur at the surface through snow fall or ablation. Expressing the accumulation, M, in metres of ice depth per unit time and unit area, the net input at the surface is

$$F_{surf} = M\Delta x\Delta y. \tag{6.41}$$

Note that $M < 0$ corresponds to surface ablation. Basal melting can be included in this term, but is usually small compared with surface accumulation or ablation, except perhaps on floating ice shelves.

When the mass fluxes into and out of the column are not in balance, the thickness of the column must change. The rate of thickness change is $\partial H/\partial t$ and the corresponding change in volume of the column is

$$\frac{\partial H}{\partial t}\Delta x \Delta y. \tag{6.42}$$

This volume change must balance the sum of equations (6.40)–(6.42). Dividing by the area of the column gives

$$\frac{\partial H}{\partial t} = -\frac{H\Delta U + U\Delta H}{\Delta x} + M. \tag{6.43}$$

Taking the limit $\Delta x \to 0$ gives the continuity equation

$$\frac{\partial H}{\partial t} = -\frac{\partial(HU)}{\partial x} + M. \tag{6.44}$$

When flow in both horizontal directions is considered, a term describing advection in the horizontal y-direction must be added to the right hand side.

6.2.6 Basal sliding and bed deformation

Glacier flow may result from two mechanisms, namely internal deformation, as described by the flow law (equation (6.20)), and sliding of the ice over its bed. Basal sliding is important where the basal temperature has reached the pressure-melting temperature and a lubricating water layer is present. While speeds associated with internal deformation are typically of the order of 10 to 100 meters per year, sliding allows for much more rapid flow with speeds of up to several kilometres per year. Consequently, the occurrence of sliding can have a dramatic impact on glacier flow and mass balance.

Two mechanisms allow basal ice to move past obstacles and irregularities in the glacier bed, namely *regelation* and *enhanced creep*. Firstly, regelation occurs because the pressure-melting temperature is lower on the upstream side of basal protuberances, where the normal stress acting on the vertical face is compressive, than on the downstream side, where the normal stress is tensile. This results in upstream melting and downstream refreezing, thus allowing the ice to move past the obstacle. The latent heat released by refreezing must be conducted through the obstacle to the upstream side to maintain the temperature difference, and this process is most effective for small obstacles. Secondly, the stress difference across the obstacle leads to enhanced deformation of ice around the obstacle. This process is more effective for larger obstacles. By considering a simplified glacier bed consisting of a smooth inclined plane on which cubical protuberances are superimposed, and assuming that the bed is frictionless so that all resistance to flow is due to normal stresses acting on the vertical faces of these obstacles, Weertman (1957a, 1964) derived the following sliding relation:

$$U_s = A_s \left(\frac{\tau_b}{R^2} \right)^{(1+n)/2},\tag{6.45}$$

in which A_s is a sliding parameter that depends on material properties of the bed and basal ice, and R represents the bed roughness parameter (the ratio of obstacle spacing and obstacle height). Subsequent studies have expanded the Weertman model to more realistic basal geometries but yield essentially similar results (Kamb, 1970; Nye, 1969, 1970).

According to the Weertman model for sliding, the only controls on sliding speed are the drag at the glacier base and the roughness of the underlying bed. However, observations on sliding glaciers suggest that the sliding speed can vary considerably over short time intervals, in particular after rain-fall events that apparently lead to increased water storage under the glacier. To account for these observations, the sliding law (equation (6.45)) must be corrected for the effect of subglacial water pressure. Bindschadler (1983) compared several sliding relations against observations and found the following relation to give the best description of the data:

$$U_s = A_s \frac{\tau_b^3}{P_e},\tag{6.46}$$

with the effective basal pressure equal to the ice overburden pressure minus the subglacial water pressure

$$P_e = \rho g H - P_w.\tag{6.47}$$

This sliding relation is most often used in numerical models, but it should be kept in mind that its validity has yet to be established unambiguously.

An important assumption in models for basal sliding is that the bed under the ice is hard and non-deformable. Many glaciers, however, rest on soft beds consisting of sediments that may be deforming, and subglacially deposited till or other unlithified sediments are exposed in regions that were formerly covered by ice. Boulton and Jones (1979) propose that, under favourable conditions, most notably high basal water pressure and thus low effective pressure, these sediments may be actively deforming, thus contributing to the motion of the glacier above. After discovering a layer of till under Whillans ice stream in West Antarctica, Alley and co-workers successfully advanced the hypothesis that the fast motion of this ice stream (and by association that of the other active ice streams draining into the Ross ice shelf), in spite of unusually small driving stresses, can be attributed to a continuous layer of actively deforming till (Alley *et al.*, 1986, 1987). However, subsequent shear tests on till samples retrieved from the bed under Whillans ice stream indicate that this till behaves more or less like a plastic material, unable to support stresses in excess of a few kilopascals (Kamb, 1991) rather than as the viscous material supposed in Alley's hypothesis. At this stage, the role of till deformation on glacier flow remains under debate (see Van der Veen, 1999a, sect. 4.5). Modellers should keep in mind that, even where such deformation may contribute significantly to ice discharge, the constitutive properties of the till are not well established, and any relation between ice speed and till viscosity and thickness, and basal drag remains more poorly constrained than the sliding relation.

The modified Weertman sliding law (equation (6.47)) exhibits great sensitivity to the effective basal pressure, in particular when the ice approaches near flotation and this pressure becomes very small. Thus it becomes necessary to add a description of subglacial hydraulics when modelling glacier flow. Several models for subglacial drainage have been proposed, including a film of water of more or less uniform thickness (Weertman, 1972), channels incised into the base of the ice (R-channels; Röthlisberger, 1972) or into the bed below (N-channels), a linked-cavity system (Fowler, 1987a; Kamb, 1987; Walder, 1986), or water flow through a subglacial acquifer (Lingle and Brown, 1987). For each model, the relation between water flux and water pressure is different to the extent that changes in the nature of the drainage system may lead to glacier instability. For example, in an R-channel the effective pressure decreases as the water flux through the channel increases, whereas in a linked-cavity system the water pressure increases as the water flux increases. Fowler (1987b) argues that the transition from drainage through R-channels to linked-cavity drainage may be responsible for periodically occurring glacier surges during which the speed of the glacier increases by an order of magnitude or so (Van der Veen, 1999a, section 10.4). Accommodating such transitions into a (numerical) model requires detailed understanding of the transient nature of the drainage system. Moreover, in nature the situation may be more complex than commonly assumed in theoretical treatments. For example, Hooke (1989) argues that observational evidence favours a model in which a linked-cavity system is transected by a few broad, shallow conduits that drain melt water more directly. For glaciers resting on soft beds, the situation may be even more complex. In view of these remarks, it appears that no single model can be applied and that the nature of subglacial drainage varies from glacier to glacier and possibly throughout the year as well as melt-water input changes.

The actual distribution of subglacial water pressure depends on many factors, as noted above, and can only be evaluated for idealized geometries. These derivations are often lengthy and involve many simplifying assumptions that may not apply directly to actual glaciers. For example, the analysis of Röthlisberger (1972) assumes a steady water flux through the tunnels, which may not be realistic at the onset of the melting season, when water input is presumably largest. Apart from such reservations, a more fundamental question is whether water pressures within subglacial conduits are indeed appropriate to use in a sliding law. These conduits typically occupy a small fraction of the glacier bed and, averaged over larger areas, the effective basal pressure may be expected to differ from that directly above one of these conduits. In numerical models, with horizontal grid spacings typically exceeding 1 km or so, the areal average pressure should be used in the sliding law, rather than very localized pressures. In other words, the fraction of the bed experiencing separation between ice and underlying substrate may be the more relevant variable determining skin roughness.

The range of possible areal-averaged water pressures can be readily determined, independent of the nature of the drainage system. The effective pressure at the bed, given by equation (6.47), cannot become negative, as this would imply a net upward force on the base of the ice. (During short-lived local events, the effective pressure may exceed the weight of the ice, but for most practical modelling applications such unusual events need not be considered.) Thus the maximum water pressure is given by the ice overburden pressure,

$$P_{\mathrm{w}}^{\max} = \rho g H. \tag{6.48}$$

The minimum water pressure follows from the condition that water underneath the glacier must be able to reach the terminus. That is, the gradient in water pressure must be sufficiently large to force the water over the sloping bed. This requirement yields for the minimum water pressure

$$P_{\mathrm{w}}^{\min} = -\rho_{\mathrm{w}} g b, \tag{6.49}$$

where ρ_{w} represents the density of water and $z = -b$ denotes the elevation of the bed. This water pressure applies to the situation where water flows unimpeded under the glacier, that is, a full and easy connection exists between water under the glacier and the open sea.

Summarizing the foregoing, an adequate and realistic description of basal sliding and water pressure under the ice remains lacking. The modified Weertman sliding relation (equation (6.46)), with the effective pressure calculated based on the assumption of full and easy access to the ocean (equation (6.49)) has been used frequently and appears to have gained status of 'empirically verified'. This status is not fully warranted, however, and what is needed is greater effort by the glaciological community to establish more realistic basal boundary conditions for use in numerical models.

6.3 Hierarchy of models

6.3.1 Introduction

The previous section discussed the tools available to glaciologists to model the dynamic response of glaciers to changes in external forcings, including surface mass balance. The challenge faced by modellers is to strike the right balance between model complexity and including the relevant physical processes. In theory, it is possible to construct a numerical model that involves no assumptions other than about the form of the flow law and sliding relation. As discussed by Van der Veen (1989), stresses throughout a glacier can be calculated from the force balance equations if a sliding relation relating sliding speed to basal drag is adopted. This model could be coupled to a temperature model to account for the temperature dependence of the rate factor. However, such a model would require computational resources that extend well beyond the limits currently available. Consequently, it becomes necessary to simplify the model by introducing certain assumptions that reduce the number of computational steps involved. While this is a legitimate procedure, one should remain cognizant of the implications. Consider as an example a numerical model driven by a balance between driving stress and basal drag, the so-called *lamellar flow* model. If basal sliding is not included, the model glacier may be expected to be notoriously sluggish, with comparatively small and slow response to large climate forcing. The reason for this is that the equilibrium profile of a glacier in which basal drag supports the driving stress is rather insensitive to the surface mass balance, and this insensitivity is reflected by the model. Thus, the *a priori* assumption of lamellar flow dictates to a large extent the type of model behaviour one can expect when running a numerical time-evolving model.

The force balance equation (6.16) identifies three sources for the resistance to flow, namely basal drag, lateral drag and gradients in longitudinal stress. Retaining any of these, or perhaps a combination of two, leads to different models that apply to different glaciers. For example, lamellar flow is appropriate when considering the flow in the interior regions of polar ice sheets, while a combination of gradients in longitudinal stress and lateral drag best describes floating ice shelves.

6.3.2 Lamellar flow

In the conventional model for glacier flow, the driving stress is taken to be balanced by drag at the glacier bed. The force balance equation (6.16) reduces to

$$\tau_{dx} = \tau_{bx}. \tag{6.50}$$

In this model, the only non-zero resistive stress is that associated with vertical shear, and this stress increases linearly with depth, from zero at the glacier surface to the maximum value – equal to the driving stress – at the bed (Van der Veen, 1999a, p. 103). Thus,

$$R_{xz}(z) = \frac{h-z}{H}\tau_{dx} = -\rho g(h-z)\frac{\partial h}{\partial x}, \tag{6.51}$$

$$R_{yz}(z) = \frac{h-z}{H}\tau_{dy} = -\rho g(h-z)\frac{\partial h}{\partial y}, \tag{6.52}$$

with $z = 0$ at sea level and $z = h$ representing the upper ice surface.

The only non-zero strain rates are the vertical strain rates, $\dot{\varepsilon}_{xz}$ and $\dot{\varepsilon}_{yz}$, and these are related to the shear stresses through the flow law (equation (6.29)). Neglecting along-flow gradients in the vertical velocity, the shear strain rate equals half the vertical gradient in ice velocity (equation (6.34)) and

$$\frac{\partial u}{\partial z} = 2\left(\frac{R_{xz}}{B}\right)^n, \tag{6.53}$$

$$\frac{\partial v}{\partial z} = 2\left(\frac{R_{yz}}{B}\right)^n. \tag{6.54}$$

Noting that $B^{-n} = A$, the vertical gradients in the two horizontal components of velocity are

$$\frac{\partial u}{\partial z} = -2A(\rho g)^n(h-z)^n\left[\left(\frac{\partial h}{\partial x}\right)^2 + \left(\frac{\partial h}{\partial y}\right)^2\right]^{n-1/2}\frac{\partial h}{\partial x}, \tag{6.55}$$

$$\frac{\partial v}{\partial z} = -2A(\rho g)^n(h-z)^n\left[\left(\frac{\partial h}{\partial x}\right)^2 + \left(\frac{\partial h}{\partial y}\right)^2\right]^{n-1/2}\frac{\partial h}{\partial y}. \tag{6.56}$$

Integrating these expressions from the bed ($z = h - H$) to some depth z in the ice gives the velocity at that depth. Introducing the vector of surface slope,

$$\nabla h = \left(\frac{\partial h}{\partial x}, \frac{\partial h}{\partial y}\right), \tag{6.57}$$

the horizontal velocity vector at depth z is

$$\mathbf{u}(z) = -2(\rho g)^n (\nabla h \cdot \nabla h)^{(n-1)/2} \nabla h \int_{h-H}^{z} (h - \bar{z})A \, d\bar{z} + U_{\mathrm{b}}, \tag{6.58}$$

in which U_{b} represents the basal slip vector. The horizontal ice flux is obtained by integrating this expression over the full ice thickness:

$$HU = -2(\rho g)^n (\nabla h \cdot \nabla h)^{(n-1)/2} \nabla h \int_{h-H}^{h} \int_{h-H}^{z} (h - \bar{z})A \, d\bar{z} \, dz + HU_{\mathrm{b}}. \tag{6.59}$$

The integral on the right hand side can be evaluated if the depth variation of the rate factor is known. This requires the vertical temperature profile to be either prescribed or calculated.

The conditions under which the lamellar flow assumptions are valid can be assessed by conducting a scaling analysis of the force balance equations (e.g. Hutter, 1983). Generally, the glacier geometry must be simple, with the surface and bed topographies resembling plane slabs. In reality, glaciers are much more complex, but nevertheless the lamellar flow solution remains a good approximation for ice flow in the interior regions of ice sheets.

6.3.3 Including lateral drag

On glaciers that are laterally bounded by rock walls or, as in the case of some of the West Antarctic ice streams, by almost stagnant ice, lateral drag may balance part or all of the driving stress, and basal drag must be smaller than the driving stress. To account for this effect, Nye (1965) introduced the *shape factor, f.* This factor represents the proportion of driving stress that is supported by drag at the glacier bed, and

$$\tau_{\mathrm{bx}} = f \tau_{\mathrm{dx}}. \tag{6.60}$$

The remainder of resistance to flow, $(1 - f)\tau_{\mathrm{dx}}$, originates at the lateral margins where the ice moves past rock walls or stagnant ice. The shape factor for a very wide glacier on a rough bed is thus unity, and it is less for narrow glaciers.

There are two methods for calculating the shape factor for a given glacier geometry. The first possibility is to calculate a value for f that produces the correct surface velocity at the glacier centre line if lamellar flow is assumed. The 'correct' velocity can be obtained by explicitly computing the stress distribution in a cross-section (Nye, 1965; Van der Veen, 1999a, sect. 5.4). The second possibility is to calculate the shape factor from the requirement that forces must balance over the entire width of the glacier. In the first method, a local value for f is calculated, whereas the second method yields a width-averaged shape factor. The two methods yield values that differ by 10 to 20% depending on the shape of the basin (Nye, 1965; Van der Veen, 1999a, section 5.4). Raymond (1980, fig. 10) uses Glen's flow law and volume flux continuity considerations to estimate that the shape factor varies between 0.4 and 0.7 along Variegated glacier, Alaska. For Vernagtferner, a small valley glacier in Austria, Kruss and Smith (1982) use the shape factor as a tuning parameter, and find that its value is about 0.5 but varying along-flow. The numbers indicate that, in these cases, lateral drag and basal drag contribute more or less equally to opposing the driving stress.

While the shape factor is an expedient technique for incorporating lateral drag into a (numerical) model, it is not fully realistic because its value is commonly taken to be independent of glacier speed. In reality, one would expect lateral drag to become more important as the glacier speeds up. This interaction can be included by relating the shape factor to the glacier speed at the centre line. To achieve this, two important assumptions need to be made about how lateral drag varies with depth and across the glacier. The first assumption is that lateral drag does not vary with depth or, equivalently, that an appropriately weighted depth-mean value may be used in the force balance equation. The second assumption is that, across the glacier, lateral drag is equally important and supports the same fraction of driving stress at any point across the glacier. Both assumptions may be challenged, in particular the second one. Full stress calculations for a rectangular cross-section show that, in the absence of basal sliding, flow resistance from lateral drag is greatest near the glacier margins (Van der Veen, 1999a, fig. 5.9). On the other hand, that calculation does not include the warming effect from concentrated lateral shear which tends to reduce the stress associated with a particular strain rate, which would make lateral drag more constant in the transverse direction.

Considering the balance equation (6.16), resistance to flow is described by the last term on the right hand side. Averaged over the half-width, W, of the glacier, lateral resistance is

$$\overline{-\frac{\partial}{\partial y}(HR_{xy})} = -\frac{1}{W} \int_{-W}^{0} \frac{\partial}{\partial y}(HR_{xy})\,dy = -\frac{1}{W}[HR_{xy}(0) - HR_{xy}(-W)]. \qquad (6.61)$$

The ice velocity is taken to be symmetric around the centre line ($y = 0$) so that the lateral shear stress is zero there. Making the further simplification that the shear stress at both margins is equal (but of opposite sign), with value τ_s, and denoting the thickness at the lateral margins by H_W, lateral resistance acting on a section of glacier of unit width is then equal to $H_W \tau_s / W$, and the width-averaged force balance equation becomes

$$\bar{\tau}_{dx} = \bar{\tau}_{bx} + \frac{H_W \tau_s}{W}. \qquad (6.62)$$

Note that in these expressions the overbar represents width-averaged values.

The next step is to link the value for the shear stress at the margins, τ_s, to the glacier speed. To do so, several simplifying assumptions are made, as noted above. Firstly, the driving stress and ice thickness are assumed to be constant in the transverse direction and lateral drag supports the same fraction of driving stress across the glacier. It then follows from the balance equation (6.16) that the shear stress, R_{xy}, varies linearly across the glacier:

$$R_{xy}(y) = ay + b. \tag{6.63}$$

The constants follow from the boundary conditions that the shear stress must be zero at the centre line (thus, $b = 0$) while the value at the margin equals R_{xy}. From the balance equation (6.62) it follows that:

$$\tau_s = \frac{W}{H}(\tau_{dx} - \tau_{bx}), \tag{6.64}$$

and the shear stress across the glacier becomes

$$R_{xy}(y) = -\frac{y}{H}(\tau_{dx} - \tau_{bx}). \tag{6.65}$$

Invoking the flow law, equation (6.29), this stress can be linked to the transverse gradient in ice velocity. Neglecting the contribution of vertical shear to the effective strain rate, the flow law reduces to

$$R_{xy} = B\left(\frac{1}{2}\frac{\partial u}{\partial y}\right)^{1/n}, \tag{6.66}$$

or

$$\frac{\partial u}{\partial y} = 2\left(\frac{R_{xy}}{B}\right)^{n}. \tag{6.67}$$

Substituting expression (6.65) for the shear stress, and integrating with respect to the transverse direction, y, gives the transverse velocity profile,

$$u(y) = u(0)\left(1 - \left(\frac{y}{W}\right)^{n+1}\right). \tag{6.68}$$

The centre-line velocity is given by

$$u(0) = \frac{2}{n+1}\left(\frac{\tau_{dx} - \tau_{bx}}{BH}\right)^{n}W^{n+1}. \tag{6.69}$$

Van der Veen and Whillans (1996) apply a model similar to the one described above to study evolution and stability of West Antarctic ice streams. To eliminate basal drag from equation (6.69) for the centre-line speed, these authors adopt a Weertman-type sliding relation, modified to account for the effective basal pressure which is approximated by the height above buoyancy. That is

$$\tau_{bx} = \mu A_s\left(H + \frac{\rho}{\rho_w}H_b\right)^{1/n}\bar{U}^{1/n}, \tag{6.70}$$

where μ is a friction parameter, A_s is a sliding parameter, ρ and ρ_w represent the densities of ice and sea-water, respectively, and H_b denotes the depth below sea level. In this expression, $\bar{U}$ represents the width-averaged ice velocity, obtained by integrating the transverse profile, equation (6.68) across the full width of the glacier. Using equation (6.70) to eliminate basal drag from the expression for the ice speed results in an equation that relates ice discharge to geometry parameters (driving stress, water depth, etc.), which can be used in the continuity equation to compute changes in ice thickness. Other forms of the sliding relation could be adopted, but the essential procedure remains the same as followed by Van der Veen and Whillans (1996).

6.3.4 Ice-shelf spreading

The third potential source of flow resistance opposing the driving stress is associated with gradients in longitudinal stress, or differential pushes and pulls in the direction of flow, described by the second term on the right hand side of the balance equation (6.16). Note that this term may act in co-operation with the driving stress if the ice is being pushed from up-glacier or pulled from down-glacier. On grounded glaciers, gradients in longitudinal stress contribute little to the balance of forces, even where the longitudinal stress itself becomes large. For example, Whillans and Van der Veen (1993a) use measurements of surface speed along a flow line extending from the head of the Whillans ice stream to the calving front of the Ross ice shelf to evaluate the partitioning of flow resistance among basal drag, lateral drag and gradients in longitudinal stress. They find that longitudinal tension or compression is unimportant along the entire length of flow line considered, except in the vicinity of the Crary ice rise. From theoretical considerations, this was to be expected. Applying scale analysis to estimate the relative magnitude of the terms in the balance equation, it can be shown that gradients in longitudinal stress – often referred to as the 'T-term' – become negligible when evaluated over horizontal distances of several ice thicknesses. Thus, it appears that, on grounded glaciers, this source of flow resistance need not be considered if the large-scale dynamical behaviour is of interest.

Several modelling studies have attempted to incorporate longitudinal tension and compression. Alley and Whillans (1984) apply a non-steady ice-flow model including longitudinal stresses to investigate the stability of the East Antarctic ice sheet to sea-level rise. A similar model was used by Van der Veen (1985, 1987) to study the effect of changes in back-stress at the grounding line on the discharge of interior ice. Greuell (1989, 1992) includes longitudinal stresses in a numerical model of the Hintereisferner, Austria. All of these modelling studies suggest that including longitudinal stresses into a numerical model has little effect on the model behaviour, thus supporting the conclusions from theoretical studies.

While, on grounded glaciers, longitudinal stress gradients play a minor role in the balance of forces, the situation is different on floating ice shelves and ice tongues, where basal drag is zero. On these glaciers, resistance to flow is partitioned between lateral drag and gradients in longitudinal stress. In the case of a free-floating ice shelf, the force balance equation reduces to

$$\tau_{dx} = -\frac{\partial}{\partial x} \int_{h-H}^{h} R_{xx} \, dz. \tag{6.71}$$

Substituting equation (6.8) for the driving stress and integrating with respect to the flow direction, x, gives, after invoking hydrostatic equilibrium to link the surface elevation to ice thickness,

$$R_{xx}^{(0)} = \frac{1}{2}\rho g \left(1 - \frac{\rho}{\rho_w}\right) H. \tag{6.72}$$

The superscript (0) indicates that this solution refers to a free-floating 'Weertman-type' ice shelf. Making the assumption that the ice shelf is spreading in the direction of flow only (that is, an ice shelf in a parallel-sided bay is considered with zero or small side-wall friction), the stretching rate is found to be (Van der Veen, 1999a, section 5.6; Weertman, 1957b)

$$\dot{\varepsilon}_{xx}^{(0)} = \left[\frac{\rho g}{4B}\left(1 - \frac{\rho}{\rho_w}\right)\right]^n H^n. \tag{6.73}$$

This expression can be modified to allow for spreading in the other horizontal direction, but this only affects the constant factor on the right hand side, and the stretching rate remains proportional to the nth power of ice thickness (Thomas, 1973; Van der Veen, 1999a, sect. 5.6).

Few, if any, floating glaciers are supported by gradients in longitudinal stress only. Instead, most ice shelves and ice tongues have formed in embayments, and lateral drag provides some additional restraint to oppose the driving stress. Again making the assumption that lateral drag is equally important across the width of the ice shelf, the balance equation becomes, analogous to equation (6.62),

$$\tau_{dx} = -\frac{\partial}{\partial x}(H R_{xx}) + \frac{H\tau_s}{W}. \tag{6.74}$$

Following the same procedure as above, the lateral shear stress at the margins can be related to the width-averaged ice velocity, and

$$\tau_s = B\left(\frac{5}{2}\frac{\bar{U}}{W}\right)^{1/n}. \tag{6.75}$$

Neglecting the contribution of lateral shear to the effective strain rate, the force balance equation becomes

$$\tau_{dx} = -\frac{\partial}{\partial x}\left[2B\left(\frac{\partial \bar{U}}{\partial x}\right)^{1/n}\right] + \frac{BH}{W}\left(\frac{5}{2}\frac{\bar{U}}{W}\right)^{1/n}, \tag{6.76}$$

and this expression can be solved iteratively to find the discharge along the ice shelf.

The ice shelf model summarized above applies to one-dimensional flow that is laterally constrained by rock walls or perhaps stagnant ice. The model can be expanded to allow for varying width, but the essential assumption is that transverse flow in the y-direction can be ignored. This may be a reasonable assumption for floating ice tongues on the periphery of the Greenland ice sheet, most of which have formed in fjords, but does not apply to the large ice shelves surrounding the Antarctic ice sheet. For example, The Ross and Ronne ice shelves are nourished by several ice streams and outlet glaciers, and flow is more complex than can be described by considering a single flow line only. To account for the two-dimensional nature of the flow of these ice shelves, the balance equations in both horizontal directions need to be considered and solved iteratively for the velocity distribution.

Neglecting vertical shear (Sanderson and Doake, 1979) and taking horizontal strain rates constant with depth, the two horizontal balance equations expressed in terms of velocity gradients are (Huybrechts, 1992; Van der Veen, 1999a, sect. 8.8)

$$\frac{\tau_{dx}}{B} = -\frac{\partial}{\partial x}\left(H\dot{\varepsilon}_e^{1/n-1}\left(2\frac{\partial U}{\partial x} + \frac{\partial V}{\partial y}\right)\right) - \frac{\partial}{\partial y}\left(H\dot{\varepsilon}_e^{1/n-1}\frac{1}{2}\left(\frac{\partial U}{\partial y} + \frac{\partial V}{\partial x}\right)\right),$$

(6.77)

$$\frac{\tau_{dy}}{B} = -\frac{\partial}{\partial x}\left(H\dot{\varepsilon}_e^{1/n-1}\frac{1}{2}\left(\frac{\partial U}{\partial y} + \frac{\partial V}{\partial x}\right)\right) - \frac{\partial}{\partial y}\left(H\dot{\varepsilon}_e^{1/n-1}\left(\frac{\partial U}{\partial x} + 2\frac{\partial V}{\partial y}\right)\right),$$

(6.78)

in which the effective strain rate is given by

$$\dot{\varepsilon}_e^2 = \frac{1}{2}\left(\frac{\partial U}{\partial x}\right)^2 + \frac{1}{2}\left(\frac{\partial V}{\partial y}\right)^2 + \frac{1}{4}\left(\frac{\partial U}{\partial y} + \frac{\partial V}{\partial x}\right)^2.$$

(6.79)

Because the balance equations contain velocity gradients in both horizontal directions, the velocity at any grid point depends on velocities at neighbouring grid points and a solution has to be found using numerical iteration techniques. For example, Huybrechts (1992) uses the point relaxation technique to solve this set of coupled equations for the two components of velocity.

6.3.5 Ice shelf/ice sheet interaction

One particular issue that has received considerable attention, and that continues to be debated, is the stability of the marine-based west Antarctic ice sheet. Many glaciologists have argued that this ice sheet, which rests on a former sea floor well below sea level, is inherently unstable and may respond drastically to changes in climate (Bentley, 1983, 1984; Hughes, 1973, 1983, 1992; Lingle, 1984, 1985; Mercer, 1968, 1978; Thomas, 1979; Thomas and Bentley, 1978; Thomas, Sanderson and Rose, 1979; Weertman, 1974). According to this view, the peripheral ice shelves exert some stabilizing *back-stress* on the inland ice. Reducing or eliminating altogether this back-stress would lead to greatly increased discharge from the interior and, because a grounding line is unstable if the bed slopes downward toward the ice-sheet interior (as is the case in west Antarctica), may lead to irreversible grounding line retreat and possibly the collapse of the entire ice sheet in west Antarctica.

The basic idea behind the marine instability hypothesis is simple indeed, and centres on the notion of *creep thinning*. Considering flow along a flow line, the continuity equation (6.45) can be written as

$$\frac{\partial H}{\partial t} = -U\frac{\partial H}{\partial x} - H\frac{\partial U}{\partial x} + M.$$

(6.80)

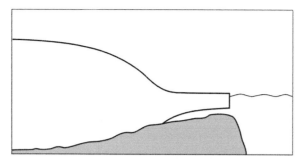

Figure 6.7. Illustrating an unstable grounding line when the bed slopes downward toward the ice-sheet interior.

The second term on the right hand side represents creep thinning or, equivalently, thinning associated with along-flow gradients in ice velocity. From the definition of strain rates, this term is equal to the product of ice thickness and stretching rate, $H\dot{\varepsilon}_{xx}$. At the grounding line, the ice becomes afloat and the stretching rate can be estimated from the ice-shelf solution. Assuming for now that the ice shelf is spreading freely, the stretching rate is given by equation (6.73) and, taking $n = 3$, creep thinning is proportional to the fourth power of thickness:

$$H\dot{\varepsilon}_{xx} = CH^4, \tag{6.81}$$

in which C includes the constants in equation (6.73). Now consider the situation where the bed below the grounded glacier slopes down toward the interior, as shown in Figure 6.7. If the grounding line advances for some reason, the thickness at the grounding line decreases, thus strongly reducing creep thinning, thereby allowing the grounding line to advance further. On the other hand, if some process makes the grounding line retreat, the thickness at the transition from grounded ice to floating ice increases, thereby greatly increasing creep thinning, causing the grounding line to retreat further. Thus, because creep thinning depends critically on the ice thickness at the grounding line, a positive feedback exists that amplifies initially small perturbations at the grounding line. In the case where the bed slopes down toward the ocean, the feedback is negative and the position of the grounding line remains more stable.

As noted above, the major Antarctic ice shelves have formed in embayments, and part of the driving stress is opposed by lateral drag originating at the lateral margins or where the ice moves past obstacles such as ice rises. The effect is to reduce the stretching rate compared with the Weertman solution, equation (6.73). Defining the back-stress as the integrated resistance to flow from lateral drag,

$$\sigma_b = \frac{1}{H} \int_x^L \frac{H\tau_s}{W} \, d\bar{x}, \tag{6.82}$$

where $x = L$ denotes the position of the calving front, the stretching rate is given by

$$\dot{\varepsilon}_{xx} = \left(\frac{R_{xx}^{(0)} - \sigma_b}{B} \right)^n.$$ (6.83)

In other words, the effect of a constrained ice shelf is to reduce the stretching rate at the grounding line. Any reduction in back-stress, for example due to shortening of the shelf or thinning of the ice shelf, leads to increased creep thinning at the grounding line and, as explained above, this may lead to irreversible collapse of the ice sheet.

The difficulty with the marine instability model outlined above is that creep thinning is calculated based on ice-shelf dynamics, but the ice velocity is not affected by the presence of the ice shelf. In other words, the first term on the right hand side of the continuity equation (6.80) is not affected by a reduction in back-stress. However, all strain rates affect the ice velocity through the effective strain rate, and one may expect an increase in speed itself as the stretching rate increases. By not including this interaction, the model becomes inherently unstable, and a stable response to changes at the grounding line is prohibited. Using a numerical model in which the stretching rate at the grounding line affects the velocity through the effective strain rate, Van der Veen (1985, 1987) found a more stable ice-sheet response in which the marine instability is, to a large extent, countered by an increase in ice discharge from the interior.

The question of the stability of the West Antarctic ice sheet remains open. Certainly, the simple feedback envisioned in earlier models may be questioned and is neither supported by observational evidence or modelling studies. Nevertheless, the flow of this ice sheet remains poorly understood, and numerical models do not adequately account for the dynamics of fast moving ice streams. These ice streams can undergo important changes in relatively short time periods (Whillans, Bentley and Van der Veen, 2001), but the causes and controls on these changes are at present not well known.

6.4 Evaluating terrestrial ice-mass models

6.4.1 Terminology

The main objective of developing numerical or analytical models describing glacier flow is to gain better understanding of the behaviour of glaciers and their possible response to changes in external forcings. These models are based on the theory outlined in the preceding sections but usually contain simplifications to make a solution tractable. This is a fully acceptable procedure, provided the model is evaluated in some way and provided the model is not applied to situations in which the simplifications may not apply. It appears that only recently have glaciologists turned their attention toward evaluating the complex models developed over the past few decades. The EISMINT model inter-comparison discussed more fully in section 6.4.3 represents a good first step, but much remains to be done in the area of evaluation of glaciological models. For this reason, and because of its obvious importance, we discuss this issue in quite some detail here, in the hope that this will be of use to modellers and may bring some uniformity in how model results are evaluated.

Van der Veen (1999b) offers a review of some of the philosophical issues underlying the evaluation of ice-mass models. A major issue is the confusing terminology used in this area. In particular, words such as 'verification', 'validation' and 'testing' are often used interchangeably. It is therefore necessary to define the terminology, based on that proposed by Van der Veen (1999a). We advocate the use of four terms applicable to the process of model evaluation, namely verification, validation, calibration and confirmation. These terms have been introduced to other branches of science and appear to describe the different levels involved in the optimum way.

A model may be said to be *verified* if its ability to describe accurately natural processes under consideration has been unambiguously established. Several studies (referred to in Van der Veen, 1999b) have convincingly argued that models of natural systems cannot be verified. Consider ice-sheet models and, in particular, the commonly used flow law (equation (6.20)). This law is based on results from laboratory experiments and interpretation of field observations, and its form can be justified to some extent by theoretical arguments, but its applicability to real ice masses has not been proven conclusively. Indeed, it is theoretically impossible to demonstrate that Glen's flow law applies to real glaciers. Ice in these glaciers has undergone deformation for periods ranging from a few years to hundreds of thousands of years under stresses that are not known but likely much larger than can be achieved in laboratory settings. Further, processes acting on the scale of individual crystals, such as rotation of the *c*-axis, or recrystallization, cannot be incorporated into the flow law except in a bulk sense. To verify the flow law would require experimental observation of all of these processes, not only throughout the entire glacier being modelled, but throughout its entire history and future as well. And, of course, these observations would have to be conducted in such a way as not to interfere with the flow of the glacier.

With verification a theoretical impossibility, not only for ice-sheet models but, indeed, for any model describing a natural system, including Newton's laws of motion, the only remaining pragmatic choice is to relax the requirements for model evaluation and to demonstrate that some level of confidence can be placed in the predictions of the model in question. To do so requires that the model be validated and confirmed.

We propose that the term *validation* should be restricted to assessing the numerical model only, and to establish that the model does not contain any obvious or detectable flaws (Van der Veen, 1999b). That is, a conceptual model for glacier flow (e.g. the laminar flow model) is adopted and computer software developed to solve the mathematical equations describing the conceptual model. Often these equations are ordinary or partial differential equations, principally involving rates of change of variables that are assumed to vary continuously through both time and space. After reliable numerical software has been constructed that solves the underlying differential equations in a consistent manner, the predictions of this numerical model can be compared with observations of the processes and features that are being modelled. This comparison with real-world observations will be termed model *confirmation*, and is complicated by the fact that the solution of most numerical models will require some information about the real system, for instance initial conditions, boundary conditions and model forcing. Further, most models contain a number of poorly constrained parameters that can be adjusted to improve the match between model predictions and observations. This process of parameter tuning is called *calibration*. It is important to note that a successful calibration does not necessarily mean that the model is a realistic description of the physical processes being studied, but rather that the model contains enough adjustable parameters to make model predictions match observations. Confirmation of the model involves comparison of model predictions against observations that were not used in the calibration process. While absolute confirmation is not possible – as this would imply establishing the truth of the model, i.e. verification – the level of confidence that can be placed in the model will increase as the model is confirmed by increasingly more independent observations.

Before discussing the issues surrounding the various steps involved in the model evaluation process, we will quickly review the main types of cryospheric model in use today.

6.4.2 Types of ice-mass models

A wide variety of models has been used to study the dynamics and evolution of glaciers and ice sheets. Although all based on the physics outlined in the preceding sections of this chapter, they differ in which of the equations are explicitly included in the model (i.e. the physics of the model), the way these equations are solved (the numerics of the model), as well as the dimensions of the model domain (whether the model is time-dependent or steady-state and whether it explicitly incorporates the vertical dimension as well as one or both of the horizontal, spatial dimensions).

Since the 1980s, two main schools of ice-mass model have developed, namely prognostic and diagnostic models. They differ primarily in the physics that are incorporated, but have coincidental differences in the numerics employed and the dimensionality and nature of the problems considered. Prognostic models developed primarily from the work of the Melbourne group led by Prof. W. F. Budd in the late 1970s, and gained significant impetus from the publication of *Ice Sheets and Climate* by Oerlemans and Van der Veen (1984) and the work of Huybrechts (1992). Diagnostic models have been developed by many researchers to understand particular processes better or to study the flow regime of individual glaciers.

Prognostic models

This type of model has principally been used to address the response of ice masses to an imposed climate change. In this context, the key interactions being investigated are between the (heavily) parameterized effects of a change in climate on the accumulation and ablation fields experienced by an ice mass, and the ice mass's response in terms of changed geometry (ice thickness and surface slope) and flow. More recently, this type of model has been used to investigate the potential for internally generated flow instability (i.e. surging) (Marshall and Clarke, 1997; Pattyn, 1996; Payne, 1995, 1999). In this application, the crucial interactions are between the thermal and flow regimes of an ice mass. The basic set of equations solved by these models is listed below.

(1) Diagnostic equations for both components of horizontal velocity as algebraic functions of local ice geometry (in particular, ice thickness and surface slope) and ice rheology (namely the rate factor in Glen's flow law; equation (6.34)) derived from the assumption that basal resistance alone balances gravitational driving forces (equations (6.57) and (6.58)).
(2) A prognostic equation for internal ice temperature evolution, given appropriate boundary conditions at the upper (mean annual air temperatures) and lower (geothermal heat flux as well as any heat generated by basal slip over an immobile substratum) ice surfaces (equation (6.44)).
(3) A diagnostic equation for vertical velocity based on the divergence of the horizontal velocity fields generated in equation (6.1).
(4) A prognostic equation for ice thickness as the difference between snow accumulation and snow/ice melt, and the convergence and divergence of horizontal ice flow (equation (6.44)).

Additional parameterizations are required to relate ice rheology (the rate factor A) to the evolving characteristics of the ice (for instance, its temperatures and crystal fabric), and to provide the horizontal velocity at the ice/bed interface (basal slip or sliding), mean annual air temperatures, and snow accumulation and snow/ice melt (i.e., net mass balance), as well as to incorporate the effects of isostatic bedrock movements under the load of the ice mass. Chapter 5 describes the development of the models necessary to determine net mass balance, and the reader is referred to that chapter for further details. It should, however, be noted here that the models incorporated into large-scale, ice-mass simulations of the type discussed here invariably tend toward the simpler, highly parameterized end of the spectrum of the model described in Chapter 5. In particular, the degree-day method is most often used to predict ablation rates, and accumulation rates are typically prescribed from observations with a simple altitudinal correction. This is principally because of the dearth of meteorological data available to drive these models.

Component (1) makes use of the most basic assumptions about the force balance within the ice mass (see section 6.3.2). Components (1) and (4) are, in practice, combined by making use of the local, algebraic form of equation (6.57) to produce a non-linear parabolic equation for ice-thickness evolution (where the non-linearity enters through an effective diffusivity coefficient that depends on ice thickness to the fifth power and the ice-surface slope squared; e.g. equation (6.58)). This type of model therefore has at its heart the solution of two evolutionary equations (for ice thickness and temperature).

Prognostic models are typically implemented using finite-difference techniques on a regular grid of nodes in the two horizontal dimensions, and using a stretched co-ordinate system in the vertical (Jenssen, 1977). Finite-element implementations exist (e.g. Hulbe and MacAyeal, 1999), although these are often performed on a regular grid (e.g. Fastook and Prentice, 1994). This school of model has been used to study the dynamics of the Greenland and Antarctic ice sheets, and to predict both their evolution through glacial–interglacial cycles (over timescales of hundreds of thousands of years) as well as to assess their response to future climate change. Chapters 11 and 13 report these results in detail. This type of model has also been used to study valley glaciers, in which case only one horizontal dimension is incorporated (parallel to the flow of the glacier), and the effects of variations in glacier width are incorporated by modifying the prognostic equation for ice-thickness evolution.

Diagnostic models

The second type of ice-mass model in wide use today differs in two main respects from that described above. Firstly, the models are normally diagnostic and do not include the time evolution of the ice mass. The second difference is that a far more detailed model of the internal stress regime (force balance) of the ice mass is used. In particular, the key assumption that basal resistance alone balances gravitational driving forces is relaxed so that a more complete force balance is considered in which contributions from lateral and longitudinal stresses are included. These two differences are not unconnected in that the amount of computer-processing power typically required to solve these extended force balances in diagnostic mode precludes the additional computational expense of incorporating time evolution (in which the force balance would have to be solved for each of thousands of time steps).

This type of approach is essential in modelling the flow of floating ice shelves, and the initial applications were limited to this area (e.g. Thomas and MacAyeal, 1982). In this case, the force balance can be simplified considerably because the underlying ocean water cannot exert any traction on the ice mass. Basal traction is therefore zero, and vertical shearing can also be assumed to be minimal. This reduces the complexity of the problem considerably because vertical averages represent the system accurately (reducing a three-dimensional problem to a two-dimensional planform one; equations (6.77)–(6.79)). More recently, this type of model has been applied to ice streams. The assumption that vertical shear is minimal is more equivocal in this case. It seems likely that this is the case for the ice streams of the Siple Coast region in West Antarctica (MacAyeal, 1989) but cannot be generally assumed (e.g. Joughin *et al.*, 2001). This type of model typically employs the finite-element technique because of the very irregular spatial extent of the features being modelled (ice shelves and streams), although finite-difference implementations do exist. In particular, the force balance approach developed by Van der Veen and Whillans (1989) is based on finite differences, and has proven to be very useful in understanding the relative importance of lateral drag and basal traction in the ice streams of the Siple Coast (Whillans and Van der Veen 1993a, 1997).

More recently, models have been developed that do not rely on the assumption of negligible vertical shear and solve the full three-dimensional problem. This type of model is still very much in its infancy, but offers the possibility of a general modelling framework that does not require the *a priori* classification of ice flow into shelf-like flow or fully grounded flow (in which a local gravity–basal traction balance can be assumed). This type of model also has applications in studying the flow of valley glaciers, where the normal assumptions about flow regime (either of a local force or of negligible vertical shear) are untenable. Examples of this approach include the work by Herterich (1987) on the grounding zone, as well as applications to valley glaciers by Blatter (1995) and Gudmundsson (1999).

It should be noted that a few 'whole ice sheet' models exist that combine the two groups of model discussed above and couple grounded ice-mass models of the first type with more complete stress balance models of the second type (in particular, the ice-stream and ice-shelf components are modelled using this approach and are also allowed to evolve prognostically).

Examples include the models of the Antarctic ice sheet by Huybrechts and De Wolde (1999) and Ritz, Rommelaere and Dumas (2001).

6.4.3 Model validation

The process of constructing a working computer-based model from continuous equations can itself be divided into a number of stages. The most widely discussed is the process by which the continuous, partial differential equations that describe the model conceptually are transformed into a form that is tractable by a computer. The two principal methods used to make this transition are the finite-element and finite-difference techniques. Both are standard methods and we will not discuss their details further, other than to note that both aim to replace the continuous forms with a form where variables are known only at distinct, finite (both temporally and spatially) locations. In the case of finite differences, this discretization process generates equations related to points in space, while the finite-element technique generates equations relating to conservation over an area. Other methods of analysis do, however, exist and are often complementary to the more traditional discretization-based approaches. For instance, considerable progress has recently been made in the application of normal-model analysis to models of the cryosphere (Hindmarsh, 1997).

Although the discretization process described above has received much attention in the literature, it should be kept in mind that it is not the only stage involved in producing a reliable computer-based model. The generation of error-free computer programs is far from trivial given the complexity of the current generation of numerical ice-mass models. Indeed, much of the work involved in the EISMINT benchmark centred around the identification of programming errors and the problems associated with defining an unequivocal description of the experiments to be performed.

The EISMINT inter-comparison

The European ice sheet modelling initiative (EISMINT) model validation exercise is the only attempt by the glaciological modelling community to validate different models in a consistent fashion. The majority of large-scale ice-flow models have been developed by individual researchers or within small institution-based groups. This 'do-it-yourself' approach is in stark contrast to the large community-based models that dominate research in the counterpart fields of meteorology and oceanography.

EISMINT was developed to address this diversity of existing models. It had two main aims. Firstly, to perform a model inter-comparison exercise so that a concensus could be reached on the basic predictions of the models. Secondly, it was hoped that this process would help to identify areas of good practice in modelling. It would also help to identify any obvious flaws in individual existing models and provide a useful benchmark to speed the development of future models. The EISMINT project was funded by the European Science Foundation in two phases which ran from May 1993 to September 1995, and from January 1996 to December 1997.

Model inter-comparisons were made at three levels. The first level was aimed solely at understanding the effects of numerical implementation on model prediction. For this reason, the physics incorporated into the models was tightly constrained, as were the values of the various parameters and boundary conditions employed in the models. The separate origins of many of the existing models has led to a wide variety of numerical techniques being employed. Several finite-element models exist, although the majority of ice-sheet models employ finite differences. Within the latter class, a wide variety of methods are employed to deal with the basic non-linearity of ice flow (see above), so that fully implicit, semi-implicit and explicit discretizations all exist, as well as models which incorporate one or both horizontal dimensions, and may or may not include the vertical.

The second level of inter-comparison allowed individual modellers to use their pre-ferred values for the various model parameters. The aim of this inter-comparison was to determine the effect that the many poorly constrained physical parameters have on the over-all prediction of the models. The final level of inter-comparison was to model 'real' ice masses through a given climate-change scenario. Examples included a long-term steady-state, the response to stepped changes in forcing, the response to glacial–inter-glacial climate change and the response to future anthropogenically driven climate change. This latter ex-ercise focussed on the two present-day ice sheets of Greenland and Antarctica, as well as the response of valley glaciers. The aim here was to test the predictions of complete models and therefore not to prescribe the details of individual experiments completely. In this way, models which had already been used to generate predictions on ice-sheet response to climate change could be compared fully. The aim of the various levels of inter-comparison therefore progressed from details of numerical implementation toward whole-model assessment. It is important to emphasize that the whole EISMINT exercise was principally driven by model inter-comparison and not by comparing against real-world data sets. This was principally because of the paucity of real-world data available for such testing.

EISMINT levels one and two

A 'level one' inter-comparison of existing ice-sheet models is presented by Huybrechts *et al.* (1996). A series of experiments was presented which could be adapted for mod-els with one and two horizontal dimensions, and which may or may not include internal temperature evolution. The boundary conditions and parameter values used in the models were prescribed as tightly as possible. A major simplification was that, although internal temperature calculations could be incorporated, their effect on ice viscosity was omit-ted. The methodology was therefore to reduce potential differences between models to a minimum so that any observed differences could be interpreted purely in terms of model numerics.

The actual inter-comparison exercise progressed through a number of stages in which in-dividual groups re-submitted their results and a consensus set of results gradually emerged. This process was found to be necessary because many of the differences initially identified between models actually arose because of trivial reasons (such as ambiguity in the experi-ment descriptions). This process was not without its dangers, in that results inevitably tended to converge toward median values as outlying models were modified. The modification process may have been genuine (for instance in finding coding errors) but may also have been driven by a perceived need to conform to the bulk of the models. In this way, genuine reasons for difference may have become obscured. This was partially avoided, in some cases, by the existence of certain analytical solutions to the equations describing ice-sheet flow. The design of the experiments recognized the existence of these solutions, which to a large extent can be taken as 'truth'. Unfortunately, the number of analytical solutions available is very limited and applies to highly simplified situations. Nonetheless, their use was an important feature of the inter-comparison.

In general, the models agreed closely in terms of how ice thickness and internal temperature evolved and in the predicted velocity patterns within the ice mass. Interpretation of the remaining scatter between models yielded useful information on the effect of spatial discretization on model prediction. Two groups of models were found to exist within the results based on the horizontal discretization used to incorporate ice flow. Models which employed a more stable discretization were found to be less accurate in comparison to a more precise formulation which is, however, less numerically stable. This difference can be interpreted in terms of the amount of spatial averaging implicit in a particular numerical solution scheme, and is discussed further by Hindmarsh and Payne (1996). The Huybrechts *et al.* (1996) inter-comparison and those discussed below provide useful benchmarks for future glaciological modelling. They also provide an effective means of testing models under development (this is important given the limited number of analytical solutions against which to test) and also highlight potential difficulties in the numerical implementation of the ice-flow equations.

The major shortcoming of the Huybrechts *et al.* (1996) inter-comparison is that the ice-flow and temperature calculations evolved separately. While this eased the complexity of the inter-comparison task, it represents an unrealistic assumption and does not reflect current model usage. A subsequent exercise (Payne *et al.*, 2000) dealt with the effects of thermo-mechanical coupling. The experimental design was similar to the Huybrechts *et al.* (1996) inter-comparison so that all boundary conditions were symmetrical. The results, however, showed much more variability between models. Most models showed signs of spatial patterns in the basal temperature distribution, which, because of the symmetry of the boundary conditions, must have been generated internally within the ice-flow equations. One likely mechanism is the interaction between ice temperature and flow, or 'creep instability' (Clarke *et al.*, 1977). The presence of the patterning is a fairly robust feature of all models, but its details vary between models. This implies that the numerical implementation is, to a certain extent, guiding the development of the patterning. The presence of high frequency noise in some of the model results also implies that some of the models are experiencing numerical difficulties.

EISMINT level three

Simulations of the Greenland and Antarctic ice sheets, as well as various valley glaciers and the Ross ice shelf, Antarctica, comprised the third level of the EISMINT exercise. The aim of this level was to compare model predictions of the past and future evolution of these ice masses with a minimum of constraints on the details of the individual models. The details of the Antarctic inter-comparisons will not be discussed here because the number of models participating in these exercises was relatively small. The ice-shelf exercise is described in MacAyeal *et al.* (1996).

Nine groups participated in the Greenland inter-comparison. Experiments consisted of simulating the past evolution of the ice sheet over the last two glacial–inter-glacial cycles (the last 250 000 years). In addition, response to various climate-change scenarios over the next 500 years was simulated. Realistic boundary conditions were employed for both the underlying bedrock topography and the climate forcing over the period of interest (obtained from the various Greenland summit ice cores). All models simulated the basic geometry of the ice sheet in a very realistic fashion, with the positions of all of the major ice domes and divides being modelled correctly.

Results highlighted the importance of the parameterization used to model ice/snow-melt (ablation) and precipitation, and their relationships to air temperature. The areal extent of the ice sheet is principally determined by the predicted ablation within the model. There was a certain degree of consensus on the timing and magnitude of areal extent changes over the last 250 000 years before present (BP), and the present-day and the last inter-glacial (the Eemian) were found to represent minima in ice extent. In contrast, the average thickness of the ice sheet is determined principally by the precipitation rate. This is because the area experiencing ablation only represents a narrow fringe around the ice sheet, while the majority of the ice sheet is affected by accumulation directly. The relationship between air temperature and ablation and precipitation is complex: increased air temperatures imply greater ablation rates but also greater accumulation rates (because of the increased moisture content of warm air). This led to the counter-intuitive result (which was confirmed by all models) that the last glacial maximum (LGM at 21 000 years BP) was a period when areal extent was at a maximum (reduced ablation) but the ice sheet was relatively thin (divide thickness 3150 ± 150 m). In contrast, the Eemian ice mass was relatively less extensive but was thicker (divide thickness 3300 ± 100 m). The present-day ice mass appears to be intermediate in terms of both area and thickness (divide thickness 3200 ± 100 m). It is encouraging that the majority of models were able to simulate the various conflicting effects consistently.

The contribution of the world's valley glaciers to future sea-level change is investigated by Oerlemans *et al.* (1998). Although this work was part of the EISMINT project, it does not address model inter-comparison but concentrates on the modelled response of 12 valley glaciers to imposed climate warming. A mass balance model for each glacier was separately calibrated using the past variation of the glacier's length. A variety of linear warming rates were then imposed with and without concurrent changes in precipitation. Differences between the predicted responses of individual glaciers are large and are created mainly by differences in their hypsometry. Results from the scenarios vary greatly from instances where few glaciers will survive past the year 2100, to a relatively minor (10 to 20%) loss in volume from 1990 levels.

Conclusions

Three main scientific conclusions sprang from the EISMINT inter-comparison exercise. Firstly, the current generation of ice-sheet models are generally consistent with one another and with available analytical solutions. The main area of concern lies in their ability to cope with thermo-mechanical coupling. The Payne *et al.* (2000) inter-comparison highlights major differences between models when ice flow and temperature are coupled. The reason for these differences has recently been addressed by Hindmarsh (2001), who (on the basis of a normal-mode analysis) suggests that the process of solving the underlying thermodynamic equations numerically may be much more complicated than previously thought, and that details of individual numerical schemes may indeed create artefacts in the results generated by these models.

The second conclusion stemming from EISMINT is that the application of existing models to the present-day ice sheets again reveals a large degree of correspondence, which is largely due to the response being mass balance driven rather than reflecting changes in internal flow regime. Similarly, models of ice shelves and ice streams show a large degree of agreement. Finally, the main areas of concern reflect the boundaries between different types of ice-flow system, for instance at grounding lines (between grounded ice sheets and floating ice shelves) and at the onset and lateral boundaries of ice streams. The stability of the grounding zone is, of course, crucial in determining the stability of marine ice sheets such as the West Antarctic ice sheet. Much recent work has focussed on the way in which this zone is incorporated into numerical models. In particular, Hindmarsh and Le Meur (2001) suggested that details of the numerical scheme employed may alter the overall predictions made by marine ice sheet models. This is clearly a major concern and must affect the confidence that we attach to models predicting future behaviour of West Antarctica.

6.4.4 Model calibration and confirmation

After a particular computer-based model has been validated and the numerical scheme shown to be consistent with the underlying differential equations and free of errors, it is necessary to confirm that the model is indeed a realistic representation of the real-world system. Before this can be done, three further actions must be taken, which all involve some interaction with real-world observations. The first is that input data sets need to be defined. The precise details of these data will inevitably depend on the details of the model, but they are likely to involve information on boundary conditions and initial conditions. The second action is to calibrate any poorly constrained parameters in the model. Finally, the data that are going to be used to confirm the veracity of the model must be defined. Normal practice in other areas of environmental modelling is to ensure that the data used in confirming the model are kept separate from those used to calibrate it. This separation is difficult to achieve when modelling the large ice sheets because of the limited availability of suitable data, and calibration and confirmation are often confused. Indeed, more often than not, all available data are needed to calibrate the model fully, and therefore most models are seldom subjected to rigorous confirmation.

One possibility that has often been used in the literature is to calibrate a model based on the past behaviour of the ice mass being modelled, to attempt to confirm the model using the present-day ice mass and to predict the ice sheet's future behaviour. Good examples of this approach are the twin papers of Huybrechts (2002) and Huybrechts and De Wolde (1999), on the past and future behaviour of, respectively, the Greenland and Antarctic ice sheets. This approach is hampered by lack of detailed knowledge of the past ice sheets and climate forcing, and by the fact that many different processes can often create the same observed behaviour (especially if this observed behaviour is very poorly constrained by data, as it often is for the past). This type of equi-finality is discussed in the context of the retreat of the West Antarctic ice sheet after the last glacial maximum by Hindmarsh and Le Meur (2001). They suggested that several different processes can cause this type of retreat in numerical ice-sheet models. Examples include changing ice dynamics at the grounding line and changes in basal sliding (which are often related to sea level on the basis of subglacial water pressures). It is therefore possible to generate the correct result (a retreat) for the wrong reason, or, more precisely, for a reason that may not be applicable in the future.

The types of data used as input (for prescribing boundary conditions, etc.) and for model confirmation differ between the two types of models discussed above. Prognostic ice-sheet models generate a time-dependent ice-mass geometry (thickness and, to a certain extent, bedrock topography) and require only initial conditions for these variables. In contrast, diagnostic models employ data on geometry as model inputs. Information on ice-surface velocity can be used to calibrate or confirm both types of model, while information on an ice mass's thermal regime and internal structure (as revealed by radio-echo sounding) are typically only used to confirm the predictions of prognostic models, although in principle they could also be used in the confirmation of diagnostic models.

Irrespective of the type of model being evaluated, attempts at model confirmation are hampered by the paucity of data available. Although the advent of satellite observations has improved this situation enormously, three main factors still conspire to restrict our ability to confirm models with observations effectively. The first is that observations are typically only available for the upper ice surface (examples being high resolution data on this surface's elevation, rate of change and horizontal velocity). Information from the interior of an ice mass or from the ice/bed interface are only available from relatively expensive and time-consuming methods, such as airborne or field radio and seismic sounding and ice drilling.

The second factor hampering effective model confirmation is the fact that the length of the observational record is only a small fraction of the natural timescales over which ice masses operate. For instance, timescales for response to changes in external forcing range from decades for small mountain glaciers to tens of thousands of years for the ice sheet in East Antarctica. In contrast, the era of extensive observation only extends back some 50 years at most (significantly less in the case of satellite observations). Although qualitative interpretations can be made that extend this record further back in time, the information so derived is often fragmentary and difficult to interpret unequivocally. Examples include the geomorphologic record left by more extensive glaciers and ice sheets in the past, as well as the record of changing flow captured by lineations on the Ross ice shelf, Antarctica (Fahnestock *et al.*, 2000).

Another major restriction to confirming model predictions is the quality of the input data available to drive these models. This is particularly true for bedrock topography, which largely determines ice thickness at regional scales, and therefore the stress, velocity and thermal regimes. There are large gaps in our information on the bedrock topography of the Antarctic ice sheet (in particular in East Antarctica). The recent BEDMAP compilation (Lythe and Vaughan, 2001) represents a major step forward, but approximately one-quarter of the continent still remains unsurveyed (see Chapter 12). The coverage of the Greenland ice sheet is far more comprehensive (thanks in part to the recent PARCA initiative; see Chapter 10). Clearly, any attempt to compare, for instance, modelled ice-surface velocity or elevation with observations in areas with little or no data on bedrock topography is not a true test of the model but a test of the extrapolation schemes used to fill the gaps in the original bedrock-data coverage. Although less restrictive, inadequacies in our knowledge of the geothermal heat flux under ice masses and their surface meteorology are also major impediments to effective confirmation.

Most parameters used in ice-flow models are fairly well known (ice density, thermal conductivity, etc.), but there are three areas where the parameters are very poorly defined and must be adjusted through calibration. The usual procedure for calibration is to adjust these parameters such that the gross form of the modelled ice mass compares favourably with reality. In particular, tuning concentrates on achieving a realistic horizontal extent for the modelled ice mass and a realistic height-to-width ratio (that is, the ice thickness at the divide is used for calibration). The three sets of parameters that are available for tuning are associated with the viscosity of ice, basal sliding and net mass balance. Chapter 5 offers an extensive discussion on mass balance modelling, and we will not dwell on this topic other than to stress the sensitivity of prognostic ice-sheet models to the parameterization of mass balance. For example, Ritz, Fabre and Letréguilly (1996) attempted to find the degree-day parameterization (in addition to ice-flow parameterizations) that produced a best-fit to the present-day volume and extent of the Greenland ice sheet after allowing the ice sheet to evolve through a glacial–inter-glacial cycle. Their best-fit values for the degree-day coefficients were several times larger than values obtained from field-based experiments. In fact, simulations that used the field-based estimates resulted in a 36% over-prediction of ice volume.

The calibration of parameters associated with the viscosity of ice (the rate factor in Glen's flow law) and with basal sliding is particularly problematic because both have very similar effects on model dynamics. Except on ice streams, the majority of ice deformation occurs close to the bed, in the lower 20% of the ice column. It is therefore difficult to differentiate between the effects of an enhanced value of the rate factor and the onset of basal sliding, especially if the only velocity data available apply to the ice surface. The relationship between the rate factor and ice temperature is reasonably well understood (see section 6.2.4), but factors such as impurity content and crystal fabric that are known to have similar effects are not incorporated in ice-flow models. Rather, an arbitrary tuning multiplier, ranging from 1 to 5, is applied to the rate factor. Models of the Greenland ice sheet normally account for the observed difference in the mechanical properties of Holocene (harder) and older (softer) ice near the base. However, it should be stressed that, at present, we have little way of knowing whether the introduction of a tuning multiplier is legitimate (reflecting real differences in ice rheology) or whether this procedure masks an under-estimation of basal sliding. This is particularly true of the West Antarctic ice sheet, which is thought to be wet-based and therefore capable of basal sliding over the vast majority of its area (Hulbe and MacAyeal, 1999).

The basic problem that one is presented with in confirming the veracity of an ice-flow model is that the problem is massively under-determined. The models have large degrees of freedom, in that the large number of processes that are typically incorporated into operational ice-mass models means that there are often many different ways of producing the same observed behaviour. For instance, thickening at an ice-sheet divide can be produced by locally colder ice (stiffer hence slower flowing), enhanced local accumulation rates, or a reflection of a general thickening in response to a larger ice-mass extent. Similarly, increases in ice-surface velocity may be attributable to either softer, more deformable, ice or to enhanced basal sliding. This problem is compounded by the paucity of input data used in numerical models (as noted above in particular for bedrock topography and surface mass balance). The general lack of test data (plus the limitations of what data there are currently available, highlighted above) combines with this great number of the degrees of freedom to make agreement between model and observation relatively easy to achieve. However, this agreement does not necessary reflect a satisfactory confirmation of the model, but is simply a consequence of the ease of tuning an ill-constrained model to inadequate data. The advent of high resolution, spatially extensive satellite data will certainly improve this situation, and giant strides toward effective confirmation have been made, especially in the area of ice-stream modelling (e.g. MacAyeal, Bindschadler and Scambos, 1995).

Faced with this under-determinacy, a logical approach would be to attempt to quantify the uncertainty in predictions through the use of an extensive sensitivity analysis. This is generally done, and most modelling papers have a section on sensitivity analysis, although the type of analysis presented is often fairly limited. In particular, results are presented either from a series of experiments in which the values of individual parameters are changed within a reasonable range (typically no more than five experiments) or from one-off experiments aimed at particular scenarios. A more rigorous approach would investigate the full parameter space using experiments in which parameter values are jointly varied. This is common practice in many other areas of environmental modelling (e.g. Chapman *et al.*, 1994; Vachaud and Chen, 2002). This approach was used by Van der Veen (2002) to evaluate how well the future contribution of the polar ice sheets to global sea level can be predicted. Unfortunately, the computational expense of many ice-flow models makes a systematic parametric sensitivity analysis a desirable goal that is unlikely to be addressed until either ice-flow models become more computationally efficient or computer resources expand significantly.

Confirming models using ice-mass geometry data

The elevation of the upper surface of an ice mass is a predicted quantity in prognostic models and a powerful confirmation tool. Recent developments in satellite and airborne altimetry have greatly expanded information on the topography of this surface, which is now available over entire ice masses and at spatial resolutions and vertical accuracies previously unattainable. Digital terrain models of the Greenland and Antarctic ice sheets are now available on 5 × 5 km spatial grids (see Chapters 4, 10 and 12). This resolution is indeed too fine for some modelling applications, in particular models that rely on the assumption of local balance between gravitational driving and basal traction, which is appropriate only at scales greater than 10 to 20 times the ice thickness (~20–40 km for the ice sheets).

Unfortunately, the modelling community still lags behind the observational community in its use of these new data. A comparison of model output with observation typically comprises a quantitative comparison of variables such as overall ice extent, volume, average thickness and, perhaps, the elevations at selected locations such as ice divides or ice-core sites. It may also include a qualitative comparison of the spatial locations of modelled and observed features, such as divides, local domes and outlet glaciers or ice streams. Clearly, this is a major under-utilization of available data. An analysis of regional differences between modelled and observed ice-surface elevations (given that the model has been calibrated to replicate the observed mean surface elevation) would highlight regional differences in flow, and perhaps aid in unravelling the combined effects of basal sliding and ice softening discussed above.

The accuracy of modern-day satellite altimetry is now such that meaningful estimates of local surface elevation change can be made. The length of the satellite record limits the accuracy of the derived rate of surface change. For example, the observed signal still falls below the level of natural variability in snow fall for much of Antarctica (Wingham *et al.*, 1998). However, it is clear that certain areas of Antarctica (e.g. the Pine Island and Thwaites glaciers and their drainage basins; Shepherd *et al.*, 2001) and Greenland (Abdalati *et al.*, 2001) are undergoing change at a rate that can only be explained in terms of changes in ice-flow regime (Van der Veen, 2001). The ice-sheet modelling community has yet to rise to the challenge that this evidence of on-going dynamic change poses in terms of model confirmation.

Confirming models of ice velocity

As with surface topography, remote sensing has revolutionized the availability of measurements of ice-surface velocity. A range of techniques is now available for obtaining surface velocities over large regions at high spatial resolution and accuracy. Remote-sensing techniques using satellite-derived information (reviewed in Chapter 4) include interferometry (based on synthetic aperture radar data, INSAR; e.g. Kwok and Fahnestock, 1996), speckle tracking (also using synthetic aperture radar data; e.g. Gray *et al.*, 2001) and the more traditional method of tracking visible features such as crevasses (based on data from a wide range of satellite-borne instruments; e.g. Bindschadler *et al.*, 1996). To complement this information, very accurate (although not spatially extensive) velocity measurements are available from ground-based global positioning system (GPS) surveys (e.g. Thomas *et al.*, 2000; Whillans and Van der Veen, 1993b).

To our knowledge, measured surface velocities have not been used to confirm prognostic ice-sheet models. In part this may be because, initially, measurements of surface velocity focussed on special regions such as the West Antarctic ice streams and fast moving outlet glaciers both in Greenland and Antarctica. These features tend to be of limited areal extent and are not well captured in most whole ice-sheet models.

Velocity maps are, however, widely used in conjunction with diagnostic models, in particular of ice streams. This use does not aim to confirm model predictions but uses the model in an inverse sense to determine what basal characteristics are consistent with the observed surface velocities and the conceptual model. Prognostic ice-sheet modelling is forward in nature, and a particular set of model equations is integrated forward in time given a set of initial and boundary conditions. The resultant predictions are then compared with available observational evidence to confirm or calibrate the model. This approach is essentially different from diagnostic models, whose aim is to determine what sets of boundary (and initial) conditions are consistent with the observations given a particular set of model equations. A classic example would be to determine the set of boundary conditions operating at the bed of a glacier that is consistent with the observed surface topography and velocities. This type of model requires very detailed observational information, which is now available for both ice-surface velocity and elevation. Early examples of this approach applied to ice streams include work by MacAyeal *et al.* (1995) using a detailed forward model and a control-methods inversion technique (MacAyeal 1993), and Gudmundsson, Raymond and Bindschadler (1998), using theoretically derived transfer functions based on a linearized model. In principle, it may be possible to invert surface information for bedrock topography as well as basal traction and ice rheology. The force budget approach of Van der Veen and Whillans (1989) has also been applied as an inverse technique to derive estimates of lateral and longitudinal resistive forces (and basal resistance by assuming that this represents the difference between the sum of these two forces and gravitational driving) for the ice streams of the Siple Coast (Whillans and Van der Veen, 1993a, 1997). All of these inverse-type applications have, so far, only been applied to ice streams, where ice-surface velocities can be assumed to be a very close approximation to basal sliding velocities (i.e., vertical shear is assumed minimal, as discussed in section 6.4.2). It is unclear whether they will be equally as successful when applied to flow regimes where vertical shear makes an important contribution to ice flow.

In recent years, a number of modelling studies have compared predicted surface speeds to so-called balance velocities, first introduced by Budd, Jenssen and Radok (1971). The concept of balance velocities has recently received much attention (Bamber, Vaughan and Joughin, 2000b; Budd and Warner, 1996). It is based on the concept of mass conservation, and uses the integrated ice accumulation along the flow line upstream of the location of interest to estimate the flux of ice passing through the point. Information on local ice thickness then allows an estimate of the vertically averaged velocity to be made. The technique assumes that ice always flows downhill (which is acceptable on large spatial scales away from ice shelves and ice streams), and that the ice mass is in equilibrium with its mass balance, so that all ice accumulated upstream of a point does indeed pass through that point (this assumption is unlikely to be fulfilled in practice). Several studies have compared INSAR-derived velocities with balance velocities (e.g., Bamber, Hardy and Joughin, 2000a), and differences between the two are often interpreted as indicating mass imbalance. It should be noted that to derive spatial maps of balance velocity, a numerical technique is required to route flow over the two-dimensional surface of the ice mass. Budd and Warner (1996) present such a technique for ice masses, and this technique is also widely used in the hydrological community to route rain water over land surfaces (e.g., Tarboton 1997). Work in the latter area indicates that details of the numerical technique can affect the pattern of routing, especially in areas of low surface slope (typical of ice masses).

The use of velocity maps to confirm the predictions of prognostic ice mass models is fairly limited. Bamber *et al.* (2000c) compare balance velocities with the predictions of a thermomechanical ice-sheet model for Greenland. The veracity of the balance velocities is first confirmed by comparison with INSAR and GPS estimates. The gross comparison between model and balance velocities is favourable; however, the two do not compare well in detail. In particular, the north-west Greenland ice stream is present in the balance velocities but not predicted by the model (this theme is discussed further below). Another example of the use of balance velocities in model confirmation is presented by Payne (1998) for West Antarctica, in which the locations of zones of fast ice flow predicted by a thermomechanical ice-sheet model are compared with balance velocity maps. The analysis is, however, purely qualitative.

Confirming models of ice-mass temperature

The internal thermal regime of glaciers and ice sheets is known to be very important in determining their flow. This is because the viscosity of ice is strongly dependent on its temperature. Paterson and Budd (1982) and Hooke (1981) proposed relationships between ice temperature and the value of the rate factor in Glen's flow law (equations (6.34) and (6.35)). Both studies adopt an exponential Arrhenius-type relationship, and find that ice can deform 1000 times more rapidly when at the melting point than at $-50\,°C$. The viscosity of ice also depends on a number of other factors, such as its impurity content and water content (water exists between crystals even at subzero temperatures).

Ice temperature near the glacier bed is thought to be more important in determining ice velocity than that of ice in the upper sections of an ice column because the vertical shear stresses, which are mostly responsible for internal deformation, are greatest near the bed of a glacier (equations (6.51) and (6.52)). The temperature of the bed itself and, in particular, whether it is at the melting point or not is also important in determining glacier flow. This is because the presence of melt water at the bed is a necessary pre-condition for slip of the glacier over the underlying bedrock, irrespective of whether slip occurs by sliding over a solid substratum or deformation of a mobile till layer (see section 6.2.6).

Ice-temperature evolution is therefore normally incorporated into models of ice masses and, in particular, is a very important component of current models of Antarctica and Greenland (Huybrechts, 2002; Huybrechts and De Wolde, 1999). The thermodynamics of ice flow are discussed in section 6.2.4. The physical constants involved (the conductivity, latent heat capacity and specific heat capacity) are well known. However, the boundary conditions, in particular the geothermal heat flux under the present-day ice sheets, is very poorly constrained by observations. The normal practice is to assume a constant value over the entire ice-sheet domain. Exceptions are Ritz (1987), who applied the geothermal heat flux to the base of the 3 km bedrock slab underlying the ice sheet, and therefore incorporated the thermal response of the bedrock to the ice sheet (this has since become normal practice amongst the ice-sheet modelling community), and Takeda, Cox and Payne (2002), who ascribe different values to West and East Antarctica on the basis of their different geological histories (West Antarctica is younger and should therefore have a higher heat flux).

Three factors in the above discussion combine to make the confirmation of modelled near-bed ice temperatures important. These are the importance of the near-bed thermal regime in determining ice flow (both via internal deformation and basal slip), the importance of geothermal heat flux in the near-bed heat budget and uncertainty concerning the spatial distribution of geothermal heat flux. Models of the basal thermal regime are therefore very important within the overall ice-mass modelling framework, but, at the same time, they are relatively easy to calibrate (because of the importance of geothermal heat flux) and very poorly constrained (because of the uncertainty of geothermal heat flux). Recent work by Fahnestock *et al.* (2001) emphasizes this point. These authors postulate the existence of a geothermal hot spot at the head of a fast-flow feature called the north-east Greenland ice stream (NEGIS). Supporting evidence is the down-warping of radar-echo sounding (RES) internal layers (presumed to be caused by enhanced basal melt) that is coincident with an observed magnetic anomaly. This suggestion is particularly intriguing because ice-sheet models have been consistently unable to simulate the NEGIS. This inability may be because of inadequacies in the input data, in particular the assumption of a spatially constant geothermal heat flux. Experiments using ice-sheet models with a series of geothermal anomalies are clearly desirable.

There are potentially three main sources of data that can be used to confirm models of the basal thermal regime. These are the existence of subglacial lakes, characteristics of the bed revealed by radar-echo sounding and ice-core temperature records. We will discuss the strengths and limitations of these data sources in turn.

Subglacial lakes exist under both the Greenland and Antarctic ice sheets. They are most easily recognized by their effect on ice-surface topography. The section of the ice sheet directly overlying the lake acts as a free-floating ice shelf and does not experience any basal drag. The ice surface in this area is therefore both very smooth and flat. These features are easily identified using satellite altimetry. Siegert *et al.* (1996) and Dowdeswell and Siegert (1999) present inventories of the subglacial lakes of Antarctica. They identify over 70 lakes, which are mostly located in the interior of the ice sheet under the deep, slow flowing ice near domes and divides. The existence of a subglacial lake implies that the ice sheet's bed must be experiencing melt. A map of the positions of subglacial lakes could therefore be used to confirm an ice-sheet model's prediction of basal thermal regime. Huybrechts (1992) attempted such confirmation.

Unfortunately, a number of practical factors make this conceptually appealing confirmation tool very difficult to use. The first is one of spatial scale. Ice-sheet models, such as the one used by Huybrechts (1992), are based on a relatively coarse spatial grid (typically using grid cells between 20 and 40 km square). Dowdeswell and Siegert (1999) indicate that the majority (approximately 75%) of lakes are less than 10 km long and their mean length is 14.2 km (excluding Lake Vostok). Lake Vostok is 230 km long and is a distant outlier to the main population of lakes. This implies that almost all lakes are sub-grid features within ice-sheet models. Further, lakes are likely to be located in bedrock trenches that will experience a very different local thermal regime compared with that of the neighbouring ice sheet. Dowdeswell and Siegert (1999) estimate that the average lake depth is between 50 and 250 m, but the later estimate of Siegert *et al.* (2001) suggests that Lake Vostok is over 1000 m deep (in comparison to the depth of overlying ice of 3750 to 4150 m). In any case, the implication is that the lake occupies a deep trench that will affect the local thermal regime by depressing the pressure-melting point and increasing the insulating effect of the overlying ice mass.

A second factor that makes the use of subglacial lakes as a confirmation tool problematic is the method used to derive the bedrock topography used in ice-sheet models. The bedrock topography is typically derived from airborne radar-echo soundings (RES). The information is used to estimate ice thickness, which is then subtracted from ice-surface elevation to obtain an estimate of the elevation of the ice bed. Normally this figure equates to the bedrock elevation, but over lakes it will correspond to the lake surface level (water is opaque at the frequencies used in airborne RES). Bedrock compilations such as the BEDMAP project do not account for this discrepancy and therefore do not incorporate the troughs occupied by subglacial lakes. Mismatch between the predicted basal thermal regime and lake locations may therefore be due to errors in the input data for bedrock topography.

In summary, the concept of using subglacial lake locations to confirm models of basal thermal regime is conceptually appealing but is not particularly effective at ice-sheet scales. An alternative may be to use RES information in another way. Gades *et al.* (2000) use ground-based RES traverses across Siple Dome in West Antarctica to infer bed properties and, in particular, to identify the presence of water in basal sediments. Copland and Sharp (2001) use a similar technique to map thermal and hydrological conditions under John Evans glacier, Ellesmere Island, using ground-based RES traverses. Both studies calculate the bed reflection power (BRP) and attempt to correlate this to the nature of the basal ice/sediment interface. The power returned depends on many factors, including the transmitted power, polarization and scattering at the ice/bed interface, antenna characteristics, as well as on the strength of the basal return. These additional factors are assumed to be either constant or a function of ice thickness.

BRP is determined empirically from the squared amplitude of the echo from the bed (in millivolts) and corrected for ice-thickness variations using a regression of BRP against ice thickness (the corrected BRPs are then the residuals of the BRP/thickness regression). Positive values of relative BRP are taken to indicate warm basal thermal conditions (i.e. water present at the bed) while negative relative values imply frozen-bed conditions. This interpretation derived from the very high dielectric constant of water relative to both rock and ice. On a more qualitative level, internal RES reflectors can also be interpreted as thermal boundaries within an ice mass; examples being cold ice overlaying warm in polythermal glaciers such as John Evans (Copland and Sharp, 2001) and Bakaninbreen in Svalbard (Murray *et al.*, 2000).

Bentley, Lord and Liu (1998) use airborne RES data to map relative BRP for a large region of the Siple Coast, West Antarctica, covering ice stream C and Whillans ice stream. They use a more sophisticated model to determine expected BRP than discussed above. Differences between returns from the underside of the Ross ice shelf, the ice streams and inter-stream ridges are evident. In particular, the results suggest that ice stream C (although flowing slowly) has higher bed return strengths (i.e. is wetter?) than the neighbouring fast moving Whillans ice stream. The ridge between the two ice streams has lower values than either ice stream, which implies that it may be frozen.

It is clear that this technique offers great potential in mapping the basal thermal regimes of ice masses, and that this information would represent a strong confirmation tool for the modelling community. On a qualitative level, it is clear that three distinctive subglacial environments can be identified from airborne RES data (Siegert, 1999). These are: frozen ice/bed interfaces with a weak, scattered return; subglacial water masses with a strong return over a horizontal plain; and water-saturated sediment that has a return almost as strong as for water masses but is not so laterally extensive. Difficulties in applying this technique over an ice sheet include variable transmission through ice with differing temperatures and impurity contents (Corr, Moore and Nicolls, 1993).

Model predictions of ice temperature can, of course, also be confirmed against records obtained from deep ice cores. The number of useful data sets that extend from the ice surface all the way to the bed is, however, very limited. Ice cores are point sources of data and suffer from the disadvantages identified above for point information. In particular, the usefulness of comparing output from large-scale ice-sheet models (with a minimum spatial resolution of 20 km) with some near-margin cores is questionable. Examples include the cores Law Dome and Terre Adelie in East Antarctica.

The records from the Vostok and dome C cores are very useful in constraining model temperature predictions. The ice near the base of the Vostok core is approximately 420 000 years old (Petit *et al.*, 1999), which means that ice-sheet models must be driven by realistic climatic records for all of this period if a reasonable comparison is to be made. Unfortunately, the only climate records that can be used to provide this forcing come from the ice-core air temperature record. This record is, however, derived as a proxy estimate from the isotopic content of the ice (e.g. Jouzel *et al.*, 1987). There is therefore some circularity involved when using the same data to force the model and confirm it.

The East Antarctic ice cores suffer from the disadvantage, in terms of model confirmation, of being taken from ice divides and domes. These sites experience (by definition) extremely slow flow, and the primary driving factors for the vertical profile of ice temperature are geothermal heat flux in the lower sections of the core and vertical advection in the upper sections, determined to a large extent by surface accumulation. Given our lack of knowledge on the spatial variation of geothermal heat (see above) and the accumulation history, it could be argued that confirmation in these cases is simply a matter of selecting a combination of geothermal heat flux and accumulation forcing that produces the observed vertical temperature gradient!

In West Antarctica, ice-core temperature records are more plentiful. They include that of the Byrd Station ice core, which is located some distance from the ice divide, as well as numerous deep cores from the Siple Coast area. These include ice cores drilled to the bed on Siple Dome, ice streams C and D, and Whillans ice stream. These data provide a far more challenging confirmation tool because they occur in a variety of glaciologically active locations. For instance, the cores taken from ice streams will experience significant horizontal advection throughout their profiles and variable amounts of frictional heating at their bed. Hulbe and MacAyeal (1999) present one of the few attempts to use these data to confirm an ice-flow model.

A good variety of useful cores also exist for Greenland including cores near to Summit (GRIP, NGRIP and GISP2) and along the ice-sheet main divide (Dye 3), as well as on the flanks of the ice sheet (Camp Century). Few attempts have been made to use this information to confirm the many ice-sheet models that have been applied to Greenland.

The use of RES data to confirm models of glacier flow

The previous section touched upon the use of airborne and ground RES data to confirm models. The strength of the return from the glacier bed/substrate interface can be used as an indicator of the presence of subglacial water because of water's very high dielectric constant. RES data typically also contain information on a number of internal reflecting horizons. These horizons are often continuous and traceable over hundreds to thousands of kilometres (Siegert, Hodgkins and Dowdeswell, 1998). There are often ten or more horizons distributed throughout an ice column, although they tend to be concentrated in the upper levels. The interest of these horizons in the context of model confirmation is that they are believed to be isochronous and thus provide information on the history of flow within an ice mass.

The internal layers are known to be associated with changes in the electrical properties of ice. While their exact mode of formation remains unclear, several mechanisms are thought to be important. The first is associated with changes in the density of ice, and this is most prevalent in the upper levels of an ice mass, where density changes are generated in the firn compaction. The second is associated with changes in the acidity of ice and, in particular, surface deposition of aerosols generated by volcanic activity. The main implication of both processes is that the layers are generated at or close to the ice surface and are therefore likely to be isochronous. The final mechanism results from the interaction of acidic layers with the ice mass's internal stress regime (especially in the presence of significant bed topography) that may result in changes in ice crystallography. If an internal reflecting horizon can be traced to an ice-core location, then the age–depth relationship for the ice core can be used to date the horizon. This then allows a dated stratigraphic horizon to be extended spatially across the ice mass.

Although the glaciological application of internal layers is still in its infancy, two main uses can be identified. The first uses the relatively shallow layers in the upper levels of an ice mass to determine past accumulation rate variations. Flow in these areas is dominated by the gradual burial and compression of ice layers by successive snow fall (the layers' thinning rate or the vertical strain rate), and vertical shearing is at a minimum. In this case, a dated stratigraphic horizon allows one to estimate past accumulation rates if an assumption is made about the vertical form of the thinning rate. Two models are commonly used. The Nye (1963) model assumes that the thinning rate is uniform with depth. The Dansgaard and Johnsen (1969) model improves on this by assuming that vertical strain is constant to a certain depth and decreases linearly below this depth to zero at the bed. In either case, vertical strain rate at the ice surface can be related to the accumulation rate (assuming the ice surface is flat or flowing very slowly). While this application has more relevance to the derivation of accumulation input data than to model confirmation, the second application of RES data is directly relevant to confirmation.

The pattern of horizontal flow divergence will affect the height of isochrones within an ice mass (Weertman, 1976). These markers may also record changes in this pattern. At present, little work has been done in this area, but information is likely to be generated in two main areas. Firstly, qualitative data will be available on the presence and form of internal layering. For instance, fast flow appears to disrupt the layering (Robin, 1983), and so the absence of layers could be used as an indicator of fast flow (whether present or past). Similarly, layers that have become buckled are often observed near the onset of fast flow (M. Siegert, personal communication). Secondly, detailed studies of the relative depths of layers in the lower levels of an ice mass may yield information on the onset of fast flow and changing dynamics of fast flow features.

In order to make best use of this new confirmation data, models of glaciers and ice sheets must be able to predict the location (in three-dimensional space) of isochronous surfaces. This involves the solution of an equation for the pure advection of a tracer around a given flow field. In this case, the passive tracer is the time at which the ice entered the model domain (as snow fall on the ice-sheet surface). Several attempts have already been made to do this based on the same finite-difference approach that is used to model the evolution of internal temperature. However, pure advection is notoriously difficult to model within a finite-difference framework and requires the addition of artificial diffusion to remain stable. This has the implication that the artificial diffusion may dominate where advection rates are slow, for instance near the bed, and produce unrealistic ages in these areas. It seems likely that more advanced techniques may be necessary (for instance semi-Lagrangian techniques; see Brasseur and Madronich (1992)).

A similar opportunity for model confirmation is the use of the isotopic content of ice. The isotopic content of ice is determined at deposition primarily by the characteristics of the atmospheric moisture source from which the snow fell, as well as the temperature at which it formed (Dansgaard *et al.*, 1973). Within an ice mass, isotopic content behaves as a passive tracer that is advected within the flow field with negligible diffusion. The depth profiles of isotopic content available from ice cores can be used to confirm the model's predictions of the depth variation of isotopic content at that location, which will be a product of the predicted flow field's variation through time (Clarke and Marshall, 2002).

6.5 List of symbols

A	rate factor in the flow law relating strain rates to stresses
b	bed elevation
B	rate factor in the flow law relating stresses to strain rates
C_p	specific heat capacity
f	shape factor accounting for lateral drag
F	ice flux
g	gravitational acceleration
G	geothermal heat flux
h	ice surface elevation
H	ice thickness
k	thermal conductivity
K	thermal diffusivity
L	glacier length or half-width
L	lithostatic stress
L_f	specific latent heat of fusion
M	surface mass balance
M_f	amount of melt water that refreezes per unit time and volume
n	exponent in Glen's flow law
P_o	ice overburden pressure
P_e	effective basal pressure
P_w	subglacial water pressure
Q	activation energy for creep
R	gas constant
R_{ij}	resistive stress components ($i, j = x, y, z$)
s	dimensionless vertical co-ordinate
t	time
T	ice temperature
T_s	surface temperature
T_b	basal temperature
u, v, w	ice velocity components in the x-, y- and z-directions
U, V	depth-averaged horizontal components of ice velocity
U_b	basal slip velocity vector
U_s, V_s	components of the sliding velocity
W	half-width of flow band
x, y, z	cartesian co-ordinate system, with the vertical z-axis positive upward

α	slope of the upper ice surface, positive in the direction of decreasing surface elevation
γ_b	vertical temperature gradient at the glacier base
Δt	time step
$\Delta x, \Delta y, \Delta z$	horizontal and vertical grid spacings
$\dot{\varepsilon}_{ij}$	components of the strain rate tensor $(i, j = x, y, z)$
$\dot{\varepsilon}_e$	effective strain rate
η	effective viscosity of ice
ρ	ice density
ρ_w	density of water
σ_{ij}	components of the full stress tensor $(i, j = x, y, z)$
σ'_{ij}	components of the deviatoric stress tensor $(i, j = x, y, z)$
τ_{ij}	shear stress $(i \neq j = x, y, z)$
τ_{bi}	basal drag in the i-direction $(i = x, y)$
τ_{di}	driving stress in the i-direction $(i = x, y)$
τ_e	effective stress
τ_s	shear stress at lateral margins

References

Abdalati, W. *et al.* 2001. Outlet glacier and margin elevation changes: near-coastal thinning of the Greenland ice sheet. *J. Geophys. Res.* **106**, 33 729–41.

Alley, R. B. 1992. Flow-law hypotheses for ice-sheet modeling. *J. Glaciol.* **38**, 245–56.

Alley, R. B. and Whillans, I. M. 1984. Response of the East Antarctic ice sheet to sea-level rise. *J. Geophys. Res.* **89**, 6487–93.

Alley, R. B., Blankenship, D. D., Bentley, C. R. and Rooney, S. T. 1986. Deformation of till beneath ice stream B, West Antarctica. *Nature* **322**, 57–9.

  1987. Till beneath ice stream B. 3. Till deformation: evidence and implications. *J. Geophys. Res.* **92**, 8921–9.

Baker, R. W. 1987. Is the creep of ice really independent of the third deviatoric stress invariant? In Waddington, E., ed., *The Physical Basis of Ice Sheet Modelling*. IAHS publ. no. 170, pp. 7–16.

Bamber, J. L., Hardy, R. J. and Joughin, I. 2000a. An analysis of balance velocities over the Greenland ice sheet and comparison with synthetic aperture radar interferometry. *J. Glaciol.* **46**, 67–74.

Bamber, J. L., Vaughan, D. G. and Joughin, I. 2000b. Widespread complex flow in the interior of the Antarctic ice sheet. *Science* **287** (5456), 1248–50.

Bamber, J. L., Hardy, R. J., Huybrechts, P. and Joughin, I. 2000c. A comparison of balance velocities, measured velocities and thermomechanically modelled velocities for the Greenland Ice Sheet. *Ann. Glaciol.* **30**, 211–16.

Bentley, C. R. 1983. The west Antarctic ice sheet: diagnosis and prognosis. In *Proceedings of Carbon Dioxide Conference: Carbon Dioxide, Science and Consensus*. Berkeley Springs, W. Va., Sept. 19–23, 1982, IV.3–IV.50.

  1984. Some aspects of the cryosphere and its role in climate change. In Hansen, J. E. and Takahashi, T., eds., *Climate Processes and Climate Sensitivity*. Geophysical Monograph 29. Washington, D.C., American Geophysical Union, pp. 207–20.

Bentley, C. R., Lord, N. and Liu, C. 1998. Radar reflections reveal a wet bed beneath stagnant ice stream C and a frozen bed beneath ridge BC, west Antarctica. *J. Glaciol.* **44**, 149–56.

Bindschadler, R. A. 1983. The importance of pressurized subglacial water in separation and sliding at the glacier bed. *J. Glaciol.* **29**, 3–19.

Bindschadler, R., Vornberger, P., Blankenship, D., Scambos, T. and Jacobel, R. 1996. Surface velocity and mass balance of ice streams D and E, west Antarctica. *J. Glaciol.* **42** (142), 461–75.

Blatter, H. 1995. Velocity and stress fields in grounded glaciers: a simple algorithm for including deviatoric stress gradients. *J. Glaciol.* **41**, 333–44.

Boulton, G. S. and Jones, A. S. 1979. Stability of temperate ice caps and ice sheets resting on beds of deformable sediments. *J. Glaciol.* **24**, 29–43.

Brasseur, G. P. and Madronich, S. 1992. Chemistry-transport models. In Trenberth, K., ed., *Climate System Modeling*. Cambridge University Press, pp. 491–518.

Budd, W. F. 1970a. The longitudinal stress and strain-rate gradients in ice masses. *J. Glaciol.* **9**, 19–27.

    1970b. Ice flow over bedrock perturbations. *J. Glaciol.* **9**, 29–48.

Budd, W. F. and Warner, R. C. 1996. A computer scheme for rapid calculations of balance-flux distributions. *Ann. Glaciol.* **23**, 21–7.

Budd, W. F., Jenssen, D. and Radok, U. 1971. Derived physical characteristics of the Antarctic ice sheet. ANARE interim report, Series A (IV), Glaciology Publ. pp. 120, 178.

Chapman, W. L., Welch, W. J., Bowman, K. P., Sacks, J. and Walsh, J. E. 1994. Arctic sea-ice variability – model sensitivities and a multidecadal simulation. *J. Geophys. Res.* **99**, 919–35.

Clarke, G. K. C. and Marshall, S. J. 2002. Isotopic balance of the Greenland ice sheet: modelled concentrations of water isotopes from 30,000 BP to present. *Quat. Sci. Rev.* **21**, 419–30.

Clarke, G. K. C., Nitsan, U. and Paterson, W. S. B. 1977. Strain heating and creep instability in glaciers and ice sheets. *Rev. Geophys. & Space Phys.* **15**, 235–47.

Copland, L. and Sharp, M. 2001. Mapping thermal and hydrological conditions beneath a polythermal glacier with radio-echo sounding. *J. Glaciol.* **47**, 232–242.

Corr, H., Moore, J. C. and Nicolls, K. W. 1993. Radar absorption due to impurities in Antarctic ice. *Geophys. Res. Lett.* **20**, 1071–4.

Dansgaard, W. and Johnsen, S. J. 1969. A flow model and a time scale for the ice core from Camp Century, Greenland. *J. Glaciol.* **8**, 215–23.

Dansgaard, W., Johnsen, S. J., Clausen, H. B. and Langway, C. C. 1973. Stable isotope glaciology. *Medd. Grønl.* **197** (2), 1–53.

Dowdeswell, J. A. and Siegert, M. J. 1999. The dimensions and topographic setting of Antarctic subglacial lakes and implications for large-scale water storage beneath continental ice sheets. *Geol. Soc. Am. Bull.* **111**, 254–63.

Fahnestock, M. A., Scambos, T. A., Bindschadler, R. A. and Kvaran, G. 2000. A millennium of variable ice flow recorded by the Ross ice shelf, Antarctica. *J. Glaciol.* **46**, 652–64.

Fahnestock, M., Abdalati, W., Joughin, I., Brozena, J. and Gogineni, P. 2001. High geothermal heat flow, basal melt, and the origin of rapid ice flow in central Greenland. *Science* **294**, 2338–42.

Fastook, J. L. and Prentice, M. 1994. A finite-element model of Antarctica – sensitivity test for meteorological mass-balance relationship. *J. Glaciol.* **40**, 167–75.

Fowler, A. C. 1987a. Sliding with cavity formation. *J. Glaciol.* **33**, 255–67.

1987b. A theory of glacier surges. *J. Geophys. Res.* **92**, 9111–20.

Gades A. M., Raymond, C. F., Conway, H. and Jacobel, R. W. 2000. Bed properties of Siple Dome and adjacent ice streams, west Antarctica, inferred from radio-echo sounding measurements. *J. Glaciol.* **46**, 88–94.

Glen, J. W. 1955. The creep of polycrystalline ice. *Proc. Roy. Soc. London, Ser. A* **228**, 519–38.

1958. The flow law of ice. A discussion of the assumptions made in glacier theory, their experimental foundations and consequences. IAHS publ. no. 147, pp. 171–83.

Gray, A. L., Short, N., Mattar, K. E. and Jezek, K. C. 2001. Velocities and flux of the Filchner ice shelf and its tributaries determined from speckle tracking interferometry. *Can. J. Remote Sensing* **27** (3), 193–206.

Greuell, W. 1989. Glaciers and climate. Ph.D. Thesis, University of Utrecht.

1992. Hintereisferner, Austria: mass-balance reconstruction and numerical modelling of the historical length variations. *J. Glaciol.* **38**, 233–44.

Gudmundsson, G. H. 1999. A three-dimensional numerical model of the confluence area of Unteraargletscher, Bernese Alps, Switzerland. *J. Glaciol.* **45**, 219–30.

Gudmundsson, G. H., Raymond, C. F. and Bindschadler, R. 1998. The origin and longevity of flow stripes on Antarctic ice streams. *Ann. Glaciol.* **27**, 145–52.

Herterich, K. 1987. On the flow within the transition zone between ice sheet and ice shelf. In Van der Veen, C. J. and Oerlemans, J., eds., *Dynamics of the West Antarctic Ice Sheet*. Dordrecht, D. Reidel, pp. 185–202.

Hindmarsh, R. C. A. 1997. Normal modes of an ice sheet. *J. Fluid Mech.* **335**, 393–413.

2001. Influence of channelling on heating in ice-sheet flows. *Geophys. Res. Lett.* **28**, 3681–4.

Hindmarsh. R. C. A. and Le Meur, E. 2001. Dynamical processes involved in the retreat of marine ice sheets. *J. Glaciol.* **47**, 271–82.

Hindmarsh, R. C. A. and Payne, A. J. 1996. Time-step limits for stable solutions of the ice-sheet equation. *Ann. Glaciol.* **23**, 74–85.

Hooke, R. Le B. 1981. Flow law for polycrystalline ice in glaciers: comparison of theoretical predictions, laboratory data, and field measurements. *Rev. Geophys. & Space Phys.* **19**, 664–72.

1989. Englacial and subglacial hydrology: a qualitative review. *Arctic & Alpine Res.* **21**, 221–33.

1998. *Principles of Glacier Mechanics*. Upper Saddle River, NJ, Prentice Hall.

Hughes, T. J. 1973. Is the West Antarctic ice sheet disintegrating? *J. Geophys. Res.* **78**, 7884–910.

1983. The stability of the West Antarctic ice sheet: what has happened and what will happen. In *Proceedings of Carbon Dioxide Conference: Carbon Dioxide, Science and Consensus*. Berkeley Springs, W. Va. Sept. 19–23, 1982. IV.51–IV.73.

1992. On the pulling power of ice streams. *J. Glaciol.* **38**, 125–51.

1998. *Ice Sheets*. Oxford University Press.

Hulbe, C. L. and MacAyeal, D. R. 1999. A new numerical model of coupled inland ice sheet, ice stream, and ice shelf flow and its application to the West Antarctic ice sheet. *J. Geophys. Res.* **104**, 25 349–66.

Hutter, K. 1983. *Theoretical Glaciology*. Dordrecht, D. Reidel.

Huybrechts, P. 1992. The Antarctic ice sheet and environmental change: a three-dimensional modelling study. *Berichte zur Polarforschung* **99**, 241 pp.

1994. The present evolution of the Greenland ice sheet: an assessment by modelling. *Global & Planetary Change* **9**, 39–51.

2002. Sea-level changes at the LGM from ice-dynamic reconstructions of the Greenland and Antarctic ice sheets during the glacial cycles. *Quat. Sci. Rev.* **21**, 203–31.

Huybrechts, P. and De Wolde, J. 1999. The dynamic response of the Greenland and Antarctic ice sheets to multiple-century climatic warming. *J. Climate* **12**, 2169–88.

Huybrechts, P. *et al.* 1996. The EISMINT benchmarks for testing ice-sheet models. *Ann. Glaciol.* **23**, 1–14.

Jenssen, D. 1977. A three-dimensional polar ice sheet model. *J. Glaciol.* **18**, 373–90.

Joughin, I., Fahnestock, M., MacAyeal, D., Bamber, J. L. and Gogineni, P. 2001. Observation and analysis of ice flow in the largest Greenland ice stream. *J. Geophys. Res.* **106**, 34 021–34.

Jouzel, J. *et al.* 1987. Vostok ice core: a continuous isotope temperature record over the last climatic cycle (160,000 years). *Nature* **329**, 402–8.

Kamb, B. 1970. Sliding motion of glaciers: theory and observations. *Rev. Geophys. & Space Phys.* **8**, 673–728.

1987. Glacier surge mechanism based on linked cavity configuration of the basal water conduit system. *J. Geophys. Res.* **92**, 9083–100.

1991. Rheological nonlinearity and flow instability in the deforming bed mechanism of ice stream motion. *J. Geophys. Res.* **96**, 16 585–95.

Kamb, W. and Echelmeyer, K. 1986. Stress-gradient coupling in glacier flow: IV. Effects of the 'T' term. *J. Glaciol.* **32**, 342–9.

Kruss, P. D. and Smith, I. N. 1982. Numerical modelling of the Vernagtferner and its fluctuations. *Zeits. Gletscherkund & Glazialgeol.* **18**, 93–106.

Kwok, R. and Fahnestock, M. A. 1996. Ice sheet motion and topography from radar interferometry. *IEEE Trans. Geosci. Remote Sensing* **23** (1), 189–200.

Lingle, C. S. 1984. A numerical model of interactions between a polar ice stream and the ocean: application to Ice Stream E, west Antarctica. *J. Geophys. Res.* **89**, 3523–49.

1985. A model of a polar ice stream, and future sea-level rise due to possible drastic retreat of the West Antarctic Ice Sheet. In *Glaciers, Ice Sheets, and Sea Level: Effect of a CO_2-induced Climatic Change.* US Department of Energy Report DOE/EV/60235–1, pp. 317–30.

Lingle, C. S. and Brown, T. J. 1987. A subglacial aquifer bed model and water pressure dependent basal sliding relationship for a west Antarctic ice stream. In Van der Veen, C. J. and Oerlemans, J., eds., *Dynamics of the West Antarctic Ice Sheet.* Dordrecht, D. Reidel, pp. 249–85.

Lythe, M. B. and Vaughan, D. G. 2001. BEDMAP: a new ice thickness and subglacial topographic model of Antarctica. *J. Geophys. Res.* **106** (B6), 11 335–51.

MacAyeal, D. R. 1989. Large-scale ice flow over a viscous basal sediment – theory and application to ice stream-B, Antarctica. *J. Geophys. Res.* **94**, 4071–87.

MacAyeal, D. R. 1993. A tutorial on the use of control methods in ice-sheet modelling. *J. Glaciol.* **39**, 91–8.

MacAyeal, D. R., Bindschadler, R. A. and Scambos, T. A. 1995. Basal friction of ice-stream-E, west Antarctica. *J. Glaciol.* **41**, 247–62.

MacAyeal, D. R., Rommelaere, V., Huybrechts, P., Hulbe, C. L., Determan, J. and Ritz, C. 1996. An ice-shelf model test based on the Ross ice shelf, Antarctica. *Ann. Glaciol.* **23**, 46–51.

McTigue, D. F., Passman, S. L. and Jones, S. J. 1985. Normal stress effects in the creep of ice. *J. Glaciol.* **31**, 120–6.

Man, C.-S. and Sun, Q.-X. 1987. On the significance of normal stress effects in the flow of glaciers. *J. Glaciol.* **33**, 268–73.

Marshall, S. J. and Clarke, G. K. C. 1997. A continuum mixture model of ice stream thermomechanics in the Laurentide ice sheet 2. Application to the Hudson Strait ice stream. *J. Geophys. Res.* **102**, 20615–37.

Mercer, J. H. 1968. Antarctic ice and Sangamon sea level. IAHS publ. no. 79, pp. 217–25.
　1978. West Antarctic ice sheet and CO_2 greenhouse effect: a threat of disaster. *Nature* **271**, 321–5.

Murray, T. *et al.* 2000. Glacier surge propagation by thermal evolution at the bed. *J. Geophys. Res.* **105**, 13491–507.

Nye, J. F. 1953. The flow law of ice from measurements in glacier tunnels, laboratory experiments and the Jungfraufirn borehole experiments. *Proc. Roy. Soc. London, Ser. A* **219**, 477–89.
　1963. Correction factor for accumulation measured by the thickness of annual layers in an ice sheet. *J. Glaciol.* **4**, 785–8.
　1965. The flow of a glacier in a channel of rectangular, elliptic or parabolic cross-section. *J. Glaciol.* **5**, 661–90.
　1969. A calculation on the sliding of ice over a wavy surface using a Newtonian viscous approximation. *Proc. Roy. Soc. London, Ser. A* **311**, 445–67.
　1970. Glacier sliding without cavitation in a linear viscous approximation. *Proc. Roy. Soc. London, Ser. A* **315**, 381–403.

Oerlemans, J. and Van der Veen, C. J. 1984. *Ice Sheets and Climate.* Dordrecht, D. Reidel.

Oerlemans, J. *et al.* 1998. Modelling the response of glaciers to climate warming. *Climate Dyn.* **14**, 267–74.

Paterson, W. S. B. 1994. *The Physics of Glaciers, 3rd edn.* Oxford, Pergamon Press/Elsevier.

Paterson, W. S. B. and Budd, W. F. 1982. Flow parameters for ice-sheet modeling. *Cold Regions Sci. Tech.* **6**, 175–7.

Pattyn, F. 1996. Numerical modelling of a fast-flowing outlet glacier: experiments with different basal conditions. *Ann. Glaciol.* **23**, 237–46.

Payne, A. J. 1995. Limit cycles in the basal thermal regime of ice sheets. *J. Geophys. Res.* **100**, 4249–63.
　1998. Dynamics of the Siple Coast ice streams, west Antarctica: results from a thermomechanical ice sheet model. *Geophys. Res. Lett.* **25**, 3173–6.
　1999. A thermomechanical model of ice flow in west Antarctica. *Climate Dyn.* **15**, 115–25.

Payne, A. J. *et al.* 2000. Results from the EISMINT Phase 2 simplified geometry experiments: the effects of thermomechanical coupling. *J. Glaciol.* **46**, 227–38.

Petit, J. R. *et al.* 1999. Climate and atmospheric history of the past 420,000 years from the Vostok ice core, Antarctica. *Nature* **399**, 429–36.

Raymond, C. F. 1980. Valley glaciers. In Colbeck, S. C., ed., *Dynamics of Snow and Ice Masses.* New York, Academic Press, pp. 79–139.

Rigsby, G. P. 1958. Effect of hydrostatic pressure on the velocity of shear deformation of single ice crystals. *J. Glaciol.* **3**, 273–8.

Ritz, C. 1987. Time dependent boundary conditions for calculation of temperature fields in ice sheets. In Waddington, E., ed., *The Physical Basis of Ice Sheet Modelling.* IAHS publ. no. 170, pp. 207–16.

Ritz, C., Fabre, A. and Letréguilly, A. 1996. Sensitivity of a Greenland ice sheet model to ice flow and ablation parameters: consequences for the evolution through the last climatic cycle. *Climate Dyn.* **13**, 11–24.

Ritz, C., Rommelaere, V. and Dumas, C. 2001. Modeling the evolution of Antarctic ice sheet over the last 420 000 years: implications for altitude changes in the Vostok region. *J. Geophys. Res.* **106**, 31 943–64.

Robin, R. de Q. 1983. Radio-echo studies of internal layering of polar ice sheets. In Robin, G. de Q., ed., *The Climatic Record in Polar Ice Sheets*. Cambridge University Press, pp. 180–4.

Röthlisberger, H. 1972. Water pressure in intra- and subglacial channels. *J. Glaciol.* **11**, 177–204.

Sanderson, T. J. O. and Doake, C. S. M. 1979. Is vertical shear in an ice shelf negligible? *J. Glaciol.* **22**, 285–92.

Schowalter, W. R. 1978. *Mechanics of Non-Newtonian Fluids.* Oxford, Pergamon Press.

Schubert, G. and Yuen, D. A. 1982. Initiation of ice ages by creep instability and surging of the east Antarctic ice sheet. *Nature* **296**, 127–30.

Shepherd, A., Wingham, D. J., Mansley, J. A. D. and Corr, H. F. J. 2001. Inland thinning of Pine Island Glacier, west Antarctica. *Science* **291**, 862–4.

Siegert, M. J. 1999. On the origin, nature and uses of Antarctic ice-sheet radio-echo layering. *Prog. Phys. Geog.* **23**, 159–79.

Siegert, M. J., Dowdeswell, J. A., Gorman, M. R. and McIntyre, N. F. 1996. An inventory of Antarctic sub-glacial lakes. *Antarctic Sci.* **8**, 281–6.

Siegert, M. J., Hodgkins, R. and Dowdeswell, J. A. 1998. A chronology for the dome C deep ice-core site through radio-echo layer correlation with the Vostok ice core, Antarctica. *Geophy. Res. Lett.* **25**, 1019–22.

Siegert, M. J. *et al.* 2001. Physical, chemical and biological processes in Lake Vostok and other Antarctic subglacial lakes. *Nature* **414**, 603–9.

Takeda, A. L., Cox, S. J. and Payne, A. J. 2002. Parallel numerical modelling of the Antarctic ice sheet. *Computers & Geosci.* **28**, 723–34.

Tarboton, D. G. 1997. A new method for the determination of flow directions and contributing areas in grid digital elevation models. *Water Resources Res.* **33**, 309–19.

Thomas, R. H. 1973. The creep of ice shelves: theory. *J. Glaciol.* **12**, 45–53.
    1979. The dynamics of marine ice sheets. *J. Glaciol.* **24**, 167–77.

Thomas, R. H. and Bentley, C. R. 1978. A model for the Holocene retreat of the west Antarctic ice sheet. *Quat. Res.* **10**, 150–70.

Thomas, R. H. and MacAyeal, D. R. 1982. Derived characteristics of the Ross ice shelf, Antarctica. *J. Glaciol.* **28**, 397–412.

Thomas, R. H., Sanderson, T. J. O. and Rose, K. E. 1979. Effect of a climatic warming on the west Antarctic ice sheet. *Nature* **227**, 355–8.

Thomas, R. H. *et al.* 2000. Substantial thinning of a major east Greenland outlet glacier. *Geophys. Res. Lett.* **27**, 1291–4.

Vachaud, G. and Chen, T. 2002. Sensitivity of a large-scale hydrologic model to quality of input data obtained at different scales; distributed versus stochastic non-distributed modelling. *J. Hydrology* **264**, 101–12.

Van der Veen, C. J. 1985. Response of a marine ice sheet to changes at the grounding line. *Quat. Res.* **24**, 257–67.
    1987. Longitudinal stresses and basal sliding: a comparative study. In Van der Veen, C. J. and Oerlemans, J., eds., *Dynamics of the West Antarctic Ice Sheet*. Dordrecht, D. Reidel, pp. 223–48.

1989. A numerical scheme for calculating stresses and strain rates in glaciers. *Math. Geol.* **21**, 363–77.

1999a. *Fundamentals of Glacier Dynamics.* Rotterdam, A. A. Balkema.

1999b. Evaluating the performance of cryospheric models. *Polar Geog.* **23**, 83–96.

2001. Greenland ice sheet response to external forcing. *J. Geophys. Res.* **106**, 34 047–58.

2002. Polar ice sheets and global sea level: how well can we predict the future? *Global & Planetary Change* **32**, 165–94.

Van der Veen, C. J. and Whillans, I. M. 1989. Force budget: I. Theory and numerical methods. *J. Glaciol.* **35**, 53–60.

1990. Flow laws for glacier ice: comparison of numerical predictions and field measurements. *J. Glaciol.* **36**, 324–39.

1996. Model experiments on the evolution and stability of ice streams. *Ann. Glaciol.* **23**, 129–37.

Walder, J. S. 1986. Hydraulics of subglacial cavities. *J. Glaciol.* **32**, 439–45.

Weertman, J. 1957a. On the sliding of glaciers. *J. Glaciol.* **3**, 33–8.

1957b. Deformation of floating ice shelves. *J. Glaciol.* **3**, 38–42.

1964. The theory of glacier sliding. *J. Glaciol.* **5**, 287–303.

1972. General theory of water flow at the base of a glacier or ice sheet. *Rev. Geophys. & Space Phys.* **10**, 287–333.

1974. Stability of the junction of an ice sheet and an ice shelf. *J. Glaciol.* **13**, 3–11.

1976. Sliding-no sliding zone effect and age determinations of ice cores. *Quat. Res.* **6**, 203–7.

Whillans, I. M. and Van der Veen, C. J. 1993a. Patterns of calculated basal drag on ice stream-B and ice stream-C, Antarctica. *J. Glaciol.* **39**, 437–46.

1993b. New and improved determinations of velocity of ice streams B and C, West Antarctica. *J. Glaciol.* **39**, 483–90.

1997. The role of lateral drag in the dynamics of ice stream B, Antarctica. *J. Glaciol.* **43**, 231–7.

Whillans, I. M., Bentley, C. R. and Van der Veen, C. J. 2001. Ice streams B and C. In Alley, R. B. and Bindschadler, R. A., eds., *The West Antarctic Ice Sheet: Behavior and Environment.* Antarctic Research Series vol. 77. Washington, D.C., American Geophysical Union, pp. 257–81.

Wingham, D. J., Ridout, A. J., Scharroo, R., Arthern, R. J. and Shum, C. K. 1998. Antarctic elevation change from 1992 to 1996. *Science* **282**, 456–8.

Yuen, D. A. and Schubert, G. 1979. The role of shear heating in the dynamics of large ice masses. *J. Glaciol.* **24**, 195–212.

Yuen, D. A., Saari, M. R. and Schubert, G. 1986. Explosive growth of shear-heating instabilities in the down-slope creep of ice sheets. *J. Glaciol.* **32**, 314–20.

7

Modelling the dynamic response of sea ice

WILLIAM D. HIBLER, III
International Arctic Research Center,
University of Alaska at Fairbanks

7.1 Introduction

The dynamical response of sea ice to climate change depends on a complex interplay between mechanical and thermodynamic processes driven by radiation, temperature, wind and oceanic forcing. Because of ice deformation, a typical $100\,\text{km}^2$ patch of sea ice will contain a variety of ice thicknesses. These thicknesses range from open water to very thick ice, including pressure ridges extending possibly 30 m or more below the surface. On top of this matrix there is often a relatively thin snow cover. Although thin, this snow cover can cause substantial insulation of the ice and reduce its growth rate.

While the main driving forces that move this ice cover come from wind and currents, ice does not just move as a passive tracer. Instead, ice has a motion and, more notably, deformation that is significantly affected by the ice interaction. Far from shore, the effects of interaction are more subtle, but still considerable in that excessive ice buildup is prevented by ice pressure. In addition, deformation typically takes place in the form of long intersecting leads; a fracturing pattern that is common in the brittle failure of many materials. From models and observation, ice pressure is comparable in magnitude to the buildup of surface pressure in the ocean by sea-surface tilt. Ice stresses averaged over the ice thicknesses that are typically used in large-scale models, for example (Hibler, 2001), are of the order of $2\text{--}5 \times 10^4\,\text{N/m}^2$, which is approximately equivalent to the bottom pressure of a 2–5 m high column of water.

The thermodynamic properties of sea ice are complicated by the fact that when sea-water freezes under quiescent conditions, much of the dissolved salt is expelled leaving newly formed (first-year) ice with a salinity of about 5–10 parts per thousand (ppt). The remaining salt is trapped in small (0.1–1 mm) brine pockets, giving sea ice thermodynamically very different properties from fresh water ice. These brine pockets can also migrate. Some of the salt in these brine pockets drains or is flushed out by melt water in subsequent summers so that multi-year sea ice in the Arctic has a salinity of only 2–4 ppt. Because of these

This work was supported by the Frontier Research System for Global Change through the International Arctic Research Center. The author would particularly like to thank Dr Tony Payne for insightful readings and considerable editorial aid, which made completion of this chapter possible; Dr Jonathan Bamber for persistent support and encouragement; and Dr Greg Flato for in-depth reviews and comments.

Mass Balance of the Cryosphere: Observations and Modelling of Contemporary and Future Changes, eds. Jonathan L. Bamber and Antony J. Payne. Published by Cambridge University Press. © Cambridge University Press 2003.

features, solidification of salt water is really a problem involving a binary system exhibiting a two-component mushy character almost everywhere. Similarly under melt conditions, the change of the conductivity and internal melting due to penetrating radiation can greatly change equilibrium thicknesses of sea ice due mainly to internal increase of brine volume.

The presence of this dynamic ice cover substantially alters heat, salt and momentum transfers between the atmosphere and ocean, and hence has the potential to alter atmospheric and oceanic circulation. In these modifications, sea-ice dynamics plays a pivotal role inasmuch as leads and ridging substantially change the air–sea heat exchanges as compared with a constant thickness ice cover. Moreover, as climate changes, the mechanical characteristics may alter the evolution of the system by modifying the deformation field for the same dynamical forcing. In addition, ice transport allows net local imbalances in salt and heat to exist.

Finally, the fact that sea ice is formed from and floats in the ocean means that modelling the growth drift and decay of sea ice is inherently an ice–ocean dynamics problem, in which the oceanic boundary layer characteristics must at least be considered. Because of these considerations, much of the current effort in large-scale ice modelling is taking place in the context of coupled ice–ocean circulation models at increasingly fine resolution.

In the modelling overview presented here, physical processes generally felt to be pivotal to the response of sea ice to past and future climate changes are emphasized. Because of their complexity, many of the processes are only tractable by means of numerical investigation. Consequently the emphasis here is on determining physical processes identified by certain archetypal model or analytical studies. A more complete complementary description of the current state of modelling contemporary and future sea-ice characteristics is found in Chapter 9.

The overall structure of this chapter is divided into three broad categories: a brief review of selected observational characteristics (section 7.2); a detailed examination of the components of coupled dynamic thermodynamic sea-ice models (sections 7.3–7.6); and a final section examining a hierarchy of model simulations of the evolution of sea ice, with special emphasis on dynamical effects. In the various sections, short analytic and numerical examples are given to illustrate certain processes or physics, and to identify current scientific questions and areas of research. In the final section, selected results from fully coupled dynamic thermodynamic models of differing complexity are given to show characteristic behaviour, limitations, timescales, predictions, and the effects of ice dynamics on climate change. While thermodynamics is treated in some detail, the emphasis is on dynamical processes; a focus reflected in the order of the sections. Overall, this chapter is designed to be a pedagogical aid in understanding sea-ice models used in investigations of contemporary and future changes in the sea-ice component of the cryosphere.

7.2 Selected observational sea-ice motion: mechanical and physical characteristics

While this chapter is focussed on modelling the dynamical response of sea ice, to set the framework for the physical treatment of the system, certain observational characteristics are critical. These observations help to establish scaling of the momentum equations and to identify essential physical processes. A much more comprehensive discussion of observational characteristics is examined in Chapter 8. Here, a few key drift and deformation features especially critical to ice mechanics and ice growth and decay are given. Physical and mechanical properties relevant to both thermodynamics and ice dynamics are also briefly discussed.

7.2.1 Sea-ice drift, deformation and pressure ridges

On short timescales, ice tends to follow the wind, with the drift approximately following the geostrophic wind with about 1/50 of the magnitude (Zubov, 1943). This feature has been quantified in greater detail by Thorndike and Colony (1982) using smoothed drifting buoy data and smoothed wind fields, which somewhat mimics the effects of ice interaction. Since wind variations are larger than currents on short timescales, this also means that fluctuations in ice drift will typically be dominated more by wind than by currents. However, on a long timescale, the more steady currents, together with ice mechanics effects (Hibler and Bryan, 1987; Steele *et al.*, 1997), can play a major role.

The general drift rates and some information on variability can be deduced from the mean monthly drifting stations positions plotted in Figure 7.1. Two notable features illustrated by this figure are the presence of substantial drift (even in winter) parallel to the Alaskan coast and a relatively small component of motion perpendicular to the Canadian archipelago near the pole. A feature of drift tracks not shown in Figure 7.1 is the meandering nature of the drift, with week-long reversals in direction being common. Also, as drifting stations approach the Greenland–Spitsbergen passage, the drift rates begin to increase and meandering decreases. This phenomenon continues after the ice exits from the basin, the drift rates being as large as 25 km per day in the East Greenland region (Vinje, 1976). In recent years, the presence of large numbers of remote buoys in the Arctic have made the ice-drift field the most precisely measured surface field of any world ocean.

Since ice drift generally follows the wind, the shifting of ice-drift patterns in response to large-scale oscillations in wind fields has been of great interest in recent years. Recent work has emphasized patterns of wind fields coincident with the North Atlantic oscillation (NAO) index and their effect on ice drift (Polyakov, Proshutinsky and Johnson, 1999; Proshutinsky and Johnson, 1997; Wang and Ikeda, 2000). The NAO is largely a measure of shifts in high latitude wind fields in the northern hemisphere (Hurrell, 1995). Under this classification, there is a weakened anti-cyclonic flow together with a more concentrated Beaufort Sea gyre under the so-called high index (Figure 7.2(a)). With a low index, the drift pattern tends to have stronger anti-cyclonic flow with a much larger and more pronounced Beaufort Gyre (Figure 7.2(b)) together with greater recirculation of ice. Differences (Tucker *et al.*, 2001) in ice thicknesses deduced from sonar data in the Beaufort Sea between the early 1980s and the late 1990s can be related to these differences. In particular, Tucker *et al.* (2001) found significantly thicker ice in the Beaufort Gyre region (86°N, 146°W) in 1986–87 during

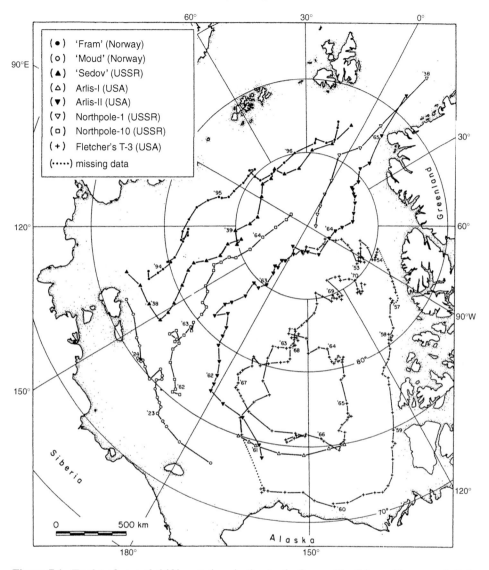

Figure 7.1. Tracks of several drifting stations in the Arctic Ocean. Monthly positions are plotted, with the year of each January 1 position indicated. (From Hastings (1971).)

the low NAO index than in the 1990s when a high index prevailed. They attributed much of this difference to longer residence ice time due to the ice circulation and hence more accumulation of ice mass through ridging.

In the Arctic, a dominant term in the ice-mass budget of the sea ice is the ice flux through the Fram Strait. Probably the most detailed observations of outflow have been made by Kwok and Rothrock (1999) using passive microwave satellite data. These results

(a) (b)

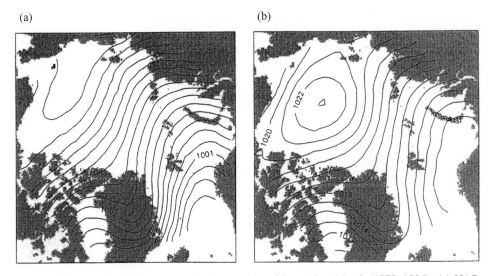

Figure 7.2. Mean sea-level pressure for the months of December–March (1978–1996): (a) NAO index >+1; (b) NAO index <−1. (Contour interval 1 mbar.) (From Kwok and Rothrock (1999).)

(Figure 7.3) show an areal outflow that varies somewhat with location in the Fram Strait. The outflow tends to correlate strongly with the average geostrophic wind through the Fram Strait and also with the North Atlantic oscillation. However, the best-fit regression coefficient between the December–March area flux and geostrophic wind is one-half of the October–May regression coefficient. Also the outflow magnitudes are significantly smaller during December to March for the same atmospheric pressure gradient across the Fram Strait. This difference is most likely due to ice mechanics effects (see section 7.4.3). Areal outflows are relatively precisely measured by this method, while ice-mass outflow is more uncertain due to lack of knowledge about the lateral thickness profile of the sea ice across the strait. Typical long-term estimates (Aagaard and Greisman, 1975; Kwok and Rothrock, 1999; Vijne and Finnekasa, 1986) of ice-area outflow and ice-mass outflow (based partially on moored upward looking sonar instruments) are ~ 0.1 Sverdrup (3154 km^3 per year) and $\sim 900 \times 10^6$ km^2 per year for ice area. There is considerable inter-annual variability. Areal ice volume fluxes from interpolated buoy drift data (Colony, 1990) are of similar magnitude. Interestingly, as technology has presumably improved over the last 15 years, volume fluxes have decreased from highs of 4–5 $\times$ 10^3 km^3 per year (Vijne and Finnekasa, 1986; Wadhams, 1983). A major issue is uncertainty in the lateral thickness profile across the strait (Vigne, Norland and Kvambekk, 1998).

Concomitant with ice drift is ice deformation. This deformation fluctuates in a manner that is not directly related to wind field, especially at high frequencies. As discussed below, ice deformation is felt to be significantly modified by ice mechanics. Typical measurements of deformation are shown in Figure 7.4. These results were obtained from an experiment (Hibler *et al.*, 1974a) in the spring of 1972 in the Beaufort Sea. In this experiment a

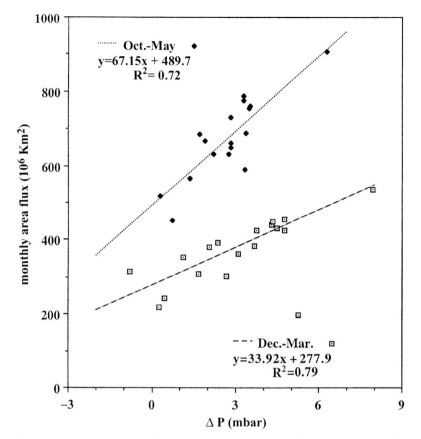

Figure 7.3. Typical areal outflow estimates over the time period 1979 to 1996 (from Kwok and Rothrock (1999)). Illustrated here are the scattergrams showing observations as well as the correlation and regression line between monthly area flux and gradient in the monthly sea-level pressure across the Fram Strait for October–May (dotted line) and December–March (dashed line).

detailed set of small-scale measurements of the temporal variability of the strain were obtained by using laser surveying techniques to measure the positions of a large number of reflectors mounted on towers on the ice. The smoothed deformation (Figure 7.4(c)) from this meso-scale array compares well with larger-scale observations. In addition to deformation occurring in response to large-scale wind fields, analysis (Hibler, 1974) indicates that there is a definite tendency for the deformation to oscillate at around 12-hour intervals (Figure 7.4(d)). In the polar regions this period coincides with the inertial period and also with the semi-diurnal tidal period. This high frequency character of ice deformation has been a general feature of deformation measurements almost everywhere, including the East Greenland marginal sea-ice zone (Lepparanta and Hibler, 1987) and the Antarctic ice pack (Heil, Lytle and Allison, 1998). In terms of ice drift there appears to be a seasonal character

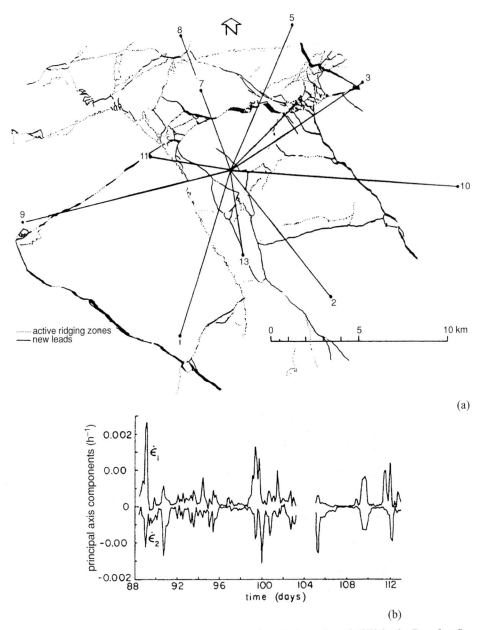

(a)

(b)

Figure 7.4. (a) Meso-scale strain array in the late winter/early spring of 1972 in the Beaufort Sea with an overlay of active leads and ridging zones on April 6, 1972. (b) Ice deformation time series from meso-scale strain array. (c) Comparison of smoothed meso-scale and macro-scale (strain triangle over ~100 km region) divergence rates and clockwise vorticity (Hibler *et al.*, 1974a). (d) Spectra of atmospheric pressure and meso-scale ice divergence rate from strain array shown in (a) (Hibler, 1974).

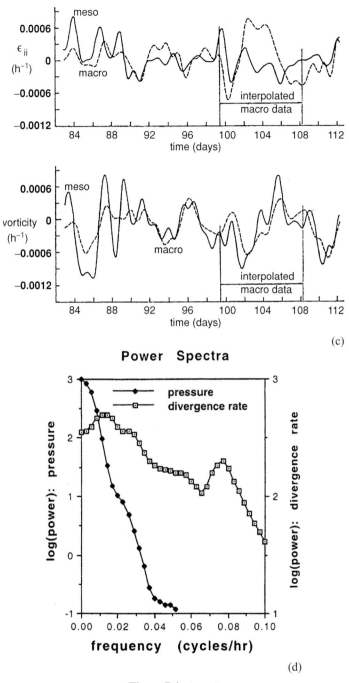

(c)

Power Spectra

(d)

Figure 7.4. (*cont.*)

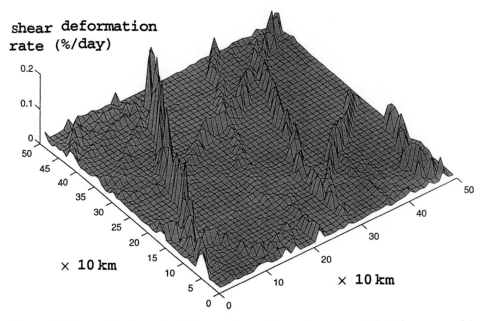

shear deformation rate (%/day)

× 10 km

× 10 km

Figure 7.5. Perspective view of spatially varying maximum shear rate obtained from sequential synthetic aperture radar (SAR) images over a three-day period in the central Arctic Basin. (R. Kwok private communication; see also Kwok (2001).)

of the oscillations in drift (Thorndike and Colony, 1980) presumably due to greater damping by ice interaction in winter.

The spatial structure of deformation, however, often tends to take place along linear oriented patterns. This is most effectively illustrated by spatial deformation patterns on a 5 km footprint obtained from processing synthetic aperture radar (SAR) imagery (Kwok, 2001; see Figure 7.5). While such deformation patterns have been apparent for some time on visual snapshots (e.g., Marko and Thompson, 1977), the actual deformation taking place concomitant with the formation of leads has not normally been directly determined. Examination of SAR and visual imagery has also led to a consistency in lead intersection angles. In particular, Cunningham, Kwok and Banfield (1994) demonstrated a remarkable tendency for leads to intersect at angles of about 40–50° with the orientation of the leads tending to be symmetric about principal axes of the strain rate tensor.

Important manifestations of deformation are large pressure ridges which routinely reach depths greater than 20 m (Hibler, Weeks and Mock, 1972; McLaren, 1989; Wadhams, 1981, 1983, 1992) and have been observed as deep as 47 m (W. Lyons, personal communication). Analysis (Hibler *et al.*, 1972) of 1960 winter-cruise data from the Sargo (Figure 7.6(a)) shows distributions (Figures 7.6(b,c)) of pressure ridges with typically at least ten ridges per kilometre deeper than 6.1 m (20 ft). The pressure ridge distributions were found to fit a theoretical distribution (Hibler *et al.*, 1972) based on the assumption that any arrangement of ridges with the same volume of deformed ice were equally probable. In the central Arctic

~10% of the ridges were deeper than 20 m, whereas in the most heavily ridged region ~10% of the ridges were below 30 m.

Observations of spatial and temporal variations of ridging in the western Arctic Basin (Hibler, Mock and Tucker, 1974b) deduced from airborne laser profilometers generally show the spatial variability of ridging to correlate well with mean thickness estimates obtained from submarine sonar data (e.g., Bourke and Garrett, 1987; Bourke and McClaren, 1992). In addition to these basin-wide variations, near-shore ridge studies (Tucker, Weeks and Frank, 1979) off the north slope of Alaska and Canada show a buildup of ridging near the coast which decays several hundred kilometres offshore. Field studies show this region to be highly deformed with substantial amounts of rubble pile-up (Kovacs, 1976). The magnitude of ridging can vary from year to year. Hibler *et al.* (1974b), for example, found the ridging intensity (a rough measure of the mean thickness of ridged ice) to be 50% larger off the Canadian archipelago in the winter of 1972 than in 1971, and to comprise (assuming

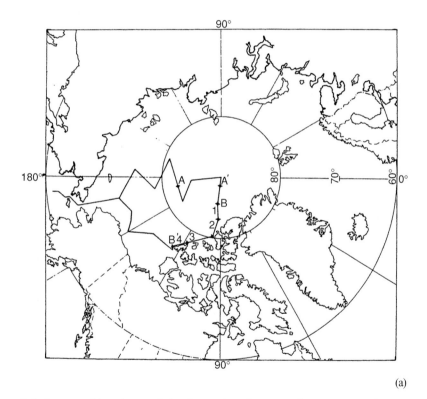

(a)

Figure 7.6. Pressure ridge characteristics from submarine sonar data. (a) Approximate locations (W. Lyons, personal communication) of sonar profile tracks. Observed and theoretical distribution for ridges deeper than 6.1 m in (b) central Arctic province A–A′ and (c) sub-province 2 of track B–B′. In (b) there were a total of 5702 ridges, and the normalized theoretical probablility density function (Npdf) was $0.346 \exp(-0.0116h^2)$. For (c) there were 3601 ridges with an (Npdf) of $0.174 \exp(-0.006h^2)$. (From Hibler *et al.* (1972).)

Central Arctic

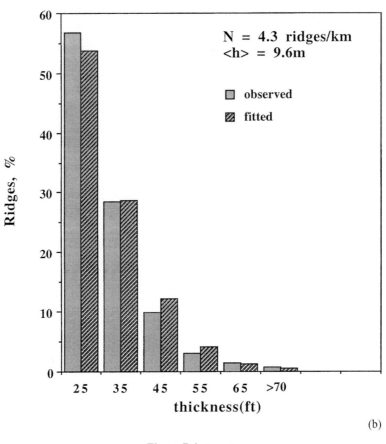

(b)

Figure 7.6. (*cont.*)

triangular ridges) of the order of $2\,m^3/m^2$ of ice in 1972. While it is difficult to determine, estimates of ridged ice area fractions from submarine transects are generally quite high: 60–80% by Williams, Swithinbank and Robin (1975), 44% by Wadhams and Horne (1980), and 38–43% by McLaren (1989).

First-year pressure ridges with block sizes up to 2 m (Tucker and Govoni, 1981) become incorporated in multi-year floes if they survive a melt season. These consolidated multi-year pressure ridges are extremely strong (Kovacs *et al.*, 1973) and consistently (Kovacs and Mellor, 1974; Weeks *et al.*, 1971) have a smaller keel to depth ratio (~3.2 m) than first-year pressure ridges (~4.5 m). Multi-year pressure ridges also ablate more slowly (Koerner, 1973) and may well comprise a large fraction of the ice volume. Observations, for example, of the physical and crystalline properties of ice in the Fram Strait region (Tucker, Gow and Weeks, 1987) found that one-third of ice cores drilled in multi-year ice (which comprised

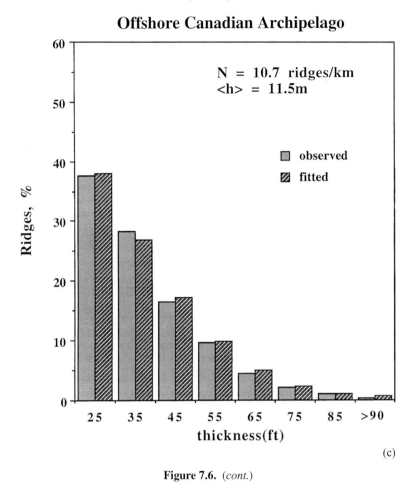

Figure 7.6. (*cont.*)

over 84% of the ice volume in the region) were identified as ridged ice; this was despite the fact that floes which appeared from the surface to be ridged were explicitly avoided.

7.2.2 Ice stress and physical properties

A key thermodynamic feature of sea ice is the presence of internal brine pockets whose size and salinity depends on local temperatures (Weeks and Ackley, 1986) in the ice. Many of the complexities of thermodynamic growth and small-scale ice strength arise from this fact. A comprehensive review of many of these properties may be found in Weeks and Ackley (1986). Here, some of the salient features relevant to treatments of thermodynamics in geophysical-scale ice models are briefly discussed.

The basic character of the average salinity of first- and multi-year ice in the Arctic is shown in Figure 7.7. Figure 7.7(a) (from Cox and Weeks, 1974) shows that the vertically

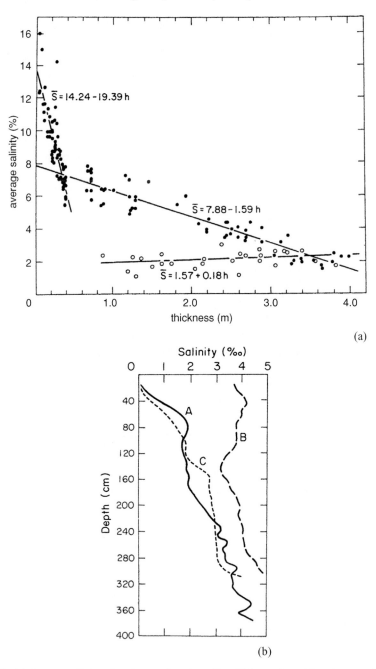

Figure 7.7. (a) Average salinity of sea ice as a function of ice thickness for cold sea ice sampled during the growth season and warm sea ice sampled during or at the end of the melt season: (solid circles) cold sea ice; (open circles) warm sea ice. (b) Average salinity profiles for multi-year ice. Curves A and B are the average hummock and depression salinity profiles obtained by Cox and Weeks (1974). Curve C is the multi-year ice average salinity profile determined by Schwarzacher (1959). (From Cox and Weeks (1974).)

averaged salinity is closely correlated with the ice thickness and depends on the overall temperature regime. As can be seen, the average salinity drops rapidly as the thickness approaches 0.4 m and then decreases more slowly for cold ice. After ice survives a melt season its salinity profile takes on a characteristic appearance. This is made possible by a 'steady-state' seasonal growth cycle of old ice, which consists of ablation in the summer and ice replacement in the winter. The multi-year salinity profile used in many thermodynamic models was developed by Schwarzacher (1959) on the basis of a large number of cores in multi-year ice (Figure 7.7(b)). However, while such a standard profile appears reasonable for hummocked ice, the study by Cox and Weeks (1974) indicates that different profiles can occur in low spots of multi-year ice where melt ponds have formed in summer. This fact emphasizes that the characteristics of old 'level' ice are not horizontally homogeneous.

The development of the capability to make stress measurements efficiently over the past few years (Cox and Johnson, 1983) has, in principle, provided a means to estimate stress buildup in pack ice directly. Heretofore, estimates of stress have been largely based on comparisons of non-linear sea-ice models with buoy-based ice-drift observations. Representative sets of ice-stress measurements were made in the eastern Arctic by Tucker and Perovitch (1992) and Tucker *et al.* (1991) and in the western Arctic (Beaufort Sea) by Richter-Menge (1997) and Richter-Menge and Elder (1998). Typical measurements of stresses (Figure 7.8) in the thick and thin ice show a high degree of variability. In practice, these changes appear to be associated with the deformation field over a region. Records from sensors very near the edges of multi-year floes are suggestive of rapid stress buildup and release, implying loading and subsequent fracturing. In the centres of the floes the stresses are smaller and have a somewhat reduced magnitude inasmuch as they represent the integrated stress around all boundaries. In the Beaufort Sea the central floe stresses tend to have maxima in the range of 50×10^3 Pa. At edge sites where stresses are highly concentrated, the stresses may rise to 200 kPa.

A dominant component of measured stresses in pack ice is a diurnal cycle (Lewis and Richter-Menge, 1998). This cycle tends to be suppressed in the absence of a daily

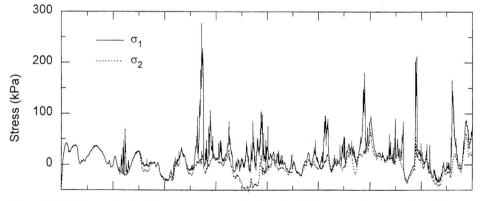

Figure 7.8. Measured principal stress components at a site ∼300 m from the edge of a multi-year floe over a 160-day interval. (From Richter-Menge and Elder (1998).)

temperature swing, and is thought to be due to thermal stresses in the ice floes due to differential heating (Lewis, 1998). Because of this strong diurnal variability, it is not clear whether semi-diurnal variations (Tucker *et al.*, 1991) are a separate phenomenon, or are harmonics of a daily thermal cycle.

An important characteristic feature of geophysical-scale ice mechanics is the stoppage of flow through narrow channels by 'arching', as analysed, for example, by Sodhi (1977) in the Bering Strait. This arching phenomenon forms a free surface downwind or downstream of a narrow passage, which can support enough stress to prevent ice flow through the channel. However, depending on the ice strength, increases in the wind or currents will break the arch, whose extent depends on the failure strength of the ice and the passage width. One of the most notable examples of recurring 'arching' is in the narrow Nares Strait region above Baffin Bay. This phenomenon normally occurs every year after an initial period of ice flow through the strait and plays a significant role in the formation of the north-water polyna by preventing the southward flow of ice into the north-water region.

Relatively stationary 'fast' ice is a mechanical/thermodynamic phenomenon prominent in shallow shelf regions of the Arctic. This phenomenon is somewhat related to the presence of arching in pack ice. Due to the large extent of shallow shelves in the Russian Arctic, fast ice is more extensive there and can extend far from the coast. Studies of the seasonal and inter-annual variability (Brigham, 2000) of fast ice in the Russian Arctic show it to extend several hundred kilometres off the coast and to have a pronounced seasonal signal. Over the period 1975 to 1995, Brigham (2000) found the maximum extents to have a range of 600 000–900 000 km^2 over the longitudinal band 58°E to 180°. Of this variability, only a small amount ($\sim$90 000 km^2) could be explained by a uniform linear reduction over two decades. Off the Alaskan coast, fast ice often contains numerous ridging bands (Weeks *et al.*, 1977) indicating that it has been formed in incremental stages becoming gradually more extensive in character. Once formed, detailed measurements (Weeks *et al.*, 1977) of the motion of the fast ice in the Prudhoe Bay area show it typically to move less than 1 metre or so.

7.3 Modelling sea-ice drift and deformation

7.3.1 Equations of motion

Sea ice moves in response to wind and water currents and the internal stress in the ice. Considering sea ice to be a two-dimensional continuum, it obeys the normal Euler equations of motion (e.g., Fung, 1977)

$$m\frac{D\mathbf{v}}{Dt} = \boldsymbol{\nabla} \cdot \boldsymbol{\sigma} + \mathbf{X}, \tag{7.1}$$

where $\mathbf{v}$ is the ice velocity, m is the ice mass per unit area, $\boldsymbol{\sigma}$ is the second-order internal stress tensor in the sea ice due to ice interaction, $\mathbf{X}$ is the total body force on the ice, and the substantial derivative of the ice velocity is

$$\frac{D\mathbf{v}}{Dt} = \frac{\partial \mathbf{v}}{\partial t} + (\mathbf{v} \cdot \nabla)\mathbf{v}. \tag{7.2}$$

The inter-facial stresses between the ice, atmosphere and ocean are part of the body forces $\mathbf{X}$ in Euler's equation. Explicitly writing out these terms in the case that inter-facial stresses are known, the equation of motion is given by

$$m\frac{D\mathbf{v}}{Dt} = -mf\mathbf{k} \times \mathbf{v} + \tau_a + \tau_w + \nabla \cdot \boldsymbol{\sigma} - mg\nabla H, \tag{7.3}$$

where, in addition to the symbols defined in equation (7.2), $\mathbf{k}$ is a unit vector normal to the surface, f is the Coriolis parameter, τ_a and τ_w are the body forces per unit area due to air and water stresses and H is the height of the sea surface. In this formulation, τ_w is assumed to include frictional drag due to the relative movement between the ice and the underlying ocean.

In the steady-state solution of the ice–ocean boundary layer (a good approximation at timescales longer than a few inertial periods), the air and water stresses are normally determined from idealized integral boundary layer theories, assuming constant turning angles (Brown, 1980; Leavitt, 1980; McPhee, 1979):

$$\tau_a = c_a'(\mathbf{V}_g\cos\phi_a + \mathbf{k} \times \mathbf{V}_g\sin\phi_a) \tag{7.4}$$

$$\tau_w = c_w'[(\mathbf{V}_w - \mathbf{v})\cos\phi_w + \mathbf{k} \times (\mathbf{V}_w - \mathbf{v})\sin\phi_w], \tag{7.5}$$

where $\mathbf{V}_g$ is the geostrophic wind, $\mathbf{V}_w$ represents the geostrophic ocean currents near the ocean surface, c_a' and c_w' are air and water drag coefficients and ϕ_a and ϕ_w are air and water turning angles. In practice, both the currents and the sea-surface tilt are estimated from geostrophic considerations by setting H equal to the dynamic height and computing currents by $\mathbf{V}_w = \frac{g}{f}\mathbf{k} \times \nabla H$. In general, both c_a' and c_w' are non-linear functions of the winds and currents. The two most commonly used formulations are (1) linear, where c_a' and c_w' are taken to be constant, and (2) quadratic, where

$$c_a' = \rho_a c_a |\mathbf{V}_g|, \tag{7.6}$$

$$c_w' = \rho_w c_w |\mathbf{V}_w - \mathbf{v}|, \tag{7.7}$$

with c_a and c_w being dimensionless drag coefficients (with typical values of 0.0012 and 0.0055, respectively; McPhee (1980)) and turning angles ϕ_a and ϕ_w of about 25° in the Arctic or −25° in the Antarctic. In the linear models, a linear drag coefficient of $\rho(fK_v)^{1/2}$ (where ρ is the density and K_v is the vertical eddy viscosity) is obtained from classical Ekman layer theory with a turning angle of 45°. In a real turbulent boundary layer the vertical eddy viscosity varies with depth and velocity (McPhee, 1982). For this reason, quadratic drag laws with turning angles different from 45° give a better fit to observations (McPhee, 1980) than the simple linear Ekman theory. However, it should be noted that the best fit to observations occurs with drag laws where the turning angle varies with the surface stress.

The dominant terms in the temporally averaged momentum balance in equations (7.3)–(7.7) are the *air* and *water* stresses and the *ice interaction*. The Coriolis term, tilt and steady current terms are about an order of magnitude smaller. Also the inertia term in this time-averaged formulation is rather small, as it would take about one hour for ice to come to a steady-state drift state after the sudden imposition of a wind field. That the ice interaction is the same order of magnitude as the wind and water drag has been verified by observation in enclosed areas, such as the Bay of Bothnia, where the ice sometimes hardly moves at all, even with significant winds (Lepparanta, 1980; Omstedt and Sahlberg, 1977). However, as verified by a large number of numerical simulations (section 7.4), ice interaction typically opposes the wind stresses and hence does not prevent very high coherence between wind fluctuations and ice drift.

While this steady-state formulation is used in almost all climate studies, typical observations of ice motion and deformation (section 7.2.1) show the motion to have considerable power at the inertial period even in the absence of wind. An integral formulation (McPhee, 1978) that retains this motion within the above framework is to consider the turbulence large so that for all timescales longer than ~ 1 hr the boundary layer transport and the ice motion are inter-dependent, with the integrated mass transport of the oceanic boundary layer ($\mathbf{M_w} = \rho_w \bar{\mathbf{U}}$ taken to be given by

$$\mathbf{M_w} = (\rho_w c_w / f) V \mathbf{V} e^{-i\beta}, \tag{7.8}$$

where $\mathbf{V} = \mathbf{v} - V_w$ is the ice velocity relative to the steady currents. In these expressions, two-dimensional vectors have been written in complex form, i.e. $\mathbf{A} = A e^{i\delta}$, with δ the counter-clockwise turning angle from the real axis. Substituting this expression into the equation of motion for the combined momentum ($\mathbf{M} = \mathbf{M_w} + m\mathbf{V}$) for the ice and oceanic boundary layer,

$$\frac{D\mathbf{M}}{Dt} = \tau_a + f\mathbf{k} \times \mathbf{M} + \nabla \cdot \sigma. \tag{7.9}$$

Expressing all momentum components in terms of the ice velocity $\mathbf{V}$, we obtain a more realistic equation of motion (in complex form) for the ice cover:

$$m\frac{D\mathbf{V}}{Dt} - (\rho_w c_w / f)\frac{D(V\mathbf{V})}{Dt}e^{-i\beta} + ifm\mathbf{V} + \rho_w c_w V\mathbf{V}e^{-i(\pi-\beta)} = \tau_a + \nabla \cdot \sigma. \tag{7.10}$$

The main feature of equation (7.9) (or equation (7.10)) as compared with equations (7.3) and (7.5) is the presence of inertial oscillations largely driven by the oceanic boundary layer. These oscillations are undamped solutions of the homogeneous portion of equation (7.9) in the absence of ice interaction. They have periods of $(12/\sin \lambda)$ hour, where λ is the latitude, and are a ubiquitous phenomenon in the upper ocean (Gill, 1982). These oscillations describe a particle motion proceeding clockwise in the northern hemisphere. In equations (7.3) and (7.5) these solutions are artificially damped out in less than an hour or so, which is unrealistic.

Note that, in the steady-state limit, this more general expression (equation (7.10)) reduces to the same quadratic drag result as expressed in equations (7.3)–(7.8). However, there is a major difference in the relative magnitude of the momentum balance components on timescales of less than a few days. The importance of the time-dependent character of the momentum equation can be demonstrated by comparing unsmoothed buoy drift characteristics with simulated values (Heil and Hibler, 2002). Some of these characteristics are discussed in sections 7.4.3 and 7.6.6.

7.3.2 Deformation scaling of momentum equations

To a large degree, the deformation or differential drift of ice may be considered to be a perturbation on the smoother ice-drift fields which, as noted above, tend to follow the wind field. This deformation is, however, critical to the dynamical evolution of the thickness of the sea-ice cover as it affects open water creation and ridging. Because it is a 'perturbation', the essential scaling, however, can be quite different.

Aspects of deformation scaling may be examined (Hibler, 1974) by using a linear version of the steady-state ice-drift equations coupled with a linear-viscous rheology, so the $\boldsymbol{\nabla} \cdot \boldsymbol{\sigma}$ term in the momentum balance is given by

$$\boldsymbol{\nabla} \cdot \boldsymbol{\sigma} = \eta \nabla^2 \mathbf{v} + \zeta \nabla(\nabla \cdot \mathbf{v}), \tag{7.11}$$

where $\mathbf{v}$ is the ice velocity, and η and ζ are constant shear and bulk viscosities, respectively. Expressing the geostrophic wind in terms of the atmospheric pressure by

$$U_g = -\frac{1}{\rho f}\frac{\partial P}{\partial y} \tag{7.12}$$

$$V_g = \frac{1}{\rho f}\frac{\partial P}{\partial x} \tag{7.13}$$

and taking the curl and divergence of the linearized momentum balance (equation (7.3) with c'_a and c'_w taken to be constant in equations (7.4) and (7.5) and $\nabla H = 0$), we obtain two linear equations for $\nabla \cdot \mathbf{v}$ and the vorticity $\omega\big(=\frac{1}{2}\big(\frac{\partial v_y}{\partial x}-\frac{\partial v_x}{\partial y}\big)\big)$:

$$[(\eta + \zeta)\nabla^2 - D\cos\theta]\nabla\cdot\mathbf{v} + [mf + D\sin\theta]\,2\omega = \frac{B\sin\phi}{\rho f}\nabla^2 P \tag{7.14}$$

$$-[mf + D\sin\theta]\nabla\cdot\mathbf{v} + [\eta\nabla^2 - D\cos\theta]2\omega = \frac{-B\cos\phi}{\rho f}\nabla^2 P. \tag{7.15}$$

Similar equations for different components of the strain rate in a given co-ordinate system could also be obtained (Geiger, Hibler and Ackley, 1997).

Independent of the linear rheology assumption, these equations may be scaled with observed deformation data. Referring back to Figure 7.4 for deformation, we find typical values of divergence and vorticity to be $\sim 4 \times 10^{-4}$ per hour and 8×10^{-4} per hour. Using the fact that the divergence of the surface wind field is ~ 0.14 per hour, the wind stress term in these deformation equations is about ten times larger than the water stress terms in the equations of motion. Consequently, in this case the essential balance on scales of $\sim 100\,\mathrm{km}$ is between the gradient of the surface stress and the gradient of the force due to the internal ice stress,

$$0 \approx \nabla(\tau_a) + \nabla (\boldsymbol{\nabla} \cdot \boldsymbol{\sigma}). \tag{7.16}$$

Within the linear-viscous approximation in equations (7.14) and (7.15), the effect of high viscosity is to smooth the atmospheric pressure field spatially in order to obtain the local deformation. This linear procedure has also been applied directly to ice drift with best fits to observations yielding substantially higher viscosities in winter than in summer (Hibler and Tucker, 1979). However, application of such linear models is limited to relatively homogeneous regions far from boundaries. As discussed in section 7.4, to explain ice mechanics successfully highly non-linear rheologies are in general required.

Limiting solutions of equations (7.14) and (7.15) for large viscosities in the infinite boundary limit (Hibler, 1974) show that the ice would be expected to converge if the surface winds were directed inward, which could occur due to a low pressure region, for example. This is in contrast to the free drift solution which applies as η, $\zeta \to 0$, where $\boldsymbol{\nabla} \cdot \mathbf{v} \sim \nabla^2 P$. The vorticity ω, however, has the same sign in either limit. Comparisons of these limiting cases with temporally smoothed observations (Figure 7.9) show the large viscosity limit to be in much better agreement with the observed deformation, indicating that the effects of ice interaction were likely to be very substantial. Moreover, least squares regression fits of large viscosity solutions upon the divergence rate and vorticity yielded η and ζ values of $\sim 10^{12}$ kg/s with $\zeta \sim 2\eta$. The physical notion here is that the ice interaction prevents large deformation from occurring and hence the divergence rate tends to follow the wind gradients, especially on shorter spatial scales. The viscous approximation may be justified for *small* deformation magnitudes if one considers sufficient spatial and temporal smoothing of the velocity field with plastic behaviour on the local scale (Hibler, 1977).

While these linear solutions are only approximate, the main point here is that ice mechanics alone can cause the ice deformation to differ substantially from the expectations based on simple linear drift rules such as that of Zubov (1943) or Thorndike and Colony (1982). Other relevant mechanisms can be repeated lead and ridging events (see section 7.4.3). This is not widely appreciated, as much interpretation of recent decreases in Arctic atmospheric pressure (Walsh, Chapman and Shy, 1996) has been based on anti-cyclonic wind forcing favouring convergent ice motion. This has also been a common interpretation in the analysis of the effect of Arctic oscillation wind patterns (Polykov *et al.*, 1999). However, the ubiquitous presence of average ice divergence was recently demonstrated by spatially dense SAR imagery (R. Kwok, private communication), which shows all regions in the Arctic (including the region close to the archipelago) to undergo net long-term divergence. Interestingly, recent detailed analysis of SAR data by Moritz and Stern (2001) yielded divergence rates having a zero correlation with the Laplacian of the atmospheric pressure field. In this author's opinion, this result emphasizes the necessity for using non-linear sea-ice mechanics in ice models to simulate deformation properly.

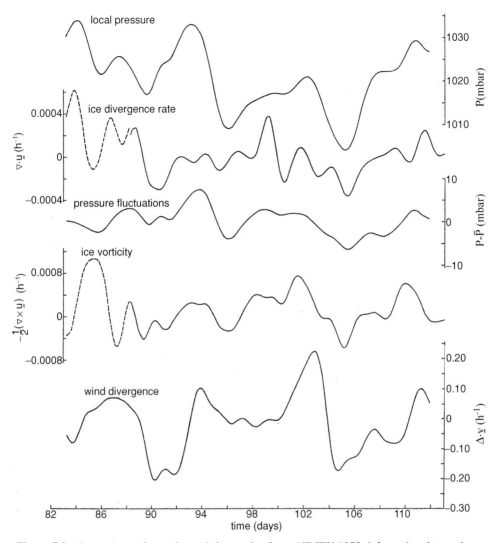

Figure 7.9. Comparison of experimental time series from AIDJEX 1972 deformation data and atmospheric observations of surface wind speed and pressure. The deformation data is taken from a densely sampled meso-scale array covering about a 20 km region, and the wind data is taken from surface observations of wind at three ice camps separated by ∼100 km in a triangle. Pressure data from land stations were also used in the pressure plots. (From Hibler (1974).)

7.4 Sea-ice mechanics

Typical non-linear formulations of ice mechanics in dynamic thermodynamic models yield a stress gradient $\nabla \cdot \sigma$ generally in opposition to the wind stress (Hibler and Bryan, 1987; Steele *et al.*, 1997). When vectors are averaged over time periods of a few months or longer, the ice stress gradient tends to have a magnitude about 50% of the average wind stress (see Table 1 from Hibler and Bryan, 1987). However, as examined below, due to non-linear ice interaction, the daily stress gradient is a smaller fraction of the wind stress gradient. Because of these considerable effects, proper treatment of ice mechanics plays a key role in modelling the evolution of sea ice.

Successful descriptions of pack-ice failure and flow have typically incorporated non-linear plastic failure models with differing treatments of shear strength and flow rules. There are two general treatments of plastic failure that generally parallel the classic Von Mises and Tresca yield curves in solid mechanics (Malvern, 1969). The essential idea in these plastic failure models is to have a rate-independent failure stress once plastic failure is initiated. Much of the initial development of plastic sea-ice models was based more on aggregate energetic arguments (Coon, 1974; Rothrock, 1975; see also section 7.6 below) than on particular physical failure processes applicable to fractures or sliding friction between floes. Recently proposed rheologies (Hibler and Schulson, 2000) have similarities to plastic and brittle failure constitutive laws developed in smaller-scale laboratory studies of ice (e.g., Schulson and Nickolayev, 1995). These rheologies also have significant similarities to granular coulombic failure models used in soils and rock mechanics (see, e.g., Dempsey, Palmer and Sodhi, 2001; Schulson, 2001).

7.4.1 Aggregate isotropic sea-ice constitutive laws

Aggregate isotropic descriptions of pack ice have typically sought to portray the collective behaviour of pack ice with a large number of failing ridges and leads comprising the pack-ice continuum. In these descriptions the failure is considered to describe statistical aggregate behaviour, with both the stress and strain comprising a reasonable statistical number of leads, fractures and ridges. Depending on the constitutive law used to determine ice stress from deformation, this continuum description does not in principle prevent the formation of long-oriented failure features in simulations similar to those seen in SAR imagery.

Considering sea ice to be a two-dimensional isotropic fluid, a general non-linear constitutive law applicable to a relation between stress and deformation rate (Malvern, 1969) is

$$\sigma_{ij} = 2\eta \, \dot{\varepsilon}_{ij} + [(\zeta - \eta)\dot{\varepsilon}_{kk} - P], \tag{7.17}$$

where repeated subscripts are summed over, σ_{ij} is the two-dimensional stress tensor, ε_{ij} is the two-dimensional strain rate tensor, and ζ, η and P are functions of the two invariants of the strain rate tensor $\dot{\varepsilon}_{ij}$. In the general Reiner–Rivlin fluid there is also an additional term $\lambda \dot{\varepsilon}_{ik}\dot{\varepsilon}_{kj}$, where repeated subscripts are summed over. However, since P can also be a non-linear function of the strain rate invariants, this additional λ-term is usually omitted in sea-ice rheologies. In constructing a law beginning with this general framework, one would like to take into account the following observational characteristics and/or intuitively reasonable assumptions: discontinuous slippage near shore; relatively coherent motion; possible lack of ice motion under considerable wind forcing; small tensile strength and high compressive strengths. These last two features are also characteristic of biaxial failure of laboratory ice (Schulson and Nickolayev, 1995).

While linear-viscous models are useful for mechanistic calculations, it became clear, from theoretical and numerical modelling studies in the early 1970s, that some of the basic phenomena of ice dynamics and ice deformation discussed above could not be explained by linear rheologies. For example, using a linear-viscous model employing both bulk and shear viscosities, it was found that best-fit viscosities of the order of 10^{11} to 10^{12} kg/s were needed to model deformation (Hibler, 1974) and drift (Hibler and Tucker, 1979) far from shore. Near shore, however, investigations of satellite imagery near coastal boundaries indicated viscosities as small as 10^8 kg/s (Hibler, 1974), while model simulations of tidal-induced drift indicated that in very small channels near-shore viscosities might be as small as 10^5 kg/s (Sodhi and Hibler, 1980). Such variations indicated that some type of non-linear behaviour was necessary.

One way to formulate a non-linear rheology consistent with the above characteristics is to make use of plasticity theory (see, e.g., Pritchard, 2001). This was first noted by Coon (1974), who based much of this proposition on the similarities between sea ice and a granular medium. The basic assumptions are plastic failure together with small or zero tensile stresses. For comparison, it is notable that the two-dimensional Von Mises failure criterion (Malvern, 1969) for a thin steel plate or shell is an ellipse centred at the origin. Some typical isotropic yield curves (Figure 7.10) proposed for sea ice along these lines are an elliptical yield curve (Hibler, 1979) and a coulombic yield curve (Overland and Pease, 1988; Smith, 1983), which has a more direct explanation in terms of oriented failure along surfaces and will be discussed below. Initial plastic pack-ice simulations utilized a circular yield curve for simplicity (Coon and Pritchard, 1974). Two-dimensional energetic arguments by Rothrock (1975) lead to a proposed 'teardrop' yield curve, and kinematic arguments by

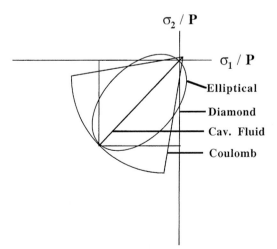

Figure 7.10. Typical 'aggregate' isotropic yield curves proposed for sea ice.

Bratchie (1984) lead to a 'sine lens' yield curve, which also can be considered 'scale-invariant' (Hibler, 2001). A 'square' or diamond yield curve was proposed by Pritchard (1978) based partially on frictional arguments. A particularly simple conceptual yield curve is the 'cavitating fluid' (Flato and Hibler, 1992), whereby ice resists compression but has no resistance to opening. This yield curve is useful where temporally averaged wind forcing is employed. With all such yield curves, to supply a complete description of the stress state in the system one also needs to have a way of obtaining stresses inside the yield curve. Moreover, if one wishes to model deformation and flow, one needs a flow rule to specify the stresses for different strain rates.

The specification of the stresses inside the yield curve has traditionally used elastic be-haviour (Pritchard, 2001). With elastic behaviour inside the yield curve and rate-independent stresses on the yield surface, energetic conditions typically lead to the so-called normal flow rule. This rule specifies that in coincident principal axes space, the ratio of the principal strain rate components is specified by a vector normal to the surface. Since energy may be stored and released by the elastic portion of the system, this constraint can be shown to necessitate a convex yield surface and a normal flow rule. If some other type of closure, such as rigid or viscous, is used, a normal flow rule is not required by this argument as there is never any storage of mechanical energy. A typical simple coulombic yield curve, for exam-ple, often assumes pure shear deformation for a variety of stress states. Numerical solution of the elastic–plastic formulation is typically by direct explicit methods (Pritchard, 2001; Pritchard, Coon and McPhee, 1977). While elastic closure may be very valuable mathemat-ically, since 'plastic' pack-ice rheologies seek to describe the collective behaviour of pack ice, there appears to be very little, if any, physical basis for necessarily assuming elastic behaviour at the large scale.

Another description of the stress states inside as well as on the plastic yield curve, is the 'viscous–plastic' (vp) rheology approach introduced by Hibler (1979, 1980a) for the purpose of modelling non-linear flow for a wide variety of yield surfaces (see, e.g., Ip, 1993; Ip, Hibler and Flato, 1991). Here, rigid flow is approximated by a state of very slow creep, a procedure which can be justified in part by considering a statistical aggregate of different deformation states undergoing rigid plastic flow in any control volume (Hibler, 1977). In practice, this description may be thought of as essentially a highly non-linear viscous fluid, as described in equation (7.17), with the viscosity adjusted to yield rate-independent stresses for large strain rates. (The 'viscous–plastic' formulation should not be confused with a 'visco-plastic' or Bingham rheology. A visco-plastic rheology is similar to rigid plasticity except that at failure viscous, rather than plastic, flow occurs; see, e.g., Shames and Cozzarelli, 1992.) This viscous–plastic formulation is widely used in a variety of ice–ocean and atmosphere–ice–ocean models. Particular rheology and solution procedures are given below.

A third method, often used in soil mechanics, is to consider the system to be rigid inside the yield surface. With this procedure one needs some type of constraint to determine the interior stresses. A procedure for numerically solving for such interior stresses in the rigid approximation was proposed by Flato and Hibler (1992) for coulombic-type rheologies with no dilatation. Here one uses the constraint that there is no deformation for stresses inside the yield curve. This general procedure was subsequently extended to a more general coulombic yield curve, including dilatation, by Tremblay and Mysak (1997).

In the case of the elliptical yield curve with the viscous–plastic approximation and a normal flow rule, a simple closed set of equations (Hibler, 1977, 1979) can be formulated relating the stress to the strain rate anywhere on or inside the yield surface. This yield curve and flow rule is widely used in many numerical sea-ice models (see, e.g., Chapter 9; Lemke, Owens and Hibler, 1990; Lemke *et al.*, 1997). Referring back to equation (7.17), for an elliptical yield curve with a normal flow rule, the non-linear bulk and shear viscosities ζ and η can easily be shown (Hibler, 1977, 1979) to be functions of the strain rate invariants according to

$$\zeta = \text{minimum} \left[\frac{P^*}{2\Delta}, \zeta_{\text{max}} \right] \tag{7.18}$$

$$\eta = \frac{\zeta}{e^2}, \tag{7.19}$$

where

$$\Delta = \left[(\dot{\varepsilon}_{11}^2 + \dot{\varepsilon}_{22}^2) \left(1 + \frac{4}{e^2} \right) + \frac{\dot{\varepsilon}_{12}^2}{e^2} + 2\dot{\varepsilon}_{11}\dot{\varepsilon}_{22} \left(1 - \frac{1}{e^2} \right) \right]^{1/2}; \tag{7.20}$$

P^* is the ice strength (equal to the pressure P for high strain rates), and e is a constant (typically $\sqrt{2}$). To approximate stress states inside the yield curve, ζ and η are capped at some large maximum values (ζ_{max} and $\eta_{\text{max}} = \zeta_{\text{max}}/e^2$) for small strain rates.

In order to ensure that there is no stress at zero strain rates, in recent formulations it is insisted (Hibler and Ip, 1995; Hibler and Schulson, 2000; Ip *et al.*, 1991) that the pressure P in equation (7.17) is given by

$$P = 2\Delta\zeta, \tag{7.21}$$

where ζ is given by equation (7.18). When plastic flow is occurring, P will be a constant (P^*). However, for $\zeta = \zeta_{\text{max}}$, P will be less than or equal to P^*, and the stress state will lie on a smaller geometrically similar yield curve, as shown in Figure 7.11, going through the origin. This condition together with the above flow rule ensures that the internal mechanical energy dissipation $\sum_{ij=1}^{2} \dot{\varepsilon}_{ij}\sigma_{ij}$ is positive definite. This result is in contrast to the earlier formulation of an elliptical viscous–plastic rheology by Hibler (1979), where P was taken to be a constant times P^*. As noted by Schulkes (1996), this energy condition is the main requirement for overall energetic stability of the viscous–plastic rheology, since it guarantees, via the continuum energy equations, that the kinetic energy will be bounded in the absence of body forces. Simulations (Hibler and Ip, 1995; and see below) show the inclusion of equation (7.21) to reduce substantially excessive stoppage of ice motion.

The numerical procedure proposed by Hibler (1979) as part of a dynamic thermodynamic sea-ice model represents probably the most direct method of solving numerically the

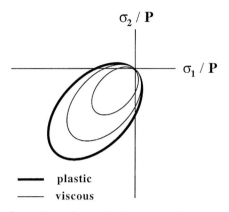

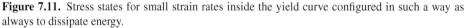

Figure 7.11. Stress states for small strain rates inside the yield curve configured in such a way as always to dissipate energy.

viscous–plastic dynamical equations. In this direct implicit solution the coupled non-linear momentum equations are linearized and solved implicitly by successive relaxation at each time step with a locally implicit solution of anti-symmetric terms. Since the linear system of equations thus formed is positive definite (but not symmetric), other more efficient numerical solvers could probably be used. Similar relaxation procedures are used, for example, by Harder and Fischer (1999). Because of this implicit solution there is no time step limitation on the momentum equation, although with time steps long compared with the forcing, the evolution of the non-linear terms (Figure 7.12) will not keep up with the forcing changes unless repeated iterations at each time step are used. Such 'updating' procedures at each time step were used, for example, by Harder and Fischer (1999), Hibler and Ip (1995), Hilmer, Harder and Lemke (1998), Ip (1993), and Kreyscher *et al.* (2000).

A subsequent more efficient 'splitting' method was proposed by Zhang and Hibler (1997) where only a portion of the coupled equations are solved at each time step. Although more efficient, this procedure solves only a portion of the momentum equations at each time step and hence is less dissipative than the Hibler (1979) method. Moreover, at large time steps the Zhang and Hibler (1997) method may have convergence problems. A more convergent 'splitting procedure' has subsequently been developed by J. K. Hutchings and colleagues. Such splitting techniques have been used in very high resolution ice–ocean simulations (e.g., Maslowski *et al.*, 2000). Hunke and Dukowicz (1997), and Hunke and Zhang (1999) developed an 'elastic–viscous–plastic' (evp) solution procedure whereby an artificial elastic term is added so explicit time steps may be taken, yielding a more efficient solution on vector and parallel computers. While not fully vectorizable, the Hibler (1979) procedure can also be concurrentized on parallel computers, but the initial set-up-time scales with the number of processors.

It should be emphasized that, given the same numerical discretization of the non-linear viscous and pressure terms, all of the numerical solutions should converge to the same plastic solution, i.e. certain cases for which analytic solutions exist (Lepparanta and Hibler,

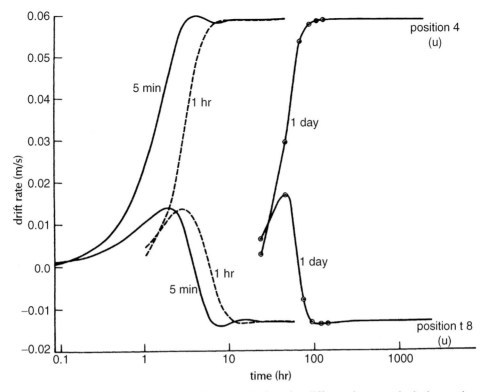

Figure 7.12. Approach to equilibrium of the ice velocity using different time steps in the integration of the viscous–plastic momentum equations. In all cases the ice was initially at rest and then a constant wind field was turned on. Position 4 is very near a boundary, whereas position 8 is near the centre of the grid. (From Hibler (1979).)

1985; see also section 4.3.1). However, different discretizations are typically used as well as time steps in comparisons (Arbetter, Curry and Maslanik, 1999; Hunke and Zhang, 1999) so precise differences are at present poorly determined. Moreover, no updating at all of the viscous–plastic (vp) solution at each time step was used by Hunke and Zhang (1999), which resulted in artificially long response times for the vp model as compared with the evp model. The original Hibler (1979) rectangular finite-difference discretization and solution procedure ensures that the viscous terms are locally and globally energy dissipative for each linear solution, and that the full rheology is dissipative if the equations are iterated to full plastic flow with the formulation of equations (7.21). Recent developments are full curvilinear energy dissipative discretizations for spatially varying bulk and shear viscosities by Hunke and Dukowicz (2002), and a spherical co-ordinate dissipative finite-difference discretization by Hibler (2001) retaining all metric terms. There is also a plethora of other numerical methods (e.g., finite-element methods) that can and have been applied to solving these equations, although the overall time stepping procedures are usually similar to one of the methods described above.

7.4.2 Coulombic and fracture-based isotropic models

While they have many similarities to aggregate rheologies based largely on energy arguments, coulombic and fracture-based rheologies are based more on either failure across faults or sliding friction between discrete particles, or floes in the case of sea ice. In this sense, the archetypal classic plastic analogue would be the Tresca failure criterion, where failure is considered to occur when shear stress across any surface reaches a critical value. The classic coulombic rheology (Overland and Pease, 1988; Overland *et al.*, 1998; Smith 1983; Tremblay and Mysak, 1997) extends this concept to make the critical failure shear stress across any surface depend on the compressive stress on that surface. Because of this emphasis on failure surfaces and faults, this type of rheology provides a more natural formulation for treating oriented fractures and failure in sea ice and for building up a material description of sea ice based on such phenomena. In the case of fracturing it can be argued that similar rheological behaviour should apply on the large and small scales (Hibler, 2001; Schulson and Hibler, 1991). Such rheologies also provide a framework for describing likely mechanisms leading to oriented fractures and leads in sea ice (Hutchings and Hibler, 2002; Schulson, 2001).

The simplest coulombic model may be conveniently described by considering an oriented failure surface in two dimensions. For the coulomb sliding friction model (Figure 7.13(a)), we imagine a flaw oriented at an angle θ relative to the most compressive stress σ_y applied to the local region. We consider this flaw and all other flaws in the region to have a local friction coefficient of μ. With the 'coulomb model' assumption the flaw will fail when the shear stress τ along the flaw is related to the compressive stress σ across the flaw according to the frictional sliding approximation: $\tau_c = \mu\sigma + b$, where b is a measure of the cohesive strength of the material and τ_c is the critical shear stress for failure. To simplify the mathematics we consider all stresses to have units of bar, so that this equation becomes

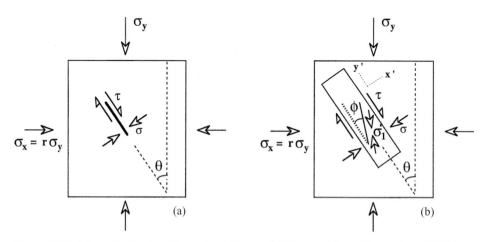

Figure 7.13. Schematic showing (a) a coulomb frictional sliding model and (b) a 'finite-width' failure model allowing for dilatation.

$\tau_c = \mu\sigma + 1$. From the orientation of flaw in Figure 7.13(a), for a given confinement ratio r less than one, the flaw will fail for an external principal stress σ_y given by

$$\sigma_y \frac{(1-r)\sin(2\theta)}{2} = \mu\frac{(1+r)\sigma_y}{2} - \mu\frac{(1-r)\cos(2\theta)\sigma_y}{2} + 1. \qquad (7.22)$$

Equation (7.22) was obtained by expressing the failure equation of the flaw in terms of the external principal stresses σ_y and $r\sigma_y$. To find the angle of flaws most likely to fail, we can consider σ_y to be a function of θ and seek a minimum value of σ_y. Differentiating this equation with respect to θ and setting $d\sigma_y/d\theta = 0$, we obtain the classic result

$$\tan(2\theta_c) = \pm\frac{1}{\mu}. \qquad (7.23)$$

Rearranging equation (7.6), it is also clear that if flaws at the fixed angle θ_c preferentially fail, the relationship between the maximum external shear stress, $\sigma_s = \sigma_y(1+r)/2$, and compressive stress, $\sigma_c = \sigma_y(1+r)/2$, is given by

$$\sigma_s = \cos(2\theta_c)\sigma_c + \sin(2\theta_c). \qquad (7.24)$$

This is the principal axes space yield curve following from the local sliding frictional model with the angle of internal friction given by $\beta = (\pi/2) - 2\theta_c = \tan^{-1}(\mu)$.

While useful, the classic coulombic rheology of Figure 17.13(a) has the drawback that only pure shear deformation is allowed. A useful physical model (Coon *et al.*, 1998; Hibler and Schulson, 2000) that can be used to extend the coulombic rheology to arbitrary dilatation is the 'finite-width' oriented flaw in Figure 7.13(b). While loosely referred to as 'finite width', the flaw may be made infinitesimally narrow without affecting the theory. In this limit it takes on features of a mathematical characteristic used in partial differential equations inasmuch as the strain and velocity derivative are still defined on the 'strip', even though its width is arbitrarily small (Mendelson, 1968). It can be shown (Hibler and Schulson, 2000) that insisting on continuity of stresses at the interface of the flaw and minimizing the stresses as before leads to a preferred angle of failure that depends on the flow rule of the rheology of the thin ice, which is in general larger than the coulombic angle for a sliding surface. In the case of a more complex rheology applying to the thin ice, the preferred orientation can be solved for numerically (see Hibler and Schulson (2000) for details). Since in numerical models the failure patterns take place along finite-width zones, these failure intersection angles are also the ones we expect from numerical simulations. As an example, a viscous–plastic formulation of a non-normal flow rule coulombic yield curve (Figure 7.14) based on observed laboratory results (Schulson and Nickolayev, 1995) was proposed by Hibler and Schulson (2000), and used in the 'lead' in the finite-width model in Figure 7.13(b). The predicted flaw angles tend to be much narrower than in the elliptical yield curve for diverging conditions. Confirmation for the theory comes from viscous–plastic boundary value simulations forced by stresses at the boundaries. In these simulations the 'modified coulombic' rheology leads to localized deformation strikes (see

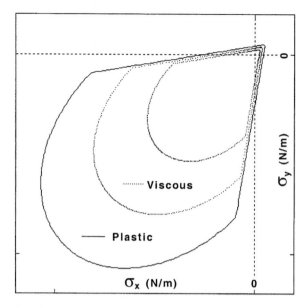

Figure 7.14. A coulombic yield curve used for a numerical calculation of composite failure of thick ice with an oriented thin ice lead as shown in Figure 7.13(b). This yield curve was based on laboratory observations of the brittle failure of ice by Schulson and Nickolayev (1995). The dotted lines are the stress states for very small strain rates. (From Hibler and Schulson (2000).)

fig. 10 of Hibler and Schulson (2000)) in good agreement with the theory. With an elliptical rheology, there are no clearly oriented failure results in this experiment.

The concept is that this type of 'coulombic' model should be much more successful (Aksenov and Hibler, 2001) at simulating oriented deformation patterns (see Figure 7.5) in the Arctic ice pack. That this is indeed the case can be shown (Hutchings and Hibler, 2002) by beginning with an initial heterogeneous ice strength field. With this initial condition, appropriate wind forcing causes the ice to weaken and fail along intersecting oriented zones (Figure 7.15) very similar to observations (Kwok, 2001). In this simulation a wind field having a constant deformation rate with a -1.4 ratio of the principal strain rate components was utilized. With this wind field and initial condition, Hutchings and Hibler (2002) find intersecting failure features (Figure 7.15(b)) form with average spacing scaling with the square root of the average ice strength over the square root of the spatial wind stress gradient. This scaling may be physically argued on the basis of allowable rigid block motion dimensions with zero stress boundary conditions at the 'lead'. Due to the strength heterogeneity, the initial plastic deformation field is rather chaotic without any clear oriented structure. However, with weakening due to divergence, local failure points nucleate the formation of intersecting failure lines, which form in only a few hours. Over this period the global averaged stress magnitude drops by about a factor of 5, while the stress in the oriented failure zones drops by several orders of magnitude. This general behaviour is consistent with stress observations (see, e.g., Figure 7.8).

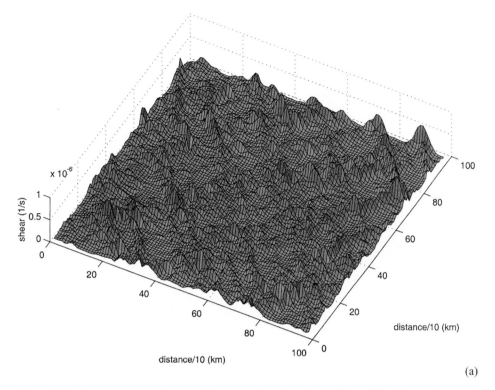

Figure 7.15. Shear strain rate simulated for a two-dimensional grid (200 × 200) of ice with strength initially varying randomly and the wind forcing having a uniform strain field (-1.4 = ratio of principal strain rate axes) everywhere. (a) Plastic equilibrium with no weakening. (b) The ice has been allowed to weaken in proportion to local divergence rate for 3.6 hours.

By considering the steady-state equations of motion of an idealized intersecting set of leads, Hutchings and Hibler (2002) argue that the orientation of the leads should be predicted by an energy minimization calculation (for opening and hence weakening flaws) with respect to orientation of an intersecting set of 'finite' failure zones as shown in Figure 7.16(a). This energy minimization solution can differ from 'characteristics' based on static solutions (e.g., Erlingsson, 1991; Pritchard, 1988) and can be used for non-normal flow rule rheologies. Applying the Hibler and Schulson (2000) procedure to the configuration of Figure 7.16(a) yields the predicted angles shown in Figure 7.16(b). The main relevant point here is that the ellipse (or any normal-flow-rule based rheology) yields much larger (and hence less realistic) intersection angles (see Figure 7.5) than fracture-based rheologies. This is especially true in the limit of close to pure shear strain rates that are often encountered in typical sea-ice deformation fields. The main physical reason for this is the cusp in the yield curve where different failure mechanisms (see, e.g., Schulson and Nickolayev (1995)) take over.

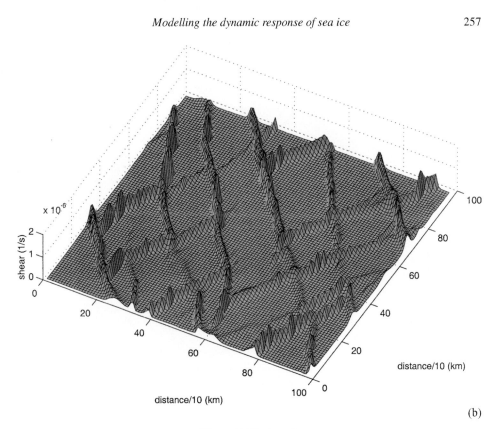

(b)

Figure 7.15. (*cont.*)

One other notable approach to modelling granular flow that is somewhat related to the continuum coulombic models is the discrete element simulation of sea ice (e.g., Hopkins, 1996). In this approach the motion of large numbers of discrete distributions of floes are solved for individually with coulombic friction at contact points. While not highly applicable to climate-scale simulations, such models are particularly useful for process simulations to estimate realistic thickness distribution ridging parameterizations in larger-scale models.

The main relevance of the 'coulombic' rheologies to the evolution of the mass budget of sea ice lies in their deformation characteristics of shearing and dilating of intersecting fractures, an interpretation which becomes more relevant in high resolution models that can resolve local scales. With this interpretation, open water formation is dictated by divergence rate except for certain deformation states (e.g., Hibler and Schulson, 2000; Hutchings and Hibler, 2002) requiring both opening and closing sets of leads. In the case of aggegate rheologies, on the other hand, it is usually hypothesized (see section 7.6.2) that both divergence and convergence take place simultaneously over a wide range of deformation states. As discussed in section 7.7.3, how this partition is made in numerical model simulations significantly affects ice-mass characteristics. It is also notable that certain coulombic rheologies are arguably 'scale'-invariant (Hibler, 2001) and hence would be expected to

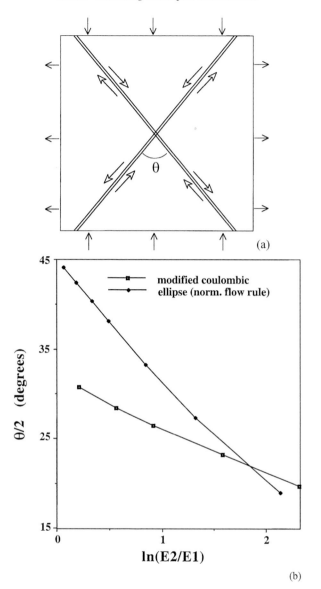

(a)

(b)

Figure 7.16. (a) Intersecting 'finite'-width faults together with a specified divergent strain field used to estimate lead orientation for different rheologies. (b) Predicted lead orientation obained by finding the minimum energy of the intersecting fault model for different rheologies vs. the natural logarithm of the ratio of the principal axis strain rate magnitudes. E2 is positive and E1 is the magnitude of the convergent principal axis value. (From Hutchings and Hibler (2002).)

apply at very high resolution as well as larger scales. This view is, however, in contrast to arguments by Overland *et al.* (1995) who argue for a scale break in rheology around ~10 km.

7.4.3 Effect of plastic ice interaction on modelled ice drift

Characteristics of plastic flow particularly relevant to the sea-ice mass balance are its tendency to yield large-scale ice drift similar to the wind forcing and the tendency for ice flowing through narrow channels to slow down and sometimes totally stop via the formation of static arches. Both these features occur because of the highly non-linear character of plastic flow. Because of the first characteristic, large-scale ice motion fields will tend to correlate highly with wind field variations, even with high ice stresses. The second 'arching' characteristic is important because of the critical control of relatively narrow outflow passages, most notably the Fram Strait, on the ice-mass budget in the Arctic basin. Some of these characteristics are examined below.

A mechanistic one-dimensional plastic system

Some physical notions regarding plastic flow and sea-ice drift may be illustrated by analysing a special one-dimensional case of the momentum balance employing only a linear water drag term cu, an external constant wind stress τ and a one-dimensional ice stress σ:

$$cu - \frac{\partial \sigma}{\partial x} = \tau. \tag{7.25}$$

Let us now consider a rigid plastic case with constant strength with rigid walls at $x = 0$ and $x = L$. For the plastic rheology assume that $\sigma = -P$ for $\partial u/\partial x < 0$ and $\sigma = 0$ for $\partial u/\partial x > 0$. These assumptions define a rigid plastic rheology with no tensile strength. A solution for this case may be constructed by noticing (a) that for any convergent deformation, $\sigma = -P$, while for no deformation the stress is in the rigid state so that $0 \geq \sigma \geq -P$, and (b) that the maximum force would be expected to take place at the right hand boundary since the wind stress has built up at this point.

With these assumptions in mind, it is easy to show (see, e.g., Hibler (1985)) that u takes on a constant value

$$u = \frac{\tau L - P}{cL}, \tag{7.26}$$

with the stress σ varying linearly over the whole length of the ice cover according to

$$\sigma(x) = -P + (cu - \tau)(x - L). \tag{7.27}$$

This analysis is also easily extended to include non-linear water drag terms. In this case, the steady-state momentum balance for this one-dimensional system is

$$\rho_a c_a u_g^2 = \rho_w c_w u^2 - \frac{\partial \sigma}{\partial x}, \tag{7.28}$$

where u_g is the geostrophic wind. By the same methods as employed in the linear case for a constant wind, the solution for this system is

$$u = \left(\frac{\rho_a c_a u_g^2 - P/L}{\rho_w c_w} \right)^{1/2} \tag{7.29}$$

if $P < (\rho_a c_a u_g^2)L$ and $u = 0$ otherwise. For a numerical comparison relevant to the Arctic basin we take $L = 2.5 \times 10^3$ km, which is the scale of the Arctic basin, and $P = 8.25 \times 10^4$ N/m, a typical model-based ice strength for 3 m thick ice in the Arctic basin (Hibler and Walsh, 1982). Keeping in mind that this ice strength may be a bit high, one obtains a comparison of free drift and ice interaction modified drift as shown in Figure 7.17. Note that for large wind speeds there is little difference between free drift and the rigid plastic result, whereas for smaller wind speeds the difference is very marked, with the rigid plastic system stopping totally.

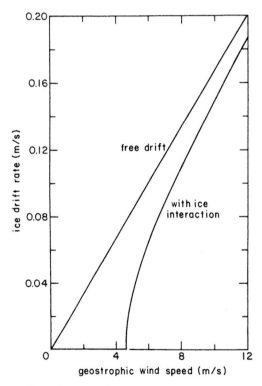

Figure 7.17. Typical one-dimensional plastic solution characteristics with constant strength and a fixed constant wind field. Quadratic boundary layer drags were used, and for the ice interaction case typical strength and length scales relevant to the Arctic Basin were assumed.

Comparison of large-scale simulated plastic drift and deformation characteristics

Comparisons of simulated and observed drift characteristics (Hibler and Ip, 1995; Ip, 1993) tend to support the basic elements of the mechanistic one-dimensional results described above: namely the stoppage of ice drift under low wind speeds and the high correlation of simulated ice drift, albeit with substantial ice stresses. A number of these features are illustrated in Figures 7.18 and 7.19, taken from a series of multi-year simulations by Ip (1993) employing a hierarchy of plastic rheologies using a two-level dynamic thermodynamic model (Hibler, 1979) with a full heat budget code described below (see sections 7.5.3 and 7.6.4). The main variables are the mean ice thickness per unit area, h, and the ice compactness, A, with the ice strength P related to h and A by $P = P^* h \, e^{-C(1-A)}$, where $P^* = 2.75 \times 10^3 \, \text{N/m}^2$ (Hibler and Walsh, 1982) and $C = 20$. Simulations were carried out over 1975 to 1985. The yield curves are basically as shown in Figure 7.10, except that the coulombic yield curve goes to the origin and has no tensile strength.

The characteristic stoppage of ice is illustrated in Figure 7.18, where observed daily distribution of drift speeds are compared with values simulated with a plastic rheology (Figure 7.18(a)) and in Figure 7.18(b) with a 'free drift' simulation with no ice interaction. What is clear from these figures is that with a rheology including shear strength the distribution of drift speeds demonstrates a higher fraction of low or zero ice motion, in general agreement with the simple one-dimensional model physics. However, the 'free' drift model, or any direct linear model correlating ice drift with wind with fixed spatially invariant coefficients, does not reproduce the observations. It should be noted, however, that this realistic correlation of modelled ice drift requires an energy dissipative rheology. Use of a fixed pressure term without the pressure being dependent on deformation (equation (7.21)) results in excessive amounts of stoppage (Figure 7.18(c)).

While this reduction of drift speeds can occur at low wind speeds, there is, however, a very strong complex correlation (Figure 7.19(a)) between wind and simulated ice drift, as we would expect from physical considerations. It is also notable that there is a strong negative correlation between wind force and the ice force arising from the gradient of the ice stress tensor. This general behaviour was also subsequently noted by Steele *et al.* (1997) in a similar model calculation, and earlier by Hibler and Bryan (1987) in an ice–ocean model calculation. Depending on the season, the average magnitude of the stress gradient is about 10–30% of the average wind force magnitude (Figure 7.19(b)). If, on the other hand, vector force averages are taken (Hibler and Bryan, 1987), then the ice force can account for a significantly higher fraction of the vector averaged wind force.

While drift characteristics are affected by ice mechanics, ice deformation is modified in a much more pronounced way. This is shown in Figure 7.19(c), where the averaged deformation time series for the Beaufort Sea region is plotted for a full rheology and the free drift case. Clearly, the effect of rheology, especially when combined with ice-thickness evolution equations, is to cause a general divergence of the ice cover. This is in contrast to the free drift case where there is always a convergent condition because of the dominant anti-cyclonic condition over the Beaufort Sea. Although part of this condition is due to the

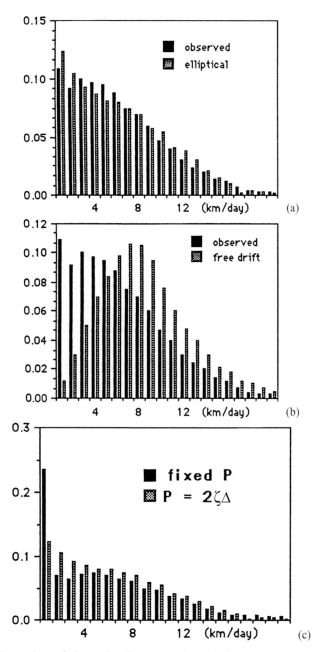

Figure 7.18. Comparison of observed and simulated buoy drift statistics over a seven-year period (1979–1985) for (a) an energy dissipative elliptical rheology (see Figure 7.10) and (b) for free drift. In part (c) the effect of taking a constant pressure rather than energy dissipative rheology (see equation (7.21) is shown. (From Ip (1993).)

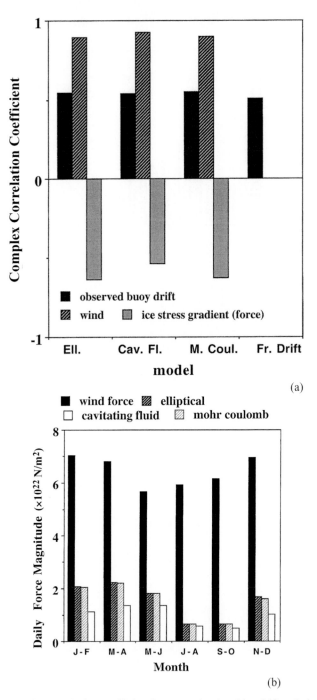

(a)

(b)

Figure 7.19. (a) Complex correlation coefficient between simulated ice drift and observed ice drift, geostrophic wind velocity and simulated ice stress gradient ($\nabla \cdot \sigma$). Three different rheologies and free drift (no ice interaction) were utilized. (b) Average daily magnitude of the wind stress and the ice stress gradient ($\nabla \cdot \sigma$) for five different rheologies as a function of time of year. (c) Seasonally averaged divergence rate for three different rheologies and a free drift model. Statistics in (a)–(c) were compiled over a seven-year period (1979–1985). (From Ip (1993).)

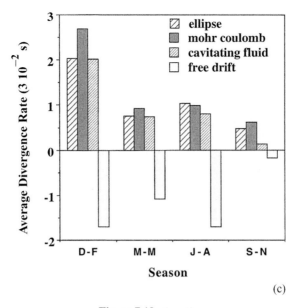

Figure 7.19. (*cont.*)

reduced magnitude of the divergence rate, as argued in idealized linearized calculations above, it is also likely to be affected by freezing of ice in leads formed by a divergent pack. This additional ice can then oppose convergence with subsequent changed wind fields.

Improvement of simulations by including 'inertial imbedding'

As noted in section 7.3.2, most models utilize the basic equation of motion specified by equations (7.3)–(7.5). However, these equations ignore the presence of considerable inertial power in the upper portions of the ocean, which significantly affect the higher frequency motion of the ice drift. Simulations (Heil and Hibler, 2002) including this inertial energy (by utilizing equation (7.10) for the equation of motion) improve simulations of buoy drift at all frequencies. This is graphically illustrated in Figure 7.20, which shows simulated and observed power spectra for the two different models. As in the previous section, the simulations were carried out in conjunction with 'two-level' ice evolution equations described in section 7.6. The simulated power spectra now much more closely follow the observed drift for this central Arctic buoy. Most important is the greatly enhanced coherence at all frequencies. This improved coherence is also reflected in the overall correlation coefficients that are substantially greater for the inertial imbedded result ($R_x = 0.87$, $R_y = 0.84$) than for the non-imbedded model ($R_x = 0.59$, $R_y = 0.48$).

These comparisons demonstrate both the importance of ice interaction in restraining the modelled ice-velocity fluctuations to values close to the observed ice velocity and the capability of non-linear plastic rheologies (together with ice evolution equations discussed below) to supply the appropriate amounts of high frequency energy. An underlying reason for

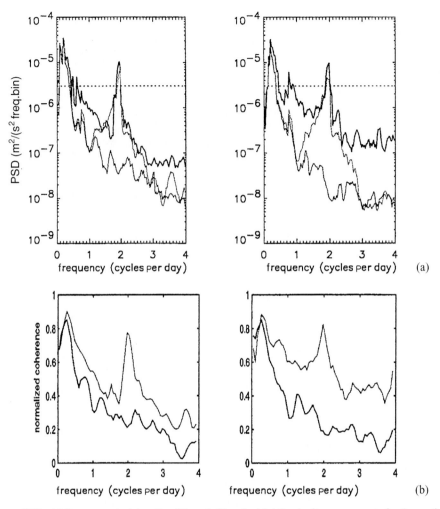

Figure 7.20. (a) Power spectral density of the x (left) and y (right) velocity components for the motion of a buoy (buoy A1; see Figure 7.32) over a several month long period in the central portion of the Arctic Basin. The heavy top curve represents the observations, the middle curve represents simulated values including 'inertial imbedding' (equation (7.10.)), and the bottom curve is with no inertial imbedding (equations (7.3)–(7.5)). (b) Normalized coherence of the x (left) and y (right) components of the observations and the two model results. The high coherence is for the model including 'inertial imbedding' and the low coherence is without inertial imbedding. (From Heil and Hibler (2002).)

this improvement is the artificial damping of power by the passive water drag if the traditional momentum equations are used (see, for example, traditional comparisons by Steele *et al.* (1997)). In the imbedded equations ((7.9) and (7.10)), the only way the combined average integrated ice/boundary layer energy may be reduced is via ice interaction dissipation or lateral viscosity in the ocean unless some small artificial damping is used. (In reality, other ocean damping mechanisms exist, such as damping by internal waves.) These imbedded simulations also greatly improve the simulation of ice deformation, as discussed in section 7.6.6. Little work to date has been carried out with this improved momentum formulation.

The effect of rheology on outflow

An important aspect of plastic flow with a rheology with some cohesive strength is the ability to form some type of static arch when the material flows through narrow channels. Even if total stoppage does not occur, the flow can be significantly reduced by ice mechanics. This is particularly important for the Arctic basin due to the control that the ice export through the Fram Strait exerts on the mass budget as well as affecting the degree to which ice may be transported through the Canadian archipelago. Indeed, to first order, the mass budget (Koerner, 1973) may be considered to be a balance between outflow and growth, with higher outflow inducing thinner ice and thus more growth. The potential of partial 'mechanical arching' across the Fram Strait to induce multiple equilibrium states in the Arctic ice pack is investigated in the context of a relatively idealized model in section 7.7.

Arching is a statically indeterminate problem in that different shapes of arches will have different strengths and breaking limits. Nonetheless, the arching phenomenon can be relatively easily analysed in planar systems. An idealized analytic analysis (Richmond and Gardner, 1962) of arching for a free arch between two vertical walls can be made with some simplifying assumptions utilizing a coulombic rheology with cohesive strength (i.e. $b \neq 0$ in equation (7.24)). Assuming (1) in the vicinity of the static arch, there is no variation of stresses in the y-direction, and (2) a Mohr–Coulomb criterion with cohesive strength, Richard and Gardner (1962) obtained the maximum channel width that could arch under a constant wind stress τ to be given by

$$\lambda_{\text{max}} = \frac{\alpha(\cos\theta_c)b}{\tau}, \tag{7.30}$$

where α contains geometrical factors of order 1, and θ_c and b are defined in equations (7.23) and (7.24). For channel widths larger than this, the arch will fail at the boundaries and the ice will flow through the channel. This analysis also yields a minimum width through which no flow can occur for given τ and b. The key scaling here is the arch width, which is proportional to the cohesive strength and inversely proportional to the wind stress. Also, as the friction increases (wider coulombic yield with θ_c decreasing), a wider arch can be formed.

The arching problem has been analysed numerically by Ip (1993) using a variety of channel configurations. These simulations were carried out both with and without thickness advection equations with similar results. While there is some effect, the actual channel configuration was found not to be highly critical on the arching initiation. For a given arch configuration the solutions were found to follow very well a non-dimensional scaling, so the numerical results can be used to examine the flow and arching through any size channel with given wind stresses and ice strengths.

Considering a tapered arch (Hibler and Hutchings, 2003), as in Figure 7.21(a), the basic parameters are the length λ of the opening, the ice strength P and the wind stress $\rho_a c_a u_g^2$.

For this system, the x momentum equation in the absence of Coriolis force is given by

$$0 = \rho_a c_a u_g^2 - \rho_w c_w u^2 + \left(\frac{\partial \sigma_{xx}}{\partial x} + \frac{\partial \sigma_{xy}}{\partial y} \right). \tag{7.31}$$

Expressing x and y in terms of λ and stress in terms of P, we have, after dividing by the wind stress, the dimensionless equation

$$1 = \beta - \gamma \left(\frac{\partial \sigma'_{xx}}{\partial'_x} + \frac{\partial \sigma'_{xy}}{\partial'_y} \right), \tag{7.32}$$

where primed values of stresses and x and y are dimensionless, and β and γ are dimensionless parameters given by

$$\beta = \frac{\rho_w c_w u^2}{\rho_a c_a u_g^2}, \quad \gamma = \frac{P}{\lambda \rho_a c_a u_g^2}. \tag{7.33}$$

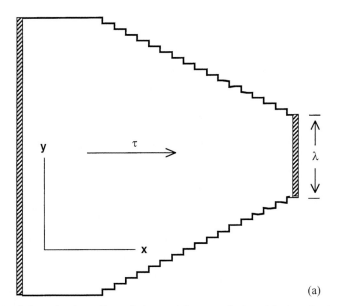

(a)

Figure 7.21. (a) Schematic view of boundaries used for numerical 'arching' study. (b) Dimensional average ice velocity and ice volume flow results versus ice thickness. In the tapered channel of (a), the shaded bands show regions where zero ice strength is assumed so the ice can freely flow into or away from these regions. A constant stress of $\tau = 0.4$ N/m^2 in the direction of the arrow is used for the body force. To simplify scaling, a linear water drag ($0.56u$) is used. For a plastic rheology the modified coulombic yield curve of Hibler and Schulson (2000) is used, and the ice strength is taken to scale linearly with thickness according to $P = 4 \times 10^4 h$ N/m, with h in metres. The ice area outflow in (b) may be put in non-dimensional form by taking the ordinate to be $\beta = 0.56u/\tau$, with τ the wind stress ($= 0.4$ N/m^2) and the abcissa to be $\gamma = P/(\lambda \tau)$, where λ is the width of opening. (From Hibler and Hutchings (2003).)

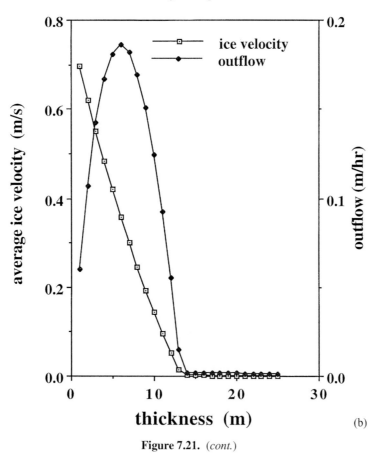

Figure 7.21. (*cont.*)

Consequently, in dimensionless form, the solution for β, which is a measure of the ice velocity, should only depend on γ unless the geometry of the boundaries changes. Investigation of the solutions for different resolutions and different geostrophic wind speeds shows this scaling to hold.

The character of the outflow is illustrated in Figure 7. 21(b), which has been taken from an investigation by Hibler and Hutchings (2003) of multiple equilibrium states induced by ice mechanics. Dimensional results for the area and volume outflow versus ice thickness (and hence ice strength) are shown. The results may be easily converted to non-dimensional form for the area outflow as noted above. For the volume outflow the appropriate ordinate will scale as $(P/\tau^2)\Delta$, where Δ is the volume outflow in dimensional form.

The basic character of the solutions is a gradual decrease of the ice velocity as the strength increases or the opening span (λ) decreases. At some point the velocity stops, and a static solution with an arch formed is obtained numerically, and the system is motionless with a free surface forming below an arch. Analysis of this system with a cavitating fluid has also been carried out by Ip (1993). In this case there is no arching, and, after a small decrease

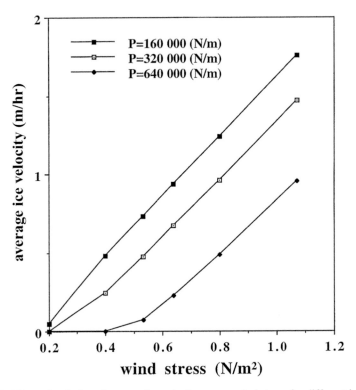

Figure 7.22. Dimensional plot of average ice velocity versus wind stress for different ice strengths. These results were obtained from the non-dimensional arching results of Figure 7.21. Note the similarity to outflow observations shown in Figure 7.3.

of ice velocity with h, the ice velocity becomes independent of h (i.e., ice strength). In the case of no ice interaction, there is some fixed velocity as a fraction of the wind speed. Analysis shows the estimated analytic strength limit to be slightly smaller ($\gamma_{\text{crit}} \approx 1.5$) than the numerical result ($\gamma_{\text{crit}} \approx 2.1$). This result is consistent with the fact that a constant channel width was assumed in the analytic calculation, whereas a tapered channel was used in the numerical experiments, so that arching was initiated at a slightly wider channel for the same forcing than the analytic limit.

Since the average ice velocity tends to scale with the areal outflow, it is instructive to plot average ice velocity in dimensional form for different ice strengths as a function of ice strength (Figure 7.22). This figure closely resembles observations of areal ice outflow through the Fram Strait shown in Figure 7.3 if we assume that during winter ice interaction effects are more pronounced. This figure also suggests that outflow results based only on correlations with atmospheric pressure gradients (Vinje, 2001) may be somewhat high in winter months. Aspects of the volume outflow results are discussed in sections 7.6 and 7.7 in conjunction with the capability of the arching mechanism to induce multiple equilibrium ice-thickness states if ice growth and ice advection are added to this highly non-linear system.

7.5 Sea-ice thermodynamics

7.5.1 Idealized growth: the Stefan problem

A wide variety of sea-ice thermodynamic models make use of steady-state linear profiles of temperature (Figure 7.23) to calculate the growth of a given thickness of sea ice. As shown by Worster (1999), given an idealization of growing sea ice as having fixed conductivity and heat capacity, for steady boundary conditions this approximation is very good provided the Stefan number S is small, where

$$S = \frac{L}{C_p(T_m - T_B)} \tag{7.34}$$

with L the latent heat of fusion, C_p the specific heat capacity, and T_m the melting point of the ice; T_B, the top boundary temperature of ice, is taken to be less than T_m.

The classic steady-state 'Stefan' problem used in many ice growth models consists of taking the growth condition to be approximated by

$$\rho L \dot{h} \approx k \frac{(T_m - T_B)}{H}, \tag{7.35}$$

which may readily be solved to give

$$H \approx \left(\frac{2Kt}{S}\right)^{1/2}. \tag{7.36}$$

Worster (1999) has compared the exact solution, beginning with $H = 0$, with the linear profile solution, and shows the steady-state approximation to have a less than 7% relative error for $S > 2$. Since for sea ice typical values of C_p and L are 2×10^6 J/(m^3 K) and 3×10^8 J/m^3,

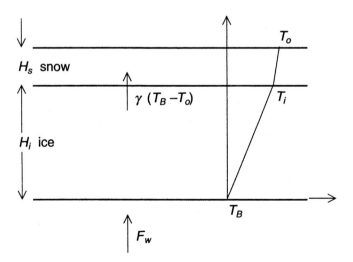

Figure 7.23. Combined snow and ice system with linear temperature profiles used in 'Stefan' steady-state thermodynamic ice models.

respectively, the Stefan number will be quite large even for small temperature differences. However, it should be emphasized that this solution only applies to constant boundary temperatures. For varying ice surface temperatures that occur regularly, heat capacity effects are substantial. In addition, the binary nature of sea ice makes these assumptions invalid in many cases, as noted below.

7.5.2 Empirical analytic sea-ice growth models for seasonal ice

For thin ice growing to a maximum thickness over one growth season, the Stefan approximation provides a framework to obtain reasonable agreement with observations. Introducing the concept of a degree day, $\theta(t)$,

$$\theta(t) = \int_0^t (T_f - T_a)\, dt,$$ (7.37)

where T_f is the freezing temperature of sea-water and T_a is the air temperature. A good empirical approximation (see fig. 12.6 in Hibler and Flato (1992)) to ice thickness, H, as a function of degree days based on data analysis, is given by Anderson (1961):

$$H^2 + 5.1H = 6.7\theta,$$ (7.38)

where H is in centimetres and θ has units of $°C$ day. Taking T_B in the Stefan approximation (equation (7.35)) equal to T leads to

$$\rho_i L H^2 = \kappa_i \theta.$$ (7.39)

This equation provides a physical basis for the squared term in the empirical Anderson equation.

This formulation is unrealistic, however, for thin ice, since, under those circumstances, the surface ice or snow temperature (Figure 7.23), T_o, can be much warmer than T_a. To account for this, some type of average transfer coefficient C_t can be assumed that describes both sensible and latent heat exchanges. In addition, to account for snow cover, a snow depth H_s and thermal conductivity κ_s may be added. These extended equations may be straightforwardly solved (see Hibler and Flato, 1992; Maykut, 1978) yielding

$$H^2 + \left(\frac{2\kappa_i}{\kappa_s}H_s + \frac{2\kappa_i}{C_t}\right)H = \frac{2\kappa_i}{\rho_i L}\theta.$$ (7.40)

With the numerical values (see Maykut, 1986) $C_t = 209$ J/(cm day K), $\kappa_i = 1758$ J/(cm day K), $\rho_i L = 272$ J/cm^3, $\kappa_s = \kappa_i/6.5$, and with H in centimetres, the above formula becomes

$$H^2 + (13.1H_s + 16.8)H = 12.9\theta.$$ (7.41)

While neither of these formulae precisely fit the Anderson empirical results, there are a number of useful conclusions that can be drawn from these analytic results. From this result it is evident that the growth slows down drastically once the ice becomes thicker than about

20 cm. It is also clear that the effect of snow is very critical for thin ice, and that it will greatly slow down the growth rate. Finally, a third interesting point is that the thicknesses are amazingly independent of the amount of heat transfer to the atmosphere, as the latter only becomes critical for very thin ice. The square term in equation (7.40), which does not involve the rate of heat transfer to the atmosphere, becomes dominant for thicker ice. Thus many of the detailed heat loss mechanisms, such as wind speed, etc., are not highly critical to the final equilibrium thickness of the ice, which is basically a function of the degree days.

7.5.3 Full heat budget thermodynamic models

To take into account the full variability of different radiative and temperature forcing as well as temporal variability of snow fall and oceanic heat flux, it is necessary to utilize a complete surface heat budget. This heat-budget computation is used together with some type of thermodynamic model which can also account for melting. The most common procedure is to make use again of the steady-state temperature profile, as in the simplified Stefan approach in conjunction with a full heat budget (e.g., Parkinson and Washington, 1979). However, without 'thermal' inertia effects, seasonal variations in ice thickness tend to be too large. As will be discussed in the next section, the two main problems with this approach are the neglect of the internal heat capacity and the effects of internal brine pockets. These effects particularly come into play as one considers the warming of ice and the problem of equilibrium thickness of multi-year ice. The steady-state temperature profile approach does, however, allow the relative roles of different heat-budget components to be analysed.

Considering an ice slab with a snow layer on top (Figure 7.23), if one assumes no melting at the snow–ice interface, by continuity the amount of heat going through this interface must be the same in the snow and the ice so that

$$(T_i - T_o)\frac{\kappa_s}{H_s} = \frac{\kappa_i}{H}(T_B - T_i), \tag{7.42}$$

where, as depicted in Figure 7.23, T_i is the surface temperature of the ice and T_o is the surface temperature of the snow. This equation allows a solution for T_i in terms of T_B and T_o. Substituting the resulting expression into the conductive flux through the ice gives

$$(T_i - T_o)\frac{\kappa_s}{H_s} = \gamma(T_b - T_o), \tag{7.43}$$

where γ is given by

$$\gamma = \frac{\kappa_i}{H + (\kappa_i/\kappa_s)H_s}. \tag{7.44}$$

Using the sign convention that fluxes into the surface are considered positive, the complete heat-budget equation at the snow surface then becomes

$$(1 - \alpha)F_s + F_1 + \rho_a C_p C_H V_g(T_a - T_o)$$
$$+ \rho_a L_v C_E(q_a(T_a) - q_s(T_o)) - \varepsilon\sigma T_o^4 + \gamma(T_B - T_o), = 0, \tag{7.45}$$

where α is the surface albedo, ρ_a is the density of air, C_p is the specific heat of air, V_g is the wind speed, q_a is the specific humidity of the air, q_s is the specific humidity at the temperature of the surface, and F_s and F_l are the incoming short-wave and long-wave radiation terms. The constants C_H and C_E are bulk sensible and latent heat transfer coefficients, L_v is the latent heat of vaporization and ε is the surface emissivity. The equation is usually solved iteratively (see, e.g., appendix B of Hibler (1980b) for details and numerical values of various constants) for the ice surface temperature. The conduction of heat through the ice is used to estimate ice growth using

$$\gamma(T_B - T_o) - F_w = \rho_i L \frac{dH}{dt}, \qquad (7.46)$$

where F_w is the oceanic heat flux. In the case that the calculated surface temperature of the ice is above melting, it is then set equal to melting and the imbalance of surface flux is used to melt ice.

To give some feeling for the relative role of the different components, Table 7.1 (taken from Hibler (1980b)) gives a breakdown of the various heat-budget components near the North Pole taken from the solution for such a balance equation without snow cover. As can be seen, the qualitative statements made earlier about the dominance of sensible heat fluxes in winter and the dominance of the short-wave fluxes in summer are borne out by the more detailed heat-budget calculations.

7.5.4 Effects of internal brine pockets and variable conductivity

Apart from the absence of dynamics, the steady-state thermodynamic models discussed above greatly over-simplify the thermodynamic response of the sea ice. The main simplifications are that the effects of internal brine pockets and internal heat capacity have not been considered. Adding this simplification to the neglect of non-steady-state internal temperature variations, the steady-state 'Stefan' solutions tend to substantially over-estimate the seasonal swing of ice thickness. Additionally, unless some corrective modification of surface albedo is made, the neglect of internal melting due to brine pockets will cause the equilibrium thickness from such models to be substantially larger than observations.

In sea ice, the density, specific heat and thermal conductivity are all functions of salinity and temperature (the dependence on temperature is indirectly also due to salinity). These dependences are caused by salt trapped in brine pockets (see Figure 7.7) that are in phase equilibrium with the surrounding ice. The equilibrium is maintained by volume changes in the brine pockets. A rise in temperature causes the ice surrounding the pocket to melt, diluting the brine and raising its freezing point to the new temperature. Because of the latent heat involved in this internal melting, the brine pockets, for example, act as a thermal reservoir, retarding the heating and cooling of the ice. Since the brine has a smaller conductivity and a greater specific heat than ice, these parameters change with temperature.

The classic treatment of these variations was the development of a time-dependent thermodynamic model for level multi-year sea ice by Maykut and Untersteiner (1971).

Table 7.1. *Calculated heat-budget components near the North Pole for different ice thicknesses in units of centimetres of ice per day.*

1 cm of ice per day is equivalent to 34.95 W/m^2.

Heat-budget component	February 18				June 10				June 26		
	Open water	Ice thickness (m)			Ice thickness (m)			Open water	Ice thickness (m)		
		0.5	1.0	4.0	0.5	1.0	4.0		0.5	1.0	4.0
Net short-wave radiation	0	0	0	0	2.1	2.1	2.1	7.5	3.2	3.2	3.2
Net long-wave radiation	−3.0	−0.4	−0.1	0.1	−0.7	−0.8	−0.8	−0.6	−0.9	−0.9	−0.9
Sensible heat flux	−12.3	−2.5	−1.5	−0.5	−0.8	−0.7	−0.7	0	−0.1	−0.1	−0.1
Latent heat flux	−3.1	−0.2	−0.1	0	−0.6	−0.5	−0.5	0	−0.2	−0.2	−0.2
Conductive flux		3.1	1.7	0.4	0	−0.1	−0.1		−0.3	−0.1	0
Growth rate	18.4	3.1	1.7	0.4	0	−0.1	−0.1	−6.9	−2.0	−2.0	−2.0

From Hibler (1980b).

This work was heavily based on a variety of observations of the physical properties of sea ice dating as far back as Malmgren (1933) together with a series of numerical calculations of the heat conduction in sea ice by Untersteiner (1964). In their comprehensive analysis, Maykut and Untersteiner (1971) allowed the product of the density and heat capacity, ρc, and the thermal conductivity κ_i to be functions of the salinity S and temperature T. In order to calculate these parameters consistently, they made use of a vertically varying salinity profile (Schwarzacher, 1959), which was assumed to be constant – an assumption that provides one of the main restrictions of their model. These parameters were then used in a time-dependent thermal diffusion equation, which also allowed for short-wave radiation penetration into the ice,

$$(\rho c)_i \frac{\partial T}{\mathrm{d}t} = k_i \frac{\partial^2 T}{\partial z^2} + K_i I_0 \exp(-K_i z), \tag{7.47}$$

where the subscripts i denote the ice, T is the temperature, K_i is the short-wave extinction coefficient and I_0 is the amount of short-wave radiation passing through the ice surface. In the standard case simulations, I_0 was assumed to be 17% of the incident radiation. For the snow cover, a similar diffusion equation was used with fixed values of the parameters $(\rho c)_s$, k_s and I_0 set equal to zero. These equations were solved as a time-dependent, initial-value problem. Incoming long-wave radiation (F_{lw}), incoming solar radiation (F_{sr}), latent and sensible heat fluxes (F_1, F_s) and outgoing short-wave radiation (σT^4) were included as driving functions. These functions are combined with the thermal diffusion (equation (7.47)) to form a surface heat budget, which is solved by iteration to determine a surface temperature. This is then used as a boundary condition for further integration of the same equation. Also included as driving functions were a specified snow-fall rate versus time and an oceanic heat flux F_w into the ice from below to parameterize the effects of boundary layer entrainment of heat from below and the effects of heat due to radiation penetration into the oceanic boundary layer through leads and thin ice.

A typical simulation (after a 38-year integration to yield an equilibrium state) is shown in Figure 7.24(a). In close agreement with observations, over an annual cycle, a bottom accretion of 45 cm occurs, balanced by a surface ablation of 40 cm and a bottom ablation of 5 cm. Note that the bottom ablation occurs in early fall, after most of the surface ablation is over. An instructive way to display the overall thickness evolution of the Maykut and Untersteiner model is the equilibrium cycle in thickness and surface temperature space (Figure 7.24(b)). Viewed in this way, it is clear that most of the growth and diminution of the ice thickness occurs during periods when the ice temperature is relatively constant, being near 0 °C during the ablation period and close to the air temperature during growth. An extreme idealization of this characteristic was proposed by Thorndike (1992a, b) for idealized climate sensitivity studies. He suggested that many features of a perennial ice cover could be approximated by a constant winter and summer forcing. The transition time between these two periods is then dictated by the time taken to either cool or warm up the ice in accordance with its heat capacity.

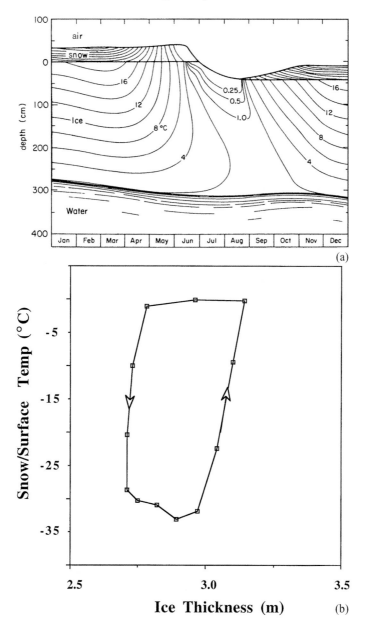

Figure 7.24. (a) Predicted values of equilibrium temperature and thickness for sea ice. (From Maykut and Untersteiner (1971).) Isotherms are labelled in negative degrees Celsius. In this simulation, an upward oceanic heat flux of $1.5\,\text{kcal}/\text{cm}^2$ per year is assumed, and 17% of the net short-wave radiation is allowed to pass through the ice surface (when it is snow-free). In addition, the albedo of snow-free, melting ice is taken to be 0.64. The snow-cover albedos are taken to vary seasonally, as specified by Marshunova (1961). (b) Equilibrium cycle of the Maykut–Untersteiner model plotted in thickness surface temperature space. The upper left-hand point is the August monthly mean.

In practice, uncertainty about the magnitude and variability of the oceanic heat flux and snow cover creates substantial uncertainty in the simulated equilibrium thickness of sea ice. In the case of the oceanic heat flux, the results are generally straightforward. Increasing the heat flux reduces the rate of accretion on the bottom of the ice and somewhat increases the bottom ablation. The processes at the upper boundary, on the other hand, are relatively insensitive to the ocean–ice heat flux. By reducing the heat flux to zero, the ice thickness almost doubles.

The effects of varying the snow cover are more complex. While a thinner snow cover allows greater cooling of the ice during winter, it is also removed earlier in the spring, which decreases the average albedo and prolongs the period of ice ablation. The effect is further complicated by internal melting due to the absorbed radiation once the ice is snow-free. Because of these effects, the snow depth has an unexpectedly small effect until 70 cm is reached. When this occurs, the decrease in ice ablation tends to dominate the ice thickness. Up to 120 cm, the snow cover in the model will totally disappear at some time in the summer.

A simplified thermodynamic model that included brine pockets was proposed by Semtner (1976) by asserting that sea ice could be thought of as a matrix of brine pockets surrounded by ice, where melting can be accomplished internally by enlarging the brine pockets rather than externally by decreasing the thickness. As a consequence, for the same forcing, sea ice can have a substantially greater equilibrium thickness than fresh-water ice. In this model, the snow and ice conductivities are fixed, and the salinity profile does not have to be specified. To account for internal melting, an amount of penetrating radiation is stored in a heat reservoir without causing ablation. Energy from this reservoir is used to keep the temperature near the top of the ice from dropping below freezing in the fall. As shown in Figure 7.25, using this 'brine damping' concept, Semtner was able to reproduce many aspects of Maykut and Untersteiner's results within a few per cent.

It is also possible to modify a steady-state 'Stefan' model to reproduce correct mean ice thicknesses, although such models still over-estimate the seasonal swing of ice thickness. For an even simpler diagnostic model, for example, Semtner proposed that a portion of the penetrating radiation I_0 be reflected away and the remainder applied as a surface energy flux. In addition, to compensate for the lack of internal melting, the conductivity is increased to allow greater winter freezing. This 'Stefan' equilibrium model does, however, produce a greater seasonal swing (Figure 7.25) of ice thickness, a flaw which Zhang and Rothrock (2001) have shown also occurs in more complete thickness distribution models that use 'Stefan'-like thermodynamics.

A variety of one-dimensional models subsequent to the Maykut and Untersteiner model (see, e.g., Steele and Flato (2000)) have focussed on reduced vertical resolution together with emphasis on different aspects of the heat budget and the effects of internal brine pockets. Recent examples include Gabison (1987), Flato and Brown (1996) and Lindsay (1998). A notable recent development has been energy-, or in certain cases 'enthalpy'-, conserving thermodynamic models (Bitz and Lipscomb, 1999; Winton, 2000) where strict energy conservation combined with changes in brine volume (*assuming* a fixed salinity profile) are insisted upon. These models appear to be able to reproduce a much more

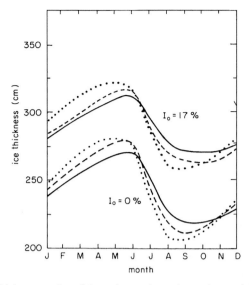

Figure 7.25. Annual thickness cycles of three thermodynamic sea-ice models for the cases of 0 and 17% penetrating radiation: (solid lines) Maykut and Untersteiner model; (dashed lines) three-layer Semtner model; (dotted lines) zero-layer Semtner model. The forcings (other than the penetrating radiation) are the same as in Figure 7.21. (Re-drafted from Semtner (1976).)

realistic seasonal cycle of ice thickness with few vertical temperature levels. The effects of the inclusion of such models in a full thickness distribution is given in section 7.6.3.

In addition to the above models, which are largely consistent reformulations of the Maykut–Untersteiner approach, there are a variety of aspects of sea-ice thermodynamics that are currently being investigated in more detail both in the laboratory and the field (see, e.g., Eicken *et al.* (1995)). One notable feature of these investigations that has received special attention in the Antarctic is the presence of 'snow ice' (Ackley, Lange and Wadhams, 1990; Eicken, 1995). Basically, in the Antarctic pack the heavy snow cover causes sea ice to be depressed below water level and flooded by sea-water. As a result of this flooding, snow ice forms through the congelation of sea-water and brine in a matrix of meteoric ice (i.e., snow). This can increase the effective ice thickness in many regions. A related problem is that of frazil ice formation (Ackley *et al.*, 1990), as opposed to plate ice as assumed in the Maykut–Untersteiner model. Many of these features are considered to be especially critical in the marginal ice zone regions, and also play a role in sediment uptake in the ice.

7.6 Ice-thickness distribution theory: dynamic thermodynamic coupling

A key coupling between sea-ice thermodynamics and ice dynamics is the ice-thickness distribution. The basic physical notion there is that the deformation causes pressure ridging and open water creation. Thermodynamic processes, on the other hand, tend to ablate pressure ridges and remove thin ice by growth in winter, and create thinner ice and open water in summer. From a typical ice-thickness distribution, we can think of deformation as affecting the higher (thick) and lower end of the thickness distribution, thus causing a spreading of the distribution. Conversely, thermodynamic processes without deformation cause a concentration of the thickness toward a centre value.

The coupling of thermodynamics with deformation is critically affected by mechanical effects, which limit the character and amount of deformation. As a consequence there can be a significant feedback between mechanics and thickness since, as the ice state changes, the deformation can also change. Assumptions on how the ridging process occurs for given deformation states can modify the temporal and spatial growth and thickness patterns of the ice pack, and hence modify its response to climatic change.

7.6.1 Evolution equations for the ice-thickness distribution

A theory of ice-thickness distribution may be formulated (Thorndike *et al.*, 1975) by postulating an areal ice-thickness distribution function and developing equations for the dynamic and thermodynamic evolution of this distribution. An underlying assumption is that all thicknesses of the ice pack move with the same velocity field. Following Thorndike *et al.* (1975), $g(h)dh$ is defined to be the fraction of area (in a region centred at position x,y at time t) covered by ice with thickness between h and $h + dh$. Neglecting lateral melting effects, it is easy to derive the following governing equation for the thickness distribution:

$$\frac{\partial g}{\partial t} + \nabla \cdot (vg) + \frac{\partial (f_g g)}{\partial H} = \Psi, \qquad (7.48)$$

where f_g is the vertical growth (or decay) rate of ice of thickness H, and Ψ is a redistribution function (depending on H and g) that describes the creation of open water and the transfer of ice from one thickness to another by rafting and ridging. An underlying assumption in this formulation is that a region statistically large enough to describe a thickness distribution is used. However, this assumption is more a constraint on observations than on the continuum model. Except for the last two terms, equation (7.48) is a normal continuity equation for g. The last term on the left hand side can also be considered a continuity requirement in thickness space since it represents a transfer of ice from one thickness category to another by the growth rates. An important feature of this theory is that growth occurs by rearranging the relative areal magnitudes of different thickness categories.

This theory was extended by Hibler (1980b) to include lateral melting by adding an additional term F_L to equation (7.48), where

$$\int_0^\infty F_L \, dh = 0.$$

This constraint follows from the fact that lateral melting will be compensated for by a change in open water extent. The physical notion here is that thick ice will have a larger vertical interface with the ocean than thin ice, and hence a constant value reducing all categories proportional to their abundance is a good first approximation. This theory may also be extended explicitly to include ridged ice (Flato and Hibler, 1995) by breaking the distribution into ridged ($g_r(h)$) and unridged ($g_u(h)$) components, each of which evolve according to an evolution equation of the form of equation (7.48).

The multi-level ice-thickness distribution theory represents a very precise way of handling the thermodynamic evolution of a continuum comprising a number of ice thicknesses. However, the price paid for this precision is the introduction of a complex mechanical redistributor. The other issue is how to keep track of the stored energy and thermal properties of a variety of ice thicknesses as they are transferred into thicker ice by ridging. Also, as noted below, how the mechanical redistribution is carried out for arbitrary deformation states can significantly affect spatial patterns and magnitudes of ice-thickness characteristics.

To describe the redistribution, one must specify what portion of the ice distribution is removed by ridging, how the ridged ice is redistributed over the thick end of the thickness distribution, and how much ridging and open water creation occur for an arbitrary two-dimensional strain field, including shearing as well as convergence or divergence. In selecting a redistributor, one can be guided by the conservation conditions on Ψ:

$$\int_0^\infty \Psi \, dh = \nabla \cdot \mathbf{v} \tag{7.49}$$

$$\int_0^\infty h\Psi \, dh = 0. \tag{7.50}$$

Equation (7.49) follows from the constraint that Ψ renormalize the g distribution to unity due to changes in area. Equation (7.50) follows from conservation of mass, and basically states that Ψ does not create or destroy ice but merely changes its distribution. An additional assumption in equation (7.50) is that the ice mass is related in a fixed manner to the thickness. Additional guidance on choosing a redistributor may be obtained from mechanical considerations as discussed below.

7.6.2 Consistency of isotropic plastic models with ridge building

It is possible to base the smooth plastic descriptions of plastic flow, as, for example, the elliptical yield curve described above, on the physics of pressure ridging. Mechanistic pressure ridge models (Hopkins, 1994; Parmeter and Coon, 1972) show the work needed to build ridges is relatively independent of how fast the ridge is formed, provided that the major work done in ridge building is due to potential energy changes caused by ice pile-up. Rothrock (1975) extended this concept to two dimensions by insisting that two-dimensional deformational work be explicitly related to the amount of work done by ridging. There is high uncertainty on how to handle the amount of energy due to friction and deformation, which are largely caused by lateral sliding along leads or faults. If one assumes that the energy dissipated in pressure ridges is proportional to the change in potential energy, and that ridging is the only energy loss, then a constraint on ψ is

$$C \int_0^\infty h^2 \Psi \, dh = \sum_{i,j=0}^{2} \sigma_{ij} \dot{\varepsilon}_{ij}, \tag{7.51}$$

where C is a constant related to the densities of water and ice and the frictional energy losses during ridging. In the case that only potential energy losses are considered, $C = C_b$ with

$$C_b = \frac{1}{2} \rho_i \left(\frac{\rho_w - \rho_i}{\rho_w} \right) g, \tag{7.52}$$

where ρ_w is the density of water, ρ_i is the density of ice and g is the acceleration due to gravity. In the special case of pure convergence it is not hard to show that equation (7.51) serves as a definition of ice strength. A key question is the magnitude of frictional losses in the ridging process. Early estimates of frictional losses put them at similar values to C_b, which have subsequently been shown to be low.

In order explicitly to ensure consistency of mechanical and ridging (Hibler, 1980b) the redistributor Ψ may be written in the form

$$\Psi = \delta(h)(M + \dot{\varepsilon}_{kk}) + M\omega_r,$$

where M is a normalized mechanical energy dissipation term and ω_r is the so-called ridging mode function. This function describes the transfer of thin ice into a distribution of thicker, ridged ice, acting simultaneously as a sink of thin ice and a source of thick ice in such a way as to conserve area and volume. If all of the energy losses result from ridging, then $M = p^{-1} \sum_{i,j=1}^{2} \sigma_{ij} \dot{\varepsilon}_{ij}$ (Hibler, 1980b), where p is the hydrostatic compressive strength and σ_{ij} and $\dot{\varepsilon}_{ij}$ are the components of the two-dimensional stress and strain rate tensors, respectively. In the other extreme, $M = 0$. More details on these equations may be found in Hibler (1980b).

There are several major issues with the energetic consistency equations that make the mechanical aspects of the ice-thickness distribution theory rather arbitrary and generally render the theory somewhat unwieldy. The most notable are: how to carry out the transfer of thin ice to thick ice by ridging; how much frictional energy losses occur during ridging; and how much ridging and concomitant open water creation occur under an arbitrary deformation state involving some shear. This latter feature can largely be dispensed with by using a fracture-based model at high resolution rather than an aggregate model (see section 7.4.2).

For the the transfer of thin ice to thick ice by ridging, Thorndike *et al.* (1975) suggested a redistributor which transfers ice into categories that are fixed multiples of the initial thickness. However, ridge observations and theoretical considerations suggest that such a redistributor is unrealistic. Hibler (1980b) proposed a scaling law for ridge building with ice redistributed uniformly up to a maximum thickness, scaling as the square root of the thickness of ice being ridged. Limited data (Figure 7.26) on the block size of ice in ridges

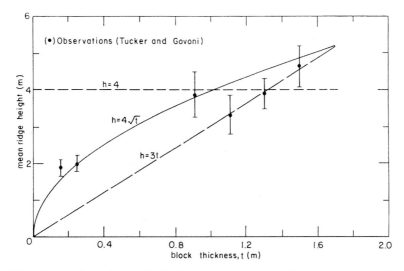

Figure 7.26. Observations of mean block thicknesses in pressure ridges versus mean ridge height. Together with strengths resulting from different redistribution scaling laws for ridging in variable thickness models. (From Tucker and Govoni, 1981.)

(Tucker and Govini, 1981) versus mean ridge height tend to support this scaling. Subsequent work (Babko, Rothrock and Maykut, 2002) has added an explicit formulation of ice rafting.

With any of these redistributors, however, discrepancies still remained for some time in that the local ridging losses were much smaller than needed in large-scale simulations to obtain realistic ice drifts (Flato and Hibler, 1995; Hibler, 1980b). A major contribution to resolving this inconsistency was made in the late 1980s by the ice–ocean dynamics group at Dartmouth through a series of large-scale simulations (Flato and Hibler, 1995) and discrete-element simulations of the ridging process by Hopkins and Hibler (1991a) and Hopkins (1994). The basic results were that frictional losses in ridging (Figure 7.27) were found to be about an order of magnitude higher in the ridge building than the potential energy losses. By increasing this energy loss in large-scale non-linear plastic models (Flato and Hibler, 1995), stresses large enough to yield realistic sea-ice drifts were obtained, provided that redistribution functions were realistic.

The other major uncertain aspect of the thickness distribution is the amount of ridging under pure shear. Within the conventional thickness distribution approach and an aggregate rheology (Rothrock, 1975; Thorndike *et al.*, 1975), energetic consistency of a rheology demands, arguably, that under shearing conditions both ridging and open water creation must occur. Variable thickness simulations (Flato and Hibler, 1995) have shown that including or excluding this energetic consistency can lead to substantial changes in the simulated mean ice thicknesses and in the relative volume of ridged ice, even though the general shape of the thickness distribution is relatively unchanged. This consistency may be formulated in terms of sets of deforming oriented leads (Ukita and Moritz, 1995) which provides a link to lead structure. However, such a procedure provides no unique way of choosing the lead

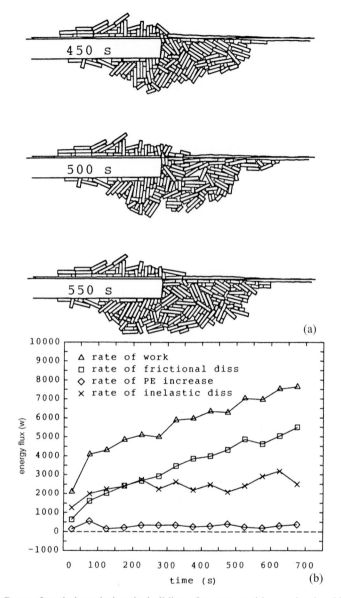

Figure 7.27. Rates of work done during the building of a pressure ridge as simulated by a discrete-element model. (a) Simulated ridge near the end of a numerical experiment. (b) Energy rates for different components in the energy budget. (From Hopkins and Hibler (1991a).)

orientation. Anisotropic models (Coon *et al.*, 1998; Hibler and Schulson, 2000), including continuity of stresses across the leads (discussed in section 7.4.2), provide a unique way based on dynamical constraints to make this choice. These models typically include a great deal of shear energy losses due to sliding of leads so that additional open water creation is not usually required. Moreover, isotropic 'fracture-based' rheologies (Hibler, 2001; Hibler and Schulson, 2000) provide a more precise theoretical determination of this partition which ultimately derives from laboratory work if the 'scale-invariant' assumption can be invoked.

7.6.3 Characteristics of thickness distribution models coupled to specified deformation

While it does not consider mechanical feedback effects, coupling of thickness distribution evolution models with thermodynamic models provides a useful way to estimate the role of a variable thickness sea-ice cover. A number of 'partially coupled' model studies of this type are reviewed by Steele and Flato (2000). One of the more comprehensive studies is that of Maykut (1986). Using relatively smoothed deformation from the Aidjex experiment, Maykut (1986) investigated the role of thinner ice in the overall heat budget as well as the different heat-budget components. These results show the dominance of thin ice in the overall heat budget. Basically, when variations in ice thickness were taken into account, the amount of heat conducted to the surface doubled. Maykut found that most of this additional heat was lost to the atmosphere via the turbulent fluxes. Notable was the fact that open water categories (0–0.1 m) did not dominate the balance; rather, it was the intermediate thicknesses (0.2–0.8 m) that were responsible for over half the heat loss through the ice. Maykut's study also emphasized the role of heat absorbed through leads, which may go to melting thick pressure ridges or possible lateral melting. Considerable experimental work has been performed on lateral melting (most notably by Perovich, Elder and Richter-Menge, 1997) which has allowed the development of formulations that can be used with the variable thickness models.

The degree to which inclusion of improved thermodynamics might change variable thickness results with specified deformation was examined by J. Zhang and colleagues. Utilizing Winton's (2000) three-level enthalpy conserving model, Zhang and Rothrock (2001) included this formulation in each separate fixed thickness category and made use of an Eulerian fixed ten-level thickness grid. Deformation and thermodynamic forcing were taken from a fully coupled dynamic thermodynamic model near the pole (Zhang *et al.*, 2000; see also section 7.7). As an intermediate model they also included a model with an additional temperature layer but without conservation of internal heat. Although without full mechanical feedback it is difficult to assess the full sensitivity, their main result (Figure 7.28) was that, mirroring thermodynamic model results, including full enthalpy conservation increased the mean ice thickness and substantially reduced the seasonal swing. Specifically, a 5–10% effect on the ice thickness with a much less reduction of thickness in summer was found.

One difficulty with all specified strain comparisons is a realistic estimation of the deformation field. Referring to Figure 7.16, portraying a typical deformation field with intersecting shearing and opening 'faults' or 'leads', the average strain field is very well defined, and can be obtained by a boundary integral using Green's theorem regardless of whether the faults

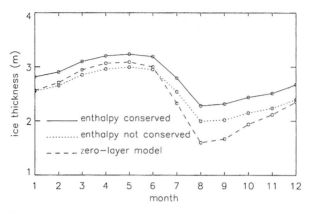

Figure 7.28. Equilibrium ice thickness results for an enthalpy conserving variable thickness model tested with specified deformation. In the solid and dotted curves, there are three temperature layers, whereas the dashed line is a zero-layer Stefan model with a uniform temperature gradient between the top and bottom of the ice.

are arbitrarily small. However, estimation of the strain by only a few velocity points on the ice can yield substantial errors, and one should minimally do a least squares analysis (e.g., Hibler *et al.*, 1974a) to estimate the error. Moritz and Stern (2001) have effectively used the boundary integral method employing SAR data at three-day intervals. Such data are not widely available, and as a consequence strain data are usually obtained from temporally and spatially smoothed velocity fields (see, e.g., Thorndike (1986)) which under-estimate high frequency motion among other measures. Consequently, the resulting thickness distributions must be viewed with caution. In practice, aspects of the energetic consistency argument mentioned above (which are at best arbitrary even with remote-sensing estimates; e.g., Stern, Rothrock and kwok (1995)) must be invoked. Even with this inclusion, high frequency variability is neglected. P. Heil and colleagues have shown, for example, that inclusion of this variability (from observations) yields a significant increase in the net sea-ice production (Heil, Allison and Lytle, 2001; Heil *et al.*, 1998).

7.6.4 Two-level ice-thickness distribution

Many features of the thickness distribution may be approximated by a two-level sea-ice model (Hibler, 1979), where the ice-thickness distribution is approximated by two categories: thick and thin. Variations of this model have been used in a large number of investigations, most recently for example by Harder and Fischer (1999), Hilmer and Lemke (2000), Holloway and Sou (2002), and Kreyscher *et al.* (2000) (see also Chapter 9). Within this two-level approach, the ice cover is broken down into an area A (often called the compactness), which is covered by thick ice, and a remaining area $1-A$, which is covered by thin ice, which, for computational convenience, is always taken to be of zero thickness (i.e., open water). The idea is to have the open water approximately represent the combined fraction of both open water and thin ice up to some cutoff thickness H_0. The remainder of the ice is considered to be distributed between zero and $2H/A$ in thickness.

For the overall mean thickness H and compactness A, the following continuity equations are used:

$$\frac{\partial H}{\partial x} = -\frac{\partial(uH)}{\partial x} - \frac{\partial(vH)}{\partial y} + S_H, \qquad (7.53)$$

$$\frac{\partial A}{\partial x} = -\frac{\partial(uA)}{\partial x} - \frac{\partial(vA)}{\partial y} + S_A, \qquad (7.54)$$

where $0 \le A \le 1$, u, v are the components of the ice velocity vector, and S_H and S_A are thermodynamic terms. While the first equation is essentially a continuity equation for ice mass (characterized by the mean ice thickness or equivalently the ice mass per unit area) with a thermodynamic source term, the second equation for the compactness implictly includes some treatment of the ridging process. By including the restriction that $A \le 1$, a mechanical sink term for the areal fraction of ice has been added to a continuity equation for ice concentration. This sink term is considered to apply when $A = 1$, and under converging conditions removes enough ice area through ridging to prevent further increases in A. Note, however, that, independent of A, ice mass is rigidly conserved by the conservation equation for H. The thermodynamic growth and decay terms are typically determined by assuming a uniform distribution of ice thicknesses between zero and $2H/A$ (see, e.g., Hibler (1979); Hibler and Walsh (1982); Walsh, Hibler and Ross (1985)).

To characterize the strength P in this two-level model, Hibler proposed that

$$P = P^* H \, e^{-C(1-A)}, \qquad (7.55)$$

where $C \sim 20$ and P^* is a constant. The most widely used value for the strength is $P^* = 2.75 \times 10^3 \, \text{N/m}^2$, which was obtained by Hibler and Walsh (1982) by fitting progressive vector plots of predicted ice drift with observations. This value is larger than the value initially proposed by Hibler (1979) due to seven-day-averaged wind being used in the initial simulations. Similar values have been used by most other investigators mentioned above in this section.

7.6.5 Relative characteristics of two-level and multi-level models in numerical simulations

To allow either the multi-level or two-level sea-ice model to be integrated over a seasonal cycle, it is necessary to include some type of oceanic boundary layer or ocean model. The simplest approach (e.g., Walsh et al., 1985) is to include a motionless fixed depth mixed layer. Another approach is to use some type of one-dimensional mixed layer, such as a Kraus–Turner-like mixed layer used by Lemke et al. (1990) for the Weddell Sea. A third approach is to utilize a complete oceanic circulation model which also allows lateral heat transport in the ocean (Hibler and Zhang, 1994; Zhang et al., 1998). This latter approach is discussed in more detail in section 7.7.1.

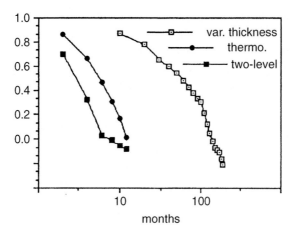

Figure 7.29. Monthly mass anomaly (deviation from the average seasonal value) auto-correlation of Arctic ice mass for three different ice models used to simulate the Arctic ice cover over several decades. The variable thickness model results are from Flato (1995), and the thermodynamic and two-level model results are from Walsh *et al.* (1985).

Results from both two-level and multi-level simulations are discussed in Chapter 9 by G. M. Flato and by Steele and Flato (2000), who emphasize ice growth and models with specified deformation. Selected multi-level simulation results are discussed below. Of particular relevance to climatic variability is the inertia that a full thickness distribution inserts into the ice-mass characteristics. This is illustrated in Figure 7.29, where we show auto-correlations of the mass anomalies for three different dynamic thermodynamic models of the Arctic ice pack with different thickness distribution formulations. Note that with a full multi-level thickness distribution, it take about eight years for the auto-correlation function of the thickness anomalies to fall off to $1/e$ of the variance. This is in sharp contrast to a two-level model where there is a very rapid balancing of the strength and thickness (via equation (7.55)) on timescales of a few months; so much so that the auto-correlation time for a two-level model is less than for a thermodynamic model. The inertia of the multi-level model is very large because the thick ridged ice plays no direct role in the ice strength, and slowly reaches equilibrium by a combination of advection, growth, melt and formation.

It should be emphasized that these temporal correlations occur even though a uniform distribution of thick ice is used in thermodynamic calculations in the two-level models. Without this thermodynamic correction, the two-level model (see, e.g., Hibler and Flato (1992)) yields ice about 20% thinner than thermodynamic thicknesses for the Arctic basin and about 35% thinner than a full variable thickness simulation. This thermodynamic correction does improve the growth rates but still does not supply a precise treatment of thin ice, mechanical coupling, nor a full thickness distribution for comparison with observations. Full thickness distribution models are used, for example, by Flato and Hibler (1995), Hibler (1980b), Holland and Curry (1999), Polykov *et al.* (1999) and Zhang *et al.* (2000).

7.6.6 Thickness strength coupling: kinematic waves and inertial variability

High frequency variability in ice motion and deformation at sub-daily timescales can have a significant impact on the mass budget of sea ice (Heil *et al.*, 1998). As noted above, the most pronounced form of high frequency variability is inertial variability, which has some seasonal signature almost universally felt to be due to enhanced damping by ice interaction in winter. The main explanation of these phenomena is a strong turbulent coupling between the ice and the oceanic boundary layer so that the ice motion is a signature of the larger inertia of the oceanic boundary layer. When coupled with evolution of the thickness distribution (e.g., Hibler, Heil and Lytle, 1998), significant variations in compactness can occur.

A separate, related mechanism that is probably related to inertial scale variability in ice deformation are waves (loosely referred to as kinematic waves) arising from coupling between the momentum equations and ice-thickness evolution equations. These waves have a more rapid propagation speed than the advection timescale, and hence at higher resolutions the use of longer time steps can result in instabilities in the coupled equations. In such cases, either coupled implicit solutions are utilized or smaller time steps employed. We briefly describe kinematic waves, the ramifications of ice mechanics on high frequency deformation and some ice arching issues.

Kinematic waves in sea ice

The basic idea of kinematic waves in sea ice is that the conservation equation in conjunction with the ice momentum equation sets up a wave due to the velocity being dependent on quantities in the conservation equation (such as the compactness). The term 'kinematic' arises from the fact that inertia plays no role in the wave.

To examine some analytic characteristics, consider ice piling up against the coast with a motion that is controlled by an ice pressure P. Considering the case where ice piles up continuously and smoothly so that the ice failure is always on the yield curve for appropriate forcing, we may take $P = f(A) = P_1 e^{-C(1-A)}$, where A is the compactness. Then with a linear water drag the equilibrium equation of motion for a one-dimensional system is

$$\frac{\partial P}{\partial x} + \tau - cu = 0. \tag{7.56}$$

In the absence of thermodynamics, the conservation equation for A is

$$\frac{\partial A}{\partial t} + \frac{\partial (uA)}{\partial x} = 0. \tag{7.57}$$

Solving for u from the momentum equation, the conservation equation may be rewritten as

$$\frac{\partial A}{\partial t} - \frac{\partial}{\partial x} \left\{ \left[\frac{\partial P(A)}{\partial x} - \tau \right] A \right\} = 0. \tag{7.58}$$

Considering P to be dependent upon x through A and τ constant, this equation may be put in the form

$$\frac{\partial A}{\partial t} + Q(A)\frac{\partial A}{\partial x} = \eta(A)\frac{\partial^2 A}{\partial x^2}, \tag{7.59}$$

where $\eta(A) = A\frac{\partial P}{\partial A}$ and $Q(A)$ represents the remainder of the terms when the differentiation is carried out. This is basically a damped hyperbolic wave equation (Whitham, 1974) with the essential non-linearity inherent in the term $Q(A)\frac{\partial A}{\partial x}$. This equation may be thought of as a series of damped propagating waves with the wave speed dependent on the value of A (see, e.g., Whitham (1974)). For example, an analytic simple case of equation (7.59) (still highly non-linear) is Burgers' equation where $Q(A) = A$ and $\eta(A) =$ constant. Burgers' equation is soluble analytically (Whitham, 1974) and tends to lead to sharp fronts propagating even though the initial conditions may be smooth.

A numerical example of ice buildup similar to the analytic example, together with attendant effects of non-linear thickness strength coupling, was examined by Hibler, Udin and Ullerstig (1983). In this study, coupled momentum and conservation equations were numerically integrated in conjunction with an ice strength parameterization. In this simulation the coupled equations were integrated for 18 hours at time steps of 15 minutes on a 9×9 grid. The initial conditions consisted of 10% open water together with a mean ice thickness of 0.5 m. To simplify analysis, the ice strength is taken to depend only on the compactness:

$$P = (0.5)P^*e^{-C(1-A)}. \tag{7.60}$$

In addition, the turning angles and coriolis parameter were set equal to zero. A constant wind speed of 9.23 m/s in the positive x-direction was used.

The behaviour of the velocity variability and ice buildup are shown in Figure 7.30. Even though the forcing is fixed, points nearest the boundary first slow down and then speed up. The initial slowing down of the ice near the coast is due to the ice becoming stronger as the compactness decreases. However, the region of low compactness eventually becomes large enough to accumulate an adequate wind fetch to overcome the differential of plastic stresses, and the near-shore ice begins to drift faster. In terms of the compactness, the wave is essentially a propagating front between high and low compactness propagating outward. The non-linear coupling between the strength and the compactness (equation (7.60)) can also lead to fluctuations in the ice velocity, as is apparent in the velocity plots in Figure 7.30 at ~6 hours. These secondary waves were removed when a linear compactness strength relationship was utilized.

Overall it is possible that continual excitation of such waves may be responsible for observed meso-scale sea-ice fluctuations which may affect the ice-mass balance. Probably the main relevance, however, of these waves to modelling the evolution of sea ice is their constraint on time step limitations in high resolution models. Since the wave propagation time is probably an order of magnitude faster than the advection speed of the ice, typical explicit integrations of the coupled equations requires a Courant, Friedrichs and Lewy

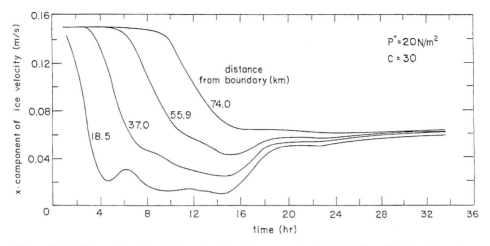

Figure 7.30. Time series of *x*-velocity at grid points progressively further from right hand boundary of an idealized ice buildup numerical experiment.

(CFL) criterion (Richtmyer and Morton, 1967) for these waves for overall stability unless an implicit solution for the momentum and thickness equations is utilized.

Inertial variability in sea-ice deformation

The basic character of simulated and observed inertial oscillations was briefly mentioned above (see Figure 7.4) in conjunction with the dynamical equations of motion. Here we focus briefly on the mechanisms for this variability, and the role of ice mechanics in damping the oscillations.

An idealized investigation of the effects of the inertial imbedding procedure proposed by McPhee (1978) was carried out by Hibler *et al.* (1998) with application to the Antarctic ice pack. This simulation was similar to the kinematic wave study mentioned above. However, when a pulse of wind is applied to this system with appropriate strengths, the outward propagating front of the high compactness couples with the inertial motion to form bands of oscillating deformation that are very slowly damped.

More relevant to the sea-ice mass budget, however, are realistic two-dimensional simulations. Such simulations (Heil and Hibler, 2002) demonstrate that with realistic temporally and spatially varying winds, non-linear mechanical damping from plastic rheologies supplies about the correct mechanical damping with resultant deformation. The statistical characteristics of the resultant deformation are illustrated in Figure 7.31, which shows simulated and observed deformation spectra from two sets of three drifting buoys with the locations shown in Figure 7.32(a). Basically, with imbedding the two deformation spectra agree very well with reasonable coherence. This improved coherence is also reflected in a very significant correlation coefficient ($r = 0.87$) between the simulated and observed divergence for the array corresponding to the left hand panels in Figure 7.31.

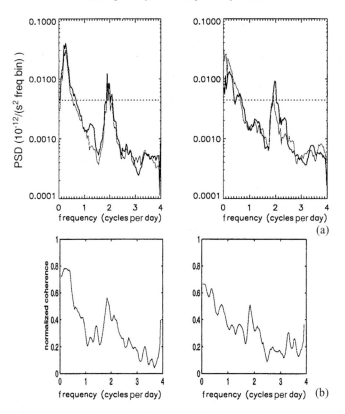

Figure 7.31. (a) Power spectral densities of observed (darker curve) and simulated (with inertial imbedding) divergence rate for two different buoy arrays. (b) Normalized coherence between simulated and observed divergence rates.

Inertial imbedding also greatly improves the variance of the simulated divergence compared with conventional simulations using, say, quadratic drag. Geiger *et al.* (1997), for example, found that while the viscous–plastic rheology performed best of all rheologies tested, modelled ice deformation components in the Weddell Sea showed a much lower variance than that yielded by the observed values, and a very low coherence with buoy deformation even at low frequencies. This was especially true of divergence, whose variance magnitude was only 10% of the observed divergence variance magnitude. Including the inertial energy of the ocean via inertial imbedding greatly ameliorates the simulation of divergence, and hence appears to be a key component in simulating realistic ice-thickness characteristics.

Interestingly, much of the inertial energy in this deformation field also arises from the non-linear ice mechanics. In particular, Heil and Hibler (2002) also examined simulations with smoothed wind fields with inertial imbedding both with and without ice mechanics

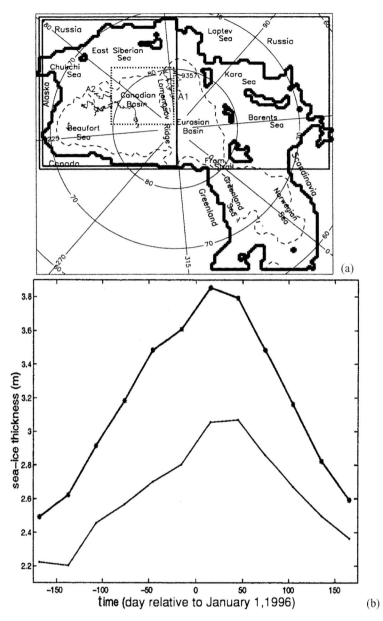

Figure 7.32. (a) Outline of model domain used for ice–ocean inertial oscillation study. The left hand solid box indicates the region where ice thicknesses were analysed. Trajectories of two arrays of three buoys each are shown: A1 near the pole and A2 in the Beaufort Sea. Buoys shown travelled from the top toward the bottom of the graph. The trajectory of IABP 9357, which forms part of A1 (see Figure 7.20), is labelled. (b) Time series of ice thickness over the central Arctic Basin from a dynamic thermodynamic sea-ice model with and without inclusion of 'inertial embedding' and the resultant increased high frequency deformation.

effects. The velocity magnitude spectra show that with unsmoothed wind the inertial imbedding alone supplies a resonance that will cause amplification of the small amount of high frequency noise in the wind field, and hence create a spectral peak in simulated buoy velocity. However, in the case of smoothed wind, the imbedding alone, which is essentially a linear operation on the wind field, produces no appreciable inertial power. These results show that the non-linear ice mechanics coupled to the thickness evolution equations supplies a cascade of energy to higher frequencies that generates inertial power. Consequently, we may think of the ice mechanics as generating energy through this cascade and then also damping this energy to realistic levels.

The effects of this variability on the mass balance are significant with up to 1 m of additional ice formation in the central Arctic region (Figure 7.32(b)). While utilizing only a two-level distribution simulation, Heil and Hibler (2002) showed that, in equilibrium, enhanced high frequency variability substantially increases (Figure 7.32(b)) the ice volume averaged over a centre portion of the Arctic basin (as noted in Figure 7.32(a)). Basically, there is a greater increase in winter in the ice mass, with a somewhat reduced mass increase in summer. The likely physics here is an increase in open water in winter, which greatly increases ice production. However, during the decay season the additional radiation absorption provided by increased open water due to variability also causes more melt, hence the seasonal asymmetry.

Ice arching with growth and advection

It is instructive to examine the highly non-linear arching problem analysed in section 7.4.3 in the presence of ice growth and ice advection. Referring back to Figure 7.21(a) for the channel configuration, when ice growth occurs in the channel the outflow can be largely stopped or freely flowing depending on the rate of growth (Hibler and Hutchings, 2003). This is illustrated in Figure 7.33, where thickness characteristics are shown for growth rated inversely dependent upon thickness according to $f = \alpha[1/(H + 0.4) - 0.005)$, with α a variable and f in units of metres per hour and H in metres. Upstream differencing is used for ice advection.

With high growth rates ($\alpha = 5/2.25$), the ice gradually forms a strong enough arching formation effectively to stop the outflow (Figure 7.33(a)). In the case of slower growth ($\alpha = 2.5$) the ice creates a narrow channel through the ice pack through which ice flow occurs relatively rapidly (Figure 7.33(b)). Such features are very characteristic of the ice-thickness patterns in the ice pack in regions near the Bering Strait (L. Shapiro, personal communication). This transition to total stoppage occurs rapidly with growth rate. Since both states are forced with the same wind, the physical notion is that there is potential for multiple equilibrium states with the same wind forcing dependent on the initial conditions. This will in fact occur for $\alpha = 2.5$. In particular, if the initial ice thickness is taken to be very large a result similar to Figure 7.33 is obtained (Hibler and Hutchings, 2003).

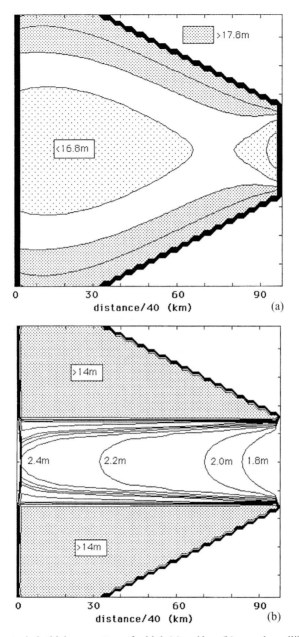

Figure 7.33. Characteristic thickness patterns for high (a) and low (b) growth equilibrium simulations for the arching problem. (From Hibler and Hutchings (2003).)

7.6.7 Ice–tide interaction and stationary shore fast ice

Two topics related to the evolution of sea ice in near-shore regions that intrinsically involve dynamic thickness coupling are the presence of relatively stationary shore 'fast' ice, and ice–tidal interaction. Since the most significant tidal effects (see, e.g., Kowalik (1981)) occur in shelf and coastal regions, these topics have a geographical relationship. Moreover, the relative stationarity of the fast ice causes an increased damping of the tides in contrast to the behaviour under a highly non-linear but mobile ice cover.

Fast ice typically forms early in the winter near coastal regions. On the Russian side of the Arctic (in the Laptev and east Siberian Seas, for example), collected observations show fast ice can extend several hundred kilometres from the coast. Off the Alaskan coast, fast ice often contains numerous ridging bands, indicating that it has been formed in incremental stages becoming gradually more extensive in character. Once formed, detailed measurements of the motion of the fast ice in the Prudhoe Bay area show it to move typically less than a metre or so. Little work has been done on simulating fast ice. However, it is notable that, in a similar manner to the formation of linear kinematic features (see Figure 7.15) with appropriate boundary conditions, non-linear 'plastic' models can simulate a boundary separating rigid and highly mobile pack ice. Pritchard (1978), for example, showed that for sufficiently high strengths, irregular boundaries can produce shaded regions of dead ice utilizing a plastic rheology. Moreover, pressure ridges incorporated in the fast ice are often grounded (Kovacs, 1972), and hence can provide a variety of anchor points as the ridging occurs and increase the subsequent stability. More detailed investigations of this phenomenon are currently underway.

With respect to tides, there are significant sea-surface height variations as well as rectified currents resulting from tides in the Arctic (and Antarctic), especially in shallow shelf regions. A prominent tidal component is the M_2 semi-diurnal tidal component, simulated values (Kowalik and Proshutinsky, 1994) of which are shown in Figure 7.34(a). In near-shore regions, such as the Barents Sea, the tidal amplitude can be over a metre as a result (verified by observations). In the central portion of the Arctic basin, the tidal variations are significantly smaller (Figure 7.34(a)), with elevation changes of less than a few centimetres. While in high latitudes inertial oscillations (discussed in section 7.6.3) have a similar temporal frequency, they are a distinct phenomenon. Most notably, the inertial oscillations have in principle no phase locking in time and are more stochastic in character. They also proceed clockwise in the northern hemisphere. Because of this similarity of frequency, it is difficult to distinguish clearly tides from inertial variability. Indeed, in cases where tidal motion sweeps out a clockwise ellipse, the two phenomena may interact strongly. Such tidal behaviour occurs (Kowalik and Proshutinsky, 1994) in many parts of the Arctic (the Laptev and east Siberian Seas, for example). This inertial–tidal interaction problem utilizing non-linear ice mechanics has been little investigated.

While not including highly non-linear ice models or 'inertial imbedding' concepts, Arctic ice–tide models (Kowalik and Proshutinsky, 1994) have robustly demonstrated the capability to simulate significant amounts of ice production in near-shore regions. This production is

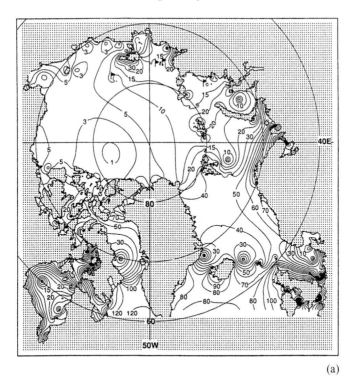

(a)

Figure 7.34. (a) Computed amplitude (cm) of surface elevation for the semi-diurnal tidal constituent M_2. Flags denote shelf wave regions. (b) Residual ice motion in the east Siberian and Laptev Seas caused by M_2 waves. Ice viscosity coefficients are taken as $\eta = \zeta = 10^7$ cm^2/s, and the ice pressure coefficient is $A_p = 10^{10}$ cm^2/s. (Note: for $A_p = 10^8$ cm^2/s there is considerably less rectification.) (From Kowalik and Proshutinsky (1994).)

largely due to rectification effects by the sea ice, whereby the ice mechanics resist converging sea ice but has little resistance to opening. To study ice–tide interaction, Kowalik and Proshutinsky (1994) used a linear-viscous model (see section 7.3.2) with constant bulk and shear viscosities (ζ and η) as in equation (7.11). However, in addition they added a pressure term (see equation (7.17)) that resists compression with a linear-viscous behaviour for convergence ($P = \Lambda_p \nabla \cdot \mathbf{v}$ for $\nabla \cdot \mathbf{v} < 0$ with Λ_p constant) and is zero ($P = 0$ if $\nabla \cdot \mathbf{v} \geq 0$). Since their shear viscosities are relatively small, for large Λ_p their model is similar to the more non-linear cavitating fluid model of Flato and Hibler (1992). In the cavitating fluid model, no convergence is allowed until the ice pressure exceeds some critical value. This ice pressure remains fixed independent of convergence rate for constant strength. For diverging flow, the ice pressure is zero.

   Near the coast, Kowalik and Proshutinsky (1994) found significant rectification by the ice interaction over a tidal period. Figure 7.34(b), for example, shows the residual ice motion over a tidal cycle for the M_2 component. Kowalik and Proshutinsky explain this residual ice

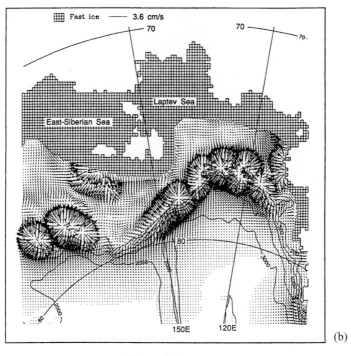

(b)

Figure 7.34. (*cont.*)

motion to be related to the ice convergence and divergence. During half of the tidal cycle, ice motion is suppressed by the high ice compactness or by a solid boundary (the fast-ice edge). During the second part of the tidal cycle, ice moves in the opposite direction, the compactness of the ice decreases and regions of open water appear. Based on typical growth rates of thin ice, the authors argue that up to one metre per year of ice could be formed in the centres of one of these rectified tidal regions, as shown in Figure 7.34(b). However, in the central Arctic region analysed by Heil and Hibler (2002) for inertial variability, Kowalik and Proshutinsky find less than 2 cm of ice produced per year by ice–tide interaction.

Kowalik and Proshutinsky (1994) also argue that the ice mechanics has little impact on the tides. However, this result may be dependent on their rheology and ice–ocean coupling methods, and is currently being investigated in more detail. In particular, their ice–ocean coupling procedure takes the stress into the ocean to be the water drag on the bottom of the ice without consideration for the convergence of the ice, which can also modify the Ekman flux into the ocean. This coupling is used in many ice–ocean model studies (e.g., Holland *et al.*, 1993; Nazarenko, Holloway and Tausner, 1998; Mellor and Kantha, 1989; see also Chapter 9). It is, however, in contrast to the coupling used by Hibler and Bryan (1987), where the combined ice and oceanic boundary layer transport is used for the Ekman flux. Under this paradigm (which is partially consistent with the inertial imbedding approach), the effective stress transferred into the ice–ocean system is the wind stress minus the ice interaction force. In cases of ice divergence or convergence, the differences between these formulations are significant.

7.7 A selected hierarchy of dynamic thermodynamic simulations of the evolution of sea ice

To examine the interplay of dynamics and thermodynamics on the evolution of the mass balance of sea ice, a selected hierarchy of model simulations is presented here. Many of these model runs are coupled ice–ocean model studies. The focus is on simulations illustrating the role of dynamical processes. For this purpose we proceed from idealized models to models employing a full variable thickness distribution. In practice there are a very large number of contemporary dynamic thermodynamic model simulations of sea-ice evolution, most emphasizing the thermodynamic characteristics. A review of different model simulations of the Arctic with an emphasis on thermodynamic response together with specified deformation may be found in Steele and Flato (2000). In addition, for a more complete description of different simulations of contemporary change and future variability with emphasis on coupled climate simulations, see Chapter 9.

7.7.1 Selected characteristics of ice–ocean circulation models

It should be emphasized that if realistic ice-margin simulations are desired in large-scale models, it is critical to utilize coupled ice–ocean models, otherwise the location of the ice margin cannot be realistically simulated. A useful mode for this purpose is a robust diagnostic model whereby a weak relaxation to observed temperature and salinity is used everywhere except in the upper boundary layer where ice is present. This is particularly graphically illustrated in simulation results (Figure 7.35(a)) from such an ice–ocean model by Hibler and Bryan (1987). In this model a weak relaxation to observed temperature and salinity (three years) was utilized. Sea ice was treated by a two-level dynamic thermodynamic sea-ice model, with a viscous–plastic rheology. The improved ice margin is due to the large oceanic heat flux from the deeper ocean into the upper mixed layer. Analysis shows that much of the heat flux occurs in winter and that much of the flux is due to deep convection. This deep convection brings up warm water and prevents ice formation in early winter.

Inter-annual variability of the ice margin in the Greenland and Barents Sea is, however, less critically affected by ocean variations. Results from a higher resolution version (Hibler and Zhang, 1994) of the Hibler–Bryan 'robust' diagnostic ice–ocean model are shown in Figure 7.35(b). A typical stream function for this model showing ocean transport into the Barents Sea and the model domain is given in Figure 7.36(a). A five-year relaxation to observed temperature and salinity was used everywhere except in the upper two levels, which essentially comprised the oceanic boundary layer. While there is some improvement in the ice-margin variations with the inclusion of variable ocean circulation, the simulations

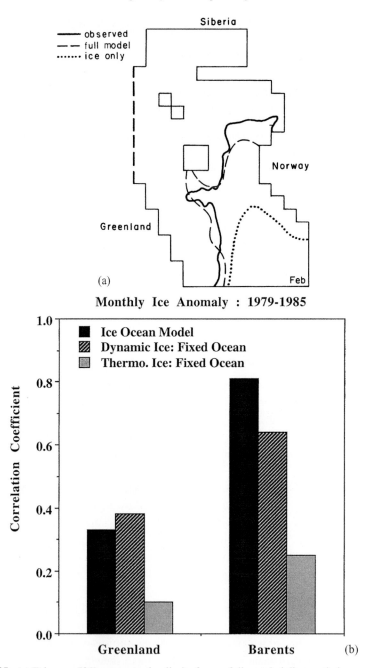

Figure 7.35. (a) February 50% concentration limits from a full coupled diagnostic ice–ocean model and for an ice-only model which includes a fixed-depth mixed layer. A one-degree resolution was used. (From Hibler and Bryan (1987).) (b) Correlation between monthly simulated and observed ice-edge anomalies using a 40 km diagnostic ice–ocean model. A two-level ice-thickness distribution was used in both coupled models.

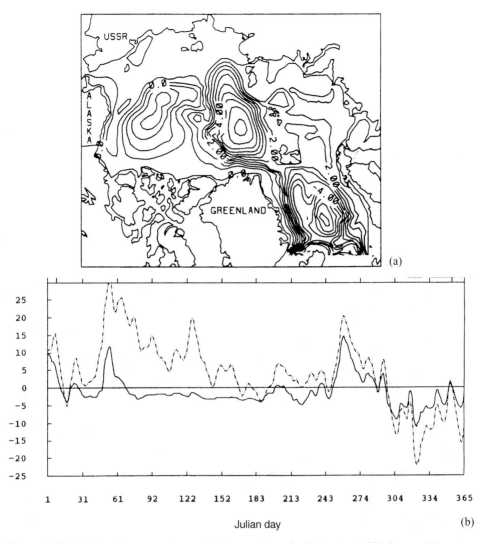

Figure 7.36. (a) Monthly mean stream function contours for September 1980 from a 40 km res-olution ice–ocean model of the Arctic, Greenland and Norwegian Seas. Contour interval is 0.5 sv $(1\,\mathrm{sv} = 10^6\,\mathrm{m}^3/\mathrm{s})$. (b) Time series of vertically integrated flow $\sim$50 km north of Point Barrow for 1979. Negative values represent eastward flow. The solid line is with the stress into the ocean modi-fied by the ice interaction, and the dashed line is the current with wind stress transferred directly into the ocean without modification.

with a fixed annual oceanic heat flux correlate essentially as well. Part of this is likely to be due to the fact that ice transport effects and oceanic circulation effects tend to vary in a similar manner in these regions. For example, stronger northerly winds advecting ice further south into the Barents Sea also tend to cause a reduction of northward transport of heat in the ocean. The only model that works extremely poorly is a thermodynamic-only ice model where ice advection and ice deformation effects are not included.

As noted above, an important aspect of the Hibler and Bryan (1987) type ice–ocean model is the coupling of the ice to the ocean. This coupling consists of taking the stress into the ocean to be given by the wind stress minus the ice interaction. Hibler and Bryan (1987) argue that this procedure rigidly conserves momentum for the ice–ocean system and properly takes into account the combined Ekman flux of both the ice and oceanic boundary layer. The difference between this coupling and taking the stress into the ocean to be the water drag (see, e.g., Nazarenko *et al.* (1998); Wang and Ikeda (2000), the Community Climate System Model model of the National Center for Atmospheric Research, and many others) can be significant. In particular, since the simulated ice drift correlates closely with the wind (see Figure 7.19), the water drag stress field closely resembles the wind. Within this assumption, an example of the effect of this coupling by Zhang (1993) on the Alaska coastal current is shown in Figure 7.36(b).

With the full Hibler and Bryan (1987) coupling, the stream function in September 1983 (from Zhang (1993)) clearly shows an Alaskan counter-current running along the coast. Analysis (Zhang, 1993) of the vertical cross-section of this current shows it to be similar to the Beaufort Sea undercurrent based on hydrographic data and current measurements (Mountain, 1974). The simulated current was two to three grid cells wide (80 to 120 km) with a typical summer mean current speed of 15 cm/s, while the observed current of Mountain (1974) was about 75 km wide with speed up to 30 cm/s. However, in a simulation with the stress into the ocean taken to be the wind stress, the eastward character (Figure 7.36(b)) of the current in 1979 almost totally disappears except in the fall. Analysis (Zhang, 1993) of the stress into the ocean system at this location shows that the modification of stress into the ocean by the ice interaction is very great, even though the ice is generally moving westward. Consequently, Zhang's explanation is that a portion of the inflow from the Bering Strait has a tendency to flow eastward along the Alaskan coast toward the Beaufort Sea if ice exists to subdue the air stress. Otherwise the often easterly wind will frequently prevent this from happening. It is of course possible that the ice drag is also reduced, but ice drift in this region is sometimes actually enhanced by the non-linear character of the ice interaction. Given current interest in understanding the effects of fast ice and coastal processes on the evolution of sea ice, these coupling issues are of relevance and bear consideration.

A particular problem with most ice–ocean circulation models is the tendency to yield too much salt content in the central Arctic if the model is run without any 'nudging' to observed temperature and salinity data (Hakkinen, 1993; Hakkinen and Mellor, 1992). This in turn tends to yield a weakened surface current circulation pattern (see Zhang *et al.*, 1998) compared with observations. An examination of this problem was methodically carried out by Zhang *et al.* (1998) by performing a series of 70-year simulations with and without various forms of 'diagnostic nudging' to observed temperature and salinity. The results (Figure 7.37) show that without relaxation of any kind the ice–ocean model tends to yield a dome of saltier surface water in the interior of the Arctic basin (Figure 7.37(a)). This is in contrast to observations indicating that the surface salinity should be increasing more or less uniformly

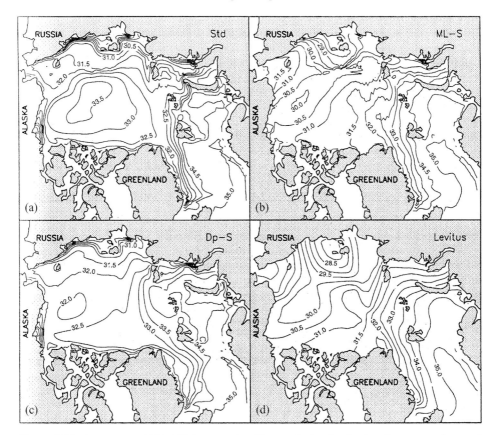

Figure 7.37. Seven-year (1979–1985) mean mixed-layer salinity distributions in parts per thousand predicted by three different ice–ocean models together with Levitus (1982) data in (d). (a) The standard (Std) model is a fully prognostic model with imbedded mixed layer. (b) The specified mixed-layer salinity (ML-S) model is the same model with a diagnostic relaxation (30-day time constant) to observed salinity only in the mixed layer. (c) The deep salinity (Dp-S) model has a five-year relaxation to observed salinity in the ocean everywhere below the mixed layer. For Dp-S there is no relaxation in the mixed layer. (A model with relaxation to both temperature and salinity was similar to Dp-S.) Contour interval is 0.5 ppt. (From Zhang *et al.* (1998).)

as one approaches the Fram Strait (Figure 7.37(d)). If, on the other hand, one utilizes either a strong (∼30-day) relaxation to surface salinity (Figure 7.37(b)) or a weak relaxation (∼three years) to salinity *below* the mixed layer (Figure 7.37(c)), then surface salinity fields in closer agreement with observation are obtained. These nudging effects also lead to more realistic ocean surface currents, and hence better simulations of ice drift. However, Zhang *et al.* (1998) emphasize that diagnostic relaxation, especially when done below the mixed layer, can lead to somewhat unrealistic convection effects. The precise reason for these differences is an active area of research. Explanations range from indirectly hypothesizing sinking of shelf water along the coasts (Aagaard and Carmack, 1994; Aagaard, Coachman and Carmack, 1981); uncertainties with regard to the precipitation in the Arctic (Ranelli and Hibler, 1991); and inadequate treatments of vertical boundary layer processes and horizontal eddies in models (McPhee, 1999; Zhang, Maslowski and Semtner 1999). Another possibility in light of this review is inadequate simulation of ridged ice. This problem and related ice–ocean issues are currently active areas of investigation.

A final interesting feature of coupled ice–ocean models on the hemispheric scale is the presence of inter-annual and inter-decadal oscillations (Hibler and Zhang, 1995; Zhang, Lin and Greatbatch, 1995) in the ice characteristics. These oscillations are related to the thermohaline overturning circulation and occur even with temporally constant atmospheric forcing in 'sector models'. They are not however present in the absence of the thermodynamic effects of a coupled ice cover (Hibler and Zhang, 1995), and hence appear to be an intrinsic ice–ocean phenomenon. While the inclusion of ice transport can change the frequency of the oscillations (Hibler and Zhang, 1995), analysis shows these variations to be mainly due to the thermal insulating effect of sea ice. Since the rate of heat transfer into the ocean depends non-linearly on the ice thickness and the heat flux from the deeper ocean, a simple set of equations describing the system and admitting non-linear oscillations can be formulated (see, e.g., Hibler and Zhang (1995)). The most notable variability found by Hibler and Zhang was in the ice thickness, which tended to vary with about a six-year cycle without ice transport, and at approximately four years with ice dynamics included. In addition there are longer-term decadal variations in the northward water and heat transport in these models (Zhang *et al.*, 1995). The main relevance of these oscillations to the dynamic response of sea ice is that they naturally exist in the system, even in the absence of atmospheric feedback. Consequently, such variability is a candidate for variations of ice thickness in a natural climate sense as opposed to climatic warming.

7.7.2 Multiple equilibrium states of mechanistic dynamic thermodynamic sea-ice models

In the Arctic basin, equilibrium ice thickness depends upon residence time, ice growth and ice outflow in a highly non-linear manner. Because of this, it is in principle possible for these coupled models to have multiple equilibrium states for the same forcing, especially under cooler climate conditions. Whether or not this has happened in the past is an interesting question that has received little attention. The key mechanical component leading to multiple equilibrium states is the arching phenomenon which can restrict the ice flow. In addition there are albedo feedback effects that can, in principle, lead to multiple equilibrium states even under present climate conditions (Thorndike, 1992a) due to physics similar to the 'small ice cap instability' inherent in Budyko–Sellers models (see, e.g., North (1988) and North, Cahalan and Coakley (1981)).

The role of mechanical effects in inducing such multiple equilibrium states has been demonstrated by Hibler and Hutchings (2003) in a two-dimensional dynamic thermodynamic model of the Arctic ice cover. In practice such states appear to be accessible to most non-linear dynamic thermodynamic sea-ice models, although little investigation of

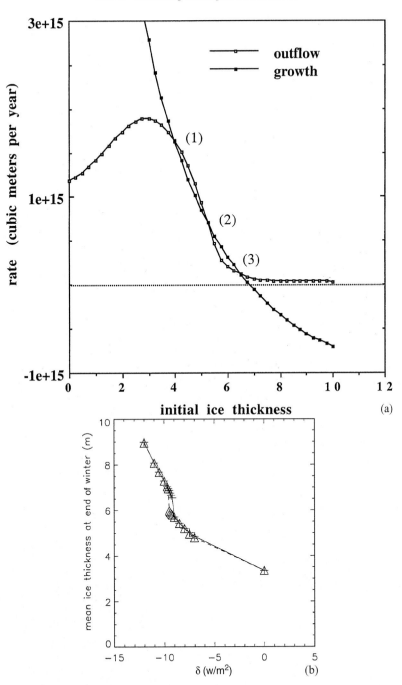

Figure 7.38. (a) Outflow and basin-averaged ice growth as a function of initial ice thickness for a specific value of $\delta = -11$ W/m^2. (b) Maximum ice thickness in seasonal equilibrium as a function of δ. States intitialized with both thick and thin ice are shown. (c) Monthly basin outflow in sverdrups (1 sv $= 10^6$ m^3/s) and (d) monthly basin-averaged ice growth (sv). (From Hibler and Hutchings (2003).)

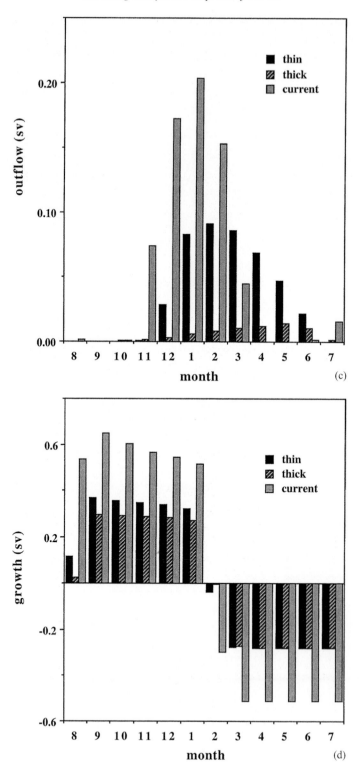

(c)

(d)

this phenomenon has been done. In order to consider a range of climate states, Hibler and Hutchings (2003) used the idealized thermodynamic ice model proposed by Thorndike (1992b). In addition to examining the overall qualitative sensitivity of sea ice to climatic perturbations, this thermodynamic model has also been used in conjunction with more traditional Budyko–Sellers energy balance models to examine multiple equilibrium states of a stationary sea-ice cover (Thorndike, 1992a, b). Although the seasonal response is unrealistic, the 'Stefan'-like thermodynamic model of Thorndike (1992a, b) reproduces salient aspects of seasonal thermodynamic models. The essential idea in this model is to divide the year into a warm and cold season each of length $Y = 0.5$ year. The climate is described by down-welling long-wave radiation ($f_{lwc} = 180\,\mathrm{W/m^2}$) in the cold season and long-and short-wave radiation during the warm season of $f_{lww} = 270\,\mathrm{W/m^2}$ and $f_{sw} = \mathrm{W/m^2}$. The final variable is the heat supplied from the ocean, which is idealized (Maykut and Untersteiner, 1971) to be a constant of the order of $2\,\mathrm{W/m^2}$ for present-day Arctic conditions. Climatic change is introduced via a uniform perturbation δ to the long-wave radiation fluxes f_{lwc} and f_{lww}. An important difference from the traditional 'Stefan' models is the introduction of a heat capacity during the transition periods from warming to cooling and vice versa.

To examine multiple equilibrium states, Hibler and Hutchings (2003) examined a full coupled dynamic thermodynamic model of the Arctic basin with a 100 km resolution and outflow only allowed at the Fram Strait. The ice strength P^* was related to the mean ice thickness h by $P^* = (4 \times 10^4)h\,\mathrm{N/m}$, with h given in metres. The mean ice thickness was the only thickness distribution variable, and evolved according to $\nabla \cdot (\mathbf{v}h) + f(h) = 0$, where $f(h)$ is the ice growth rate calculated from the idealized thermodynamic model. In the thermodynamic calculations the conductivity was increased to $k = 3.4\,\mathrm{W/m^2}$ to account for the modification to ice growth by deformation. Mean monthly wind forcing was used together with linear wind and water drag formulations. The growth period was considered to be from October to March.

The basic character of potential multiple equilibrium states is illustrated in Figure 7.38. Part (a) of this figure shows outflow and basin-averaged ice growth at the end of two years. To the degree that the initial thickness is a good indicator of the basin-averaged growth and outflow, three possible equilibrium states are identified by the intersection of the outflow and growth conditions in Figure 7.38(a). In particular the rate of change of ice thickness dh/dt is given by

$$\frac{dh}{dt} = G(h) - O(h),\tag{7.61}$$

where $G(h)$ is the growth rate and $O(h)$ is the outflow rate. Considering h_0 to be a solution of this equation and $h_1 = h - h_0$ to be a small thickness perturbation relative to the solution, one obtains to lowest order the rate of change of the perturbation to be given by

$$\frac{dh_1}{dt} = [G'(h) - O'(h)]h_1.\tag{7.62}$$

Clearly, the small perturbation will grow unless $[G'(h) - O'(h)] < 0$, a condition which, by inspection of Figure 7.38(a), is met for solutions (1) and (3) but not for solution (2).

Because solution (2) is not stable, as one proceeds from a cold climate to a warm climate (or vice versa) there will be a rapid jump to a lower thickness state which then changes less rapidly with warming as the outflow is much more significant and yields a lower ice thickness with higher growth rates. This is shown in Figure 7.38(b), which shows mean ice thickness for actual multiple equilibrium states obtained by 300-year simulations initialized with very thin or very thick ice. The change of slope is significant. If one were to interpolate linearly the trend before the drop, an approximately zero ice thickness under present conditions would be expected as compared with the $\sim$2 m that is simulated. In addition, because of the multiple equilibrium states, the transition between thick and thin states will depend upon whether one is gradually warming or cooling.

Seasonal outflow and growth characteristics for two equilibrium states for $\delta = -9.5\,\text{W/m}^2$ are shown in Figures 7.38(c) and (d). Also shown is a control solution for $\delta = 0$. The main feature of the 'thick' solution is a greatly reduced outflow, with the annual average outflow of the thick case being 16% of the thin solution and 12% of the standard simulation. In the thick solution the outflow occurs mainly during the melt season with little correlation with the wind stress in the Fram Strait region. In the standard case ($\delta = 0$), for example, because of the weaker strengths the outflow tends to correlate strongly with the wind causing the outflow to peak in late winter, even though the strength is largest then. However, with the thick solution, and, to a lesser degree, the thin multiple equilibrium solution, the outflow tends to be seasonal with much less outflow in the growth period. In terms of magnitude, the outflow is small compared with the growth rates, although outflow is of course comparable to the difference between the growth and melt rate.

While the above model is very idealized, the fact that such multiple equilibrium states may exist in full sea-ice circulation models indicates that such states may be latent in more complex thickness distribution simulations, especially under climatic cooling conditions. In the above analysis these states ultimately depend on shear strength in the ice rheology and should not, in principle, be present for a 'cavitating fluid' rheology. In addition to paleoclimate studies, including sea-ice dynamics (e.g., Vavrus and Harrison (2003), the effect of non-linear mechanics on the evolution of sea ice over the twentieth century should be considered. Moreover, even if not possible for present climate conditions, the existence of such states can affect the response of the models to major changes in the wind fields related to inter-decadal variability. Consequently, interpretations of the overall historical circulation of the Arctic ice cover based only on correlations with winds (Vijne, 2001) are questionable.

7.7.3 Arctic Basin variable thickness simulations

Arctic Basin simulations employing non-linear plastic rheologies together with full multiple level thickness distribution and a heat-budget-based thermodynamic code provide a useful mechanism for examining the interplay between dynamics and thermodynamics in sea-ice and ice–ocean models. In the highly selected simulations presented here, we focus on models with full thickness distributions and examine results using observed forcing, in some cases over 18-year periods. Natural variability in the forcing, or removal of variability, then provides the means to assess sensitivity.

The variable thickness model simulations discussed below (Flato and Hibler, 1995; Zhang and Rothrock, 2001; Zhang *et al.*, 2000) are largely based on the variable thickness model solution framework of Hibler (1980b). This numerical framework represents the most direct solution of the equations of motion in that fixed Eulerian grids were utilized in both space and ice-thickness space. These fixed thickness and space grids facilitate the maintenance of conservation properties of the thickness distribution, but for full resolution of the thickness distribution a large number of levels are typically required. Other formulations exist. Bitz *et al.* (2001) for example have emphasized fewer categories, with both the mean thickness and ice area in each category as variables. Following Bratchie (1984) and Pritchard (1981), Polyakov *et al.* (1999) utilized fewer thickness categories up to some thick ice level where both thickness and area for a thick ice category is utilized. Probably the most useful procedure, however is that of Lipscomb (2001), who utilizes a fixed grid in thickness space together with a remapping procedure (Dukowicz and Baumgardner, 2000) to treat the thickness advection term $\mathrm{d}f_g/\mathrm{d}H$ term in equation (7.48). Using this procedure with five to seven fixed thickness categories, Lipscomb (2001) is able to produce results commensurate with 20 to 30 categories (e.g., Hibler, 1980b) without remapping. In most cases, for the non-linear ice dynamics, investigators have utilized a variation on the viscous–plastic constitutive law formulation with an elliptical yield curve of Hibler (1979), possibly with a numerical solution using the elastic–viscous–plastic method (Hunke and Dukowicz, 1997).

While a short simulation (five years) with relatively smooth wind fields (seven-day averaged wind) the initial simulation by Hibler (1980b) demonstrated the characteristic ice buildup along the Canadian archipelago (Figure 7.39(a)). This buildup is largely caused by the ice velocity field (Figure 7.39(b)) advecting ice into this region as well as episodic ice convergence in this region. Since the ice convergence is episodic, it is difficult to tell whether there is a net convergence. This overall buildup is characteristic of variable thickness sea-ice models, and is in general agreement with contours of ice thickness from submarine sonar data (Bourke and Garrett, 1987) as well as with ridge statistics (see section 7.2). Analysis of actual ridge production over the year (not shown – see Hibler (1989)) shows much higher intensity of ridging near the outflow region. Consequently, the ice buildup is more of a complex combination of ice advection and local deformation. This is in contrast to results obtained with the two-level thickness distribution (sections 7.6.4 and 7.6.5), which tend to be more of a short-term balance of deformation and advection. Sensitivity studies (not shown – see Hibler (1980b)) with greater strength do, however, show a decrease in the maximum thickness in the archipelago region, so that strength parameterizations in the variable thickness model are important here.

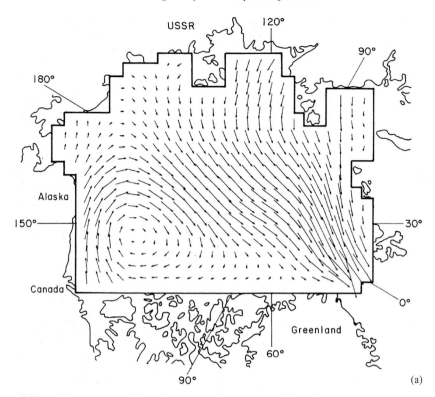

USSR 120°
90°
180°
Alaska
150° 30°
Canada
0°
Greenland
60°
90° (a)

Figure 7.38. (a) Average annual ice velocity and (b) average April ice-thickness contours (metres) at the end of a five-year variable thickness dynamic thermodynamic sea-ice model simulation. Seven-day averaged winds from May 1962 to May 1963 were used for dynamical forcing. Thermodynamic forcing similar to that used in Parkinson and Washington (1979) utilized monthly averaged atmospheric climatological air temperatures, dew points and daily long-and short-wave radiative forcing. (In (a) a velocity vector one grid space long represents 0.02 m/s.) (From Hibler (1980b).)

Strength sensitivity simulations carried out by Hibler (1980b) also demonstrate the effects of ice mechanics on ice outflow. In particular, increases of the strength by an order of magnitude over the standard case resulted in a total stoppage of flow for several of the winter months. The result supports the arching analysis carried out above and amplifies the importance of having as strong a variability in the wind forcing as possible. However, use of smoothed wind fields with the highly non-linear rheologies employed in these models will likely result in rather sluggish ice flow and certainly poorer renditions of ice deformation.

Ridged ice and sensitivity to mechanical parameters

Using daily atmospheric forcing over the time period 1979 to 1985, Flato and Hibler (1995) carried out a series of variable thickness simulations including an explicit treatment of ridging. The focus was largely on determining appropriate and most critical parameters in

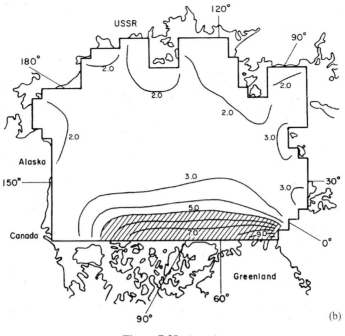

Figure 7.39. (*cont.*)

the mechanical redistributor and the buildup of ridged ice. In the standard version of this model, the energy dissipation during ridging is taken to be 17 times the potential energy of ridge building via the redistribution function. This value was obtained by insisting that the computed and average monthly drift magnitudes of several buoys were within about 2%. The value of 17 so obtained compares well with the range of 10–17 determined by discrete-element simulation of individual ridge building events (Hopkins, 1994; see also Figure 7.27).

A notable characteristic (Figure 7.40(a)) of the buildup to equilibrium of this model is the long time (approximately ten years) taken for ridged ice fraction to come to equilibrium. This is opposed to the level ice fraction that comes into equilibrium more rapidly. Timescales by different thickness categories were subsequently investigated more methodically by Holland and Curry (1999); see also Chapter 9. It is also notable that in quasi-equilibrium the ridged ice accounts for more than half the ice volume, a feature that Flato and Hibler (1995) argue is qualitatively consistent with a wide variety of data. The long timescale (approximately seven years) is in contrast to a two-level model where the strength is directly coupled to ice thickness (see, e.g., Figure 7.29). This history effect means that wind patterns may take a number of years to affect the full thickness distribution. Basically, ridging is a complex balance between ridge formation, advection and decay, with the actual ridging occurring in a given year being significantly smaller and differently distributed than the full distribution of ridged ice.

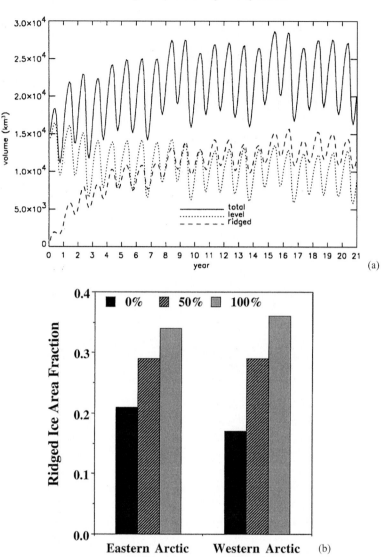

Figure 7.40. (a) Time series of total, ridged and undeformed ice area over the model domain for a multi-level sea-ice simulation (Flato and Hibler, 1995), with explicit accounting for the distribution of ridged ice. A 21-year simulation was obtained by repeating 1979–1985 forcing three times. (b) Sensitivity of ridged ice volume per unit area to energetic consistency assumption. Three cases were examined: convergence only, where ridges are only created by ice convergence proportional to the convergence rate; 50% consistency, where 50% of the mechanical energy is accounted for by additional ridging (and concomitant open water creation) under shearing conditions; and 100% consistency. The western Arctic point was located at 76°N, 160°W and the data were compiled for October 31, 1982. The eastern Arctic point was located at 84°N, 5°W, with data compiled for June 30, 1985. (From Flato and Hibler (1995).)

Investigation of different energetic consistency formulations by Flato and Hibler (1995) showed that much of the ridging was arising from the 50% energetic consistency assumption in the standard simulation. When either 100% or 0% energetic consistency was employed (in the latter case open water can only be created by divergence and ridged by convergence – not simultaneously by open water creation and ridging during shearing events) 10–14% changes in mean ice thickness were observed with the volume of ridged ice changing even more. Even larger changes in the modelled ice area fraction (Figure 7.40(b)) occurred. Although not shown, these energetic consistency variations also change the spatial pattern of thickness, especially in the Beaufort Sea. Clearly, variations on how this feature is treated can significantly change results.

The relative role of dynamics and thermodynamics in historical variability

Changes in wind patterns can supply a very substantial change in the ice buildup. An example of how wind patterns can dominate, especially as one shifts between different Arctic oscillation patterns, was provided by Zhang *et al.* (2000) using a multi-level thickness sea-ice model with atmospheric forcing over the time period 1979 to 1996. The formulation of this model was essentially the same as in Hibler and Flato (1992) and Hibler (1980b) and was coupled to an ocean circulation model (see, e.g., Zhang *et al.*, 1998). Over this time period there was a general shift of the wind patterns and ice velocity patterns (see, e.g., Figure 7.2) from low North Atlantic oscillation (NAO) patterns with a strong gyre-like pattern to high NAO with a weakened and shrunken gyre. Some of the thickness patterns are shown in Figure 7.41. These thickness characteristics are qualitatively similar to most dynamic thermodynamic simulations. The main differences are the presence of significantly different thickness anomalies in response to changes in the wind forcing. Moreover, almost all the thickness changes were due to dynamical changes as opposed to thermodynamic changes (Figure 7.42). This figure shows simulations for the ice mass in the eastern and western Arctic (based on the Greenwich meridian as demarcation) with fixed and variable thermal forcing.

The authors explain this difference to be largely due to a reduced transport of ice from the western to the eastern Arctic under a low NAO pattern. Conversely, during the high NAO pattern, there is much less transport of thick ice into the eastern Arctic. When no variation in thermal forcing was employed, there was little change in the results. This occurred even though there was a substantial increase in mean air temperature over the Arctic Ocean of about 4 °C between 1980 and 1996. It should be cautioned in this simulation that the spatial pattern of the thickness increase in the western Arctic is not in agreement with recently reported submarine thickness observations by Tucker *et al.* (2001) over the time period 1986 to 1994. In fact, Tucker *et al.* find a decrease in thickness along the transect in disagreement with the results of Zhang *et al.* (2000). Consequently, although the overall shifts are clearly consistent between time periods, there are some apparent discrepancies in the anomaly pattern. Clearly there is motivation for improvement of this model.

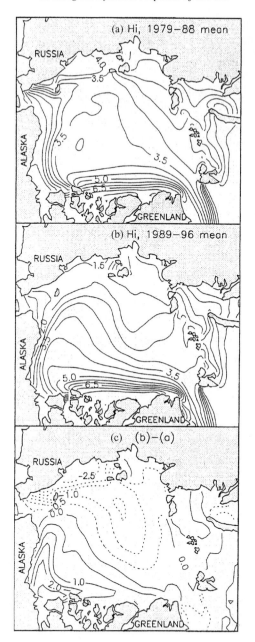

Figure 7.41. Simulated mean ice-thickness fields (Zhang *et al.*, 2000) for (a) 1979 to 1988 and (b) 1989 to 1996. Part (c) shows their difference field. The contour interval is 0.5 m.

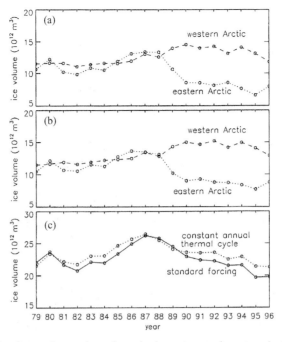

Figure 7.42. Simulated annual mean ice volume in the eastern and western Arctic ((a) for standard forcing; (b) for constant annual thermal cycle), and for the whole Arctic ((c) for both types of forcing). The prime meridian divides the two regions. (From Zhang *et al.* (2000).)

7.7.4 The response of sea ice to climate change: the effect of ice dynamics

Early dynamic thermodynamic sea-ice modelling studies (Hibler, 1984) have suggested that sea-ice motion might exert a negative climate feedback (Vavrus and Harrison, 2003) on ice-thickness changes. This negative feedback stems from two processes absent in a stationary ice pack: ice advection and local thickness variations due to ice deformation. Under this paradigm, regions characterized by sea-ice convergence tend to resist thinning caused by atmospheric thermal perturbations. Local thickness variations resulting from deformation are important because they lead to differential growth rates. Since columnar ice growth is inversely proportional to ice thickness, sea ice tends to restore itself toward its original thickness, while ice motion causes thickness changes. Likewise, because the strength of ice is proportional to its thickness, mobile ice tends to resist thermodynamic perturbation through easier ridging as the floes thin and more 'difficult' ridging as they thicken. Overall, these considerations suggest that models including dynamical sea-ice processes should tend to be less sensitive to changes in atmospheric forcing, at least with regard to sea-ice thickness (Curry, Schramm and Ebert, 1995; Hibler, 1989; Holland *et al.*, 1993; Pollard and Thompson, 1994).

In addition to ice modifying ice-thickness changes, ice dynamics and deformation can be argued to have a mitigating effect on high latitude atmospheric warming. This characteristic was indirectly indicated by idealized warming experiments (Hibler, 1984) of a two-level dynamic thermodynamic model of the Weddell Sea ice pack (Hibler and Ackley, 1983). In the warming experiments, climatic change was crudely approximated by a 4 °C temperature change, which affects incoming long-wave radiation and sensible heat losses in summer and winter. Warming was found to change thicknesses by similar factors in the dynamic and thermodynamic cases, with only a small negative feedback in the dynamic case. However, the heat exchange into the atmosphere (Figure 7.43(b)) due to ice growth and oceanic heat loss to the atmosphere over the whole grid were much less sensitive to warming in the dynamics case. Moreover, heat gained by the atmosphere and ice growth in the current climate case (Figure 7.43(a)) are both substantially less for the thermodynamic-only model than for the full dynamics case. Based on these current climate characteristics, one would expect that with full ice dynamics the heat transfer into the atmosphere would be enhanced and hence cause a local warming anomaly compared to thermodynamics alone. Moreover, because of the reduced change of sea-to-air heat exchange (Figure 7.43(b)), with dynamics one would expect a reduced polar temperature amplification under increased CO_2.

A methodical investigation of ice dynamical effects on the response to climatic change of a fully coupled atmosphere–sea-ice model has been carried out by Vavrus (1999) and Vavrus and Harrison (2003). In addition to reduced polar amplification mentioned above, Vavrus (1999) identified several new ice dynamic feedback mechanisms unique to coupled atmosphere–ice models. The atmosphere–ocean model used for this study was the GENESIS2 coupled atmosphere–mixed layer model that was originated from the NCAR community climate model, version 1 (Thompson and Pollard 1997). Sea-ice dynamics was treated with a cavitating fluid rheology (Flato and Hibler, 1992) together with a two-level ice-thickness distribution. Ocean currents for sea-ice drift were specified and oceanic north-ward transport of heat was parameterized by certain prognostic heat fluxes into the boundary layer.

Of particular interest for this review are the different atmospheric characteristics simulated with and without sea-ice dynamics. The results for the present climate (Figure 7.44) show that even small amounts of leads, which are created by moving ice, cause substantial warming at the Arctic Ocean surface and in the lower troposphere. This result is consistent with earlier GCM-based studies of both polar regions (Simmonds and Budd, 1991; Vavrus, 1995). In comparison to the simulation with thermodynamics only (TI) (Vavrus, 1999; see also Maslanik (1997)) finds that inclusion of dynamics (DI) causes reduced ice coverage (not shown here) from the Laptev Sea westward to Svalbard, due to the mean offshore ice drift along the Eurasian shelf and much more diffuse ice margin in the North Atlantic. These ice coverage differences cause higher temperatures in the lower troposphere than TI in the vicinity of the reduced ice concentrations (Figure 7.44(a)) and allow generally warmer conditions to spread over the entire Arctic Ocean. Conversely, ice dynamics force

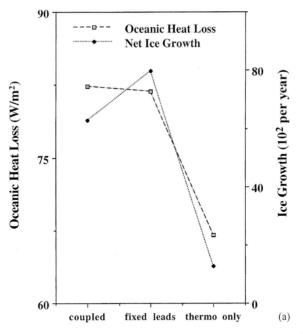

Figure 7.43. (a) Net sea-ice growth and oceanic heat loss to the atmosphere over the growth period (Julian day 60–240) from a dynamic thermodynamic model (Hibler and Ackley, 1983) of the Weddell Sea ice cover. Results are for a full dynamic thermodynamic model (coupled), a thermodynamic only model (thermo only) and a thermodynamic model with a specified seasonal fraction of leads (fixed leads). (b) Change in net ice growth and oceanic heat loss over the same time period between a 5 °C warming simulation and the control simulation for the same models as in part (a). Atmospheric forcing data from 1979 were used to drive the model. (Results from Hibler (1984).)

much greater ice coverage just east of Greenland, where only a minimal amount of sea ice grows thermodynamically, and these higher ice concentrations extend equator-ward along the southern Greenland coasts. As a result, annual air temperatures are up to 2 °C colder than in TI along the east Greenland coast. These surface temperature differences propagate aloft into the middle troposphere, causing differences in geopotential height (Figure 7.44(b)) almost directly above the core surface anomalies. This results in anomalous higher geopotential heights in the Svalbard–Norwegian sea region, for example, where the DI simulation produces warmer temperatures.

The sensitivity of the individual models with (DI) and without (TI) dynamics to climatic warming (Vavrus and Harrison, 2003) is shown in Figure 7.45. In the thermodynamic case, the spatial structure of the ice fraction (Figure 7.45(a)) changes consists of a roughly annular contraction of the ice pack in TI, accompanied by a pronounced melt back along the east Greenland coast and at the boundary of the Arctic Ocean with the Greenland–Barents Seas. Conversely, the ice fraction anomaly pattern in DI features a dipole structure within the Arctic Ocean, with the largest decreases along and north of the Siberian coast and smallest decreases along and north of the Greenland–Canadian archipelago. Vavrus and

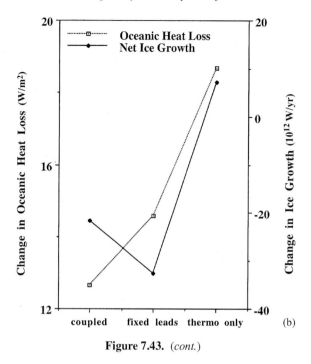

Figure 7.43. (*cont.*)

Harrison (2003) note that this type of spatial anomaly resembles the trend of ice-cover changes in recent decades (see section 7.7.3). This general pattern in the DI simulations facilitates the melting of thin, less compact ice in the divergent Eurasian sector and hampers melting in the convergent Canadian sector, where sea ice is relatively thick and compact.

These differences in ice fraction cause corresponding heating anomalies (Figure 7.45(b)) in the lower troposphere. As in the case with the present climate, these heating anomalies also cause substantial pressure anomalies (not shown). In both pairs of experiments, the largest anomalies occur over regions of maximum retreat. Particularly notable is the muted and reduced warming over the central basin in the DI experiment. This basic result is that change in oceanic heat loss in the Weddell Sea causes a reduced sensitivity of the temperatures over the basin as well as a reduced response in ice thickness in that region. This reduced sensitivity in ice thickness persists over the whole range of warming and paleoclimate scenarios, as shown in Figure 7.46(a). Vavrus and Harrison (2003) attribute the change in thickness to two different processes. Smaller thickness anomalies under positive radiative perturbations appear to result from easier ridging partially compensating for the thermodynamic thinning. The muted thickness response under negative forcing, on the other hand, is due to equator-ward spreading of the ice pack offsetting the thermally induced thickening. The differences in ice thickness between DI and TI constitute a negative feedback in every pair of simulations; all of which are statistically significant at the 99% confidence level, based on a Student's t-test.

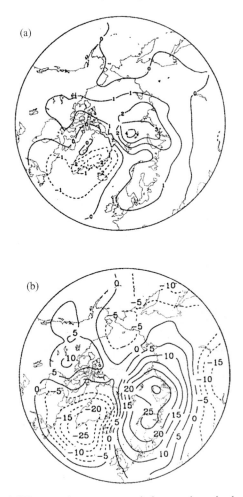

Figure 7.44. Mean annual differences between a coupled atmosphere–ice boundary layer circulation model with and without dynamic sea ice (model with dynamics − model without dynamics) for current climate conditions. Variables are (a) surface air temperature (kelvin), (b) geopotential height (metres) at the $\sigma = 0.695$ model level ($\sim$750 Pa level). (From Vavrus (1999).)

The changes in ice concentration (Figure 7.46(b)) due to dynamics induced mobility demonstrate differing feedbacks under warming and cooling. Nonetheless, the results are amenable to consistent interpretation based on ice rectification effects. In particular under positive radiative perturbations, the *decrease* in ice coverage is much smaller when dynamic ice is included, ranging from 43% of the thermodynamic response at 6000 years to 68% at 10 000 years. Conversely, the *increase* in ice fraction at 115 000 years is more than twice as large with ice dynamics than without. Sea-ice dynamics thus represents a negative feedback on ice coverage when the Arctic climate is warmer than present, but a positive feedback when the Arctic climate is colder. Vavrus and Harrison (2003) attribute this

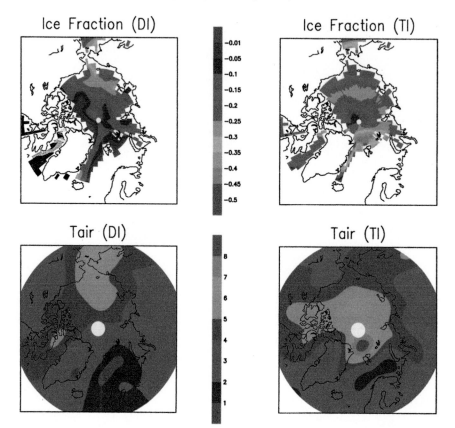

Figure 7.45. Mean annual anomalies of Arctic ice fraction (ice concentration) and surface air temperature for $2 \times CO_2$ simulations with dynamic ice (DI) and thermodynamic only ice (TI). The anomalies are computed as differences from the DI or TI control simulation (Figure 7.44). (From Vavrus (1999).)

non-uniform response to the tendency of ice motion to spread ice equator-ward. This feature opposes a contraction of the pack in a warm scenario, but reinforces an expansion in a cool scenario.

The impact of dynamic ice on ice concentration is reflected strongly in the response of surface air temperature (Figure 7.46(c)). Hence, ice motion always promotes cooling in the interior Arctic relative to the corresponding TI experiment. While much of this effect can be attributed to the ice fraction feedback discussed above, further analysis by Vavrus and Harrison (2003) shows modified circulation patterns to supply an important role. In particular they note that the geostrophic surface wind anomalies associated with the sea-level pressure changes induced by the warming anomalies are such that the DI simulation produces enhanced outflow of cold Arctic air and ice over the Norwegian and Barents Seas. The TI simulation, on the other hand, generates enhanced inflow of warm and moist air of Atlantic origin into the central Arctic. Consequently the shifting of the atmospheric

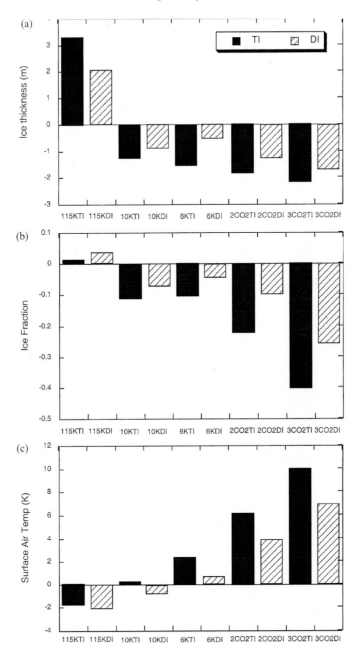

circulation, and not just the one-dimensional response, aids in the negative feedback effects under warming expected on the basis of air–sea heat exchange alone.

While the details of the atmospheric circulation changes and different anomaly patterns are complex, this sequence of simulations graphically emphasizes the difference that sea-ice dynamics can make to the response of the climate system to climate warming or cooling. Indeed, the base state of circulation is itself modified. It is of course possible that some of this response is model-dependent. Simulations with the Canadian climate model, for example, yield little difference in sensitivity between dynamic and thermodynamic cases under climate warming (Flato *et al.*, 2000). However, the other possibility is that the base state (Maslanik *et al.*, 1996) of the GENESIS model used by Vavrus and Harrison (2003) is more realistic in the Arctic region to begin with, so that the sensitivities are more realistic. It is also notable that shear strength effects not included in the cavitating fluid model may affect some of the sensitivity analysis, especially in cases of climatic cooling. Moreover, inclusion of full thickness distribution formulations (see, e.g., Flato and Hibler (1995) and Lipscomb (2001)), and high frequency variability (Heil and Hibler, 2002) may modify aspects of the sensitivity.

7.8 Concluding remarks

The main thrust of this chapter has been to elucidate the interplay between dynamics and thermodynamics in understanding and interpreting the dynamic response of sea ice. Developing the capability to model this response has necessitated the creation of sea-ice circulation models that bear many resemblances to atmospheric and oceanic circulation models. Indeed, as these models become utilized at higher resolution and more coupled processes begin to be resolved, they will likely begin to take on more of the chaotic behaviour that characterizes atmospheric circulation models and eddy-resolving ocean circulation models. Consequently, while a number of the dynamical processes, such as 'kinematic' waves, examined in this chapter have received little attention in the climate community, as more detailed investigations of the role of sea ice in climatic change are initiated it will become more important to be aware of such phenomena.

Figure 7.46. Mean annual anomalies of central Arctic. (a) Sea-ice thickness; (b) sea-ice concentration; (c) surface air temperature for a variety of paleoclimate and warming experiments. The anomalies represent averages over the whole central portion of the Arctic Basin. The DI (TI) anomalies are computed as differences from the DI (TI) control simulation. TI anomalies are in black and DI anomalies are hatched. In addition to the $2 \times CO_2$ and $3 \times CO_2$ simulations, there are paleoclimate simulations using radiative and ice sheet reconstruction information from 6000 (6KDI and 6KTI), 10 000 (10KDI and 10KTI) and 115 000 years (115KDI and 115KTI) before present. Briefly, the 6K period represents a 'climatic maximum' with warmer temperatures than today. The 10K period has stronger radiation forcing compared to today, but the Laurentide ice sheet was still present. The 115K period, which represents the beginning of the most recent glacial stage, has much lower radiative forcing, but no ice sheets were present. (From Vavrus and Harrison (2003).)

It has recently become more commonplace to utilize some level of physically based sea-ice dynamics model in numerical investigations of climate. With this fact in mind, this chapter has examined the major features of these models with some extra emphasis on the dynamical and mechanical component of the models. The hope here is that this information will aid in a more discerning interpretation of different model-based predictions of sea-ice change. This detail also aids in identifying aspects of the dynamic response of sea ice requiring further attention or research.

The topics covered in this chapter may be briefly outlined as follows. Scale analysis of the momentum equation generally indicates that the drift of ice is largely a balance between the wind stress, water stress, and ice stress gradient with inertia and Coriolis terms playing less of a role in the time-averaged response. On timescales less than a day, however, the strong coupling via turbulence with the oceanic boundary layer leads to significant inertial motions, with the oceanic boundary layer supplying a greatly enhanced ice inertia effect. Observations, together with theoretical work, indicate that ice mechanics tends to supply a damping of this inertial motion, especially in winter. In contrast to ice drift, analysis of winter ice deformation suggests the essential scaling to be a balance between internal ice stress and wind forcing. As in ice drift, inertial motions do, however, induce sub-daily deformation variations.

Successful descriptions of sea-ice mechanics almost all involve some type of plastic failure model. However, the paradigm in recent years has shifted from aggregate granular models, with many ridges and sharp slip lines forming between defined floes, to 'fracture'-based models where the focus is on finite zones of failure. This paradigm is supported by recent advances in processing of synthetic aperture radar imagery, whose dominant spatial signature is the presence of actively shearing linear kinematic features. Because of the ostensible fractal character of sea ice viewed from the air, and the strong similarities to laboratory failure characteristics, this view lends support to scale-invariant rheologies, some of which are discussed above.

Use of either aggregate or fracture-based plastic rheologies leads to a simulated ice-velocity field that, although correlating well with wind field variations, has significantly different deformation properties than linear models following the wind. Chief among these is a persistent divergent field in almost all model studies using plastic rheologies. The strengths used in these plastic models are substantial and involve very significant amounts of friction for their justification in addition to the potential energy due to ridging. These models also tend to produce stress gradients, generally, but not always, opposing the ice motion with magnitudes typically about 25% of the wind stress. Vector averages over longer time periods of these terms yield stress gradient to wind stress ratios of about double this value. Plastic flow also leads to a particular scaling of outflow through narrow passages with possible total stoppages under either very high ice strength conditions or low body forces (wind and water stresses).

While related, thermodynamic properties of sea ice tend to be divided into the growth problem for seasonal ice and the more steady-state problem of sea ice going through repeated thaw and decay cycles. In the case of growth, exact calculations plus observations tend to support the use of the 'Stefan' approximation, where a steady-state linear temperature profile is used to estimate ice thicknesses up to a metre or so in thickness. This type of Stefan model is relatively widely used in thermodynamic models of sea ice together with full surface heat-budget formulations, but has the chief drawback that it overestimates the seasonal swing of ice thicknesses. This arises both because of the neglect of heat capacity effects and internal modifications of the brine volume.

Models for the seasonal ice are still largely dominated by the fixed salinity profile models of sea ice, whereby observational results for internal brine volume and light penetration may be used to calculate precisely internal changes in the ice properties. These models do a good job on the seasonal cycle of ice thickness, and there have been a number of recent adaptations of this general concept to thermodynamic models for climate study. The idea is to take into account both the internal change of brine volume and heat capacity effects with relatively few temperature layers. However, this approach still relies on the somewhat unrealistic assumption of constant salinity profile.

While there are simplifications, ice thickness evolution equations coupling the momentum equations to the thermodynamics tend to be largely based on characterization using the fraction of area covered by each thickness category. While this formulation, with proper numerics, provides a precise treatment of the growth of ice, there are considerable uncertainties in the treatment of ridging and the amount of concomitant ridging and open water creation to allow for under arbitrary deformation states. An additional problem until recently was the mismatch in ridging energy and large-scale mechanical energy losses needed to obtain reasonable large-scale simulations. Many of these issues were satisfactorily addressed in the 1990s by a combination of seasonal models together with discrete-element ridge building models showing that frictional losses in ridge building were ~20 times as large as potential energy losses. However, uncertainty about simultaneous ridging and open water formation remain. Fracture-based rheologies provide a more precise theoretical answer to this partition, which ultimately derives from laboratory work if the scale-invariant assumption can be invoked.

Simpler two-level formulations of the ice-thickness distribution have also been proposed and are quite useful in examining aspects of ice buildup and growth in dynamic thermodynamic models. Variations of these formulations are widely used in large-scale studies. However, as noted above, the essential timescale induced by such conservation equations in dynamic thermodynamic models is only a few months for ice buildup as opposed to at least several years for variable thickness models. The latter case is more consistent with ridge and thickness observations mentioned at the beginning of the chapter. When interpreting different model results, these timescale differences should be kept in mind.

While not examined much in climate-based studies, at shorter time- and space-scales, coupling between the ice conservation equations and the dynamical equations leads to wave-like phenomena with considerable physical and numerical interest. Probably the most physically relevant of these phenomena is inertial and tidal variability in the deformation.

Simulations discussed in sections 7.6.6 and 7.6.7 show the importance of such sub-daily deformation variations on the sea-ice mass balance. The ramifications of these largely overlooked processes are, as yet, not fully understood.

In the final section, selected longer-term coupled dynamic thermodynamic simulations either responding to climate or historical variability are briefly examined. In terms of a simple archetypal climate sensitivity study, a dynamic thermodynamic box model of the Arctic is used to show how even rudimentary knowledge of the effect of ice mechanics on outflow can lead to multiple equilibrium states of the Arctic ice-cover thickness. This simple example also shows how modified growth by lead formation and outflow tends to reduce sensitivity to thermodynamic warming or cooling.

In terms of full physics variable thickness models, simulations taken from the literature demonstrate sensitivity to mechanical parameterizations and the capabilities of these models to simulate major changes in ice cover due to circulation changes in the atmosphere. Most notable in the last example is the greatly reduced sensitivity of these more complete models to changes in thermodynamic forcing and the much greater change to dynamic forcing. This has long been expected from early model studies, but as yet does not seem to appear in fully coupled climate simulations.

Overall it is clear from this review that, while considerable progress has been made over the last several decades on modelling the dynamic response of sea ice, considerable uncertainties and research still remain. Many of these aspects were alluded to in this chapter, and should hopefully motivate further research in this field.

References

Aagaard, K. and Carmack, E. C. 1994. The Arctic Ocean and climate: a perspective. In Johannessen, O. M., Muench, R.D. and Overland, J.E., eds., *The Polar Oceans and Their Role in Shaping the Global Environment*. Geophysical Monograph 85. American Geophysical Union, pp. 5–20.

Aagaard, K. and Greisman, P. 1975. Toward new mass and heat budgets for the Arctic Ocean. *J. Geophys. Res.* **80**, 3821–7.

Aagaard, K., Coachman, L. K. and Carmack, E. C. 1981. On the halocline of the Arctic Ocean. *Deep-Sea Res. Part A* **28**, 529–45.

Ackley, S. F., Lange M. A. and Wadhams, P. 1990. Snow cover effects on Antarctic sea ice thickness, sea ice properties and processes. In Ackley, S. F. and Weeks, W. F., eds., *Proceedings of the W. F. Weeks Sea Ice Symposium*. USACREL Monograph 90–1, pp. 16–21.

Aksenov, Ye. and Hibler, W. D. III. 2001. Failure propagation effects in an anisotropic sea ice dynamics model. In Dempsey, J. P. and Shen, H. H. eds., *IUTAM Conference on Scaling Laws in Ice Mechanics and Ice Dynamics*. Dordrecht, Kluwer Academic Publishers, pp. 363–72.

Anderson, D. L. 1961. Growth rate of sea ice. *J. Glaciol.* **3**, 1170–2.

Arbetter, T. E., Curry, J. A. and Maslanik, J. A. 1999. Effects of rheology and ice thickness distribution in a dynamic-thermodynamic sea ice model. *J. Phys. Oceanogr.* **29**, 2656–70.

Babko, O., Rothrock, D. A. and Maykut, G. A. 2002. Role of rafting in the mechanical redistribution of sea ice thickness. *J. Geophys. Res.* **107** (c8), paper 27.

Bitz, C. C. and Lipscomb, W. H. 1999. An energy-conserving thermodynamic model of sea ice. *J. Geophys. Res.* **104**, 15 669–77.

Bitz, C. M., Holland, M. M., Weaver, A. J. and Eby, M. 2001. Simulating the ice-thickness distribution in a coupled climate model. *J. Geophys. Res.* **106**, 2441–64.

Bourke, R. H. and Garrett, R. P. 1987. Sea ice thickness distribution in the Arctic Ocean. *Cold Regions Sci. Technol.* **13**, 259–80.

Bourke, R. H. and McLaren, A. S. 1992. Contour mapping of Arctic Basin ice draft and roughness parameters. *J. Geophys. Res.* **97**, 17 715–728.

Bratchie, I. 1984. Rheology of an ice-floe field. *Ann. Glaciol.* **5**, 23–8.

Brigham, L. 2000. Sea ice variability in Russian Artic coastal seas: influences on the northern sea route. Ph. D. dissertation, Scott Polar Research Institute, University of Cambridge (unpublished).

Brown, R. A. 1980. Planetary boundary layer modeling for AIDJEX. In Pritchard, R. S., ed., *Sea Ice Processes and Models*. Seattle, University of Washington Press, pp. 387–401.

Colony, R. 1990. Seasonal mean ice motion in the Arctic Basin. *Proceedings of the International Conference on the Role of the Polar Regions in Global Change*. Fairbanks, University of Alaska, p. 290.

Coon, M. D. 1974. Mechanical behavior of compacted arctic floes. *J. Petrol. Technol.* **26**, 466–70.

Coon, M. D. and Pritchard, R. 1974. Application of an elastic-plastic model of arctic pack ice. In Reed, J. C. and Sater, J. E., eds., *The Coast and Shelf of the Beaufort Sea*. Arctic Institute of North America, pp. 173–94.

Coon, M. D., Knoke, G. S., Echert, D. C. and Pritchard, R. S. 1998. The architecture of an anisotropic elastic-plastic sea ice mechanics constitutive law. *J. Geophys. Res.* **103**, 21 915–25.

Cox, G. F. N. and Johnson, J. B. 1983. Stress measurements in ice. CRREL Report 83–23.

Cox, G. F. N. and Weeks, W. F. 1974. Salinity variations in sea ice. *J. Glaciol.* **13**, 109–20.

Cunningham, F. F., Kwok, R. and Banfield, J. 1994. Ice lead orientation characteristics in the winter Beaufort Sea. Paper presented at the International Geoscience and Remote Sensing Symposium, IEEE, Pasadena, CA.

Curry, J. A., Schramm, J. L. and Ebert, E. E. 1995. Sea ice-albedo climate feedback mechanism, *J. Climate* **8**, 240–7.

Dempsey, J. P., Palmer, A. C. and Sodhi, D. S. 2001. High pressure zone formation during compressive ice failure. *Engng. Fracture Mechanics* **68**. 1961–74.

Dukowicz, J. K. and Baumgardner, J. R. 2000. Incremental remapping as a transport/advection algorithm. *J. Comput. Phys.* **160**, 318–35.

Eicken, H. 1995. Effects of snow cover on Antarctic sea ice and potential modulation of its response to climate change. *Ann. Glaciol.* **21**, 369–76.

Eicken, H., Lensu, M., Lepparanta, M., Tucker, W. B. III, Gow, A. J. and Salmela, O. 1995. Thickness, structure and properties of level summer multiyear sea ice in the Eurasian sector of the Arctic Ocean. *J. Geophys. Res.* **100** (11), 22 697–710.

Erlingsson, B. 1991. The propagation of characteristics in sea-ice deformation fields. *Ann. Glaciol.* **15**, 73–80.

Flato, G. M. 1995. Spatial and temporal variability of Arctic ice thickness. *Ann. Glaciol.* **21**, 323–9.

Flato, G. M. and Brown, R. D. 1996. Variability and climate sensitivity of landfast Arctic
    sea ice. *J. Geophys. Res.* **101**, 25 767–77.
Flato, G. M. and Hibler, W. D. III. 1992. On modeling pack ice as a cavitating fluid.
    J. Phys. Oceanography **22**, 626–51.
    1995. Ridging and strength in modeling the thickness distribution of Arctic sea ice.
    J. Geophys. Res. **100**, 18 611–26.
Flato, G. M. *et al.* 2000. The Canadian Centre for Climate Modeling and Analysis global
    coupled model and its climate. *Climate Dyn.* **16** (6), 451–67.
Fung, Y. C. 1977. *A First Course In Continuum Mechanics*, 2nd edn. Englewood Cliffs,
    NJ, Prentice Hall.
Gabison, R. 1987. A thermodynamic model of the formation, growth, and decay of
    first-year sea ice. *J. Glaciol.* **33**, 105–19.
Geiger, C. A., Hibler, W. D., III and Ackley, S. F. 1997. Large-scale sea ice drift and
    deformation: comparison between models and observations in the western Weddell
    Sea during 1992. *J. Geophys. Res.* **103**, 21 893–913.
Gill, A. E. 1982. *Atmosphere-Ocean Dynamics.* New York, Academic Press.
Hakkinen, S. 1993. An Arctic source for the great salinity anomaly: a simulation of the
    Arctic ice-ocean system for 1955–1975. *J. Geophys. Res.* **98**, 16 397–410.
Hakkinen, S. and Mellor, G. L. 1992. Modeling the seasonal variability of the coupled
    Arctic ice-ocean system. *J. Geophys. Res.* **97**, 20 285–304.
Harder, M. and Fischer, H. 1999. Sea ice dynamics in the Weddell Sea simulated with an
    optimized model. *J. Geophys. Res.* **104**, 11 151–62.
Hastings, A. D. Jr. 1971. Surface climate of the Arctic basin. Selected climatic elements
    related to surface-effects vehicles. US Army Topographic Laboratory, Fort Belvoir,
    VA. ETL-TR-71-5.
Heil, P. and Hibler, W. D. III. 2002. Modeling the high-frequency component of Arctic
    sea-ice drift and deformation. *J. Phys. Oceanography* **32** (11), 3039–57.
Heil, P., Lytle, V. I. and Allison, I. 1998. Enhanced thermodynamic ice growth by sea-ice
    deformation. *Ann. Glaciol.* **27**, 493–7.
Heil, P., Allison, I. and Lytle, V. I. 2001. Effect of high-frequency deformation on sea-ice
    thickness. In Dempsey, J. P. and Shen, H. H., eds., *IUTAM Symposium on Scaling
    Laws in Ice Mechanics and Ice Dynamics.* Dordrecht, Kluwer Academic Publishers,
    pp. 417–26.
Hibler, W. D. III. 1974. Differential sea ice drift II: comparison of mesoscale strain
    measurements to linear drift theory predictions. *J. Glaciol.* **13** (69), 457–71.
    1977. A viscous sea ice law as a stochastic average of plasticity. *J. Geophys. Res.* **82**
    (27), 3932–8.
    1979. A dynamic thermodynamic sea ice model. *J. Phys. Oceanography* **9** (4), 815–46.
    1980a. Modeling pack ice as a viscous-plastic continuum: some preliminary results. In
    Pritchard, R. S., ed., *Sea Ice Processes and Models.* University of Washington Press,
    pp. 163–76.
    1980b. Modeling a variable thickness sea ice cover. *Monthly Weather Rev.* **108**, 1943–73.
    1984. The role of sea ice dynamics in modeling CO_2 increases. In Hansen, J. E. and
    Takahoshi, T., eds., *Climate Processes and Climate Sensitivity.* Geophysical
    Monograph 29. American Geophysical Union, pp. 238–53.
    1985. Modeling sea ice dynamics. In Manabe, S., ed., *Issues in Atmospheric and
    Oceanic Modeling, Part A: Climate Dynamics.* Advances in Geophysics 28,
    pp. 549–78.

1986. Ice dynamics. In Untersteiner, N., ed., *Geophysics of Sea Ice.* New York, Plenum Press, pp. 577–640.

1989. Arctic sea-ice dynamics. In Hermann, Y., ed., *Climatology, Oceanography and Geology of the Arctic Seas.* Van Nostrand, pp. 47–92.

2001. Sea ice fracturing on the large scale, *Engng. Fracture Mech.* **68**, 2013–43.

Hibler, W. D. III and Ackley, S. F. 1983. Numerical simulation of the Weddell Sea pack ice. *J. Geophys. Res.* **88**, 2873–87.

Hibler, W. D., III and Bryan, K. 1987. A diagnostic ice-ocean model. *J. Phys. Oceanography* **17** (7), 987–1015.

Hibler, W. D. III and Flato, G. M. 1992. Sea ice modeling. In Trenberth, K. E., ed., *Climate Systems Modeling.* Cambridge University Press.

Hibler, W. D. III and Hutchings, J. 2003. Multiple equilibrium Arctic ice cover states induced by ice mechanics. 16th IAHR Conference on Sea Ice Processes, Dec. 2002, Dunedin, New Zealand, vol. 3.

Hibler, W. D. III and Ip, C. F. 1995. The effect of sea ice rheology on Arctic buoy drift. In Dempsey, J. and Rajapakse, Y. D. S., eds., *Ice Mechanics.* ASME AMD vol. 207. New York, American Society of Mechanical Engineering, pp. 255–63.

Hibler, W. D. III and Schulson, E. M. 2000. On modeling the anisotropic failure and flow of flawed sea ice. *J. Geophys. Res.* **105** (C7), 17 105–19.

Hibler, W. D. III and Tucker, W. B. III. 1977. Seasonal variations in apparent sea ice viscosity on the geophysical scale. *Geophys Res. Lett.* **4** (2), 87–90.

1979. Some results from a linear viscous model of the Arctic ice cover. *J. Glaciol.* **22** (87), 293–304.

Hibler, W. D. III and Walsh, J. E. 1982. On modeling seasonal and interannual fluctuations of Arctic sea ice. *J. Phys. Oceanography* **12**, 1514–23.

Hibler, W. D. III and Zhang, J. 1994. On the effect of ocean circulation on Arctic ice-margin variations. In Johannessen, O. M., Muench, R. D. and Overland, J. E., eds., *The Polar Oceans and Their Role in Shaping the Global Environment.* Geophysical Monograph 85. American Geophysical Union, pp. 383–98.

1995 On the effect of sea-ice dynamics on oceanic thermohaline circulation. *Ann. Glaciol.* **21**, 361–8.

Hibler, W. D. III, Udin, I. and Ullerstig, A. 1983. On forecasting mesoscale ice dynamics and buildup. *Ann. Glaciol.* **4**, 110–15.

Hibler, W. D., III, Weeks, W. F. and Mock, S. J. 1972. Statistical aspects of sea ice ridge distributions. *J. Geophys. Res.* **70** (30), 5954–70.

Hibler, W. D. III, Weeks, W. F., Kovacs, A. and Ackley, S. F. 1974a. Differential sea ice drift I: spatial and temporal variations in sea ice deformation. *J. Glaciol.* **31** (69), 437–55.

Hibler, W. D., III, Mock, S. J. and Tucker, W. B. III. 1974b. Classification and variation of sea ice ridging in the western Arctic Basin. *J. Geophys. Res.* **79** (18), 2735–43.

Hibler, W. D. III, Ackley, S. F., Crowder, W. K., McKim, H. W. and Anderson, D. M. 1974c. Analysis of shear zone ice deformation in the Beaufort Sea using satellite imagery, In Reed, J. C. and Sater, J. E., eds., *The Coast and Shelf of the Beaufort Sea.* Arctic Institute of North America, pp. 285–96.

Hibler, W. D. III, Heil, P. and Lytle, V. I. 1998. On simulating high frequency variability in Antarctic sea-ice dynamics models. *Ann. Glaciol.* **27**, 443–8.

Hilmer, M. and Lemke, P. 2000. On the decrease of Arctic sea ice volume. *Geophys. Res. Lett.* **27**, 3751–54.

Hilmer, M., Harder, M. and Lemke, P. 1998. Sea ice transport: a highly variable link between Arctic and North Atlantic. *Geophys. Res. Lett.* **25**, 3359–62.

Holland, D. M., Mysak, L. A., Manak, D. K. and Oberhuber, J. M. 1993. Sensitivity study of a dynamic-thermodynamic sea ice model. *J. Geophys. Res.* **98** (C2), 2561–8.

Holland, M. M. and Curry, J. A. 1999. The role of physical processes in determining the interdecadel variability of central Arctic sea ice. *J. Climate* **12**, 3319–30.

Holloway, G. and Sou, T. 2002. Has Arctic sea-ice rapidly thinned? *J. Climate* **15**, 1691–1701.

Hopkins, M. A. 1994. On the ridging of intact lead ice. *J. Geophys. Res.* **99**, 16 351–60.
  1996. On the mesoscale interaction of lead ice and floes. *J. Geophys. Res.* **101**, 18 315–26.

Hopkins, M. A. and Hibler, W. D. III. 1991a. On the ridging of a thin sheet of lead ice, *Ann. Glaciol.* **15**, 81–6.
  1991b. Numerical simulations of a compact convergent system of ice floes. *Ann. Glaciol.* **15**, 26–30.

Hunke, E. C. and Dukowicz, J. K. 1997. An elastic-viscous-plastic model for sea ice dynamics. *J. Phys. Oceanography* **27**, 1849–67.
  2002. The elastic-viscous-plastic sea ice dynamics model in general orthogonal curvilinear co-ordinates on a sphere–incorporation of metric terms. *Month. Weather Rev.* **130** (7), 1848–65.

Hunke, E. C. and Zhang, Y. 1999. A comparison of sea ice dynamics models at high resolution. *Month. Weather Rev.* **127**, 396–408.

Hurrell, J. W. 1995. Decadel trends in the north Atlantic oscillation: regional temperatures and precipitation. *Science* **269**, 676–9.

Hutchings, J. and Hibler, W. D. III. 2002. Modeling sea ice deformation with a viscous-plastic isotropic rheology. *Proceedings of the 16th IAHR International Symposium on Ice*, vol. 2, pp. 358–66.

Ip, C. F. 1993. Numerical investigation of the effect of rheology on sea ice drift. Ph.D. thesis, Dartmouth College, Hanover, N. H.

Ip, C. F., Hibler, III W. D. and Flato, G. M. 1991. On the effect of rheology on seasonal sea ice simulations. *Ann. Glaciol.* **15**, 17–25.

Koerner, R. M. 1973. The mass balance of the sea ice of the Arctic Ocean. *J. Glaciol.* **12**, 173–85.

Kovacs, A. 1972. On pressured sea ice. In Karlsson, T., ed., *Sea Ice: Proceedings of an International Conference*. Reykjavik, National Research Council, Iceland, pp. 276–95.
  1976. CRREL Report 76–32. Hanover, NH, US Army Cold Regions Research and Engineering Laboratory.

Kovacs, A. and Mellor, M. 1974. Sea ice morphology and ice as a geologic agent in the southern Beaufort Sea. In Reed, J. C. and Sater, J. E., eds., *The Coast and Shelf of the Beaufort Sea*. Arlingtion, Virginia, Arctic Institute of North America, pp. 113–62.

Kovacs, A., Weeks, W. F., Ackley, S. F. and Hibler, W. D. III. 1973. Structure of a multi year pressure ridge. *Artic* **26** (1), 22–31.

Kowalik, Z. 1981. A study of the M_2 tide in the ice-covered Arctic Ocean. *Modeling, Identification & Control* **2** (4), 201–23.

Kowalik, Z. and Proshutinsky, A. 1994. The Arctic Ocean tides. In Johannessen, O. M., Muench, R. D. and Overland, J. E., eds., *The Polar Oceans and Their Role in*

Shaping the Global Environment. Geophysical Monograph 85. American Geophysical Union, pp. 137–58.

Kreyscher, M., Harder, M., Lemke, P. and Flato, G. M. 2000. Results of the sea ice model intercomparison project: evaluation of sea-ice rheology schemes for use in climate simulations. *J. Geophys. Res.* **105**, 11 299–320.

Kwok, R. 2001. Deformation of the Arctic Ocean sea ice cover between November 1996 and April 1997: a qualitative survey. In Dempsey, J. P. and Shen, H. H., eds., *Proceedings of the IUTAM Conference on Scaling Laws in Ice Mechanics and Ice Dynamics.* Dordrecht, Kluwer Academic Publishers, pp. 315–22.

Kwok, R. and Rothrock, D. A. 1999. Variability of Fram Strait ice flux and north Atlantic oscillation. *J. Geophys. Res.* **104** (C3), 5177–89.

Leavitt, E. 1980. Surface-based air stress measurements made during AIDJEX. In Pritchard, R. S., ed., *Sea Ice Processes and Models.* Seattle, University of Washington Press, pp. 419–29.

Lemke, P., Owens, B. and Hibler, W. D. III. 1990. A coupled sea-ice mixed layer-pynocline model for the Weddell Sea. *J. Geophys. Res*. **95**, 9513–25.

Lemke, P, Hibler, W. D., III, Flato, G., Harder, M. and Kreyscher, M. 1997. On the improvement of sea-ice models for climate simulations: the sea ice model intercomparison project. *Ann. Glaciol.* **25**, 183–7.

Lepparanta, M. 1980. On the drift and deformation of sea ice fields in the Bothnian Bay. Winter Navigation Research Board, Helsinki, Finland, Research Report 29.

Lepparanta, M. and Hibler, W. D. III. 1985. The role of plastic ice interaction in marginal ice zone dynamics. *J. Geophys. Res.* **90**, 11 899–909.

   1987. Mesoscale sea ice deformation in the east Greenland marginal ice zone. *J. Geophys. Res.* **92**, 7060–70.

Levitus, S. 1982. *Climatological Atlas of the World.* NOAA Publ. 13. Washington, D. C., US Department of Commerce.

Lewis, J. K. 1998. Thermomechanics of pack ice. *J. Geophys. Res.* **103**, (C10), 21 869–82.

Lewis, J. K. and Richter-Menge, J. A. 1998. Motion-induced stresses in pack ice. *J. Geophys. Res.* **103**, C10, 21831–21844.

Lindsay, R. W. 1998. Temporal variability of the energy balance of thick Arctic pack ice. *J. Climate.* **11**, 313–33.

Lipscomb, W. H. 2001. Remapping the thickness distribution in sea ice models. *J. Geophys. Res.* **106** (c7), 13 989–14 000.

McLaren, A. S. 1989. The under-ice thickness distribution of the Arctic Basin as recorded in 1958 and 1970. *J. Geophys. Res.* **94**, 4971–83.

McPhee, M. 1978. A simulation of inertial oscillation in drifting pack ice. *Dyn. Atmos. & Oceans* **2**, 107–22.

   1979. The effect of the oceanic boundary layer on the mean drift of pack ice: application of a simple model. *J. Phys. Oceanography* **9**, 388–400.

   1980. Analysis of pack ice drift in summer. In Pritchard, R. S., ed., *Sea Ice Processes and Models.* Seattle, University of Washington Press, pp. 62–75.

   1982. Sea ice drag laws and simple boundary layer concepts, including application to rapid melting. CRREL Report 82–4. Hanover, NH, US Cold Regions Research and Engineering Laboratory.

   1999. Parameterizing of mixing in the oceanic boundary layer. *J. Marine Syst.* **21**, 55–65.

Malmgren, F. 1933. On the properties of sea ice. *The Norwegian Polar Expedition 'Maud' Sci. Res.* **1a** (5), 1–67.

Malvern, L. E. 1969. *Introduction to the Mechanics of a Continuous Media*, Englewood Cliffs, NJ, Prentice Hall.

Marko, J. R. and Thompson, R. E. 1977. Rectilinear leads and internal motions in the pack ice of the western Arctic Ocean. *J. Geophys. Res.* **82**, 7787–802.

Marshunova, M. S. 1961. Principal characteristics of the radiation balance of the underlying surface and of the atmosphere in the Artic. *Proc. Arc. Antarc. Res. Inst.* **229**. Translated by the Rand Corporation, RM-5003-PR (1969).

Maslanik, J. 1997. On the role of sea ice transport in modifying polar responses to global climate change. *Ann. Glaciol.* **25**, 102–6.

Maslanik J., McGinnis, D., Serreze, M., Dunn, J. and Law-Evans, E. 1996. An assessment of GENESIS version 2 GCM performance for the Arctic. In *Modeling the Arctic System: A workshop on the state of modeling in the Arctic System Science Program.* The Arctic Research Consortium of the US, Fairbanks, pp. 70–2.

Maslowski, W., Newton, B., Schlosser, P., Semtner, A. and Martinson, D. 2000. Modeling recent climate variability in the Arctic Ocean. *Geophys. Res. Lett.* **27**, 3743–6.

Maykut, G. A. 1978. Energy exchange over young sea ice in the central Arctic. *J. Geophys. Res.* **83**, 3646–58.

Maykut, G. A. 1986. The surface heat and mass balance. In Untersteiner, N., ed., *Geophysics of Sea Ice.* New York, Plenum Press, pp. 395–464.

Maykut, G. A. and Untersteiner, N. 1971. Some results from a time dependent, thermodynamic model of sea ice. *J. Geophys. Res.* **76**, 1550–75.

Mendelson, A. 1968. *Plasticity: Theory and Application.* Malabar, FL, Krieger Publishing Company, p. 290.

Mellor, G. L. and Kantha, L. H. 1989. An ice-ocean coupled model. *J. Geophys. Res.* **94**, 10 937–54.

Moritz, R. E. and Stern, H. L. 2001. Relationships between geostrophic winds, ice strain rates and the piecewise rigid motions of pack ice. In Dempsey, J. P. and Shen. H. H., eds., *Scaling Laws in Ice Mechanics and Ice Dynamics.* Dordrecht, Kluwer Academic Publishers, pp. 335–48.

Mountain, D. G. 1974. Preliminary analysis of Beaufort shelf circulation in summer. In Reed, J. C. and Sater, J. E., eds., *The Coast and Shelf of the Beaufort Sea.* Arctic Institute of North America.

Nazarenko, L., Holloway, G. and Tausner, N. 1998. Dynamics of transport of 'Atlantic signature' in the Artic Ocean. *J. Geophys. Res.* **103** (13), 31 003–15.

North, G. R. 1988. Lessons from energy balance models. In Schlesinger, M. E., ed., *Physically-Based Modelling and Simulation of Climate and Climatic Change.* Dordrecht, Kluwer Academic Publishers, pp. 627–52.

North, G. R., Cahalan, R. F. and Coakley, J. A. 1981. Energy balance climate models. *Rev. Geophys. Space Phys.* **19**, 91–121.

Omstedt, A. and Sahlberg, J. 1977. Some results from a joint Swedish-Finnish sea ice experiment, March 1977. Winter Navigation Research Board, Norrkoping, Sweden, Research Report 26.

Overland, J. E. and Pease, C. H. 1988. Modeling ice dynamics of coastal seas. *J. Geophys. Res.* **93**, 15 619–37.

Overland, J. E., Walter, B. A., Curtin, T. B. and Turet, P. 1995. Hierarchy and sea ice mechanics: a case study from the Beaufort Sea. *J. Geophys. Res.* **100**, 4559–71.

Overland, J. E., McNutt, S. L., Salo, S., Groves, J. and Li, S. 1998. Arctic sea ice as a granular plastic. *J. Geophys. Res.* **103**, 21 845–68.

Parkinson, C. L. and Washington, W. M. 1979. A large-scale numerical model of sea ice. *J. Geophys. Res.* **84**, 311–37.

Parmeter, R. R. and Coon, M. D. 1972. Model of pressure ridge formation in sea ice. *J. Geophys. Res.* **77**, 6565–75.

Perovich, D. K., Elder, B. C. and Richter-Menge, J. A. 1997. Observations of the annual cycle of sea-ice temperature and mass balance. *Geophys. Res. Lett.* **24** (5), 555–8.

Pollard, D. and Thompson, S. L. 1994. Sea-ice dynamics and CO_2 sensitivity in a global climate model. *Atmos.-Ocean*, **32**, 449–67.

Polyakov, I. V., Proshutinsky, A. Y. and Johnson, M. A. 1999. Seasonal cycles in two regimes of Arctic climate. *J. Geophys. Res.* **104**, 25 761–88.

   1975. An elastic-plastic constitutive law for sea ice. *J. Appl. Mech.* **42** (E2), 379–84.

Pritchard, R. S. 1978. Effect of strength on simulations of sea ice dynamics. *Proceedings of the International Conference on Port and Ocean Engineering Under Arctic Conditions*, 1977, pp. 494–505.

   1981. Mechanical behaviour of pack ice. In Selvadurai, A. P. S., ed., *Mechanics of Structured Media*. Amsterdam, Elsevier, pp. 371–405.

   1988. Mathematical characteristics of sea ice dynamics models. *J. Geophys. Res.* **93**, 15 609–18.

   2001. Sea ice dynamics models. In Dempsey, J. P. and Shen, H. H., eds., *Scaling Laws in Ice Mechanics and Ice Dynamics. Proceedings of IUTAM Conference*, University of Alaska at Fairbanks, June 2000. Dordrecht, Kluwer Academic Publishers, pp. 265–88.

Pritchard, R., Coon, M. D. and McPhee, M. G. 1977. Simulation of sea ice dynamics during AIDJEX. *J. Pressure Vessel Technol.* **99J**, 491–7.

Proshutinsky, A. and Johnson, M. 1997. Two circulation regimes of the wind-driven Arctic Ocean. *J. Geophys. Res.* **102**, 12 493–514.

Ranelli P. H. and Hibler, W. D. III. 1991. Seasonal Arctic sea-ice simulations with a prognostic ice-ocean model. *Ann. Glaciol.* **15**, 45–53.

Richmond, O. and Gardner, G. C. 1962. Limiting spans for arching of bulk materials in vertical channels. *Chem. Engng. Sci.* **17**, 1071–8.

Richter-Mange, J. A. 1997. Towards improving the physical basis for ice-dynamics models. *Ann. Glaciol.* **25**, 177–82.

Richter-Menge, J. A. and Elder, B. C. 1998. Characteristics of pack ice stress in the Alaskan Beaufort Sea. *J. Geophys. Res.* **103**, 21 817–29.

Richtmyer, R. D. and Morton, K. W. 1967 *Difference Methods for Initial Value Problems*, 2nd edn. New York, Wiley Interscience.

Rothrock, D. A. 1975. The energetics of the plastic deformation of pack ice by ridging. *J. Geophys. Res.* **80**, 4514–19.

Schulkes, R. M. S. M. 1996. Asymptotic stability of the viscous-plastic sea ice rheology. *J. Phys. Oceanography* **26**, 279–83.

Schulson, E. M. 2001. Brittle failure of ice. *Engng. Fracture Mechanics* **68**, 1839–88.

Schulson, E. M. and Hibler, W. D. III. 1991. The fracture of ice on scales large and small: Arctic leads and wing cracks. *J. Glaciol.* **37**, 319–22.

Schulson, E. M. and Nickolayev, O. Y. 1995. Failure of columnar saline ice under biaxial compression: failure envelopes and the brittle-to-ductile transition. *J. Geophys. Res.* **100**, 22 383–400.

Schwarzacher, W. 1959. Pack ice studies in the Arctic Ocean. *J. Geophys. Res.* **64**, 2357–67.

Semtner, A. J. Jr. 1976. A model for the thermodynamic growth of sea ice in numerical investigations of climate. *J. Phys. Oceanography* **6**, 379–89.

Shames, I. H. and Cozzarelli, F. A. 1992. *Elastic and Inelastic Stress Analysis.* Englewood Cliffs, NJ, Prentice Hall, P. 121.

Simmonds, I. and Budd, W. F. 1991. Sensitivity of the southern hemisphere circulation to leads in the Antarctic pack ice. *Quart. J. Roy. Meteor. Soc.* **117**, 1003–24.

Smith, R. B. 1983. A note on the constitutive law for sea ice. *J. Glaciol.* **29**, 191–5.

Sodhi, D. S. 1977. Ice arching and the drift of pack ice through restricted channels. CRREL Report, 77–18. Hanover, NH, Cold Regions Research and Engineering Laboratory.

Sodhi, D. S. and Hibler, W. D. III. 1980. Non steady ice drift in the Strait of Belle Isle. In Pritchard, R. S, ed., *Sea Ice Processes and Models.* Seattle, University of Washington Press, pp. 177–86.

Steele, M. and Flato, G. M. 2000. Sea ice growth melt and modeling: a survey. In Lewis, E. L. *et al.*, eds., *The Freshwater Budget of the Arctic Ocean.* Dordrecht, Kluwer, pp. 549–87.

Steele, M., Zhang, J., Rothrock, D. and Stern, H. 1997. The force balance of sea ice in a numerical model of the Arctic Ocean. *J. Geophys. Res.* **102**, 21 061–79.

Stern, H. I., Rothrock, D. A. and Kwok, R. 1995. Open water production in Arctic sea ice: satellite measurements and model parameterizations. *J. Geophys. Res.* **100**, 20 601–12.

Thompson, S. L. and Pollard, D. 1997. Greenland and Antarctic mass balances for present and doubled atmospheric CO_2 from the GENESIS version 2 global climate model. *J. Climate* **10**, 871–900.

Thorndike, A. S. 1986. Kinematics of sea ice. In Untersteiner, N., ed., *Geophysics of Sea Ice.* New York, Plenum Press, pp. 489–550.

1992a. Toy model linking atmospheric thermal radiation and ice growth, *J. Geophys. Res.* **97** (6), 9401–10.

1992b. A toy model of sea ice growth. In Ojima, D., ed., *Modeling the Earth System.* Office for Interdisciplinary Earth Studies Global Change Institute vol. 3, Boulder, UCAR, pp. 225–38.

Thorndike, A. S. and Colony, R. 1980. Large-scale ice motion in the Beaufort Sea during AIDJEX, April 1975-April 1976. In Pritchard, R. S., ed., *Sea Ice Processes and Models.* Seattle, University of Washington Press, pp. 249–60.

1982. Sea ice motion in response to geostrophic winds. *J. Geophys. Res.* **87**, 5845–52.

Thorndike, A., Rothrock, D. A., Maykut, G. A. and Colony, R. 1975. The thickness distribution of sea ice. *J. Geophys. Res.* **80**, 4501–13.

Tremblay, L. B. and Mysak, L. A. 1997. Modeling sea ice as a granular material, including the dilatancy effect. *J. Phys. Oceanography* **27**, 2342–60.

Tucker, W. B., III and Govoni, J. W. 1981. Morphological investigations of first-year sea ice pressure ridge sails. *Cold Regions. Sci. Technol.* **5**, 1–12.

Tucker, W. B. III and Perovich, D. K. 1992. Stress measurements in drifting pack ice. *Cold Regions Sci. & Technol.* **20**, 119–39.

Tucker, W. B. III, Weeks, W. F. and Frank, M. 1979. Sea ice ridging over the Alaskan continental shelf. *J. Geophys. Res.* **84**, 4885–97.

Tucker, W. B., III, Gow, A. J. and Weeks, W. F. 1987. Physical properties of summer sea ice in the Fram Strait. *J. Geophys. Res.* **92**, 6787–803.

Tucker, W. B. III, Perovich, D. K., Hopkins, M. A. and Hibler, W. D. III. 1991. *Ann. Glaciol.* **15**, 265–70.

Tucker, W. B., Weatherly, J. W., Eppler, D. T., Farmer, D. and Bentley, D. L. 2001. Evidence for the rapid thinning of sea ice in the western Arctic Ocean at the end of the 1980s. *Geophys. Res. Lett.* **28** (9), 2851–4.

Ukita, J. and Moritz, R. E. 1995. Yield curves and flow rules of pack ice. *J. Geophys. Res.* **100**, 4545–57.

Untersteiner, N. 1964. Calculations of temperature regime and heat budget of sea ice in the Central Arctic. *J. Geophys. Res.* **69**, 4755–66.

Vavrus, S. J. 1995. Sensitivity of the Arctic climate to leads in a coupled atmosphere/mixed-layer ocean model. *J. Climate* **8**, 158–71.

  1999. The response of the coupled Arctic sea ice-atmosphere system to orbital forcing and ice motion at 6 kyr and 115 kyr BP. *J. Climate* **12**, 873–96.

Vavrus, S. J. and Harrison, S. P. 2003. The impact of sea ice dynamics on the Arctic climate system. *Climate Dyn.* **20**, 741–57.

Vinje, T. E. 1976. Sea ice conditions in the European sector of the marginal seas of the Arctic, 1966–1975. In *Norsk Polar-Institutt Arbok 1975*. Oslo, pp. 163–74.

  2001. Fram Strait ice fluxes and atmospheric circulation: 1950–2000. *J. Climate* **14**, (16), 3508–17.

Vinje, T. E. and Finnekasa, O. 1986. The ice transport through the Fram Strait. Report 186, 3900. Oslo, Norsk Polarinstitutt, pp. 1–39.

Vinje, T. E. Norland, N. and Kvambekk, A. 1998. Monitoring ice thickness in Fram Strait. *J. Geophys. Res.* **103**, 10 437–49.

Wadhams, P. 1981. Sea ice topography of the Arctic Ocean in the region 70 °W to 25 °E. *Roy. Soc. Phil. Trans.* A302, 1464–504.

Wadhams, P. 1983. Sea ice thickness distribution in Fram Strait. *Nature* **305**, 108–11.

  1992. Sea ice thickness distribution in the Greenland Sea and Eurasian Basin, May 1987. *J. Geophys. Res.* **97**, 5331–48.

Wadhams, P. and Horne, R. J. 1980. An analysis of ice profiles obtained by submarine sonar in the Beaufort Sea. *J. Glaciol.* **25**, 401–24.

Walsh, J. E., Hibler, W. D. III and Ross, B. 1985. Numerical simulation of northern hemisphere sea ice variability, 1951–1980. *J. Geophys. Res.* **90**, 4847–65.

Walsh, J. E., Chapman, W. L. and Shy, T. L. 1996. Recent decrease of sea level pressure in the central Arctic. *J. Climate* **9**, 480–6.

Wang, J. and Ikeda, M. 2000. Arctic oscillation and Arctic sea-ice oscillation. *Geophys. Res. Lett.* **27** (9), 1287–90.

Weeks, W. F. and Ackley, S. F. 1986. The growth, structure, and properties of sea ice. In *The Geophysics of Sea Ice*. NATO ASI Series B, vol. 146, pp. 9–164.

Weeks, W. F., Kovacs, A. and Hibler, W. D. III. 1971. Pressure ridge characteristics in the Arctic coastal environments. *Proc. 1st Intl Conf. on Port Ocean Engineering under Arctic Conditions*, vol. 1, pp. 152–83.

Weeks, W. F., Kovacs, A., Mock, S. J., Hibler, W. D. III and Gow, A. J. 1977. Studies of the movement of coastal sea ice near Prudhoe Bay, Alaska, USA. *J. Glaciol.* **19** (81), 533–46.

Whitham, G. B. 1974. *Linear and Nonlinear Waves*. New York, John Wiley and Sons.

Williams, E., Swithinbank, C. and Robin, G. De Q. 1975. A submarine sonar study of Arctic pack ice. *J. Glaciol.* **15**, 349–62.

Winton, M. 2000. A reformulated three-layer sea ice model. *J. Atmos. Oceanic Technol.* **17**, 525–31.

Worster, M. G. 1999. Solidification of fluids. In Batchelor, G. K., Moffatt, H. K. and Worster, M. G., eds., *Perspectives in Fluid Dynamics*. Cambridge University Press.

Zhang, J. 1993. A high resolution ice-ocean model with imbedded mixed layer. Ph. D.
    thesis, Thayer School of Engineering, Dartmouth College.
Zhang, J. and Hibler, W. D., III. 1997. On an efficient numerical method for modeling sea
    ice dynamics. *J. Geophys. Res.* **102**, 8691–702.
Zhang, J. and Rothrock, D. 2001. A thickness and enthalpy distribution sea-ice model.
    J. Phys. Oceanography **31**, 2986–3001.
Zhang, J., Hibler, W. D. III, Steele, M. and Rothrock, D. A. 1998. Arctic ice-ocean
    modeling with and without climate restoring. *J. Phys. Oceanography* **28**, 191–217.
Zhang, J., Rothrock, D. and Steele, M. 2000. Recent changes in Arctic sea ice: the
    interplay between ice dynamics and thermodynamics. *J. Climate*, **13** 3099–114.
Zhang, S., Lin, C. A. and Greatbatch, R. J. 1995. A decadal oscillation due to the coupling
    between an ocean circulation model and a thermodynamic sea-ice model. *J. Marine
    Res.* 53, 79–106.
Zhang, Y., Maslowski, W. and Semtner, A. J. 1999. Impact of mesoscale ocean currents on
    sea ice in high-resolution Arctic ice and ocean simulations. *J. Geophys. Res.* **104**
    (C8), 18 409–30.
Zubov, N. N. 1943. *Arctic Ice*. English Translation NTIS no. AD 426972, Springfield, VA
    (1979). Translators: Naval Oceanographic Office and the American Meteorological
    Society.

Part III

The mass balance of sea ice

8

Sea-ice observations

SEYMOUR W. LAXON
Centre for Polar Observation and Modelling, University College London
JOHN E. WALSH
Department of Atmospheric Sciences, University of Illinois
PETER WADHAMS
Department of Applied Mathematics and Theoretical Physics,
University of Cambridge
OLA M. JOHANNESSEN AND MARTIN MILES
Nansen Environmental and Remote Sensing Center, Bergen

8.1 Introduction

The Earth's climate system is presently undergoing an uncontrolled experiment as a result of man's increasing emissions of carbon dioxide (CO_2), methane (CH_4), nitrous oxide (N_2O) and other greenhouse gases – gases that exert a positive radiative forcing of climate – into the atmosphere, as well as anthropogenic aerosols (microscopic particles) that have a negative radiative forcing. As a net result of these forcings and associated dynamics, changes in global mean temperature are predicted to exceed their natural variability between the decades 1980 and 2010 (Cubasch *et al.*, 1995). An assessment by the Intergovernmental Panel on Climate Change (IPCC) concluded cautiously that the balance of observational evidence already suggests a discernible human influence on the global climate (IPCC, 1995).

As a complement to observational studies, numerical models are used to understand better climate and climate change, including the effect of anthropogenic emissions of greenhouse gases and aerosols. The most advanced climate models are coupled oceanic and atmospheric general circulation models (GCMs). These models simulate the climate system based on physical laws describing the dynamics and physics of the ocean and atmosphere, and include representations of land–surface processes and other complex processes including those related to sea ice. Model runs include changes in external forcings such as those from increasing greenhouse gases and aerosols. A consensus from the numerical modelling community is that greenhouse warming will be enhanced in the polar regions, especially the Arctic (Figure 8.1). The predicted warming for the Arctic is $\sim$3–4 °C during the next 50 years (Mitchell *et al.*, 1995) with a substantial retreat of the Arctic sea-ice cover (Manabe, Spelman and Stouffer, 1992).

The reasons for enhanced Arctic warming foreseen in models are found in atmosphere–ocean–ice interactions or feedbacks within the climate system (Randall *et al.*, 1998). A

Mass Balance of the Cryosphere: Observations and Modelling of Contemporary and Future Changes, eds. Jonathan L. Bamber and Antony J. Payne. Published by Cambridge University Press. © Cambridge University Press 2003.

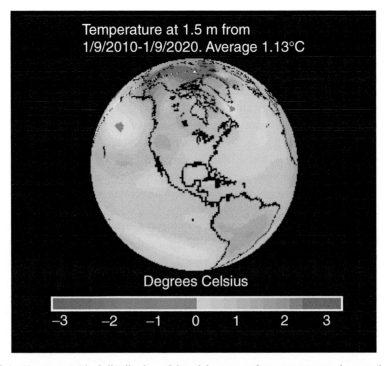

Figure 8.1. The geographical distribution of decadal mean surface temperature changes (in degrees Celsius) for the period 2010 to 2020 with respect to the period 1880 to 1920, as predicted using a coupled ocean–atmosphere general circulation model (GCM). The modelled scenario assumes gradually increasing greenhouse gases (CO_2 and others) and sulphate aerosols. (Courtesy of Hadley Centre, UK.)

pre-dominant mechanism is the temperature–ice–albedo feedback, which has a positive or amplifying effect. For example, higher temperatures can reduce the extent of highly reflective sea ice (or snow), which in turn increases the absorption of energy at the surface, which leads to reduced sea ice, and thus the positive feedback loop perpetuates itself.

The predominant feature of the Arctic's physical environment is the presence of a sea-ice cover, which is perennial in the central Arctic and at least seasonal in the marginal seas. The sea-ice cover is an important component of the climate system, modulating the exchange of heat, moisture and momentum between the ocean and atmosphere, as well as ocean stratification and deep-water formation in winter. Sea-ice variability can thus both reflect and affect climate change. Sea ice's relatively straightforward (compared to land snow cover), and fast (compared to land-ice sheets and glaciers), response to atmospheric forcing suggests that the observations of the sea-ice cover may provide early strong evidence of global greenhouse warming in the Arctic (Walsh, 1995). Moreover, the sea-ice cover is a spatially integrated indicator of climate change, in contrast to spatially sparse temperature records available for the interior of the Arctic Ocean.

8.2 Sea-ice observations

Sea-ice data are derived from various sources, each with their particular spatial and temporal sampling and other limitations. Sea-ice charts extend back over 100 years in many regions, and gridded data sets of Arctic ice extent since 1901 have been produced, albeit with some inherent uncertainties. The most reliable, homogeneous part cover the period since 1953 (Chapman and Walsh, 1993) and are derived from operational ice charts, including those based at least partly on satellite images since the early 1970s. This data set provides quantitative information on monthly Arctic ice extent, a parameter defined as the area with the ocean–ice boundary. Antarctic sea-ice data sets are considerably shorter and less complete.

The most consistent, quantitative means of studying the global sea-ice cover are satellite passive microwave remote-sensor data, available without interruption since 1978. Passive microwave sensors measure low level microwave radiation emitted from the Earth's surface and atmosphere. The measured radiance is calibrated to brightness temperature (T_B) and represents a linear function of the emissivity (i.e., radiative efficiency) and the physical temperature (T) of a substance. Even though sea ice has a lower T than open water, it has a much greater microwave emissivity and therefore a higher T_B than open water, for the microwave frequencies used for sea-ice retrieval. Algorithms applied to multi-frequency microwave T_B data are used to filter out atmospheric and other effects and then to calculate total ice concentration (the percentage of ice-covered ocean within an image pixel) (Figure 8.2), from which total ice extent (the area within the ice–ocean margin), total ice area (the area of ice-covered ocean) and open water area (the difference between extent and area) are derived. As presented later, additional parameters may be retrieved under certain conditions.

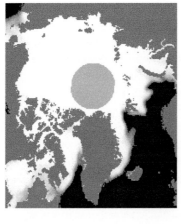 

(a)   (b)

Figure 8.2. Arctic total sea-ice concentration in winter (a) and summer (b) as derived from satellite-borne passive microwave sensor data. In these image maps, ice concentrations (percentage of ice-covered ocean in an image pixel) less than 15% are assigned as zero (open water).

Sea-ice time series derived from multi-channel passive microwave data are among the longest continuous satellite-derived geophysical records, extending over two decades. The Nimbus-7 scanning multi-channel microwave radiometer (SMMR) provided data from 1978 to 1987, and the follow-up special sensor microwave/imager (SSM/I) onboard defence meteorological satellite program (DMSP) satellites has provided data since 1987. The data are optimally re-sampled into 25 × 25 km grid cells and then issued by the National Snow and Ice Data Center (USA), providing a standard data set for the research community to analyse sea-ice variability and trends.

Most of the available data on the statistical properties of the sea-ice thickness distribution and other statistical measures of sea-ice roughness in the Arctic come from submarines. The fundamental attribute of the ice cover derivable from submarine sonar profiles is the *ice-draft distribution*, or probability density function of ice draft. In addition, the continuous profiles obtainable from a submarine (and some moored upward sonar) permit spatial statistics of the ice cover to be generated, such as the distribution of *pressure ridge* depths and spacings, *lead* widths and spacings, and *spectral* and *fractal* properties of the underside.

8.3 Sea-ice observations: the pre-satellite era

Widely cited sea-ice variations of the past few decades include the decrease of Arctic ice coverage, particularly in summer, as shown by the satellite passive microwave record (Bjorgo, Johannessen and Miles, 1997; Parkinson *et al.*, 1999); the decrease of multi-year sea-ice coverage by approximately 7% per decade (Johannessen and Miles, 2000); and the reduction of the Arctic Ocean's ice thickness, as inferred from 29 nearly coincident submarine measurements, by approximately 30% from the 1960s to the 1970s compared with the 1990s (Rothrock, Yu and Maykut, 1999). The variations of the past few decades are sufficiently large that they represent potentially significant changes in the climate system. However, actual significance can only be assessed in the context of a broader temporal record of sea-ice variability. The challenge in this regard is to utilize historical information that, unlike the gridded hemispheric fields available from satellites at one- to two-day intervals since the late 1970s, is of variable quantity and quality. Local observations from scattered sites date back several hundred years, permitting the compilation of indices that provide long-term temporal context on a local basis. Examples include (1) the Icelandic sea-ice index (e.g. Ogilvie, 1998), the spatial representativeness of which was addressed by Kelly *et al.* (1987), and (2) the index of winter ice severity in the western Baltic Sea, extending back to 1701 (Koslowski and Glaser, 1995).

Sea-ice charts for specific sectors of the Arctic date back only to the late 1800s. These charts often suffer from spatial and temporal gaps, especially during the dark months of autumn and winter. Nevertheless, the data compilations for at least some regions are sufficient to permit assessments of the broad patterns of the inter-annual to decadal variations extending back about a century. Indeed, a consolidation of the available charted data can provide a framework for an assessment of the recent (post-1978) trends summarized above. Such a framework is the subject of this section.

A twentieth-century sea-ice data consolidation can draw upon several extensive data bases that are now available in digital form. The first is the annual series of sea-ice summaries compiled from 1890 onward by the Danish Meteorological Institute (e.g. DMI, 1936). These summaries include northern hemisphere charts, which depict the ice distribution for the spring and summer months (April–August) on the basis of coastal and ship reports. The annual reports also include narrative summaries of coastal and ship-based reports for each of the remaining months of the year. The monthly charts in this series have been digitized by Kelly (1979). It should be noted that, while the Danish charts cover the northern hemisphere, they include little information for regions other than the north Atlantic.

A recent synthesis of North Atlantic sea-ice data extending back to the mid 1800s is described by Vinje (2000). Prior to about 1950, these depictions of the ice edge were based primarily on ship reports. Much of these data were also provided to the DMI for inclusion in the yearbooks cited above. In the more recent decades, aircraft and satellite observations were included in the Norwegian data set, which has recently been digitized by the Norwegian Polar Institute.

Historical sea-ice data for the western North Atlantic, spanning the period 1810 to 1958, have recently been synthesized, quality controlled and digitized by Hill (1998). This data set, which includes information on icebergs as well as sea ice, is the most comprehensive depiction of sea-ice variations in the Canadian maritime region, especially the waters around Newfoundland and Labrador. Because the ice season is relatively short in this region, the digitized ice charts for most years are limited to February, March and April.

Additional data sets that are useful in documenting sea-ice variations over the pre-satellite era include the Russian data set of Zakharov (1997), the chart series of the UK Meteorological Office and the weekly chart series of the US National Ice Center (Martin and Dedrick, 1998). A synthesis of these data into a monthly gridded data set spanning the twentieth century is described by Walsh and Chapman (2001). Vinnikov *et al.* (1999) used this data set in a comparison of recent model-derived and observationally derived trends of sea ice; they also compared time series of sea-ice coverage obtained from several of the different data sets mentioned above. While the definitions of 'ice coverage' varied among the various data sets, Figure 8.3 shows that the variations depicted by the different data sets are highly correlated, at least for the past few decades. Furthermore, because the various compilations for the first half of the twentieth century drew upon the same raw data in many instances, the variations are generally robust across the data sets even in decades prior to the satellite era.

In order to provide examples of twentieth-century sea-ice variations in a longer-term context, we show the annual index of sea-ice severity along the northern Iceland coast (Figure 8.4, from Kelly, Goodess and Cherry, 1987) and in the western Baltic Sea (Figure 8.5; Koslowski and Glaser, 1995). The Icelandic coast clearly experienced more severe ice conditions in the nineteenth century, particularly in the second half, than during most of the twentieth century (Figure 8.4). The decadal-scale excursion toward severe ice conditions

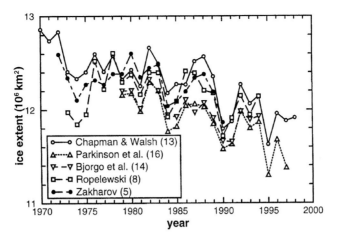

Figure 8.3. Time series, 1970s onward, of hemispheric ice extent from several observational sources. (From Vinnikov *et al.* (1999).)

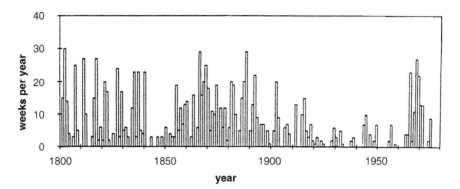

Figure 8.4. The number of weeks per year when ice affected the coast of Iceland. This is a modified version of an earlier index of sea-ice severity near Iceland, defined as the product of the number of weeks with ice per year and the number of coastal areas near where it was observed. (Adapted from Kelly *et al.* (1987) and Lamb (1977).)

in the late 1960s and early 1970s was associated with the well documented 'great salinity anomaly' (Dickson *et al.*, 1988), which followed an apparent enhancement of ice export from the Arctic Ocean through the Fram Strait (Hakkinen, 1993). A similar conclusion about the relative severity of nineteenth- and twentieth-century ice conditions emerges from the time series for the western Baltic Sea (Figure 8.5). In this case, however, the strongest twentieth-century pulse of severe ice conditions occurred in the early-to-middle 1940s, which coincided with the Second World War and its well documented cold winters in eastern Europe. Since the long period of relatively heavy ice conditions the eighteenth and nineteenth centuries had ended by 1900, there is little discernible trend toward less ice during the twentieth century in the western Baltic.

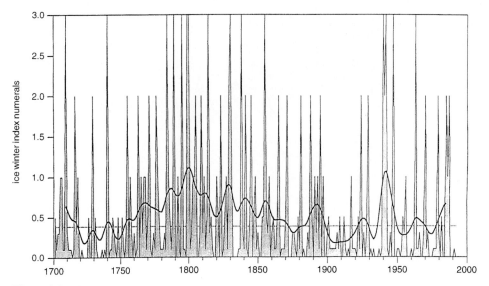

Figure 8.5. The ice winter index numerals from 1701 to 1993. The horizontal dotted line denotes the climatological mean of 0.385 for the period 1901 to 1960. The heavy solid curve represents smoothed ice winter index numerals obtained by applying a Gaussian low-pass filter with a 20-year cutoff. (From Koslowski and Glaser (1995).)

In contrast to the absence of a twentieth-century trend in the western Baltic, Figure 8.6 shows a twentieth-century decrease of sea ice in the Nordic seas, both in the western (30°W–10°E) and the eastern (10°E–70°E) sectors studied by Vinje (2000). The time series show even larger decreases when the final three to four decades of the nineteenth century are included in the plots. While the time series in Figure 8.6 are for April, the August trends are also negative (see fig. 8.4 of Vinje, 2000).

Finally, Figure 8.7 (Hill, 1998) shows the time series of winter ice anomalies in the western North Atlantic/Labrador Sea region offshore of Newfoundland. Also shown in Figure 8.7 is the corresponding time series of the North Atlantic oscillation. The winter ice anomalies show a general decrease from the late 1800s to the 1960s, followed by an irregular increase to the late 1990s. Because of the increase since the 1970s, the overall trend since 1900 is small. Figure 8.7 shows that the twentieth-century ice variability in this region is highly correlated with the North Atlantic oscillation (NAO), which has generally been in its positive phase since the 1970s; the five-year moving average NAO reached a new period of record highs during the 1990s.

The NAO is also correlated with sea-ice variations in the Nordic seas and the western Baltic (Figures 8.5 and 8.6), although the correlations are of opposite sign from those for the western North Atlantic. This opposition is consistent with the 'seesaw' nature of the wind and temperature anomalies associated with the NAO (Rogers and Van Loon, 1979). Indeed, a major emphasis of the studies by Koslowski and Glaser (1995) and Vinje (2000) is the close association between the NAO and North Atlantic sea-ice anomalies, pointing

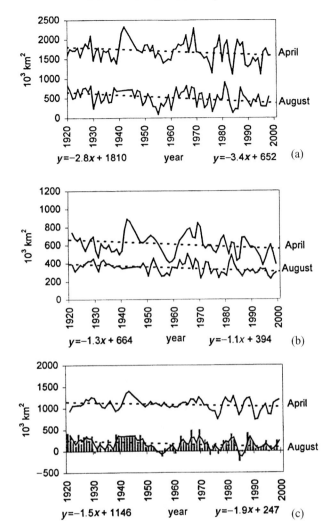

Figure 8.6. Time series of April and August ice extent given by two-year running means for (a) the Nordic seas, (b) western area and (c) eastern area, for the period 1920 to 1998. The linear regression equations used for trend estimations are given below the abscissas ($y = 10^3 \, km^2$, and $x = $ (year − 1920).)

to the importance of the atmospheric circulation in forcing inter-annual to decadal sea-ice anomalies. (The lack of correlation between the NAO and the Icelandic ice index in Figure 8.4 is probably due to the fact that Iceland is essentially the 'fulcrum' of the NAO.) The role of the NAO in the broader decrease of Atlantic sea ice since 1800 (Figures 8.5 and 8.6) is less clear because the NAO time series, derived from station pressure measurements, generally extend back only to the 1860s or 1870s. It should also be kept in mind that global temperatures appear to have increased by about 1 °C since the 1880s (IPCC, 2000), although the warming is regional and at least partially circulation-driven.

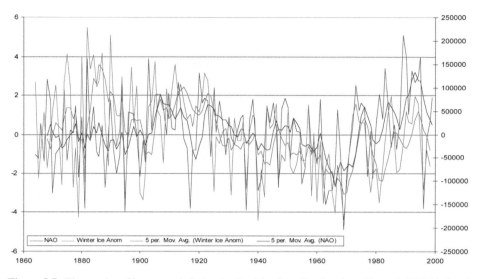

Figure 8.7. Time series of ice extent in Labrador Sea/Newfoundland region. (From B. Hill, National Research Council of Canada; see Hill (1998).)

When the various regional data sets described previously are combined into a hemispheric data base, the twentieth-century time series of annual and seasonal total ice coverage vary as shown in Figure 8.8. The negative trends of the past two to three decades are consistent with the findings based on satellite data. However, Figure 8.8 shows that the trends of the satellite period are continuations, or even accelerations, of trends that span much of the latter half of the twentieth century. By contrast, the trends during the first half of the century are small.

An intriguing feature of Figure 8.8, also found in the satellite-based studies, is that the negative trends are larger in summer than in winter. Global climate models, on the other hand, project the strongest greenhouse-induced warming in the Arctic during winter, while the projected Arctic warming during summer is very small (IPCC, 2000). Part of the explanation for this apparent discrepancy lies in the regionality of recent sea-ice trends in the North Atlantic, where the major portion of the marginal ice zone is located during winter. Offsetting trends in the western and eastern North Atlantic (Figures 8.5 and 8.6), attributable to the NAO, reduce the magnitude of the winter-time trend. By contrast, the summer ice margin is largely in the Arctic Ocean during summer, when the NAO is much weaker and displaced southward from the ice margin. Hence circulation-induced dipoles are not key characteristics of the summer sea-ice distribution.

As noted previously, the Antarctic sea-ice record is far more fragmentary than the Arctic record. While the whaling data analysed by De La Mare (1997) suggested a decrease of Antarctic sea-ice extent in early January (Figure 8.9), direct comparisons with the summer and winter results for the Arctic are precluded by the fact that January is a month of rapid sea-ice retreat in the Antarctic. Other recent comparisons with the historical record, allowing for the retreat behaviour of the ice cover in the modern record,

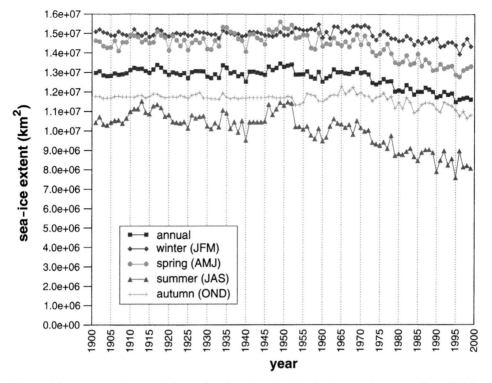

Figure 8.8. Annual and seasonal time series of total northern hemisphere sea-ice extent. (From Walsh and Chapman (2001).)

suggest that the maximum and minimum extents of Antarctic sea ice have changed little. The identification of any Antarctic changes is further complicated by interpretative difficulties in comparing ship-based observations with the satellite-derived estimates for recent decades.

8.4 Sea-ice cover: the post-satellite era

Observation studies of microwave-derived sea-ice concentration and derived parameters indicate that variability on timescales of days to weeks and longer is usually organized into geographical patterns that are associated with synoptic-scale pressure systems and larger-scale structures of atmospheric circulation variability. In the context of climate-change detection, this high frequency variability is considered to be essentially background noise, such that climate-change analyses based on sea-ice data tend to be based on monthly averages. The predominant variability in Arctic (or Antarctic) sea-ice time series is seasonal, with typical late winter (March) maximum ice extent of $\sim 15 \times 10^6 \, \text{km}^2$, compared with a late summer (September) minimum of $\sim 5 \times 10^6 \, \text{km}^2$ in the Arctic, though the absolute values may vary from study to study due to operational differences. Figure 8.10(a) shows

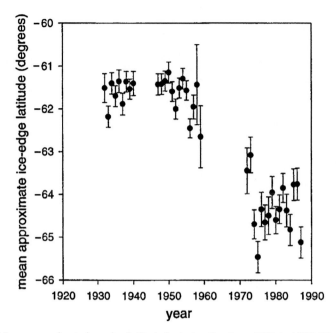

Figure 8.9. Mean approximate ice-edge latitude in Antarctica from 1931 to 1987. The estimates are from a linear model fitted to whale-catch records, and are standardized to the first 10 day period of January and the longitudinal sector 20°–30°E. (From De La Mare (1997).)

the seasonal cycle in Arctic sea-ice area, 1978–98. The seasonal variability in the Antarctic (not shown) is even greater, with the late austral summer being nearly ice-free. The seasonal cycle can be readily removed statistically, leaving a series of departures or anomalies from which remaining irregular variability and trends can be determined (Figure 8.10(b)).

The first trend analysis based on SMMR data from 1978 to 1987 found a slight negative trend in Arctic sea-ice extent (Gloersen and Campbell, 1991). The $0.032 \times 10^6 \, km^2$ per year decrease (2.4% per decade) was found to be statistically significant, with no significant trend found in the Antarctic. Subsequent data from the SSM/I provided the basis to follow up on the SMMR-derived trends. An independent analysis of SMMR and SSM/I records taken separately revealed a greater reduction in Arctic sea-ice area and extent during the SSM/I period. The decreases from 1987 to 1994 were ~4% per decade compared with ~2.5% per decade from 1978 to 1987, again with no significant trends found in the Antarctic (Johannessen, Miles and Bjørgo, 1995). The Arctic's apparently shrinking sea ice attracted considerable attention in the international climate change community. However, the high degree of inter-annual variability, coupled with the brevity of the individual time series, compelled researchers to produce longer time series and more robust trend estimates.

Since then, merged SMMR–SSM/I time series have been produced and analysed, establishing the trends more firmly. The merging of SMMR and SSM/I involves inter-comparison and adjustments (i.e., 'inter-calibration') based on the six-week overlap period when both sensors operated. The first published merged SMMR–SSM/I analysis established the trend

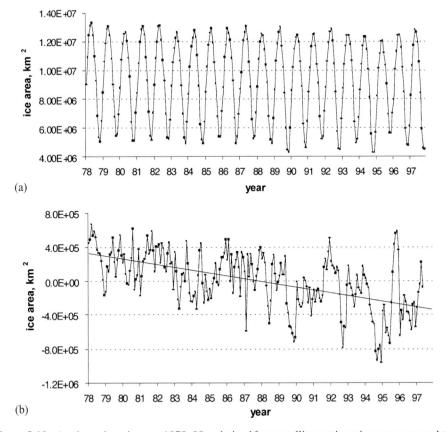

Figure 8.10. Arctic total sea-ice area 1978–98 as derived from satellite passive microwave sensor data. Shown are monthly averages (a) and anomalies (b). The linear regression in (b) indicates a negative trend $\sim 0.03 \times 10^6$ km^2 per year, which represents a nearly 6% decrease during the observation period.

in Arctic ice area and extent (1978–95) to be about -0.3×10^6 km^2 per decade, corresponding to $\sim 3\%$ per decade, with no significant change in the Antarctic. The $\sim 3\%$ per decade decrease in the Arctic ice extent (1978–97) was subsequently corroborated in other analyses (Parkinson *et al.*, 1999). The latter also identified the seasonal and regional patterns of trends, while the former confirmed the hemispheric asymmetry seen earlier (Bjorgo *et al.*, 1997; Johannessen *et al.*, 1995). The Antarctic sea-ice cover may have even increased slightly ($\sim 1.5\%$) (Cavalieri *et al.*, 1997), though there is disagreement as to the statistical significance of such a small trend (Bjorgo *et al.*, 1997; Cavalieri *et al.*, 1997). The hemispheric ice covers fluctuate quasi-periodically, with predominant periods between three and five years, though their variability is apparently not correlated (Cavalieri *et al.*, 1997). The cause of this quasi-periodic behaviour may be related to large-scale atmospheric teleconnection patterns such the El Niño southern oscillation (ENSO) (Gloersen, 1995).

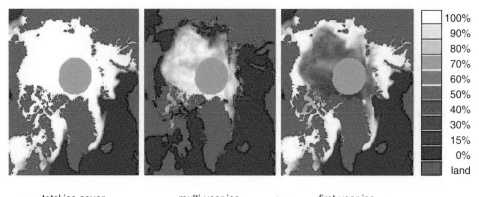

total ice cover　　=　　multi-year ice　　+　　first-year ice

Figure 8.11. Arctic total sea-ice cover and its two components, multi-year (MY) and first-year (FY) ice, as derived from satellite passive microwave sensor data in winter. The scale indicates the concentration (percentage) of each ice type in each image pixel.

The seasonality and forcing mechanisms behind the decreases in Arctic ice extent in the 1990s were then analysed using SMMR–SSM/I data (1979–95) together with meteorological data fields (Masalanik, Serreze and Barry, 1996). The ice reductions were found to be most pronounced in the Siberian sector in the summer, with record low Arctic ice minima in 1990, 1993 and 1995, apparently linked to atmospheric circulation anomalies – in particular, an increase in low pressure systems and associated advection of warm air from the Eurasian land mass in the 1990s. The pronounced summer reductions suggested consequential changes in other aspects (e.g., perennial ice pack) of the ice cover.

Perennial, multi-year ice (i.e. that having survived the summer melt) is approximately three times thicker than first-year or seasonal ice (~1–2 m), such that changes in its distribution could also both reflect and affect climate change. As mentioned earlier, multi-year (MY) and first-year (FY) ice have different radiative properties, permitting discrimination using multi-channel passive microwave T_B data during the winter months when their signatures are relatively stable (Figure 8.11). In summer, the effects of melt ponds and melting snow on the ice confound the signal.

The possibility of monitoring inter-annual variations in MY ice area was explored using SMMR data (Comiso, 1990), but its potential has remained under-realized until recently. In a recent study, 20 years of the SMMR and SSM/I data were used to produce and analyse spatially integrated time series of MY and FY ice areas in winter, revealing the ice cover's changing composition (Johannessen, Shalina and Miles, 1999). The methods used in the analysis were based on the approach used previously for merging SMMR–SSM/I sea-ice time series (Bjorgo *et al.*, 1997), with additional methods for robust estimation of MY and FY ice areas. Because the SMMR–SSM/I sensor overlap occurred during the boreal summer (i.e. during a period when ice signatures are unstable), it was unreasonable to inter-calibrate directly the SMMR- and SSM/I-derived MY and FY ice areas, as was done for total ice concentration. Instead, the estimates were made by: (i) fine-tuning assumed MY ice emissivities based on Arctic field measurements; (ii) fine-tuning the weather filters to reduce false ice signatures off the main ice edge; (iii) analysing spatial and temporal variations in the MY ice distribution for coherence; (iv) comparing with independent field and aerial data; and (v) analysing each summer's minimum ice area in conjunction with the following winter's MY ice area, which should correspond (Comiso, 1990). The last-mentioned procedure, (v), confirmed a close correspondence ($r \sim 0.82$), with the winter-averaged MY ice area only 13% less than the summer minimum – substantially closer than the 25–40% obtained previously (Comiso, 1990). The difference is partially explained by the metamorphism of second-year ice into mature MY ice, such that the microwave-derived MY ice area generally increases during the winter – the MY ice area in late winter (March) is only 9% less than the summer minimum. The remaining difference may be largely explained by ice outflow through the Fram Strait (between Greenland and Svalbard) during winter. Therefore, the time series of MY and FY ice areas could reasonably be considered to represent the character of the winter ice cover.

The winter-averaged MY area decreased by $0.031 \times 10^6 \, \text{km}^2$ per year (Figure 8.12) compared with a total ice area decrease of $0.024 \times 10^6 \, \text{km}^2$ per year averaged over the same five months. The difference ($0.007 \times 10^6 \, \text{km}^2$ per year) represents replacement by FY ice, such that the proportion of MY to FY ice changed accordingly. The $0.031 \times 10^6 \, \text{km}^2$ per year decrease in MY ice area represents a proportionally large ($\sim$7% per decade) reduction in the MY ice area during 1978 to 1998 (Figure 8.12), compared with an $\sim$2% per decade decrease in the total ice area in winter (Johannessen *et al.*, 1999). The apparent 14% reduction in MY ice area over two decades is corroborated by other analyses, such as SMMR-SSM/I data analysis that found an 8% increase (5.3 days) in the length of the sea-ice melt season in the Arctic from 1978 to 1996 (Smith, 1998). It is also supported by an analysis of oceanographic data that has revealed changes in Arctic water masses since the 1970s that are reasoned to stem from a substantial ($\sim$2 m) melting of perennial MY ice (McPhee *et al.*, 1998).

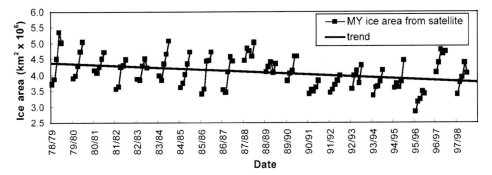

Figure 8.12. Arctic multi-year (MY) sea-ice area in winter (November to March) as derived from satellite passive microwave sensor data. The linear regression indicates that the MY ice area decreased by $\sim$610 000 km^2 during the 20-year observation period, which represents a decrease of $\sim$14%.

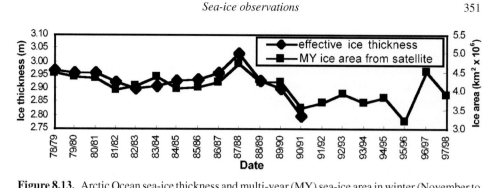

Figure 8.13. Arctic Ocean sea-ice thickness and multi-year (MY) sea-ice area in winter (November to March), as derived from ice surface measurements (1978–91) and satellite passive microwave sensor data (1978–98). The strong correlation ($r \sim 0.88$) between the ice parameters during the overlap suggests that observed areal decreases in MY ice area are associated with decreases in ice thickness, rather than representing only peripheral changes.

In order to assess the significance of the observed reductions in MY ice area in terms of the ice cover's mass balance, spatially and temporally coincident data on ice-thickness are needed. Spatially averaged ice-thickness estimates (Nagurnyi, Korostelev and Abaza, 1994; Nagurnyi, Korostelev and Ivanov, 1999) from Russian drifting stations have been compared with MY ice area during the common observation period 1978 to 1991 (Figure 8.13), finding a close correspondence ($r \sim 0.88$) between these independently derived parameters. The strength of the correlation was unexpected – after all, MY ice could become substantially thinner and still count toward the MY ice area. However, the observed relationship with ice thickness suggests that the observed decrease in MY ice area (1978–98) represents a substantial (i.e., mass balance) rather than a peripheral effect.

8.5 Mean ice thickness and its variability

Our knowledge of the regional variability of the ice-thickness distribution, $g(h)$, in the Arctic comes almost entirely from upward sonar profiling by submarines. Therefore our level of knowledge of $g(h)$, and of the mean regional thickness h_m, depends on whether submarines have been able to operate in the area concerned. To date, data have been obtained mainly from British submarines operating in the Greenland Sea and Eurasian Basin, and from US submarines operating in the Canada and Eurasian Basins, with some earlier cruises in the Canadian Arctic. Very large data sets were more recently obtained from annual cruises during the US SCICEX civilian submarine programme of 1993–97 (Rothrock *et al.*, 1999).

Moving northward from sub-polar regions, the ice in Baffin Bay is largely thin first-year ice with a modal thickness of 0.5–1.5 m (Wadhams, McLaren and Weintraub, 1985). In the southern Greenland Sea, too, the ice, although composed largely of partly melted multi-year ice, also has a modal thickness of about 1 m (Vinje, 1989; Wadhams, 1992), with the decline in mean thickness from the Fram Strait giving a measure of the fresh-water input to the Greenland Sea at different latitudes. Over the Arctic Basin itself there is a gradation in mean ice thickness from the Soviet Arctic, across the Pole and toward the coasts of north Greenland and the Canadian Arctic archipelago, where the highest mean thicknesses of some 7–8 m are observed (Bourke and McLaren, 1992; LeSchack, 1980; Wadhams, 1981). These overall variations are in accord with the predictions of numerical models (see Chapter 9) (Hibler, 1979, 1980), which take account of ice dynamics and deformation as well as ice thermodynamics.

The temporal variability of $g(h)$ has been much less extensively measured. Only the advent of SCICEX allowed a comparison between extensive modern data and a more temporally and spatially sparse assortment of earlier data sets. Pre-SCICEX results show that in the Eurasian Basin far from land, in the region between the Pole and Svalbard, the mean ice draft seemed remarkably stable between different seasons and different years (McLaren, 1989; Wadhams, 1989, 1990a), although McLaren found considerable inter-annual variability at the Pole itself. In the region between the Pole and Greenland, where a buildup of pressure ridging normally occurs due to the motion of the ice cover toward a downstream land boundary, very considerable differences in $g(h)$ and in mean draft (more than 15%) were observed between October 1976 and May 1987 (Wadhams, 1990b). In the Beaufort Sea in summer, also, considerable differences were observed between records taken several years apart (McLaren, 1989). Bourke and Garrett (1987) used all available data to attempt to construct seasonal contour maps of mean ice draft. Figure 8.14 shows Bourke and Garrett's maps for summer and winter, and Figure 8.15 shows the variability observed by Wadhams between 1976 and 1987. It should be noted that the data used for the Bourke and Garrett map did not include open water, so these maps are over-estimates of the mean ice draft. The maps were updated by Bourke and McLaren (1992), who also gave contour maps of standard deviation of draft, and mean pressure ridge frequencies and drafts for summer and winter, based on 12 submarine cruises.

In order to assess reliably whether ice-thickness changes are occurring in the Arctic, it is necessary to obtain area-averaged observations of mean ice thickness over the same region using the same equipment at different seasons or in different years. Ideally the region should be as large as possible, and the measurements should be repeated annually in order to distinguish between a fluctuation and a trend. Early approaches toward this goal included the work of McLaren (1989), who compared data from two US Navy submarine transects of the Arctic Ocean in August 1958 and August 1970, stretching from the Bering Strait to the North Pole and down to the Fram Strait. He found similar conditions prevailing in each year in the Eurasian Basin and North Pole area, but significantly milder conditions in the Canada Basin in 1970. The difference is possibly due to anomalous cyclonic activity as observed in the region in recent summers (Serreze, Barry and McLaren, 1989).

Wadhams (1989) compared mean ice thicknesses for a region of the Eurasian Basin lying north of the Fram Strait, from British Navy cruises carried out in October 1976, April–May 1979 and June–July 1985. All three data sets were recorded using similar sonar equipment. It was found that a box extending from 83°30'N to 84°30'N and from 0° to 10°E had an especially high track density from the three cruises (400 km in 1976, 400 km in 1979 and

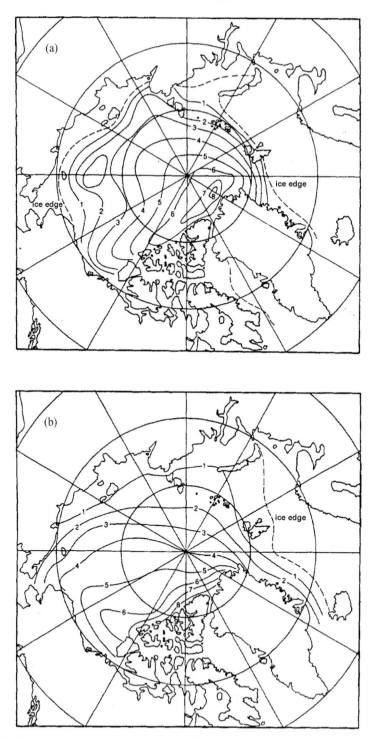

Figure 8.14. Contour maps of estimated mean ice drafts for (a) summer and (b) winter in the Arctic Basin. (After Bourke and Garrett (1987).)

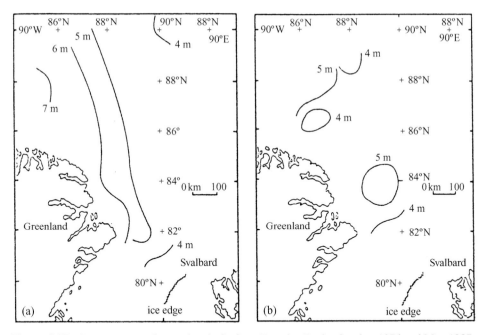

Figure 8.15. Contour maps of mean ice drafts from Eurasian Basin, October 1976 and May 1987. (After Wadhams (1990b).)

1800 km in 1985), and this was selected for the comparison. It is a region far from any downstream boundary, and represents typical conditions in the Trans Polar Drift Stream prior to the acceleration and narrowing of the ice stream which occurs as it prepares to enter the Fram Strait. The mean thicknesses from the three cruises were remarkably similar: 4.60 m in 1976; 4.75 m in 1979; and 4.85 m in 1985. It should be remembered that these data sets were recorded in different seasons as well as different years.

Later, Wadhams compared data from a triangular region extending from north of Greenland to the North Pole, recorded in October 1976 and May 1987 (Wadhams, 1990b). Mean drafts were computed over 50 km sections, and each value was positioned at the centroid of the section concerned; the results were contoured to give the maps shown in Figure 8.15. There was a decrease of 15% in mean draft averaged over the whole area (300 000 km^2), from 5.34 m in 1976 to 4.55 m in 1987. Profiles along individual matching track lines (Figure 8.16) show that the decrease was concentrated in the region south of 88°N and between 30° and 50°W. From Figure 8.7 it appears that the buildup of pressure ridging which gave the high mean drafts near the Greenland coast in 1976 was simply absent in 1987, but in fact the situation is not that simple. Table 8.1 shows a comparison between the probability density functions of ice thickness from the pairs of profiles shown in Figure 8.16 (strictly, two 300 km sections from the southern part of the N–S transect and two 200 km sections from 40–50°W in the E–W transect). In 1987 there was more ice present in the form of young ice in refrozen leads (coherent stretches of ice with draft less than 1 m) and

Table 8.1. *Comparisons of ice statistics from 1976 and 1987 data sets.*

	N–S transect		E–W transect	
	1976	1987	1976	1987
Mean draft (m)	6.09	5.31	6.32	4.07
Ice <2 m draft (%)	11.6	16.7	7.9	29.9
Ice 2–5 m draft (%)	48.7	38.6	46.5	39.6
Ice <5 m draft (%)	39.7	44.7	46.0	30.5
Refrozen leads (%)	4.0	7.9	3.7	15.6

as FY ice (draft less than 2 m). There was less MY ice (interpreted as ice 2–5 m thick) and less ridging (ice more than 5 m thick) in 1987 in the E–W transect, although slightly more ridged ice in the N–S transect. The main contribution to the loss of volume appears to be, then, the replacement of MY and ridged ice by young and FY ice.

For instance, taking ice of 2–5 m thickness as an indicator of undeformed MY ice fraction, this declined from 47.6% in 1976 to 39.1% in 1987, a relative decline of 18%. This is in agreement with recent results of Johannessen *et al.* (1999), who found that MY ice fraction in the Arctic (estimated from passive microwave data) suffered a 14% decrease during the period 1978 to 1998.

A final regional comparison, which the 1987 data set made possible, occurs in the region immediately north of the Fram Strait, between 82°N and 80°N (Figure 8.17). Here it was possible to compare data from the four years 1976, 1979, 1985 and 1987. This is a region which is ice-covered in most years, and where mixing occurs between the various ice streams preparing to enter the Fram Strait, notably the streams of old, deformed ice moving south from the North Pole region and south-east from the region north of Greenland; and the stream of younger, less heavily deformed ice moving south-west from the seas north of Russia. Figure 8.17 shows all available mean drafts from 50 km sections (from Wadhams, 1981, 1983, 1989, and unpublished analyses). There is very good consistency among these four data sets, regardless of year or season of generation; fluctuations appear to be random in character, and where centroids from different experiments lie close to one another, the mean drafts are usually similar. Only the 1976 data points appear somewhat thicker than their neighbours.

McLaren *et al.* (1992) analysed 50 km and 100 km sections of ice profile centred on the North Pole from six cruises from 1977 to 1990. They found that the mean ice draft from 50 km sections in the late 1970s (1977, 1979) was 4.1 m (4.2 m, 4.0 m, respectively), while the mean draft for the late 1980s was 3.45 m (2.8 m, 4.1 m, 3.3 m, 3.6 m for 1986, 1987, 1988, 1990 respectively). The difference of 0.65 m is 15%. Using a t-test they showed that this difference of means is significant only at the 20% level, i.e. non-significant although possibly indicative.

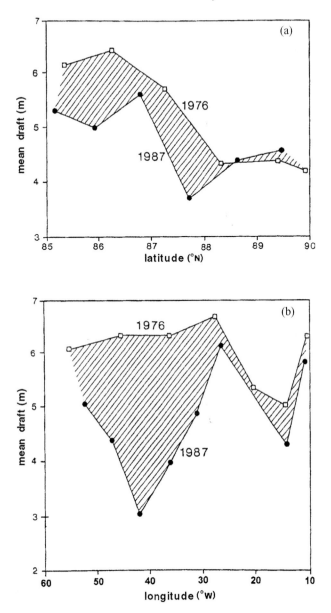

Figure 8.16. (a) Comparison of mean ice drafts from 1976 and 1987 along a N–S transect from the North Pole to 85°N. (b) Comparison of mean ice drafts for transect across north of Greenland from 60°W to 10°W.

In the most recent study by Rothrock *et al.* (1999), sonar data obtained from the US civilian SCICEX submarine programme in September–October of 1993, 1996 and 1997 were compared with data obtained during six summer cruises during the period 1958 to 1976: 29 crossing places were identified, where a submarine track from the recent period

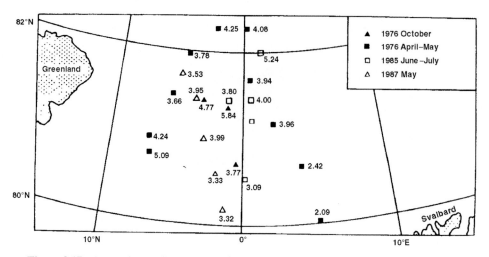

Figure 8.17. Comparisons of mean ice drafts measured in the region north of the Fram Strait.

crossed one from the early period, and the corresponding tracks (of average length 160 km) were compared in thickness. In each case the mean thicknesses obtained were adjusted to a standard date of September 15 using an ice–ocean model to account for seasonal variability. The 29 matched data sets were divided into six geographical regions (Figure 8.18). The decline in mean ice draft was significant for every region and increased across the Arctic from the Canada Basin towards Europe – it was 0.9 m in the Chukchi Cap and Beaufort Sea, 1.3 m in the Canada Basin, 1.4 m near the North Pole, 1.7 m in the Nansen Basin and 1.8 m in the eastern Arctic. Overall, the mean change in draft was from 3.1 m in the early period to 1.8 m in the recent period, a decline of 42%.

Rothrock *et al.* (1999) commented that the decline in mean draft could arise thermodynamically from any of the following flux increases:

(1) a 4 W/m^2 increase in ocean heat flux;
(2) a 13 W/m^2 increase in pole-ward atmospheric heat transport; or
(3) a 23 W/m^2 increase in down-welling short-wave radiation during summer.

Clearly a change in ice dynamics can also produce a change in mean ice draft, both via changes in deformation and redistribution within the Arctic Basin (this is discussed further in Chapter 9).

This is the most extensive comparison so far, but it should be noted that all data sets involved are from summer, mostly late summer, so that the reported decline refers to only one season of the year, and that most track comparisons occur over the North Pole region and Canada Basin, with few in the Eurasian Arctic and none south of 84° in the Eurasian Basin.

Complementary to the Rothrock *et al.* study are comparisons from the Eurasian Basin and Greenland Sea made using data from British submarine cruises (one British cruise was also used in the Rothrock study). Wadhams did not correct for seasonal variability

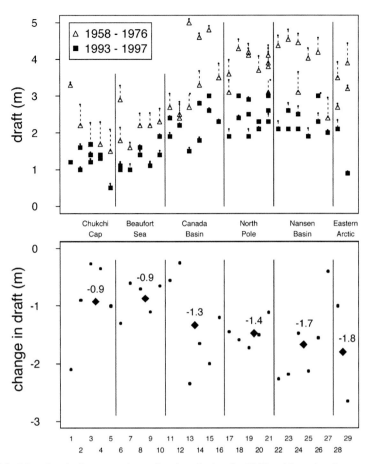

Figure 8.18. Mean ice drafts at crossings of cruises during the 1990s with cruises between 1958 and 1976. From Rothrock *et al.* (1999).)

between the 1976 measurements, made in October, and those of 1987, made in April–May. If this is done in the manner used by Rothrock *et al.* (1999), the decrease in mean ice draft (standardized to September 15) becomes much greater at 42%, since April–May is the time of greatest ice thickness. This is in excellent agreement with the Rothrock *et al.* results for the entire overall data set that they analysed, yet occurring within a period of only 11 years. This indicates either that thinning occurs faster in the Eurasian Basin than elsewhere in the Arctic (which is suggested by the geographical trend of the Rothrock *et al.* data) or that it is invalid to compare data sets from different times of year simply by standardizing to 'summer' through use of a model.

The latter problem is largely overcome in an analysis of the most recent British data set, obtained in September 1996 by Wadhams (1990b) aboard HMS *Trafalgar*. Wadhams and Davis (2000) includes a comparison of these data with results from October 1976. The two

Table 8.2. *Mean sea-ice drafts in Eurasian Basin in vicinity of the Greenwich meridian, 1976 to 1996.*

Latitude range (°N)	Mean draft, 1996	Mean draft, 1976	1996 as a percentage of 1976
81–82	1.57	5.84	26.9
82–83	2.15	5.87	36.6
83–84	2.88	4.90	58.7
84–85	3.09	4.64	66.6
85–86	3.54	4.57	77.4
86–87	3.64	4.64	78.5
87–88	2.36	4.60	51.2
88–89	3.24	4.41	73.4
89–90	2.19	3.94	55.5
Overall	2.74	4.82	56.8

submarines followed similar courses between 81°N and 90°N on about the 0° meridian, and it was found that about 2100 km of track from each submarine, when divided into 100 km sections, were close enough in correspondence to count as 'crossing tracks' in the sense used by Rothrock *et al.* (1999). The overall decline in mean ice thickness between 1976 and 1996 was 43%, in remarkably close agreement with Rothrock *et al.* The mean drafts in 1° bins of latitude are given in Table 8.2.

It can be seen from Table 8.2 that there was a significant decrease of mean draft at every latitude, but that the decline is largest just north of the Fram Strait and near the North Pole itself. A characteristic of the ice cover observed from below was the large amount of completely open water present at all latitudes. A seasonality correction to the 1976 data for the slight difference in mean draft between October and September brings the ratio to 59.0% for September, a decline of 41%. Thus the British and the US data are in remarkably good agreement in describing a very significant thickness decrease in Arctic Ocean sea ice.

A cautionary note must be sounded in that these significant thickness decreases from spatially averaged data conceal large random variabilities at given locations. Time series of ice draft at fixed locations have been obtained from moored upward sonar systems, of which the most comprehensive set spans the Fram Strait. Vinje, Nordlund and Kvambekk (1998), in an analysis of data from 1991 to 1998, show that inter-seasonal and inter-annual variability in thickness far exceed any trend, although of course the length of the data set is only seven years.

8.6 Current evidence for change

The balance of observational evidence indicates a sea-ice cover in transition, which could eventually lead to a different ice–ocean–atmosphere regime in the Arctic, altering heat and mass exchanges as well as ocean stratification. However, 20 years of microwave satellite data may be inadequate to establish that this is a long-term trend rather than reflecting decadal-scale atmosphere–ocean variability such as the ENSO and NAO (Hurrell, 1995). The NAO is the tendency for simultaneous strengthening or weakening of the Icelandic low and the Azores high, the two semi-permanent atmospheric pressure centres in the North Atlantic. The NAO is strongest in winter, though it remains the predominant mode of atmospheric circulation in the region year round. In its positive mode, it leads to enhanced westerlies across the North Atlantic, associated with positive anomalies in storminess, precipitation and temperature in northern Europe and beyond. Recently, another apparent mode of high latitude atmospheric variability, the so-called Arctic oscillation (AO) (Thompson and Wallace, 1998), has been introduced. The AO describes the dominant spatial pattern of variability in mean sea-level pressure (SLP) north of 20°N, encompassing the regional NAO. Although its proponents consider it to be a more fundamental structure than the NAO, their temporal correlation in the Atlantic sector is 0.95, such that it may be essentially indistinguishable from the NAO (Deser, 2000).

Inter-annual variability in the NAO is known to be strongly coupled to fluctuations in Arctic sea-ice motion (Kwok and Rothrock, 1999) and ice export through the Fram Strait (Kwok, 2000), as well as regional sea-ice extent (Deser, 2000; Mysak *et al.*, 1996). The NAO winter index has also been found to be lag-correlated with the Arctic minimum ice area following summer, and hence the following winter MY ice area ($r = -0.54$), such that the NAO index explains $\sim$25% (r^2) of the MY ice variability (Johannessen *et al.*, 1999). The connection to atmospheric circulation anomalies underscores the need to produce and analyse consistently longer-term sea-ice data sets. Indeed, it is possible that should atmospheric circulation anomalies seen in, e.g., NAO and AO indices during recent decades return to 'normal', the Arctic sea-ice cover would probably rebound accordingly. On the other hand, these atmospheric circulation anomalies themselves may be part of global warming. For example, some greenhouse warming modelling analyses indicate increasingly positive states of the NAO and AO. Furthermore, a recent comparison of GCM-simulated sea-ice extent and observed sea-ice trends concludes that there is less than a 2% chance that the 1978–98 sea-ice trends arise from natural variability, and less than 0.1% for 1953–98 sea-ice trends (Vinnikov *et al.*, 1999).

8.7 Consequences of change

Extrapolating the estimated (Rothrock *et al.*, 1999) thinning rate of the ice cover of 4 cm per year indicates that the Arctic Ocean could be essentially ice-free in summer in the next 50 years. Present uncertainties about recent and future melting rates notwithstanding, there is a range of potential consequences – some 'negative' and some 'positive' – of a disappearing ice cover that can be hypothesized.

(i) Dramatic reductions in albedo would have significant effects on radiation and energy balances and atmospheric and oceanic circulation in the high and mid latitudes, hence altering weather patterns and storm tracks, frequency and intensity.

(ii) Exposure of vast areas of the Arctic Ocean with cold open water, which has a high capacity for CO_2 absorption, could become an important sink of atmospheric CO_2, thus mitigating global warming. For example, first-order estimates based on measurements of carbon fluxes in the Greenland Sea and a box model indicate that $0.3–0.6 \times 10^{15}$ g of carbon could be absorbed each year by an ice-free Arctic Ocean, a 15–30% increase above present levels.

(iii) Broad changes in the marine ecosystem (e.g., less plankton in the North Atlantic due to less ice and greater inflow of melt water (Reid *et al.*, 1998)) would have a negative impact on Arctic and sub-Arctic marine biodiversity, though with an increased area for fisheries in previously ice-covered areas, as well as increased offshore and gas activities and marine transportation, including the northern sea route north of Siberia.

(iv) Changes in the pathways and spreading of melt water and in the stratification in the northern North Atlantic Ocean, and the effects of reduced deep-water formation on the global thermohaline circulation (Mauritzen and Hakkinen, 1999) (the so-called 'conveyer belt'), which could diminish or otherwise change the Gulf Stream and transatlantic drift stream (Samiento *et al.*, 1998), thereby greatly altering the climate of Europe.

It is therefore of utmost importance that monitoring and research on the Arctic Ocean and surrounding seas be prioritized as one of the 'grand challenges' for the international science community. Further systematic monitoring and analysis of sea ice and other high latitude environmental parameters such as ice sheets, glaciers, snow cover and permafrost, together with oceanographic and atmospheric data as well as modelling simulations, are needed to understand better the patterns, processes and consequences of climate change in the Arctic.

8.8 Future prospects

The measurements of sea-ice thickness from space has previously been described as an unachievable goal. However, recent research has demonstrated the feasibility of obtaining direct measurements of sea-ice elevation from space-borne altimetry (Peacock *et al.*, 1997). Despite the fact that the Earth resources satellite (ERS) instruments were not designed for operation over sea ice, validation against submarine measurements has shown that errors in ice-thickness estimation are as little as 0.5 m. At this level these data can start to provide the necessary means to validate model predictions of thickness and to place longer-term observations, such as those from submarines, in context. Figure 8.19 shows contour plots of the mean end of summer and end of winter ice thickness estimated from ERS-2 altimeter measurements of ice elevation. Comparison of the altimeter thickness estimates with those of Bourke and Garrett (Figure 8.14) shows that ice during the 1997–98 winter was considerably thinner than suggested in the climatology derived from data gathered in the 1960s. This agrees with observations of much thinner ice than expected during the surface heat budget of the Arctic Ocean (SHEBA) transit and the larger-scale submarine study by Rothrock showing thinner ice during the 1990s than previously (Rothrock *et al.*, 1999).

Retrieval of ice elevations will become an increasingly important goal for future altimeter missions. The RA-2 instrument on-board Envisat, launched in March 2002, employs a new tracking and echo recording system which is anticipated to overcome many of the

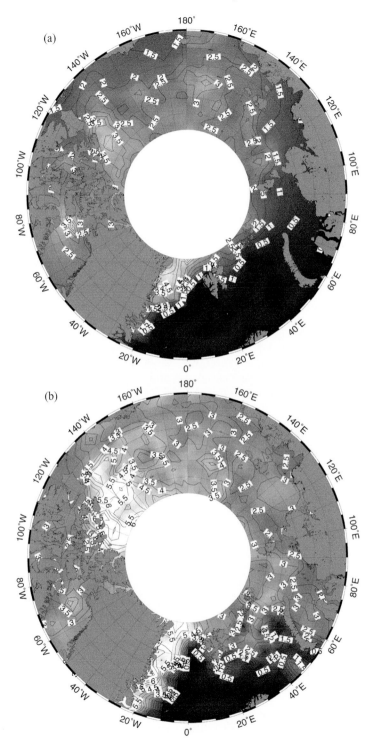

Figure 8.19. Mean altimeter-derived ice thickness for (a) September–October, 1997, (b) February–March, 1998.

problems encountered in using ERS data over sea ice. The ICESat laser mission, launched in January 2003, is the first mission designed to employ a space-borne laser for ice elevation measurements. This has the potential to provide ice freeboard measurements in clear sky conditions over a footprint of only 160 m and to latitudes of 86°N. Finally, the CryoSat satellite, due for launch in late 2004, will employ novel Doppler processing of radar signals to limit the illuminated footprint to ~1 km, providing much better sampling than current space-borne radar altimeters. The latitudinal coverage, to 88°N, will provide almost complete altimetric coverage of the Arctic Ocean for the first time.

References

Bjorgo, E., Johannessen, O. M. and Miles, M. W. 1997. Analysis of merged SMMR-SSMI time series of Arctic and Antarctic sea ice parameters 1978–1995. *Geophys. Res. Lett.* **24** (4), 413–16.

Bourke, R. H. and Garrett, R. P. 1987. Sea ice thickness distribution in the Arctic Ocean. *Cold Regions Sci. & Technol.* **13**, 259–80.

Bourke, R. H. and McLaren, A. S. 1992. Contour mapping of Arctic basin ice draft and roughness parameters. *J. Geophys. Res.* **97** (C11), 17715–28.

Cavalieri, D. J., Gloersen, P., Parkinson, C. L., Zwally, H. J. and Comiso, J. C. 1997. Observed hemispheric asymmetry in global sea ice changes. *Science* **278**, 1104–6.

Chapman, W. L. and Walsh, J. E. 1993. Recent variations of sea ice and air temperatures in high latitudes. *Bull. Am. Meteorol. Soc.* **74**, 33–47.

Comiso, J. C. 1990. Arctic multiyear ice classification and summer ice cover using passive microwave satellite data. *J. Geophys. Res.* **95** (C8), 13411–22.

Cubasch, U. *et al.* 1995. A climate change simulation starting at an early time of industrialisation. *Climate Dyn.* **11**, 71–84.

De La Mare, W. K. 1997. Abrupt mid-twentieth-century decline in Antarctic sea ice extent from whaling records. *Nature* **389**, 57–9.

Deser, C. 2000. On the teleconnectivity of the 'Arctic oscillation'. *Geophys. Res. Lett.* **27**, 779–82.

Dickson, R. R., Meincke, J., Malmberg, S. A. and Lee, A. J. 1988. The 'great salinity anomaly' in the northern North Atlantic, 1968–1982. *Prog. Oceanography* **20**, 103–51.

DMI. 1936. Isforholdene I de artiske Have 1936. *Nautisk-Meteorologisk Aarbog*. Copenhagen, Danish Meteorological Institute.

Gloersen, P. 1995. Modulation of hemispheric sea-ice cover by ENSO events. *Nature* **373**, 503–4.

Gloersen, P. and Campbell, W. J. 1991. Recent variations in Arctic and Antarctic sea-ice covers. *Nature* **352**, 33–6.

Hakkinen, S. 1993. An Arctic source for the great salinity anomaly: a simulation of the Arctic ice ocean system for 1955–1975. *J. Geophys. Res.* **98** (C9), 16397–410.

Hibler, W. D. III. 1979. A dynamic-thermodynamic sea ice model. *J. Phys. Oceanography*, **9**, 815–46.

1980. Modelling a variable thickness sea ice cover. *Month. Weather Rev.* **108**, 1943–73.

Hill, B. T. 1998. Historical record of sea ice and iceberg distribution around Newfoundland and Labrador, 1810–1958. *Proceedings of the Workshop on*

Operational Sea Ice Charts of the Arctic (Seattle, WA), World Climate Research Programme/Arctic Climate System Study, pp. 12–13.

Hurrell, J. 1995. Decadal variability in the North Atlantic oscillation regional temperatures and precipitation. *Science* **269**, 676–9.

IPCC. 1995. *The Science of Climate Change*. Cambridge University Press.
   2000. *Climate Change: The Scientific Basis*. IPCC Third Assessment Report, Cambridge University Press.

Johannessen, O. M. and Miles, M. W. 2000. Arctic sea ice and climate change – will the ice disappear in this century? *Sci. Prog.* **83** (3), 209–33.

Johannessen, O. M., Miles, M. and Bjørgo, E. 1995. The Arctic's shrinking sea ice. *Nature* **376**, 126–7.

Johannessen, O. M., Shalina, E. V. and Miles, M. W. 1999. Satellite evidence for an Arctic sea ice cover in transformation. *Science* **286** 1937–9.

Kelly, P. M. 1979. An Arctic sea ice dataset, 1901–1956. *Glaciol. Data* **GD-5**, 101–6.

Kelly, P. M., Goodess, C. M. and Cherry, B. S. G. 1987. The interpretation of the Icelandic sea ice record. *J. Geophys. Res.* **92**, (C10), 10 835–43.

Koslowski, G. and Glaser, R. 1995. Reconstruction of the ice winter severity since 1701 in the western Baltic. *Climatic Change* **31**, 79–98.

Kwok, R. 2000. Recent changes in Arctic Ocean sea ice motion associated with the North Atlantic oscillation. *Geophys. Res. Lett.* **27**, 775–8.

Kwok, R. and Rothrock, D. A. 1999. Variability of Fram Strait ice flux and North Atlantic oscillation. *J. Geophys. Res.* **104** (C11), 5177–89.

Lamb, H. H. (1977). *Climate: Present, Past and Future*, vol. 2. London, Methuen.

LeSchack, L. A. 1980. Arctic Ocean sea ice statistics derived from the upward-looking sonar data recorded during five nuclear submarine cruises. Technical Report, LeSchack Associates Ltd., University Blvd. W., Silver Spring, MD.

McLaren, A. S. 1989. The under-ice thickness distribution of the Arctic basin as recorded in 1958 and 1970. *J. Geophys. Res.* **94** (C4), 4971–83.

McLaren, A. S., Walsh, J. E., Bourke, R. H., Weaver, R. L. and Wittmann, W. 1992. Variability in sea ice thickness over the North Pole from 1977–1990. *Nature* **358**, 224–6.

McPhee, M. G., Stanton, T. P., Morison, J. H. and Martinson, D. G. 1998. Freshening of the upper ocean in the Arctic: is perennial sea ice disappearing? *Geophys. Res. Lett.* **25** (10), 1729–32.

Manabe, S., Spelman, M. J. and Stouffer, R. J. 1992. Transient responses of a coupled ocean-atmosphere model to gradual changes in atmospheric CO2. *J. Climate* **5**, 105–26.

Martin, D. L. and Dedrick, K. R. 1998. US National Ice Center sea ice charts in a digital GIS-based time series and climatology. *Proceedings of the Workshop on Operational Sea Ice Charts of the Arctic* (Seattle, WA) World Climate Research Programme/Arctic Climate System Study, pp. 17–19.

Masalink, J., Serreze, M. C. and Barry, R. G. 1996. Recent decreases in Arctic summer ice cover and linkages to atmospheric circulation anomalies. *Geophys. Res. Lett.* **23**, 1677–80.

Mauritzen, C. and Hakkinen, S. 1999. On the relationship between dense water formation and the 'meridional overturning cell' in the North Atlantic Ocean. *Deep-Sea Res. I* **46**, 877–94.

Mitchell, J. F. B., Johns, T. C., Gregory, J. M. and Tett, S. F. B. 1995. Climate response to increasing levels of greenhouse gases and sulphate aerosols. *Nature* **376**, 501–4.

Mysak, L. A., Ingram, R. G., Wang, J. and van der Baaren, A. 1996. The anomalous sea-ice extent in Hudson Bay, Baffin Bay and the Labrador Sea during three simultaneous NAO and ENSO episodes. *Atmos.-Ocean* **34**, 313–43.

Nagurnyi, A. P., Korostelev, V. G. and Abaza, P. A. 1994. Wave method for evaluating the effective ice thickness of sea ice in climate monitoring. *Bull. Russ. Acad. Sci. Phys. Suppl. Phys. Vibr.* **58**, 168–74.

Nagurnyi, A. P., Korostelev, V. G. and Ivanov, V. V. 1999. Multiyear variability of sea ice thickness in the Arctic basin measured by elastic-gravity waves on the ice surface. *Meteor. Hydrol.* **3**, 72–8. (In Russian; English translation available from the Nansen Environmental and Remote Sensing Center, Bergen, Norway.)

Ogilvie, A. E. J. 1998. Historical sea-ice records from Iceland ca. A.D. 1145 to ca. 1950. *Proceedings of the Workshop on Operational Sea Ice Charts of the Arctic* (Seattle, WA). World Climate Research Programme/Arctic Climate System Study, pp. 1–2.

Parkinson, C. L., Cavalieri, D. J., Gloersen, P., Zwally, H. J. and Comiso, J. C. 1999. Spatial distribution of trends and seasonality in the hemispheric sea ice covers: 1978–1996. *J. Geophys. Res.* **104** (C9), 20 837–56.

Peacock, N. R., Laxon, S. W., Scharoo, R. and Maslowski, W. 1997. Improving the signal to noise ratio of altimetric measurements in ice covered seas. *EOS (Suppl.)* **78** (46), F140 (abstract).

Randall, D. *et al.* 1998. Status and outlook for large-scale modelling of atmosphere-ice-ocean interaction in the Arctic. *Bull. Am. Meteor. Soc.* **79**, 197–219.

Reid, P. C., Edwards, M., Hunt, H. G. and Warner, A. J. 1998. Plankton change in the North Atlantic. *Nature* **391**, 546.

Rogers, J. C. and Van Loon, H. 1979. The seesaw in winter temperatures between Greenland and northern Europe. Part II: Some oceanic and atmospheric effects in middle and high latitudes. *Month. Weather Rev.* **107** (5), 509–19.

Rothrock, D. A., Yu, Y. and Maykut, G. A. 1999. Thinning of the Arctic sea-ice cover. *Geophys. Res. Lett.* **26** (23), 3469–72.

Samiento, J. L., Hughes, T. M. C., Stouffer, R. J. and Manabe, S. 1998. Simulated response of the ocean carbon cycle to anthropogenic climate warming. *Nature* **393**, 245–9.

Serreze, M. C., Barry, R. G. and McLaren, A. S. 1989. Seasonal variations in sea ice motion and effects on ice concentration in the Canada basin. *J. Geophys. Res.* **94** (C8), 10 955–70.

Smith, D. M. 1998. Recent increase in the length of the melt season of perennial Arctic sea ice. *Geophys. Res. Lett.* **25**, 655–8.

Thompson, D. J. W. and Wallace, J. M. 1998. The Arctic oscillation signature in wintertime geopotential height and temperature fields. *Geophys. Res. Lett.* **25**, 1297–300.

Vinje, T. 1989. An upward looking sonar ice draft series. In Axelsson, K. B. E. and Franssom, L. A., eds., *Proceedings of the 10th International Conference on Port & Ocean Engineering Under Arctic Conditions*, vol. 1. Luleå University of Technology, pp. 178–87.

  2000. Anomalies and trends of sea ice extent and the atmospheric circulation in the nordic seas during the period 1864–1998. *J. Climate* **14** (3), 255–67.

Vinje, T., Nordlund, N. and Kvambekk, A. 1998. Monitoring ice thickness in the Fram Strait. *J. Geophys. Res.* **103** (C5), 10 437–49.

Vinnikov, K. *et al.* 1999. Global warming and northern hemisphere sea ice extent. *Science* **286**, 1934–37.

Wadhams, P. 1981. Sea ice topography of the Arctic Ocean in the region 70°W to 25°E. *Phil. Trans. Roy. Soc., Lond.* **A302** (1464), 1445–85.

    1983. Sea ice thickness distribution in the Fram Strait. *Nature* **305**, 108–11.

    1989. Sea-ice thickness in the trans polar drift stream. *Rapp. P-v Reun Cons. Int. Explor. Mer* **188**, 59–65.

    1990a. Ice thickness distribution in the Arctic Ocean. In Murthy T. K. S. *et al.*, eds., *Ice Technology for Polar Operations.* Southampton, Computational Mechanics Publications.

    1990b. Evidence for thinning of the Arctic ice cover north of Greenland. *Nature* **345**, 795–7.

    1992. Sea ice thickness distribution in the Greenland Sea and Eurasian basin, May 1987. *J. Geophys. Res.* **97** (C4), 5331–48.

Wadhams, P. and Davis, N. R. 2000. Further evidence for ice thinning in the Arctic Ocean. *Geophys. Res. Lett.* **27** (24), 3973–5.

Wadhams, P. A., McLaren, S. and Weintraub, R. 1985. Ice thickness distribution in Davis Strait in February from submarine sonar profiles. *J. Geophys. Res.* **90** (C1), 1069–77.

Walsh, J. E. 1995. Long-term observations for monitoring of the cryosphere. *Climatic Change* **31**, 369–94.

Walsh, J. E. and Chapman, W. L. 2001. Twentieth-century sea ice variations from observational data. *Ann. Glaciol.* **33**, 444–8.

Zakharov, V. F. 1997. Sea ice in the climate system. World Climate Research Programme/Arctic Climate System Study, WMO (TD-no. 782).

9

Sea-ice modelling

GREGORY M. FLATO
Canadian Centre for Climate Modelling and Analysis,
Meteorological Service of Canada

Representing the mass balance of sea ice involves solving a coupled, non-linear set of equations describing ice motion (momentum balance), thermodynamic growth and melt (energy balance) and the transport and redistribution of ice thickness (area and volume conservation). A detailed description of the theory underlying sea-ice models is provided in Chapter 7, and so only a brief review is provided here. It is the case that approximations employed to represent many important processes, both in stand-alone ice models and in more comprehensive global climate models, lead to errors or uncertainties in simulated sea-ice behaviour. Nonetheless, because of the scarcity of direct observations related to sea-ice mass balance, models provide valuable insight into the mean state of the ice cover and its historical and projected future changes.

9.1 Brief overview of sea-ice models

The sea-ice mass balance involves local growth and melt, horizontal transport and deformation. These processes alter the local mean thickness (ice volume per unit area) and involve exchanges of mass (fresh water) and energy with the atmosphere and ocean. Figure 9.1 provides an illustration, for the Arctic, based on results of a stand-alone sea-ice model (Hilmer, Harder and Lemke, 1998). In this figure the annual mean ice transport is indicated by the vectors, the annual mean ice thickness by the solid contours and the net freezing rate (net ice growth minus ice melt – directly proportional to the salt flux delivered to the ocean surface) by the coloured shading. The general pattern is anticyclonic circulation within the Arctic basin and outflow of ice through the Fram Strait that is balanced by net ice growth over much of the basin. The mean transport pattern leads to convergent deformation and thickening along the Canadian Arctic islands and Greenland, with divergence and correspondingly thin ice along the central Eurasian coast. For Antarctica, the pattern (not shown)

I thank Jonathan Gregory for providing the data used to construct Figures 9.9 and 9.10. I also thank Michael Hilmer for providing Figures 9.1, 9.5 and 9.8, Markus Harder for Figure 9.2, and Greg Holloway for Figure 9.7. I also thank the contributors to CMIP for providing the data used to construct Figures 9.3 and 9.4. Finally I thank Francis Zwiers, François Primeau and John Walsh for helpful comments.

Mass Balance of the Cryosphere: Observations and Modelling of Contemporary and Future Changes, eds. Jonathan L. Bamber and Antony J. Payne. Published by Cambridge University Press. © Cambridge University Press 2003.

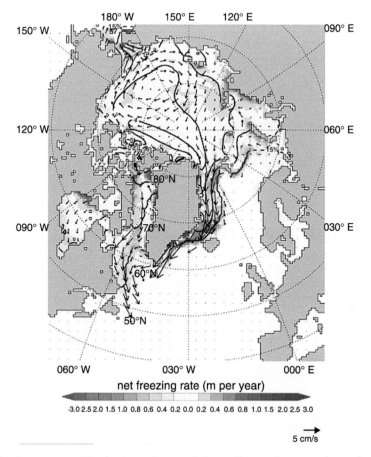

Figure 9.1. Components of the Arctic sea-ice mass balance. Vectors show annual mean ice transport; the solid contours (contour interval of 1 m) show annual mean ice thickness; and the coloured shading indicates net freezing rate (net ice growth minus melt – positive values correspond to a net salt flux to the ocean). (Figure provided by M. Hilmer; Hilmer *et al.* (1998).)

is one of net divergence, and hence net ice growth, around much of the continent (with net melt further out in the Southern Ocean), and rather thinner ice in general.

Representing features of the sort illustrated in Figure 9.1 in either a stand-alone sea-ice model or in a global climate model involves the following:

- solution of a momentum equation to obtain the ice velocity field;
- solution of a thermodynamic equation to obtain net ice growth or melt;
- solution of conservation equations that includes ice transport and deformation along with the thermodynamic sources and sinks.

These model components will be discussed briefly in turn. More comprehensive reviews are available in Chapter 7, Fichefet, Goosse and Morales Maqueda (1998), Hibler (1986), Hibler and Flato (1992) and Steele and Flato (2000).

9.1.1 Momentum equation

The balance of horizontal forces on a parcel of ice may be represented as

$$\rho_i h \frac{\partial u}{\partial t} = -\rho_i h \mathbf{k} \times \mathbf{u} + \tau_a + \tau_o - \rho_i g h \nabla \eta + \nabla \cdot \sigma, \tag{9.1}$$

where ρ_i is the ice density, h is the mean thickness (ice volume per unit area), $\mathbf{u}$ is the ice velocity vector, t is time, $\mathbf{k}$ is a unit vertical vector, τ_a is the wind stress applied on the ice surface, τ_o is the ocean stress applied on the ice underside, g is the acceleration due to gravity, η is the ocean surface elevation and σ is the internal ice stress tensor (a non-linear function of the ice state and strain or strain rate). Note that in this equation the non-linear momentum advection term has been neglected (on the basis of a scaling argument). In addition, the ice is considered to behave as a two-dimensional continuum. This assumption is harder to justify, especially at spatial scales of tens of kilometres or less where pack ice behaves more as a collection of semi-rigid 'floes' or 'plates' (e.g. Moritz and Stern, 2001; Moritz and Ukita, 2000; Thorndike, 1987); however, as in soil mechanics, the continuum assumption is plausible at scales much larger than a typical ice floe, which is the case for the numerical grids employed by most of the models whose results will be discussed here.

The acceleration term on the left hand side of equation (9.1) is typically negligible for timescales longer than a few hours. It is, of course, crucial if one wishes to resolve the inertial oscillations apparent in high frequency ice motion, especially in the Antarctic (e.g. Geiger, Hibley and Ackley, 1998). Heil and Hibler (2002) describe a modification to equation (9.1) to account for the 'added mass' of the upper ocean in these oscillations, but this has not been included in the large-scale models considered here. In some circumstances, the deformation associated with inertial oscillations may contribute to the sea-ice mass balance (via additional opening and rafting/ridging), but this contribution is likely to be small at the basin scale.

The first three terms on the right hand side of equation (9.1) are the Coriolis force and the body forces due to wind drag on the upper surface and water drag on the ice underside. These latter terms are typically represented by a bulk boundary layer parameterization. The fourth term is the (generally small) force associated with tilt of the sea surface.

The last term in equation (9.1) is the force arising from gradients in the internal ice stress field; that is, the force resulting from the resistance of sea ice to deformation. This term poses one of the main challenges in sea-ice modelling. In general, the stress state depends on the ice state (e.g. thicker ice is stronger and so able to support larger stresses than thin ice) and on the extent to which the ice is being deformed. The relation between stress and strain or strain rate is termed 'rheology', and there have been several rheologies used in sea-ice modelling.

Early efforts represented the ice as a linear viscous fluid wherein stress is linearly dependent on strain rate (e.g. Campbell, 1965; Hibler, 1974); however, such an approximation fails to reproduce the differences in deformation between the central pack and the near-shore regions. An elastic–plastic rheology (in which the stress depends linearly on strain until it reaches a specified failure or 'yield' strength) was proposed by Pritchard (1975) and used successfully by Pritchard, Coon and McPhee (1977) in a short-term, regional application. The necessity of resolving elastic waves and employing a Lagrangian grid renders the elastic–plastic scheme somewhat undesirable for long-term climate studies, although some of these shortcomings have been addressed recently (Polyakov *et al.*, 1998). The viscous–plastic model, introduced by Hibler (1979), grew out of these earlier attempts. Like the elastic–plastic rheology, it requires a specification of the plastic yield strength which, for mathematical convenience, was represented by an elliptical yield curve (see Hibler (1979) for a definition and further details). The difference is that pre-yield stress states are assumed to be linearly related to strain rate, as in the viscous rheology. It should be noted that there is no implication that sea ice behaves as a viscous fluid – indeed the pre-yield viscosities are somewhat arbitrary and are chosen in practice so that non-yielding ice is 'nearly rigid' (much as in the elastic–plastic case). The advantage of the viscous closure is that it permits an efficient numerical scheme well suited to large-scale, long-term simulations. The recent 'elastic–viscous–plastic' scheme of Hunke and Dukowicz (1997) is a variation on Hibler's scheme, and, while not really a different rheology (the elastic term is added strictly as a numerical artifice to allow an alternative and more efficient time-stepping scheme), it is an approach that is becoming widely used. Other authors have explored alternative yield formulations (alternative yield curve shapes or assumptions regarding the relative orientation of stress and strain rate vectors – the so-called 'flow rule'). Examples here include the cavitating fluid rheology and its extension to a Mohr–Coulomb triangular yield curve (Flato and Hibler, 1992), and the later generalization of the Mohr–Coulomb case to include dilation effects (Tremblay and Mysak, 1997). All of the rheologies discussed so far (and indeed all of those currently in use in large-scale models) assume that the ice cover is isotropic (i.e. material properties do not depend on direction). However, deformation features such as leads and ridges typically do have a preferred orientation, and so recent work by Coon *et al.* (1998) and Hibler and Schulson (2000) has focussed on the development of anisotropic rheologies that account for their effects.

9.1.2 Thermodynamics

Thermodynamic growth and melt involves the energy balance at the upper surface and underside of the ice cover, along with internal heat conduction and storage. Ice growth or melt at the ice underside results from the difference between heat conducted away from the boundary into the ice and the heat flux supplied from the ocean. At the upper surface, ice can form by submergence when the snow burden is sufficiently large; however, this surface more typically determines the heat lost from the ice in the winter (in which case the insulating effect of snow plays an important role) and the heat gained in summer (which directly causes melt when the surface reaches the melting temperature).

Because of the vast difference between typical horizontal and vertical length-scales, sea-ice thermodynamics is generally regarded as a one-dimensional (vertical) problem. Maykut and Untersteiner (1971) provided the first comprehensive numerical model of one-dimensional sea-ice thermodynamics, and, although their numerical method was somewhat unwieldy, their model set the stage for those that followed. In particular, the simplifications introduced by Semtner (1976) are still widely used today. One-dimensional models, expanded to include multiple ice types and sophisticated parameterizations of surface processes, have been employed by Björk (1997), Ebert and Curry (1993) and Schramm *et al.* (1997) in a variety of processes using periodic, climatological forcing. Flato and Brown (1996) employed a one-dimensional thermodynamic model to study inter-annual variability in land-fast ice thickness, while Bitz and Lipscomb (1999) developed a model with a refined representation of internal heat storage (in brine pockets). Several studies have shown that assuming a linear temperature profile through the ice (i.e. not resolving the internal temperature) leads to a distortion of the seasonal cycle; however, they also show that rather few vertical layers are required to resolve the temperature profile adequately. What has become clear from work like that of Shine and Henderson-Sellers (1985) is the importance of properly representing the evolution of surface albedo during the melt season – a process still represented rather crudely in most models.

There is a large body of literature on sea-ice thermodynamics that cannot be covered adequately here; however, a recent review of thermodynamic models, their results and sensitivity analysis is available in Steele and Flato (2000).

9.1.3 Conservation equations

Sea ice moves over the surface of the ocean as a compressible material with sources/sinks of volume provided by thermodynamic evolution, and sources/sinks of area provided by divergence or convergence in the motion field. The minimum requirement is therefore a conservation equation for volume (usually in terms of mean ice thickness, h) and one for area (usually in terms of concentration, A), i.e.

$$\frac{\partial h}{\partial t} = -\nabla \cdot (\mathbf{u}h) + S_h, \tag{9.2}$$

$$\frac{\partial A}{\partial t} = -\nabla \cdot (\mathbf{u}A) + S_A, \tag{9.3}$$

where S_h and S_A are thermodynamic source terms for mean thickness and concentration, respectively. In equation (9.3) one must also enforce the constraint $A \leq 1$ (or include an additional term to guarantee this, e.g. Gray and Morland (1994)) to represent crudely the thickness buildup by ridging during convergent deformation. This is the 'two-category' representation of sea ice that was used in the model of Hibler (1979) and many others (in the literature this is often referred to as a 'two-level' model). In this case the ice strength must be parameterized in terms of h and A only, and an approach introduced by Hibler (1979) remains the most widely used, namely

$$p = p^*h \exp[-C(1 - A)], \tag{9.4}$$

where p^* and C are parameters often taken to be 27.5 kN/m^2 and 20, respectively.

The two-category approach ignores several important features of sea ice that are important in its mass balance, such as the non-linear effect of small-scale variations in ice thickness on thermodynamic growth (e.g. Maykut, 1978). Indeed, observations such as those made from submarine sonar indicate that a range of thickness, from open water through nearly level first- and multi-year ice, to ridged ice 10 m thick or more, may be encountered in a transect of only one kilometre in length. A framework for describing this thickness variability was proposed by Thorndike *et al.* (1975), making use of the so-called thickness distribution function, $\hat{g}(h)$. This function is analogous to the probability density function for ice thickness that evolves according to

$$\frac{\partial \hat{g}(h)}{\partial t} = -\nabla \cdot (\mathbf{u}\hat{g}) - \frac{\partial(F\hat{g})}{\partial h} + L + \psi, \tag{9.5}$$

where F is the thermodynamic growth rate, L represents the lateral melt of ice and ψ is a 'redistribution function' that converts thin ice to thick, ridged ice during deformation. In such a formulation, energy losses associated with ridging can be accounted for in a more physically based parameterization of ice strength (Flato and Hibler, 1995; Rothrock, 1975) that is reasonably consistent with detailed ridging models (e.g. Hopkins, 1994; Hopkins and Hibler, 1991).

In practice, $\hat{g}(h)$ is typically discretized into a fixed number of thickness intervals, and so models of this sort are termed 'multi-category' models – an example of which appears in Hibler (1980). One of the principal uncertainties involves the ridge redistribution term that was explored in some detail by Flato and Hibler (1995). There are also some numerical difficulties that have been addressed recently by Bitz *et al.* (2001) and Lipscomb (2001). As a result, some global climate models are now implementing models based on this 'multi-category' formulation.

9.2 Mean thickness

Estimates of the mean spatial pattern of sea-ice thickness are available from many stand-alone models for both the Arctic and Antarctic. The details depend directly on the atmospheric forcing fields, dynamic and thermodynamic parameterizations, and the time period simulated. Figure 9.2 shows the mean March Arctic ice-thickness fields produced by four different models subjected to the same observationally based atmospheric forcing over the period 1986 to 1992. These results, obtained as part of the Sea Ice Model Intercomparison Project (SIMIP; Kreyscher, Harder and Lemke, 1997; Kreyscher *et al.*, 2000) illustrate the sensitivity of the simulated ice-thickness field to the representation of ice rheology. Three of the four rheologies (viscous–plastic, cavitating fluid and 'stoppage') are used in

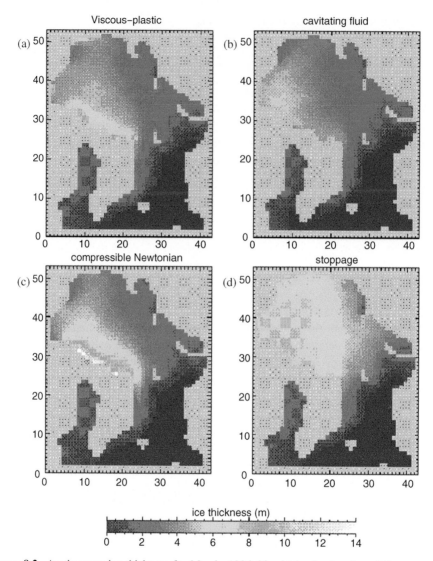

Figure 9.2. Arctic mean ice thickness for March, 1986–92, obtained using four different sea-ice models: (a) viscous–plastic; (b) cavitating fluid; (c) compressible Newtonian fluid; (d) 'stoppage' scheme, in which internal ice stress is zero when the ice thickness is less than 3 m and the ice is motionless when its thickness exceeds 3 m. See Kreyscher *et al.* (1997, 2000) for further details.

contemporary global climate models. Although the differences are substantial, all of the models exhibit a pattern of thicker ice in the North American portion of the basin, thinning toward Eurasia, especially western Eurasia, in qualitative accord with submarine observations. Kreyscher *et al.* (2000) performed a number of comparisons to available observations (thickness, velocity, etc.) and found that the viscous–plastic model performs best, followed by the cavitating fluid model. However, it should also be noted that ice thickness, and other sea-ice quantities, are sensitive to a variety of rather poorly known parameters, allowing some opportunity for optimization ('tuning'). Such sensitivity will be discussed further in section 9.2.3.

Global climate models that include prognostic sea-ice calculations suffer not only from errors and approximations in the sea-ice component, but also from errors in the forcing provided by their atmospheric and ocean components. Many of the global climate models used to date have ignored ice motion and have treated only sea-ice thermodynamics. The Coupled Model Intercomparison Project (CMIP) affords the opportunity to compare the sea-ice simulations with several current climate models (Meehl *et al.*, 1997). Figures 9.3 and 9.4 show composite results from a subset of the CMIP model sea-ice simulations. Figures 9.3(a) and 9.4(a) display winter-time ice-cover results. The lightest shading indicates the region where less than half of the models have ice; the darkest shading indicates where more than 90% of the models have ice. In both hemispheres the 50% contour (the median ice boundary) agrees reasonably well with the observed ice edge (e.g. Gloersen *et al.*, 1992), although it should be noted that two models whose output indicated no ice in the southern hemisphere were not included. The difference between the 10% and 90% contours gives an indication of the range of modelled ice coverage. Figures 9.3(b) and 9.4(b) display the winter-time mean ice thickness averaged over twelve models in the case of the northern hemisphere and ten models in the case of the southern hemisphere. Figures 9.3(c) and 9.4(c) show the inter-model standard deviation as a measure of the range of thicknesses produced by these climate models. The northern hemisphere mean thickness is somewhat less than that inferred from observations, with maximum values of just over 2.5 m. In addition, the spatial pattern does not exhibit the pronounced increase toward North America and Greenland that is seen in Figure 9.2. This may be largely a consequence of the fact that only seven of the CMIP models include any representation of ice motion, and only four of these include a prognostic solution of the sea-ice momentum equation (the remaining three use a variant of the 'stoppage' scheme mentioned above). In addition, recent work by Bitz, Fyfe and Flato (2002) illustrates the important role of errors in GCM wind fields in simulated ice-thickness patterns for models that include ice dynamics.

9.2.1 Spatial and temporal variability

Sea ice is predominantly a 'driven' system. Although its dynamics include a non-linear rheology, with important consequences (see Chapter 7), the resulting mass balance components are largely deterministic given prescribed atmospheric and oceanic forcing. Therefore, sea-ice variability is primarily a consequence of the relatively high frequency variability imparted by the atmosphere and the relatively lower frequency variability imparted by the ocean. Of these, atmospherically driven variability has been much better studied.

Spatial variations in ice motion result in deformation which, over timescales up to a few days, dominate the local opening and closing so important to ice growth. This high

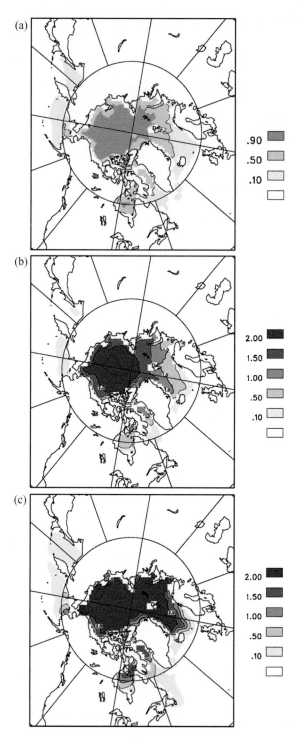

Figure 9.3. Arctic winter-time (December–February) sea-ice results from the coupled model inter-comparison project (CMIP). (a) 10, 50 and 90 percentile ice-cover boundaries obtained from 14 of the participating models. (b) Model ensemble mean ice thickness and (c) inter-model thickness standard deviation obtained from 12 of the participating models.

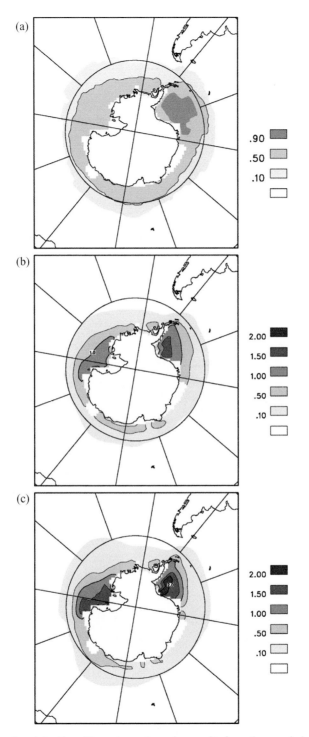

Figure 9.4. Antarctic winter-time (June–August) sea-ice results from the coupled model intercomparison project (CMIP). (a) 10, 50 and 90 percentile ice-cover boundaries obtained from 12 of the participating models. (b) Model ensemble mean ice thickness and (c) inter-model thickness standard deviation obtained from ten of the participating models.

frequency motion includes synoptic forcing from the atmosphere and, in some locations, important contributions from tidal and inertial oscillations (e.g. Geiger *et al.*, 1998; Heil and Hibler, 2002). At longer timescales, ice transport variations lead to large-scale variability in ice thickness and net freezing rate. The forcing in this case is related to large-scale modes of variability in the atmosphere (such as the North Atlantic oscillation, NAO, and the Arctic Oscillation, AO), and, for the Arctic at least, this kind of ice-motion variability has received considerable attention recently (e.g. Maslowski *et al.*, 2000; Polyakov, Proshutinsky and Johnson, 1999; Proshutinsky and Johnson, 1997; Proshutinsky, Polyakov and Johnson, 1999; Zhang, Rothrock and Steele, 2000). The picture that emerges is one of variations between a large Beaufort gyre (much as in Figure 9.1) that spans most of the western Arctic, and a much smaller gyre confined to the Beaufort Sea proper.

Arctic ice-thickness variability, forced primarily by variations in wind-driven deformation and transport, appears to have a characteristic pattern, illustrated in Figure 9.5 (from Hilmer and Lemke (2000)). This figure shows the inter-annual standard deviation of ice thickness

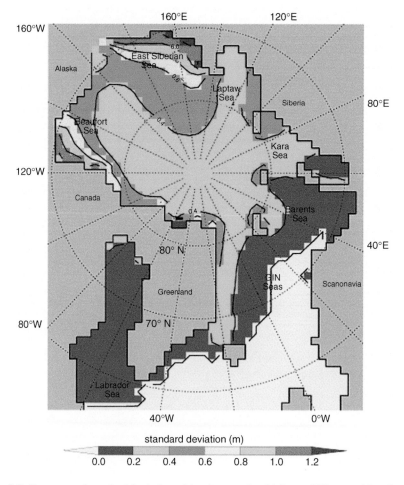

Figure 9.5. Inter-annual standard deviation of Arctic mean ice thickness (Hilmer and Lemke, 2000).

obtained from a 40-year ice model simulation. A similar result, using a different model, was obtained by Flato (1995). The variability is largest in the Beaufort and East Siberian Seas, and appears to result primarily from simultaneous thick anomalies in one region and corresponding thin anomalies in the other. Zhang *et al.* (2000) term this the 'east–west Arctic anomaly pattern' (EWAAP).

Temporal variability of Arctic ice volume cannot, so far, be estimated from observations, but it has been examined in various model studies. Arctic basin-scale studies, using observationally based forcing fields spanning several decades, have been performed (e.g. Chapman *et al.*, 1994; Flato, 1995; Häkkinen, 1993; Hilmer and Lemke, 2000; Walsh, Hibler and Ross, 1985), as have century or longer simulations with one-dimensional models (e.g. Bitz *et al.*, 1996; Häkkinen and Mellor, 1990; Holland and Curry, 1999). In either case, the simulated ice volume varies predominantly on inter-decadal and longer timescales. Some studies have concluded that variations in thermodynamic forcing (pole-ward heat transport by the atmosphere) dominate, while others have pointed toward variability in wind-driven deformation. Thermodynamic forcing is amplified by positive albedo feedbacks, which are rather sensitive to details of the model parameterizations (Battisti, Bitz and Moritz, 1997). On the other hand, ridging processes, creating thick ice that survives for many years, may act to damp out forced variability (Holland and Curry, 1999). Figure 9.6 illustrates this effect with the temporal auto-correlation function of simulated Arctic ice volume from a model that included an explicit representation of the thickness distribution function (Flato, 1995). One sees that the correlation timescale for thin ice (<1 m) is approximately two months, for medium ice (2 to 5 m) the timescale is about 20 months, whereas for the total ice volume, the timescale is roughly seven years. The reason is that thick, ridged ice (which accounts for nearly half of the simulated ice volume) is able to survive many summer melt seasons before ultimately being exported from the Arctic.

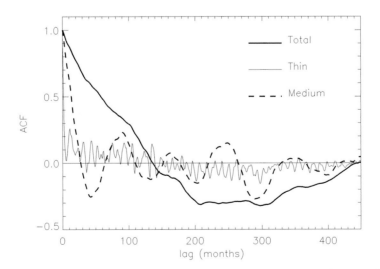

Figure 9.6. Auto-correlation functions (ACF) of total, medium (2 to 5 m) and thin (<1 m) ice Arctic sea-ice volume anomaly time series (Flato, 1995).

Knowledge of temporal and spatial variability in ice thickness is important in interpreting rather sparse observations from submarines. Wadhams (1990) compared submarine observations from 1976 and 1987 which showed a 15% decrease in ice thickness in a region north of Greenland. From Figure 9.5 one sees that a model-based estimate of the inter-annual standard deviation near the north Greenland coast is in excess of 0.4 m (where the modelled and observed mean thickness is roughly 6 m). So, sampling variability provides one 'explanation' for the observed decrease and is, in any case, the null hypothesis to be rejected. In a more recent comparison of submarine observations over much of the central Arctic, Rothrock, Yu and Maykut (1999) found a roughly 40% decrease in thickness between the 1960s and 1990s averaged over 29 locations where a comparison could be made. In a recent study (Holloway and Sou, 2002), the ice thickness obtained from a model run over the same period was sampled in the same way as the observations. The thickness change at the sample locations agreed well with the observed change, but, interestingly, the overall volume change was only about 12% (broadly commensurate with the observed 3% per decade decline in Arctic ice extent). Figure 9.7 displays the modelled ice-thickness

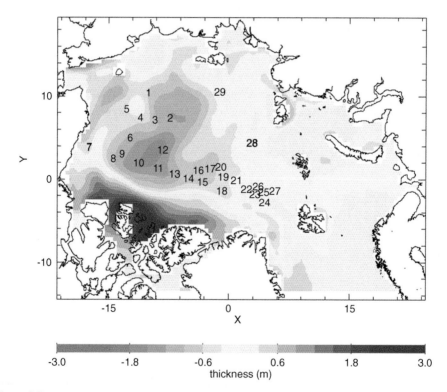

Figure 9.7. Change in modelled ice thickness obtained as the average of years 1993, 1996 and 1997, minus 1958, 1960, 1962, 1970 and 1976. The numbers indicate the locations of the submarine-derived differences used by Rothrock *et al.* (1999). (Figure provided by G. Holloway (Holloway and Sou, 2002).)

change over the period compared by Rothrock *et al.* (1999); it shows that the spatial distribution of thickness changed substantially (in a manner consistent with Figure 9.5), and that the submarine observations coincidentally sampled only the region that experienced thinning.

9.2.2 Ice export

In both the Arctic and the Antarctic, net ice growth is, over the long term, balanced by transport. In the Antarctic ice is generally transported northward away from the continent, and thus the export is not localized (e.g. Stössel, Lemke and Owens, 1990). Antarctic model studies have primarily focussed on ice export in the Weddell Sea (e.g. Harder and Fischer, 1999). In the Arctic most export passes through the Fram Strait, and several model studies have focussed on Fram Strait export and its variability. An example, based on Hilmer *et al.* (1998), is provided in Figure 9.8 and shows a time series of simulated annual mean ice export anomalies from 1958 to 1997. Studies like this have shown that ice export variability is primarily driven by local wind forcing which, by virtue of proximity to the Icelandic low, is connected to the North Atlantic oscillation. Indeed, Kwok and Rothrock (1999) demonstrate a significant correlation between the NAO index and observationally based estimates of Fram Strait ice area flux. However, a recent model study by Hilmer and Jung (2000) indicates that the correlation between the NAO and Fram Strait ice export was considerably weaker in the past, owing to rather subtle shifts in NAO-related wind patterns.

Because the export of sea ice from the Arctic is such an important source of fresh water to the North Atlantic, large anomalies in ice export may have profound impacts on ocean stratification and convection. A particular episode of anomalously large export, associated with the 'great salinity anomaly' of the late 1960s, has received particular attention (e.g. Dickson *et al.*, 1988). Model studies such as those of Häkkinen (1993) and Hilmer *et al.* (1998) have illustrated the role of local and regional atmospheric forcing in events such as this, and in low frequency outflow variability in general.

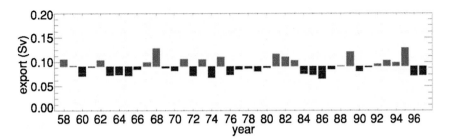

Figure 9.8. Simulated annual mean Fram Strait ice export anomalies. Export is measured in 'sverdrups' (sv), defined as 10^6 m^3/s. (From Hilmer *et al.* (1998).)

9.2.3 Sensitivity to model parameterizations

The modelled sea-ice mass balance components depend directly on assumptions, approximations and parameterizations of unresolved physical processes. Because many of the relevant parameters are poorly constrained by observations (and some, like geophysical-scale ice strength, are virtually impossible to measure directly), the uncertainties imbued on model variables have typically been assessed by conducting sensitivity studies (e.g. Holland, Mysak and Manak, 1993). In the case of ice or ice–ocean models, the results are of course also sensitive to errors in the prescribed atmospheric (and oceanic) forcing.

Perhaps the least constrained parameterizations involve representation of internal ice stresses. The resistance to deformation afforded by large-scale ice dynamics is a significant term in the sea-ice momentum balance (e.g. Steele *et al.*, 1997), and thus plays an important role in simulated ice motion (e.g. Flato and Hibler, 1992) and the resulting thickness buildup pattern illustrated in Figure 9.2. Resistance of sea ice to deformation is characterized by a 'yield curve' representing the locus of stress states separating nearly rigid behaviour from plastic flow (see Chapter 7 for details). The yield curve must be specified in a model, and the dependence of simulated thickness and motion statistics on yield curve shape has been investigated by Ip *et al.* (1991) and Kreyscher *et al.* (1997, 2000). The results so far have indicated that a plastic rheology with a yield curve similar to the ellipse proposed by Hibler (1979) compares most favourably with a range of observations. A closely related issue is the parameterization of ice strength. For the widely used 'two-category' representation of ice thickness, a parameterization based on mean thickness and concentration is used (equation (9.4)); in 'multi-category' models, a parameterization based on energy losses during ridging is used (Rothrock, 1975). The parameter values are typically chosen by comparison to buoy drift observations (e.g. Hibler and Walsh, 1982), but the two parameterizations produce significantly different results when compared directly in the same model (Flato, 1996). In multi-category models, the strength parameterization is closely coupled to the representation of ice ridging, further complicating efforts to optimize parameter values (Flato and Hibler, 1995).

Thermodynamic parameterizations also have a significant impact on modelled sea-ice mass balance components, particularly as they directly affect ice growth and melt. Various authors have performed sensitivity studies related to parameterizations involving surface albedo (e.g. Curry, Schramm and Ebert, 1995), thermal conductivity (e.g. Fichefet, Tartinville and Goosse, 2000), heat capacity (Bitz and Lipscomb, 1999), snow-cover effects (e.g. Fichefet and Morales Maqueda, 1999), the inclusion of multiple thickness categories (e.g. Schramm *et al.*, 1997) and the potential role of two-dimensional heat conduction effects in ridged ice (Schramm, Flato and Curry, 2000). Representation of lead thermodynamics, frazil ice formation,[1] multi-year ice metamorphosis and other potentially important processes remain rather poorly explored. A detailed review of all such studies is beyond the scope of this chapter; however, results from a recent study using

---

[1] Frazil ice forms in turbulent supercooled water and consists of small platelets roughly 1 mm in diameter. These platelets ultimately clump together and consolidate as cooling continues.

Table 9.1. *Importance of sea-ice processes in each hemisphere.*

Process	Northern hemisphere	Southern hemisphere
Internal heat storage	unimportant	unimportant
Brine pocket storage	important	unimportant
Penetrating short-wave radiation	important	unimportant
Sub-grid-scale variability in heat conduction	important	important
Snow cover	unimportant	important
Snow ice formation (by submergence)	unimportant	important
Lead formation	important	important
Ice motion/deformation	important	important
Resistance to shear	intermediate	unimportant

Adapted from Fichefet and Morales Maqueda (1997).

a global ice–upper ocean model (Fichefet and Morales Maqueda, 1997), summarized in Table 9.1, are instructive. Although by no means definitive, the use of a global model allows the relative importance of various processes (judged in a qualitative manner) to be compared for the northern and southern hemispheres.

Although only a subjective assessment from one model, the results in Table 9.1 indicate that sub-grid-scale variability in heat exchange, due to variations in ice thickness and snow cover, play an important role in both hemispheres, as do processes related to ice motion, deformation and lead formation. Because of the larger snow-fall rates and thinner ice, snow processes play a bigger role in the southern hemisphere ice cover. On the other hand, thicker, longer-lived ice in the northern hemisphere is more strongly affected by processes involving penetration of short-wave radiation into the ice, and the latent heat exchange between brine pockets and the surrounding ice.

9.3 Modelling future changes in sea-ice mass balance

Climate model simulations and recent observations agree that surface air temperature increases due to enhanced greenhouse gas forcing are largest at high latitudes. This pole-ward amplification of warming is apparent in projections of future climate change (e.g. Kattenberg *et al.*, 1996), and is in large part a consequence of sea-ice-related feedbacks (e.g. Rind *et al.*, 1995). Corresponding reductions in sea-ice thickness and extent are therefore anticipated. As an example, Figure 9.9 compares northern and southern hemisphere annual mean sea-ice extent simulated by two global climate models over the period 1900 to 2100 using the same historical and 'business as usual' future scenario for greenhouse gas and aerosol forcing. The curve labelled CGCM2 refers to the second-generation model of the Canadian Centre for Climate Modelling and Analysis (Flato and Boer, 2001); the curve

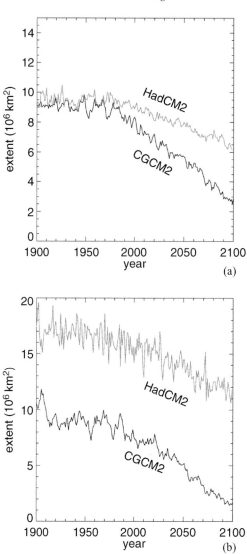

Figure 9.9. Simulated annual mean sea-ice extent for the northern (a) and southern (b) hemispheres. Results from CGCM2 and HadCM2 are shown.

labelled HadCM2 refers to the second-generation model of the Hadley Centre for Climate Prediction and Research (Johns *et al.*, 1997). Both models include a representation of sea-ice motion (cavitating fluid in the case of CGCM2 and 'stoppage' in the case of HadCM2). Both models under-estimate the historical northern hemisphere sea-ice extent (observed to be roughly 13×10^6 km^2), but are split on southern hemisphere extent (observed to be roughly 14×10^6 km^2). The results shown in Figures 9.3 and 9.4 indicate that errors of this magnitude are not uncommon in current-generation climate models, owing largely to the

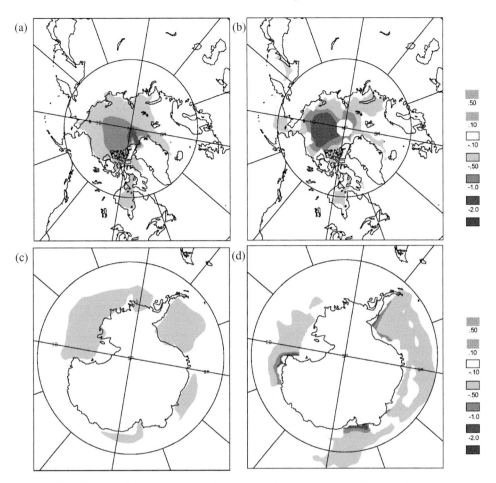

Figure 9.10. Simulated change in winter-time sea-ice thickness in the northern and southern hemi-spheres. (a), (c) Results from CGCM2. (b), (d) Results from HadCM3.

sensitivity of ice extent to the subtle balance of heat in the atmosphere and ocean near the ice edge. Nevertheless, both models are in general agreement with the observed northern hemisphere decline of 3% per decade since 1978 (e.g. Cavalieri *et al.*, 1997), although CGCM2 projects a much more rapid future decline than does HadCM2. Vinnikov *et al.* (1999) provide more detailed analysis of results like these for the northern hemisphere. In the southern hemisphere, the two models agree more closely with regard to the future trend if one removes their respective historical biasses.

Projected changes in ice thickness can also be obtained from global models. Figure 9.10 displays the change in thickness between 1971 and 1990 and 2041 and 2060 for winter-time in both hemispheres (December to February in the northern hemisphere and June to July in the southern hemisphere). Results on the left hand side of the figure are from CGCM2, described above, while those on the right are from a more recent version of the Hadley

Centre model, HadCM3 (Gordon *et al.*, 2000). The two models agree reasonably well in terms of both the magnitude and spatial pattern of the thickness change, with values greater than 0.5 m over the central Arctic and less than 0.5 m over almost all of the Antarctic ice pack.

The model results discussed above project a continuing decline in ice thickness and extent as the climate warms – persistent negative mass balance. Of course, these are only two of many available climate models, and, while they agree qualitatively, the magnitude of the projected decline and its spatial pattern differ. The results shown earlier in Figures 9.3 and 9.4 indicate substantial variation in coupled climate model simulations of contemporary sea-ice extent and thickness, and this inter-model spread reflects the cumulative effect of differences in a whole range of model details (atmosphere, ocean and sea ice). These differences also affect the response of modelled sea ice to climate forcing. That is, a model projection of future climate change depends in part on the response of the sea-ice component to climatic perturbations (as distinct from the response of the mean state to model parameters).

Various model studies have attempted to estimate the potential effect of sea-ice parameterizations on climate sensitivity (e.g. Arbetter, Curry and Maslanik, 1999; Flato, 1998; Hibler, 1984; Holland, 1998; Pollard and Thompson, 1994). Although the experiments differ substantially in terms of model domain, applied perturbation and variable(s) analysed, the general impression is that including dynamics tends to decrease sea-ice sensitivity to perturbations, whereas including a representation of the thickness distribution increases or decreases the sensitivity depending on the details of the experiment performed and the particular diagnostic quantity used to measure the model's response. In any case, it remains difficult to anticipate the impact of a change in sea-ice parameterization once all the feedbacks of a coupled ocean–ice–atmosphere model are included. Results of model inter-comparison studies will continue to be valuable in this regard.

9.4 Summary and conclusions

Numerical models encapsulate much of what is known about the physics of sea ice, and thereby provide a means of investigating the role of various processes. In addition, when driven by observationally based atmospheric and oceanic forcing, such models provide estimates of quantities not easily observable. Finally, when used in the context of global climate simulations, such models provide projections of future changes in sea-ice conditions and behaviour.

Stand-alone models are able to produce credible estimates of the spatial pattern and temporal evolution of sea-ice thickness, its transport and deformation, and net growth and melt. In many cases, these mass balance components are poorly observed, and so model results may aid in the interpretation of the sparse observations that are available. The potential for a more formal combination of observations and models, via data assimilation techniques, has not so far been realized, although some initial attempts have been made (e.g. Meier, Maslanik and Fowler, 2000; Thomas *et al.*, 1996). Sea-ice mass balance estimates using data assimilation methods are likely to become more widespread in the future. Data assimilation could also be used to improve estimates of various model parameters, particularly those related to geophysical-scale mechanical properties which are otherwise virtually impossible to measure directly.

Global climate models necessarily include a sea-ice component, although often employing parameterizations that are less sophisticated than those in 'state-of-the-art' stand-alone models. In the case of climate models, the discrepancies between modelled and observed sea-ice features, and differences from one model to another, are due not only to errors and uncertainties in the sea-ice component, but also to errors in the atmospheric, oceanic and even terrestrial components of the model. Many global modelling centres have or are introducing substantial improvements into the sea-ice component of their climate models, in particular through the use of more sophisticated treatments of sea-ice dynamics and multi-layer formulations of vertical heat conduction. The on-going atmospheric and coupled model inter-comparison projects will allow the impact of these improvements to be assessed.

Climate model projections generally indicate enhanced warming at high latitudes, with a corresponding decline in ice extent and thickness. However, the rate of decline and the strength of the various feedbacks that enhance or damp the warming response are highly uncertain. Observations aimed at testing and improving sea-ice parameterizations, and at evaluating modelled trends and variability, will continue to be important in refining projections of future changes in the sea-ice mass balance.

References

Arbetter, T. E., Curry, J. A. and Maslanik, J. A. 1999. Effects of rheology and ice thickness distribution in a dynamic-thermodynamic sea ice model. *J. Phys. Oceanography* **29**, 2656–70.

Battisti, D. S., Bitz, C. M. and Moritz, R. E. 1997. Do general circulation models underestimate the natural variability in the Arctic climate? *J. Climate* **10**, 1909–20.

Bitz, C. M. and Lipscomb, W. H. 1999. An energy-conserving thermodynamic model of sea ice. *J. Geophys. Res.* **104**, 15 669–77.

Bitz, C. M., Fyfe, J. C. and Flato, G. M. 2002. Sea ice response to wind forcing from AMIP models. *J. Climate* **15**, 522–36.

Bitz, C. M., Battisti, D. S., Moritz, R. E. and Besley, J. A. 1996. Low-frequency variability in the Arctic atmosphere, sea ice, and upper-ocean climate system. *J. Climate* **9**, 394–408.

Bitz, C. M., Holland, M. M., Weaver, A. J. and Eby, M. 2001. Simulating the ice-thickness distribution in a coupled climate model. *J. Geophys. Res.* **106**, 2441–64.

Björk, G. 1997. The relation between ice deformation, oceanic heat flux, and the ice thickness distribution in the Arctic Ocean. *J. Geophys. Res.* **102**, 18 681–98.

Campbell, W. J. 1965. The wind-driven circulation of ice and water in a polar ocean. *J. Geophys. Res.* **70**, 3279–301.

Cavalieri, D. J., Gloersen, P., Parkinson, C. L., Comiso, J. C. and Zwally, H. J. 1997. Observed hemispheric asymmetry in global sea ice changes. *Science* **278**, 1104–6.

Chapman, W. L., Welch, W. J., Bowman, K. P., Sacks, J. and Walsh, J. E. 1994. Arctic sea ice variability: model sensitivities and a multi-decadal simulation. *J. Geophys. Res.* **99**, 919–35.

Coon, M. D., Knoke, G. S., Echert, D. C. and Pritchard, R. S. 1998. The architecture of an anisotropic elastic-plastic sea ice mechanics constitutive law. *J. Geophys. Res.* **103**, 21 915–25.

Curry, J. A., Schramm, J. L. and Ebert, E. E. 1995. Sea ice-albedo climate feedback mechanism. *J. Climate* **8**, 240–7.

Dickson, R. R., Meinke, J., Malmberg, S. A. and Lee, A. J. 1988. The 'great salinity anomaly' in the northern North Atlantic 1968–1982. *Prog. Oceanography* **20**, 103–51.

Ebert, E. E. and Curry, A. J. 1993. An intermediate one-dimensional thermodynamic sea ice model for investigating ice-atmosphere interactions. *J. Geophys. Res.* **98**, 10 085–110.

Fichefet, T. and Morales Maqueda, M. A. 1997. Sensitivity of a global sea ice model to the treatment of ice thermodynamics and dynamics. *J. Geophys. Res.* **102**, 12 609–46.
   1999. Modelling the influence of snow accumulation and snow-ice formation on the seasonal cycle of the Antarctic sea-ice cover. *Climate Dyn.* **15**, 251–68.

Fichefet, T., Goosse, H. and Morales Maqueda, M. A. 1998. On the large-scale modelling of sea ice and sea ice–ocean interactions. In Chassignet, E. P. and Verron, J., eds., *Ocean Modeling and Parameterization*. Dordrecht, Kluwer, pp. 399–422.

Fichefet, T., Tartinville, B. and Goosse, H. 2000. Sensitivity of the Antarctic sea ice to thermal conductivity. *Geophys. Res. Lett.* **27**, 401–4.

Flato, G. M. 1995. Spatial and temporal variability of Arctic ice thickness. *Ann. Glaciol.* **21**, 323–9.
   1996. Parameterizing the strength of Arctic sea ice. *Proceedings of the ACSYS Conference on the Dynamics of the Arctic Climate System*, Göteborg, Sweden. November 7–10, 1994. World Climate Research Programme, WMO/TD no. 760, pp. 278–82.
   1998. The sensitivity of models in the SIMIP hierarchy to thermodynamic perturbations. *Proc of the ACSYS Conference on Polar Processes and Global Climate*, Orcas Island, USA, November 3–6, 1997. World Climate Research Programme, WMO/TD no. 908, pp. 51–3.

Flato, B. M. and Boer, G. J. 2001. Warming asymmetry in climate change simulations. *Geophys. Res. Lett.* **28**, 195–8.

Flato, G. M. and Brown, R. D. 1996. Variability and climate sensitivity of landfast Arctic sea ice. *J. Geophys. Res.* **101**, 25 767–77.

Flato, G. M. and Hibler, W. D. III. 1992. Modeling pack ice as a cavitating fluid. *J. Phys. Oceanography* **22**, 626–51.
   1995. Ridging and strength in modeling the thickness distribution of Arctic sea ice. *J. Geophys. Res.* **100**, 18 611–26.

Geiger, C. A., Hibler, W. D. III and Ackley, S. F. 1998. Large-scale sea ice drift and deformation: comparison between models and observations in the western Weddell Sea during 1992. *J. Geophys. Res.* **103**, 21 893–913.

Gloersen, P., Campell, W. J., Cavalieri, D. J., Comiso, J. C., Parkinson, C. L. and Zwally, H. J. 1992. Arctic and Antarctic sea ice, 1978–1987: satellite passive microwave observations and analysis. NASA SP-511, 290 pp.

Gordon, C. *et al.* 2000. The simulation of SST, sea ice extents and ocean heat transports in a version of the Hadley Centre coupled model without flux adjustments. *Climate Dyn.* **16**, 147–68.

Gray, J. M. N. T. and Morland, L. W. 1994. A two-dimensional model for the dynamics of sea ice. *Phil. Trans. Roy. Soc. London A* **347**, 219–90.

Häkkinen, S. 1993. An Arctic source for the great salinity anomaly: a simulation of the Arctic ice-ocean system for 1955–1975. *J. Geophys. Res.* **98**, 16 397–410.

Häkkinen, S, and Mellor, G. L. 1990. One hundred years of Arctic ice cover variations as simulated by a one-dimensional, ice-ocean model. *J. Geophys. Res.* **95**, 15 959–69.

Harder, M. and Fischer, H. 1999. Sea ice dynamics in the Weddell Sea simulated with an optimized model. *J. Geophys. Res.* **104**, 11 151–62.

Heil, P. and Hibler, W. D. III. 2002. Modeling the high-frequency component of Arctic sea-ice drift and deformation, *J. Phys. Oceanography* **32**, 3039–57.

Hibler, W. D. III. 1974. Differential sea ice drift. II: comparison of mesoscale strain measurements to linear drift theory predictions. *J. Glaciol.* **13**, 457–71.

    1979. A dynamic thermodynamic sea ice model. *J. Phys. Oceanography* **9**, 817–46.

    1980. Modeling a variable thickness sea ice cover. *Month. Weather Rev.* **108**, 1943–73.

    1984. The role of sea ice dynamics in modeling CO_2 increases. *Climate Processes and Climate Sensitivity*. Geophysical Monograph 29. American Geophysical Union, pp. 238–53.

    1986. Ice dynamics. In Untersteiner, N., ed., *Geophysics of Sea Ice*. New York, Plenum Press, pp. 577–640.

Hibler, W. D. III and Flato, G. M. 1992. Sea ice models. In Trenberth, K. E., ed., *Climate System Modeling*. Cambridge University Press, pp. 413–36.

Hibler, W. D. III and Schulson, E. M. 2000. On modelling the anisotropic failure and flow of flawed sea ice. *J. Geophys. Res.* **105**, 17 105–20.

Hibler, W. D. III and Walsh, J. E. 1982. On modeling seasonal and interannual fluctuations of Arctic sea ice. *J. Phys. Oceanography* **12**, 1514–23.

Hilmer, M. and Jung, T. 2000. Evidence for a recent change in the link between the North Atlantic oscillation and Arctic sea ice export. *Geophys. Res. Lett.* **27**, 989–92.

Hilmer, M. and Lemke, P. 2000. On the decrease of Arctic sea ice volume. *Geophys. Res. Lett.* **27**, 3751–4.

Hilmer, M., Harder, M. and Lemke, P. 1998. Sea ice transport: a highly variable link between Arctic and North Atlantic. *Geophys. Res. Lett.* **25**, 3359–62.

Holland, D. M., Mysak, L. A. and Manak, D. K. 1993. Sensitivity study of a dynamic thermodynamic sea ice model. *J. Geophys. Res.* **98**, 2561–86.

Holland, M. M. 1998. The impact of the ice thickness distribution on simulated Arctic budgets and climate. *Proceeding of the ACSYS Conference on Polar Processes and Global Climate*, Orcas Island, USA, November 3–6, 1997. World Climate Research Programme, WMO/TD no. 908, pp. 93–5.

Holland, M. M. and Curry, J. A. 1999. The role of physical processes in determining the interdecadal variability of central Arctic sea ice. *J. Climate* **12**, 3319–30.

Holloway, G. and Sou, T. 2002. Is Arctic sea ice rapidly thinning? *J. Climate* **15**, 1691–1701.

Hopkins, M. A. 1994. On the ridging of intact lead ice. *J. Geophys. Res.* **99**, 16 351–60.

Hopkins, M. A. and Hibler, W. D. III. 1991. On the ridging of a thin sheet of lead ice. *Ann. Glaciol.* **15**, 81–6.

Hunke, E. C. and Dukowicz, J. K. 1997. An elastic-viscous-plastic model for sea ice dynamics. *J. Phys. Oceanography* **27**, 1849–67.

Ip, C. F., Hibler, W. D. III and Flato, G. M. 1991. On the effect of rheology on seasonal sea-ice simulations. *Ann. Glaciol.* **15**, 17–25.

Johns, T. C. *et al.* 1997. The second Hadley Centre coupled ocean-atmosphere GCM: model description, spinup and validation. *Climate Dyn.* **13**, 103–34.

Kattenberg, A. *et al.* 1996. Climate models – projections of future climate. In Houghton, J. *et al.*, eds., *Climate Change 1995. The Science of Climate Change*. Cambridge University Press, pp. 285–358.

Kreyscher, M., Harder, M. and Lemke, P. 1997. First results of the sea-ice model intercomparison project (SIMIP). *Ann. Glaciol.* **25**, 8–11.

Kreyscher, M., Harder, M., Lemke, P. and Flato, G. M. 2000. Results of the sea ice model intercomparison project: evaluation of sea-ice rheology schemes for use in climate simulations. *J. Geophys. Res.* **105** (C5), 11 299–320.

Kwok, R. and Rothrock, D. A. 1999. Variability of Fram Strait ice flux and North Atlantic oscillation. *J. Geophys. Res.* **104**, 5177–89.

Lipscomb, W. H. 2001. Remapping the thickness distribution in sea ice models. *J. Geophys. Res.* **106**, 13 989–14 000.

Maslowski, W., Newton, B., Schlosser, P., Semtner, A. and Martinson, D. 2000. Modeling recent climate variability in the Arctic Ocean. *Geophys. Res. Lett.* **27**, 3743–6.

Maykut, G. A. 1978. Energy exchange over young sea ice in the central Arctic. *J. Geophys. Res.* **83**, 3646–58.

Maykut, G. A. and Untersteiner, N. 1971. Some results from a time-dependent thermodynamic model of sea ice. *J. Geophys. Res.* **76**, (6) 1550–75.

Meehl, G. A., Boer, G. J., Covey, C., Latif, M. and Stouffer, R. J. 1997. Intercomparison makes for a better climate model. *EOS, Trans. Am. Geophys. Union* **78**, 445–6.

Meier, W. N., Maslanik, J. A. and Fowler, C. W. 2000. Error analysis and assimilation of remotely sensed ice motion within an Arctic sea ice model. *J. Geophys. Res.* **105**, 3339–56.

Moritz, R. E. and Stern, H. L. 2001. Relationships between geostrophic winds, ice strain rates and the piecewise rigid motions of pack ice. In Dempsey, J. P. and Shen, H. H., eds., *Proceedings of the IUTAM Symposium on Scaling Laws in Ice Mechanics and Ice Dynamics*, Fairbanks, June 13–16, 2000. Dordrecht, Kluwer.

Moritz, R. E. and Ukita, J. 2000. Geometry and the deformation of pack ice, Part I: a simple kinematic model. *Ann. Glaciol.* **31**, 313–22.

Pollard, D. and Thompson, S. L. 1994. Sea-ice dynamics and CO_2 sensitivity in a global climate model. *Atmos.-Ocean* **32**, 449–63.

Polyakov, I. V., Proshutinsky, A. Y. and Johnson, M. A. 1999. Seasonal cycles in two regimes of Arctic climate. *J. Geophys. Res.* **104**, 25 761–88.

Polyakov, I. V. *et al.* 1998. Coupled sea ice-ocean model of the Arctic Ocean. *J. Offshore Mech. & Arctic Engrng.* **120**, 77–84.

Pritchard, R. S. 1975. An elastic-plastic constitutive law for sea ice. *J. Appl. Mech.* **43E** 379–84.

Pritchard, R. S., Coon, M. D. and McPhee, M. G. 1977. Simulation of sea ice dynamics during AIDJEX. *J. Pressure Vessel Technol.* **99J**, 491–7.

Proshutinsky, A. Y. and Johnson, M. 1997. Two circulation regimes of the wind-driven Arctic Ocean. *J. Geophys. Res.* **102**, 12 493–514.

Proshutinsky, A. Y., Polyakov, I. V. and Johnson, M. A. 1999. Climate states and variability of Arctic ice and water dynamics during 1946–1997. *Polar Res.* **18**, 135–42.

Rind, D., Healy, R., Parkinson, C. and Martinson, D. 1995. The role of sea ice in $2 \times CO_2$ climate model sensitivity. Part I: The total influence of sea ice thickness and extent . *J. Climate* **8**, 449–63,

Rothrock, D. A. 1975. The energetics of plastic deformation of pack ice by ridging, *J. Geophys. Res.* **80**, 4514–19.

Rothrock, D. A., Yu, Y. and Maykut, G. A. 1999. Thinning of Arctic sea-ice cover. *Geophys. Res. Lett.* **26**, 3469–72.

Schramm, J. L., Flato, G. M. and Curry, J. A. 2000. Toward modeling of enhanced basal melting in ridge keels. *J. Geophys. Res.* **105**, 14 081–92.

Schramm, J. L., Holland, M. M., Curry, J. A. and Ebert, E. E. 1997. Modeling the thermodynamics of a sea ice thickness distribution, 1. Sensitivity to ice thickness resolution. *J. Geophys. Res.* **102**, 23 079–91.

Semtner, A. J. Jr. 1976. A model for the thermodynamic growth of sea ice in numerical investigations of climate. *J. Phys. Oceanography* **6**, 379–89.

Shine, K. P. and Henderson-Sellers, A. 1985. The sensitivity of a thermodynamic sea ice model to changes in surface albedo parameterization. *J. Geophys. Res.* **90**, 2243–50.

Steele, M. and Flato, G. M. 2000. Sea ice growth melt and modeling: a survey. In Lewis, E. L. *et al.*, eds., *The Freshwater Budget of the Arctic Ocean*. Dordrecht, Kluwer, pp. 549–87.

Steele, M., Zhang, J., Rothrock, D. and Stern, H. 1997. The force balance of sea ice in a numerical model of the Arctic Ocean. *J. Geophys. Res.* **102**, 21 061–79.

Stössel, A., Lemke, P. and Owens, W. B. 1990. Coupled sea ice – mixed layer simulations for the Southern Ocean. *J. Geophys. Res.* **95**, 9539–55.

Thomas, D., Martin, S., Rothrock, D. and Steele, M. 1996. Assimilating satellite concentration data into an Arctic sea ice mass balance model, 1979–1985. *J. Geophys. Res.* **101**, 20 849–68.

Thorndike, A. S. 1987. A random discontinuous model of sea ice motion. *J. Geophys. Res.* **92**, 6515–30.

Thorndike, A. S., Rothrock, D. A., Maykut, G. A. and Colony, R. 1975. The thickness distribution of sea ice. *J. Geophys. Res.* **80**, 4501–13.

Tremblay, L.-B. and Mysak, L. A. 1997. Modeling sea ice as a granular material, including the dilatancy effect. *J. Phys. Oceanography* **27**, 2342–60.

Vinnikov, K. Y. *et al.* 1999. Global warming and northern hemisphere sea ice extent. *Science* **286**, 1934–7.

Wadhams, P. 1990. Evidence for thinning of the Arctic ice cover north of Greenland. *Nature* **345**, 795–7.

Walsh, J. E., Hibler, W. D. III and Ross, B. 1985. Numerical simulation of northern hemisphere sea ice variability, 1951–1980. *J. Geophys. Res.* **90**, 4847–65.

Zhang, J., Rothrock, D. and Steele, M. 2000. Recent changes in Arctic sea ice: the interplay between ice dynamics and thermodynamics. *J. Climate* **13**, 3099–114.

Part IV
The mass balance of the ice sheets

10

Greenland: recent mass balance observations

ROBERT H. THOMAS

EG&G Services, NASA Wallops Flight Facility, Virginia

10.1 Introduction

Tide-gauge measurements indicate that sea level has risen by about 15 ± 5 cm over the past century, with perhaps 7.5 ± 5 cm of this rise caused by effects other than changes in the polar ice sheets (IPCC, 2001). The missing 7.5 ± 7 cm was most probably caused by net losses from the Greenland and Antarctic ice sheets at an average rate of about 300 ± 280 km^3 of ice per year. This is equivalent to $12 \pm 11\%$ of their combined annual snow accumulation, and would represent a thinning rate of 2 ± 2 cm per year averaged over the entire area of both ice sheets. Although the uncertainty of this estimate is quite large, it is far lower than that resulting from many decades of glaciological observations, which have yet to yield even the sign of the collective mass balance of these two ice sheets, with an uncertainty equivalent to about 20% of their combined snow-accumulation rate. Until recently, this level of uncertainty applied equally to both ice sheets, and provided the prime motivation for NASA's Program for Arctic Regional Climate Assessment (PARCA) which had, as its initial goal, measurement of the mass balance of the Greenland ice sheet. To a large extent, this goal has been achieved, and PARCA has significantly improved our knowledge of many of the factors that determine the mass balance. Consequently, this chapter presents a description of the PARCA measurements and a summary of major results, many of which incorporate measurements from other programs, such as ice coring by the Greenland ice sheet programme and the north Greenland traverse, and coastal studies of ablation by Danish and Dutch scientists. Much of the material presented here is a summary of results described in the PARCA special section of *J. Geophys. Res. (Atmos.)*, where readers can find more detail covering the topics discussed here.

PARCA was formally initiated in 1995 by combining into one co-ordinated programme various investigations associated with efforts, started in 1991, to assess whether airborne laser altimetry could be applied to measure ice-sheet elevation changes (Thomas *et al.*, 2001a). The prime result is an order-of-magnitude improvement in our estimates for the

This work was compiled with help from scientists involved with PARCA, who contributed to the PARCA special section in the *Journal of Geophysical Research (Atmospheres)* **106** (D24), 2001.

Mass Balance of the Cryosphere: Observations and Modelling of Contemporary and Future Changes, eds. Jonathan L. Bamber and Antony J. Payne. Published by Cambridge University Press. © Cambridge University Press 2003.

mass balance of the entire ice sheet, with quite detailed assessments of the behaviour of smaller regions within the ice sheet. Higher elevation parts of the ice sheet have been very close to balance for the past few decades, with thinning predominating over the last few years nearer the coast. Total sea-level rise associated with this thinning is very small, indicating that Antarctica is probably responsible for much of the unexplained increase in sea level. Significant progress has also been made both in the development of new techniques for glaciological research, and in process studies. Taken as a whole, results from PARCA represent a major advance in our knowledge and understanding of the total mass balance of the Greenland ice sheet, and they form a baseline set of measurements for comparison with precise surface-elevation measurements to be acquired by NASA's geoscience laser altimeter system (GLAS) aboard ICESat, launched in January 2003, and by ESA's CryoSat radar altimeter with a planned launch a year or two later. Here, we present major results from the programme to date, with an assessment of our current understanding of the mass balance of the ice sheet. First, however, we briefly describe the Greenland ice sheet and identify major differences between it and the Antarctic ice sheet.

10.1.1 The polar ice sheets

The Greenland ice sheet occupies a latitude band extending from 60°N to 80°N, and covers an area of $1.7 \times 10^6 \, \text{km}^2$. With an average thickness of 1600 m, it has a total volume of approximately $3 \times 10^6 \, \text{km}^3$ – equivalent to a sea-level rise of about 7 m. It consists of a northern dome and a southern dome, with maximum surface elevations of approximately 3200 m and 2850 m, respectively, linked by a long saddle with elevations around 2500 m. Bedrock beneath the central part of the ice sheet is remarkably flat and close to sea level, but the ice sheet is fringed almost completely by coastal mountains, through which it is drained by many glaciers.

The ice sheet in Greenland differs significantly from that in Antarctica, which is almost ten times larger in volume. Antarctica straddles the South Pole, and has a dominant influence on its own climate and on the surrounding ocean, with cold conditions even during the summer and around its northern margins. Away from the coast, much of Antarctica is a cold desert, with very low precipitation rates. There is little surface melting, even near the coast, and most melt water soaks into underlying snow and refreezes. Because of the cold conditions, vast floating ice shelves exist around much of the continent. Ice drainage is primarily by glaciers and ice streams, some of which penetrate deep into the heart of the ice sheet, moving at maximum speeds of a few hundred metres per year, apart from a few exceptions. Most glaciers and ice streams flow into ice shelves, which are also fed by snow accumulation on their surfaces. The ice shelves thin toward their seaward ice fronts, partly by ice creep and partly by basal melting, with melting rates generally of a few tens of centimetres per year. Thus, ice loss from the Antarctic ice sheet is primarily by basal melting and iceberg calving from ice shelves.

By contrast, the Greenland climate is strongly affected by its proximity to other land masses and to the North Atlantic, with the Gulf Stream to the south and regions of North Atlantic deep water production to the east and west. Ice-core data from the summit of the Greenland ice sheet indicate that Greenland temperatures and accumulation rates can increase significantly over periods of a few years to decades (Alley *et al.*, 1993). Other major contrasts with Antarctica include widespread summer melting, higher accumulation rates, very few ice shelves, and faster glaciers. Summer melting occurs over about 50% of the ice-sheet surface, depending on summer temperatures, with much of the resulting melt water flowing into the sea, either along channels cut into the ice surface or by draining to the bed via crevasses. The average accumulation rate (approximately 0.3 m water per year) is more than double that for Antarctica, so, although the ice sheet is far smaller, its total annual exchange of water with the ocean is about 30% that for Antarctica. Higher coastal temperatures in Greenland do not favour ice shelves, but there are a few along the north and north-east coasts. Basal melting rates from these reach values exceeding 10 m per year, and they give an indication of the potential effect of warmer ocean temperatures around Antarctica. Most Greenland outlet glaciers are narrower, by an order of magnitude, than their Antarctic counterparts, but some reach speeds that are an order of magnitude larger. Consequently, these glaciers drain very large volumes of ice, with discharge rates strongly determined by fast-glacier dynamics, which are poorly understood.

10.1.2 *Greenland and sea-level change*

Although the Greenland ice sheet is an order of magnitude smaller in volume than the Antarctic ice sheet, it is particularly important to the study of sea-level change in a warming climate for two reasons. Firstly, it is likely to respond most rapidly to warmer temperatures because surface melting occurs already over almost half its surface and in all coastal regions. This means that small increases in air temperatures would result in large inland migrations of summer melt zones up the comparatively gentle slopes of interior parts of the ice sheet. Increasing melt would reduce ice-sheet volume directly by drainage into the ocean, and indirectly by lubricating the base of outlet glaciers. By contrast, there is very little summer melting in Antarctica, and substantial warming would be required for summer melting to migrate up the steeper coastal slopes of the ice sheet. Secondly, Greenland may give an indication of Antarctic conditions in a warmer climate. In particular, markedly higher ice velocities on most Greenland outlet glaciers probably result from the lack of the buffering effect of large ice shelves. If so, Antarctic ice discharge could increase significantly if climate warming were to weaken or remove key ice shelves. This may already be happening on Thwaites and Pine Island glaciers, which have little ice-shelf restraint, are moving faster than other Antarctic glaciers at speeds approaching those of Greenland glaciers and are thinning quite rapidly (Shepherd *et al.*, 2001). Consequently, investigation of Greenland outlet glaciers should help in assessing the likelihood of substantial changes in Antarctica under conditions of a warming climate.

10.1.3 Program for Arctic Regional Climate Assessment (PARCA)

PARCA's inception owed most to the global positioning system of satellites – GPS. By the late 1980s, these were sufficiently well established, and data-analysis techniques sufficiently mature, to offer the possibility of providing very precise knowledge of aircraft trajectories via kinematic solutions of GPS data acquired aboard the aircraft (Krabill and Martin, 1987). Until then, the lack of such knowledge had seriously limited measurement of absolute surface topography from aircraft, and measurement of small changes associated with ice-sheet thickening or thinning was out of the question. Consequently, starting in 1991, NASA's Polar Research Program sponsored an exhaustive series of tests in Greenland to assess the accuracy of an existing scanning laser altimeter flown aboard a GPS-equipped aircraft over the ice sheet (Krabill *et al.*, 1995). Soon after, this programme was enhanced by development and airborne testing of a coherent ice depth-sounding radar (Gogineni *et al.*, 1998). Results were sufficiently promising to warrant aircraft surveys in 1993 and 1994 of both surface elevation and ice thickness over all the major ice drainage basins on the ice sheet, with the intention of repeating these surveys in the future to determine elevation changes.

During this period, other events occurred to help establish a basis for PARCA: synthetic aperture radar (SAR) data from the ERS-1 satellite were used to compile a mosaic map of Greenland (Fahnestock *et al.*, 1993), and to infer ice velocities by interferometric analysis of time-separated image pairs (Goldstein *et al.*, 1993); research was on-going on the interpretation of time series of satellite microwave data in terms of ice-surface characteristics; analysis of the effects of inter-annual variability of snow-accumulation rates indicated that these alone could cause quite large rates of ice-surface elevation change over periods of several years (Braithwaite, 1993; Van der Veen, 1993); and ice-core analyses were providing more precise annual resolution due to use of continuous flow techniques, identification of multiple seasonally varying parameters, and a strong commitment to high-precision dating (Anklin *et al.*, 1998; Mosley-Thompson *et al.*, 2001). Consequently, in 1995, PARCA was formally established, comprising the following key elements.

  (i) Direct measurement of ice-surface elevation change from time series of satellite radar and aircraft laser altimeter data.
 (ii) Localized measurements of long-term ice-thickness change from measurements of the vertical motion of markers buried in the ice.
(iii) Indirect mass-budget estimates of long-term changes in ice-sheet volume by comparing total snow accumulation with total ice discharge.
(iv) Historical ice-thickness change inferred from precise GPS and gravity measurements of crustal motion at coastal sites on each side of the ice sheet.
 (v) Ice-core acquisition and analysis to improve estimates of total snow accumulation and its long-term trends, and to assess the impacts on surface-elevation change of inter-annual variability in snow-accumulation rates.
(vi) Estimation of snow-accumulation rates at high temporal resolution, by model assimilation of analysis fields provided by global weather models, and validation by analysis of high resolution snow-pit profiles.

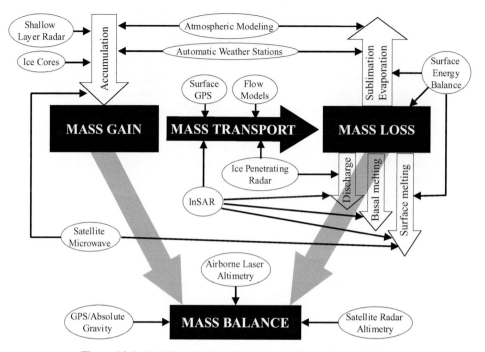

Figure 10.1. PARCA, showing links between the various activities.

(vii) Establishment of a network of automatic weather stations (AWS) to monitor weather conditions, local energy balance, and snow-fall events in different climatic zones on the ice sheet.

(viii) Investigation of processes associated with surface melting, from measurements taken along a transect through the ablation zone near the west coast of the ice sheet, and from satellite estimates of albedo.

(ix) Investigation of near-surface processes associated with surface melting, ice-layer formation, hoar-frost formation, and total snow-accumulation rate in order to improve our ability to infer these parameters from satellite data.

(x) Investigations of individual glaciers and ice streams using satellite SAR interferometry for ice velocities, core data for snow-accumulation rates, and aircraft measurements of ice thickness and surface topography.

Figure 10.1 shows how these activities relate to the PARCA goal of quantifying and understanding the mass balance of the ice sheet, and Figure 10.2 shows locations on the ice sheet where most of the associated measurements were made. 1999 was a milestone year for PARCA, with completion of aircraft re-surveys to provide very accurate measurements of changes in surface elevation over a five-year period for the entire ice sheet, completion of the ice coring, and significant advances in other aspects of the programme. Since then, emphasis has shifted to analysis of the data, with new measurements focussed on addressing important problems revealed by the analysis.

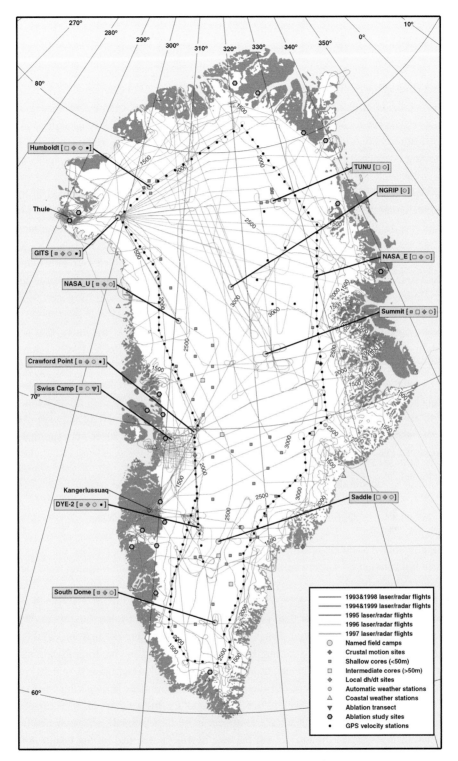

Figure 10.2. Greenland, showing the locations of PARCA activities (Thomas *et al.*, 2001a), and of European ablation and surface mass balance investigations near the coast. (After Weidick (1995).)

10.2 Components of ice-sheet mass balance

Here, 'mass balance' refers to the total mass balance of large catchment regions within the ice sheet and of the entire ice sheet, as defined by equation (2.2) in Chapter 2. Mass is added by snow fall, sublimation and occasional rain, and is lost by melting, evaporation and ice discharge. The net accumulation rate can be readily inferred from stake, snow-pit, and ice-core measurements, with the prospect of future large-area coverage by remote sensing. Loss by melting is generally estimated from stake measurements, and is more complex because some of the melt water refreezes after draining into near-surface snow (Pfeffer, Meier and Illangasekare, 1991). Melt loss from beneath the ice sheet is usually neglected in mass-budget calculations because it is considered to be very small compared with other losses, but this is certainly not the case for floating glacier tongues and ice shelves, where melting rates can reach several metres per year. However, past estimates of total Greenland ice-sheet mass balance have implicitly neglected these losses by assigning mass loss to either surface ablation or iceberg calving. This problem can be avoided by measuring ice discharge across the grounding line, and this is made possible by new techniques for measuring ice velocities over very large regions from SAR data.

Glaciological observations of the Greenland ice sheet began in the early 1900s, and our state of knowledge in the late 1980s was well reviewed by Reeh (1989). Estimates of surface elevation had been inferred over much of the ice sheet away from the coast by satellite radar altimetry (Bindschadler *et al.*, 1989), and extensive radar ice-depth sounding during the 1970s (Gudmandsen, 1976) provided a first mapping of most of the bed, but with the limitations of potentially large aircraft-navigation errors, and poor data acquisition over areas of warmer ice. Components of the total mass balance of the ice sheet were summarized by Weidick (1985) as:

$$\text{accumulation: } 500 \pm 100 \, \text{km}^3 \text{ of water equivalent per year,}$$
$$\text{melting: } 295 \pm 100 \, \text{km}^3 \text{ of water equivalent per year,}$$
$$\text{calving: } 205 \pm 60 \, \text{km}^3 \text{ of water equivalent per year.}$$

No significance should be attached to the exact mass balance implied by the numbers, because information on iceberg calving was so poor that the calving estimate was inferred assuming the ice sheet to be in steady state, implicitly including basal melting from floating glacier extensions. Indeed, Reeh (1985) estimated a total calving flux of $310 \, \text{km}^3$ of water equivalent per year –50% larger than given above – but he stressed that only about 45% of this estimate was based on actual data. Consequently, it is surprising that the calving uncertainty was assumed to be lower than those for accumulation and melting. Recent results described below have reduced the uncertainty in total accumulation (to less than $\pm 20 \, \text{km}^3$ of water equivalent per year), but not substantially for melting, and the true calving uncertainty may be even larger than that given above. Consequently, most of the error estimates for these quantities listed by the IPCC (2001) appear to be far too small, although we should note that the associated estimate of total balance is almost identical to that resulting from repeat laser altimetry surveys of the ice sheet by PARCA (Krabill *et al.*, 2000). Despite this entirely fortuitous agreement, we do not include their assessment of the current mass balance here because it gives a totally false impression of the accuracy of most components of the ice-sheet mass balance. Instead, we shall summarize results from various PARCA investigations that show the ice sheet at higher elevations to be in total mass balance to within an equivalent thickness change of about ± 1 cm per year ($10\,\mathrm{km}^3$ of ice per year), with thinning predominating at lower elevations to yield a net loss of about $50\,\mathrm{km}^3$ of ice per year.

Knowledge of surface topography significantly improved during the 1990s, when information from all available sources, including PARCA laser surveys over the ice sheet and Danish photogrammetric surveys of the coast, were merged to yield a high resolution digital elevation map of almost the entire island (Ekholm, 1996). In addition, ice-thickness measurements made during all PARCA airborne surveys (Gogineni *et al.*, 2001) have been analysed in conjunction with the earlier measurements to produce an improved map of bedrock topography (Bamber, Layberry and Gogineni, 2001).

10.2.1 Accumulation

Snow-accumulation estimates over the ice sheet were inferred by Bender (1984) and by Ohmura and Reeh (1991), based on measurements from more than 200 pits and cores in the accumulation zone and precipitation observations made at coastal weather stations. These measurements referred to a range of different time periods, and more than half were based on only one to two years of snow fall (Bales *et al.*, 2001b). Average net snow fall over the entire ice sheet was estimated to be about 30 cm water equivalent per year. More recently, in addition to results from PARCA, there are three other reports of accumulation on the ice sheet: the Summit region (Bolzan and Strobel, 1994); the EGIG line (Anklin *et al.*, 1994; Fischer *et al.*, 1998); and the North GRIP traverse (Fischer, 1997; Fischer *et al.*, 1998; Friedmann *et al.*, 1995). When these data are included and short-term data excluded (Bales, Mosley-Thompson and McConnell, 2001a), total accumulation on the ice sheet, averaged over ten years or more, is very close to the value inferred by Ohmura and Reeh (1991), suggesting that this estimate is accurate to about 3%. However, the new data show significant regional differences from the earlier analysis. Moreover, it is clear that there is very large inter-annual variability in accumulation rates, particularly in areas of high accumulation.

10.2.2 Surface ablation

Surface ablation rates were first measured more than 100 years ago (Drygalski, 1897), but available information is sparse and primarily from the south-west coast (Figure 10.2), with few intensive investigations of energy balance until recent work by European groups (e.g. Braithwaite, 1995a, 1996; Greuell *et al.*, 2001; Konzelmann *et al.*, 1994; Oerlemans and Vugts, 1993; Van de Wal, 1996; van de Wal and Oerlemans, 1994). Surface melting is generally measured using an array of stakes, limiting the size of the region from which data can be acquired. However, it may be possible to use remotely sensed data to extrapolate this information over far larger areas. Melt zones, and possibly melt intensity, derived from satellite passive microwave data, can be used with estimated or measured melt rates, to infer total melting from large regions (Mote, 2000). A second approach is to estimate ablation rates as the difference between SAR-derived vertical ice velocity at the surface and elevation-change rates measured by repeat laser altimetry (Reeh *et al.*, 2002). Both of these approaches are in their infancy, and are the subject of on-going research, but they offer the potential for substantial improvement in our knowledge and understanding of surface ablation.

Ablation depends strongly on temperature and therefore implicitly on elevation. But unfortunately the ablation–elevation relationship is different at different locations, and it can vary from year to year. To overcome this, models are used to estimate melt rates from observations of surface temperature or, more elaborately, from energy balance calculations. The simplest of these is the 'positive degree-day (PDD) model', which assumes that total summer melting is proportional to local positive air temperatures integrated over the days of the ablation season (Braithwaite, 1984, 1995b), with local temperatures inferred from weather station data and assumed lapse rates. Although this is an intuitively attractive approach, and very simple to apply in models of past and future ice sheets (eg. Reeh, 1991), reality is more complex, with both lapse rates and the PDD proportionality factor changing both regionally and temporally. In view of the poor coverage of measurements, and uncertainties in existing models, our knowledge of total melting and its probable response to changing climate is still poor, and this is a research topic identified by PARCA as a high priority for future work.

10.2.3 Ice discharge

Calving of icebergs represents a large fraction of the ice lost from the ice sheet, but direct observations are too limited to estimate the calving flux from the entire ice sheet (Reeh, 1985). Instead, calving is usually estimated by assuming balance and subtracting other losses from the total accumulation. Even if the ice sheet is in balance, this actually gives an estimate of total calving plus melting from the upper and lower surfaces of ice shelves and floating ice tongues, and it has become clear that total losses by basal melting are far larger than those by calving for some glaciers. Accurate calving estimates would require the measurement of ice thickness and velocity along all major calving ice fronts – a formidable task – and they would still reflect only part of the total volume of ice discharged from the ice sheet. Basal melting from ice shelves and floating glacier tongues, at rates of 10 m per year or more, would also have to be estimated.

Fortunately, recent advances in SAR interferometry over ice provide an alternative approach, which is described in more detail below. In brief, the SAR analysis yields spatially detailed estimates of both ice velocity and grounding-line location for outlet glaciers where appropriate data have been acquired. The discharge flux of ice across the grounding line can then be calculated, using ice-thickness measurements from airborne surveys or estimates based on surface-elevation measurements and hydrostatic equilibrium. In addition, repeated SAR mapping gives an indication of grounding-line migration over time.

10.3 PARCA measurements

Before PARCA began in 1995, NASA research on the Greenland ice sheet was centred around the development and testing of new techniques for glaciological research: aircraft laser altimetry; ice depth sounding using a coherent radar; velocity mapping from SAR interferometry; and ice-surface characteristics from satellite microwave data. This work continued within PARCA, and this section presents brief summaries of some of these techniques.

10.3.1 Snow-accumulation rates

Three approaches were applied by PARCA to determine snow-accumulation rates: shallow ice coring; inference from satellite microwave data; and estimation by meteorological modelling.

Shallow ice coring

Firn and ice cores provide annual-layer thicknesses that are the best available representations of net annual accumulation. These annual accumulation data reflect a combination of both the larger-scale climate signal and local glaciological noise at the drill site. The PARCA ice-coring initiative has resulted in an impressive collection of shallow to intermediate-length cores (totaling roughly 2100 m) from about 60 Greenland locations (Figure 10.2). All cores cover recent decades, eight records extend back at least 250 years, and one record includes annually resolved accumulation for the past 800 years. Table I in Mosley-Thompson *et al.* (2001) contains the basic information for each PARCA core. The strength of this data set is the dating precision. The development of a continuous flow analysis system (Anklin *et al.,* 1998; Fuhrer *et al.*, 1993) has made it possible to measure simultaneously multiple chemical species such as hydrogen peroxide, calcium, ammonium and nitrate. Some of these species exhibit seasonal concentration variations, and, when coupled with the seasonal variations of insoluble dust and oxygen isotopic ratios ($\delta^{18}O$), both measured in discrete samples, the dating precision is increased significantly. In many cases, the PARCA cores have no dating error. This is often confirmed by the identification of beta radioactivity horizons (see fig. 5 in Mosley-Thompson *et al.*, 2001) associated with known atmospheric thermonuclear tests (e.g., 1952 and 1963) and/or identification of explosive volcanic events such as Laki (1783–84) or Tambora (1815) by excess sulphate concentration (see figs. 2 and 4 in Mosley-Thompson *et al.*, 2001 and fig. 10 in Mosley-Thompson *et al.*, 1993). Identification of the annual layers, coupled with the density, allows construction of records of net annual mass accumulation (McConnell *et al.*, 2001), providing critical information to other investigations designed to address the past and present mass balance of the Greenland ice sheet.

Accumulation rates from satellite microwave data

Accumulation rates in dry snow zones can now be estimated in several different ways using satellite observations of microwave emission and backscattering. An empirical method has been proposed by Bolzan and Jezek (2000), who showed that annually averaged brightness temperatures at 19 and 37 GHz correlate very strongly with annual accumulation (and likewise for multi-year-averaged measurements) in a 150 km × 150 km region around the Greenland summit. A second empirical method, proposed by Drinkwater, Long and Bingham (2001) uses radar scatterometer image data from Seasat, ERS-1/2, NSCAT and Quikscat. On the basis of a strong Ku-band relationship between the rate of decrease in backscatter coefficient with incidence angle, snow-accumulation retrievals are made possible in the dry snow zone. The observed relationship appears to be determined by the covariance of density, grain size, and accumulation rate, and incidence-angle, and thus path-length integrated volume scattering effects in the upper layers of firn. Retrieval errors using this technique are less than 5% when compared with maps compiled from *in situ* data. Decadal changes in dry snow zone accumulation have been inferred using differences between scatterometer data collected in 1978 and 1996.

A more physical method by Zwally and Giovinetto (1995) is based on the covariance of emissivity at 19 GHz with accumulation rate, and judicious blending of historical ground-based observation. The emissivity/accumulation-rate covariance is explained as being due to physical relationships between snow-grain size and both accumulation rate and emission, but regionally dependent (empirical) adjustments in the emissivity/accumulation-rate parameterization are required. Root-mean-square retrieval errors for this method in the dry snow zone on Greenland are roughly 10%. A fourth method has been developed by Winebrenner, Arthern and Shuman (2001) based on covariance of fine-scale (order 0.5 cm) density layering and accumulation rate, and the effect of layering on the polarization of emission at 6.7 GHz (Figure 10.3). While the latter effect is quantitatively understood, the layering/accumulation-rate covariance is not, and must be specified with the aid of ground-based data. In the case of Greenland, retrieval errors for this method are approximately 10%.

Accumulation rates from atmospheric analyses

Observations of precipitation over Greenland are limited to the coastal regions and have large uncertainties. By contrast, the analysed wind, geopotential height and moisture fields for the free atmosphere over the ice sheet are available for recent years from atmospheric analyses, and this information has been used to retrieve precipitation over Greenland using a dynamic method. Results for 1985 to 1996 and their seasonal and inter-annual variations were initially related to the North Atlantic oscillation (NAO) index (Bromwich *et al.*, 1999; Chen, Bromwich and Bai, 1997). If the NAO index increases, the precipitation over southern

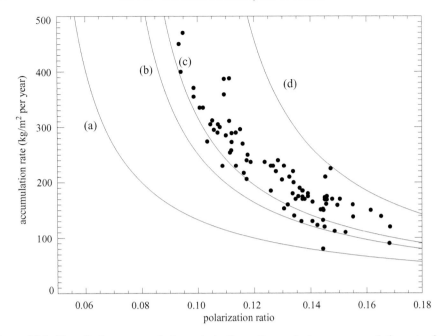

Figure 10.3. Plot of microwave emission polarization ratios against snow-accumulation rates from Winebrenner *et al.* (2001). Polarization ratios are from SMMR data for 4.5 cm wavelength (1979–85), and the observed accumulation rates (black dots) are from *in situ* measurements compiled by Ohmura and Reeh (1991). The four theoretical curves reflect increasing crust density and frequency from (a) to (d).

and western Greenland decreases, and vice versa. The correlation coefficient between these two series for the winter of 1985–95 is −0.60 to −0.80. Newly measured time series of recent accumulation from ice cores reported for 11 sites located near the 2000 m contour of the Greenland ice sheet have been compared with retrieved precipitation covering the same time range of 1985 to 1995. Good agreement in their inter-annual variations has been found if the retrieved precipitation values at individual sites are scaled up by a fixed amount (McConnell *et al.*, 2000a). For subseasonal observations, the modelling data appear to require a similar fix, estimating approximately 56% of the observed net accumulation but with good agreement in annual accumulation pattern (Shuman *et al.*, 2001a). More troubling is the introduction of accumulation gradients in the model output where none appear to exist in reality. However, Hanna, Valdes and McConnell (2001) found excellent agreement between the PARCA ice-core measurements and annual snow accumulation derived from European Centre for Medium-range Weather Forecasting (ECMWF) analyses without the need for scaling.

Based on the evaluation of recent Greenland precipitation studies (Bromwich *et al.*, 1998), several of the deficiencies in the precipitation spatial distributions are probably related to the topographic data initially employed in the modelling. The impact of using the modern topographic data set of Ekholm (1996) to replace the inaccurate US Navy 10 arc minute data set used initially, shows that some deficiencies of the retrieved precipitation are improved. The topographic effect on precipitation can be more accurately modelled by using a separation of the horizontal pressure gradient force in sigma-coordinates into its irrotational and rotational parts, which are expressed by the equivalent geopotential and geo-streamfunction, respectively (Chen and Bromwich, 1999). This procedure is used to enhance the dynamic method, yielding improved agreement between the inter-annual variations observed in accumulation from ice cores and those found from the dynamically retrieved precipitation. No adjustment of retrieved precipitation amounts is now needed.

Accumulation estimates from the ice cores provide detailed information on temporal variability, but this is partly obscured by spatial noise from surface irregularities (Van der Veen and Bolzan, 1999), such as sastrugi and undulations (see figs. 6 and 7 in Mosley-Thompson *et al.*, 2001). This spatial noise must be estimated and removed from ice-core records for reliable descriptions of temporal variability to be obtained. Collection of several duplicate cores separated by tens of kilometers allows this spatial noise to be averaged out, and provides spatially averaged temporal variability on a length-scale that approaches that resolved by the precipitation calculations. The best set of such cores was collected at Humboldt (Mosley-Thompson *et al.*, 2001), where low accumulation enhances the significance of spatial variability. It is desirable to collect such information in a contemporary sampling of all four major ice-sheet environments (south, western, central and north) for testing and calibration of retrieved precipitation amounts. The required data could be collected from multiple cores, as at Humboldt, or by radar mapping of the depth of shallow, isochronous layers, as described in the following section. Once completed, such an effort would not need to be repeated, and only a very limited coring programme would be desirable to check on-going precipitation calculations.

10.3.2 Ice depth sounding and layer tracking

A knowledge of ice thickness is required both for estimating mass balance and for investigating its causes. Consequently, PARCA supported development and improvement of the University of Kansas coherent radar depth sounder operating at a centre frequency of 150 MHz (Gogineni *et al.*, 1998). The transmitter generates a pulse of 1.6 μs duration and about 100 W peak power, which is frequency-modulated over a bandwidth of 17 MHz. It uses two antennas that are mounted under the left and right wings of the aircraft: one for transmission and the other for reception. Each antenna is a four-element dipole array with two-way beamwidths of 18° and 66° in planes perpendicular and parallel to the flight path, respectively. The receiver amplifies and compresses the received signals in a weighted surface acoustic wave (SAW) compressor to an effective pulse length of about 60 ns, resulting in a depth resolution of about 4.5 m in ice. The compressed signal is coherently detected and integrated by summing consecutive pulses. This serves as a low-pass filter and reduces the along-track antenna beamwidth. The pulse compression and coherent processing are the features that make this system unique and capable of sounding outlet glaciers and ice-sheet margins. Good ice-thickness data were collected over 90% of the flight-lines shown in Figure 10.2 (Gogineni *et al.*, 2001), with radar-determined ice thicknesses within ±10 m of those measured at the Greenland Ice Sheet Project (GISP) and Greenland Ice-coring Project (GRIP) core sites. All the PARCA ice-thickness data were processed, and are available for use by the scientific community. These data also show continuous internal layers to a depth of about 2.5 km for 3 km thick ice over distances of several hundred kilometres, which provide information on ice dynamics and are useful for validating ice-sheet models. Radar data have also been used to determine surface roughness and echo strength at the ice–bedrock interface (Allen *et al.*, 1997).

    PARCA ice-core results have highlighted the importance of quantifying the temporal and spatial variability of snow fall, and this can be achieved over very large regions only by accurately mapping the depth of shallow isochronous layers. In order to meet this long-term objective, PARCA supported development of an ultra wideband radar operating over the frequency range from 170 to 2000 MHz for high resolution mapping of internal layers in the top 200 m of ice. A surface-based test of this system was performed at the North GRIP drill site during August 1999, and data were collected over a 10 km transect. The results show that radar-determined depths to internal layers were within ±2 m of those from an ice core (Kanagaratnam *et al.*, 2001). Future plans call for testing of an airborne version of the sounder, toward development of an operational system by about 2002.

10.3.3 Ice velocities and glacier grounding lines from SAR interferometry

Since it was first used to measure ice motion by Goldstein *et al.* (1993), satellite radar interferometry (SRI) has emerged as a major new tool for measuring ice motion over large, featureless areas, at high resolution and with a uniform sampling scheme. Many advances in SRI have been made as part of PARCA investigations, including the first measurements of ice-sheet flow (Joughin, Winebrenner and Fahnestock, 1995; Joughin, Kwok and Fahnestock, 1996a; Rignot, Jezek and Sohn, 1995) ice-sheet surface topography (Joughin *et al.*, 1996b; Kwok and Fahnestock, 1996), vector motion from crossing orbits (Joughin, Kwok and Fahnestock, 1998; Mohr, Reeh and Madsen, 1998), and grounding-line position and migration with time (Rignot, 1996, 1998). These investigations benefited from the large volume of SAR data collected by the European Space Agency's ERS-1 and ERS-2 satellites. SRI-derived velocities, combined with ice-thickness data from the University of Kansas radar sounder, have provided significantly improved estimates of ice discharge from the large outlet glaciers that fringe the perimeter of Greenland and control its mass discharge (Joughin *et al.*, 1999a; Rignot *et al.*, 1997). In particular, these measurements showed that basal melting of floating glacier tongues is a dominant process of mass loss in the north (Rignot *et al.*, 1997), and that discharge of the north-east sector of the ice sheet is controlled by a previously unmapped 600 km long ice stream (Figure 10.4) reminiscent of some in Antarctica (Fahnestock *et al.*, 1993; Joughin *et al.*, 1999b, 2001). The SRI techniques

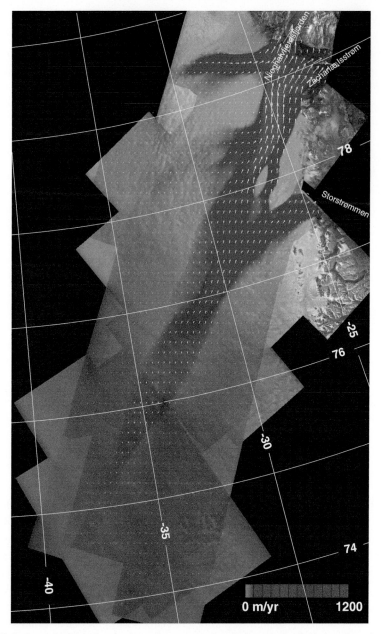

Figure 10.4. Ice velocities, for the large ice stream that drains much of the ice sheet in northern Greenland, inferred from interferometric analysis of ERS SAR data. (Joughin *et al.* (2001). Copyright ESA.)

developed under PARCA for Greenland are now employed to study the Antarctic ice sheet, and have already led to a number of important discoveries.

10.3.4 Ice-surface characteristics from satellite data

Early PARCA airborne and *in situ* observations better defined relationships between space-borne microwave data and glaciological processes. This has led to significant advances in mapping zones of summer melt and of associated snow facies, and estimation of snow-accumulation rates. Visible and infra-red data have also been used to infer surface temperatures and snow albedo. In addition, space-borne imagery has been used to map margins of the western Greenland ice sheet over time (Sohn, Jezek and Van der Veen, 1998).

Summer melt zones

A marked change in microwave emission characteristics during the onset of melt allows for the classification of wet and dry snow based on passive microwave brightness temperatures from the scanning multi-spectral microwave radiometer (SMMR) and special sensor microwave/imager (SSM/I) instruments. A single-channel approach using 37 GHz horizontal brightness temperatures provides details on the day-to-day variability of surface melt characteristics (Mote and Anderson, 1995). A multi-channel approach, using the 19 GHz horizontal polarization combined with the 37 GHz vertical polarization (Figure 10.5) has also been used (Abdalati and Steffen, 1997a). This is a higher inertia signal that dampens some of the day-to-day variability, and it is well suited to longer-term monitoring such as on seasonal and inter-annual scales. Currently, the methods are binary in nature, in that the firn can be classified either as wet or dry, but the different temporal inertia of the two methods shows potential for more quantitatively describing the degree of wetness. Moreover, progress is being made in combining these binary melt estimates with positive degree-day probability distributions to estimate ablation rates for some parts of the ice sheet (Mote, 2000). In addition, the annual extent of melt, the melt duration, and the length of the melt season for 1979 to 1997 (Joshi, 1999) were inferred using an edge-detection algorithm applied to passive microwave time series. The new information on melt duration and length of melt season is better related to global temperature trends than melt extent alone, largely because of the ephemeral nature of melt along transition zones between percolation and dry snow facies. Such analyses are preliminary, but they provide a significant first step toward quantitatively estimating ablation rates from passive microwave satellite data. These methods will be applicable to the advanced microwave scanning radiometer (AMSR), which has a spatial sampling interval (12.5 km) half that of SSM/I.

Other instruments that can be used for melt zone mapping include synthetic aperture radar (SAR) (Fahnestock *et al.*, 1993), wind scatterometer (Drinkwater and Long, 1998; Long and Drinkwater, 1999) and visible sensors such as the advanced very high resolution radiometer (AVHRR). SAR and scatterometer methods rely on differing radar backscatter characteristics between wet, dry, and refrozen snow, while methods using visible sensors are based on reflectance properties and provide additional information about the radiant energy exchanges. Both SAR and AVHRR offer means of mapping melt/ablation characteristics at higher resolution, but SAR has limited temporal coverage, and visible sensors are limited by cloud cover, which can result in substantial gaps in coverage. These data can also be expensive and require considerable computational resources. Scatterometers such as SeaWinds onboard QuikSCAT and ADEOS-2 are proposed by Long and Drinkwater (1999) as the most suitable alternative, as they provide daily, ice-sheet-wide coverage at a spatial resolution of about 4.5 km. The potential also exists for combining observations from passive microwave, scatterometer, SAR and visible imagery for a more complete description of the ice-sheet melt characteristics.

Surface temperature and albedo

It is important to monitor accurately the surface albedo because it controls the amount of solar energy available for absorption by the snow pack and is therefore closely linked with surface melting. Snow-surface temperature and albedo can be derived from thermal and visible satellite imagery under clear skies, provided a correction for the intervening atmosphere is made. Further corrections are typically needed for the derivation of surface albedo, such as a correction for directional effects and conversion from narrow band visible and near-infra-red albedos to a broad-band albedo. Surface temperature and albedo maps over Greenland have been computed using data from the AVHRR aboard weather satellites from 1989 to 1993 (Stroeve and Steffen, 1998; Stroeve, Nolin and Steffen, 1997). More extensive time series of surface albedo (Figure 10.6) and temperature for the polar regions are now available from the AVHRR polar pathfinder data (APP), at the National Snow and Ice Data Center in Boulder, Colorado. Spanning the period from 1981 to 1998, gridded maps of surface albedo and temperature at 5 km spatial resolution are available for both hemispheres. Data from the moderate resolution imaging radiometer (MODIS) can also be used for monitoring the spatial and temporal variations of surface albedo and temperature over Greenland, with improved accuracy in surface temperature and albedo, along with increased accuracy in cloud detection.

Extensive inter-comparison between surface albedo measured at automatic weather stations and that from the APP data set for 1997–98 indicate that APP-derived surface albedos are, on average, 10% less than those measured *in situ* (Stroeve, 2001). However, a positive bias of about 4% is observed in the *in situ* measurements, which would reduce the difference between the two measurements to approximately 6%. Limited comparisons between AVHRR-derived surface temperatures and those measured at the equilibrium line altitude (69.5° N, 49.3° W) indicate that surface temperature can be derived to an accuracy of less than 1 K during summer. However, the *in situ* measurements of both albedo and temperature refer to single points, whereas the satellite-derived values are averages over quite large areas, and further comparisons in time and space are needed. Meanwhile, Greuell and Knap (2000) used AVHRR-derived albedo over a section of the Greenland ice sheet to determine the slush limit. The method is based on a strong gradient in spatial variability of the albedo

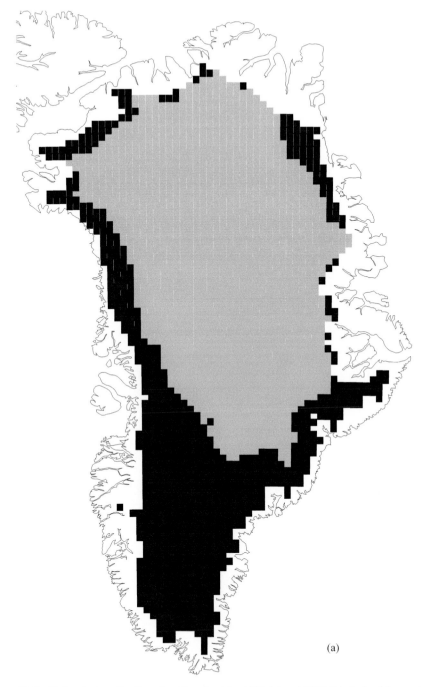

(a)

Figure 10.5. Melt extent on the Greenland ice sheet for 1991 (a) and 1992 (b), derived from satellite passive microwave data using the cross-polarized gradient ratio method of Abdalati and Steffen (2001). The black pixels indicate portions of the ice sheet that experienced melt during the summer, and the grey pixels depict

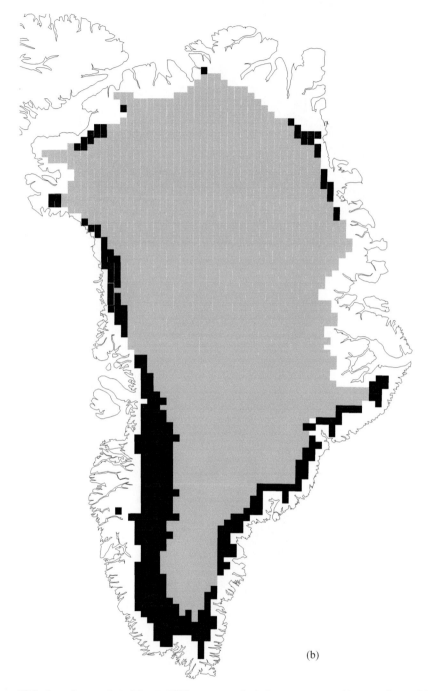

(b)

Figure 10.5. (*cont.*) areas that did not. 1991 was a particularly warm year with extensive melting; 1992 was unusually cool with very relatively little melt. The low melt of 1992 is believed to be attributable to the stratospheric loading caused by the eruptions of Mt Pinatubo (Abdalati and Steffen, 1997b).

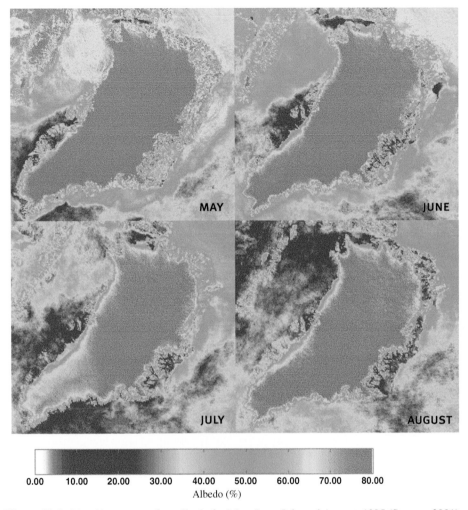

Figure 10.6. Monthly mean surface albedo for May, June, July and August, 1995 (Stroeve, 2001).

at the slush limit, with low variability in the snow zone and high variability in the slush zone, and consequently does not depend on the absolute accuracy of albedo estimates.

Observations of microwave emission at 6.7 GHz (4.5 cm wavelength) can also be used to estimate mean snow surface temperature in dry snow zones. The relatively long, 4.5 cm wavelength, emission originates from depths down to roughly 80 m, and thus is little affected by seasonal temperature variations. Fine-scale (order 0.5 cm) density layering in the snow affects both the polarization of emission (which is directly observable) and effective emissivity (which is not) in such a way as to allow accurate estimation of emissivity from polarization (Winebrenner *et al.*, 2001). The combination of emissivity and observed brightness temperature thus yields an estimate of the 10–80 m firn temperature, which closely approximates the mean surface temperature.

Snow facies

Long and Drinkwater (1994, 1999) find Ku-band and C-band satellite scatterometer images to be extremely effective for delimiting distinctive characteristics, or 'facies', of near-surface snow. This is because microwave backscatter is highly sensitive to physical characteristics of the snow and firn in the top few metres. The primary characteristics apparent in microwave radar data are indicative of the occurrence, intensity and duration of summer melting as well as the seasonal metamorphosis of the firn. Summer surface melting creates stratification and distinctive changes that have a dominant impact on scattering characteristics, particularly in the percolation zone. Refreezing of downward percolating melt water leads to permanent buried ice lenses and ice pipes which produce extremely intense backscattering in winter. This typically leads to seasonal backscatter coefficients close to and exceeding 0 dB (Drinkwater *et al.*, 2001). Within central Greenland the dry snow zone is one of much lower backscatter, and a sharp spatial gradient in backscatter marks the boundary between the percolation zone and dry snow. At higher elevations the surface experiences no melting, and this region exhibits the least seasonal variability in backscatter. At the lowest elevations, within the ablation zone, the snow cover is entirely melted to produce runoff in summer, and little of the incident radar energy is reflected back to the satellite. Drinkwater *et al.* (2001) determine the extent of this zone by summing the areas of the percolation and dry snow zone and subtracting them from the total area of the ice sheet. The results of facies mapping (Figure 10.7) appear consistent with previous findings of other investigators using SAR data Fahnestock *et al.*, 1993; Jezek, Gogineni and Shanableh, 1994). However, by comparison with SAR data, time series data from C- and Ku-band radar scatterometers are far more compact and more easily analysed, and they currently provide the most convenient data source for studies of snow facies and their changes with time.

10.3.5 Automatic weather station (AWS) network and meteorological observations

The Greenland Climate Network (GC-Net) was established in spring 1994 with the objective of monitoring climatological and glaciological parameters at various locations on the ice sheet over a time period of at least ten years (Steffen, Box and Abdalati, 1996). The Summit AWS in this network continues observations begun in support of the GISP2 project in 1987 (Shuman *et al.*, 2001b). The network consists of 19 stations with a distributed coverage over the Greenland ice sheet (Figure 10.2). Four stations are located at high elevations (~3000 m) along a north–south direction; 11 stations lie close to the 2000 m contour line; and four stations are in the ablation region near Swiss Camp. In addition to these PARCA stations, several stations have been installed in ablation regions near the coast by Danish and Dutch investigators. Each AWS measures 32 parameters, including temperature, humidity, wind speed and wind directions at two levels, short-wave incoming and reflected radiation, net radiation, snow height, pressure, snow-temperature profile from 0 to 10 m depth, GPS time and location. These measurements routinely provide meteorological conditions, including

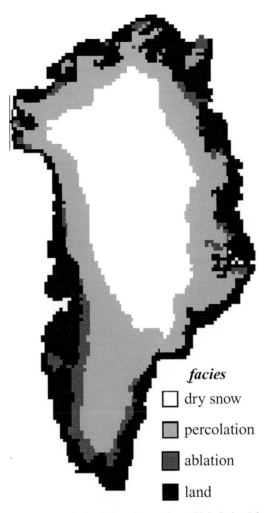

Figure 10.7. Map of snow and ice facies in late September, 1996, derived from satellite radar scatterometer data (Drinkwater *et al.*, 2001). The upper (i.e. the dry snow line) and lower boundaries of the percolation zone were delineated using the -1.8 dB contour in a frequency ratio image (C band/Ku band) computed by differencing collocated weekly mean images of vertically-polarized ERS-2 and NSCAT backscatter normalized to $40°$ incidence (i.e. $\Delta\sigma°$ ($40°$) $= -1.8$ dB). The dry snow zone is marked by negligible seasonality in backscatter coefficient, $\sigma°$, while the percolation zone exhibits the largest seasonal amplitude on Earth. The area remaining up to the outer margin of the ice sheet is assumed to be the ablation zone.

accumulation rate in near-real time to many scientists studying a broad spectrum of research problems, such as: verification of satellite-derived surface albedo; boundary conditions for ice-core chemical transfer models; comparison with analyses from weather models; coastal versus inland climate comparison; calculation of blowing-snow flux, evaporative mass flux, and surface energy balance; and logistic support for research camps on the ice sheet (e.g.

Steffen, 1995). These data will also be important for the interpretation of ICESat and other satellite measurements.

Based on the AWS data, a new annual mean air temperature map was produced (Steffen and Box, 2001), showing the annual mean air temperature to be approximately 2 °C warmer for the central part of Greenland for the time period 1995 to 1999, compared with the standard decade 1951 to 1960 (Ohmura, 1987). This temperature increase is significant, but the timing of the warming cannot be assessed because of the limited AWS time series. The warming decreases with elevation to about 1 °C for the elevation range 1000 to 2000 m, and this pattern of warming was predicted with high resolution global circulation model runs for CO_2-doubling by Ohmura, Wild and Bengtsson (1996). The net sublimation was also derived from AWS data for the entire ice sheet; the total loss is of the order of 0.5×10^{14} kg, or 10% of the annual precipitation (Box and Steffen, 2001). Regions most sensitive to sublimation in terms of the surface mass balance are the ablation zone and sites with small accumulation rates in the north of the ice sheet.

10.3.6 Total mass balance

Four different approaches were applied by PARCA to infer the mass balance of the Greenland ice sheet: a conventional comparison of total snow accumulation with ice discharge; repeated surveys by precise airborne laser altimetry; time series of satellite radar altimetry measurements; and point measurements of elevation change made down shallow bore holes.

Traverse around the ice sheet

During 1993 to 1997, ice discharge velocities were measured by repeat GPS surveys at stations approximately 30 km apart (Figure 10.8), completely circumnavigating the ice sheet close to the 2000 m contour, and ice thicknesses were measured by low frequency radar along the same route (Thomas *et al.*, 1998, 2000a). Seaward ice discharge across this traverse was calculated as the product of thickness and velocity, after correction for the estimated ratio between column-averaged and surface velocities (Huybrechts, 1996). This was compared with total upstream snow accumulation to give estimates of average thickening/thinning rates for individual large catchment basins and for the entire central part of the ice sheet. Assuming that ice velocity at the traverse stations changes slowly with time, the results apply to the period over which snow-accumulation rates are averaged – generally a few decades. In addition to the first accurate mass-budget assessment of any large ice sheet, this work provides a wealth of information on ice velocities and their spatial variation around the entire central portion of the ice sheet.

Aircraft laser altimetry

Airborne laser altimetry is an application of GPS, inertial navigation, and laser ranging to produce accurate and dense measurements of topography (Krabill *et al.*, 1995). Precise laser ranges to the surface below the aircraft are converted to vectors via attitude

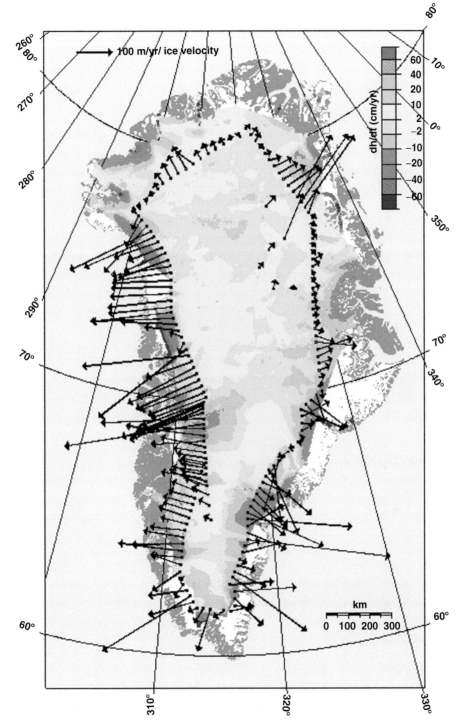

Figure 10.8. Ice velocities, measured approximately along the 2000 m contour line (Thomas *et al.*, 2000a), and Greenland ice-sheet thickening/thinning rates measured by repeat laser altimetry surveys between 1993/4 and 1998/9 (Krabill *et al.*, 2000).

measurements from an inertial navigation system. These are referenced to Earth centre-of-mass co-ordinates through post-flight kinematic GPS determination of the trajectory of the aircraft. Over Greenland, a conically scanning laser was used to provide a swath of data, typically 140–250 m wide, facilitating repeated measurements separated by days or years. Aircraft guidance is derived from real-time GPS positioning. During the past ten years, overall accuracy has improved, and recent data show sub-ten-centimetre repeatability over flight-lines of several hundred kilometres that were re-surveyed after a few days.

Estimates of elevation change can be made by comparing the elevations of individual footprints, from different flights, that lie within a prescribed horizontal separation (typically 1 m), and this is done for areas where the surface is rough or crevassed. However, most of the ice sheet is sufficiently smooth and flat to be well described as a series of plane surfaces, or 'platelets', fitted to the data. Data from different flights are compared by seeking any platelet from the second flight with its centre lying close (typically within 75 m) to the centre of a platelet from the first flight, and comparing heights that have been interpolated to the point midway between the two platelet centres by taking account of the platelet slopes. The root-mean-square fit of the numerous separate laser shots to these platelets is typically 5 cm or better, and, for most purposes, the platelets adequately represent the information contained within the laser data, and they enormously reduce the data volume.

Laser altimeter surveys were completed over all major drainage basins on the Greenland ice sheet in 1993 and 1994 (Figure 10.2), with re-surveys along the same flight-lines in 1998 and 1999. Comparison of results from the two surveys gives estimates of surface-elevation changes over the five-year interim (Figure 10.8).

Satellite radar altimetry

Changes in the surface elevation of the polar ice sheets have been measured using time series of surface elevations obtained from satellite radar altimeters (Zwally *et al.*, 1989). Data from Seasat (1978) and Geosat (1985–89) provided coverage up to ±72° latitude. Radar altimeters aboard the ERS-1/2 satellites extended coverage to ±81.5°, with a continuous time series of data since July 1991. The NASA ice sheet altimeter pathfinder programme has re-processed these data using a consistent set of environmental corrections and orbit solutions. Early ice-sheet change studies using Seasat and Geosat altimeter data suffered from a variety of problems because these instruments were never intended for this application. With progressive improvements in orbit computation (Tapley *et al.*, 1996), orbit-error analysis and reduction (e.g. Davis *et al.*, 2000; Yoon, 1998), ice-sheet re-tracking (Davis, 1997), and identification of measurement-system biasses (e.g. Davis *et al.*, 2000), ice-sheet satellite radar altimetry has now evolved to a sufficient state of maturity that regional changes in surface elevation can be inferred with an accuracy of the order of a few centimetres per year over periods of five to ten years (Davis, Kluever and Haines, 1998; Davis *et al.*, 2000; Wingham *et al.*, 1998; Zwally, Brenner and Dimarzio, 1998). However, these error estimates do not include possible effects of temporal changes in the radar backscatter characteristics of the surface snow (Davis *et al.*, 2001), and of range biasses that might be associated with off-nadir pointing of the radar beam (Zwally and Brenner, 2001). Moreover, because of the large radar 'footprint' and poor tracking near the ice-sheet margin, reliable elevation-change estimates can be acquired only over ice sheets with surface slopes less than about one degree.

Local rates of ice-sheet-thickness change

Local rates of ice-thickness change were measured at 11 sites on the ice sheet (Figure 10.2) using precise GPS surveys. The method entails comparing the vertical component of ice velocity, as measured using GPS surveys of markers anchored at various depths in firn, with the local, long-term rate of snow accumulation from core stratigraphy (Hamilton and Whillans, 2000). Adjustments are made to the vertical velocities to account for vertical motion due to along-slope flow and firn compaction. The difference between the adjusted vertical velocity and the accumulation rate is the local rate of ice-sheet-thickness change, and is approximately equal to the local mass balance as defined by equation (2.1) in Chapter 2. The results apply to timescales defined by the length of the accumulation records, with the assumption that ice dynamics do not change over the same time interval. This is probably reasonable for much of the inland ice sheet where measurements were made, but may not apply in areas of faster moving ice with significant basal sliding. While the results apply locally, they represent averages over longer periods, and comparison with regional mass-budget estimates from the traverse gives an indication of spatial variability of ice-thickening/thinning rates.

10.4 Results

10.4.1 Accumulation and its variability

Accurate spatial estimates of snow accumulation are a central ingredient in essentially all studies of ice-sheet mass balance. Recent PARCA, GISP and European cores have approximately doubled the number of years of accumulation records over the ice sheet, and have addressed many of the previous uncertainties in accumulation. This is because most recent cores cover at least two decades, and most older point measurements were for fewer than ten years. Analysis of these results (Figure 10.9) shows that accumulation in the west–central part of the ice sheet is considerably lower than the Ohmura and Reeh (1991) estimates, with regions of both higher and lower values for the southern portion (Bales *et al.*, 2001b). Although the total accumulation for the entire ice sheet is almost identical to that based on the earlier sparser data, there are still areas of uncertainty, particularly in the south, where spatial variability is very high.

The spatially distributed records of accumulation from the PARCA cores also provide an excellent index of how large-scale atmospheric circulation influences the Greenland climate. Over the past 130 years, approximately one-third of the observed variability in annual accumulation in the north-west can be linked to the North Atlantic oscillation (NAO) (Bales *et al.*, 2001a). The deeper PARCA cores from west-central and north-west Greenland, which cover several centuries, offer the potential for development of a proxy NAO history and the opportunity to improve the initial efforts by Appenzeller, Stocker and Anklin *et al.* (1998) based on a single ice-core record. Preliminary data suggest that cores further inland show a weaker correlation with NAO indices. Perhaps the best promise for understanding how NAO and related large-scale phenomena influence regional accumulation on the ice sheet will come from more detailed analysis of the accumulation records, establishing spatially averaged accumulation variations on a year-by-year basis (McConnell *et al.*, 2001).

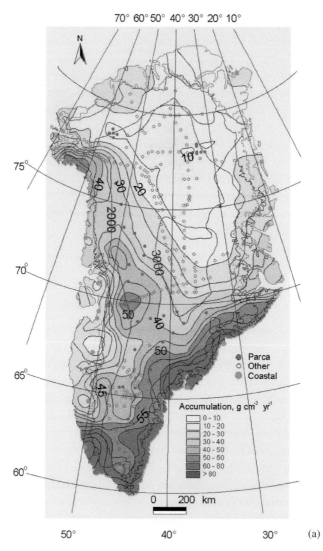

Figure 10.9. (a) Estimated snow-accumulation rates derived from both historical and recent data after rejection of those considered dubious or which referred to very short time periods (Bales *et al.*, 2001b). (b) Map showing differences between these estimates and those from Ohmura and Reeh (1991).

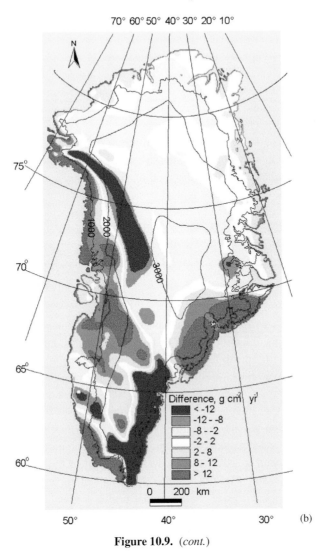

70° 60° 50° 40° 30° 20° 10°

Difference, g cm² yr⁻¹
< -12
-12 - -8
-8 - -2
-2 - 2
2 - 8
8 - 12
> 12

0   200 km

50°   40°   30°   (b)

Figure 10.9. (*cont.*)

Temporal variability and its effect on surface-elevation changes

While repeat altimetry surveys over southern Greenland show little overall elevation change at higher elevations, they show large spatial variability (Davis *et al.*, 1998; Krabill *et al.*, 1999; Zwally *et al.*, 1998). In particular, substantial thickening was observed to the west of the ice-sheet elevation divide below $\sim 68°$N with strong thinning to the east. Taken alone, these altimetry studies were unable to determine the underlying geophysical processes (e.g. changing snow accumulation, increased basal slip) driving current ice-sheet elevation change, nor to assess if present rates of change exceed the natural variability. To address these issues, McConnell *et al.* (2000b) used widely distributed records of annual snow accumulation from ice cores and a physically based model of snow densification (Arthern and Wingham, 1998) to derive the 1978–88 ice-sheet elevation change (dH/dt) resulting solely from measured variability in snow accumulation (dA/dt). While error analyses indicate that uncertainty in accumulation-driven dH/dt arises primarily from well documented spatial variability in snow accumulation, overall uncertainties were generally smaller than the estimated dH/dt. Because accumulation-forced dH/dt was in close agreement with observed dH/dt from Seasat/Geosat satellite altimetry at 11 of 12 locations studied, the majority of the 1978–88 change in ice-sheet elevation above 2000 m was attributed to temporal and spatial variability in accumulation. Similar analyses using seven longer ice-core accumulation records indicate that the observed ten-year changes in ice-surface elevation for central parts of the ice sheet are typical of accumulation-driven dH/dt over the past few centuries. However, areas of very rapid change, particularly in the south-east, cannot be explained in this way, and appear to be undergoing dynamic changes (Thomas *et al.*, 2001b).

10.4.2 Surface ablation

Most investigations of surface ablation have been concentrated in west Greenland in connection with the mapping of hydropower potential. More recently, measurements have been made in connection with climatology studies (for example, Braithwaite, 1983; Ohmura *et al.*, 1994; Van de Wal *et al.*, 1995). These studies indicate that ice-ablation rates of 3–4 m per year are common at lower elevations on outlet glaciers in south Greenland. The AWS located at 900 m elevation in the ablation region near Swiss Camp (Figure 10.2) measured an average ablation of 1 m per year for 1996 and 1997, with summer air temperatures near the melting point and clear-sky conditions. However, along much of the western slope the net ablation is difficult to assess due to the wide extent of the slush zone (30–50 km wide). Near the Swiss Camp, PARCA installed four AWS along a profile through the ablation zone across all glacier facies, with a similar profile planned near Kangerlussuaq to complement the long-term observations started in 1990 by the Institute for Marine and Atmospheric Research at Utrecht University (summarized by Greuell *et al.* (2001)). Data from these should provide crucial information in the future. Strong albedo feedback and consequently the increase in absorbed solar radiation is the major driver for surface ablation in summer. Melt-water lakes, which are common in this elevation band all around the perimeter of Greenland, enhance the surface ablation further, owing to their lower surface albedo.

Melt trends

Analysis of the passive-microwave-derived melt characteristics of the Greenland ice sheet has been extended to include data from 1995 to 1999 acquired by the SSM/I F-13 instrument, increasing the melt record to 20 years. Extending the time series required matching of the F-13-derived melt characteristics to those simultaneously derived with the F-11 sensor

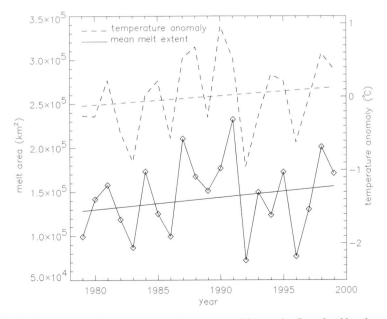

Figure 10.10. Time series of the average area of summer melting on the Greenland ice sheet for 1979 to 1999, derived from satellite passive microwave data (see Figure 10.5). Also shown are the average temperature anomalies for the same time period, inferred from coastal weather station data (Abdalati and Steffen, 2001). The melt area shows an increasing trend of almost 1% per year, coinciding with a total warming of 0.24 °C over the 21-year period.

during a several-month period of overlap. Such geophysical-parameter matching is essential to account for instrumental differences, and to assure consistency within the time series. Results from the longer time period spanning four instruments show an increasing melt trend for the period 1979–1999 of about 1% per year (Figure 10.10), driven primarily by the behaviour of the western side of the ice sheet. The melt demonstrates a high temporal variability and shows signs of being influenced by the eruption of Mt Pinatubo and other climatological phenomena.

10.4.3 Outlet glaciers

Perhaps the most dramatic result from PARCA is the observation of widespread, rapid thinning of most near-coastal ice (Figure 10.8) surveyed by the ATM (Krabill *et al.*, 2000). Most of these surveys were made along outlet glaciers (Figure 10.2), many of which thinned by more than 5 m between 1993/4 and 1998/9 (Abdalati *et al.*, 2001). More detailed studies of many glaciers have been made by satellite radar interferometry (SRI), to detect glacier surges, to infer grounding-line positions and migration rates, and to estimate ice-discharge and ice-shelf basal melt rates. The mini-surge of Ryder Gletscher (Joughin *et al.*,

1996c) confirms that surging glaciers are not uncommon in northern Greenland (Higgins and Weidick, 1988), even though summer melt-water production is limited. Glacier grounding lines, mapped for the first time in north Greenland at a resolution of a few tens of metres (Figure 10.11), and associated ice velocities and thicknesses, indicate that 3.5 times more ice flows out of this part of the ice sheet than previously estimated (Rignot *et al.*, 1997). The discrepancy between prior and new estimates is due to extensive basal melting beneath ice shelves. The inferred basal melt rates (10–20 m per year) are one to two orders of magnitude larger than estimates for most Antarctic ice shelves. Comparison of coastal ice discharge with mass accumulation suggests that slightly more ice is being discharged from northern Greenland glaciers than accumulates in the interior (Rignot *et al.*, 2001), but the errors associated with snow accumulation and runoff remain large. On a shorter timescale, but with greater certainty, SRI measured grounding-line retreat of major north Greenland glaciers (Humboldt, Petermann, Ryder, Ostenfeld and Nioghalvfjerdbrae) between 1992 and 1996 (Rignot, 1998; Rignot *et al.*, 2000, 2001). Retreat rates of several hundred metres per year suggest associated ice thinning at the metre-per-year level, which is unlikely to be caused by temporal changes in accumulation and/or ablation, implying that ice near the grounding line must be thinning by anomalously high creep.

Along the south-east coast of Greenland, ice thinning is even more dramatic (Krabill *et al.*, 1999, 2000). SRI velocity estimates show Helheim Gletscher to be moving at 8 km per year, which is faster than Jakobshavn Isbrae in west Greenland. Kangerdlugssuaq Gletscher, another fast-moving glacier, is thinning at rates increasing to 10 m per year at the coast, but with stable ice-front positions since the early 1960s and steady ice velocities until the late 1990s when the glacier apparently accelerated, probably causing the rapid thinning (Thomas *et al.*, 2000b). In addition to these and other glaciers draining east Greenland, glaciers in the north-west are also undergoing rapid thinning (Abdalati *et al.*, 2001).

North-east ice stream

Initial work on a SAR mosaic and visible imagery revealed the presence of a large ice stream in north-east Greenland which discharges ice into three major outlet glaciers along the coast. This feature reaches well into the interior, with a catchment that includes the northern side of the summit dome. A combination of interferometric analysis of ERS-1 and ERS-2 SAR data, controlled by balance velocities and GPS measurements, and ice-penetrating radar measurements, have allowed us to characterize the ice flow and basal configuration of this ice stream (Joughin *et al.*, 2001). The velocity map makes this ice stream the best documented in terms of surface motion, and provides a rigorous boundary condition for understanding the mechanics of rapid ice flow (Figure 10.4). There are similarities between this ice stream and extensive ice streams in the interior of east Antarctica. However, it differs significantly from the extensively studied ice streams of west Antarctica; most notably, surface slopes are considerably larger in Greenland, indicating a stronger interaction with the bed.

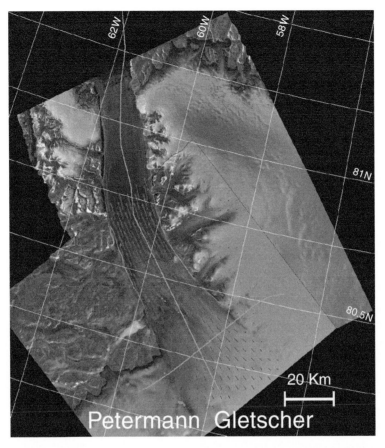

Figure 10.11. Ice velocities and grounding-line position for the Petermann glacier, derived from interferometric analysis by I. Joughin and E. Rignot from a combination of ERS InSAR ascending and descending tracks. The 1996 grounding-line position is shown in blue. Flow vectors are shown in red. Velocity contours, in metres per year, are shown in black. (Copyright ESA 1996.)

10.4.4 Ice-sheet mass balance

PARCA results have significantly improved our understanding of the mass balance of the Greenland ice sheet. Prior to the programme, we could not determine even whether the ice sheet was increasing or decreasing in volume, and mass balance errors were equivalent to about ±10 cm per year thickness change for the entire ice sheet (Reeh, 1989). Since then, repeat surveys by satellite radar altimeter (1978–88 and 1992–99) and by aircraft laser altimeter (1993/4–1998/9), and volume balance estimates from comparison of total snow accumulation with total ice discharge, all show that the entire region of the ice sheet above about 2000 m elevation has been close to balance (within 1 cm per year) for at least the past few decades, but with smaller areas of quite rapid change that can largely be explained by temporal variability in snow-accumulation rates. Some areas, however, appear to be

Table 10.1. *Average ice-sheet thickening/thinning rates*
(dH/dt mm per year).

Region	dH/dt (traverse)	dH/dt (radar)	dH/dt (laser)
Time period	1970–95[a]	1978–88	1993/4–1998/9
North	−3 ± 6	–	+18 ± 7
South	+4 ± 17	+10 ± (10–20)	+8 ± 7
Total	0 ± 7	–	+10 ± 5

[a] PARCA accumulation estimates are predominantly for approximately this period, but accumulation estimates for the entire ice sheet are based on measurements referring to periods ranging back for many decades, so the traverse results probably apply to a time period considerably longer than 25 years, assuming no change in ice-discharge velocities during the period.

undergoing large changes, which may be ongoing adjustments to events since the last glacial maximum, or they may be indicative of changes that began only recently (Thomas *et al.*, 2001b). In particular, most surveyed outlet glaciers are thinning in their lower reaches (Abdalati *et al.*, 2001), and a large area of ice sheet in the south-east has thinned significantly over the past few decades, at rates that increase to more than 1 m per year near the coast. Only part of this thinning can be explained by increased melting associated with recent warmer summers, indicating that ice discharge velocities must also have increased.

PARCA results give three independent estimates of thickening/thinning rates (dH/dt) at higher elevations on the ice sheet. Results from the velocity traverse around the ice sheet represent comparison between total snow fall onto, and total ice discharge out of, large zones on the ice sheet, wheras radar and laser altimetry results provide far more spatial detail. Moreover, the traverse results refer to longer time periods than the altimeter results, which refer only to the period of repeat surveys (1978–88 for radar and 1993/4–1998/9 for laser). Nevertheless, when the altimetry results are averaged over the zones for which traverse dH/dt estimates were made (Figure 10.12), most zones show agreement within estimated errors (Thomas *et al.*, 2001b). This suggests that the effects of temporal variability in snowfall, averaged over large areas of ice sheet, are substantially damped over periods of five years or more. This is further supported by close agreement between the different estimates of average thickening/thinning rate for the northern and southern parts of the ice sheet and for the entire ice sheet.

The results in Table 10.1 indicate that higher-elevation parts of the ice sheet are almost exactly in balance when taken as a whole. But it is clear from Figure 10.12 that some of the individual zones were not in balance during the survey periods and they show apparent large temporal variability. Figure 10.12 includes local estimates of dH/dt inferred from GPS surveys of markers anchored at various depths in shallow bore holes (Hamilton and Whillans, 2000). These results apply to timescales equal to those covered by local measurements of

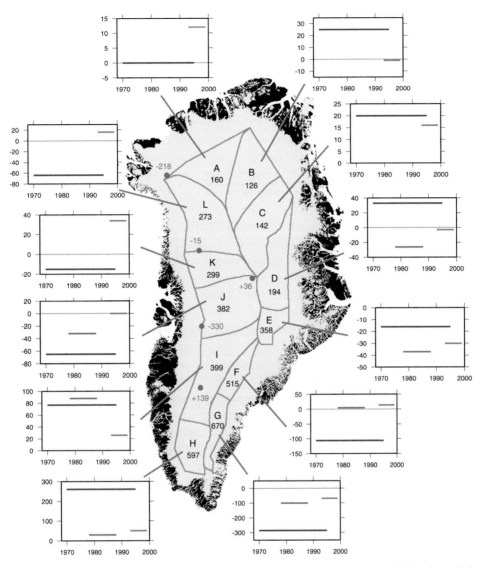

Figure 10.12. Estimates of ice-thickening rates (mm water equivalent per year) during the period 1970 to 1999 inferred from the velocity traverse, satellite radar altimetry time series and repeat aircraft laser altimetry surveys (Thomas *et al.*, 2001b). Traverse results are shown in blue, radar altimetry in green and laser in red. The average accumulation rate (mm water per year) is shown by the number in each zone, and local estimates of thickening rate (mm per year) from bore-hole measurements (Hamilton and Whillans, 2000) are shown in magenta on the map.

accumulation rates, generally decades to more than 100 years. Thus, they complement the traverse results toward providing an indication of the longer-term behaviour of the ice sheet.

The two zones with highest values of dH/dt and with the largest differences between the three different estimates are zones G and H in the southernmost part of the ice sheet

(Figure 10.12). Traverse results show zone G to be thinning by almost 30 cm per year and zone H to be thickening at about the same rate, with far smaller values of dH/dt from the shorter-term laser and radar results. Accumulation rates in both zones have the largest uncertainty of any within our entire study area. Consequently, thickening in zone H could feasibly result from the use of anomalously high accumulation rates for the traverse calculations, either because of errors or because accumulation was unusually high during the last few decades. Indeed, there is evidence for progressive, large increases in accumulation rate since the 1950s (Davis *et al.*, 2001) from a deep ice core at Dye 2 station (66.4° N; 313.8° E) that is situated to the north of zone H. If similar trends apply to zone H as well, then our assumed accumulation rates, which are based largely on PARCA cores giving average accumulation rates since the 1970s, could be as much as 20% higher than the average for the past 100 years. This could explain about half of the rapid thickening in the traverse results, but since the increasing trend continued into the 1990s we would expect the laser results also to show high values of thickening.

A possible explanation for this discrepancy is melting. Zone H is at the south-west corner of the ice sheet; it experiences summer melting over much of its surface, with very large inter-annual variability, and significant surface lowering during a single warm summer. For instance, laser surveys along the traverse line in 1997 and 1998 showed a local surface lowering of about 1 m more than the annual average inferred from the 1993 and 1998 surveys. This was probably caused by the extremely warm summer of 1998, with extensive surface melting over much of southern Greenland. One such event would reduce altimetry-derived thickening rates measured over a five-year interval by 20 cm per year. Although this would apply only to lower-elevation parts of the zone, with less impact further inland, part of the difference between altimetry and traverse dH/dt estimates could be explained by such melting. However, the traverse estimates of dH/dt in this zone are large enough to suggest significant imbalance between ice discharge and accumulation rates averaged over the past century, with resulting long-term thickening. Support for longer-term thickening in the south-west is provided by the local measurement of dH/dt in zone I (Figure 10.12) which shows a thickening rate of 139 ±111 mm per year (Hamilton and Whillans, 2000), compared with the traverse average for this zone of 77 ± 27 mm per year. Moreover, long-term thickening in this area is suggested by a model simulation of the ice sheet forced by the past temperature record (Huybrechts, 1994), archaeological and historical data (Weidick, 1993), and observations of crustal depression (Wahr *et al.*, 2001).

Continuous GPS measurements by a receiver installed by PARCA on bedrock near Kangerlussuaq, show that the Earth's crust is subsiding by about 6 ± 1 mm per year. This result is qualitatively consistent with archaeological and historical evidence, from along the south-west coast of Greenland, that indicate varying rates of subsidence over the last 3000 years or so. The observed rate of subsidence is too large to be caused by the Earth's elastic response to present-day variations in nearby ice. Furthermore, models of the Earth's visco-elastic response to ice-sheet thinning prior to 4000 years ago predict a present-day uplift, rather than subsidence, of 3 ± 2 mm per year (Wahr *et al.*, 2001). The GPS results thus show about 9 ± 2 mm per year of unexplained subsidence. PARCA modelling studies indicate that this subsidence rate is consistent with independent suggestions (Van Tatenhove, Van des Meer and Huybrechts, 1995; Weidick, 1993) that the western ice-sheet margin in this region may have advanced by about 50 km during the past 3000–4000 years. If this advance did occur, and if the crustal subsidence it induces is not included when removing effects of post-glacial rebound from altimetry measurements of Greenland ice-sheet elevation change, then the altimetry results for zones H and I shown in Figure 10.12 could under-estimate ice-thickening rates by as much as 10 mm per year.

Zone G, in the south-east of the ice sheet, displays an opposite trend, with the traverse thinning rate approximately five times larger than the laser estimate. An ice core within this zone (Dye 3 at 65.2°N; 316.1°E) shows a significant decrease in accumulation during the 1980s, consistent with locally high thinning rates inferred from radar data (Davis *et al.*, 2001), but the record ends in 1987, prior to the laser time period. Moreover, this would not explain why the traverse estimate is much larger than both the altimeter estimates. Possible explanations for the very high traverse thinning rates compared to all other estimates include:

(i) Under-estimation, by almost 50%, of accumulation rates in zone G. This appears unlikely, and an independent comparison of balance and measured traverse velocities using different estimated accumulation rates shows that the traverse velocities are dramatically higher than balance velocities in this area (Davis *et al.*, 2001).

(ii) Errors and/or inadequate coverage by the altimetry data. Radar coverage in zone G is less than 50%, and in zone H is less than 70%. Moreover, radar results could have been affected by changes in surface conditions, as suggested by Davis *et al.* (2001) to explain discrepancies further north in the radar data. The laser estimates are based on just a few transects, with interpolation between flight lines, and errors could be larger than the assumed ±3 cm per year, particularly if there are regions of rapid, local thinning that were not overflown.

(iii) Substantial local thinning, with thinning rates decreasing with time. This would require either progressively increasing snow accumulation or progressively decreasing ice discharge. However, the Dye 3 ice-core data indicate decreasing accumulation rates, and traverse velocities were measured between 1996 and 1997, so if ice velocities were decreasing they would be lower than the average for the past few decades, resulting in under-estimated thinning rates.

(iv) Recent increases in ice discharge. If velocities at some of the traverse stations increased very recently, our measurements would not be representative of conditions over the past few decades, as has been assumed to derive the traverse estimates. Associated thinning would be restricted to the region immediately upstream of stations with increasing ice velocities, which might not be covered by the altimetry surveys. However, the 22 cm per year difference in dH/dt estimates between traverse and laser results would require a recent increase in the average discharge velocity from all of zone G of 30% or more. For velocity increases along just one or two outlet glaciers, which is more likely, an even larger percentage increase would be required, with very localized, rapid thinning that could have been 'missed' by the laser surveys.

Although this final option appears highly unlikely, we should note that there are few, if any, measurements of time series of ice velocity in regions such as this. Consequently, we cannot confirm the common assumption that, far from the coast, ice-sheet velocities change only very slowly with time. Ice velocities at traverse stations in this zone reach higher values than at any other traverse station, and are high enough to indicate that basal ice is at the melting point, allowing ice to slide over its bed, and under these conditions ice velocities are more likely to undergo rapid changes. Moreover, this is an area of large near-coastal thinning rates (>1 m per year) inferred from most laser data seaward of the traverse (Krabill *et al.*, 1999) at rates too large to be explained by increased melting alone (Abdalati *et al.*, 2001), so local ice velocities and creep rates probably increased above their balance values to cause the thinning. Inland migration of this velocity increase could explain our results if the migration reached the traverse only recently.

By contrast to the overall balance at high elevations, the laser surveys reveal significant thinning along 70% of the ice-sheet periphery below 2000 m elevation (Abdalati *et al.*, 2001). Thinning rates of more than 1 m per year are common along many outlet glaciers, in some cases at elevations up to 1500 m. Warmer summers along parts of the coast may have caused a few tens of centimetres per year additional melting, but most of the observed thinning probably results from increased glacier velocities and associated creep rates. Three glaciers in the north-east all show patterns of thickness change indicative of surging behaviour, and one has been independently documented as a surging glacier. However, we have no explanation for the widespread thinning, at all latitudes, of most surveyed glaciers. There are a few areas of significant thickening (up to 50 cm per year), and these may be related to higher than normal local accumulation rates during the observation period. The total net thinning of the ice sheet is equivalent to a sea-level rise of approximately 0.13 mm per year, or about 10% of the observed signal (Krabill *et al.*, 2000).

10.5 Future research

Approximately half of the ice lost from the Greenland ice sheet is by surface melting and runoff into the sea, with most of the rest by ice discharge and basal melting from the floating portions of outlet glaciers. By contrast, there is very little ice loss by surface melting in Antarctica. Consequently, near-coastal ice in Greenland is likely to respond most rapidly and sensitively to warming climate, with an increase in both the area and intensity of summer melting. In addition to increasing the rate of melt-water discharge into the ocean, this would also increase melt-water flow into crevasses and thence to the beds of outlet glaciers, potentially leading to more rapid basal sliding. It has been suggested that Greenland was responsible for rapid sea-level rise during the last inter-glacial (Cuffey and Marshall, 2000), so understanding the mechanisms that control the ice-sheet balance, and how they may respond to climate change, is a high priority.

PARCA has completed the first overall assessment of the mass balance of the Greenland ice sheet, but this work represents only the reconnaissance phase of a complete study. It has shown that significant changes are taking place in coastal regions – precisely where the ice sheet should be most sensitive to climate warming. But observed warming in these regions is quite modest, and there is a clear need for well focussed research aimed at identifying the causes of this thinning, with the long-term goal of increasing understanding sufficiently for reliable prediction of Greenland ice-sheet response to prescribed changes. This will require research programmes addressing near-coastal accumulation and ablation and their temporal variability, and the dynamics of outlet glaciers with emphasis on those that are thinning rapidly. The accumulation/ablation programme should aim to improve understanding sufficiently for development of reliable parameterizations of precipitation and melt rates in terms of weather parameters that are observed and/or derived from models; and the outlet-glacier programme should identify and quantify the controls over glacier motion, and determine the sensitivity of glacier speed to perturbations in these controls.

Results from PARCA to date, together with those from studies of ablation and outlet glaciers, should provide a solid foundation for the interpretation and analysis of Greenland data from the ICESat mission. The aircraft laser altimeter measurements represent baseline data extending back to the early 1990s for comparison with ICESat measurements of surface elevations wherever orbit tracks cross the PARCA flight-lines. This will significantly extend the temporal coverage of our estimates of elevation change over most of the ice sheet. In addition, PARCA ice-core results will help isolate regions of long-term thickness change from those where change is most likely caused by temporal variability in accumulation rates, and interpretation of observed patterns of elevation change will benefit from satellite time series of surface characteristics, AWS measurements, and the results of process studies initiated by PARCA. We strongly recommend continued efforts to maintain these activities: satellite time series should be extended to include data from new sensors, such as MODIS, AMSR, Seawinds, and ICESat; the AWS network should be optimized, based on analysis of acquired data, to remove redundant stations and perhaps to establish new stations in ablation regions; and studies addressing coastal precipitation and ablation, and the dynamics of thinning glaciers, should be emphasized.

References

Abdalati, W. and Steffen, K. 1997a. Snowmelt on the Greenland ice sheet as derived from passive microwave satellite data. *J. Climate* **10**, 165–75.
  1997b. The apparent effect of the Mt. Pinatubo eruption on the Greenland ice sheet melt conditions. *Geophys. Res. Lett.* **24**, 1795–7.
  2001. Greenland ice sheet melt extent: 1979–1999. *J. Geophys. Res. Atmos.* **106** (D24), 33 983–8.
Abdalati, W. *et al.* 2001. Outlet glacier and margin elevation changes: near coastal thinning of the Greenland ice sheet. *J. Geophys. Res. Atmos.* **106** (D24), 33 729–42.
Allen, C. T., Ghandi, M., Gogineni, P. and Jezek, K. C. 1997. Feasibility study for mapping the polar ice bottom topography using interferometric synthetic-aperture radar techniques. Remote Sensing Laboratory Technical Report 11680–1, University of Kansas, January 1997.
Alley, R. B. *et al.* 1993. Abrupt increase in Greenland snow accumulation at the end of the Younger Dryas event. *Nature* **362**, 527–9.
Anklin, M., Stauffer, B., Geis, K. and Wagenbach, D. 1994. Pattern of actual snow accumulation along a west Greenland flow line: no significant change observed during recent decades. *Tellus* **46B**, 294–303.

Anklin, M., Bales, R. C., Mosley-Thompson, E. and Steffen, K. 1998. Annual accumulation at two sites in northwest Greenland during recent decades. *J. Geophys. Res.* **103**, 28 775–83.

Appenzeller, C., Stocker, T. F. and Anklin, M. 1998. North Atlantic oscillation dynamics recorded in Greenland ice cores. *Science* **282**, 446–9.

Arthern, R. J. and Wingham, D. J. 1998. The natural fluctuations of firn densification and their effect on the geodetic determination of ice sheet mass balance. *Climatic Change* **40**, 605–24.

Bales, R. C., Mosley-Thompson, E. and McConnell, J. R. 2001a. Variability of accumulation in northwest Greenland over the past 250 years. *Geophys. Res. Lett.* **28** (14), 2679–82.

Bales, R. C., McConnell, J. R., Mosley-Thompson, E. and Csatho, B. 2001b. Accumulation over the Greenland ice sheet from historical and recent records. *J. Geophys. Res. Atmos.* **106** (D24), 33 813–26.

Bamber, J. L., Layberry, R. and Gogineni, S. 2001. A new ice thickness and bedrock dataset for the Greenland ice sheet, 1, measurement, data reduction, and errors. *J. Geophys. Res. Atmos.* **106** (D24), 33 773–80.

Bender, G. 1984. The distribution of snow accumulation on the Greenland ice sheet. Masters Thesis, University of Alaska.

Bindschadler, R. A., Zwally, H. J., Major, J. A. and Brenner, A. C. 1989. Surface topography of the Greenland ice sheet from satellite altimetry. Washington, D.C., NASA Special Publication-503.

Bolzan, J. F. and Jezek, K. C. 2000. Accumulation rate changes in central Greenland from passive microwave data. *J. Geophys. Res. Oceans, Polar Geog.* **24** (2), 98–112.

Bolzan, J. F. and Strobel, M. 1994. Accumulation rate variations around Summit, Greenland. *J. Glaciol.* **40**, 56–66.

Box, J. and Steffen, K. 2001. Sublimation estimates for the Greenland ice sheet using automated weather stations observations. *J. Geophys. Res. Atmos.* **106** (D24), 33 965–82.

Braithwaite, R. J. 1983. Detection of climate signal by interstake correlations on annual ablation data, Qamanånarssûp Sermia, west Greenland. *J. Glaciol.* **35**, 253–9.

   1984. Calculation of degree-days for glacier-climate research. *Z. Gletscherkd. Glazialgeol.* **20**, 1–8.

   1993. Is the Greenland ice sheet getting thicker? *Climatic Change* **23**, 379–81.

   1995a. Aerodynamic stability and turbulent sensible-heat flux over a melting ice surface, the Greenland ice sheet. *J. Glaciol.* **41**, 562–71.

   1995b. Positive degree-day factors for ablation on the Greenland ice sheet studied by energy-balance modeling. *J. Glaciol.* **41**, 153–60.

   1996. Models of ice-atmosphere interactions for the Greenland ice sheet. *Ann. Glaciol.* **23**, 149–53.

Bromwich, D. H., Cullather, R. I., Chen, Q.-S. and Csatho, B. M. 1998. Evaluation of recent precipitation studies for Greenland ice sheet. *J. Geophys. Res.* **103**, 26 007–24.

Bromwich, D. H., Chen, Q.-S., Li, Y. and Cullather, R. I. 1999. Precipitation over Greenland and its relation to the North Atlantic oscillation. *J. Geophys. Res.* **104**, 22 103–15.

Chen, Q.-S., Bromwich, D. H. and Bai, L. 1997. Precipitation over Greenland retrieved by a dynamic method and its relation to cyclonic activity. *J. Climate* **10**, 839–70.

Chen, Q.-S. and Bromwich, D. H. 1999. An equivalent isobaric geopotential height and its application to synoptic analysis and generalized omega-equation in sigma-coordinates. *Month. Weather Rev.* **127**, 145–72.

Cuffey, K. M. and Marshall, S. J. 2000. Substantial contribution to sea-level rise during the last interglacial from the Greenland ice sheet. *Nature* **404**, 591–4.

Davis, C. H. 1997. A robust threshold retracking algorithm for measuring ice-sheet surface elevation change from satellite radar altimeters. *IEEE Trans. Geosci. Remote Sensing* **35**, 974–9.

Davis, C. H., Kluever, C. A. and Haines, B. J. 1998. Elevation change of the southern Greenland ice sheet. *Science* **279**, 2086–8.

Davis, C. H., Kluever, C. A., Haines, B. J., Perez, C. and Yoon, Y. 2000. Improved elevation change measurement of the southern Greenland ice sheet from satellite radar altimetry. *IEEE Trans. Geosci. Remote Sensing* **38**, 1367–78.

Davis, C. H., McConnell, J. R., Bolzan, J., Bamber, J. L., Thomas, R. H. and Mosley-Thompson, E. 2001. Elevation change of the southern Greenland ice sheet from 1978 to 1988: interpretation. *J. Geophys. Res. Atmos.* **106** (D24), 33 743–54.

Drinkwater, M. R. and Long, D. G. 1998. Seasat, ERS-1/2 and NSCAT scatterometer-observed changes on the large ice sheets. In *Proceedings of the Joint ESA-EUMETSAT Workshop on Emerging Scatterometer Applications – From Research to Operations*. ESA SP-424. ESTEC, Noordwijk, The Netherlands, ESA Publications Division, pp. 91–6.

Drinkwater, M. R., Long, D. G. and Bingham, A. W. 2001. Greenland snow accumulation estimates from scatterometer data. *J. Geophys. Res. Atmos.* **106** (D24), 33 935–50.

Drygalski, E. von. 1897. *Gronland-Expedition der Gesellschaft fur Erdkunde zu Berlin 1891–1893*, vol. 1. Berlin, W. H. Kuhl.

Ekholm, S. 1996. A full coverage, high-resolution, topographic model of Greenland computed from a variety of digital elevation data. *J. Geophys. Res.* **101**, 21 961–72.

Fahnestock, M., Bindschadler, R., Kwok, R. and Jezek, K. 1993. Greenland ice-sheet surface properties and ice dynamics from ERS-1 SAR imagery. *Science* **262**, 1530–4.

Fischer, H. 1997. Raumliche Variabilitat in Eiskernzeitreihen Nordostgronlands: Rekonstruktion klimatischer und luftchemischer Langzeittrends seit 1500 A. D. Ph.D. Thesis, Ruprecht-Karls-Universitat, Heidelberg.

Fischer, H. M. *et al.* 1998. Little Ice Age clearly recorded in northern Greenland ice cores. *Geophys. Res. Lett.* **25** (10), 1749–52.

Friedmann, A., Moore, J. C., Thorsteinsson, T., Kipfstuhl, J. and Fischer, H. 1995. A 1200 year record of accumulation from northern Greenland ice cores. *Ann. Glaciol.* **21**, 19–25.

Fuhrer, K., Neftel, A., Anklin, M. and Maggi, V. 1993. Continuous measurements of hydrogen peroxide, formaldehyde, calcium and ammonium concentrations along a new GRIP ice core from Summit, central Greenland. *Atmos. Environ.* **27**, 1873–80.

Gogineni, S., Chuah, T., Allen, C., Jezek, K. and Moore, R. K. 1998. An improved coherent radar depth sounder. *J. Glaciol.* **44**, 659–69.

Gogineni, S. *et al.* 2001. Coherent radar ice thickness measurements over the Greenland ice sheet. *J. Geophys. Res. Atmos.* **106** (D24), 33 761–72.

Goldstein, R. M., Engelhardt, H., Kamb, B. and Frolich, R. M. 1993. Satellite radar interferometry for monitoring ice sheet motion: application to an Antarctic ice stream. *Science* **262**, 1525–30.

Greuell, W. and Knap, W. H. 2000. Remote sensing of the albedo and detection of the slush line on the Greenland ice sheet. *J. Geophys. Res.* **105**, 15 567–76.

Greuell, W., Denby, B., Van de Wal, R. S. W. and Oerlemans, J. 2001. Ten years of mass-balance measurements along a transect near Kangerlussuaq, central west Greenland. *J. Glaciol.* **47**, 157–8.

Gudmandsen, P. 1976. Studies of ice by means of radio echo soundings. Technical University of Denmark, Laboratory of Electromagnetic Theory, R 162, Lyngby, Denmark.

Hamilton, G. S. and Whillans, I. M. 2000. Point measurements of mass balance of the Greenland ice sheet using precision vertical global positioning system (GPS) surveys. *J. Geophys. Res.* **105** (B7), 16 295.

Hanna, E., Valdes, P. and McConnell, J. 2001. Patterns and variations of snow accumulation over Greenland, 1979–98, from ECMWF analyses, and their verification. *J. Climate* **14**, 3521–35.

Higgins, A. K. and Weidick, A. 1988. The world's northernmost surging glacier? *Zeits. Gletscherkunde Glazialgeol.* **24**, 111–23.

Huybrechts, P. 1994. The present evolution of the Greenland ice sheet: an assessment by modeling. *Global & Planetary Change* **9**, 39–51.

  1996. Basal temperature conditions of the Greenland ice sheet during the glacial cycles. *Ann. Glaciol.* **23**, 226–36.

IPCC 2001. *Climate Change 2001– the Scientific Basis*. Contribution of Working Group I to the Third Assessment Report. Cambridge University Press.

Jezek, K. C., Gogineni, P. and Shanableh, M. 1994. Radar measurements of melt zones on the Greenland ice sheet. *Geophys. Res. Lett.* **21**, 33–6.

Joshi, M. 1999. Estimation of surface melt and absorbed radiation on the Greenland ice sheet using passive microwave data. Ph. D. Thesis, The Ohio State University, Columbus.

Joughin, I., Kwok, R. and Fahnestock, M. 1996a. Estimation of ice sheet motion using satellite radar interferometry: method and error analysis with application to the Humboldt Glacier, Greenland. *J. Glaciol.* **42**, 564–75.

  1998. Interferometric estimation of the three-dimensional ice-flow velocity vector using ascending and descending passes. *IEEE Trans. Geosci. Remote Sensing* **36**, 25–37.

Joughin, I., Winebrenner, D. P. and Fahnestock, M. 1995. Observations of complex ice sheet motion in Greenland using satellite radar interferometry. *Geophys. Res. Lett.* **22**, 571–4.

Joughin, I., Winebrenner, D. P., Fahnestock, M., Kwok, R. and Krabill, W. 1996b. Measurement of ice-sheet topography using satellite radar interferometry. *J. Glaciol.* **42**, 231–41.

Joughin, I., Tulaczyk, S., Fahnestock, M. and Kwok, R. 1996c. A mini-surge on the Ryder Glacier, Greenland, observed via satellite radar interferometry. *Science* **274**, 228–30.

Joughin, I., Fahnestock, M., Kwok, R., Gogineni, P. and Allen, C. 1999a. Ice flow of Humboldt, Petermann and Ryder Gletscher, northern Greenland. *J. Glaciol.* **45**, 231–41.

Joughin, I. *et al.* 1996b. Tributaries of West Antarctic ice streams revealed by RADARSAT interferometry. *Science* **286**, 283–6.

Joughin, I., Fahnestock, M., MacAyeal, D., Bamber, J. and Gogineni, P. 2001. Observation and analysis of ice flow in the largest Greenland ice stream. *J. Geophys. Res. Atmos.* **106** (D24), 34 021–34.

Kanagaratnam, P., Gogineni, S. P., Gundestrup, N. and Larson, L. 2001. High-resolution radar mapping of internal layers at the north Greenland ice core project. *J. Geophys. Res. Atmos.* **106** (D24), 33 799–812.

Konzelmann, T., Van de Wal, R. S. W., Greuell, W., Bintanja, R., Henneken, E. and Abe-Ouchi, A. 1994. Parameterization of global and longwave incoming radiation for the Greenland ice sheet. *Global & Planetary Change* **9**, 143–64.

Krabill, W. B., and Martin, C. F. 1987. Aircraft positioning using global positioning system carrier phase data. *J. Inst. Nav.* **34**, 1–21.

Krabill, W., Thomas, R., Martin, C., Swift, R. and Frederick, E. 1995. Accuracy of airborne laser altimetry over the Greenland ice sheet. *Int. J. Remote Sensing* **16**, 1211–22.

Krabill, W. *et al.* 1999. Rapid thinning of parts of the southern Greenland ice sheet. *Science* **283**, 1522–4.

　2000. Greenland ice sheet: high-elevation balance and peripheral thinning. *Science* **289**, 428–30.

Kwok, R. and Fahnestock, M. A. 1996. Ice sheet motion and topography from radar interferometry. *IEEE Trans. Geosci. Remote Sensing* **34**, 189–200.

Long, D. G. and Drinkwater, M. R. 1994. Greenland ice sheet surface properties observed by the Seasat-A scatterometer at enhanced resolution. *J. Glaciol.* **40**, 213–30.

　1999. Cryosphere applications of NSCAT data. *IEEE Trans. Geosci. Remote Sensing* **37**, 1671–84.

McConnell, J. R., Mosley-Thompson, E., Bromwich, D. H., Bales, R. C. and Kyne, J. D. 2000a. Interannual variations of snow accumulation on the Greenland ice sheet (1985–1996): new observations versus model predictions. *J. Geophys. Res.* **105**, 4039–46.

McConnell, J. R. *et al.* 2000b. Changes in Greenland ice sheet elevation attributed primarily to snow accumulation variability. *Nature* **406**, 877–9.

　2001. Annual net snow accumulation over southern Greenland from 1975 to 1998. *J. Geophys. Res. Atmos.* **106** (D24), 33 827–38.

Mohr, J. J., Reeh, N. and Madsen, S. N. 1998. Three-dimensional glacial flow and surface elevation measured with radar interferometry. *Nature* **391**, 273–6.

Mosley-Thompson, E., Thompson, L. G., Dai, J., Davis, M. and Lin, P. N. 1993. Climate of the last 500 years: high resolution ice core records. *Quat. Sci. Rev.* **12**, 419–30.

Mosley-Thompson, E. *et al.* 2001. Local to regional-scale variability of annual net accumulation on the Greenland ice sheet from PARCA cores. *J. Geophys. Res. Atmos.* **106** (D24), 33 839–52.

Mote, T. L. 2000. Ablation rate estimates over the Greenland ice sheet from microwave radiometric data. *Prof. Geog.* **52**, 322–31.

Mote, T. L., and Anderson, M. R. 1995. Variations in snowpack melt on the Greenland ice sheet based on passive microwave measurements. *J. Glaciol.* **41**, 51–60.

Oerlemans, J. and Vugts, H. 1993. A meteorological experiment in the melting zone of the Greenland ice sheet. *Bull. Am. Meteorol. Soc.* **74**, 355–65.

Ohmura, A. 1987. New temperature distribution maps for Greenland. *Zeits. Gletscherkunde Glazialgeol.* **21** (1), 1–45.

Ohmura, A. and Reeh, N. 1991. New precipitation and accumulation maps for Greenland. *J. Glaciol.* **37**, 140–8.

Ohmura, A., Wild, M. and Bengtsson, L. 1996. A possible change in mass balance of Greenland and Antarctic ice sheets in the coming century. *J. Climate* **9**, 2124–35.

Ohmura, A. *et al.* 1994. Energy balance for the Greenland ice sheet by observation and model computation. In *Snow and Ice Cover: Interactions with the Atmosphere and Ecosystem*. IAHS Publ. 233, pp. 85–94.

Pfeffer, W. T., Meier, M. F. and Illangasekare, T. H. 1991. Retention of Greenland runoff by refreezing; implications for projected sea-level rise. *J. Geophys. Res.* **96** (C12), 22 117–24.

Reeh, N. 1985. Greenland ice-sheet mass balance and sea-level change. In *Glaciers, Ice Sheets and Sea-Level: Effects of a CO_2-induced Climatic Change.* Washington, D.C., National Academy Press, pp. 155–71.

1989. Dynamic and climatic history of the Greenland ice sheet. In Fulton, R. J., ed., *The Geology of North America*, vol. K1. Quaternary Geology of Canada and Greenland, chap. 14. Boulder, CO, Geological Society of America.

1991. Parameterization of melt rate and surface temperature on the Greenland ice sheet. *Polarforschung* **59**, 113–28.

Reeh, N. *et al.* 2002. Glacier specific ablation rate derived by remote sensing measurements, *Geophys. Res. Lett.* **29**, 10.1–10.4.

Rignot, E. 1996. Tidal flexure, ice velocities and ablation rates of Petermann Gletscher, Greenland. *J. Glaciol.* **42**, 476–85.

1998. Hinge-line migration of Petermann Gletscher, north Greenland, detected using satellite radar interferometry. *J. Glaciol.* **44**, 469–76.

Rignot, E., Jezek, K. C. and Sohn, H. G. 1995. Ice flow dynamics of the Greenland ice sheet from SAR interferometry. *Geophys. Res. Lett.* **22**, 575–8.

Rignot, E., Gogineni, S., Krabill, W. and Ekholm, S. 1997. Ice discharge from north and northeast Greenland as observed from satellite radar interferometry. *Science* **276**, 934–7.

Rignot, E., Buscarlet, G., Csatho, B., Gogineni, S., Krabill, W. and Schmeltz, M. 2000. Mass balance of the northeast sector of the Greenland ice sheet: a remote sensing perspective. *J. Glaciol.* **46**, 265–73.

Rignot, E., Gogineni, S. P., Joughin, I. and Krabill, W. B. 2001. Contribution to the glaciology of northern Greenland from satellite radar interferometry. *J. Geophys. Res. Atmos.* **106** (D24), 34 007–20.

Shepherd, A., Wingham, D. J., Mansley, J. A. D. and Corr, H. F. J. 2001. Inland thinning of Pine Island glacier, west Antarctica. *Science* **291**, 862–4.

Shuman, C. A., Bromwich, D. H., Kipfstuhl, J. and Schwager, M. 2001a. Multiyear accumulation and temperature history near the North Greenland ice core project site, north central Greenland. *J. Geophys. Res. Atmos.* **106** (D24), 33 853–66.

Shuman, Fahnestock, M., Alley, R. and Stearns, C. R. 2001b. A dozen years of temperature observations at the Summit: central Greenland automatic weather stations 1987–1999. *J. App. Meteorol.* **40** (4), 741–52.

Sohn, H. S., Jezek, K. C. and Van der Veen, C. J. 1998. Jakobshavn Glacier, west Greenland: 30 years of spaceborne observations. *Geophys. Res. Lett.* **25**, 2699–702.

Steffen, K. 1995. Surface energy exchange during the onset of melt at the equilibrium line altitude of the Greenland ice sheet. *Ann. Glaciol.* **21**, 13–8.

Steffen, K. and Box, J. 2001. Surface climatology of the Greenland ice sheet: Greenland climate network 1990–1999. *J. Geophys. Res. Atmos.* **106** (D24), 33 951–64.

Steffen, K., Box, J. and Abdalati, W. 1996. Greenland climate network: GC-Net. CRREL Report on Glacier, Ice Sheets and Volcanoes, no. 96-27, 98–103.

Stroeve, J. 2001. Assessment of Greenland albedo variability from the AVHRR polar pathfinder data set. *J. Geophys. Res. Atmos.* **106** (D24), 33 989–4006.

Stroeve, J. and Steffen, K. 1998. Variability of AVHRR-derived clear sky surface temperature over the Greenland ice sheet. *J. Appl. Meteorol.* **37**, 23–31.

Stroeve, J., Nolin, A. and Steffen, K. 1997. Comparison of AVHRR-derived and in-situ surface albedo over the Greenland ice sheet. *Remote Sensing Environ.* **62**, 262–76.

Tapley, B. D. *et al.* 1996. The joint gravity model 3. *J. Geophys. Res.* **101**, 28 029–49.

Thomas, R. H., Csatho, B. M., Gogineni, S., Jezek, K. C. and Kuivinen, K. 1998. Thickening of the western part of the Greenland ice sheet. *J. Glaciol.* **44**, 653–8.

Thomas, R. *et al.* 2000a. Mass balance of the Greenland ice sheet at high elevations. *Science* **289**, 426–8.

2000b. Substantial thinning of a major east Greenland outlet glacier. *Geophys. Res. Lett.* **27**, 1291–4.

2001a. Program for Arctic regional climate assessment (PARCA): goals, key findings, and future directions. *J. Geophys. Res. Atmos.* **106** (D24), 33 691–706.

2001b. Mass balance of higher-elevation parts of the Greenland ice sheet. *J. Geophys. Res. Atmos.* **106** (D24), 33 707–16.

Van der Veen, C. J. 1993. Interpretation of short-term ice-sheet elevation changes inferred from satellite altimetry. *Climatic Change* **23**, 383–405.

Van der Veen, C. J. and Bolzan, J. F. 1999. Interannual variability in net accumulation on the Greenland ice sheet: observations and implications for mass balance measurements. *J. Geophys. Res.* **104**, 2009–14.

Van de Wal, R. S. W. 1996. Mass-balance modeling of the Greenland ice sheet: a comparison of an energy-balance and a degree-day model. *Ann. Glaciol.* **23**, 36–45.

Van de Wal, R. S. W. and Oerlemans, J. 1994. An energy balance model of the Greenland ice sheet. *Global & Planetary Change* **9**, 115–31.

Van de Wal, R. S. W. *et al.* 1995. Mass-balance measurements in the Søndre Strømfjord area in the period 1990–1994. *Z. Gletscherk. Glazialgeol.* **31**, 57–63.

Van Tatenhove, F. G. M., Van der Meer, J. J. M. and Huybrechts, P. 1995. Glacial/geomorphological research in Greenland used to test an ice-sheet model. *Quat. Res.* **44**, 317–27.

Wahr, J. M., Van Dam, T., Larson, K. and Francis, O. 2001. GPS and absolute gravity measurements in Greenland. *J. Geophys. Res. Solid Earth* **106**, 16 567–82.

Weidick, A. 1985. Review of glacier changes in west Greenland. *Z. Gletscherkd. Glacialgeol.* **21**, 301–9.

1993. Neoglacial change of ice cover and the related response of the earth's crust in west Greenland. *Rapp. Gronlands Geol. Unders.* **159**, 121–6.

1995. Satellite image atlas of glaciers of the world. Greenland, US Geological Survey Professional Paper 1386-C, Washington, D.C. US Government Printing Office.

Winebrenner, D. P., Arthern, R. J. and Shuman, C. A. 2001. Mapping ice sheet accumulation rates using observations of thermal emission at 4.5 cm wavelength. *J. Geophys. Res. Atmos.* **106** (D24), 33 919–34.

Wingham, D. J., Ridout, A. J., Scharroo, R., Arthern, R. and Shum, C. K. 1998. Antarctic elevation change from 1992 to 1996. *Science* **282**, 456–8.

Yoon, Y. T. 1998. Global orbit-error analysis using stochastic filtering methods. M. S. Thesis, University of Missouri, Columbia.

Zwally, H. J. and Giovinetto, M. B. 1995. Accumulation in Antarctica and Greenland derived from passive-microwave data: a comparison with contoured compilations. *Ann. Glaciol.* **21**, 123–30.

Zwally, H. J. and Brenner, A. C. 2001. Ice sheet dynamics and mass balance. In Fu, L. and Cazenave, A., eds., *Satellite Altimetry and Earth Sciences*. New York, Academic Press, pp. 351–69.

Zwally, H. J., Brenner, A. C. and DiMarzio, J. P. 1998. Technical comment: Growth of the southern Greenland ice sheet. *Science* **281**, 1251.

Zwally, H. J., Brenner, A. C., Major, J. A., Bindschadler, R. A. and Marsh, J. G. 1989. Growth of Greenland ice sheet: measurement. *Science* **246**, 1587–9.

11

Greenland: modelling

RODERIK S. W. VAN DE WAL
Institute for Marine and Atmospheric Research, Utrecht University

11.1 Introduction

Modelling the mass balance of the Greenland ice sheet is a way to improve our understanding of the processes that are important for the behaviour of the ice sheet. Models are tools to find out whether we can explain the observations and extrapolate them to areas for which no observations are available. The purpose of mass balance models is to relate mass balance to the prevailing or changing climate. This offers the possibility to predict how the ice sheet responds to climatic change. Changes in the ice flow have response times of the order of 10^4 years and are determined by isostasy and thermodynamics. Changes in the specific mass balance can be much faster. For the Greenland ice sheet, under the present-day climate, the long-term dynamic imbalance is probably small (Church *et al.*, 2001; Huybrechts and De Wolde, 1999). For this reason, the main focus of this chapter will be on modelling the specific mass balance. Changes in accumulation and ablation due to climate changes can contribute significantly to sea-level changes on 100-year timescales. To study this, several mass balance models for the Greenland ice sheet are used. We can distinguish three categories of models:

- general circulation models;
- parameterized models;
- boundary layer models.

General circulation models (GCMs) take into account changes in the atmospheric circulation in a realistic manner, which is why they are particularly useful for calculating (changes in) accumulation. They are, however, not yet very appropriate for ablation calculations, as will become clear later in this chapter.

Parameterized models, such as surface energy balance models and degree-day models, are particularly useful for sensitivity experiments that focus on changes in ablation resulting from climatic changes. These models can easily be included in ice dynamical models that

The contribution of N. Reeh to the section on calving and bottom melting is highly appreciated. I thank the reviewers and editors for their helpful comments that significantly improved this chapter.

Mass Balance of the Cryosphere: Observations and Modelling of Contemporary and Future Changes, eds. Jonathan L. Bamber and Antony J. Payne. Published by Cambridge University Press. © Cambridge University Press 2003.

Table 11.1. *Physical characteristics of the Greenland ice sheet.*

Mass balance components are averaged over the area of the ice sheet.

Area ($10^6 \, km^2$)	1.71
Volume ($10^6 \, km^3$)	2.85
Sea-level rise equivalent (m)	7.2^a
Accumulation (water equivalent per year)	0.304 ± 0.015
Ablation (water equivalent per year)	0.174 ± 0.019
Iceberg discharge (water equivalent per year)	0.137 ± 0.019
Bottom melting (water equivalent per year)	0.019 ± 0.002
Balance (water equivalent per year)	-0.026 ± 0.031

^{a} After isostatic rebound and sea-water replacing grounded ice and an oceanic area
of $3.62 \times 10^8 \, km^2$.

Source: Church *et al.* (2001).

address longer timescales. They are not very useful for the calculation of accumulation (changes), because they do not solve the atmospheric circulation at all.

Boundary layer models are used for local studies of ablation and yield information on relevant processes. Knowledge obtained from this type of model can be used to improve either parameterized models or GCMs.

These three types of mass balance models are used for studies of (parts) of the Greenland ice sheet. Mass balance modelling focussing on the entire ice sheet was initiated during the 1980s and 1990s. A compilation of the physical characteristics of the ice sheet, including a compilation of observed and modelled mass balance terms, is presented in Table 11.1.

A particular problem in modelling the mass balance of Greenland is the ablation zone, as will be discussed in more detail later in this chapter. The ablation zone is estimated to cover only 10% of the ice sheet (Reeh, 1989), which makes it a marginal zone of 30 km, on average. Highly parameterized models are capable of explicitly solving this marginal zone, but GCMs need downscaling techniques or crude parameterizations to enhance the resolution near the surface.

In this chapter, we will use the term (total) mass balance as defined in Chapter 2. The mass balance is found by integrating annual values of accumulation, loss by runoff and loss by iceberg discharge over the entire ice sheet. However, melt loss from beneath the ice sheet is neglected, since it is assumed to be small (see Table 11.1). The term specific (mass) balance is used for only the sum of local accumulation and ablation (see Chapter 2). The chapter starts with a discussion of models for the different components of the surface mass balance with emphasis on the modelling of ablation. The chapter concludes with a discussion of the dynamical imbalance of the ice sheet as calculated by ice flow models, and of how modelling should proceed.

11.2 Modelling the specific mass balance

Ablation depends strongly on temperature and therefore implicitly on elevation. This means that observations of the specific mass balance as a function of elevation are crucial for the validation of mass balance models, in particular if such observations are continued for several years. These profiles can only be used for validating models if climate data for the same period are available. Unfortunately, very few mass balance data have been published to date, but a compilation of all unpublished data is in progress (C. Bøggild, personal communication). Due to the limited information on ablation, it is impossible to present a compilation of the ablation data for the entire ice sheet. Therefore, we restrict ourselves here to specific mass balance profiles used in modelling studies of the ice sheet as presented in Figure 11.1. All profiles show, on average, a linear relation with elevation. Data from year to year, however, do not necessarily exhibit a linear dependence. Only a few specific mass balance records for longer than one year exist, and they are mainly from West Greenland, a climatologically homogeneous region (dry and warm). Few observations are available for other parts of the ice sheet. As a consequence, the evaluation of the performance of specific mass balance models has, until now, been restricted to crude comparisons, as presented in Figure 11.1

11.2.1 Modelling accumulation

Parameterized models and boundary layer models do not provide information on changes in the accumulation. The only possibility of acquiring that information is through the use of GCMs or numerical weather prediction models (no distinction is made here). It should, however, be kept in mind that GCMs calculate precipitation rather than accumulation. A GCM usually considers accumulation equal to precipitation minus evaporation. Local differences between precipitation minus evaporation on one hand and accumulation on the other hand, for instance due to snow drift, are not considered here. In this paragraph, we restrict ourselves to the contribution of GCMs to the calculation of the accumulation (precipitation minus evaporation) over Greenland. Several studies focus on accumulation over the Greenland ice sheet (e.g. Bromwich *et al.*, 1993; Chen, Bromwich and Bai, 1997; Genthon and Braun, 1995; Glover, 1999). A serious problem is that the models include parameterizations that are optimized for middle and low latitudes. Superimposed on this is the notorious problem of obsolete ice sheet topography data in models (e.g. Genthon and Braun, 1995; Glover, 1999). Genthon, Jouzeland and Déqué (1994) and Ohmura, Wild and Bengtsson (1996) demonstrated that simulations for Greenland improve with higher resolutions. Connolly and King (1996) pointed out that, in models for Antarctica, a substantial part of the moisture transport to the ice sheet results from exaggerated horizontal diffusion through smoothing, introduced to maintain numerical stability and hence a limited resolved moisture transport. The numerical diffusion might be important for Greenland as well. The quality of GCM results suffers from this. Nevertheless, most models for Greenland capture the main

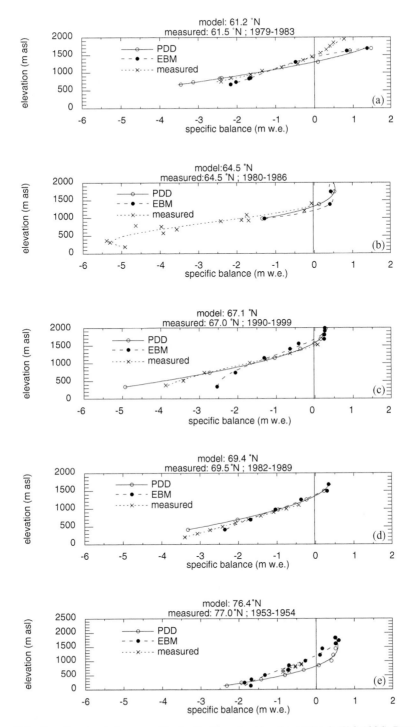

Figure 11.1. Specific mass balance profiles for the Greenland ice sheet (Van de Wal, 1996). Data from 61.5°N (Clement, 1981, 1982, 1983, 1984), 64.5°N (Braithwaite and Olesen, 1989), 67°N (Greuell *et al.*, 2001), 69.5°N (H. H. Thomsen, personal communication), 76°N (Nobles, 1960). EBM are results from an energy balance model and PDD are results from a degree-day model.

characteristics of accumulation observations. The models predict large-scale features, such as a precipitation minimum in the north-east (sometimes over-estimated) and maximum values near the south-eastern and southern coasts. The models do not always predict the band of high accumulation trending north–south along the western flank to be as pronounced as the observations show it to be, but they do at least predict it. The best approach at present is probably to use a GCM with a high resolution, although the necessarily limited length of the runs, due to the high computational costs of the climatology runs, might introduce uncertainties related to the inter-annual variability. As an example of GCM-produced accumulation, we show results of ECHAM4 T106 (1.1°) for the present-day climate in Figure 11.2(a) (Wild and Ohmura, 2000). Results are presented on an equidistant grid with a spacing of 20 km as used by many parameterized mass balance models for Greenland. Figure 11.2(a) can be compared with the actual observations in Figure 11.2(b) as presented by Ohmura and Reeh (1991). We use this climatology as an example because it has been widely used as input for mass balance modelling studies in the 1990s. However, any more up-to-date climatology (such as Bales *et al.*, 2001; Calanca *et al.*, 2000; Ohmura *et al.*, 1999) will show a similar large-scale resemblance between model results and observations.

Table 11.2 presents an overview of some of the recent model results for the accumulation. Despite the fact that the large-scale spatial distribution of the accumulation is reasonable, most models tend to over-estimate mean accumulation. The values for the central parts of the ice sheet are often too high, which is probably at least partly related to obsolete topography and limited horizontal resolution.

Until now, we have been discussing the capability of modelling the present-day distribution. Not only GCMs are used to predict accumulation changes. Attempts have been made to link accumulation changes to temperature changes (or to assume that accumulation is independent of temperature changes). This is based on the fact that the saturation vapour pressure increases with temperature at a constant relative humidity. Higher temperatures enable larger moisture contents of air, and the idea is that this leads to more precipitation. This idea appeared to be supported by a comparison of accumulation and temperature in ice cores from central parts of the ice sheet (e.g. Dahl-Jensen *et al.*, 1993). However, more and more evidence points to a more complicated dependence of accumulation changes over Greenland (e.g. Wild and Ohmura, 2000). Changes in the general circulation are probably also very important for changes in the accumulation in Greenland. This means that the use of high resolution GCMs at timescales of centuries is the only serious alternative for future work concerning the modelling of accumulation. At present, it is doubtful whether regular lower resolution GCMs are suitable for this purpose, given the over-estimation of the mean accumulation.

Despite their limitations, several GCMs have been used to quantify the accumulation change for a doubled-CO_2 climate. Wild and Ohmura (2000) present an increase of 28% in accumulation (0.9 m water equivalent per year) for the ECHAM4 T106 model. This differs substantially from the earlier estimates by the ECHAM3 T106 model (Ohmura *et al.*, 1996), which exhibited a decrease of 5% in accumulation. The large increase in ECHAM4 is mainly

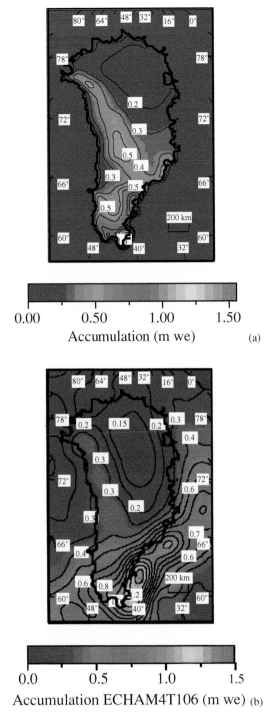

Figure 11.2. (a) Annual accumulation for the Greenland ice sheet (Ohmura and Reeh, 1991). (b) Accumulation distribution modelled by ECHAM4 T106 for the present-day situation (Wild and Ohmura, 2000).

Table 11.2. *A compilation of modelled accumulation by different techniques and models.*

Source	Accumulation (m water equivalent per year)	Scaled accumulation[a]
Genthon and Braun (1995)[b]	0.32	104%
Bromwich *et al.* (1993) 1963–1989[c]	0.41	132%
Ohmura *et al.* (1996)[d]	0.43	137%
Thompson and Pollard (1997)[e]	0.44	
Glover (1999)[f]	0.39	126%
Wild and Ohmura (2000)[g]	0.33	107%

[a] This presents the modelled accumulation scaled by the mean, 0.304 m water equivalent per year, of a compilation of observations and models by Church *et al.* (2001).

[b] ECMWF, six years of re-analyses.

[c] Advection of relative geostrophic vorticity.

[d] ECHAM3 T106, five years.

[e] Values from Thompson and Pollard are for precipitation and not for accumulation. The difference between accumulation and precipitation varies from model to model but is typically 5–15% of the precipitation. Genesis-2 T31 atmosphere and $2° \times 2°$ for the surface.

[f] UGCM T42 re-analyses downscaled by a factor of 4.

[g] ECHAM 4 T106, ten years.

due to enhanced winter accumulation, which was not found in ECHAM3. The increase in winter accumulation is associated with the northward displacement of the Icelandic low and the increased pressure gradient along the southern parts of the east and west coasts (Wild and Ohmura, 2000). These changes are favourable for more accumulation along the south-eastern coast and south-western coast. Thompson and Pollard (1997) also present an increase of 27% (equivalent to 0.12 m water equivalent per year in their model) for the doubled-CO_2 climate. These large changes justify the need for a better understanding of accumulation changes in Greenland. Application of high resolution obtained by a nested regional climate model or by using a stretched-grid GCM (Krinner *et al.*, 1997) might be a more promising approach for future work. Much improvement is also expected from the ECMWF re-analysis project (ERA40), the results of which are not yet available.

11.2.2 Modelling ablation

This section discusses the three categories of mass balance model. Refreezing is also discussed. The section concludes with a compilation of ablation model results for the entire ice sheet.

General circulation models

Despite the limited material available for validation, several ablation models have been developed in order to estimate the effects of climate change. To a large extent, the energy balance of the surface in summer-time determines the ablation in Greenland (e.g. Ambach, 1963). A rational approach is, therefore, to start by calculating the surface energy budget. GCMs yield the information needed to calculate the entire surface energy budget, but have not yet been used very much. This is firstly due to the fact that the spatial resolution is limited. The highest resolution afforded for climate studies by GCMs is currently T106 with a grid size of 120 km in the meridional direction and, in Greenland, 20–60 km in the longitudinal direction, whereas the ablation zone is on average only 30 km wide. Hence, GCM calculations are hampered by limited resolution. GCM runs with high resolution are currently too expensive to simulate periods of hundred years or longer. Alternatives are to use downscaling techniques (e.g. Glover, 1999; Thompson and Pollard, 1997). Secondly, and probably less importantly, GCMs do not take the change in geometry over the course of years into account, although there would be no objection to this.

Attempts have been made to use GCM temperature output to calculate ablation in a parameterized way, independent of the energy fluxes (Wild and Ohmura, 2000), and to use GCM output (temperature, cloudiness and accumulation) as input for a parameterized surface energy balance model (Van de Wal, Wild and De Wolde, 2001). Currently, the merits of using GCMs for specific mass balance studies of the Greenland ice sheet arise mainly from the capability of accounting for spatial and temporal changes in accumulation for which a resolution of T106 seems to be sufficient (Wild and Ohmura, 2000). This means that the present ablation models cannot take full advantage of GCMs.

Parameterized models

Alternative models for calculating the ablation are either surface models or boundary layer models. By 'surface models' we mean models that calculate the ablation without taking the vertical structure of the atmosphere (and top layer of the ice sheet) into account. Within the class of surface models we distinguish surface energy balance models and degree-day models. A surface energy balance model needs information on temperature, cloudiness and accumulation to calculate surface heat fluxes, radiation and turbulent heat exchange. Degree-day models calculate ablation based on temperature only (Chapter 5). Here, we discuss a degree-day model and a surface energy balance model.

We start with the degree-day approach presented by Reeh (1991). Ice-sheet modellers have been widely using slightly modified forms of this type of ablation model (e.g. Greve, 1995, 1997; Huybrechts, Letréguilly, and Reeh, 1991; Letréguilly, Huybrechts and Reeh, 1991; Ritz, Fabre and Letréguilly, 1997; Van de Wal and Oerlemans, 1997). The basic assumption is that ablation is linearly related to the sum of positive degree-days over the entire year (see Chapter 5). Therefore, this type of model needs a mean daily temperature distribution as input. For the Greenland ice sheet, temperature can, to a reasonable approximation, be parameterized as a function of elevation and latitude. This temperature parameterization is based on a compilation of all available monthly and annual temperature data (Ohmura, 1987). GCMs can contribute here, because they calculate temperature fields in time and space for a given geometry of the ice sheet with reasonable accuracy. Based on the monthly temperature distribution in space and time, the positive degree-day sum is calculated. Consequently, ablation is evaluated by assigning degree-day factors for snow and ice. Application of monthly mean temperatures would mean that if the average July temperature is below zero, ablation would be zero. In reality, ablation occurs if the July temperature is close to zero, due to the daily cycle and to daily weather-related temperature variations that are both excluded in the parameterization. Therefore, Reeh (1991) followed a proposal by Braithwaite (1985) and introduced a stochastic term (σ), which accounts for these random variations and allows some ablation, even with the July temperature below zero. The positive degree-day sum is used for estimating ablation (Braithwaite and Thomsen, 1984). Reeh (1991) validated his model against four specific mass balance profiles near the margins of the ice sheet.

A second surface model for ablation is the surface energy balance model as proposed by Van de Wal and Oerlemans (1994). This model can be considered a follow-up of the energy balance model presented by Oerlemans, Van de Wal and Conrads (1991). Ablation is calculated from the amount of energy available at the surface, based on a parameterization of the radiation budget and the turbulent transfer between surface and atmosphere. Input variables are the 2 m temperature and accumulation, as for degree-day models, and, in addition, cloudiness. This model neither takes into account the vertical structure of the atmosphere nor subsurface processes. We therefore use the term 'surface energy balance model' and not simply 'energy balance model' to prevent confusion among climatologists or climate modellers working with GCMs. Cloudiness is based on observations from coastal stations and an assumed reduction in cloud cover between the ice margin and the central parts of the ice sheet. Increased cloud cover reduces short-wave radiation at the surface, but increases net long-wave radiation. Turbulent energy is proportional to the difference between the 2 m temperature and surface temperature. Temperature changes have a three-fold effect in the model: turbulent heat exchange changes, long-wave radiation changes and refreezing changes. Since the model lacks vertical structure, changes in free-atmosphere temperature or a change in chemical composition of the atmosphere cannot be prescribed. Figure 11.3 depicts the specific mass balance distribution as calculated by Van de Wal and Oerlemans (1994). The figure basically shows large negative values around the margin on top of the accumulation distribution (Figure 11.2(b)).

A weakness of this type of energy balance model is the lack of detail in the subsurface energy budget. Surface temperature and refreezing are parameterized very crudely, and are validated using only Ambach's data for one location (Ambach, 1963). Since no more data were available, a more sophisticated approach was impossible. The modelling of refreezing will be discussed separately, later in this chapter.

In this energy balance model, the albedo parameterization is important because the snow and ice albedo differ considerably, and short-wave radiation is often the most important term of the energy balance. The large difference between the albedo of snow and ice implies a

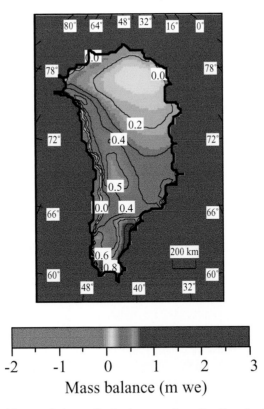

Figure 11.3. The specific mass balance distribution over the entire Greenland ice sheet, calculated by means of the surface energy balance model presented by Van de Wal and Oerlemans (1994).

strong feedback mechanism. Less snow leads to more ablation due to the lower albedo of ice. One can argue that degree-day models also include an albedo feedback mechanism. In degree-day models, the number of degree-days is linearly coupled to snow and ice melt, by degree-day factors of 0.003 for snow and 0.007 for ice (Reeh, 1991). This means that the same number of degree-days yields only 40% of the ablation for a snow surface compared to an ice surface. However, the ratio between the ablation of a snow surface and ice surface is not only determined by the albedo, but also by the other components of the surface energy budget. Van de Wal (1996) demonstrated the importance of the albedo feedback and showed that the specific mass balance sensitivity is halved if the albedo is kept constant in time and space.

Theoretically, energy balance approaches should yield better results. Several studies, however, have pointed out that energy balance models suffer from a lack of knowledge and validation of the input data. A crucial point is that the temperature distribution and variations above the Greenland ice sheet are not very well known, which means that degree-day models suffer from a similar lack of accurate input data as surface energy models. One might argue that the choice of the type of model to calculate ablation is arbitrary. Chapter 5 provides arguments for adopting either a degree-day model or an energy balance model. Both types of model, applied to the Greenland ice sheet, are capable of describing the available ablation data reasonably well (Van de Wal, 1996). This is probably due to the limited data available to test the models. A second reason is that temperature and accumulation input are similar for both types of models. Another way to obtain insight into the performance of ablation models would be to test whether they can explain the inter-annual variability in ablation instead of the mean profile. Results of this approach have not yet been published, as far as we know.

Interestingly enough, whereas both types of models simulate the present-day observations of ablation reasonably well, they show considerable differences in sensitivity. There is no obvious explanation for this discrepancy. For a perturbation of 1 K, the sensitivity of the ablation calculated by the degree-day model is 20% greater than that of the energy balance model (Van de Wal, 1996). For larger perturbations, the sensitivities diverge. This results in a 32% difference after a 100-year run for a temperature increase of 0.3 K per decade. Still, one should realize that both models exhibit an increasing sensitivity for larger perturbations. Typically, this increase is a factor of 2 for an increase in the temperature perturbation from 1 to 4 K. A constant sensitivity parameter that does not depend on the perturbation magnitude is therefore bound to fail. Consequently, rather than constant sensitivity parameters, mass balance models should be used for climate studies.

A common problem for both types of models is that the calculations are restricted to a grid of 20 km. This is not considered to represent elevation accurately enough. This problem has posed a serious limitation on the validation of specific mass balance calculations from the beginning of the 1990s (Reeh and Starzer, 1996; Van de Wal and Ekholm, 1996). It can probably be solved by the use of high resolution data sets (Janssens and Huybrechts, 2000), which are currently available (Ekholm, 1996) with a claimed accuracy of 13 m for surface slopes smaller than 1 degree.

Boundary layer models

It should be kept in mind that the ablation calculations are still crude. Major improvements in the meteorological aspects are urgently required. Surface ablation models as described in the previous section are highly parameterized. At present, insight into turbulent fluxes is limited and temperature changes in the free atmosphere are still assumed to be equal to near-surface temperature changes. Denby (2001) developed a three-dimensional boundary layer model, with a horizontal resolution of 20 km to simulate the 1998 ablation season. The model is forced by ECMWF analysis, and is capable of describing the vertical structure of the boundary layer. Sensitivity experiments with this model for a 2 K temperature increase in free-atmosphere temperature indicate that increased turbulent fluxes account for 41% of the increased ablation. Increase of the net long-wave radiation is responsible for 17% of the increased ablation and 42% is due to the albedo feedback mechanism. Changes in wind speed are of minor importance for changes in the turbulent heat flux compared with the temperature effect. The model indicates that, for a 1 K increase in free-atmosphere temperature, ablation increases by 35%, which happens to be nearly identical to results obtained with a surface energy balance model described earlier in this chapter (Van de Wal and Oerlemans, 1994). A strong aspect of the model is the result that the inter-annual variability in temperature can explain the inter-annual variability in observed mass balance along the Kangerlussuaq transect over the period 1991 to 2000. The model has been used to derive parameterizations, based on the free-atmosphere temperature, for incoming long-wave radiation and turbulent heat. These parameterizations can be used in surface energy balance models aiming at a timescale of 10 to 100 years. The study by Denby (2001) represents a major step forward in the simulation of ablation. This eliminates one of the major uncertainties arising from how the vertical structure of the atmosphere should be accounted for. It seems more important to focus on the strength of the albedo feedback, which this boundary layer model did not treat in great detail.

With respect to modelling the albedo, it is worthwhile mentioning a study by F. Lefebre and colleagues, who studied snow and ice melt at ETH Camp during the summers of 1990 and 1991 (Lefebre *et al.*, 2003). They used a multi-layered one-dimensional thermodynamic snow–ice model including physical snow metamorphism. They showed that this model is capable of simulating the ablation over two widely different years. In one year, the mass balance was negative, and the following year ended with a positive balance. Albedo was simulated correctly for these widely varying conditions. From these detailed calculations at one site, it is a major step to calculate albedo and ablation over the entire ice sheet. However, given the importance of the albedo feedback mechanism, the effort is justified. This approach also implicitly yields insight into the role of refreezing for ablation, as discussed next.

Refreezing

So far, we have restricted the discussion mainly to the calculation of surface ablation. When surface ablation occurs, melt water penetrates in the snow or firn layer and may refreeze. As a result, snow or firn density increases, but no mass is lost from the ice sheet. Later in the season, runoff occurs if the firn layer is saturated. Details of this process depend on melt rate, firn temperature, accumulation rate and density. By refreezing, energy is transported downward. This energy is not available for melting. Detailed subsurface models do exist (see Chapter 5) and work well for the simulation of refreezing for specific cases on Greenland (Greuell and Konzelmann, 1994). Unfortunately, few measurements are available to validate these models. This means that modelling these processes in detail for the entire ice sheet has not been attempted and that, so far, parameterized solutions have been used instead. A useful schematic attempt was made by Pfeffer, Meier and Illangasekare (1991). These authors defined a runoff limit, above which melt-water production is too low to saturate the firn layer and eliminate the temperature cooling during the winter. In that case, melt-water production can be neglected in specific mass balance studies. Below the runoff limit, the amount of melt water formed is enough to saturate the entire snow pack and heat it up, so all melt water runs off. Applying the model concept proposed by Pfeffer *et al.* (1991) yields a runoff limit that is situated where the annual melt is 70% of the annual accumulation (Janssens and Huybrechts, 2000). This is close to the value that Reeh (1989) had already proposed. Neglecting refreezing would lead to an over-estimation of ablation by approximately 40% for present-day conditions for a degree-day model (as described in the previous sections). This is only 10% for the energy balance model described in the previous section. Comparing the various results is not easy because of the many differences in model set-up and simulation of the refreezing process (see Chapter 5), but the differences between the results from these models warrant further study. Neglecting refreezing also leads to an over-estimation of specific mass balance sensitivity. A comparison of several parameterization schemes for retention of melt water, in specific mass balance models of the positive degree-day type, shows that overall results for the entire ice sheet are fortuitously quite similar for the various refreezing concepts. On the other hand, large spatial differences occur (Janssens and Huybrechts, 2000). More observations are apparently needed to improve this important aspect of ablation models for the Greenland ice sheet. The effect of refreezing is of primary importance for ablation calculations. Improvements can also be accomplished through a comparison of simple refreezing schemes with more complex subsurface models.

A compilation of ablation model results for the ice sheet

Ablation is an important term in the mass balance for the Greenland ice sheet. Only a few observations exist, and therefore numerical models have to be used despite the poor possibilities for calibrating the models. Here, we discuss a compilation of model results in terms of mean ablation and sensitivity of the ablation to changes in climate. We distinguish two categories of models, depending on the degree of parameterization. The first group consists of models directly derived from GCMs and using downscaling techniques to calculate the surface energy fluxes that determine ablation. We reckon the results from Thompson and Pollard (1997) and Glover (1999) to this group. All other results in Table 11.3 belong to models of the second group, which is characterized by a high degree of parameterization. Results from ECHAM (Ohmura *et al.*, 1996; Wild and Ohmura, 2000) fall into the second group since ablation is calculated from a parameterization between July temperature and melt and not from the energy fluxes of the GCM. Table 11.3 shows that the GCM studies by Thompson and Pollard (1997) and Glover (1999) are typically 50% higher in terms of mean ablation. As the parameterized models are calibrated to what is believed to be realistic, it indicates that GCMs are not yet capable of predicting the mean ablation reasonably accurately. Poor resolution, in combination with tuning based on mid-latitudes, is the most likely explanation for the over-estimation. Parameterized models do not show significant differences in the mean ablation as they are tuned to a value of roughly half the mean accumulation. Interestingly enough, significant differences in sensitivity show up. In Table 11.3, we have tried to standardize the sensitivity by the magnitude of the temperature perturbation. For some models, these results are derived from temperature perturbation experiments (those with a superscript *e*) and for others the sensitivity was calculated from doubled-CO_2 experiments (those with a superscript *h*). The final group was normalized by dividing the results by the mean local temperature increase between the doubled-CO_2

Table 11.3. *A compilation of modelled ablation by different techniques and models.*

A compilation of observations and models presented by Church *et al.* (2001) indicates a mean value of 297×10^{12} kg per year equivalent to 0.17 m per year, averaged over the ice-sheet surface area. The sensitivity dA/dT is the sensitivity of the ablation to a local change temperature expressed as global sea-level equivalent (SLC).

Source	Ablation (m per year)	Sensitivity, dA/dT (mm SLC/K) per year	$2 \times CO_2 - 1 \times CO_2$ (m per year)
Thompson and Pollard (1997)[a]	0.31		0.38
Glover (1999)[b]	0.31	0.12[c]	
Huybrechts *et al.* (1991)[d]	0.15	0.33[e]	
Van de Wal (1996)[f]	0.19	0.31[e]	
Ohmura *et al.* (1996)[g]	0.15	0.37[h]	0.21
Wild and Ohmura (2000)[i]	0.20	0.22[h]	0.16
Janssens and Huybrechts (2000)[j]	0.17	0.35[e]	
Van de Wal *et al.* (2001)[k]	0.19	0.16[h]	0.11

[a] Genesis-2 T31 atmosphere and $2° \times 2°$ for the surface.
[b] UGCM T42 re-analyses downscaled by a factor of 4.
[c] Calculated for a +2 K perturbation experiment.
[d] Degree-day model 20 km grid.
[e] Constant accumulation.
[f] Surface energy balance 20 km grid.
[g] ECHAM3 T106 five years.
[h] A mean temperature increase of 3.4 K has been used for Wild and Ohmura (2000) and Van de Wal *et al.* (2001) and 2.9 K for Ohmura *et al.* (1996).
[i] ECHAM 4 T106 ten years.
[j] Updated degree-day model includes geometry changes 5 km grid.
[k] Surface energy balance forcing ECHAM4 T106 includes geometry changes 20 km grid.

climate state and the control run. Whether these two subsets can be compared is doubtful, since a uniform temperature increase in time and space obviously results in different ablation rates compared with a large winter-time increase in central areas in combination with a small summer-time increase along the margins. A second complication in the interpretation of the sensitivity results is the role of accumulation changes. The model results derived from only a temperature change neglect the effect of variability and changes in accumulation influencing the albedo feedback, and hence ablation. Higher accumulation rates yield less ablation for identical temperatures: more snow prevents melt. For this reason it might be worthwhile comparing results from doubled-CO_2 climate runs instead of simple perturbation experiments. But, even then, large differences in sensitivity are observed that depend on how ablation is parameterized and on whether changes in the geometry are included or not. Given the limited constraints for the models, the data in Table 11.3 suggest a wide range for doubled-CO_2 experiments.

11.3 Calving and bottom melt of floating glacier tongues

Another aspect of the mass balance is the iceberg calving process, which is still poorly understood. Calving of icebergs constitutes roughly half the mass loss from the Greenland ice sheet (Table 11.1). This estimate is based on assuming a balance between total accumulation, ablation and calving. Direct observations of the calving fluxes by measuring ice velocity and thickness at the calving front are too limited in number to allow an independent estimate of the calving flux for the entire ice sheet. Reeh (1994a) and Weidick (1995) published compilations of observations of thickness, width and velocity at Greenland calving glacier fronts. These analyses show that no universal calving law can be formulated (Reeh, 1994a). This explains why relatively little effort has been made toward modelling of the calving flux from the Greenland ice sheet. An additional complication is the high resolution needed to resolve the outlet glaciers in models. So far, ice dynamical models for the entire ice sheet have had a resolution of no better than 20×20 km. This means that, for instance, even Jakobshavn Isbrae, the outlet glacier with the largest discharge in Greenland, is not resolved properly. This limitation can be partly conquered through the availability of higher resolution data sets of the topography. However, if one assumes a linear relationship between velocity and ice thickness at the front, one can estimate the contribution of calf ice from the outlet glaciers (Reeh, 1994b). From this analysis, it appears likely that outlet glaciers will become thinner in a warmer climate and that iceberg discharge will decrease. During the retreat of the ice, a transient increase of iceberg discharge may occur, depending on the timescale of the dynamic adjustment of the glaciers and the timescale of climate change. No attempts have yet been made to model the basal melt of floating glacier tongues, but recent observations justify the effort.

11.4 The dynamical imbalance

11.4.1 Introduction

Until this point, we have restricted ourselves to ablation, accumulation and calving. In order to address the question of future volume changes, it is necessary to consider the dynamic response of the ice sheet. For short periods, one tends to focus on the specific mass balance. However, the specific mass balance strongly depends on elevation, and changes in elevation should therefore be included. Several studies, e.g. Huybrechts and De Wolde (1999) and Van de Wal and Oerlemans (1997), indicate that, even on a timescale of 100 years, the dynamic response should be included. This can be achieved with a three-dimensional thermodynamic ice-sheet model (Chapter 7). For Greenland, several model studies have been performed with this type of model. These models produce a geometry in reasonable agreement with the present-day observations of elevation (e.g. Huybrechts *et al.*, 1991; Van de Wal, 1999) and the extent of the ice sheet (Van Tatenhove, Van der Meer and Huybrechts, 1995). The models, however, do not simulate the fast flowing Jakobshavn Isbrae, and neither do they simulate the Greenland ice stream in the north-east. Van der Veen (1999) is therefore justified in questioning the usefulness of these models for the long-term evolution (more than 100 years).

11.4.2 The present-day imbalance

The present-day imbalance, defined as the volume change over the last 200 years as calculated by Huybrechts (1994), may play an important role in future estimates of volume change. Even a constant future climate can lead to volume changes for some time. The imbalance has to be estimated from thermodynamic ice-sheet models that include bedrock adjustment. Results from Huybrechts (1994) and Huybrechts and De Wolde (1999) are based on such a three-dimensional time-dependent ice-sheet model that includes thermo-mechanical coupling as well as the effect of the stiffness contrast between Holocene and Pleistocene ice. The results indicate that the ice sheet as a whole is currently thickening at a mean rate of 0.1 cm per 100 years, but large spatial differences occur. The slightly lower value from the most recent calculations is due to a difference in temperature forcing, but does not contradict the basic conclusion that the Greenland ice sheet is close to equilibrium. This means that volume changes on a timescale of 100 years are most likely dominated by changes in ablation, accumulation and calving, and not by the imbalance. Although the imbalance is probably not the dominant factor, opportunities to validate the spatial distribution should be used to test the applicability of ice-sheet models for calculating volume changes. Testing the applicability of ice-sheet models can be attempted by using ice-thickness estimates from repeated airborne radar depth measurements, in combination with ice-motion data inferred from global positioning systems (GPS). Remote sensing also provides higher resolution data sets of ice thickness (Bamber, Layberry and Gogineni, 2001) and surface elevation (Ekholm, 1996). This will enable the use of higher resolution ice-sheet models, which could be important, because the narrow marginal ablation zone and outlet glaciers would be resolved in more detail.

11.4.3 Volume changes in the near future

The simplest approach to calculating volume changes would be to use mass balance sensitivities derived from ablation models. Degree-day models, surface energy balance models and boundary layer models all yield values equivalent to a sea-level rise of approximately 0.3 mm K per year. This might suggest that predicting volume changes for Greenland is simple. Problems arise from the facts that the sensitivity depends on the magnitude of the perturbation, which already varies in time and space, and that the geometry of the ice sheet and the accumulation are not constant in time either.

   Several studies have been performed to estimate the change of volume of the Greenland ice sheet in the near future with more detail than the constant sensitivity approach. Van de

Wal and Oerlemans (1997) and Huybrechts and De Wolde (1999) used temperature changes derived from a coupled zonal mean energy balance model with an annual cycle (De Wolde *et al.*, 1997). This type of model finally enabled the incorporation of seasonal and some crude latitudinal changes in temperature over the ice sheet. This was a step forward, compared with previous studies based on uniform changes in time over the entire ice sheet. These model studies predicted a sea-level increase of 80–100 mm by the year 2100 for the IS92a scenario (Warrick *et al.*, 1996). A drawback of most coupled zonal mean energy balance models is that they are incapable of simulating accumulation changes. Such changes can only be accurately derived from GCMs. Results by Wild and Ohmura (2000) show that it is very important to include changes in accumulation rate. Moreover, other results (Van de Wal and Oerlemans, 1997; Huybrechts and De Wolde, 1999) make clear that the adjustment of the ice sheet due to ice dynamics cannot be neglected on a timescale of 100 years. This means that climate models and ice-sheet models should be fully coupled, because climate variables (particularly temperature) depend on altitude. This has not yet been attempted with a GCM. In section 11.2.2, we also made clear that the direct use of GCMs for calculating ablation is still in its early stages, though major progress has been achieved in recent years. An intermediate approach is to use GCM output decoupled from changes in the elevation of the ice sheet to drive a mass balance model coupled to an ice-sheet model. Van de Wal *et al.* (2001) applied this approach in a surface energy balance model forced by ECHAM4 T106 output. This intermediate approach was also taken in IPCC 2001, through the use of a degree-day model forced by seven different GCMs (Church *et al.*, 2001). The results indicate that only a small volume change (1 to 5 cm of sea-level rise) will occur during the twenty-first century, depending on the climate-change scenario and model formulation. The reason that the change is small is that enhanced ablation is largely offset by increased accumulation. This result should be considered critically. While high resolution GCMs have the capacity of predicting fairly accurately the present-day distribution of the accumulation, whether the predictions of the accumulation changes are correct still has to be confirmed by information from firn cores and remote sensing. A comparison can then be made between the inter-annual variability of precipitation simulated by the GCM and the observations. These small changes over the next century do not imply that Greenland is not important at all for sea-level studies. Firstly, the uncertainties in these calculations are large, and, secondly, the contribution of Greenland will increase rapidly in the future. Within a greenhouse warming scenario, melting will most likely dominate the volume change on a timescale of a few hundred years. Huybrechts and De Wolde (1999) imposed a stabilization of the temperature around 2150 AD and calculated the response of the ice sheet until 3000 AD. The results show a continuous volume reduction equivalent to 9–60 cm of global sea-level rise per 100 years.

11.5 Outlook

As discussed before, volume changes are not only due to changes in specific mass balance but also to changes in geometry resulting from the ice flow (Chapter 7). Calculations with a very simple calving model indicate that the effect of changes in calving rate may be comparable to the effect of changes in accumulation or ablation (Reeh, 1994b). Model-derived patterns of ice velocity agree with surface velocities derived from synthetic aperture radar interferometry satellite data for parts of the ice sheet (Bamber *et al.*, 2000), but large local differences occur. Numerical ice-flow models do not generate the large ice stream in the north-east (Fahnestock *et al.*, 1993), due to the limited resolution and limitations in the model physics. Numerical flow models for the ice sheet are based on the shallow-ice approximation (Chapter 7) and do not include the evolution of ice streams, as this was not recognized to be important for Greenland until recently. One may therefore wonder to what degree poorly accounted for ice dynamics in the presently available ice-sheet models may further increase the uncertainty in estimates of future ice volume changes. Enhanced basal sliding of major outlet glaciers, due to increased amounts of melt water reaching the bed, may also be important on timescales of 10 to 100 years. As a result, the geometry might change significantly in marginal regions of the ice sheet, and this would have a significant impact on ablation and thus on volume changes. On the other hand, one could argue that the shallow-ice approximation is sufficiently detailed to be used for volume changes on a timescale of 100 years because it provides a reasonable approximation of the expected elevation changes which are crucial for the ablation calculations. Previous studies by Van de Wal and Oerlemans (1997) and Huybrechts and De Wolde (1999) show that it is important to include these changes in the geometry, as a fixed geometry would over-estimate the specific mass balance sensitivity. This complicates modelling efforts because it means that GCMs and ice-sheet models must be fully coupled to account for the effect of elevation changes on the GCM and on the ablation calculations of the ice sheet. This coupling is the challenge for the next generation of specific mass balance models, which should include the necessary physics and a low degree of parameterization.

References

Ambach, W. 1963. Untersuchungen zum Energieumsatz in der Ablationszone des Grønlandischen Inlandeises (Camp IV-EGIG, 69°40′05″N, 49°37′58″W). *Medd. Grønl.* **174** (4),

Bales, R. C., McConell, J. R., Mosley-Thompson, E. and Csatho, B. 2001. Accumulation maps for the Greenland Ice sheet: 1971–1990. *Geophys. Res. Lett.* **28** (15), 2967–70.

Bamber, J. L., Hardy, R. J., Huybrechts, P. and Joughin, I. 2000. A comparison of balance velocities, measured velocities and thermodynamically modeled velocities for the Greenland ice sheet. *Ann. Glaciol.* **30**, 211–16.

Bamber, J. L., Layberry, R. and Gogineni, S. 2001. A new ice thickness and bedrock dataset for the Greenland ice sheet. *J. Geophys. Res.* **106** (D24), 33 773–80.

Braithwaite, R. J. 1985. Calculation of degree-days for glacier-climate research. *Z. Gletscherkd. Glazialgeol.* **20**, 1–8.

Braithwaite, R. J. and Olesen, O. B. 1989. Detection of climate signal by inter-stake correlations of annual ablation data, Qamanârssûp Sermia, West Greenland, *J. Glaciol.* **35** (120), 253–9.

Braithwaite, R. J. and Thomsen, H. H. 1984. Runoff conditions at Paakitsup Akuliarusersua, Jakobshavn, estimated by modelling. *Grønlands Geologiske Undersøgelse Gletscher-hydrol. Meddl.* **84/3**, 22 pp.

Bromwich, D. H., Robasky, F. M., Keen, R. A. and Bolzan, J. F. 1993. Modeled variations of precipitation over the Greenland ice sheet. *J. Climate* **6**, 1253–68.

Calanca, P., Gilgen, H., Ekholm, S. and Ohmura, A. 2000. Gridded temperature and accumulation distributions for Greenland for use in cryospheric models. *Ann. Glaciol.* **31**, 118–20.

Chen, Q.-S., Bromwich, D. H. and Bai, L. 1997. Precipitation over Greenland retrieved by a dynamic method and its relation to cyclonic activity. *J. Climate* **10**, 839–70.

Church, J. A. *et al.* 2001. Changes in sea-level. In Houghton, J. T. and Yihui, D., eds., *IPCC Third Scientific Assessment of Climate Change.* Cambridge University Press.

Clement, P. 1981. Glaciological activities in the Johan Dahl Land 1980, South Greenland. *Grønlands Geologiske Undersøgelse, Ser. Report* **105**, 62–4.

1982. Glaciological investigations in connection with hydropower. *Grønlands Geologiske Undersøgelse, Ser. Report* **110**, 91–5.

1983. Mass balance measurements on glaciers in South Greenland. *Grønlands Geologiske Undersøgelse, Ser. Report* **115**, 118–23.

1984. Glaciological activities in the Johan Dahl Land area, South Greenland. *Grønlands Geologiske Undersøgelse, Ser. Report* **120**, 113–21.

Connolly, W. M. and King, J. C. 1996. A modelling and observational study of East-Antarctic surface mass balance. *J. Geophys. Res.* **101**, 1335–43.

Dahl-Jensen, D., Johnsen, S. J., Hammer, C. U., Clausen, H. B. and Jouzel, J. 1993. Past accumulation rates derived from observed annual layers in the GRIP ice core from Summit, Central Greenland. In Peltier W. R., ed., *Ice in the Climate System.* NATO ASI Series 112. Berlin, Springer, pp. 517–31.

De Wolde, J. R., Huybrechts, P., Oerlemans, J. and Van de Wal, R. S. W. 1997. Projections of global mean sea level rise calculated with a 2D energy-balance climate model and dynamic ice sheet models. *Tellus* **49A**, 486–502.

Denby, B. 2001. Ph.D. thesis, Institute for Marine and Atmospheric Research Utrecht, Utrecht University, The Netherlands.

Ekholm, S. 1996. A full coverage, high resolution, topographic model of Greenland, computed from a variety of digital elevation data. *J. Geophys. Res.* **101**, 21 961–72.

Fahnestock, M., Bindschadler, R., Kwok, R. and Jezek, K. 1993. Greenland ice sheet surface properties and ice dynamics from ERS-1 SAR imagery. *Science* **262** (5139), 1485–1616.

Genthon, C. and Braun, A. 1995. ECMWF analysis and predictions of the surface climate of Greenland and Antarctica. *J. Climate* **8**, 2324–32.

Genthon C., Jouzel, J. and Déqué, M. 1994. Accumulation at the surface of polar ice sheets: observation and modeling for global climate change. In Desbois, M. and Desalmand, F., eds., *Global Precipitations and Climate Change.* NATO ASI Series I, vol. 26, pp. 53–76.

Glover, R. W. 1999. Influence of spatial resolution and treatment of orography on GCM estimates of the surface mass balance of the Greenland ice sheet. *J. Climate* **12**, 551–63.

Greuell W. and Konzelmann, T. 1994. Numerical modelling of the energy balance and the englacial temperature of the Greenland ice sheet. Calculations for the ETH-camp location (West Greenland m a.s.l.). *Global & Planetary Change* **9** (1/2), 91–114.

Greuell, W., Denby, B., Van de Wal, R. S. W. and Oerlemans, J. 2001. Ten years of mass-balance measurements along a transect near Kangerlussuaq, Greenland. *J. Glaciol.* **47** (156), 157–8.

Greve, R. 1995. Thermomechanisches Verhalten polythermer Eisschilde, Theorie, analytik, numerik. Ph.D. Thesis, University of Darmstadt, Germany.

  1997. Application of a polythermal three-dimensional ice sheet model to the Greenland ice sheet: response to steady-state and transient climate scenarios. *J. Climate* **10**, 901–18.

Huybrechts, P. 1994. The present evolution of the Greenland ice sheet: an assessment by modelling. *Global & Planetary Change* **9**, 39–51.

Huybrechts, P. and de Wolde, J. 1999. The dynamic response of the Greenland and Antarctic ice sheets to multiple-century climate warming. *J. Climate* **12**, 2169–88.

Huybrechts, P., Letréguilly, A. and Reeh, N. 1991. The Greenland ice sheet and greenhouse warming. *Global & Planetary Change* **3** (4), 399–412.

Janssens, I. and Huybechts, P. 2000. The treatment of meltwater retardation in mass-balance parameterizations of the Greenland ice sheet. *Ann. Glaciol.* **31**, 133–40.

Krinner, G., Genthon, C., Li, Z-X. and Le Van, P. 1997. Studies of the Antarctic climate with a stretched-grid general circulation model. *J. Geophys. Res.* **10** (13), 731–45.

Lefebre, F., Gallee H., van Ypersele J. P. and Greuell W. 2003. Modeling of snow and ice melt at ETH Camp (West Greenland): a study of surface albedo. *J. Geophys. Res.* **108** (D8) art. no. 4231.

Letréguilly, A., Huybrechts, P. and Reeh, N. 1991. Steady-state characteristics of the Greenland ice sheet under different climates. *J. Glaciol.* **37** (125) 149–57.

Nobles, L. H. 1960. Glaciological investigations, Nunatarssuaq ice ramp, northwestern Greenland. SIPRE Technical Report 66, 57 pp.

Oerlemans, J., Van de Wal, R. S. W. and Conrads, L. A. C. 1991. A model for the surface balance of ice masses. Part II: application to the Greenland ice sheet. *Z. Gletscherkd. Glazialgeol.* **27/28**, 85–96.

Ohmura, A. 1987. New temperature distribution maps for Greenland. *J. Glaciol.* **37**, 140–8.

Ohmura, A. and Reeh, N. 1991. New precipitation and accumulation maps for Greenland. *J. Glaciol.* **37**, 140–8.

Ohmura, A., Wild, M. and Bengtsson, L. 1996. A possible change in mass balance of Greenland and Antarctic ice sheets in the coming century. *J. Climate* **9**, 2124–35.

Ohmura, A., Calanca, P., Wild, M. and Anklin, M. 1999. Precipitation, accumulation and mass balance of the Greenland ice sheet. *Z. Gletscherkd. Glazialgeol.* **35**, 1–20.

Pfeffer, W. T., Meier, M. F. and Illangasekare, T. H. 1991. Retention of Greenland runoff by refreezing: implication for projected sea level change. *J. Geophys. Res.* **96** (12), 22 117–24.

Reeh, N. 1989. Dynamic and climatic history of the Greenland ice sheet. In Fulton, R. J., ed., *Quaternary Geology of Canada and Greenland*. Geological Survey of Canada (1), pp. 793–822.

  1991. Parameterization of melt rate and surface temperature on the Greenland ice sheet. *Polarforschung* **59** (3), 113–28.

  1994a. Calving from Greenland glaciers: observations, balance estimates of calving rates, calving laws. In Reeh, N., ed., *Report on the Workshop on the Calving Rate of West Greenland Glaciers in Response to Climate Change*. Copenhagen, Danish Polar Center, pp. 85–102.

1994b. Sensitivity to climate change of the calf-ice production from Greenland glaciers. In Reeh, N., ed., *Report on the Workshop on the Calving Rate of West Greenland Glaciers in Response to Climate Change*. Copenhagen, Danish Polar Center, pp. 103–9.

Reeh, N. and Starzer, W. 1996. Spatial resolution of ice-sheet topography: influence on Greenland mass-balance modelling. GGU report 1996/53, pp. 85–94.

Ritz, C., Fabre, A. and Letréguilly, A. 1997. Sensitivity of a Greenland ice sheet model to ice flow and ablation parameters: consequences for the evolution through the last climate cycle. *Climate Dyn.* **13**, 11–24.

Thompson, S. L. and Pollard, D. 1997. Greenland and Antarctic mass balances for present and doubled atmospheric CO_2 from the GENESIS version-2 global climate model. *J. Climate* **10**, 871–900.

Van de Wal, R. S. W. 1996. Mass balance modelling of the Greenland ice sheet: a comparison of energy balance and degree-day models. *Ann. Glaciol.* **23**, 36–45.

1999. The importance of thermodynamics for modeling the volume of the Greenland ice sheet. *J. Geophys. Res.* **104** (D4), 3887–98.

Van de Wal, R. S. W. and Ekholm, S. 1996. On elevation models as input for mass balance calculations of the Greenland ice sheet. *Ann. Glaciol.* **23**, 181–6.

Van de Wal, R. S. W. and Oerlemans, J. 1994. An energy balance model for the Greenland ice sheet. *Global & Planetary Change* **9**, 115–31.

1997. Modelling the short-term response of the Greenland ice sheet to global warming. *Climate Dyn.* **13**, 733–44.

Van de Wal, R. S. W., Wild, M. and de Wolde, J. R. 2001. Short-term volume changes of the Greenland ice sheet in response to doubled CO_2 conditions. *Tellus* **53B**, 94–102.

Van der Veen, C. J. 1999. *Fundamentals of Glacier Dynamics*. Rotterdam/Brookfield, Balkema.

Van Tatenhove, F. G. M., Van der Meer, J. J. M. and Huybrechts, P. 1995. Glacial-geological/geomorphological research in West Greenland used to test an ice-sheet model. *Quat. Res.* **44**, 317–27.

Warrick, R. A., Le Provost, C., Meier, M. F., Oerlemans, J. and Woodworth, P. L. 1996. Changes in sea level. In Houghton, J. T. *et al.*, eds., *Climate Change 1995 – The Science of Climate*. Cambridge University Press, pp. 359–405.

Weidick, A. 1995. *Satellite Image Atlas of Glaciers of the World: Greenland*. US Geological Survey Professional Paper 1386-C. Washington, United States Government Printing Office.

Wild, M. and Ohmura, A. 2000. Changes in mass balance of the polar ice sheets and sea level under greenhouse warming as projected in high resolution GCM simulations. *Ann. Glaciol.* **30**, 197–203.

12

Mass balance of the Antarctic ice sheet: observational aspects

CHARLES R. BENTLEY
Department of Geology and Geophysics,
University of Wisconsin-Madison

12.1 Introduction

The Antarctic ice sheet (Figure 12.1) contains sufficient ice to raise the world-wide sea level by more than 60 m if melted completely. The amount of snow deposited annually on the ice sheet is equivalent to about 5 mm of global sea level, as is the mean annual discharge of ice back into the ocean. Thus, a modest imbalance between the input and output of ice might be a major contributor to the present-day rise in sea level (1.5–2 mm per year), but the uncertainty is large.

Despite all available measurements of snow accumulation, ice velocities, surface and basal melting, and iceberg discharge, it is still not known for certain whether the ice sheet is growing or shrinking. The uncertainty in the estimate of the total mass balance is at least 20% of the mass input, equivalent to a global sea-level change of about 1 mm per year. Furthermore, the fact that rates of discharge from some of the Antarctic ice streams have changed markedly in recent decades and centuries suggests that the mass balance also may be rapidly changeable.

The volume and geographic extent of the Antarctic ice sheet certainly have undergone major changes over geological time. The ice sheet was significantly larger during the last glacial maximum (LGM), some 20 000 years ago, and retreated to near its present extent within the last several thousand years; that retreat is probably on-going at present (see Chapter 13, section 13.3.1). Several mechanisms in the ice-sheet–lithosphere system, notably post-glacial isostatic uplift and the effect of temperature on the viscosity of the ice in the deeper layers, have long response times. It is likely that the ice sheet is still reacting dynamically to the glacial–inter-glacial transition between 20 000 and 10 000 years ago and to the subsequent increase in the snow accumulation rate. Consequently, the present Antarctic contribution to sea-level change (whatever it may be) could reflect changes in the accumulation rate over the past 100 years, or the long-term dynamical response of the ice sheet, or both.

Mass Balance of the Cryosphere: Observations and Modelling of Contemporary and Future Changes, eds. Jonathan L. Bamber and Antony J. Payne. Published by Cambridge University Press. © Cambridge University Press 2003.

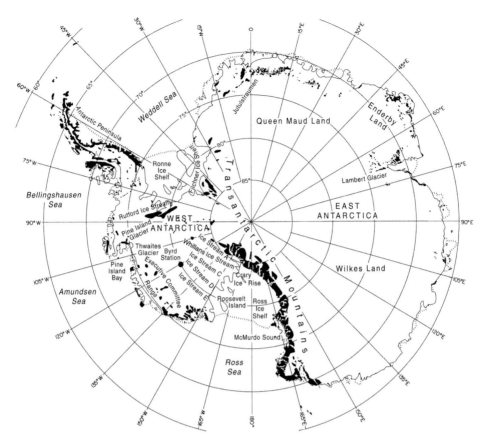

Figure 12.1. Location map of Antarctica and its ice sheet. Black areas denote mountains and other mostly exposed rock. Broken lines mark the seaward boundary of ice shelves. (Some coastline modifications are courtesy of D. G. Vaughan, British Antarctic Survey. For more detail in the Ross embayment, see Figures 12.8, 12.11 and 12.12.)

Future enhanced greenhouse warming will affect the mass balance of the ice sheet. Oceanic warming could increase basal melting of the floating ice shelves, whose thinning could result in faster flow of the ice into the ocean, contributing to sea-level rise. On the other hand, atmospheric warming and reduction in sea-ice cover almost surely will give rise to increased precipitation over the continent, thereby contributing to sea-level lowering (Chapter 13).

Realistic predictions of the response of Antarctica to enhanced greenhouse warming and the resulting sea-level contribution will not be possible until the present mass balance is determined and a better understanding of the atmosphere–ice–ocean processes and ice dynamics is developed.

In this chapter we first examine the approaches to Antarctic mass balance evaluation, including techniques to be applied in the future, and then survey the results of measurements and calculations leading to estimates of the present-day mass balance.

12.2 Measurement approaches

12.2.1 General

Two different forms of mass balance are commonly of interest to Antarctic glaciologists. (1) The mass balance of the entire ice sheet, including the ice shelves. Since the ice shelves are integral parts of the ice sheet, it is important from the standpoint of understanding the dynamics of the ice sheet as a whole to include them in mass balance analyses. (2) The mass balance of the 'inland' ice sheet, that portion that rests on a solid bed. It is only changes in the mass of this portion of the ice sheet, which is bounded by the 'grounding line' (not really a line but a zone of variable width between grounded and floating ice), that have any effect on sea level. In accordance with Archimedes' principle, ice-shelf melting or calving of icebergs leaves sea level unchanged.

There are two basic approaches to determining present-day mass balance – the integrated approach and the component approach. The integrated approach involves the determination of the change in mass of the ice sheet by measuring changes in its surface height or gravitational attraction using instruments mounted in satellites. This approach is only effective in determining the mass balance of the inland ice, since changes in thickness of the ice shelves have only a 10% effect on surface height and have no effect on gravity at all (Archimedes' principle again).

The component approach involves determining separately the mass input onto the ice sheet and the mass flux lost from the ice sheet into the surrounding ocean (Figure 12.2). This approach is applicable to either form of mass balance, although the measurements needed to determine the mass output are different in the two cases.

12.2.2 Integrated approach

Because of the inaccessibility and vast size of Antarctica, only satellite-borne instruments can approach continent-wide coverage. The only satellite-borne tool available until recently for mass balance studies was the radar altimeter (see Chapter 4). However, now there are two NASA satellite systems in orbits that reach to 86°S. One, ICESat (ice, cloud, and land elevation satellite; see Figure 12.3), launched in January 2003, bears a laser altimeter, and the other, launched in March 2002, carries a high-precision gravity-measuring system (comprising two satellites that continuously measure the distance between them). The laser altimeter, called GLAS (geoscience laser altimeter system), should yield surface elevations over the entire ice sheet (Zwally *et al.*, 2002) with a regional mean accuracy of 0.1 m or better north of 86°S, diminishing south of 86°S to about 1 m at the Pole. The gravity mission, known as GRACE (gravity recovery and climate experiment) (Wahr, Molenaar and Bryan, 1998), will be sensitive to changes in the overall mass of the ice sheet to an

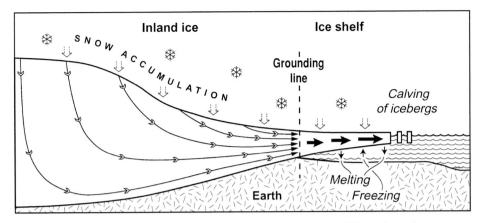

Figure 12.2. Diagrammatic sketch of ice-sheet input and output. Snow falls on the surface, is transformed to ice, which follows curved paths on the inland ice to the grounding line, where it enters the ice shelf. Flow speed is uniform with depth in the ice shelf, increasing seaward because of ice-shelf spreading and additional snow fall on the surface. The base of the ice shelf interacts with the underlying ocean, causing melting in some places and freezing in others. Pieces of the ice shelf 'calve' into the ocean to form icebergs.

Figure 12.3. The official NASA ICESat logo, including an artist's sketch of ICESat. (Courtesy NASA's GLAS science team.)

astonishing accuracy (in terms of layer thickness) of 0.1 mm per year water equivalent (National Research Council, 1997). Both satellites are expected to yield data for three to five years.

However, producing good determinations of the changing surface height and gravity field does not in itself yield changes in ice-sheet mass. Two non-trivial steps must follow to convert surface elevation changes to changes in mass. (1) A correction must be applied for height changes that simply mirror changes in the height of the bed, such as may arise from isostatic or tectonic uplift. (2) Corrected height (volume) changes must then be multiplied by the proper snow or ice density to obtain the mass change. The appropriate density to use depends upon how the change is distributed between short-term changes in surface snow fall and long-term changes associated with the flow of solid ice, which is not easily known.

The first correction applies *a fortiori* to satellite gravity measurements, because isostatic uplift is equivalent to adding a layer of the density of the Earth's mantle to the Earth column below the satellite. The density of the mantle is more than three times that of ice, so isostatic uplift of a given height will affect gravity more than three times as strongly as the same change in ice thickness. The second correction, however, is absent in the case of gravity measurements, because the gravity responds directly to mass, not volume.

There will be a strong complementarity between the GLAS and GRACE measurements precisely because one measures changes in volume and the other changes in mass. Bentley and Wahr (1998) conclude that together the two missions will attain an accuracy of $3 \, \mathrm{kg/m^2}$ in the mean accumulation rate over the continent, equivalent to 0.1 mm per year in sea-level change. However, that will not in itself reveal whether changes measured over the expected five-year duration of the two missions primarily reflect long-term dynamic changes in the ice sheet or simply inter-annual variability in precipitation over the continent.

Wahr, Wingham and Bentley (2000) have carried out an analysis that specifically looks at the problem of separating secular ice-sheet change from meteorological variability and isostatic uplift using data only from the two satellites. They conclude that the error in mean trend from a combined GLAS/GRACE analysis over a five-year mission would be about 0.2 mm per year in sea-level equivalent; they go further and estimate that the accuracy in estimating the true century-scale trend will be about 0.3 mm per year.

The accuracy over the mission lifetime is only half that predicted by Bentley and Wahr (1998), but the Wahr *et al.* (2000) analysis does not take into account the probability that independent information can reduce the errors from both of these sources. For example, the calculations of integrated moisture flux divergence around the continent (see sections 12.2.3 and 12.3.2) can be carried out annually, yielding a direct determination of the yearly mass input. Furthermore, annual variations can be measured by GLAS and GRACE themselves. In regard to isostatic uplift, analyses of gravity changes over places such as the Canadian Shield, where post-glacial rebound is known to be occurring and there is now no ice sheet, can be expected to yield evaluations of viscosity in the Earth's mantle that will be important in modelling post-glacial rebound in Antarctica.

Because of the importance of the satellite measurements to understanding sea-level change, there will be much attention paid over the next decade to the most effective solution to these problems of interpretation. Nevertheless, these problems are serious ones, and virtually assure that at least a decade of satellite measurements will be needed to obtain a quantitatively significant figure for the Antarctic contribution to sea-level change. Thus, satellite measurements of surface height and gravity will be needed beyond the five-year lifetimes of ICESat and GRACE.

12.2.3 Component approach

Mass input

There are two principal approaches to determining the mass input to the ice sheets: (a) measuring (or calculating) 'moisture flux divergence', i.e. the difference between total amounts of water vapour passing inward and outward across the ice-sheet margins, and (b) integrating local measurements of 'surface mass balance', the amount of snow that accumulates on the surface.

The moisture flux divergence technique has a major advantage in determining the overall mass input to the ice sheet, in that it does not require knowledge of snow accumulation rates in particular places. Any moisture passing southward across the continental margin that is not matched by moisture passing northward across the same boundary must represent net mass added to the ice sheet somewhere. But it is supportive of the method that recent applications that include atmospheric modelling over the ice sheet (Bromwich, Cullather and Von Woert, 1998; Van Lipzig, 1999) have reproduced the known spatial distribution of accumulation rates rather well. This technique rests heavily on meteorological models calibrated by atmospheric soundings around the continent.

On the surface, mass balance rates can be measured in several different ways. The oldest way, and traditionally the least reliable, is by the interpretation of stratigraphic variations in snow pits to identify annual layers. This technique has a troubled history in Antarctica, because many of the early measurements (i.e. in the 1950s and 1960s) were made by people who necessarily had little previous experience in interpreting the Antarctic stratigraphy. The absence of melt phenomena during the summers and the fact that in the regions of low accumulation some years may be totally missing, either for lack of significant precipitation or because of wind erosion, made interpretation difficult and mistakes commonplace. Accumulation rates were frequently greatly over-estimated (Giovinetto, Bentley and Bull, 1989).

The introduction of geochemical techniques, such as the identification of seasonal variations in oxygen isotope ratios in the firn, eventually improved the situation, although errors in interpretation of isotopic signals were also made. The discovery that the radioactive fallout from atmospheric hydrogen bomb tests on known dates in the mid 1950s and mid 1960s could be identified unequivocally as marker horizons in the firn finally provided a definitive means of determining mean accumulation rates over one to several decades. This not only removed the uncertainty in interpretation, but also greatly diminished the error associated with the inter-annual variability of precipitation. At this writing (2003), approximately 400 accumulation measurements by the fallout-horizon method exist, irregularly spread over the ice sheet.

A newer method of great potential value, but one that has not yet been fully developed and tested, is to calculate the accumulation rate from the microwave emissivity of the surface. The grain size and shape in the uppermost ten metres of snow depend upon the accumulation rate (but also on other parameters); the characteristics of the grains, in turn, affect the emissivity (Zwally, 1977). Unfortunately, these effects are not fully understood;

the best to hope for at present is an effective means of interpolating between points where the accumulation rate has been measured on the surface (Zwally and Giovinetto, 1995). For this, as well as for validation of the moisture flux technique, a good distribution of surface measurements is essential.

Alternatively, Arthern and Winebrenner (1998) have recently shown that the polarization of the microwave emissions is related to accumulation rates in dry snow zones because the length-scales of density layering are determined by the accumulation rate. Theoretical and observed polarization/accumulation rate relationships compare well over widely distributed locations in Antarctica and Greenland. A first-generation map of accumulation rates over Antarctica based on microwave polarization data agrees qualitatively on large scales with that of Zwally and Giovinetto (1995) and earlier maps (based solely on ground data), but indicates greater topographic control of accumulation rates in, for example, the Lambert Glacier drainage system, in agreement with recent Australian ground observations (Cunde *et al.*, 2001).

Mass output

The mass output is calculated differently in the two forms of mass balance estimates. For the entire ice sheet, the primary contributors are melting beneath the ice shelves and calving of icebergs (seasonal melting at the surface of the ice, followed by runoff, is properly a term in the surface mass balance). Both of these quantities are difficult to measure. Calving takes place around almost the entire vast perimeter of Antarctica and involves events that range over many orders of magnitude in size and frequency. Satellite sensors cannot see the smaller icebergs, but there is no other effective way to monitor the extensive coastline. Until recently, melting at the inaccessible bases of the ice shelves could be quantified only indirectly by glaciological measurements on the surface or from oceanographic measurements near the ice front. Since 1998, the British Antarctic Survey has been making direct measurements of bottom melt, using a phase-sensitive radar system, with a cited accuracy of 4 mm (Corr *et al.*, 2002). This exciting new technique is applicable to all ice shelves.

For the inland ice alone, it is the mass flux across the grounding lines that must be calculated. The output flux is the product of two quantities whose values must be known around the entire perimeter of the grounded ice sheet: the mean columnar velocity across the grounding line and the ice thickness, adjusted for low density near-surface layers, along the grounding line. Inland, the mean column velocity is between 80% and 100% of the surface velocity, depending on the temperature profile and the amount of basal sliding (Paterson, 1994). If measured at or just seaward of the grounding line, velocities will be essentially constant with depth in the ice. Velocities at the ice-sheet surface can be determined by a variety of techniques that are discussed in Section 12.3.2.

The principal means of measuring ice thickness is airborne radar sounding (see Chapter 4). Mean snow and ice density can be estimated with reasonably good accuracy from widely scattered deep ice-core measurements. There is no direct way to sound ice thickness from satellites, but there is an indirect means. If surface heights above sea level can be measured

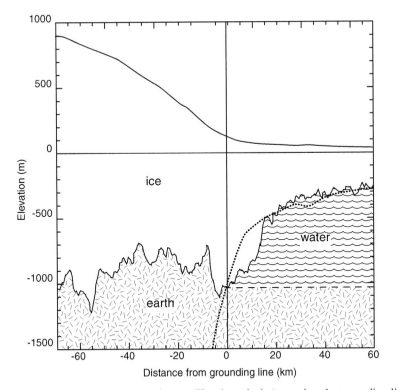

Figure 12.4. Diagram of a glacier (in this case Thwaites glacier) crossing the grounding line. Ice flow is from left to right. The dotted line shows where the base of the floating ice would be if it were in hydrostatic equilibrium. There is a large discrepancy near the grounding line, but elsewhere the equilibrium and mean measured thicknesses mostly agree closely. A major source of inaccurate ice thicknesses near the grounding line may be navigation error in old radar-sounding flights (E. Rignot, personal communication, 2002); Rignot has found better agreement between hydrostatic and measured thicknesses on several glaciers in Greenland (Rignot *et al.*, 2001). The deviation around 35 km is unexplained. (Redrawn from Rignot (2001). Reprinted from the *Journal of Glaciology* with the permission of the International Glaciological Society.)

accurately at or just seaward of the grounding line (Figure 12.2), then the ice thickness can be calculated from measured surface heights and Archimedes' principle (e.g. Rignot, 1998, 2001; see Figure 12.4). There are complications, however. Firstly, on an ice shelf within a distance from the grounding line equal to several times the ice thickness, hydrostatic balance is typically not attained (Sanderson, 1979). This effect may partly explain the discrepancy between 0 and 15 km in Figure 12.4 (but see also the caption to this figure). Thus, hydrostatic balance right at the grounding line is not assured, and an error of tens, or even hundreds, of metres could be introduced by assuming that it does.

  Secondly, measuring the height above sea level involves an intermediate step. Satellite measurements are referred to a smoothed, mean approximation to the Earth (the *ellipsoid*), which locally differs in height, generally by some tens of metres, from the real, irregular, equi-potential surface of the Earth (the *geoid*). The sea surface follows the geoid, the height of which can be determined either by direct sea-level measurements at a coastal station on the ground (which are few and far between) or by analysis of satellite orbits, most of which are poorly known over Antarctica. As a result, geoidal heights are not well known along most grounding lines. Furthermore, the small density contrast between ice and water magnifies the errors associated with uncertainties in the geoidal height and the mean density of the ice column in calculating ice thicknesses.

Thirdly, melt rates from the bottom of ice shelves can be very large near the grounding line (Jacobs, Hellmer and Jenkins, 1996; Rignot and Jacobs, 2002). Consequently, the hydrostatic balance method must be applied very close to the grounding line, the position of which is not always easy to determine. Thus, for several reasons it remains to be ascertained whether, in general application, this technique can attain the accuracy needed to make it widely effective in mass output flux calculations.

12.3 Measurement results

12.3.1 Integrated approach

The radar-altimeter-bearing satellites of the 1970s and 1980s (Seasat and Geosat) had orbits that carried them at best only to a latitude of 72°S, so they covered only a small fraction of the ice sheet. The situation is much improved now, with the ESA satellites ERS-1 and ERS-2 in orbits that reach to 81.5°S, although that still leaves a substantial portion of the ice sheet uncovered.

Five years of spatially continuous ERS-1 and ERS-2 satellite measurements (1992–1996) have been used to estimate the rate of change of surface elevation of nearly two-thirds (63%) of the grounded ice sheet (Wingham *et al.*, 1998; see Figure 12.5). The largest excluded regions are: south of the orbital limit of the satellites; a coastal strip 100 or 200 km wide where the slope is too steep for good radar altimeter returns; mountainous areas, including the entire Antarctic peninsula; and three swathes where data were lost because of tape-recorder limitations on the ERS satellites. Wingham *et al.* (1998) made corrections for instrumental effects, atmospheric conditions, surface slope and tidal variations. They also corrected for isostatic rebound, but they employed models of past and present changes in ice thickness that were cited by Wahr, Han and Trupin (1995) as examples, not firmly known quantities. Wingham *et al.* (1998) compared average rates of elevation change, calculated within 19 different drainage systems, with the variability in surface height to be expected from inter-annual snow-fall variability (taken uniformly as 15% of the mean annual accumulation), assuming a snow density of $350 \, \text{kg/m}^3$. In only one of those drainage basins was the rate of change in elevation greater than the variability – in the basin that feeds Pine Island Bay the elevation dropped an average of $117 \pm 10 \, \text{mm}$ per year, compared with a variability of 93 mm per year. Furthermore, the changes were concentrated over Thwaites

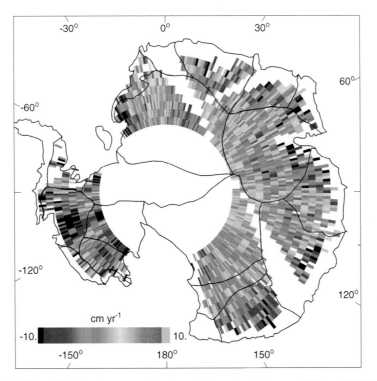

Figure 12.5. Map of height changes on the Antarctic ice sheet between 1992 and 1996 from ERS radar altimetry. Height changes have been averaged over boxes measuring 1° of latitude by 1° of longitude and are indicated by the colour scale. Black lines within the continental outline denote boundaries of drainage basins. (Reprinted with permission from Wingham *et al.* (1998). Copyright 1998, American Association for the Advancement of Science.)

glacier, which other observations show to be thinning (Rignot, 2001; see section 12.4.2). However, a detailed time series of the surface elevation shows no change for the first 2.5 years and then an abrupt switch to a drop of double the average rate. That is a clear sign of large changes in snow fall, whose presence obscures the dynamic lowering of the glacier as a whole that probably is present. But more recent analyses also show surface lowering over the entire Pine Island Bay drainage basin (Zwally *et al.*, 2002) and, particularly rapidly, over the portion of Pine Island glacier near its grounding line (Shepherd *et al.*, 2001) (see section 12.4.2).

For the two-thirds of the ice sheet sampled by Wingham *et al.* (1998), the indicated imbalance is -60 ± 76 Gt per year, or 0.17 ± 0.21 mm per year in sea-level change. Thus there is no significant evidence of a mass imbalance, and it appears from the error limits that recently the interior of the ice sheet has been at most only a modest source or sink of oceanic mass. The error limits from this study are several times smaller than those from the component approach (see section 12.3.2).

12.3.2 Component approach

Mass input

Continent-wide maps of the mass input only began to be produced after the extensive measurements of the International Geophysical Year (1957–58). Assessments have been given by (among others) Bull (1971), Giovinetto and Bentley (1985), Giovinetto and Zwally (2000; see Figure 12.6), Kotlyakov (1961), Kotlyakov, Losev and Loseva (1978), and Vaughan *et al.* (1999). The integrated net surface mass input based on the analysis by Giovinetto and Zwally (2000) is 1673 Gt per year for the grounded ice sheet and 2093 Gt per year for the entire ice sheet, including the ice shelves. (The values given by Giovinetto and Zwally (2000) have been adjusted upward to include the accumulation on the Antarctic peninsula, from Frolich (1992), as per Vaughan *et al.* (1999).) Those figures are 195 Gt per year less than the corresponding estimates by Vaughan *et al.* (1999) from essentially the same data set.

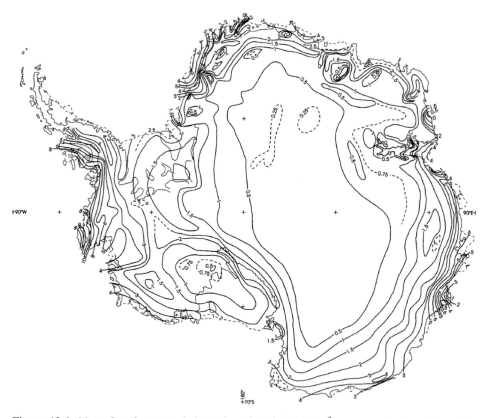

Figure 12.6. Map of surface mass balance in units of 100 kg/m^2 per year. Contour interval is 50 kg/m^2 per year up to 200 kg/m^2 per year and 100 kg/m^2 per year for larger values. Intermediate values of 25 kg/m^2 per year and 75 kg/m^2 per year are shown by dashed lines in some places. (From Giovinetto and Zwally (2000). Reprinted from the *Annals of Glaciology* with permission of the International Glaciological Society.)

It is instructive to examine the reasons for this 10% difference, especially as Vaughan *et al.* (1999) estimate the uncertainty in the mass input rate onto Antarctica from their study as ±5%. There are three principal reasons, cited in both papers.

(1) As a guide to interpolation where data are missing, Vaughan *et al.* (1999) use a background field based on a single, continent-wide correlation between microwave brightness temperatures, physical temperatures and accumulation rates (Zwally, 1977; Zwally and Giovinetto, 1995). Giovinetto and Zwally (2000), on the other hand, interpolate subjectively, based on meteorological knowledge and interpretation; furthermore, they believe that the functional fit to the microwave data used by Vaughan *et al.* (1999) was over-generalized. It is not clear whether the different modes of interpolation lead to a noticeable bias in mass balance.

(2) At some sites, particularly in two specific areas, Giovinetto and Zwally (2000), following Giovinetto and Bentley (1985) and Giovinetto *et al.* (1989), have chosen between incompatible data sets, whereas Vaughan *et al.* (1999) have averaged them. This yields a bias, because, in their selection process, Giovinetto and Zwally (2000) discounted old stratigraphic data analysed by inexperienced early observers, which were later shown by more modern techniques to be gross over-estimates of accumulation rates. The cited authors do not specifically evaluate the integrated quantitative significance of the data selections, but it appears to this writer to be of the order of 50 Gt per year.

(3) Giovinetto and Zwally (2000) apply a correction for deflation (snow blown entirely off the ice sheet) and ablation (mass loss to the ice sheet by liquid runoff) in coastal zones around the continent, whereas Vaughan *et al.* (1999) do not, on the grounds that its value is poorly known. The estimate by Giovinetto and Zwally (2000), 130 Gt per year, is indeed subject to a large uncertainty, but it is supported by a separate meteorological study of snow and water vapour transport across 70°S, which yields an estimate of 120 Gt per year, albeit also with a large uncertainty (Giovinetto, Bromwich and Wendler, 1992). Ignoring a known effect that can have only one sign introduces an automatic bias of the opposite sign.

The essential difference between these approaches is that Giovinetto and Zwally (2000) choose to apply their glaciological and meteorological knowledge to the fullest extent possible, whereas Vaughan *et al.* (1999) opt instead for computerized interpretation to avoid subjective bias. The difference of about 10% between the two calculations is a quantitative indicator of the importance of that difference in philosophy.

Recent analyses based on the moisture flux divergence have been discussed by Bromwich *et al.* (1998). When those authors exclude one set of analyses that appears to be of questionable accuracy, they find an average mean accumulation rate of 148 mm per year, with an uncertainty of about 10%. In a similar type of analysis, Van Lipzig (1999) developed and applied a regional atmospheric model to Antarctica to calculate a mean accumulation rate of 156 mm per year. The corresponding average accumulation rates from Giovinetto and Zwally (2000) and Vaughan *et al.* (1999) are 153 mm per year and 165 mm per year, respectively. Thus, on the face of it, the agreement between the results from fundamentally different techniques is very good. It is difficult, though, to ascertain to what extent the atmospheric analyses may have been influenced by the surface measurements, which are cited by Bromwich *et al.* (1998) as 'an important benchmark for comparison of new

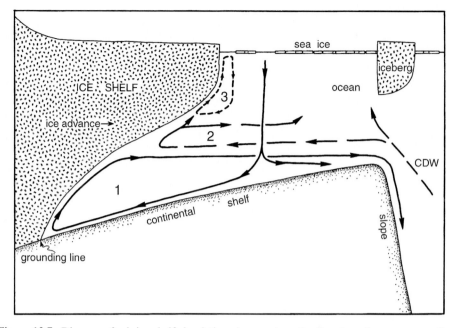

Figure 12.7. Diagram of sub-ice-shelf circulation. Arrows show the direction of ocean current flow. The circulations 1, 2 and 3 are explained in the text. CDW = circumpolar deep water. Slopes on the continental shelf and slope are greatly exaggerated. (From Jacobs *et al.* (1992). Reprinted from the *Journal of Glaciology* with permission of the International Glaciological Society.)

data sets', and as having been used 'for validation in numerous studies'. Nevertheless, the meteorological analyses at the very least show that interpolation between widely separated surface measurements has not been a large source of error. The agreement between the separate approaches thus suggests that the overall input to the ice sheet may not require major adjustments in the future. It is worth noting, however, that the fractional discrepancies within individual drainage basins, both between different analyses of surface data and between surface and atmospheric analyses, are much larger than the differences between overall means (Bromwich *et al.*, 1998; Giovinetto and Zwally, 2000; Vaughan *et al.*, 1999).

Mass output: melting of ice shelves

Jacobs *et al.* (1992) provide a good summary of oceanographic measurements and analyses leading to a calculation of the total mass of ice that is melted from the Antarctic ice shelves. They describe three modes of melting (Figure 12.7). One is generated deep beneath the interior of the shelves, near the grounding line, by the intrusion of high salinity (high density) water that is formed as sea ice freezes in front of the ice shelves (circulation 1 in Figure 12.7). This water carries the freezing temperature of the ocean surface downward beneath the thick ice to where it is as much as 1 °C warmer than the *in situ* melting point, which is lowered by the elevated pressure. The second mode (circulation 2 in Figure 12.7) involves the intrusion of relatively warm circumpolar deep water (CDW) from the continental shelf margin, where that margin is close enough to the ice shelf to allow it. CDW may be as much as 3 °C warmer than the *in situ* melting temperature in places when it comes in contact with an ice shelf. The third mode (circulation 3 in Figure 12.7) occurs near the ice fronts from tidal pumping and the seasonally warmed waters of the coastal currents.

Jacobs *et al.* (1992) estimated a total melt rate from Antarctic ice shelves of 544 Gt per year. Jacobs *et al.* (1996) update this figure to include the important contribution of very high melt rates in Pine Island Bay, extrapolated to other ice shelves of the south-east Pacific coast of the continent. Their new melt-rate total was 756 Gt per year. That figure may still be an under-estimate in light of the even higher melt rates (up to 50 m per year beneath the ice in front of Pine Island and Thwaites glaciers found by Rignot (1998, 2001).

Mass output: iceberg flux

The determination of iceberg flux is made difficult by several factors. One is the fact that extremely large icebergs calve from a particular ice front at such long time intervals that the total history of human observation may cover only a single calving event. On the other end of the size-scale is the observational difficulty in counting and measuring the very large numbers of icebergs that are too small to be observed easily, either from ships or from a satellite. Then there is the usual problem in Antarctica of interpolating and extrapolating from a small number of observations – in this case to the vast circum-Antarctic ocean. Finally, there is a large uncertainty in knowing how long icebergs survive so as to make proper allowance for counting the same icebergs more than once in multi-year surveys.

Jacobs *et al.* (1992) calculated an iceberg production rate of 2016 Gt per year from the volume of large icebergs found on satellite images since 1978 and the results of an international iceberg census project (Orheim, 1985, 1990). They suggest that their calculation could be in error by as much as one-third in either direction.

Mass output: flow speed

Surveys of output speed have been made in particular drainage basins, but not around the whole perimeter of the ice sheet. Earlier measurements based on relocation of points on the surface required reoccupation of sites after at least a year had passed. On the fast moving ice shelves, solar navigation sufficed (Budd, 1966; Dorrer, Hoffman and Seufert, 1969); elsewhere, velocity measurements were limited to triangulation from points on bedrock, and consequently were extremely rare.

Measurements aided by satellites revolutionized the determination of surface velocities. In fact, it has been a two-stage revolution, the first involving the advent of high precision point positioning by a surface observer using satellite signals for location, and the second introducing the determination of high density velocity fields from satellite imagery alone.

The first point positioning was conducted in the 1970s using Doppler techniques and the US Navy Navigational Satellite System, also known as the TRANSIT system. Velocities on a 50 km grid covering most of the Ross ice shelf were determined in this way (Thomas *et al.*, 1984), as were some velocities in the East Antarctic interior (Young, 1979) and, later, on the Ronne ice shelf (Hinze and Seeber, 1988; Jenkins and Doake, 1991) and on the West Antarctic inland ice (Whillans, Bolzan and Shabtaie, 1987). The measurements still required occupation of a station and then reoccupation in a different field season, and also necessitated many hours on-site for precise positioning. In the Ross ice shelf work, an error of about 8 m was estimated for positions occupied for three to five hours, long enough to track four satellite passes (Thomas *et al.*, 1984).

In the late 1980s and 1990s, global positioning system (GPS) observations became extensive (Chen, Bindschadler and Vornberger, 1998; Hinze and Seeber, 1988; Vaughan, 1994; Whillans and Van der Veen, 1993). GPS techniques not only increased the accuracy – typically to better than 0.1 m horizontally (Chen *et al.*, 1998) – but even made possible velocity determinations in a single season on the rapidly moving ice streams and ice shelves (Hulbe and Whillans, 1994).

The ability to use satellite imagery has vastly increased the density of measurements and the ease of obtaining them (GPS measurements are still important for ground control). The first work involved relocating such features as crevasses on repeated Landsat (or equivalent) images of the same area after the passage of some months or years (e.g. Lucchitta, Rosanova and Mullins, 1995). This was effective, but limited in accuracy by the resolution of the satellite imager. Furthermore, it could only be used where there were large features easily identifiable from satellite height.

A major advance was the advent of feature tracking using cross-correlation techniques. Scambos and Bindschadler (1993) took advantage of the broad irregularities in surface topography that are caused by the interaction of the moving ice with the basal topography and therefore are fixed in position relative to the bed. Using these for registration, they then tracked the movement of a vast array of smaller features fixed in the ice to yield the velocity field. Accuracy of one to two pixels is typical in such work. Like earlier techniques, however, feature tracking is effective only over fast moving features of the grounded ice sheet, such as ice streams. The technique is fundamentally inapplicable to floating ice; furthermore, it does not work on slow moving ice, which does not develop clear surface topography tied to the irregularities of the bed.

Finally, the accuracy and types of surfaces on which velocities could be obtained was dramatically extended by the development of synthetic aperture radar interferometry (InSAR) (Goldstein *et al.*, 1993; Joughin, *et al.*, 1999; Rignot, 1998). InSAR, in which small displacements of the surface between different satellite passes are measured by radar phase shifts they produce, is particularly effective where the ice is moving slowly. That makes it an ideal complement to the techniques used for fast moving ice. However, InSAR measurements are not limited to slow moving areas; what is needed on fast moving ice is only the appropriately short time interval between successive imagings. Consequently, maps of the velocity field of entire drainage systems, from divide to grounding line, are now becoming available (Figure 12.8).

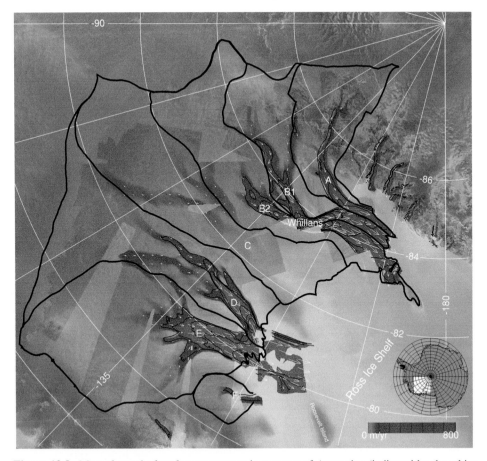

Figure 12.8. Map of speed of surface movement in a sector of Antarctica (indicated by the white box on the inset map) from InSAR data. Flow velocity is contoured with thin black lines at 100 m per year intervals. White arrows show velocity vectors in fast moving areas. Catchment boundaries for individual ice streams are plotted with thick black lines. The background map is a mosaic of imagery from the Radarsat Antarctic Mapping Project (Jezek, 1999). (Reprinted with permission from Joughin and Tulaczyk (2002), with modifications (I. Joughin, personal communication, 2003). Copyright 2002, American Association for the Advancement of Science.)

Ice thickness

A few attempts have been made to calculate ice thickness just past the grounding line from the surface elevation of the ice there (e.g. Rignot, 1998). However, by far the bulk of ice-thickness data are still from airborne radar (radio-echo) sounding. A compilation of all the ice-thickness soundings collected in Antarctica (including the old seismic soundings of pre-radar days) is available through the SCAR-sponsored BEDMAP project (Lythe *et al.*, 2001) at http://www.antarctica.ac.uk/aedc/bedmap/ (see Figure 12.9).

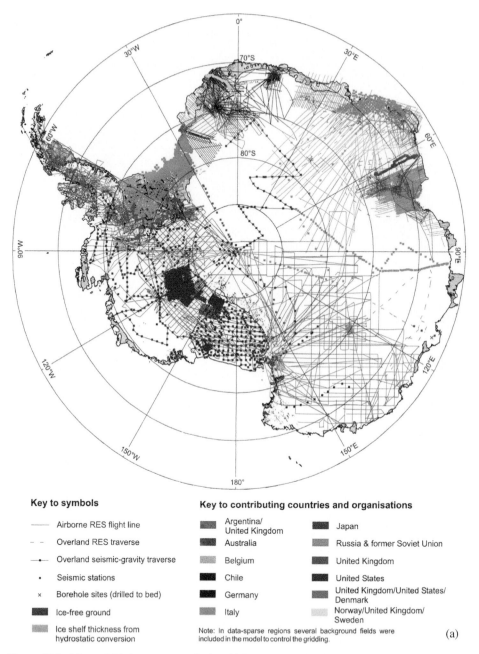

Key to symbols

— Airborne RES flight line

– – Overland RES traverse

–•– Overland seismic-gravity traverse

· Seismic stations

× Borehole sites (drilled to bed)

▮ Ice-free ground

▨ Ice shelf thickness from hydrostatic conversion

Key to contributing countries and organisations

▨ Argentina/United Kingdom

▨ Australia

▨ Belgium

▮ Chile

▮ Germany

▨ Italy

▮ Japan

▨ Russia & former Soviet Union

▮ United Kingdom

▮ United States

▨ United Kingdom/United States/Denmark

▨ Norway/United Kingdom/Sweden

Note: In data-sparse regions several background fields were included in the model to control the gridding.

(a)

Figure 12.9. Maps of (a) data coverage for ice-thickness determinations and (b) ice thickness. Both are from the British Antarctic Survey BEDMAP project (http://www.antarctica.ac.uk/aedc/bedmap/) (Lythe *et al.*, 2001).

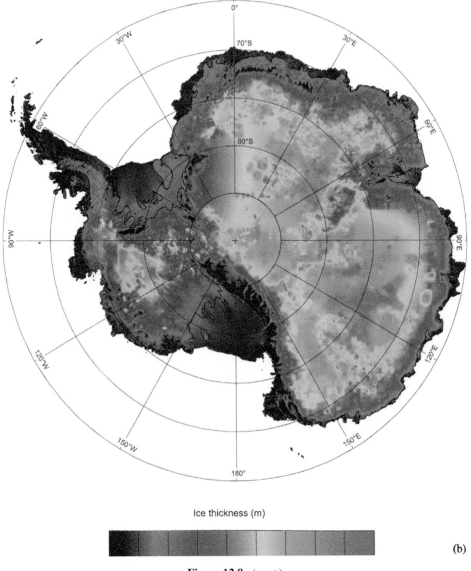

Ice thickness (m)

(b)

Figure 12.9. (*cont.*)

Examination of the BEDMAP flight-coverage map (Figure 12.9(a)) reveals that there are relatively few measurements along the ice-sheet grounding lines where output fluxes ideally should be calculated – most coastal flight-lines run normal to the ice-sheet margin. Consequently, many calculations of output flux have been made through inland gates, such as the 2000 m elevation contour. If such a gate is near the coast, it can still serve as a good proxy for the grounding line. Nevertheless, and despite the extensive ice-thickness information that does exist, uncertainty in ice thickness is probably the greatest source of error in calculating overall mass balance from the component approach at the present time.

Overall mass balance

There has been no general agreement on the overall mass balance of the inland ice sheet, and hence the Antarctic contribution to sea-level change, not even on its sign. Bentley and Giovinetto (1991) compiled all the component information they could find to calculate mass balances in about half of about two dozen drainage systems and then extrapolated to the others based on several alternative assumptions. That led to estimates of a positive mass balance of various sizes for the inland ice and a marked suggestion that a negative mass balance would not easily be compatible with glaciological data. However, more recent work has eliminated the strongly positive mass balances previously found for two major drainage systems, the Lambert glacier (Fricker, Warner and Allison, 2000; Rignot, 2002) and Pine Island glacier basins (Rignot, 1998; Rignot *et al.*, 2002; Stenoien and Bentley, 2000). In addition, Rignot (2001) and Rignot *et al.* (2002) have shown that the Thwaites glacier system, listed as in balance by Bentley and Giovinetto (1991), in fact has a decidedly negative mass balance. On the other hand, the West Antarctic drainage system into the Ross Sea, thought by Bentley and Giovinetto (1991) to have a negative mass balance, has now been shown to be growing (Joughin and Tulaczyk, 2002). Rignot (2002) also calculates mass balances for half a dozen drainage basins not previously evaluated. The new numbers combined with the balances for the other drainage systems given by Bentley and Giovinetto (1991) lead to an overall mass balance for the Antarctic ice sheet of about +7%, which is not significantly different from zero (Table 12.1).

Interestingly enough, three of the four measured systems in West Antarctica are significantly out of balance, with the Pine Island and Thwaites glacier systems shrinking and the Ross drainage system growing. The negative balance in the Pine Island and Thwaites systems accords nicely with the satellite radar altimetry of Wingham *et al.* (1998) (section 12.3.1) (the altimetry does not extend far enough south to detect the positive balance in the Ross system). Since the mass balances of all three systems are known to be changing with time (Joughin and Tulaczyk, 2002; Rignot *et al.*, 2002), the negligible overall balance for West Antarctica shown in Table 12.1 (−1% of the mass input) presumably is coincidental and transient. It is clear that the West Antarctic ice sheet is dynamically changing.

In East Antarctica, ten of the 12 measured systems show a positive or zero mass balance, although only in the system that flows through the Transantarctic Mountains into the Ross ice shelf is it significantly positive (Table 12.1). There is no indication of any East Antarctic imbalance from the satellite altimetry (Wingham *et al.*, 1998), which includes coverage of about half of the system that flows into the Ross ice shelf. The net East Antarctic balance of +11% from Table 12.1, if representative of the entire East Antarctic ice sheet, corresponds to a sea-level drop of 0.3 mm per year.

Table 12.1. *Mass balance of Antarctic drainage systems.*

System	Reference	Mass			Imbalance	
		Accumulation (Gt per year)	Outflow (Gt per year)	Net (Gt per year)	Fraction %	Significantly different from 0?
Stancomb-Wills glacier	2	14	15	−1	−7	no
Jutulstraumen	1	16	11	+5	+31	no
Eastern Queen Maud Land	1	35	35	0	0	no
Eastern Enderby Land	1	13	10	+3	+23	no
Lambert glacier	2	50	52	−2	−4	no
Denman glacier	2	34	32	+2	+6	no
Scott glacier	2	10	8	+2	+20	no
Western Wilkes Land	1	79	75	+4	+5	no
Mertz glacier	2	19	18	+1	+5	no
Ninnis glacier	2	22	20	+2	+9	no
David glacier	2	15	14	+1	+7	no
East Antarctica into Ross ice shelf	1	77	51	+26	+34	yes
East Antarctic totals		*384*	*341*	*+43*	*+11*	*?*
West Antarctica into Ross ice shelf	5	99	72	+27	+27	yes
Thwaites glacier	3	55	73	−18	−22	yes
Pine Island glacier	3	64	76	−12	−17	yes
Rutford ice stream	4	17	17	0	0	no
West Antarctic totals		*235*	*238*	*−3*	*−1*	*no*
Overall totals		*619*	*579*	*+40*	*+7*	*no*

References: 1, Bentley and Giovinetto (1991); 2, Rignot (2002); 3, Rignot *et al.* (2002); 4, Doake *et al.* (2001); 5, Joughin and Tulaczyk (2002).

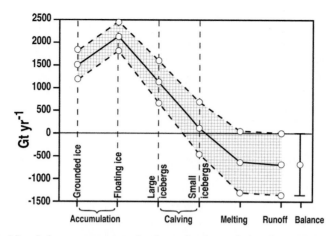

Figure 12.10. Mass balance components for the entire Antarctic ice sheet. Mass totals are added sequentially from left to right. The solid and dashed lines connect the 'best' and upper and lower limit estimates. Thus, the first set of points represents accumulation on the grounded ice and the second set represents accumulation on the grounded and floating ice together; accumulation on the floating ice is the difference between the two. The right hand error bar denotes the net result of all input and output components. (Redrawn from Jacobs *et al.* (1996). Reproduced and modified with the permission of American Geophysical Union.)

An approach that focusses first on the entire ice sheet yields a contrasting result for the Antarctic contribution to sea-level change. Jacobs *et al.* (1992; updated by Jacobs *et al.*, 1996) combine mass input, iceberg calving, and ice-shelf melting and find a negative mass balance of 700 ± 700 Gt per year (Figure 12.10). This is a large number – if all of it were attributed to the inland ice, it would yield a sea-level rise of just under 2 mm per year, i.e. enough to produce the entire observed twentieth-century rise (Church *et al.*, 2001). Although it is probable that a major part of the ice loss comes from the ice shelves (Rignot, 2002) and thus does not affect sea level at all, a significant contribution to sea-level rise is still strongly suggested. There is also the potential for an important Antarctic contribution to future sea-level rise if thinning ice shelves allow accelerated outflow of the inland ice.

The disagreement between the measurements on the ice sheet, which suggest a slightly growing ice sheet, and those around the ice-sheet margins, which suggest instead a considerable shrinkage, is intriguing and unresolved. Measurements from GLAS and GRACE are expected to provide the resolution.

12.4 The West Antarctic ice sheet – a special case?

12.4.1 General

Glaciologists generally agree that a marine ice sheet, one that rests on a bed well below sea level, is much more likely to undergo a rapid change than one lying on a higher bed. Although there are portions of the East Antarctic ice sheet that are marine, the West Antarctic ice sheet (WAIS), which comprises the bulk of the marine ice, also appears to be the most vulnerable, being open to the ocean on three sides. If the entire WAIS were discharged into the ocean, sea level would rise by five or six metres.

In fact, the WAIS surely has not been constant in size throughout the Pleistocene ice ages. Marine seismic studies show that it expanded across the continental shelf many times, (Alonso *et al.*, 1992; Anderson, 1999), whereas the subglacial occurrence of algal remains suggests that at other times the main WAIS inland ice was gone (Scherer, 1991; Scherer *et al.*, 1998). From a practical standpoint, it is crucially important to know, however, not only whether large changes in ice mass have occurred, but how rapidly they can occur and how likely it is that a large, rapid change will occur soon.

Theoretical work on marine ice-sheet stability is discussed in Chapter 13. Here we consider the field evidence relating to the WAIS, collected largely since about 1970.

12.4.2 Field evidence

The field evidence regarding WAIS stability is inconclusive but intriguing. In the 'Ross embayment' (comprising the Ross Sea, the Ross ice shelf, and the adjacent sector of West Antarctica; Figure 12.11 shows most of the Ross embayment), the distribution and depths of buried crevasses reveals that sudden reorganizations of the ice streams have occurred in the last 1000 years. Ice stream C abruptly stagnated (Retzlaff and Bentley, 1993), one of the boundaries of neighbouring Whillans ice stream (formerly called ice stream B) underwent a sudden lateral jump of 10 km or so (Clarke *et al.*, 2000), and a portion of ice stream A became quiescent (S. Shabtaie, personal communication, 1997). In addition, portions of the system are grossly out of balance today – Whillans ice stream is hyperactive (it has been losing mass but at a rapidly decreasing rate (Joughin and Tulaczyk, 2002)), ice stream C is stagnant (strongly positive mass balance), there are pronounced historical changes in the margins of Whillans ice stream near its mouth (Bindschadler and Vornberger, 1998), and Crary ice rise (an island of grounded ice in the Ross ice shelf near the mouth of Whillans ice stream) is growing and changing the regional velocity field (Bindschadler, 1993; MacAyeal *et al.*, 1987).

If the 'Ross embayment' system were unstable, one might expect these major dislocations to have caused some large changes in the outflow of ice into the Ross ice shelf. Observations of the Ross ice shelf itself can be used to test this possibility for the last 1500 years because tracers of past flow preserved in the shelf (Figure 12.12) can be compared with the present-day flow of the ice (Bentley, 1981; Bentley *et al.*, 1979; Fahnestock *et al.*, 2000; Jezek, 1984; Neal, 1979). Large surges of the inland ice would be recorded as distortions of the flow tracers. Indeed, one striking deformation of the flow tracers has been found in the form of a 'bulge' north-west of Crary ice rise (Casassa *et al.*, 1991; Fahnestock *et al.*, 2000) (Figure 12.12). Fahnestock *et al.* (2000) attribute this to a northward jump in the southern margin of ice stream C more than 700 years ago. They also show a complex history of flow

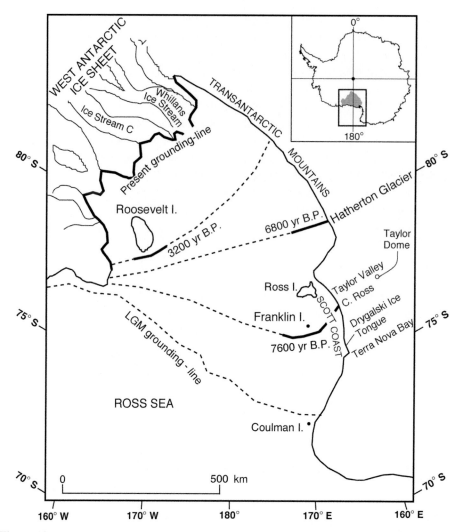

Figure 12.11. Retreat of the grounding line in the Ross embayment from its position at the last glacial maximum (LGM) to the present, based on glacial deposits and other features along the western boundary of the Ross ice shelf and on features in the ice on Roosevelt Island (solid line segments). Extrapolations from the data are dashed. Dates of various positions are indicated in years before the present (BP). (Reprinted with permission from Conway *et al.* (1999). Copyright 1999, American Association for the Advancement of Science.)

since then. Nevertheless, there is no sign of a major change in the total mass outflow in the last millennium or so.

Taken as a whole, the WAIS Ross embayment system is nearly in mass balance (Table 12.1); although measurably changing (Joughin and Tulaczyk, 2002). Furthermore, the proportion of ice flowing into the Ross ice shelf from East Antarctica and West Antarctica,

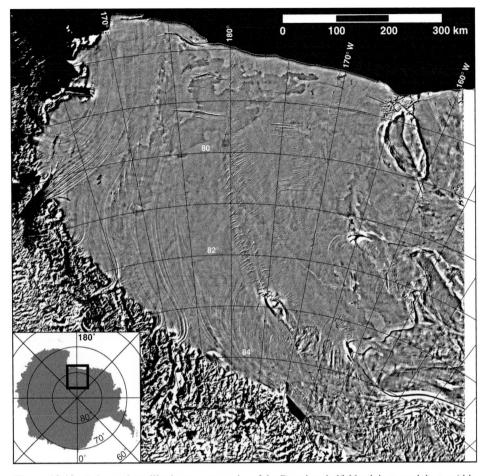

Figure 12.12. Enhanced satellite image composite of the Ross ice shelf. North is toward the top (this orientation is inverted relative to the other maps in this chapter). The Transantarctic Mountains with their outlet glaciers lie on the west and south, the complex flow region of West Antarctica (see Figure 12.8) is on the east, and the Ross Sea is to the north. Crary ice rise lies across 83°S between 170° and 175°W. The 'bulge' referred to in the text is centered on 82.5°S, 180°W, and is about 100 km across. Rifts, crevasse zones, flow stripes and surface mottling due to show conditions show prominently. (From Fahnestock *et al.* (2000). Reprinted from the *Journal of Glaciology* with the permission of the International Glaciological Society.)

respectively, has remained approximately constant over the full 1500 years. This is shown by the concordance between flow lines and flow tracers in the western portion of the ice shelf, where the flow is from East Antarctica (Bentley and Jezek, 1981). This concordance implies that the total outflow in the Ross embayment has remained relatively unchanged, despite the large internal perturbations – a large change in that outflow at some time within the last 1500 years would have caused a corresponding distortion in the flow lines from

East Antarctica, the total outflow from which presumably has been fairly steady over time. Nearly constant total flow from the WAIS at times of large internal disturbances points to a stable, not an unstable, system in the Ross embayment.

Glaciological studies of several types have all indicated that there has been no drastic change over the last 30 000 years in the height or flow of the ice sheet at Byrd Station, which lies in the West Antarctic interior within the Ross embayment (Raynaud and Whillans, 1982; Whillans, 1976, 1983). Field measurements on the surface indicate that the ice there and elsewhere in the interior Ross embayment is now thinning slowly (by a few to a 100 millimetres per year), perhaps partly in response to climatic warming about 10 000 years ago at the end of the last ice age (Hamilton and Whillans, 1997; Whillans, 1983). Ackert *et al.* (1999) suggest that a 45 m drop in ice level occurred in the Executive Committee Range north of Byrd Station (Figure 12.1) over the last 10 000 years (a mean rate of 5 mm per year. White and Steig (1997) interpret diverging trends in oxygen isotope ratios in ice-core records from Byrd Station and Taylor Dome (in the Transantarctic Mountains near Taylor glacier; Figure 12.11) as indicating a much larger drop in surface elevation at Byrd Station of about 400 m over the last 6000 years (a mean rate of 67 mm per year. Although there is quantitative disagreement, the concept of all these results is in accord with glacial geologic studies, mostly along the Scott Coast, which indicate that the grounding line in the Ross embayment retreated slowly to its present position over the last 7000 years or so (Hall and Denton, 1999). Conway *et al.* (1999) show that the grounding line passed Roosevelt Island in the north-east Ross ice shelf about 3200 years ago (Figure 12.11). Bindschadler (1998) extrapolates that retreat into the inland ice and suggests that it could extend to the ice divide in 4000–7000 years (Figure 12.13). If that were to happen, and the grounding lines in the other WAIS drainage systems were to retreat similarly, it would imply a consequent mean (but probably not steady) rate of sea-level rise of about 1 mm per year.

Study of the major drainage from the WAIS into the Ronne ice shelf also suggests that there is no gross discordance between the present velocity vectors and flow tracers in the ice shelf, although the evidence is limited (C. S. M. Doake, personal communication, 1997). Rutford ice stream, one of the major ice streams feeding the Ronne ice shelf, appears to be in balance at present (Doake *et al.*, 2001).

Observations relating to the drainage systems that flow into Pine Island Bay are more dramatic. This might indeed be the 'weak underbelly' of the WAIS (Hughes, 1980), a particularly likely site for accelerated flow because of indications that an ice shelf filling the bay may have disappeared in recent geologic time (Kellogg and Kellogg, 1986), thus perhaps reducing the restraint on the inflowing ice streams (Pine Island glacier and Thwaites glacier). Measurements in the catchment of Pine Island glacier indicate a mass balance not significantly different from zero (Stenoien and Bentley, 2000). However, a recent retreat of the grounding line of some 4 km has been revealed by InSAR measurements for this glacier (Rignot, 1998). Although the observations cover a period of only four years, refer to only part of the grounding line and are strongly non-linear in time, they are suggestive of rapid glacial recession. In this they are strikingly supported by recent work. From ERS radar altimetry, Shepherd *et al.* (2001) demonstrate substantial lowering of the surface of Pine Island glacier

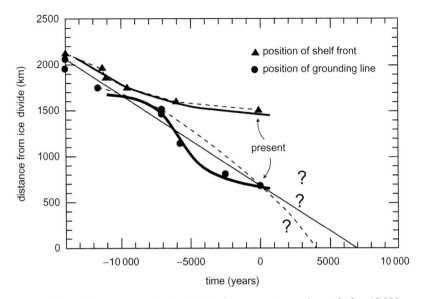

Figure 12.13. Grounding-line retreat in the Ross embayment measured over the last 13 000 years and projected into the future. Solid and dashed lines show alternate methods of projection. (Redrawn from *Science* (Bindschadler, 1998), with some additional information from R. A. Bindschadler (personal communication, 2000).)

inland from the grounding line, and, from ERS InSAR work, Rignot *et al.* (2002) reveal an acceleration of the glacier near the grounding line between 1996 and 2000. The acceleration has caused the mass balance of Pine Island glacier to turn significantly negative (Table 12.1). Equally striking are measurements and calculations for Thwaites glacier, which, while not accelerating, is growing wider and so also increasing its discharge flux. Thwaites glacier exhibited in 2000 a negative mass balance of some 22% (Rignot *et al.*, 2002).

Interesting as these accelerating negative mass balances are from a glaciological standpoint, they do not threaten a dangerous 'collapse' of the ice sheet. At the current rate of imbalance, it will take more than 20 000 years for the ice in the two drainage systems to be discharged fully, with the consequent rise in sea level of 1.25 m.

12.5 Summary

The overall mass balance of the entire Antarctic ice sheet is almost surely negative, but most of the negative balance is probably attributable to melting of floating ice, which has no effect on sea level. The Antarctic contribution to sea-level rise still remains uncertain in sign, but is apparently small in magnitude. From current evidence, it is unlikely that much of the 'unexplained' portion of world-wide sea-level rise is supplied by Antarctica. However, the evidence is becoming increasingly good that the West Antarctic ice sheet is still responding to the end of the last Ice Age. A continuing shrinkage that adds several metres to sea level over the next several millennia is a distinct possibility. The natural evolution of the West Antarctic ice sheet may gradually lead to its demise, but human activity is likely neither to accelerate nor delay that evolution significantly.

References

Ackert, R. P. Jr. *et al.* 1999. Measurements of past ice sheet elevations in interior West Antarctica. *Science* **286** (5438), 272–6.

Alonso, B., Anderson, J. B., Diaz, J. I. and Bartek, L. R. 1992. Pliocene-Pleistocene seismic stratigraphy of the Ross Sea: evidence for multiple ice sheet grounding episodes. In Elliot, D. H., ed., *Contributions to Antarctic Research III* (Antarctic Research Series). American Geophysical Union, pp. 93–103.

Anderson, J. B. 1999. *Antarctic Marine Geology*. Cambridge University Press.

Arthern, R.J. and Winebrenner, D. P. 1998. Satellite observations of ice sheet accumulation rate. In Stein, T., ed., *1998 International Geoscience and Remote Sensing Symposium (IGARSS 98) on Sensing and Managing the Environment*. Seattle, WA, July 6–10, 1998. Piscataway, NJ, IEEE Service Center, pp. 2249–51.

Bentley, C. R. 1981. Variations in valley glacier activity in the Transantarctic Mountains as indicated by associated flow bands on the Ross ice shelf. *Sea Level, Ice and Climatic Change*, IAHS publ. no **131**, pp. 247–61.

Bentley, C. R. and Giovinetto, M. B. 1991. Mass balance of Antarctica and sea level change. In Weller, G., Wilson, C. L. and Severin, B. A. B., eds., *Proceedings of an International Conference on the Role of the Polar Regions in Global Change*. June 11–15, 1990, University of Alaska Fairbanks, pp. 481–8.

Bentley, C. R. and Jezek, K. C. 1981. RISS, RISP, and RIGGS: post-IGY glaciological investigations of the Ross ice shelf in the U.S. program. *J. Roy. Soc. New Zealand* **11** (4), 355–72.

Bentley, C. R. and Wahr, J. M. 1998. Satellite gravity and the mass balance of the Antarctic ice sheet. *J. Glaciol.* **44** (147), 207–13.

Bentley, C. R., Clough, J. W., Jezek, K. C. and Shabtaie, S. 1979. Ice thickness patterns and the dynamics of the Ross ice shelf. *J. Glaciol.* **24** (90), 287–94.

Bindschadler, R. 1993. Siple Coast project research of Crary ice rise and the mouths of ice streams B and C, West Antarctica: review and new perspectives. *J. Glaciol.* **39** (133), 538–52.

1998. Future of the West Antarctic ice sheet. *Science* **282** (5388), 428–9.

Bindschadler, R. and Vornberger, P. 1998. Changes in the West Antarctic ice sheet since 1963 from declassified satellite photography. *Science* **279** (5351), 689–92.

Bromwich, D. H., Cullather, R. I. and Van Woert, M. L. 1998. Antarctic precipitation and its contribution to the global sea-level budget. *Ann. Glaciol.* **27**, 220–6.

Budd, W. F. 1966. The dynamics of the Amery Ice Shelf. *J. Glaciol.* **6** (45), 335–58.

Bull, C. 1971. Snow accumulation in Antarctica. In Quam, L. O. ed., *Research in the Antarctic*. Washington, D.C., American Association for the Advancement of Science Publ. **93**, pp. 367–421.

Casassa, G., Jezek, K. C., Turner, J. and Whillans, I. M. 1991. Relict flow stripes on the Ross ice shelf. *Ann. Glaciol.* **15**, 132–8.

Chen, X., Bindschadler, R. A. and Vornberger, P. L. 1998. Determination of velocity field and strain-rate field in West Antarctica using high precision GPS measurements. *Surv. Land Info. Syst.* **58** (4), 247–55.

Church, J. A. *et al.* 2001. Changes in sea level. In Houghton, J. T. *et al.*, eds., *Climate Change 2001: The Scientific Basis*. Cambridge University Press, pp. 639–93.

Clarke, T. S., Liu, C., Lord, N. E. and Bentley, C. R. 2000. Evidence for a recently abandoned shear margin adjacent to ice stream B2, Antarctica, from ice-penetrating radar measurements. *J. Geophys. Res.* **105** (B6), 13 409–22.

Conway, H., Hall, B. L., Denton, G. H., Gades, A. M. and Waddington, E. D. 1999. Past and future grounding-line retreat of the West Antarctic ice sheet. *Science* **286** (5438), 280–3.

Corr, H. F. J., Jenkins, A., Nicholls, K. W. and Doake, C. S. M. 2002. Precise measurement of changes in ice-shelf thickness by phase-sensitive radar to determine basal melt rates. *Geophys. Res. Lett.* **29** (8), 10.1029/2001GL014618.

Cunde, X., Jiawen, R., Qin, D., Hongqui, L., Weizhen, S. and Allison, I. 2001. Complexity of the climatic regime over the Lambert Glacier basin of the East Antarctic ice sheet: firn-core evidences. *J. Glaciol.* **47** (156), 160–2.

Doake, C. S. M. *et al.* 2001. Rutford ice stream, Antarctica. In Alley, R. and Bindschadler, R., eds., *The West Antarctic Ice Sheet, Behavior and Environment*. American Geophysical Union, pp. 221–36.

Dorrer, E., Hoffman, W. and Seufert, W. 1969. Geodetic results of the Ross Ice Shelf survey expeditions, 1962–1963 and 1965–1966. *J. Glaciol.* **8** (52), 67–90.

Fahnestock, M., Scambos, T., Bindschadler, R. and Kvaran, G. 2000. A millennium of variable ice flow recorded by the Ross ice shelf, Antarctica. *J. Glaciol.* **46** (155), 652–64.

Fricker, H. A., Warner, R. C. and Allison, I. 2000. Mass balance of the Lambert Glacier-Amery Ice Shelf system, East Antarctica: a comparison of computed balance fluxes and measured fluxes. *J. Glaciol.* **46** (155), 561–70.

Frolich, R. M. 1992. The surface mass balance of the Antarctic peninsula ice sheet. 1992. The contribution of the Antarctic peninsula ice to sea level rise. Report for the Commission of the European Communities Project EPOC-CT90–0015.

Giovinetto, M. B. and Bentley, C. R. 1985. Surface balance in ice drainage system of Antarctica. *Antarctic J. US.* **20** (4), 6–13.

Giovinetto, M. B. and Zwally, H. J. 2000. Spatial distribution of net surface accumulation on the Antarctic ice sheet. *Ann. Glaciol.* **31**, 171–8.

Giovinetto, M. B., Bentley, C. R. and Bull, C. B. B. 1989. Choosing between some incompatible data sets in Antarctica. *Antarctic J. US.* **24** (1), 7–13.

Giovinetto, M. B., Bromwich, D. H. and Wendler, G. 1992. Atmospheric net transport of water vapor and latent heat across 70°S. *J. Geophys. Res.* **97** (D1), 917–30.

Goldstein, R. M., Engelhardt, H., Kamb, B. and Frolich, R. M. 1993. Satellite radar interferometry for monitoring ice sheet motion: application to an Antarctic ice stream. *Science* **262** (5139), 1525–30.

Hall, B. L. and Denton, G. H. 1999. New relative sea level curves for the southern Scott Coast, Antarctica: evidence for Holocene deglaciation of the western Ross Sea. *J. Quat. Sci.* **14** (7), 641–50.

Hamilton, G. S. and Whillans, I. M. 1997. GPS glaciology in Antarctica. *EOS, Trans. Am. Geophys. Union* **78** (17), 100.

Hinze, H. and Seeber, G. 1988. Ice-motion determination by means of satellite positioning systems. *Ann. Glaciol.* **11**, 36–41.

Hughes, T. J. 1980. The weak underbelly of the West Antarctic ice sheet. *J. Glaciol.* **27** (97), 518–25.

Hulbe, C. L. and Whillans, I. M. 1994. Evaluation of strain rates on ice stream B, Antarctica, obtained using GPS phase measurements. *Ann. Glaciol.* **20**, 254–62.

Jacobs, S. S., Hellmer, H. H., Doake, C. S. M., Jenkins, A. and Frolich, R. 1992. Melting of ice shelves and the mass balance of Antarctica. *J. Glaciol.* **38** (130), 375–87.

Jacobs, S. S., Hellmer, H. H. and Jenkins, A. 1996. Antarctic ice sheet melting in the southeast Pacific. *Geophys. Res. Lett.* **23** (9), 957–60.

Jenkins, S. S. and Doake, C. S. M. 1991. Ice-ocean interaction on Ronne ice shelf. *J. Geophys Res.* **96** (C1), 791–813.

Jezek, K. C. 1984. Recent changes in the dynamic condition of the Ross ice shelf, Antarctica. *J. Geophys. Res.* **89** (B1), 409–16.

  1999. Glaciological properties of the Antarctic ice sheet from RADARSAT-1 synthetic aperture radar imagery. *Ann. Glaciol.* **29**, 286–90.

Joughin, I. and Tulaczyk, S. 2002. Positive mass balance of the Ross Ice Streams, Antarctica. *Science* **295** (5554), 476–80.

Joughin, I. *et al.* 1999. Tributaries of West Antarctic ice streams revealed by RADARSAT interferometry. *Science* **286** (5438), 283–6.

Kellogg, D. E. and Kellogg, T. B. 1986. Biotic provinces in modern Amundsen Sea sediments: implications for glacial history. *Antarctic J. US* **21**, 154–6.

Kotlyakov, V. M. 1961. The intensity of nourishment of the Antarctic ice sheet. *Symposium on Antarctic Glaciology*. International Association of Scientific Hydrology, Publ. **55**, 100–10.

Kotlyakov, V. M., Losev, K. S. and Loseva, I. A. 1978. The ice budget of Antarctica. *Polar Geog. Geol.* **24**, 251–62.

Lucchitta, B., Rosanova, C. and Mullins, K. 1995. Velocities of Pine Island Glacier, West Antarctica, from ERS-1 SAR images. *Ann. Glaciol.* **21**, 277–83.

Lythe, M. B. *et al.* 2001. BEDMAP: a new ice thickness and subglacial topographic model of Antarctica. *J. Geophys. Res.* **106** (B6), 11 335–51.

MacAyeal, D. R., Bindschadler, R. A., Shabtaie, S., Stephenson, S. N. and Bentley, C. R. 1987. Force, mass, and energy budgets of the Crary ice rise complex, Antarctica. *J. Glaciol.* **33** (114), 218–30.

National Research Council. 1997. *Satellite Gravity and the Geosphere: Contributions to the Study of the Solid Earth and Its Fluid Envelope*. Washington, D. C., National Academy Press.

Neal, C. S. 1979. Dynamics of the Ross ice shelf as revealed by radio echo sounding. *J. Glaciol.* **24** (90), 295–307.

Orheim, O. 1985. Iceberg discharge and the mass balance of Antarctica. In *Glaciers, Ice Sheets, and Sea Level: Effects of a CO_2-induced Climatic Change*. Washington, D.C., National Academy Press, 210–5.

  1990. Extracting climatic information from observations of icebergs in the Southern Ocean (abstract), *Ann. Glaciol.* **14**, 352.

Paterson, W. S. C. 1994. *The Physics of Glaciers*, 3rd ed. Oxford, Pergamon Press.

Raynaud, D. and Whillans, I. M. 1982. Air content of the Byrd core and past changes in the West Antarctic ice sheet. *Ann. Glaciol.* **3**, 269–73.

Retzlaff, R. and Bentley, C. R. 1993. Timing of stagnation of ice stream C, West Antarctica, from short-pulse radar studies of buried surface crevasses. *J. Glaciol.* **39** (133), 553–61.

Rignot, E. J. 1998. Fast recession of a West Antarctic glacier. *Science* **281** (5376), 549–52.

2001. Evidence for rapid retreat and mass loss of Thwaites Glacier, West Antarctica. *J. Glaciol.* **47** (157), 213–22.

2002. Mass balance of East Antarctic glaciers and ice shelves from satellite data. *Ann. Glaciol.* **34**, 217–27.

Rignot, E. and Jacobs, S. S. 2002. Rapid bottom melting widespread near Antarctic ice sheet grounding lines. *Science* **296** (5575), 2020–3.

Rignot, E., Krabill, W. B., Gogineni, S. P. and Joughin, I. 2001. Contribution to the glaciology of northern Greenland from satellite radar interferometry. *J. Geophys. Res.* **106** (D24), 34 007–20.

Rignot, E., Vaughan, D. G., Schmeltz, M., Dupont, T. and MacAyeal, D. 2002. Acceleration of Pine Island and Thwaites Glaciers, West Antarctica. *Ann. Glaciol.* **34**, 189–201.

Sanderson, T. J. O. 1979. Equilibrium profile of ice shelves. *J. Glaciol.* **22** (88), 435–60.

Scambos, T. A. and Bindschadler, R. 1993. Complex ice stream flow revealed by sequential satellite imagery. *Ann. Glaciol.* **17**, 177–82.

Scherer, R. P. 1991. Quaternary and tertiary microfossils from beneath ice stream B: evidence for a dynamic West Antarctic ice sheet history. *Palaeogeog., Palaeoclim., Palaeoecol.* **90**, 395–412.

Scherer, R. P., Aldahan, A., Tulaczyk, S., Possnert, G., Engelhardt, H. and Kamb, B. 1998. Pleistocene collapse of the West Antarctic ice sheet. *Science* **281** (5373), 82–5.

Shepherd, A., Wingham, D. J., Mansley, J. A. D. and Corr, H. F. J. 2001. Inland thinning of Pine Island Glacier, West Antarctica. *Science* **291** (5505), 862–4.

Stenoien, M. D. and Bentley, C. R. 2000. Pine Island Glacier, Antarctica: a study of the catchment using interferometric synthetic aperture radar measurements and radar altimetry. *J. Geophys. Res.* **105** (B9), 21 761–80.

Thomas, R. H., MacAyeal, D. R. and Eilers, D. H. and Gaylord, D. R. 1984. Glaciological studies on the Ross Ice Shelf, Antarctica, 1973–1978. In Bentley, C. R. and Hayes, D. E., eds., *The Ross Ice Shelf: Glaciology and Geophysics.* Antarctic Research Series, vol. 42, no. 2. Washington, D.C., American Geophysical Union, pp. 21–53.

Van Lipzig, N. P. M. 1999. The surface mass balance of the Antarctic ice sheet: a study with a regional atmospheric model. Ph.D. Dissertation, University of Utrecht. Veenendaal, Universal Press.

Vaughan, D. G. 1994. Glacier geophysical fieldwork on Ronne Ice Shelf in 1992/93. Filchner-Ronne Ice Shelf Programme (FRISP) Report 7, pp. 37–9.

Vaughan, D. G., Bamber, J. L., Giovinetto, M., Russell, J. and Cooper, A. P. R. 1999. Reassessment of net surface mass balance in Antarctica. *J. Climate* **12** (4), 933–46.

Wahr, J., Han, D. Z. and Trupin, A. 1995. Predictions of vertical uplift caused by changing polar ice volumes on a viscoelastic earth. *Geophys. Res. Lett.* **22** (8), 977–80.

Wahr, J., Molenaar, M. and Bryan, F. 1998. Time-variability of the Earth's gravity field: hydrological and oceanic effects and their possible detection using GRACE. *J. Geophys. Res.* **103** (B12), 30 205–30.

Wahr, J., Wingham, D. and Bentley, C. R. 2000. A method of combining ICESat and GRACE satellite data to constrain Antarctic mass balance. *J. Geophys. Res.* **105** (B7), 16 279–94.

Whillans, I. M. 1976. Radio-echo layers and the recent stability of the West Antarctic ice sheet. *Nature* **264** (5582), 152–5.

1983. Ice movement. In Robin, G. de Q., ed., *The Climatic Record in Polar Ice Sheets.* Cambridge University Press, pp. 70–7.

Whillans, I. M. and Van der Veen, C. J. 1993. Patterns of calculated basal drag on ice streams B and C., Antarctica. *J. Glaciol.* **39** (133), 437–46.

Whillans, I. M., Bolzan, J. and Shabtaie, S. 1987. Velocity of ice streams B & C, Antarctica. *J. Geophys. Res.* **92** (B9), 8895–902.

White, J. W. C. and Steig, E. J. 1997. Climate change in the southern hemisphere Holocene as viewed from high latitude isotopic records. *EOS, Trans. Am. Geophys. Union* **78** (17), 183.

Wingham, D. J., Rideout, A. L., Scharoo, R., Arthern, R. J. and Shum, C. K. 1998. Antarctic elevation change from 1992 to 1996. *Science* **282** (5388), 456–8.

Young, N. 1979. Measured velocities of interior East Antarctica and the state of mass balance within the I.A.G.P. area. *J. Glaciol.* **24** (90), 77–87.

Zwally, H. J. 1977. Microwave emissivity and accumulation rate of polar firn. *J. Glaciol.* **18** (79), 195–215.

Zwally, H. J. and Giovinetto, M. B. 1995. Accumulation in Antarctica and Greenland derived from passive-microwave data: a comparison with contoured compilations. *Ann. Glaciol.* **21**, 123–30.

Zwally, H. J., *et al.* 2002. ICESat's laser measurements of polar ice, atmosphere, ocean, and land. *J. Geodyn.* **34** (3–4), 405–45.

13

Antarctica: modelling

PHILIPPE HUYBRECHTS

Alfred Wegener Institute for Polar and Marine Research, Bremerhaven

13.1 Introduction

Mathematical modelling represents a vital tool for understanding and predicting the current and future behaviour of the Antarctic ice sheet. Above all, modelling tries to overcome the limitations of space and time associated with making direct observations. The dynamical timescales associated with many components of the Antarctic ice sheet are far larger than the limited period for which measurements are available. Models also generate information over the entire ice sheet and can yield insight into many processes that are often inaccessible for direct observation such as at the ice-sheet base. In addition, models are the only tools we have at our disposal to forecast the future evolution of the ice sheet.

Today, the Antarctic ice sheet contains 89% of global ice volume, or enough ice to raise sea level by more than 60 m (Table 13.1). Hence, only a small fractional change of its volume would have a significant effect on the global environment. The average annual solid precipitation falling onto the ice sheet is equivalent to 5.1 mm of sea level, this input being approximately balanced by ice discharge into floating ice shelves, which experience melting and freezing at their underside and eventually break up to form icebergs.

Changes in ice discharge generally involve response times of the order of 10^2 to 10^4 years. These timescales are determined by isostasy, the ratio of ice thickness to yearly mass turnover, processes affecting ice viscosity and physical and thermal processes at the bed. Hence it is likely that the Antarctic ice sheet is still adjusting to its past history, in particular to those changes associated with the last glacial–inter-glacial transition. Its future behaviour therefore has a component resulting from past climate changes as well as one related to present and future climate changes. To assess the future response of the ice sheet correctly, it is thus necessary to be able to distinguish between the long-term background trend and the anthropogenically induced signal due to recent and future climate changes.

Several methods have been used to assess the current evolution of the Antarctic ice sheet. The traditional method is to estimate the individual mass balance terms and make the budget. According to the Third Assessment Report of the Intergovernmental Panel on Climate Change (IPCC) (Church *et al.*, 2001), presently available data do not allow us to

Mass Balance of the Cryosphere: Observations and Modelling of Contemporary and Future Changes, eds. Jonathan L. Bamber and Antony J. Payne. Published by Cambridge University Press. © Cambridge University Press 2003.

Table 13.1. *Physical characteristics of the Antarctic ice sheet.*

Area (10^6 km^2)	12.37
Volume (10^6 km^3)	25.71
Volume (sea-level equivalent, m)^{a,b}	61.1
Accumulation^c (10^{12} kg per year)	1843 ± 76
Accumulation (sea-level equivalent, mm per year)^c	5.1 ± 0.2
Runoff^c (10^{12} kg per yr)	10 ± 10
Iceberg discharge (10^{12} kg per year)	2072 ± 304
Mass turnover time (years)	$\approx 15\,000$

^{a} Assuming an oceanic area of 3.62×10^8 km^2.
^{b} After isostatic rebound and sea-water replacing grounded ice.
^{c} Grounded ice only.
Sources: Huybrechts *et al.* (2000) and Church *et al.* (2001).

constrain this budget to better than $-376 \pm 384 \times 10^{12}$ kg per year ($-16.7\% \pm 17.1\%$ of total mass input), not significantly different from zero. Satellite altimetry has great potential to estimate the current trend, but is presently hampered by incomplete coverage and records that are too short to distinguish confidently between a short-term mass balance variation and the longer-term ice-sheet dynamic imbalance.

Because of the inaccuracy of the budget method and the short duration of satellite records, modelling of the entire ice-sheet–bedrock system over time may fill an important gap. It can provide an alternative approach to the balance problem by simulating the evolution of the ice sheet and its underlying bed over a sufficiently long time to remove transient effects, and subsequently analysing the imbalance patterns which result for the present day. Apart from yielding more insight into the role of various ice dynamic and climatic processes controlling the evolution of ice sheets, coupled ice-sheet–bedrock modelling can also help to remove the isostatic component from surface measurements to obtain ice-thickness changes, which are the relevant quantity for sea-level variations. The quality of such a calculation will depend on how well the model deals with ice and bedrock dynamics and on how well past mass balance changes can be described.

When studying the response of the Antarctic ice sheet to future climate changes, a further distinction needs to be made between surface mass balance changes and the dynamic response of the ice sheet. That is because a changing mass balance will significantly affect the distribution of ice thickness and surface slope. The resulting changes in driving stress will influence the ice flow, and thus the shape of the ice sheet, and this can in turn be expected to feed back on the mass balance components. In Antarctica, there is the additional effect of changes in ice discharge from the grounded ice sheet into the ice shelves, and the possibility of grounding-line migration. The latter directly affects the volume of grounded ice above floating, which is the relevant quantity controlling changes of ocean mass and sea level. A related aspect is the potential occurrence of unstable behaviour in the West Antarctic ice sheet (WAIS), with its buttressing ice shelves and bed so far below sea level, and proven record of ice-stream variability.

These issues are addressed here from a modelling perspective. The chapter is composed of two main parts plus a concluding summary. The first part discusses the type of models used to examine the Antarctic ice sheet. The emphasis is on three-dimensional whole ice-sheet studies as these are the tools required to investigate the overall response. It is discussed how such models are constructed and what they have taught us about the behaviour of the ice sheet. This is demonstrated with model results dealing with the current and future evolution of the Antarctic ice sheet. Much of this discussion relies on the author's own work, as to date only a few studies with complete ice-sheet/ice-shelf/lithosphere models have been published. In the second part of this chapter, the scope widens to process model studies and the potential for internally generated instability by both thermal and grounding-line mechanisms. The concluding section summarizes current knowledge gained from numerical models regarding the possible evolution of the Antarctic ice sheet during the third millennium.

13.2 Models of the Antarctic ice sheet

13.2.1 Types of models

Different types of numerical models have been applied to investigate the Antarctic ice sheet. A distinction is usually made between how these models embody horizontal space: either they study the dynamics of selected one-dimensional flow lines within the ice sheet, or they study the ice sheet in the full two-dimensional horizontal plane. The former type is often referred to as a flow-line or flow-band model and the latter as a planform model (cf. Hulbe and Payne (2001) for a recent review). Flow-line models have the principal advantage that they vastly reduce the amount of computation. Such models are therefore often used in an exploratory fashion to study the effects of a particular physical process or for situations where flow lines are strongly constrained laterally (e.g. for outlet glaciers). The disadvantage of flow-band models is that they cannot deal with situations in which the direction of ice flow or the geometry of a catchment area exhibits spatial and temporal variability. In those cases, planform modelling is more appropriate.

Planform models are distinguished by the way in which they incorporate the vertical dimension. Either processes operating through the vertical extent are incorporated by the use of a vertical average (e.g. isothermal models), in which case these models are referred to as two-dimensional planform or vertically integrated models; or these models incorporate vertical processes explicitly. Examples of such vertical processes include ice temperature, stress and velocity components, as well as fabric and water content. This class of models has been termed three-dimensional, quasi-three-dimensional, or 2.5-dimensional.

The latter terms emphasize the fact that a different set of assumptions is often employed in the vertical compared with the horizontal dimensions (a consequence of the great difference between the horizontal and vertical length-scales associated with ice masses). Very few models which incorporate the vertical dimension are truly three-dimensional because in most cases the direction of flow is determined by the surface gradient and does not vary with depth. For simplicity, however, we will continue to refer to planform models that resolve the vertical direction as three-dimensional.

Planform time-dependent modelling of ice sheets largely stems from early work by Jenssen (1977) and Mahaffy (1976). These papers develop work by Nye (1957) on what has become known as the shallow-ice approximation (Hutter, 1983). This approximation recognizes the disparity between the vertical and horizontal length-scales of ice flow, and implies flow by simple shear. This means that the gravitational driving stress is balanced by shear stresses, and that longitudinal strain rate components are neglected. The assumption requires slopes to be smoothed over a distance an order of magnitude greater than ice thickness to circumvent problems associated with small-scale bedrock irregularities. Although the assumption breaks down at the margin (large longitudinal and transverse flow gradients) and at the centre (longitudinal stress gradients prevailing), it has shown general applicability in large-scale ice-sheet modelling. The model by Mahaffy (1976) was vertically integrated and was developed as a computer program to find the heights of an arbitrary ice sheet on a rectangular grid. It incorporated Glen's flow law (Glen, 1955) for ice deformation by dislocation creep. However, in polar ice sheets the flow is also, to a large extent, a temperature-dependent problem. The first model that dealt with the flow–temperature coupling in a truly dynamic fashion was developed by Jenssen (1977). Jenssen introduced a scaled vertical co-ordinate, transformed the relevant continuity and thermodynamic equations, and presented a framework to solve the system numerically.

13.2.2 Vertically integrated whole ice-sheet models

Historically, large-scale modelling of the Antarctic ice sheet was pioneered by Budd, Jenssen and Radok (1971) in their *Derived Physical Characteristics of the Antarctic Ice Sheet*. This landmark work introduced many concepts and techniques that are still used in glaciology today. The study assumed that the ice sheet is in steady-state and that the ice always flows downhill (a consequence of the shallow-ice approximation). This allowed the identification of flow lines, along which a moving-column model was advected downstream to investigate two-dimensional (vertical-plane) temperature and velocity distributions. The flux of ice was determined by integrating snow accumulation along these flow lines. This so-called balance flux can then be used to determine the vertically averaged horizontal ice velocity, if ice thickness is known. A host of additional characteristics were also calculated by Budd *et al.* (1971), including ice residence times, vertical and horizontal strain rates, and gravitational stresses.

The first time-dependent ice-dynamical models of the Antarctic ice sheet were published in the early 1980s by Oerlemans (1982a,b) and Budd and Smith (1982). They were vertically integrated planform models coupled with simplified models of isostasy and ice-shelf formation. Both models had a coarse 100 km grid, considered isothermal ice deformation and did not accommodate for any special treatment of the flow between grounded and floating ice. Nevertheless, these models were able to reproduce the present-day distribution of ice thickness reasonably well. When submitted to sea-level and air-temperature forcing, these models highlighted the sensitivity of, in particular, the West Antarctic ice sheet.

13.2.3 Three-dimensional models

An important step forward in the numerical modelling of the Antarctic ice sheet came with the development of models which coupled the temporal evolution of ice flow and temperature (Herterich, 1988; Huybrechts and Oerlemans, 1988), and from models which considered ice flow, not only in the grounded ice sheet, but also across the grounding line and in the ice shelf (Böhmer and Herterich, 1990; Budd *et al.*, 1994; Huybrechts, 1990a,b; Huybrechts, 1992; Huybrechts and Oerlemans, 1990). This allowed one to address the evolution of the Antarctic ice sheet, without recourse to overly restrictive assumptions, and it set the stage for many of the large-scale modelling studies of the late 1980s and 1990s. Figure 13.1 shows the structure of one such model as it was described in Huybrechts (1992) and further refined in Huybrechts and De Wolde (1999) and Huybrechts (2002). The core of this model is a set of thermomechanically coupled equations for ice flow that are solved in three subdomains, namely the grounded ice sheet, the floating ice shelf and a stress transition zone in between, at the grounding line. This involves the simultaneous solution of conservation laws for momentum, mass and heat, supplemented with Glen's flow law. The flow within the three subdomains is coupled through a continuity equation for ice thickness, from which the temporal evolution of ice-sheet elevation and ice-sheet extent can be calculated. The latter is done by applying a flotation criterion, meaning that the ice-sheet geometry is entirely internally generated. The model equations are solved on a numerical grid resting on realistic bedrock topography with the finite-difference method.

In the Huybrechts model, grounded ice flow is assumed to result both from internal deformation and from basal sliding over the bed in those areas where the basal temperature is at the pressure melting point. Ice deformation in the ice-sheet domain results from shearing in horizontal planes. For the sliding velocity, a generalized Weertman relation is adopted, taking into account the effect of the subglacial water pressure. Ice shelves are included by iteratively solving a coupled set of equations for ice-shelf spreading, including the effect of lateral shearing induced by side walls and ice rises. At the grounding line, longitudinal stresses are taken into account in the effective stress term of the flow law. These additional stress terms are found by iteratively solving three coupled equations for depth-averaged horizontal stress deviators. Adjustment of the bed to changes of the ice load is taken into account by a variety of models (Le Meur and Huybrechts, 1996). In more recent versions, the bedrock model consists of a rigid elastic plate (lithosphere) that over-lies a viscous asthenosphere. This means that the isostatic compensation not only considers the local load, but integrates the contributions from more remote locations, giving rise to deviations from local isostasy. For an appropriate choice of the viscous relaxation time,

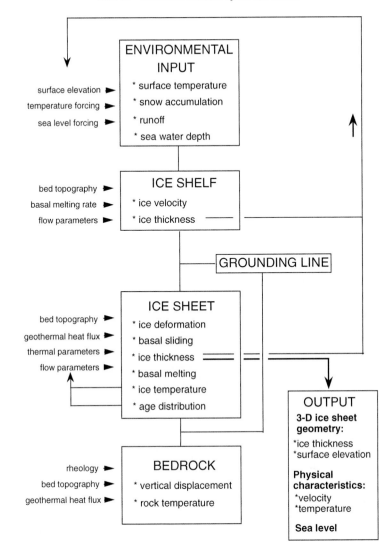

Figure 13.1. Structure of a three-dimensional ice-sheet model applied to the Antarctic ice sheet. The inputs are given at the left hand side. The model is driven by prescribed changes in environmental boundary conditions and has ice shelves, grounded ice and bed adjustment as major components. The position of the grounding line is not prescribed, but internally generated. Ice thickness feeds back on surface elevation, an important parameter for the calculation of the mass balance. The model essentially outputs the time-dependent ice-sheet geometry and the coupled temperature and velocity fields. (After Huybrechts (1992).)

this treatment produces results close to those from a sophisticated self-gravitating spherical visco-elastic Earth model, while at the same time being much more efficient in terms of computational overhead (Le Meur and Huybrechts, 1996). The horizontal grid resolution of the model is 40 km, and it has 11 layers in the vertical. This resolution was increased in recent work (Huybrechts, 2002) to 20 km and 31 layers, respectively, in an effort to better model concentrated flow in ice streams and outlet glaciers. This at the same time allowed the use of upgraded data sets for bedrock elevation and precipitation rate (Huybrechts *et al.*, 2000). Including a calculation of heat conduction in the bedrock, this gives rise to about 3×10^6 grid nodes. The numerical model used by Budd *et al.* (1994) and Warner and Budd (1998) also has a 20 km spatial resolution and was developed along similar lines, but with a somewhat simpler treatment of ice-shelf flow and no transition zone at the grounding line.

Interaction with the atmosphere and the ocean is, in these models, effected by prescribing the climatic forcing, consisting of the surface mass balance (accumulation minus ablation), surface temperature and the basal melting rate below the ice shelves. Changes in these parameters are usually parameterized in terms of air temperature. Precipitation rates are based on its present distribution and are perturbed in different climates according to sensitivities derived from ice cores or climate models. Melt-water runoff, if any, is obtained from the positive degree-day method (Braithwaite and Olesen, 1989; Reeh, 1991). This type of model is usually driven by time series of regional temperature changes (available from ice-core studies) and by the eustatic component of sea-level change. They have been used to address two main issues: the expansion and contraction of the Antarctic ice sheet during the glacial–inter-glacial cycles and the likely effects of greenhouse-induced polar warming. The answers to these questions rely on the complex interactions between grounding-line migration, interior ice-thickness changes and varying accumulation rates, as well as isostatic response, basal melting rate, ice temperature and ice viscosity.

Recent model developments have concentrated on ways of incorporating the dynamics of key areas that occur at the sub-grid scale and are therefore not well represented on 20 to 40 km grids. These principally concern the grounding line and areas of concentrated flow in outlet glaciers and ice streams, as well as details of the flow at ice divides. Such areas are characterized by large stress gradients, in which the approximations made in shallow ice-sheet models are known to break down. Ritz, Rommelaere and Dumas (2001) introduced the concept of a 'dragging ice shelf' to incorporate ice-stream dynamics, which is particularly important for the Siple Coast area of the West Antarctic ice sheet. In their model, inland ice is differentiated from an ice-stream zone by the magnitude of basal drag. This is based on the observation that ice-stream zones are characterized by low surface slopes, and thus low driving stresses, but have fast sliding. Ritz *et al.* (2001) treat these zones as semi-grounded ice shelves, and replace the shallow-ice approximation by a set of equations for ice-shelf flow to which basal drag is added. Gross model behaviour turns out to be quite similar to the whole ice-sheet models of Huybrechts, except that, in the Ritz *et al.* (Grenoble) model, the West Antarctic ice sheet has a lower surface slope near the grounding line and the break in the slope occurs further upstream at the place where the dragging ice shelf joins the inland ice subject to the shallow-ice approximation. One consequence is that grounding-line retreat in the Grenoble model occurs more readily in response to rising sea levels (Huybrechts *et al.*, 1998). Another way of incorporating smaller-scale features is to nest detailed higher-order models at higher resolution within a whole ice-sheet model. First attempts in this direction for limited inland areas near ice divides were presented in Greve *et al.* (1999) and Savvin *et al.* (2000).

Three-dimensional models have also been applied to the WAIS separately. Payne (1999) used the shallow-ice approximation in a model with fixed grounding line to investigate the interaction between thermomechanical coupling and basal sliding to generate steady-state oscillations in the ice flow. Hulbe and MacAyeal (1999) developed a numerical model which stands apart from other models in that it uses finite elements rather than finite differences to discretize the equations of motion. Their dynamic/thermodynamic model couples inland ice flow with flow in ice streams and ice shelves. The finite-element method eliminates the need for special parameterizations at flow regime boundaries, so that the ice flows smoothly from one regime to another. The method also allows for variable model resolution so one can concentrate computational effort on features of particular interest. However, the changing spatial patterns inherent in systems evolving over time pose a challenge in the application of finite-element models. Adaptive mesh generation is the best solution to that challenge, but has rarely been used in glaciology. Another limitation of the finite-element method is the associated computational burden. Therefore, principally due to their ease of use, finite differences have proven the more popular technique for ice-sheet modelling, in particular when the interest is in whole ice-sheet behaviour over longer periods of time.

13.3 Modelling the response of the Antarctic ice sheet

13.3.1 Modelling the present ice-sheet evolution

Modelling the quaternary evolution of the Antarctic ice sheet and its underlying bed is a way to obtain an estimate of the present-day ice-dynamic evolution unaffected by recent twentieth-century mass balance effects. The simulation requires time-dependent boundary conditions over a period long enough for the model to forget its initial start-up conditions. Long integrations over the last glacial cycles were analysed in Budd, Coutts and Warner (1998), Huybrechts (2002), Huybrechts and De Wolde (1999), Huybrechts and Le Meur (1999), and Ritz *et al.* (2001). Figure 13.2 shows the evolution of key glaciological variables in a typical run with the Huybrechts model over the last two glacial cycles, with forcing derived from the Vostok ice core (Petit *et al.*, 1999) and the SPECMAP sea-level stack (Imbrie *et al.*, 1984). Changing ice volume is principally a consequence of the areal expansion and contraction of the grounded ice sheet. Regional changes in ice thickness arise from these fluctuations in the location of the grounded ice sheet, and are further modulated by changes in ice temperature (cooler temperatures during a glacial result in a higher viscosity and thicker ice) and accumulation rate (reduced in cooler climates).

In the Huybrechts models, ice-sheet expansion during a glacial period mainly occurs over the Ronne-Filchner, Ross and Amery basins, and along the Antarctic peninsula. Around the East Antarctic perimeter, grounding-line advance is limited and is constrained by the proximity of the present-day grounding line to the continental shelf edge. During the last glacial maximum (LGM), modelled surface elevations over most of West Antarctica and the

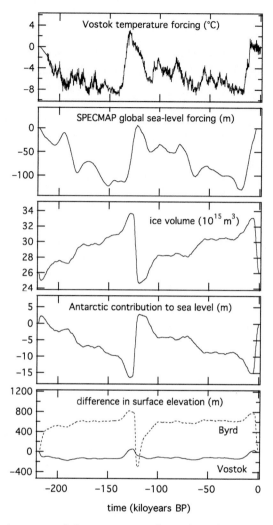

Figure 13.2. Forcing (mean annual air temperature and eustatic sea level) and predicted evolution of key glaciological variables (ice volume, contribution to sea level and local surface elevation changes at Byrd and Vostok stations) in a typical three-dimensional model experiment over the last two glacial cycles. Long spin-up times are required to model the current evolution of the ice sheet, which depends on its past history back to the last glacial period. (Based on the ice-sheet model described in Huybrechts (1992).)

Antarctic peninsula were up to 2000 m higher than present in direct response to grounding-line advance. Over central East Antarctica, surface elevations at the LGM were 100–200 m lower because of the lower accumulation rates. Holocene grounding-line retreat lags the eustatic forcing by some 10 000 years. This behaviour is related to the existence of thresholds for grounding-line retreat, and to the offsetting effect of late-glacial warming leading to enhanced accumulation rates and a thickening at the margin. In the model, most of the

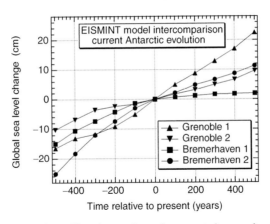

Figure 13.3. Modelled evolution of ice-sheet volume (represented as sea-level equivalent) centred at the present time resulting from on-going adjustments to climate change over the last glacial cycle. Data are from the Antarctic models that participated in the EISMINT intercomparison exercise. The Grenoble results were obtained with a model similar to the one in Ritz *et al.* (2001), and the Bremerhaven results are from the Huybrechts model. (From Huybrechts *et al.* (1998).)

retreat in West Antarctica occurs after 10 000 years BP. This timing is in line with recent geological evidence (Conway *et al.*, 1999; Ingolffson *et al.*, 1998) and is supported by some interpretations of relative sea-level data (Tushingham and Peltier, 1991). Nevertheless, small phase shifts in the input sea-level time series, inadequate representation of ice-stream dynamics and uncertainties in the Earth's rheological parameters may all have a significant effect on the model outcome.

Nonetheless, an important conclusion from this work is that the Antarctic ice sheet on the whole is still shrinking in response to grounding-line retreat during the Holocene. Experiments conducted as part of the European ice sheet modelling initiative (EISMINT) intercomparison exercise (Huybrechts *et al.*, 1998) confirm that the average Antarctic evolution at present is negative. Four different Antarctic models yield a sea-level contribution of between +0.1 and +0.5 mm per year averaged over the last 500 years (Figure 13.3), which corresponds to an average thinning of the ice sheet of between −0.3 and −1.6 cm of ice equivalent per year. Similar values of between +0.39 and +0.54 mm per year of sea-level change over the last 200 years have been found in Huybrechts and De Wolde (1999) and Huybrechts and Le Meur (1999). The geographical distribution of the local trend of Antarctic ice thickness, surface elevation and bed elevation in the latter study are displayed in Figure 13.4. The most pronounced feature is related to on-going grounding-line retreat along the Ross and Weddell Sea margins, which affects most of the West Antarctic ice sheet and nearby parts of the East Antarctic ice sheet. Local evolution rates range between −300 mm per year and +100 mm per year in West Antarctica, in contrast to the on-going slow thickening of several millimetres per year in interior East Antarctica. The latter is a direct consequence of the roughly doubled accumulation rates following the last glacial–inter-glacial transition between 15 000 and 10 000 years BP.

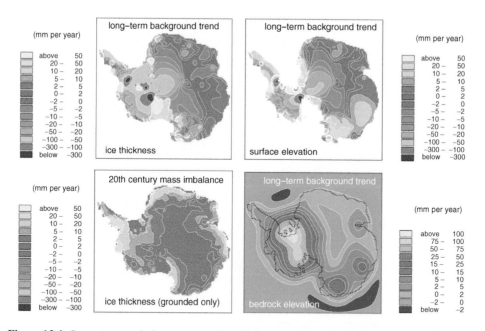

Figure 13.4. Long-term evolution patterns of ice thickness, surface elevation and bedrock elevation predicted by a coupled ice-sheet/visco-elastic bedrock model applied to the glacial cycles. The long-term patterns represent the mean evolution over the last 200 years and exclude the effects of twentieth-century climate change (Huybrechts and Le Meur, 1999). The lower left panel shows ice-thickness changes resulting from mass balance changes during the twentieth-century as predicted by scaling the ECHAM4/OPYC3 T106 patterns by the underlying ECHAM4 T42 resolution base trend (cf. Figures 13.6 and 13.7).

Glacio-isostatic modelling of the solid earth beneath the Antarctic ice sheet with pre-scribed ice-sheet evolution (James and Ivins, 1998) gives similar uplift rates to those pre-sented in Figure 13.4, indicating that the underlying ice-sheet scenarios and bedrock models are similar, but observations are lacking to validate the generated uplift rates. By contrast, Budd *et al.* (1998) find that Antarctic ice volume is currently increasing at a rate of about 0.08 mm per year of sea-level lowering because in their modelling the Antarctic ice sheet was actually smaller during the LGM than it is today (for which there is, however, little independent evidence) and the effect of the higher accumulation rates during the Holocene dominates over the effects of grounding-line changes. Model simulations of this kind do not include the possible effects of changes in climate during the twentieth century. Simulations described further below, in which an ice-sheet model is integrated using changes in tem-perature and precipitation derived from AOGCM experiments, suggest that anthropogenic climate change could have produced an additional contribution of between –0.2 to 0.0 mm per year of sea level from increased accumulation in Antarctica over the last 100 years. Figure 13.4 shows a typical example of the corresponding distribution of ice-thickness changes, which are generally positive and of the order of 0–2 cm per year.

13.3.2 Modelling the mass balance and its changes

Fundamental to the understanding of the response of the Antarctic ice sheet to climate change is its surface mass balance. Except in local regions near the coast, and in isolated blue-ice areas with their own micro-climate, surface ablation is at present negligible on the Antarctic ice sheet. That is because of the very low air temperatures, which remain well below freezing throughout the year, even in the summer at low elevation. The only exception is the northern tip of the Antarctic peninsula, but this part is characterized by steep local glaciers and covers only a very small fraction of the total Antarctic area. Thus, the primary mechanism by which climate changes affect the evolution of the Antarctic ice sheet is surface accumulation. Any deviation from its longer-term average will have an immediate effect on the total ice volume. The effect is dominant on decadal to century timescales, because ice discharge changes over longer periods. Climate changes also affect the basal melting rate below the ice shelves, but these can only influence the grounded ice sheet indirectly through weakening of the ice shelves, causing possible grounding-line retreat or enhanced outflow. Other mass balance components include blowing snow, sublimation and basal melting below the grounded ice sheet, but these components are either very small or are already accounted for in surface accumulation estimates.

Several approaches have been taken to estimate the annual snow-fall distribution over the ice sheet, ranging from classical compilations of *in situ* observations (Giovinetto and Zwally, 2000; Huybrechts *et al.*, 2000; Vaughan *et al.*, 1999), to atmospheric moisture convergence analysis based on meteorological data (Bromwich, Cullather and Van Woert, 1998; Budd, Reid and Minty, 1995; Turner *et al.*, 1999), remotely sensed brightness temperatures of dry snow (Giovinetto and Zwally, 1995a) and studies with general circulation models (Krinner *et al.*, 1997). Recent accumulation estimates over all the ice sheet (including the ice shelves) display a tendency for convergence within a range of 2100–2400×10^{12} kg per year (Church *et al.*, 2001), suggesting a remaining error of perhaps less than 10%. Nevertheless, the coverage of *in situ* data for validation of model results is still very poor over large areas, especially over the East Antarctic plateau.

To first order, it can be assumed that changes in surface accumulation are related to air temperature, because this controls the amount of water vapour that can be advected inland. Robin (1977) and Lorius *et al.* (1985) have suggested that accumulation is proportional to the saturated water vapour pressure of the air circulating above the surface inversion layer. Such a relationship appears to be particularly strong over the inland plateau and has been verified on the glacial–inter-glacial timescale from the ^{10}Be content in the Vostok ice core (Yiou *et al.*, 1985). Together with a degree-day model for melt-water runoff, this relation allows us to gain useful insight into the dependence of the surface mass balance on temperature change (Figure 13.5). As expected, accumulation increases steadily with temperature and runoff is essentially zero for present conditions. The total surface mass balance increases by 5.9% for a uniform temperature rise of $1\,°C$. The combined effect of changes in accumulation and melt-water runoff is an increase in net surface balance for a warming up to $5.3\,°C$; only temperature increases in excess of $8.3\,°C$ would lead to

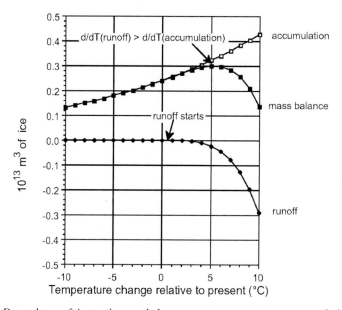

Figure 13.5. Dependence of Antarctic mass balance components on temperature relative to present. Runoff is calculated with a degree-day model, and accumulation changes are derived proportional to the saturated vapour pressure at the temperature above the surface inversion layer. For the temperature range relevant to future greenhouse warming, the relation between surface mass balance and mean air temperature is double-valued. (From Huybrechts and Oerlemans (1990).)

a reduced surface mass balance, any smaller perturbation leading to a net mass balance increase (Huybrechts and Oerlemans, 1990). These numbers are similar to those obtained by Fastook and Prentice (1994), underlining how a moderately warmer climate is likely to lead to Antarctic ice-sheet growth, and, hence, a sea-level fall from this source.

More sophisticated sensitivity analyses of Antarctica's surface mass balance have used multiple regression analyses (Fortuin and Oerlemans, 1990; Giovinetto and Zwally, 1995b), regional atmospheric models (Van Lipzig, 1999) and GCMs (Ohmura, Wild and Bengtsson, 1996; Smith, Budd and Reid, 1998; Thompson and Pollard, 1997; Wild and Ohmura, 2000). Recent progress has particularly been made with several coupled AOGCMs, especially in the 'time-slice' mode in which a high resolution model is driven by output from a low resolution transient experiment for a limited duration of time. Model resolution of typically 100 km allows for a more realistic topography crucial to resolve temperature gradients and orographic forcing of precipitation better along the steep margin of the Antarctic ice sheet. Figure 13.6 shows predicted patterns of climate change from the time-slice experiments conducted with the ECHAM4 model at T106 resolution (Wild and Ohmura, 2000). Both the simulated temperature and precipitation patterns for the present climate are generally close to compilations based on observations. In line with theoretical arguments, the model generally predicts increases of precipitation and temperature for doubled atmospheric CO_2

conditions, except over parts of Wilkes Land, for which there is no clear explanation. It can also be inferred from the summer temperature plots shown in Figure 13.6 that even in the $2 \times CO_2$ scenario, melting remains insignificant as the $0\,°C$ isoline does not reach the Antarctic coast.

Table 13.2 summarizes the mass balance sensitivity for a $1\,°C$ temperature rise from recent climate studies, expressed in equivalent global sea-level change. All data yield a precipitation increase corresponding to a fall of sea level of some $0.5\,mm/°C$ per year. The sensitivity for the case that the change in accumulation is set proportional to the relative change in saturation vapour pressure is at the lower end of the sensitivity range, suggesting that in a warmer climate changes in atmospheric circulation and increased moisture advection can become equally important to enhance accumulation, in particular close to the ice-sheet margin (Bromwich, 1995; Steig, 1997; Van Lipzig, 1999). Both ECHAM3 and ECHAM4/OPYC3 give a similar specific balance change over the ice sheet for doubled versus present atmospheric CO_2 to that found by Thompson and Pollard (1997).

Very little is known about basal melting rates below the ice shelves and on how or how fast these could respond to climatic changes. Large-scale estimates put the total melting below ice shelves at between 320×10^{12} kg per year (Kotlyakov, Losev and Loseva, 1978) and 756×10^{12} kg per year (Jacobs, Hellmer and Jenkins, 1996), or an average rate of between 0.25 and 0.55 m per year. However, the data, limited as they are, point to a large spatial variation: melt rates have been inferred in excess of 10 m per year below Pine Island glacier (Jenkins *et al.*, 1997), whereas evidence has been found for large areas of basal accretion below the Filchner-Ronne ice shelf (Oerter *et al.*, 1992). Available oceanographic studies do not permit the establishment of a clear relation between climate change, oceanic circulation, oceanic temperature and basal melting or freezing rates (Nicholls, 1997; Williams, Jenkins and Determann 1998). Factors such as summer ocean warming, length of period with open water, thermohaline properties of the source water and the details of the water circulation below the ice shelves have all been mentioned to play a role. A model study by Williams, Warner and Budd (1998) shows a quadrupling of the basal melt rate below the Amery ice shelf for an adjacent sea warming of 1 °C, but another study by Nicholls (1997) claims that climatic warming would reduce melting rates below the Ronne-Filchner ice shelf through alteration to sea-ice formation and the thermohaline circulation. Larger melt rates would thin the ice shelves, but additional feedback loops may be involved. For instance, selective removal of the warmest and thus softest ice from the ice column would decrease the depth-averaged ice temperature and thus stiffen the remaining ice, thereby decreasing ice-shelf strain rates (MacAyeal and Thomas, 1986). Many of these routes to ice-shelf changes are not represented in current ice-sheet models. Nevertheless, when rapid thinning occurs close to the grounding line, grounding-line retreat can be induced. In large-scale ice-sheet models, this occurs in two ways: steeper gradients across the grounding zone cause larger driving stresses and higher deviatoric stress gradients across the grounding zone which lead to increased strain rates and, hence, a speed-up of the grounded ice and subsequent thinning (Huybrechts and De Wolde, 1999; Warner and Budd, 1998).

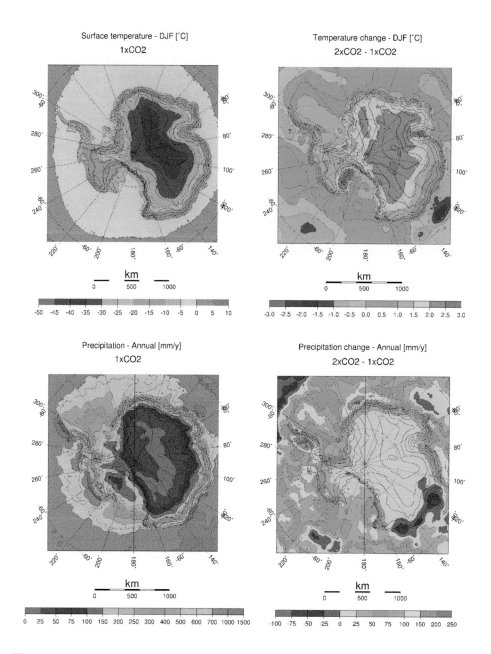

Figure 13.6. Patterns of climatic change over the Antarctic ice sheet from high resolution time-slice experiments with the ECHAM4/OPYC3 T106 model. Shown are the modelled mean annual precipitation and summer surface temperature (DJF) for the present climate (decade 1971–1980) as well as their changes at the time of doubled atmospheric CO_2 (decade 2041–2050). Over the ice sheet, total precipitation (expressed in water equivalent) exceeds total accumulation by the fractions of evaporation and liquid precipitation, but these latter components are usually small (the sublimation) or negligible (the rain fraction). (Redrawn from the information published in Wild and Ohmura (2000).)

Table 13.2. *Sensitivity of Antarctic mass balance to a 1 °C climatic warming.*

dB/dT is the mass balance sensitivity to local surface temperature change expressed as sea-level equivalent.

Source	dB/dT (mm/°C per year)	Method
Huybrechts and Oerlemans (1990)	−0.36	change in accumulation proportional to saturation vapour pressure
Giovinetto and Zwally (1995b)	−0.80[a]	multiple regression of accumulation to sea-ice extent and temperature
Ohmura *et al.* (1996)	−0.41[b]	ECHAM3/T106 time-slice ($2 \times CO_2 - 1 \times CO_2$)
Smith *et al.* (1998)	−0.40	CSIRO9/T63 GCM forced with SSTs 1950–99
Van Lipzig (1999)	−0.47[c]	regional atmospheric climate model forced with ECMWF re-analysis data
Wild and Ohmura (2000)	−0.48[b]	ECHAM4/OPYC3/T106 time slice ($2 \times CO_2 - 1 \times CO_2$)

[a] Assuming sea-ice extent decrease of 150 km/°C.
[b] Estimated from published data and the original time-slice results.
[c] Derived from an experiment with +2 °C external temperature forcing.

13.3.3 Modelling the future evolution of the Antarctic ice sheet

Response during the twenty-first century

Three-dimensional modelling studies all indicate that the dynamic response of the Antarctic ice sheet can be neglected on a century timescale, except when melting rates below the ice shelves are prescribed to rise by in excess of 1 m per year (Budd *et al.*, 1994; Huybrechts and De Wolde, 1999; Huybrechts and Oerlemans, 1990; O'Farrell *et al.*, 1997; Warner and Budd, 1998). This means that the response is essentially static, and thus that the ice flow on this timescale hardly reacts to changes in surface mass balance. Depending on the warming scenario and on how accumulation rates change with temperature, increased ice volume during the twenty-first century typically leads to global sea-level drops of the order of some 10 cm. For instance, when forced with temperature changes from a two-dimensional climate and ocean model forced by greenhouse gas rises of, respectively, 0.5% (low scenario), 1% (mid scenario) and 1.5% (high scenario) per year, Huybrechts and

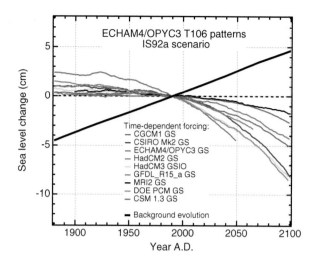

Figure 13.7. Volume changes of the Antarctic ice sheet during the twentieth and twenty-first centuries using temperature and precipitation changes from AOGCM experiments following the IS92a scenario, including the direct effect of sulphate aerosols, to derive boundary conditions for a three-dimensional ice-sheet model. The acronyms refer to the AOGCMs providing the base trend used to scale the ECHAM4/OPYC3 T106 climate-change patterns, cf. Figure 13.6. The thick black line shows the modelled long-term background trend. These experiments served as a base for the sea-level predictions of the IPCC Third Assessment Report (Church *et al.*, 2001).

De Wolde (1999) find Antarctic sea-level contributions between 1990 and 2100 to vary between −4.3 and −11.4 cm. In these runs, there is minimal surface melting, reaching only 5% of the total accumulation by AD 2100 in the high scenario. These numbers exclude the long-term background term, which was found in this study to be +5.0 cm during the same time interval. Similar changes during the twenty-first century due to accumulation changes are reported in Budd *et al.* (1994), O'Farrell *et al.* (1997) and Warner and Budd (1998).

The ice-sheet model of Huybrechts and De Wolde (1999) was used to make projections of Antarctic ice-sheet mass changes for the IPCC Third Assessment Report (Church *et al.*, 2001). Boundary conditions of temperature and precipitation were derived by perturbing present-day climatologies according to the geographically and spatially dependent patterns predicted by the T106 ECHAM4 model (Wild and Ohmura, 2000) for a doubling of CO_2 under the IS92a scenario. To generate time-dependent boundary conditions, these patterns were scaled with the area-average changes over the ice sheets as a function of time for available AOGCM results. The result is shown in Figure 13.7, suggesting global sea-level changes during the twenty-first century between −2 and −8 cm. These results were subsequently regressed against global mean temperature to enable further scaling to take into account the complete range of IPCC temperature predictions for the most recent SRES emission scenarios. Taking into account the background evolution and various sources of uncertainties, it yielded a predicted Antarctic contribution to global sea-level change between 1990 and 2100 of between -19 and $+5$ cm, which range can be considered as a 95% confidence interval (Church *et al.*, 2001).

Response during the third millennium and beyond

On centennial to millennial timescales, predictions should be based on dynamic rather than static simulations. The experiments discussed in Huybrechts and Oerlemans (1990) and Huybrechts and De Wolde (1999) demonstrate that ice dynamics counteract the accumulation-only response. Several mechanisms can be distinguished depending on the strength of the warming. For warmings below about 5 °C, runoff remains insignificant and there is hardly any change in the position of the grounding line. Under these circumstances, the inclusion of ice dynamics causes less growth, mainly because of an increase of the ice flux across the grounding line, which in part counteracts the thickening effect due to the increased accumulation rates. This increase of the flux across the grounding line is a result of both the local thickening at low elevations near to the grounding line, producing higher shear stresses and thus higher velocities, and of an increased ice-mass discharge on the ice shelves that pulls the ice out of the grounded ice sheet and is effective some distance inland. The effect becomes progressively stronger in time, and counteracts the static effect by 6% after 100 years, 30% after 500 years and more than 50% after 1000 years, reflecting response timescales at the margin (Huybrechts and De Wolde, 1999). According to this model study, the dynamic response leads to a sea-level change (relative to the background effect) of -50 cm for a surface warming of $+3$ °C and of -80 cm for a warming of $+5.5$ °C by the end of the third millennium (Figure 13.8).

For warmings that exceed 5–6 °C, on the other hand, significant grounding-line retreat sets in. In this model experiment, that is largely due to increased surface melting around the ice-sheet edge, leading to a thinning of grounded ice, but also to the effect of a warming ice shelf. The latter causes the ice shelf to deform more easily, which leads to larger flow velocities and equally a thinning. In Huybrechts and De Wolde (1999), the result is a sea-level rise of as much as $+80$ cm after 1000 years of simulated time for a warming of 8 °C. Grounding-line retreat is centred along the Antarctic peninsula and the northern-most parts of the East Antarctic perimeter (Wilkes Land in particular), with little change predicted to occur around the Ronne-Filchner and Ross ice shelves, because, even with a warming of 8 °C, no runoff takes place at their grounding zones. This response is apart from any changes in the oceanic circulation that may affect basal melting rates.

In the model studies performed by the Australian group (Budd *et al.*, 1994; O'Farrell *et al.*, 1997; Warner and Budd, 1998), large increases in bottom melting are the dominant factor in the longer-term response of the Antarctic ice sheet, even for moderate climate warmings of a few degrees. Budd *et al.* (1994) found that, without increased accumulation, the increased basal melt of 10 m per year would greatly reduce ice shelves and contribute to a sea-level rise of over 0.6 m after 500 years, but no drastic retreat of the grounding line. Reduction of grounded ice volume was mainly by high strain rates and thinning rates

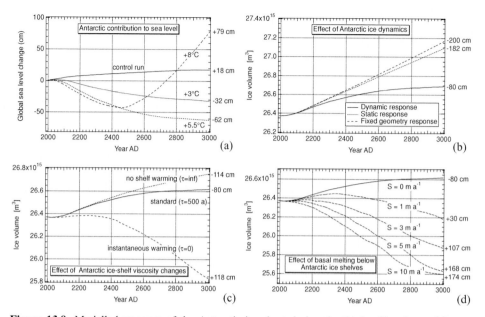

Figure 13.8. Modelled response of the Antarctic ice sheet during the third millennium subject to climate-warming scenarios predicted by a two-dimensional climate and ocean model forced by greenhouse gas concentration rises between two and eight times the present CO_2 by AD 2130, and kept constant after that (a). Also shown are the effects of ice dynamics (b) and the role of viscosity changes (c) and basal melting rates below the ice shelves (d) for the middle scenario ($4 \times CO_2$, $+5.5\,°C$). S is the applied basal melting rate and τ is the characteristic e-folding timescale for temperature adjustment in the ice shelf. The numbers in the right hand margins refer to the equivalent sea-level change between AD 1990 and 3000. (From Huybrechts and De Wolde (1999).)

near grounding lines. With a similar ice-sheet model, but different forcing derived from simulations with the CSIRO9 Mk1 fully coupled AOGCM, O'Farrell *et al.* (1997) find a sea-level rise of 0.21 m after 500 years for a transient experiment with basal melt rates evolving up to 18.6 m per year. In the study by Warner and Budd (1998), a bottom melt rate of 5 m per year causes the demise of WAIS ice shelves in a few hundred years and would remove the marine portions of the West Antarctic ice sheet and a retreat of coastal ice toward more firmly grounded regions elsewhere over a time period of about 1000 years. Predicted rates of sea-level rise are, in this study, up to between 1.5 and 3.0 mm per year depending on whether accumulation rates increase together with the warming. Similar volume reductions under conditions of high bottom melting were obtained in Huybrechts and De Wolde (1999). Allowing for runoff in addition to increased accumulation, they found a sea-level rise of more than 150 cm after 1000 years for a scenario involving a warming of $+5.5\,°C$ and a basal melting rate of 5 m per year (Figure 13.8). Though these are large shrinking rates, obtained under severe conditions of climate change, they cannot be considered to support the concept of a catastrophic collapse or strongly unstable behaviour of the WAIS, which

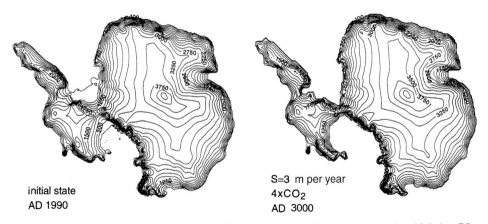

Figure 13.9. Snapshots of Antarctic surface elevation evolution for an experiment in which $4 \times CO_2$ and a uniform melting rate of 3 m per year below the ice shelves are applied. The plots demonstrate that changes are mainly restricted to grounding-line recession in West Antarctica, whereas the East Antarctic ice sheet thickens slightly. The righthand plot corresponds to a global sea-level rise of 1.07 m with respect to the 1990 initial state. Contour interval is 250 m; the lowest contour is close to the grounding line. (After Huybrechts and De Wolde (1999).)

is usually defined to mean its demise within several centuries, implying sea-level rises in excess of 10 mm per year (Mercer, 1978; Oppenheimer, 1998; Vaughan and Spouge, 2002).

The study by Huybrechts and De Wolde (1999) also highlighted the effect of reduced ice-shelf viscosity following a climatic warming to enhance strain rates and grounding-line retreat. The ice-sheet geometries shown in Figure 13.9 show that bottom melting causes most grounding-line retreat for the Ross ice shelf. Changes in the geometry of the East Antarctic ice sheet, on the other hand, are hardly distinguishable on the scale of the plots, but involve a thickening of up to 100 m on the plateau and some grounding-line retreat along the most over-deepened outlet glaciers in the lower right quadrant of the plotted ice sheet (in particular Totten, Ninis and Mertz glaciers). However, it should be stressed that the mechanics of grounding-line migration are not fully understood, and so these model results depend on how ice-sheet flow is coupled with ice-shelf flow. Also, none of these three-dimensional models adequately includes ice streams, and it may be that the ice streams are instrumental in controlling the behaviour and future evolution of the ice sheet in West Antarctica.

Independent of bottom melting below the ice shelves and ice-stream dynamics, surface melting sets an upper temperature limit on the viability of the Antarctic ice sheet, because runoff would eventually become the dominant wastage mechanism. For warmings of more than 10 °C, the mass balance at sea level around the Antarctic coast would be sufficiently negative so that the grounded ice sheet is no longer able to feed an ice-shelf. Also the WAIS ice shelves would disintegrate to near to their inland limits as summer temperatures rise above the thermal limit of ice-shelf viability believed to be responsible for the recent collapse of ice shelves at the northern tip of the Antarctic peninsula (Doake *et al.*, 1998;

Skvarca *et al.*, 1998; Vaughan and Doake, 1996). Disintegration of WAIS would, in that case, result because the WAIS cannot retreat to higher ground once its margins are subjected to surface melting and begin to recede (Huybrechts, 1994). Depending on the strength of the warming, such a disintegration would take at least a few millennia. Thresholds for disintegration of the East Antarctic ice sheet by surface melting involve warmings above 20 °C, a situation that has not occurred for at least the last 15 million years (Barker *et al.*, 1999), and which is far more than thought possible under any scenario of climatic change currently under consideration.

13.4 Potential sources of instability

The modelling studies reviewed above did not specifically address particular mechanisms that could cause runaway behaviour of the Antarctic ice sheet in the future. Such mechanisms have been identified in some glaciers, and a basic question is whether they could also occur in large ice sheets. Examples include the surging of valley glaciers and rapid retreat of tide-water glaciers. The search for instabilities in the Antarctic ice-sheet system has directed a great deal of research over the last two decades. Two mechanisms are potentially relevant for the Antarctic ice sheet. Grounding-line instability refers to an ice sheet based on a bed below sea level and is often mentioned in association with the West Antarctic ice sheet. Thermomechanically induced instabilities could occur in any type of grounded ice and have been discussed for both the East and West Antarctic ice sheets. An excellent review of potential sources of instability relevant to the West Antarctic ice sheet was given in Hulbe and Payne (2001), which is expanded upon in the discussion below.

13.4.1 Grounding-line instability

The marine nature of the West Antarctic ice sheet has led many glaciologists to believe that it may be inherently unstable and may respond drastically and irreversibly to a warming climate (Hughes, 1975; Lingle, 1984; Mercer, 1978; Thomas, 1979; Thomas and Bentley, 1978; Thomas, Sanderson and Rose, 1979; Weertman, 1974). This view is based on the idea that the large ice shelves surrounding West Antarctica control the discharge of inland ice and that a grounding-line tends to be unstable if the sea depth is greater than a critical depth and the sea floor slopes down toward the ice-sheet interior. The field evidence regarding WAIS (in-)stability is discussed in Chapter 12.

The original idea of the marine ice-sheet instability lies in the Weertman (1957) analysis of ice-shelf spreading. In that work, an expression was derived for unidirectional spreading of a confined ice shelf that leads to a creep thinning rate in the floating ice, and, thus, near the grounding line, which depends on the fourth power of ice thickness. Flow at the grounding line should therefore increase rapidly as a retreating grounding line progressively encounters deeper water and thus thicker ice, as is typically the case in West Antarctica.

This creates a positive feedback that will result in further retreat of the grounding line and a complete collapse, unless there is a sufficiently high bedrock sill on which the grounding line can achieve equilibrium (Thomas, 1979). The WAIS owes its existence, in an extension of that concept, to frictional drag from a combination of grounded 'pinning points' in the ice shelf and shear resistance along the sides of their enclosing embayments. That drag creates a 'back-pressure' from the ice shelves, which restrains the outflow of the grounded inland ice (Thomas *et al.*, 1979). If the effect of a warming were to weaken the ice shelves through increased calving and bottom melting, the ensuing reduction in back stress would strongly diminish their buttressing effect, and could initiate a collapse in as little as a century, causing sea level to rise at a mean rate of some 50 mm per year (Mercer, 1978).

However, the anchor upon which the instability hypothesis rests is the transmission of stress from floating to grounded ice. Van der Veen (1986) pointed out the inconsistency in incorporating ice-shelf dynamics at the grounding line, mainly because the ice-sheet/ice-shelf feedback is taken into account at one point only and a discontinuity is introduced when calculating creep thinning rates and ice velocities. In the Thomas *et al.* (1979) analysis, creep thinning at the grounding line is calculated from an expression based on ice-shelf spreading, but the movement at the grounding line is controlled by basal shear stresses. That excludes a negative feedback whereby increased advection from the ice sheet can compensate for the thinning. When calculating ice velocities explicitly along an entire flow line, Van der Veen (1985) found that the position of the grounding line was only little influenced by changes in ice-shelf back-pressure, and, furthermore, that an ice shelf is not needed to stop grounding-line retreat. In a further development, Hindmarsh (1993) argues that the transition zone between grounded ice and floating ice is of such limited extent that it is unlikely for any stress transmission to take place. Numerical simulations (Herterich, 1987; Lestringant, 1994) and an analytical analysis of stick-slip transitions (Barcilon and MacAyeal, 1993) support this view. Interpretation of strain rates across the grounding zone of Ekström ice shelf arrives at the same conclusion (Mayer and Huybrechts, 1999). Hence, the original marine ice-sheet instability hypothesis can be refuted on the grounds of the simplified model it was based on.

In reality, however, the situation of the WAIS is complicated by the presence of fast flowing ice streams whose characteristics blend gradually into those of the ice shelves and whose response times to changes at the grounding line appear to be very rapid (Alley and Whillans, 1991). There is a considerable body of evidence for ice-stream variability (Retzlaff and Bentley, 1993; Stephenson and Bindschadler, 1988), but the mechanisms for these oscillations are not well understood. They have been ascribed to processes such as basal water diversion to a neighbouring ice stream (Anandakrishnan and Alley, 1997) or thermomechanical interactions between competing catchment areas (Payne, 1999). Other mechanisms for dynamic changes have also been proposed at the interface of ice that is lubricated by sediment and water (Anandakrishnan *et al.*, 1998; Bell *et al.*, 1998; Blanken-ship *et al.*, 1986), the transverse shear zone where fast moving ice meets relatively static ice (Echelmeyer *et al.*, 1994; Jacobson and Raymond, 1998) and the ice-stream onset regions where slowly flowing inland ice accelerates into the ice streams. Just to what extent the ice streams contribute to the stability of the WAIS is in dispute. On the one hand, the inherent ability of ice streams to transport ice rapidly from the interior to the ocean indicates, in the view of some glaciologists, an enhanced capability for a drastically accelerated output flux. A contrary view is that the rapid response time of ice streams removes the flux imbalance at the grounding line that was the basis of the instability model, and that the purported grounding-line instability very well may not exist (Hindmarsh, 1993).

13.4.2 Thermomechanical instability

Three types of thermomechanical instability are identified in ice-sheet models. These relate to creep instability; the downstream transition from frozen to melting basal conditions; and the occurrence of warm-based ice encircled by frozen bed conditions. The first two processes are related to the temperature dependence of the flow law for ice. The third relies on the geometry of the basal temperature field. These instabilities are internal, in that they depend on flow-dependent thermal evolution of the ice sheet. Under suitable conditions, these instabilities can lead to free oscillations, i.e. cyclical behaviour without external forcing.

The temperature dependence of the flow properties of ice contains the potential for positive feedback between predicted ice velocity and temperature fields. As ice temperature increases, ice deformation rates will also increase. This leads to enhanced flow, increased dissipation and further warming. Clarke, Nitsan and Paterson (1977) introduced the term 'creep instability' for the process whereby an initial temperature anomaly leads to a runaway increase in ice velocity. Initial numerical studies of creep instability employed vertical, one-dimensional models of constant thickness resting on a bed with constant slope. In the study by Clarke *et al.* (1977), it was demonstrated that the feedback between temperature and ice flow can lead to bifurcation. Under specific conditions, multiple steady-states occur: either the ice sheet is thick and cold with low velocities, or it is thin and warm with high velocities, implying that the ice sheet can switch suddenly from the slow mode to the fast mode if external conditions change. Clarke *et al.* also carried out an analysis of linear stability, revealing that typical e-folding times for the runaway increase are in the 10^3 to 10^4 years range for large ice sheets. The study by Schubert and Yuen (1982) elaborates further on this problem. Their scenario calls on climatically enhanced accumulation to increase the thickness of the East Antarctic ice sheet above a critical value for the onset of an explosive shear heating instability. This would cause massive melting at the base and initiate a surge. In Yuen, Saari and Schubert (1986), conditions for the onset of this explosion were investigated. For certain values of ice rheological parameters, they found that a sudden increase in ice-sheet thickness by 1–2 km could lead to melting of the basal shear layer in only thousands of years.

One objection that can be made against these analyses, however, is that they are 'local': horizontal temperature advection and driving stress (ice-sheet geometry) are not allowed to react to the changing temperature and velocity fields. These processes would tend to dampen temperature perturbations. Numerical modelling studies conducted with a thermomechanical flow-band model suggest that, for the East Antarctic ice sheet, this horizontal heat advection prohibits the development of any runaway warming (Huybrechts and Oerlemans, 1988). Hulbe (1998) discusses the positive feedback in the context of a quasi-three-dimensional thermomechanical finite-element model that includes both horizontal and vertical temperature advection and diffusion. In that analysis, the tendency toward excessive heating in deep ice is mitigated firstly by corresponding large vertical strain rates, which thin the ice and thus increase upward diffusion of temperature, and secondly by enhanced downstream advection.

Payne (1995), Pattyn (1996) and Greve and MacAyeal (1996) studied a related form of thermomechanical instability. In these models, an instability arises because of an assumed abrupt increase in sliding velocity with the onset of basal melting. The sudden transition leads to a pronounced step in the ice-surface profile above the warm–cold ice transition. The steep surface slope in turn increases the gravitational driving stress and deformational velocity, and thus viscous dissipation also increases dramatically. Payne (1995) estimates a 16-fold increase in dissipation for a doubling of ice surface slope. The location of the warm–cold ice transition point can migrate rapidly upstream as a consequence of this localized heating and associated enhanced flow, causing a surge. Eventually, reduced ice thicknesses and enhanced cold-ice advection lead to stagnation. The extent of the bed area at the melting temperature depends on the air temperature and accumulation rate. For low temperatures, most of the glacier remains frozen to its bed, and oscillations are restricted to the downstream area. The period of the oscillations also changes as the accumulation rate changes: for low rates the dominant period is 6000–7000 years, for intermediate accumulation the period is 3000–4000 years, while for larger accumulation the oscillations become irregular with periods in excess of 10 000 years. Applying the same physics to the WAIS in three dimensions, Payne (1999) finds large spatial variability concentrated in the ice streams along the Siple Coast. This is attributed to competition between adjacent ice streams for increased drainage area, and leads to internally generated cycles of ice-stream growth and stagnation. Typical periods of these oscillations are between 5000 and 10 000 years. Despite large variability in output fluxes of individual ice streams by up to a factor of 5, the overall volume of the WAIS, however, hardly changed, supporting the suggestion that the ice streams may act to remove the imbalance of individual drainage basins (Hindmarsh, 1993).

A third form of internal instability is discussed by Oerlemans (1983) and MacAyeal (1992) in studies that seek cyclic behaviour in ice sheets. Both employ thermomechanical models that associate the presence of basal melt water with enhanced basal sliding. The latter model is applied to the WAIS only, uses ice-stream specific stress balance equations, but ignores horizontal temperature advection. Oerlemans (1983) uses a constant climate forcing, while MacAyeal (1992) specifies a climate cycle according to the Vostok ice-core record. The result in both cases is a cycle of slow ice-sheet growth and rapid discharge. Self-sustained oscillations were also found by Oerlemans and Van der Veen (1984) when employing a two-dimensional model of the Antarctic ice sheet. Their model contains a simplified calculation of the basal ice temperature as well as the feedback between ice flow and basal water. Also here, for certain values of the model parameters, ice thickness over

East Antarctica varied periodically with a time interval of typically a few thousand years. This was caused by the accumulation of basal water under the thicker parts. West Antarctica, on the other hand, remained close to a steady-state with extensive melting and high sliding velocities. Nevertheless, much also seems to depend on how sliding is parameterized in terms of basal water and on how effectively melt water is removed.

MacAyeal's (1992) inclusion of subglacial till dynamics leads to episodic ice-sheet fluctuations that are out of phase with the climate forcing. Collapses occur at times irregularly spaced relative to the ice-age cycle, because of the long time constant of the subglacial system. His modelling yielded three collapses in 10^6 years; it also produced iceberg pulses, with fluxes of 1000 to 3000 Gt per year (3–10 mm per year sea-level rise) lasting a few decades, about ten times during the million-year model run. These too occurred pseudo-randomly. The periodic behaviour relies on the development of basal ice at its melting point in the interior of the ice sheet, where ice is thick, while ice nearer the margins remains frozen to the bed. Eventually, the pool of warm-based ice breaks through the encircling cold-based ice, leading to a large, rapid surge. The thin, post-surge ice sheet refreezes to its bed, thickens over time and the cycle repeats. However, some approximations in MacAyeal's model remain crude, particularly those relating to the bed and to horizontal advection, which is not considered, thus precluding thermodynamic feedback with the flow from upstream. The spatial pattern of warm basal temperature in the interior and cold basal temperature near the margins is also the opposite to that predicted by the majority of WAIS thermomechanical models (Budd *et al.*, 1971; Huybrechts, 1992; Payne, 1999). The main process favouring warm-based divides is the increased thermal insulation afforded by thick ice. Processes favouring cold-based divides and warm-based margins are enhanced cold-ice advection at the divide and increased dissipation as ice discharge increases towards the margin. The models which predict cold-based interiors are physically more realistic because horizontal temperature advection is fully incorporated. Also, West Antarctic ice streams are already today known to be melting at their beds, yet none of them seem to be in a 'collapsing' mode.

13.5 Conclusions and further outlook

This chapter has discussed some of the models in use today to study the behaviour of the Antarctic ice sheet and their application in understanding the current and future response of the ice sheet, and has identified potential sources of internal instability. Although many aspects of large-scale ice-sheet models still remain rather crude and need to be developed further, a number of key results are starting to emerge. From the modelling evidence reviewed in this chapter, it appears that the Antarctic ice sheet is late in the cycle of ice-sheet growth and retreat and is still responding to the end of the last Ice Age. Most of the ongoing shrinking is concentrated in the West Antarctic ice sheet as a delayed response to post-glacial sea-level rise, whereas the East Antarctic ice sheet may be close to a stationary state, or growing slightly in response to the increased accumulation rates. Short and yet incomplete satellite altimetry records indeed support such a broad picture, although such a comparison should be reserved because of the different time periods involved (Wingham *et al.*, 1998). The background trend may well dominate the response of the ice sheet during the twenty-first century, and is probably of the order of an equivalent global sea-level rise of a few centimetres per century (Church *et al.*, 2001).

A climatic warming during the third millennium will probably lead to an increase in precipitation which has an immediate effect on Antarctic ice volume. Increased precipitation will increase the mass input onto the ice sheet, but any balancing increase in outflow will take centuries to millennia to compensate for the gain. So, in the short-term, the ice sheet will probably grow slowly relative to its current state of balance. There is, indeed, evidence, both from moisture flux modelling (Bromwich *et al.*, 1998) and glaciological observations (Mosley-Thompson *et al.*, 1999; Peel and Mulvaney, 1988), that that is already happening. As long as the warming remains moderate (<5 °C), surface melting is likely to remain insignificant for the mass balance, and the increase in accumulation will dominate over any increase in melting. Typical ice growth rates are predicted to be of the order of 5–10 cm per century of equivalent sea-level lowering. For larger warming, however, surface melting rates are predicted to become more important, especially at the Antarctic peninsula and around the East Antarctic perimeter. As a consequence, ice-sheet models exhibit thinning at the ice-sheet margin, and eventually grounding-line retreat results. It would take up to several centuries before these effects become apparent, but, depending on the strength of the warming, volume reductions equivalent to sea-level rises of several tens of centimetres per century are a distinct possibility (Huybrechts and De Wolde, 1999).

Faster changes could only arise from instability mechanisms able to increase the outflow by a multiple of the present amount. Thermomechanical instabilities have been identified in ice-sheet models, but may not be very realistic as most of these models address processes in isolation and often neglect crucial mechanisms shown to counteract any runaway behaviour. Also, these instability mechanisms operate over time periods of millennia and are therefore of less relevance on a human timescale. And, if they should occur, they would probably not be linked to anthropogenic climatic warming, but to past climatic changes (MacAyeal, 1992). That is because ice sheets take thousands of years to respond to changes in surface temperature, as it takes that long for temperature changes to penetrate to the bed, and only there could increasing temperatures affect the flow rates. The grounding-line instability hypothesis as originally formulated in the 1970s is demonstrably a model artefact and can therefore not be invoked as a serious candidate to induce a collapse of the West Antarctic ice sheet. The dynamic connection between ice shelves and the Siple Coast ice streams may modify that view, but to date no credible model has been put to work that is able to produce an average speed-up of the total outflow by at least the factor of 10 required for a sea-level rise in excess of 10 mm per year (Bentley, 1997, 1998a,b).

On a multiple-century timescale, oceanic warming is another crucial factor that could affect the grounded ice sheet because of its potentially weakening effect on ice shelves. However, GCM studies have suggested that oceanic warming in the far Southern Ocean would be delayed by centuries compared with the rest of the world because of the large-scale sinking of surface waters around Antarctica (Manabe *et al.*, 1991). Recent spectacular breakups of the Larsen ice shelves in the Antarctic peninsula (Doake *et al.*, 1998) demonstrate the existence of an abrupt thermal limit on ice-shelf viability associated with regional atmospheric warming (Skvarca *et al.*, 1998). But the WAIS ice shelves are not immediately threatened by this mechanism, which would require a further warming of 10 °C before the −5 °C mean annual isotherm reached their ice fronts (Vaughan and Doake, 1996). Although atmospheric warming would increase the rate of deformation of the ice, causing the ice shelf to thin, response timescales are equally of the order of several hundred years (Rommelaere and MacAyeal, 1997). Furthermore, it can be questioned whether ice-shelf thinning would have any drastic effect on the inland ice. Models indicate that, in any case, very high average melting rates in excess of 10 m per year would be required to produce ice-sheet retreats equivalent to sea-level rises of the order of 1–3 mm per year (Huybrechts and De Wolde, 1999; Warner and Budd, 1998). Based on the current model evidence, it is therefore very unlikely that Antarctica would undergo major volume losses during the next few centuries, and perhaps even during most of the third millennium.

Even though much progress has been made in understanding and modelling of the Antarctic ice sheet, several lines of future investigation are apparent. These can be roughly divided into two groups: improved boundary and test data, and incorporation of more appropriate physics. Climate and mass balance related boundary conditions are obviously vital to simulate correctly ice-sheet evolution. Present-day atmospheric boundary conditions, such as mean annual air temperature and snow accumulation, are known to a level of accuracy commensurate with that required by ice-sheet models, but their patterns of change in past, as well as future, climates are poorly constrained. Palaeoclimatic data show that storm strengths and trajectories have changed in the past and have greatly affected accumulation, especially at the ice-sheet margin. This raises the possibility that future circulation changes will occur and will also affect precipitation. It is important that atmospheric and snow-surface processes be understood well enough so that model-based predictions of snow accumulation can be made directly, rather than using predictions of temperature and assumed temperature sensitivity or perturbation methods to estimate precipitation. Even more troublesome is the melt rate from the underside of the ice shelves, which may affect grounding lines but for which we have very limited data. The same is also true of the geothermal heat warming at the ice-sheet base, which exerts a crucial control on the spatial extent of basal melting, but for which there is virtually no data. In addition, the process of iceberg calving, and more generally the disintegration of ice shelves, is currently not well understood and therefore impossible to model with confidence.

Model validation and testing is important. A series of validation experiments were developed within the framework of the European Ice Sheet Modelling Initiative (EISMINT) to provide tests for ice-shelf and ice-sheet numerics, thermomechanical coupling and planform models applied to the Antarctic and Greenland ice sheets (Huybrechts *et al.*, 1996, 1998; MacAyeal *et al.*, 1996; Payne *et al.*, 2000). These have enabled modellers to discover errors and numerical instabilities and to upgrade individual models. However, models should also be tested against field data, but this is hampered by a paucity of data at the appropriate spatial and temporal scales. In particular, models of basal thermal regime, basal hydrology and subglacial sediment deformation remain untested except at a very limited number of bore holes and by broad comparison with geophysical inferences from seismics or radio-echo sounding. A linked aspect concerns the incorporation of fast flowing outlet glaciers and ice streams, which are not described well in contemporary ice-sheet models. That is partly a resolution problem, but also the proper physics of ice-sheet dynamics at bases and lateral margins of ice streams and at grounding lines are not yet well known. In particular, treatment of basal sliding, together with the appropriate stresses involved, needs to be incorporated in a more realistic way. Many of these challenges pertain in particular to modelling of the West Antarctic ice sheet, as models based on the shallow-ice approximation perform more satisfactorily for the largely continental-based East Antarctic ice sheet. Clearly, there is much room for improvement of numerical models of the Antarctic ice sheet.

References

Alley, R. B. and Whillans, I. M. 1991. Changes in the West Antarctic ice sheet. *Science* **254**, 959–63.

Anandakrishnan, S. and Alley, R. B. 1997. Stagnation of ice stream C, West Antarctica by water piracy. *Geophys. Res. Lett.* **24** (3), 265–8.

Anandakrishnan, S., Blankenship, D. D., Alley, R. B. and Stoffa, P. L. 1998. Influence of subglacial geology on the position of a West Antarctic ice stream. *Nature* **394**, 62–5.

Barcilon, V. and MacAyeal, D. R. 1993. Steady flow of a viscous ice stream across a no-slip/free-slip transition at the bed. *J. Glaciol.* **39** (131), 167–85.

Barker, P. F., Barrett, P. J., Cooper, A. F. K. and Huybrechts, P. 1999. Antarctic glacial history from numerical models and continental margin sediments. *Palaeogeog., Palaeoclimatol., Palaeoecol.* **150**, 247–67.

Bell R. E. *et al.* 1998. Influence of subglacial geology on the onset of a West Antarctic ice stream from aerogeophysical observations. *Nature* **394**, 58–62.

Bentley, C. R. 1997. Rapid sea-level rise soon from West Antarctic ice sheet collapse? *Science* **275**, 1077–8.

    1998a. Ice on the fast track. *Nature* **394**, 21–2.

    1998b. Rapid sea-level rise from a West-Antarctic ice-sheet collapse: a short-term perspective. *J. Glaciol.* **44** (146), 157–63.

Blankenship, D. D., Bentley, C. R., Rooney, S. T. and Alley, R. B. 1986. Seismic measurements reveal a saturated porous layer beneath an active Antarctic ice stream. *Nature* **322**, 54–7.

Böhmer, W. J. and Herterich, K. 1990. A simplified 3-D ice sheet model including ice shelves. *Ann. Glaciol.* **14**, 17–19.

Braithwaite, R. J. and Olesen, O. B. 1989. Calculation of glacier ablation from air temperature, west Greenland. In Oerlemans, J., ed., *Glacier Fluctuations and Climatic Change*. Dordrecht, Kluwer Academic Publishers, pp. 219–33.

Bromwich, D. H. 1995. Ice sheets and sea level. *Nature* **373**, 18–19.

Bromwich, D. H., Cullather, R. I. and Van Woert, M. L. 1998. Antarctic precipitation and its contribution to the global sea-level budget. *Ann. Glaciol.* **27**, 220–6.

Budd, W. F. and Smith, I. N. 1982. Large-scale numerical modelling of the Antarctic ice sheet. *Ann. Glaciol.* **3**, 42–9.

Budd, W. F., Jenssen, D. and Radok, U. 1971. *Derived Physical Characteristics of the Antarctic Ice Sheet*. ANARE Interim Report, Series A (IV), Glaciology Publication no. 18, University of Melbourne.

Budd, W. F., Jenssen, D., Mavrakis, E. and Coutts, B. 1994. Modelling the Antarctic ice sheet changes through time. *Ann. Glaciol.* **20**, 291–7.

Budd, W. F., Reid, P. A. and Minty, L. J. 1995. Antarctic moisture flux and net accumulation from global atmospheric analyses. *Ann. Glaciol.* **21**, 149–56.

Budd, W. F., Coutts, B. and Warner, R. C. 1998. Modelling the Antarctic and northern-hemisphere ice-sheet changes with global climate through the glacial cycle. *Ann. Glaciol.* **27**, 153–60.

Church, J. A. *et al.* 2001. Changes in sea level. In Houghton, J. T. *et al.*, eds., *Climate Change 2001: The Scientific Basis*. Cambridge University Press, pp. 639–94.

Clarke, G. K. C., Nitsan, U. and Paterson, W. S. B. 1977. Strain heating and creep instability in glaciers and ice sheets. *Rev. Geophys. & Space Phys.* **15**, 235–47.

Conway, H. W., Hall, B. L., Denton, G. H., Gades, A. M. and Waddington, E. D. 1999. Past and future grounding-line retreat of the West Antarctic ice sheet. *Science* **286**, 280–6.

Doake, C. S. M., Corr, H. F. J., Rott, H., Skvarca, P. and Young, N. W. 1998. Breakup and conditions for stability of the northern Larsen ice shelf, Antarctica. *Nature* **391**, 778–80.

Echelmeyer, K. A., Harrison, W. D., Larsen, C. and Mitchell, J. E. 1994. The role of the margins in the dynamics of an active ice stream. *J. Glaciol.* **40** (136), 527–38.

Fastook, J. L. and Prentice, M. L. 1994. A finite-element model of Antarctica: sensitivity test for meteorological mass balance relationship. *J. Glaciol.* **40** (134), 167–75.

Fortuin, J. P. F. and Oerlemans, J. 1990. Parameterisation of the annual surface temperature and mass balance of Antarctica. *Ann. Glaciol.* **14**, 78–84.

Giovinetto, M. B. and Zwally, H. J. 1995a. An assessment of the mass budgets of Antarctica and Greenland using accumulation derived from remotely sensed data in areas of dry snow. *Zeits. Gletscherkunde & Glazialgeol.* **31**, 25–37.

1995b. Annual changes in sea ice extent and of accumulation on ice sheets: implications for sea level variability. *Zeits. Gletscherkunde & Glazialgeol.* **31**, 39–49.

2000. Spatial distribution of net surface accumulation on the Antarctic ice sheet. *Ann. Glaciol.* **31**, 171–8.

Glen, J. W. 1955. The creep of polycrystalline ice. *Proc. Roy. Soc. London Series B* **228**, 519–38.

Greve, R. and MacAyeal, D. R. 1996. Dynamic/thermodynamic simulations of Laurentide ice sheet instability. *Ann. Glaciol.* **23**, 328–35.

Greve, R., Mügge, B., Baral, D. R., Albrecht, O. and Savvin, A. 1999. Nested high-resolution modelling of the Greenland Summit Region. In Hutter, K., Wang, Y. and Beer, H., eds., *Advances in Cold-Region Thermal Engineering and Sciences*. Berlin, Springer Verlag, pp. 285–306.

Herterich, K. 1987. On the flow within the transition zone between ice sheet and ice shelf. In Van der Veen, C. J. and Oerlemans, J., eds., *Dynamics of the West Antarctic Ice Sheet*. Dordrecht, D. Reidel, pp. 185–202.

1988. A three-dimensional ice-sheet model of the Antarctic ice sheet. *Ann. Glaciol.* **11**, 32–5.

Hindmarsh, R. C. A. 1993. Qualitative dynamics of marine ice sheets. In Peltier, W. R., ed., *Ice in the Climate System*. NATO ASI Series I12, pp. 68–99.

Hughes, T. J. 1975. The West Antarctic ice sheet: instability, disintegration, and initiation of ice ages. *Rev. Geophys. & Space Phys.* **13**, 502–26.

Hulbe, C. L. 1998. Heat balance of West Antarctic ice streams, investigated with a numerical model of coupled ice sheet, ice stream and ice shelf flow. Ph.D. thesis, University of Chicago.

Hulbe, C. L. and MacAyeal, D. R. 1999. A new numerical model of coupled inland ice sheet, ice stream, and ice shelf flow and its application to the West Antarctic ice sheet. *J. Geophys. Res.* **104** (B11), 25 349–66.

Hulbe, C. L. and Payne, A. J. 2001. The contribution of numerical modelling to our understanding of the West Antarctic ice sheet. In Alley, R. B. and Bindschadler, R. A., eds. *The West Antarctic Ice Sheet: Behaviour and Environment.* Antarctic Research Series, 77, Washington D. C., American Geophysical Union, pp. 201–19.

Hutter, K. 1983. *Theoretical Glaciology.* Dordrecht, D. Reidel.

Huybrechts, P. 1990a. A 3-D model for the Antarctic ice sheet: a sensitivity study on the glacial–interglacial contrast. *Climate Dyn.* **5**, 79–92.

1990b. The Antarctic ice sheet during the last glacial-interglacial cycle: a three dimensional experiment. *Ann. Glaciol.* **11**, 52–9.

1992. *The Antarctic Ice Sheet and Environmental Change: A Three-Dimensional Modeling Study.* Berichte zur Polarforschung 99, Bremerhaven, Alfred-Wegener-Institut für Polar- und Meeresforschung.

1994. Formation and disintegration of the Antarctic ice sheet. *Ann. Glaciol.* **20**, 336–40.

2002. Sea-level changes at the LGM from ice-dynamic reconstructions of the Greenland and Antarctic ice sheets during the glacial cycles. *Quat. Sci. Rev.* **21** (1–3), 203–31.

Huybrechts, P. and De Wolde, J. 1999. The dynamic response of the Greenland and Antarctic ice sheets to multiple-century climatic warming. *J. Climate* **12** (8), 2169–88.

Huybrechts, P. and Le Meur, E. 1999. Predicted present-day evolution patterns of ice thickness and bedrock elevation over Greenland and Antarctica. *Polar Res.*, **18** (2), 299–308.

Huybrechts, P. and Oerlemans, J. 1988. Evolution of the East Antarctic ice sheet: a numerical study of thermo-mechanical response patterns with changing climate. *Ann. Glaciol.* **11**, 52–9.

1990. Response of the Antarctic ice sheet to future greenhouse warming. *Climate Dyn.* **5**, 93–102.

Huybrechts, P. *et al.* 1996. The EISMINT benchmarks for testing ice-sheet models. *Ann. Glaciol.* **23**, 1–12.

Huybrechts, P. *et al.* 1998. *Report of the Third EISMINT Workshop on Model Intercomparison.* Strasbourg, European Science Foundation.

Huybrechts, P., Steinhage, D., Wilhelms, F. and Bamber, J. L. 2000. Balance velocities and measured properties of the Antarctic ice sheet from a new compilation of gridded datasets for modeling. *Ann. Glaciol.* **30**, 52–60.

Imbrie, J. Z. *et al.* 1984. The orbital theory of Pleistocene climate: support from a revised chronology of the marine δ18O record. In Berger, A. *et al.*, eds., *Milankovitch and Climate.* Dordrecht, D. Reidel, pp. 269–305.

Ingolfsson, O. *et al.* 1998. Antarctic glacial history since the last glacial maximum: an overview of the record on land. *Antarctic Sci.* **10** (3), 326–44.

Jacobs, S. J., Hellmer, H. H. and Jenkins, A. 1996. Antarctic ice sheet melting in the Southeast Pacific. *Geophys. Res. Lett.* **23** (9), 957–60.

Jacobson, H. P. and Raymond, C. F. 1998. Thermal effects on the location of ice stream margins. *J. Geophys. Res.* **103** (B6), 12 111–22.

James, T. S. and Ivins, E. R. 1998. Predictions of Antarctic crustal motions driven by present-day ice sheet evolution and by isostatic memory of the last glacial maximum. *J. Geophys. Res.* **103** (B3), 4993–5017.

Jenkins, A., Vaughan, D. G., Jacobs, S. J., Hellmer, H. H. and Keys, J. R. 1997. Glaciological and oceanographic evidence of high melt rates beneath Pine Island Glacier, West Antarctica. *J. Glaciol.* **43** (143), 114–21.

Jenssen, D. 1977. A three-dimensional polar ice sheet model. *J. Glaciol.* **18** (80), 373–89.

Kotlyakov, V. M., Losev, K. S. and Loseva, I. A. 1978. The ice budget of Antarctica. *Polar Geog. & Geol.* **2** (4), 251–62.

Krinner, G., Genthon, C., Li, Z. X. and Le Van, P. 1997. Studies of the Antarctic climate with a stretched-grid general circulation model. *J. Geophys. Res.* **102** (D12), 13 731–45.

Le Meur, E. and Huybrechts, P. 1996. A comparison of different ways of dealing with isostasy: examples from modeling the Antarctic ice sheet during the last glacial cycle. *Ann. Glaciol.* **23**, 309–17.

Lestringant, R. 1994. A 2D finite element study of the flow in the transition zone between an ice sheet and an ice shelf. *Ann. Glaciol.* **20**, 67–72.

Lingle, C. S. 1984. A numerical model of interactions between a polar ice stream and the ocean: application to ice stream E, West Antarctica. *J. Geophys. Res.* **89**, 3524–49.

Lorius, C. *et al.* 1985. A 150 000-year climatic record from Antarctic ice. *Nature* **316**, 591–6.

MacAyeal, D. R. 1992. Irregular oscillations of the West Antarctic ice sheet. *Nature* **359**, 29–32.

MacAyeal, D. R. and Thomas, R. H. 1986. The effects of basal melting on the present flow of the Ross ice shelf, Antarctica. *J. Glaciol.* **32** (110), 72–86.

MacAyeal, D. R., Rommelaere, V., Huybrechts, P., Hulbe, C. L., Determann, J. and Ritz, C. 1996. An ice-shelf model test based on the Ross ice shelf. *Ann. Glaciol.* **23**, 46–51.

Mahaffy, M. A. W. 1976. A three-dimensional numerical model of ice sheets: tests on the Barnes ice cap, Northwest Territories. *J. Geophys. Res.* **81** (6), 1059–66.

Manabe, S., Stouffer, R. J., Spelman, M. J. and Bryan, K. 1991. Transient response of a coupled ocean-atmosphere model to gradual changes of atmospheric CO_2. Part I: Annual mean response. *J. Climate* **4** (8), 785–818.

Mayer, C. and Huybrechts, P. 1999. Ice-dynamic conditions across the grounding zone, Ekströmisen, East Antarctica. *J. Glaciol.* **45** (150), 384–93.

Mercer, J. H. 1978. West Antarctic ice sheet and CO_2 greenhouse effect: a threat of disaster. *Nature* **271**, 321–5.

Mosley-Thompson, E., Paskievitch, J. F., Gow, A. J. and Thompson, L. G. 1999. Late 20th century increase in South Pole accumulation. *J. Geophys. Res.* **104** (D4), 3877–86.

Nicholls, K. W. 1997. Predicted reduction in basal melt rates of an Antarctic ice shelf in a warmer climate. *Nature* **388**, 460–2.

Nye, J. F. 1957. The distribution of stress and velocity in glaciers and ice sheets. *Proc. Roy. Soc. London Series A* **239**, 113–33.

Oerlemans, J. 1982a. A model of the Antarctic ice sheet. *Nature* **297** (5967), 550–3.
    1982b. Response of the Antarctic ice sheet to a climatic warming: a model study. *J. Climatol.* **2**, 1–11.
    1983. A numerical study on cyclic behaviour of polar ice sheets. *Tellus* **35A**, 81–7.

Oerlemans, J. and Van der Veen, C. J. 1984. *Ice Sheets and Climate.* Dordrecht, D. Reidel.

Oerter, H. *et al.* 1992. Evidence for basal marine ice in the Filchner-Ronne ice shelf. *Nature* **358**, 399–401.

O'Farrell, S. P., McGregor, J. L., Rotstayn, L. D., Budd, W. F., Zweck, C. and Warner, R. C. 1997. Impact of transient increases in atmospheric CO_2 on the accumulation and mass balance of the Antarctic ice sheet. *Ann. Glaciol.* **25**, 137–44.

Ohmura, A., Wild, M. and Bengtsson, L. 1996. Present and future mass balance of the ice sheets simulated with GCM. *Ann. Glaciol.* **23**, 187–93.

Oppenheimer, M. 1998. Global warming and the stability of the West Antarctic ice sheet. *Nature* **393**, 325–32.

Pattyn, F. 1996. Numerical modelling of a fast flowing outlet glacier: experiments with different basal conditions. *Ann. Glaciol.* **23**, 237–46.

Payne, A. J. 1995. Limit cycles in the basal thermal regime of ice sheets. *J. Geophys. Res.* **100** (B3), 4249–63.

1999. A thermomechanical model of ice flow in West Antarctica. *Climate Dyn.* **15**, 115–25.

Payne, A. J. *et al.* 2000. Results from the EISMINT Phase 2 simplified geometry experiments: the effects of thermomechanical coupling. *J. Glaciol.* **46** (153), 227–38.

Peel, D. A. and Mulvaney, R. 1988. Air temperature and snow accumulation in the Antarctic Peninsula during the past 50 years. *Ann. Glaciol.* **11**, 206–7.

Petit, J. R. *et al.* 1999. Climate and atmospheric history of the past 420 000 years from the Vostok ice core, Antarctica. *Nature* **399**, 429–36.

Reeh, N. 1991. Parameterisation of melt rate and surface temperature on the Greenland ice sheet. *Polarforschung* **59**, 113–28.

Retzlaff, R. and Bentley, C. R. 1993. Timing of stagnation of ice stream C, West Antarctica, from short-pulse radar studies of buried surface crevasses. *J. Glaciol.* **39** (133), 553–61.

Ritz, C., Rommelaere, V. and Dumas, C. 2001. Modeling the evolution of the Antarctic ice sheet over the last 420000 years: implications for altitude changes in the Vostok region. *J. Geophys. Res.* **106** (D23), 31 943–64.

Robin, G. de Q. 1977. Ice cores and climatic change. *Phil. Trans. Roy. Soc. Lond. A.* **280**, 143–68.

Rommelaere, V. and MacAyeal, D. R. 1997. Large-scale rheology of the Ross ice shelf, Antarctica, computed by a control method. *Ann. Glaciol.* **24**, 43–8.

Savvin, A., Greve, R., Calov, R., Mügge, B. and Hutter, K. 2000. Simulation of the Antarctic ice sheet with a three-dimensional polythermal ice-sheet model, in support of the EPICA project. II: Nested high-resolution treatment of Dronning Maud Land, Antarctica. *Ann. Glaciol.* **30**, 69–75.

Schubert, G. and Yuen, D. A. 1982. Initiation of ice ages by creep instability and surging of the East Antarctic ice sheet. *Nature* **296**, 127–30.

Skvarca, P., Rack, W., Rott, H. and Ibarzabal y Donangelo, T. 1998. Evidence of recent climatic warming on the eastern Antarctic Peninsula. *Ann. Glaciol.* **27**, 628–32.

Smith, I. N., Budd, W. F. and Reid, P. 1998. Model estimates of Antarctic accumulation rates and their relationship to temperature changes. *Ann. Glaciol.* **27**, 246–50.

Steig, E. J. 1997. How well can we parameterize past accumulation rates in polar ice sheets? *Ann. Glaciol.* **25**, 418–22.

Stephenson, S. N. and Bindschadler, R. A. 1988. Observed velocity fluctuations on a major Antarctic ice stream. *Nature* **334**, 695–7.

Thomas, R. H. 1979. The dynamics of marine ice sheets. *J. Glaciol.* **24** (90), 167–77.

Thomas, R. H. and Bentley, C. R. 1978. A model for Holocene retreat of the West Antarctic ice sheet. *Quat. Res.* **10**, 150–70.

Thomas, R. H., Sanderson, T. J. O. and Rose, K. E. 1979. Effect of climatic warming on the West Antarctic ice sheet. *Nature* **277**, 355–8.

Thompson, S. L. and Pollard, D. 1997. Greenland and Antarctic mass balances for present and doubled atmospheric CO_2 from the GENESIS version-2 global climate model. *J. Climate* **10**, 871–900.

Turner, J., Connolley, W. M., Leonard, S., Marshall, G. J. and Vaughan, D. G. 1999. Spatial and temporal variability of net snow accumulation over the Antarctic from ECMWF re-analysis project data. *Int. J. Climatol.* **19**, 697–724.

Tushingham, A. M. and Peltier, W. R. 1991. Ice-3G: a new global model of Late Pleistocene deglaciation based upon geophysical predictions of post-glacial relative sea level change. *J. Geophys. Res.* **96** (B3), 4497–523.

Van der Veen, C. J. 1985. Response of a marine ice sheet to changes at the grounding line. *Quat. Res.* **24**, 257–67.

   1986. Ice sheets, atmospheric CO_2 and sea level. Ph.D. Thesis, University of Utrecht.

Van Lipzig, N. P. M. 1999. The surface mass balance of the Antarctic ice sheet: a study with a regional atmospheric model. Ph.D. thesis, University of Utrecht.

Vaughan, D. G. and Doake, C. S. M. 1996. Recent atmospheric warming and retreat of ice shelves on the Antarctic Peninsula. *Nature* **379**, 328–31.

Vaughan, D. G. and Spouge, J. 2002. Risk estimation of collapse of the West Antarctic ice sheet. *Climatic Change* **52**, 65–91.

Vaughan, D. G., Bamber, J. L., Giovinetto, M. B., Russell, J. and Cooper, A. P. R. 1999. Reassessment of net surface mass balance in Antarctica. *J. Climate* **12**, 933–46.

Warner, R. C. and Budd, W. F. 1998. Modelling the long-term response of the Antarctic ice sheet to global warming. *Ann. Glaciol.* **27**, 161–8.

Weertman, J. 1957. Deformation of floating ice shelves. *J. Glaciol.* **3**, 38–42.

   1974. Stability of the junction of an ice sheet and an ice shelf. *J. Glaciol.* **13** (67), 3–11.

Wild, M. and Ohmura, A. 2000. Changes in mass balance of the polar ice sheets and sea level under greenhouse warming as projected in high resolution GCM simulations. *Ann. Glaciol.* **30**, 197–203.

Williams, M. J. M., Jenkins, A. and Determann, J. 1998. Physical controls on ocean circulation beneath ice shelves revealed by numerical models. In Jacobs, S. J. and Weiss, R. F., eds., *Ocean, Ice, and Atmosphere: Interactions at the Antarctic Continental Margin*. Antarctic Research Series 75. Washington D.C., American Geophysical Union, pp. 285–99.

Williams, M. J. M., Warner, R. C. and Budd, W. F. 1998. The effects of ocean warming on melting and ocean circulation under the Amery ice shelf, East Antarctica. *Ann. Glaciol.* **27**, 75–80.

Wingham, D. J., Ridout, A. J., Scharroo, R., Arthern, R. J. and Shum, C. K. 1998. Antarctic elevation change from 1992 to 1996. *Science* **282**, 456–8.

Yiou, F., Raisbeck, G. M., Bourles, D., Lorius, C. and Barkov, N. I. 1985. ^{10}Be in ice at Vostok Antarctica during the last climatic cycle. *Nature* **316**, 616–17.

Yuen, D. A., Saari, M. R. and Schubert, G. 1986. Explosive growth of shear-heating instabilities in the down-slope creep of ice sheets. *J. Glaciol.* **32** (112), 314–20.

Part V

The mass balance of ice caps and glaciers

14

Arctic ice caps and glaciers

JULIAN A. DOWDESWELL
Scott Polar Research Institute, University of Cambridge
JON OVE HAGEN
Department of Physical Geography, University of Oslo

14.1 Introduction

The ice caps and glaciers outside the Antarctic and Greenland ice sheets account for only about 4% of the area and 0.5% of the volume of ice on land, and would yield a global rise in sea level of about 0.5 m if they were to melt completely (Dyurgerov and Meier, 1997a; Meier and Bahr, 1996). However, these ice masses, individually of up to 10^4 km^2, may be more significant contributors than the great ice sheets to sea-level rise today, and may remain so over the next century at least (Meier, 1984; Warrick *et al.*, 1996). This is a function of both the climatic sensitivity of the geographical areas in which they are located and of their relatively rapid response time to environmental changes (Johannesson, Raymond and Waddington, 1989).

The ice caps and glaciers of the Arctic islands make up about 45% of the 540 000 km^2 or so of ice outside Antarctica and Greenland (Dyurgerov and Meier, 1997a, b). If the small glaciers and ice caps on Greenland and Antarctica are included, there is a global total of 680 000 km^2 and about 180 000 km^3 of ice, excluding the great ice sheets (Meier and Bahr, 1996). In either case, the glaciers and ice caps of the Arctic islands form a significant area and volume of the world's ice (Figure 14.1). In addition, atmospheric general circulation models (GCMs) predict that the Arctic will warm preferentially over the next 100 years or so, relative to lower latitudes, thus making the mass balance of Arctic glaciers and ice caps critical to future projections of global sea level in a warming world (e.g. Cattle and Crossley, 1995). However, relatively few systematic and sustained field measurement programmes of glacier mass balance are available for Arctic glaciers, and most of those that have been undertaken are for relatively small glaciers (10 km^2) which terminate on land (Dowdeswell *et al.*, 1997; Jania and Hagen, 1996). Few field measurements of larger ice caps (10^3 km^2) have taken place (cf. Koerner, 1996; Reeh *et al.*, 2001). The relatively large footprint of the current generation of satellite radar altimeters makes them unsuitable for calculating

This is a contribution to the work of the NERC Centre for Polar Observation and Modelling (CPOM). The paper draws on work supported by several UK Natural Environment Research Council (NERC) and European Union grants. We thank Robin Bassford, Toby Benham, Evelyn Dowdeswell and Martin Sharp for providing figures, and Mark Dyurgerov for commenting on the manuscript.

Mass Balance of the Cryosphere: Observations and Modelling of Contemporary and Future Changes, eds. Jonathan L. Bamber and Antony J. Payne. Published by Cambridge University Press. © Cambridge University Press 2003.

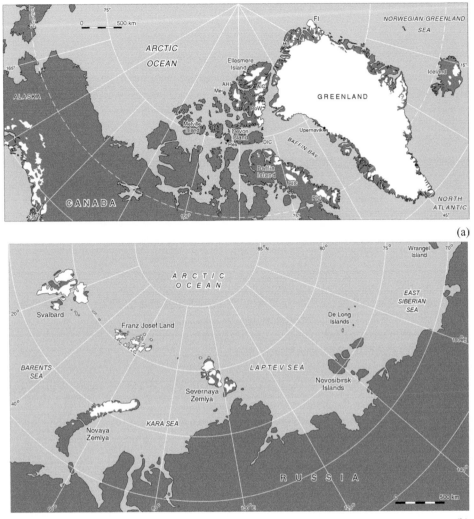

Figure 14.1. The distribution of glaciers and ice caps in the Arctic islands. (a) Canadian Arctic and Greenland. DIC is Devon ice cap; AIC is Agassiz ice cap; PoW is Prince of Wales ice cap; BIC is Barnes ice cap; PIC is Penny ice cap; FI is Flade Isblink, HT is Hans Tausen ice cap; AHI is Axel Heiberg Island, Mei is Meighen Island; Res is Resolute Bay. (b) The Eurasian Arctic, including Svalbard and the Russian Arctic archipelagos. The ice caps and glaciers from which mass balance data are taken are shown (for Svalbard, see Figure 14.6).

elevation change through time on ice caps of this size, although the laser altimeters of the ICESat and forthcoming Cryosat missions should yield accurate ice-surface elevation data. In addition, very little is known about the magnitude of mass loss through iceberg production at the extensive marine margins of many of these larger Arctic ice caps, and this parameter has often been excluded from mass balance calculations (e.g. Dowdeswell *et al.*, 1997).

In this chapter, the distribution and mass balance of Arctic glaciers and ice caps are discussed in the context of recent and projected future environmental change. The distribution of Arctic ice masses today is described, together with the modern meteorological settings in which they occur. Field measurements of the mass balance of Arctic ice caps and glaciers, available for the past few decades, are then summarized, together with model predictions of glacier responses to climatic warming and cooling. The ice caps and glaciers of Svalbard are then used as a case study of how the net balance of a whole archipelago can be assessed using mass balance gradients specified for a number of areas within the islands. Some of the problems raised through this analysis are then discussed, including those of iceberg production and the potential utilization of forthcoming satellite radar altimeters for measuring the volume change of relative small ice caps is emphasized.

14.2 Distribution, extent and volume of ice

Almost $250\,000\,\mathrm{km}^2$ of ice exists on land in the High Arctic archipelagos of Canada, Norwegian Svalbard and Russia (Dowdeswell, 1995; see Figure 14.1 and Table 14.1), with a further $70\,000\,\mathrm{km}^2$ or so from the ice caps and small glaciers beyond the margins of the Greenland ice sheet (Weidick and Morris, 1998). Canadian Ellesmere Island contains over $80\,000\,\mathrm{km}^2$ of ice, and ice extends over $151\,000\,\mathrm{km}^2$ of the Canadian Arctic islands as a whole (Table 14.1). Glaciers and ice caps cover $36\,600\,\mathrm{km}^2$ of the Svalbard archipelago, and over $55\,000\,\mathrm{km}^2$ of the Russian Arctic islands (Table 14.1).

The distribution of these ice masses within each archipelago is a function of the climate and climate gradients across each, together with the large-scale topography of each area. In the Canadian Arctic islands, the largest ice caps are present on the eastern sides of Ellesmere, Devon and Baffin islands, adjacent to the moisture sources represented by Baffin Bay and the North Water polynya. In Greenland, about 15% of local glaciers and ice caps occur in West Greenland, 50% in the north and 35% in the east (Weidick, 1995). Within Svalbard, the eastern islands of the archipelago, Edgeøya, Nordaustlandet and Kvitøya, have the largest ice coverage, linked to the coldest temperatures combined with moisture from the Barents Sea. In the Russian Arctic islands, the percentage of ice cover declines southward in Novaya Zemlya and Severnaya Zemlya, and there is also a strong eastward trend toward very cold and dry conditions, which is reflected in an almost complete absence of glaciers east of Severnaya Zemlya in the low-lying islands of the Laptev and East Siberian seas (Koryakin, 1988).

The bulk of the ice in the High Arctic islands is held within a series of relatively large ice caps of up to $10^4\,\mathrm{km}^2$, although large numbers of independent glaciers from 0.1 to $100\,\mathrm{km}^2$ are also present (e.g. Koerner, 2002; Liestøl, 1993; Weidick, 1995). The largest ice caps in the Canadian Arctic, and indeed in the Arctic as a whole, include the Agassiz ice cap ($17\,300\,\mathrm{km}^2$) in Ellesmere Island, the Devon ice cap (about $12\,000\,\mathrm{km}^2$ and $4000\,\mathrm{km}^3$), and the Barnes and Penny ice caps (each almost $6000\,\mathrm{km}^2$) on Baffin Island. In Greenland,

Table 14.1. *The area covered by ice caps and glaciers in the Arctic islands, excluding the Greenland ice sheet.*

Region of the Arctic (Figure 14.1)	Ice-covered area (km^2)
Svalbard, Arctic Norway	36 600
Franz Josef Land, Russia	13 700
Novaya Zemlya, Russia	23 600
Severnaya Zemlya, Russia	18 300
Ellesmere Island, Canada	80 500
Axel Heiberg Island, Canada	11 700
Devon Island, Canada	16 200
Baffin & other Arctic Canada	43 400
Greenland (excluding the ice sheet)[a]	70 000
Total	314 000

[a]The value of 70 000 km^2 for Greenland should be treated as an estimate (Weidick and Morris, 1998).

outside the 1 756 000 km^2 ice sheet, the largest independent ice cap is the 7000 km^2 Flade Isblink in the north-east, while the more northerly Hans Tausen ice cap, at 3975 km^2 and 763 km^3, is probably the best studied (Reeh *et al.*, 2001). In the Eurasian Arctic, Austfonna on Nordaustlandet in eastern Svalbard is the largest ice cap at 8120 km^2 and 1900 km^3 (Dowdeswell, 1986; Hagen *et al.*, 1993). The largest independent Russian ice cap is the 5575 km^2 and 2180 km^3 Academy of Sciences ice cap on Severnaya Zemlya (Dowdeswell *et al.*, 2002), although the spine of the northern island of Novaya Zemlya represents a larger ice-covered area, cut by many nunataks and outlet glaciers. The numerous ice caps on Franz Josef Land contain about 2100 km^3 of ice (J. A. Dowdeswell *et al.*, unpublished data).

A comprehensive calculation of the volume of Arctic ice caps and glaciers is difficult because, whereas the areas of almost all Arctic ice masses have been measured quite accurately from satellite images and aerial photographs, measurements of ice thickness using ice-penetrating radar instruments operating at megahertz frequencies are not always available. Estimates of ice volume in these areas are based on empirical power-law relationships between ice area and volume (e.g. Bahr, Meier and Peckham, 1997). Weidick (1995), for example, estimated that the glaciers and ice caps beyond the main ice sheet on Greenland probably have a volume of about 20 000 km^3. For Svalbard, Hagen *et al.* (1993) calculated a total ice volume of approximately 7000 km^3.

14.3 Recent climate of the Arctic

The mass balance of ice masses is related strongly to the regional climate in which the ice masses are located. The climate of the Arctic over the past few decades to centuries, and indeed changes over that period, provide an important perspective through which to view records of the recent mass balance of Arctic ice caps and glaciers. This is because a time lag exists between climate change and glacier-terminus response to that change. Mass balance is perturbed throughout the length of a glacier, but is transferred down-glacier at finite velocities over a range of distances. The effect of a shift in climate therefore arrives at a glacier margin over a period, and the terminus position is then a weighted mean of past climate changes over the time interval (T_m) beyond which there is no memory of former climate (Johannesson *et al.*, 1989). T_m is the time constant in an exponential, asymptotic approach to a new steady-state after a given shift in climate. Johannesson *et al.* (1989) proposed that $T_m = h/-b_t$, where h is maximum glacier thickness and b_t is the mass balance at the terminus, which is a negative value. This equation suggests that adjustments to changing mass balance may be of the order of decades for many smaller Arctic valley and outlet glaciers. The larger ice caps in the High Arctic, where ice may be over 500 m thick and mass loss at the margins is relatively slow, may have a longer adjustment time of a few hundred years.

Oxygen-isotope and melt-layer records from several deep ice cores in High Arctic ice caps have allowed the reconstruction of climate change over the past few hundred years. Where little or no surface melting takes place to complicate core stratigraphy and chemistry, a temporal resolution of a few years is possible over the past few centuries. However, in the Eurasian High Arctic, where mean annual temperatures are significantly higher, melting and refreezing effects dominate, and chronology is more difficult to establish (e.g. Dowdeswell, Drewry and Simões, 1990; Koerner, 1997). Some broad similarities are present in the oxygen-isotope records from Devon Island, Canada (Paterson *et al.*, 1977), through Camp Century in North Greenland (Johnsen *et al.*, 1970), to Lomonosovfonna in Svalbard (Gordiyenko *et al.*, 1981) and the Vavilov ice cap in Severnaya Zemlya (Kotlyakov, Zagorodnov and Nikolayev, 1990). The previous two to three centuries have markedly more negative oxygen isotopic ratios than the twentieth century, marking the cold period sometimes known as the Little Ice Age (Grove, 1988).

The isotopic and stratigraphic evidence from High Arctic ice cores indicates that the cold Little Ice Age began to ameliorate from about 1860 to 1880 in the Canadian Arctic, Greenland and the eastern Eurasian Arctic, but that warming took place somewhat later in Svalbard. Changes toward less negative ratios over the last 100–150 years are about 1.5 ppt on the Canadian Arctic ice caps, representing a temperature rise of about 2.5 °C over Little Ice Age conditions. The recent records of melt layers in Arctic ice cores also show similarities over space and with the isotopes. On Devon Island, the period since about 1860 has shown a rise in the number and thickness of melt layers relative to the preceding 300 years, with summer warming intensifying since the 1920s (Koerner, 1977; Koerner and Paterson, 1974). The very low frequency of melt layers about 150 years ago indicates that this was the interval of coldest summers over the whole Holocene (Koerner and Fisher, 1990). In Svalbard ice cores, the interval interpreted to represent the period from about 1550 to 1920 contains up to 30–40% less refrozen ice layers than that since 1920 (Tarussov, 1992). The melt-layer signal from the large ice caps in the Russian archipelago of Severnaya

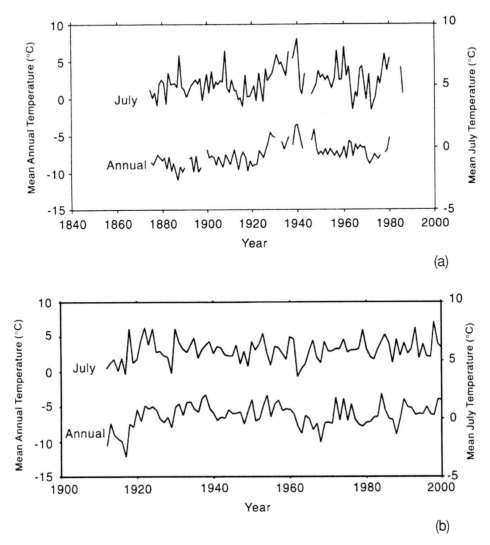

Figure 14.2. Meteorological data from (a) Upernavik, West Greenland (72°47′N, 56°10′W), and (b) Isfjord Radio/Longyearbyen, Svalbard (78°04′N, 13°38′E).

Zemlya indicates a warming trend from about 120 to 140 years ago (Kotlyakov *et al.*, 1989; Tarussov, 1992).

   Temperature and precipitation records from the High Arctic prior to the Second World War are available from only a very limited number of stations (Dowdeswell, 1995; Dowdeswell *et al.*, 1997). The longest meteorological records are those from Upernavik at 73°N in West Greenland and Isfjord Radio (now relocated to Svalbard Airport) at 78°N in Spitsbergen (Figure 14.2). Temperature measurements from 1875 at Upernavik show a warming of

2 °C in mean annual temperature between the last quarter of the nineteenth century and the twentieth, with a particularly rapid rise of about 3.5 °C if the ten years around 1920 are taken alone (Figure 14.2). The period from about 1920 to 1950 was particularly warm in West Greenland, and was followed by a cooling of approximately 1.5 °C in mean annual temperature since then. The mean annual temperature in Svalbard also rose by 4–5 °C between 1912 and 1920 (Førland, Hanssen-Bauer and Nordli, 1997). Precipitation also appears to have increased from the early twentieth century, although it should be noted that solid precipitation is difficult to measure accurately, and is often highly variable both spatially and with altitude (e.g. Cogley *et al.*, 1995; Woo *et al.*, 1983). Indications of an upward shift can be seen in the record from Svalbard and in the incomplete time series from Upernavik. For these High Arctic locations, the absolute increase in precipitation is about 100 mm per year. The relatively large shifts to warmer conditions and increased precipitation are inferred to indicate the end of the Little Ice Age in these areas of the Arctic.

Most other High Arctic meteorological stations have records of little more than 50 years at best. They show a high degree of inter-annual variability, but also some changes on the scale of decades. For example, the interval from 1940 to the mid 1950s was relatively warm in the northern Taymyr Peninsula, close to Severnaya Zemlya in the Russian Arctic. There was a period of relative cooling in summer temperature, in particular from about 1964 to 1977 recorded at Resolute, Canada (Bradley and England, 1978), followed by warmer temperatures. It is difficult to pick out significant trends in climate from the shorter records from most other High Arctic stations, and the minor excursions in temperature that have taken place in these areas are considerably less pronounced than the more significant changes which mark the end of the Little Ice Age in the longer data sets from Upernavik and Svalbard.

Recently, variations in the pressure field of the North Atlantic and Arctic regions, with associated effects on both temperature and precipitation, have been identified as an important source of inter-annual variability in the climate. Changes in the sea-level pressure difference between the Azores and Iceland are known as the North Atlantic oscillation (NAO) (Hurrell, 1995; Hurrell and Van Loon, 1997). When the NAO index is positive, that is when the pressure gradient is relatively high, more major winter storms cross the North Atlantic, resulting in wet and relatively warm winters in Europe and cold and dry conditions in northern Canada and Greenland. The NAO can be calculated back to 1864 using meteorological observations, and has been reconstructed back several hundred years using proxy palaeo-climate indicators (e.g. Black *et al.*, 1999).

The Arctic oscillation (AO), which is correlated strongly with the NAO, is calculated using factor analytic methods from winter-time data on the equivalent height of the 1000 hPa pressure surface north of 20°N, and the spatial structure of this index can be mapped over the Arctic (Baldwin and Dunkerton, 1999; Thompson and Wallace, 1998). Variations in these indices may assist in explaining both the short-term variability in Arctic climate and mass balance and, importantly, the way in which different parts of the Arctic may respond differently to the same forcing function due to their particular synoptic climatology.

14.4 Field observations of mass balance on Arctic ice masses

Observations of glacier and ice-cap mass balance have been undertaken at a number of sites, distributed widely through the islands of the High Arctic (Figure 14.1). The longest observational records, of 40 years, come from the Canadian Arctic, but the largest number of ice masses whose mass balance has been measured are on Svalbard. All mass balance data are presented in water equivalent units. Details of the mass balance measurement methods, and their associated sources of uncertainty, are given in Chapter 2 of this volume.

14.4.1 Canadian Arctic islands

Mass balance measurements have been undertaken at several glaciers and ice caps within the Canadian Arctic since about 1960, providing the longest time series of field measurements anywhere in the Arctic islands (Figure 14.1). Records from the Devon and Melville ice caps ($12\,000\,km^2$ and $68\,km^2$), located about $800\,km$ apart, provide data for the eastern and western ends of the North-west Passage (75°N), near Baffin Bay and the Arctic Ocean margin, respectively. Further north, at about 80°N, long series of measurements are also available for the White Glacier ($39\,km^2$) on Axel Heiberg Island and the Meighen Island ice cap ($85\,km^2$). Shorter series of observations have also been made on several other Canadian Arctic glaciers (Koerner, 1996), but the above provide the most comprehensive records.

The net mass balance records for these four Canadian Arctic ice masses are illustrated in Figure 14.3. Cumulating all these annual measurements, 72% record negative net mass balances. In addition, the mean net balance for each ice mass indicates an overall loss of mass: $-0.06\,m$ per year (standard deviation $\sigma = 0.12\,m$ per year) for the Devon ice cap; $-0.12\,m$ per year ($\sigma = 0.27$) for the Meighen ice cap; $-0.14\,m$ per year ($\sigma = 0.25$) for

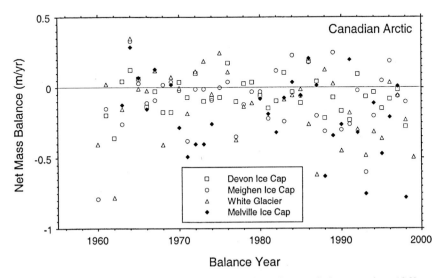

Figure 14.3. Net mass balance records for four Canadian Arctic ice caps since 1960.

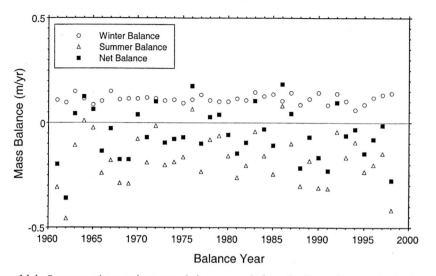

Figure 14.4. Summer, winter and net mass balance records from the Devon ice cap, Arctic Canada.

White Glacier; and −0.17 m per year ($\sigma = 0.25$) for the Melville ice cap. This conforms to the observation that many Canadian Arctic ice masses show signs of retreat from a maximum position related to the colder Little Ice Age (e.g. Jacobs, Simms and Simms, 1997). However, the inter-annual variability in net balance records is relatively high at each ice mass (Figure 14.3), as indicated by the high standard deviations for each of the four records. This variability is associated mainly with changes in summer melting. Winter precipitation is much more consistent from year to year (Figure 14.4).

14.4.2 Greenland glaciers and ice caps

The major focus for mass balance studies in Greenland has been the Greenland ice sheet itself (Chapter 11). However, a small number of mass balance investigations have also been made at ice caps and glaciers beyond the margins of the ice sheet in south-west and north Greenland. In the south-west, field measurements of winter, summer and net balance were made at three ice masses between the inland ice and the coast for between three and five years. Valhaltindegletscher and Narssaq Brae are cirque glaciers of area less than 2 km², located between 60 and 62°N. Qapiarfiup sermia is a 21 km² ice cap located between 65 and 66°N. The Hans Tausen ice cap is north of the main ice sheet in Peary Land at 82.5°N. It is considerably larger, at almost 4000 km², and here snow-pit and stake measurements were made for the balance year 1994–95 on the 15 km long Hare outlet glacier, and calculations were then generalized to the whole ice cap (Reeh *et al.*, 2001).

The net mass balance records for these four Greenland ice masses, a total of 13, are shown in Figure 14.5; 64% of these data points show a negative net balance. The two glaciers at 61°N both have average net balances that are negative: −0.17 m per year for Valhaltindegletscher

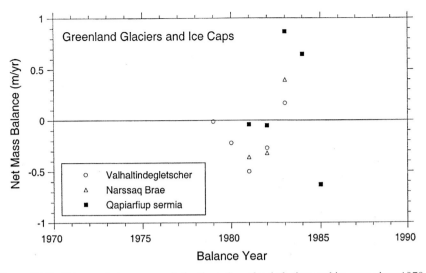

Figure 14.5. Net mass balance records for three Greenland glaciers and ice caps since 1979.

and −0.09 m per year at Narssaq Brae over five-and three-year periods, respectively. Further north-west, the Qapiarfiup sermia ice cap has a five-year record with a positive mean net balance of 0.16 m per year. Finally, the single net balance value at Hans Tausen ice cap is −0.08 m per year. However, it should be noted that the standard deviation for all three of the south-west Greenland ice masses is high, and in none of the cases can the mean net mass balance be regarded as significantly different from zero.

14.4.3 Svalbard

More mass balance investigations have taken place on the ice masses of Spitsbergen in the Svalbard archipelago than in other parts of the Arctic (Figure 14.6) (Dowdeswell *et al.*, 1997; Hagen, 1996a; Hagen and Liestøl, 1990). This allows us to look not only at the results of these studies, but also to consider differences in the records from glaciers of different sizes and from those that end on land compared with tide-water glaciers terminating in fjords.

A total of about 170 measurement years of net mass balance have been made on more than ten different ice masses within Spitsbergen (Figure 14.7(a)). Of these, only 10% record a positive balance year. The number of annual observations, the net balance and its standard deviation are shown for each Spitsbergen glacier in Table 14.2. Almost all the observed glaciers have a negative mean net balance, and a number have means that are more than one standard deviation below the level of zero net balance. Only one, Kongsvegen, has a mean mass balance greater than zero, although this is not statistically significant.

It has been observed from time series of aerial photographs that most Svalbard glaciers have been retreating and thinning at least since the 1930s, and most likely since the end of the Little Ice Age some decades earlier (Werner, 1993). The finding that most glaciers have a negative balance is in accord with this general recession. In addition, Lefauconnier and Hagen (1990) have calibrated meteorological records for Svalbard against observed

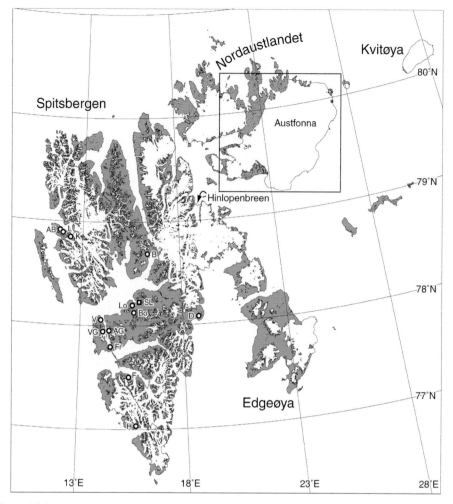

Figure 14.6. Map of the glaciers and ice caps on Svalbard in the Eurasian Arctic (see Figure 14.1(b)). The box over Nordaustlandet indicates the location of Figure 14.9. The sites of glacier mass balance measurements are shown by circles, and the meteorological station at Svalbard Lufthavn is indicated by a square labelled SL. Some glaciers are labelled by initials: AB is Austre Brøggerbreen; L is Midre Lovenbreen; K is Kongsvegen; B is Bertilbreen; V is Vøringbreen; VG is Vestre Grønfjordbreen; AG is Austre Grønfjordbreen; Fr is Fridtjovbreen; Bo is Bogerbreen; Lo is Longyearbreen; D is Daudbreen; F is Finsterwalderbreen; and H is Hansbreen.

net balance data, and have then used the regression equations generated to extrapolate the mass balance time series back to the beginning of meteorological data collection in 1912. They predict that Austre Brøggerbreen has thinned by an average of almost 40 m over this period. This is a calculation again consistent with a general observed reduction in glacier dimensions within the archipelago.

Most of the Spitsbergen glaciers at which mass balance measurements have been undertaken are less than 10 km² in area, and a number are less than 5 km² (Figure 14.7(b)

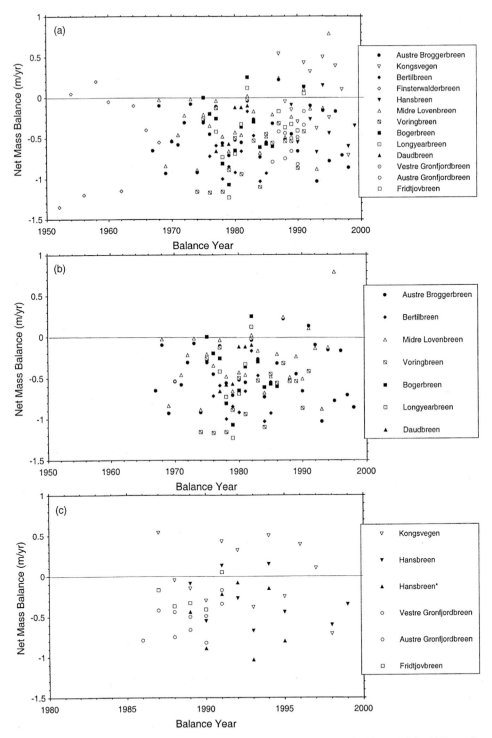

Figure 14.7. Records of net mass balance for Svalbard glaciers (located in Figure 14.6). (a) Records for 13 Svalbard glaciers. (b) Records for a subset of seven glaciers of area less than 6 km² (Table 14.2). (c) Records for Svalbard glaciers ending in marine waters.

Table 14.2. *Mean net mass balance data (in water equivalent units) and ancilliary information for the Spitsbergen glaciers at which field observations have taken place.*

N of obs. is the number of balance years for which measurements have been made. For marine margins, Y indicates that all or part of the glacier terminus is marine, and N that it ends on land.

Glacier (Figure 14.1)	Area (km²)	N of obs.	Mean net balance (m per year)	Standard dev. (m per year)	Marine margin
A. Brøggerbreen	6	32	−0.45	0.32	N
A. Grønfjordbreen	38[a]	6	−0.63	0.20	Y
Bertilbreen	5	11	−0.72	0.29	N
Bogerbreen	5	12	−0.43	0.36	N
Daudbreen	2	6	−0.36	0.27	N
Finsterwalderbreen	11	9	−0.51	0.59	N
Fridtjovbreen	49	5	−0.25	0.19	Y
Hansbreen	57	7	−0.52[b]	0.39	Y
Kongsvegen	105	12	0.04[b]	0.40	Y
Longyearbreen	4	6	−0.55	0.45	N
M. Lovenbreen	6	28	−0.30	0.36	N
V. Grønfjordbreen	38[a]	4	−0.46	0.16	Y
Vøringbreen	2	18	−0.64	0.37	N

[a] Area of Grønfjordbreen as a whole.
[b] Includes losses by iceberg calving.

and Table 14.2). Only one ice mass, Konsgvegen, is above 100 km². This is also the only Spitsbergen glacier to exhibit a positive mass balance over the period of observations, albeit at a value that is probably insignificant. Hagen (1996b) has suggested that larger Svalbard ice masses, which are likely to have higher elevation accumulation areas, may be closer to a steady-state than the small valley and cirque glaciers at which most measurements have been made. This is an important point concerning the representativeness of the glaciers in both Svalbard and elsewhere that have been selected for study (Cogley and Adams, 1998). Most have been chosen for logistical, rather than scientific, reasons.

Five of the Spitsbergen ice masses investigated are tide-water glaciers, where all or part of their terminus ends in the sea (Figure 14.7(c) and Table 14.2). For both Hansbreen and Kongsvegen, the mass loss through iceberg calving has been measured and deducted from the winter accumulation along with summer melting. Calving losses at Kongsvegen are small at 0.005 km³ per year (0.05 m per year averaged over the glacier area), both because only a small part of its margin reaches to the adjacent fjord and because the glacier is in the quiescent phase of the surge cycle (Melvold and Hagen, 1998). At Hansbreen, observations averaged over a number of years suggest a calving rate of 0.02 km³ per year (0.4 m per year

averaged over the glacier area), as this is a more active glacier with an ice–ocean interface of
about 1.5 km (Jania and Kaczmarska, 1997). It is clear from these investigations that, if mass
loss through iceberg production is unknown, then conventional mass balance observations
represent only a minimum value for the summer component of the annual net balance.
However, problems in the field measurement of internal accumulation in mass balance
studies, where melt water has refrozen below the previous year's summer surface, may
offset this effect because neglect of this form of accumulation would have the opposite sign
to the calving term.

14.4.4 Russian Arctic archipelagos

Only very few measurements of mass balance have been made in the Russian Arctic islands (Glazovskiy, 1996). The only comprehensive data are those from the Vavilov ice cap
(1820 km^2) on October Revolution Island at 79°N in the Severnaya Zemlya archipelago
(Figure 14.1). To supplement these observations, we also include 24-year records from two
small cirque glaciers in the Russian Polar Ural Mountains, at IGAN Glacier (0.7 km^2) and
Obruchev Glacier (0.3 km^2) (Figure 14.8).

The net mass balance of Vavilov ice cap, averaged over the ten years of measurement, is
−0.03 m per year ($\sigma = 0.12$ m per year). This mean value is, because of the very high inter-
annual variability, not significantly different from zero balance (Figure 14.8). Comparing
aerial photographs acquired in 1952 and 1985 shows that the ice cap has advanced on its
south and west margins, but the northern margin has retreated: overall its area expanded by
3.5 km^2 over this period (Glazovskiy, 1996). However, no measurements of, for example,
changing elevation at the ice-cap crest are available, and the ice cap may be close to balance.

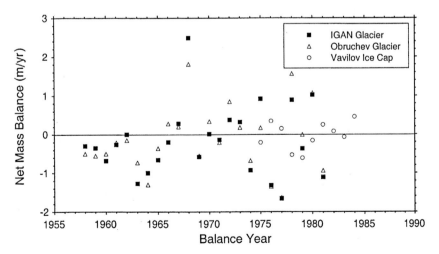

Figure 14.8. Net mass balance records for the Vavilov ice cap in Severnaya Zemlya and two small
glaciers in the Siberian Ural Mountains (Figure 14.1(b)).

Observations on changing ice-front positions in the Russian Arctic islands, summarized by Glasovskiy (1996), suggest that they have generally been retreating during the twentieth century, providing a qualitative indication that the mean mass balance in the Russian Arctic archipelagos has been negative over this period.

The two glaciers in the Polar Urals also show a mean net balance for the period 1958 to 1981 that is negative: -0.19 m per year ($\sigma = 0.91$ m per year) for IGAN Glacier and -0.13 m per year ($\sigma = 0.86$ m per year) for Obruchev Glacier. In contrast to most of the mass balance records for the Arctic islands, both winter and summer balance records show high variability at these glaciers, accounting for the very high standard deviations calculated (Figure 14.8).

14.4.5 Trends in Arctic glacier mass balance

Linear regression analysis of the mass balance time series for the glaciers shown in Figures 14.3–14.8 yielded over 70% with positive slope coefficients. Given that over 80% of the glaciers have a negative mean annual balance, this could be taken to imply a tendency toward a less negative state. However, when the correlation coefficients (R) for the fit of least squares regression lines are examined, almost all are very low: R values are less than 0.3 in most cases. The vast majority of Arctic glaciers for which mass balance time series are available therefore show no statistically significant trend toward less negative balances. This finding is in agreement with the work of Cogley *et al.* (1995). It further implies that the mass balance of most Arctic glaciers continues to be negative.

Cogley *et al.* (1995) suggest that there is some evidence (at the 68% level of significance) for a trend toward less negative mass balance in the records from a small number of Arctic glaciers: from the Meighen ice cap in Arctic Canada and from the Vavilov ice dome in the Russian High Arctic. Two Spitsbergen glaciers also show significant positive trends. It should be emphasized, however, that each of these glaciers nonetheless has a negative mean annual balance over the period of observation.

14.5 Modelling the response of Arctic glaciers and ice caps to climate change

The mass balance response of several Arctic glaciers and ice caps to climatic warming and cooling has been predicted using two forms of model. An energy balance approach was used by Fleming, Dowdeswell and Oerlemans (1997) to model Midre Lovenbreen and Austre Brøggerbreen, 5.5 and 6 km^2 glaciers in north-west Spitsbergen (Figure 14.6). For the 3975 km^2 Hans Tausen ice cap in North Greenland, Reeh *et al.* (2001) used a positive degree-day model to calculate ice-surface melting. In both studies, winter, summer and net mass balance observations were available to calibrate the models, for a ten-year period for the Spitsbergen glaciers and for a single year for the North Greenland ice cap. Average accumulation vales for the period 1975 to 1995 were also derived from ice cores on the Hans Tausen ice cap (Reeh *et al.*, 2001). At present, Austre Brøggerbreen and Midre Lovenbreen, Spitsbergen, have net balances of −0.45 and −0.30 m per year, respectively, averaged over the past 30 or so years of observations. For Hans Tausen ice cap, the value is −0.14 m per year.

The models were then used to predict the effects of recent climate change on net mass balance. In Spitsbergen, the sensitivity of net balance to climate warming was −0.6 m/K per year (Fleming *et al.*, 1997). For the Hans Tausen ice cap, the value was −0.14 m/K per year (Reeh *et al.*, 2001). This difference in response to warming probably reflects the greater sensitivity of Svalbard ice masses to climate change relative to those in climates with a lower mass turnover and significantly colder mean annual temperatures, although hyposometric variability between ice masses may also play an important part. A summer warming of about 3.5 °C in north-west Spitsbergen, and of approximately 5 °C at Hans Tausen, would result in ablation over the entire surface of each ice mass and their rapid disappearance. Neither model incorporates a dynamic treatment of surface-elevation change and ice flow, and so the incremental effects of surface lowering on the rates of melting through time are not calculated. If ice dynamics were included in modelling, the time each ice mass would take to disappear could be calculated for different temperature and precipitation-change scenarios (e.g. Raper, Brown and Braithwaite, 2000).

Energy balance modelling also predicts that a cooling of about 0.6 °C, or a precipitation increase of around 25%, would be required to give a net mass balance close to zero for the two Spitsbergen glaciers (Fleming *et al.*, 1997). For the Hans Tausen ice cap, a cooling of 1.3 °C would bring it into balance with the mean snow accumulation for the period 1975 to 1995 (Reeh *et al.*, 2001). This gives some indication of the magnitude of climate change that these ice masses may have experienced during the Little Ice Age, although, given that most Arctic ice masses grew significantly during that period and that precipitation was probably lower, the magnitude of modelled temperature change is likely to represent minimum values for cooling.

14.6 Overall mass balance of the Svalbard archipelago: a case study

The discussion so far has concerned the mass balance of individual glaciers from the Arctic islands and the geographical and temporal patterns that they exhibit. We now consider the more difficult problem of calculating the overall mass balance of one of the Arctic archipelagos – Svalbard, with its 36 600 km^2 of ice caps and glaciers. This is clearly a major aim of glacier mass balance investigations, but the lack of suitable satellite altimetric tools to allow synoptic coverage of ice masses as small as 0.1 to 100 km^2 means that the spatial distribution of available data is limited. The following discussion, based on an analysis by Hagen *et al.* (2003), derives an initial estimate of the overall mass balance of the archipelago, but it is also instructive as an indicator of the shortcomings of the available evidence and of the key problems that remain to be tackled effectively.

The overall net balance of Svalbard ice masses, or the annual volume change ($\partial V/\partial t$), can be calculated from the mass balance equation,

$$\partial V/\partial t = M_{\mathrm{a}} - M_{\mathrm{m}} - M_{\mathrm{c}} - M_{\mathrm{b}},$$

where M_a is the annual surface accumulation, M_m is the mass loss by annual surface runoff, M_c is loss by iceberg production and M_b is the bottom melting or freezing-on under any floating ice margins. There are no floating glacier fronts in Svalbard (Dowdeswell, 1989), and M_b therefore equals zero. Total ice-volume change can thus be found from the net surface mass balance $(M_a - M_m)$ and the iceberg calving term, M_c.

14.6.1 Surface mass balance

Svalbard was divided into 11 regions, conforming to the major drainage basins in the archipelago defined from the glacier inventory of Svalbard (Hagen *et al.*, 1993). For each region, a mass balance gradient was estimated, representing a mean of the spatial variability within that area. The gradients are also time-averaged values, in that they include mass balance and snow accumulation measurements acquired over the last 30 years or so. Net mass balance data from individual glaciers in Svalbard suggest that no temporal trend is present (Figure 14.7(a)), and thus the measurements should represent the present climate.

The evidence used to construct the mass balance gradients in each area of the archipelago is derived from several sources: field measurements of winter accumulation, summer melting and net mass balance at individual glaciers (Figure 14.7 and Table 14.2); snow accumulation from shallow cores and snow-distribution maps obtained from depth-probing and ground-penetrating radar (e.g. Winther *et al.*, 1998); and equilibrium-line altitude distribution maps derived from aerial photographs and satellite imagery. The set of glaciers where mass balance has been measured systematically includes only about 0.5% of the total ice-covered area of Svalbard, but shallow cores have been drilled in a number of places throughout the archipelago to detect radioactive reference horizons from the fallout of 1962–63 nuclear bomb tests and from the 1986 Chernobyl accident (Pinglot *et al.*, 1999). The reference layers could be detected in all cores in the accumulation area of the glaciers. Cores were taken at varying altitudes, thus providing average net balance values and the net balance gradient for the ice-mass accumulation areas.

It is important to specify the accumulation pattern on at least one of the large ice caps in eastern Svalbard, as well as for the smaller ice masses in Spitsbergen. This work was carried out on Austfonna in Nordaustlandet, the largest ice cap in the archipelago (Figure 14.6), where 29 shallow ice cores were retrieved (Figure 14.9). The mean annual net accumulation was calculated from the detection of radioactive layers in the cores. The Chernobyl layer was located in 19 ice cores, all drilled in the accumulation area of Austfonna, and the nuclear test layer was located in two deeper ice cores (Pinglot *et al.*, 2001). The spatial variation of the winter snow cover was mapped using snow-probing, ground-penetrating radar methods and global positioning system (GPS). The altitudinal gradient of the mean net mass balance and the altitude of the mean equilibrium line was then reconstructed for five transects radiating from the crest of the ice cap (Figure 14.10), which were used to produce estimates of net balance gradient $(\delta b_n/\delta z)$ in the eastern part of the archipelago.

Hypsometric data for each region were taken from digital elevation models (DEMs) with a 100 m horizontal resolution, produced from Norsk Polarinstitutt 1:100 000 maps

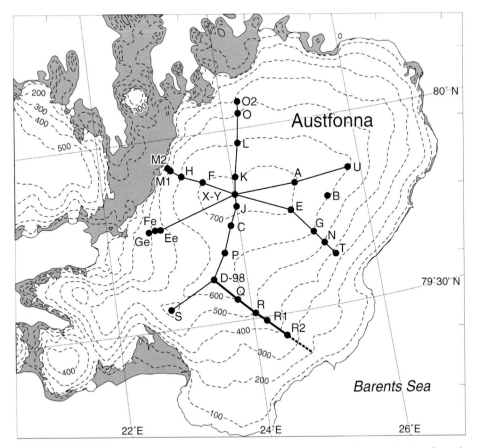

Figure 14.9. The network of shallow ice cores used to measure accumulation over the Austfonna ice cap in eastern Svalbard (Figure 14.6). Mass balance data from the transect of cores labelled D to R2 are shown in Figure 14.10. Ice-surface elevation contours are in metres.

(contour interval 50 and 100 m). The DEMs covered both glaciers and ice-free areas, but the latter were masked out in further analysis. The net surface mass balance gradients for each region were then applied to the DEM information on $\delta A/\delta z$ for that region. $\delta b_n/\delta z$ was then combined with $\delta A/\delta z$ to give the total net balance (M_n). M_n therefore equals $\Sigma b_{ni} A_i$, where $i = 1, \ldots, n$ is each 100 m altitude interval and A is the surface area in that vertical interval. Combining these values for each region gives the volume of mass either lost or gained in the archipelago, and the specific average net balance for Svalbard can then be derived by dividing by the total ice-covered area. The use of regionally derived net balance gradients and hypsometries is of particular importance because the overall mass balance of Svalbard can now be calculated in regionally representative height intervals of 100 m instead of using mean values given for a small number of individual glaciers (Figure 14.11).

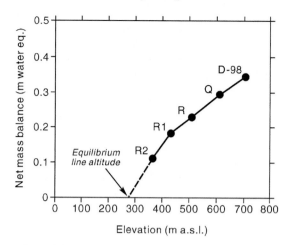

Figure 14.10. The gradient of net mass balance of the Austfonna ice cap derived from accumulation measurements in shallow ice cores (Figure 14.9). The equilibrium-line altitude is inferred from the mass balance gradient along this transect, located in Figure 14.9.

14.6.2 Iceberg production

The ice flux at the marine glacier margins of Svalbard is the product of terminus velocity, ice thickness at the calving front and the horizontal width of the ice front. Advance or retreat of the tide-water ice front also adds a negative or positive component to calving. The average thickness of the 1000 km long calving front in the archipelago is estimated at about 100 m, although many outlet glaciers have no ice-penetrating radar or marginal bathymetric data. Many of these ice fronts are relatively slow moving or stagnant with velocities of less than 10 m per year, although some relatively fast flowing outlet glaciers flow at 50 to 100 m per year and sometimes a little more (e.g. Dowdeswell and Collin, 1990; Lefauconnier, Hagen and Rudant, 1994). An initial estimate of the average velocity of calving fronts through the archipelago is about 20 to 40 m per year. Thus, the ice flux at the calving front is estimated to be about 3 ± 1 km^3 per year. In addition, the annual retreat was estimated to produce a further 1 km^3 per year. The total calving loss, M_c, is thus about 4 ± 1 km^3 per year.

An additional complication to the assessment of the rate of iceberg calving within Svalbard is that a number of glaciers and ice-cap drainage basins undergo periodic surges (Hagen *et al.*, 1993; Hamilton and Dowdeswell, 1996; Jiskoot, Murray and Boyle, 2000). Between this activity, the termini of these ice masses are largely stagnant, contributing little to mass loss through iceberg production. In years when major Svalbard tide-water outlet glaciers surge, several cubic kilomteres of ice can be released to the adjacent seas (Liestøl, 1969; Schytt, 1969). For example, the 1250 km^2 Hinlopenbreen surged in 1970 and calved about 2 km^3 of icebergs in a single year (Liestøl, 1973). Synoptic satellite radar altimetric investigations of ice-surface velocities on Svalbard ice caps and glaciers will eventually provide, together with airborne ice-penetrating radar measurements of ice thickness, a more complete data set on iceberg calving (Dowdeswell *et al.*, 1999).

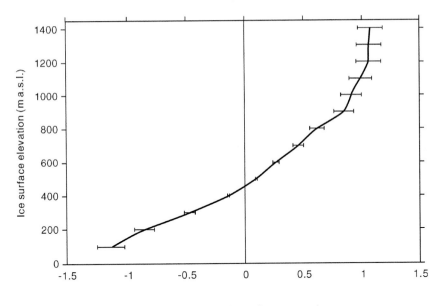

Figure 14.11. The relationship between specific net balance and ice-surface elevation, averaged over all regions of Svalbard.

14.6.3 Net mass balance of Svalbard

The total net surface balance, M_n, combining all regions, was found to be slightly negative at $-0.5\,\mathrm{km}^3 \pm 0.05\,\mathrm{km}^3$ per year. The specific surface net balance is then $-0.014\,\mathrm{m}$ per year. The calving loss, M_c, is estimated as $-4\,\mathrm{km}^3 \pm 1\,\mathrm{km}^3$ per year. The overall net balance of Svalbard ice massses, $\partial V/\partial t$, is then $-4.5\,\mathrm{km}^3 \pm 1\,\mathrm{km}^3$ per year, giving a specific net balance of $-0.12\,\mathrm{m} \pm 0.05\,\mathrm{m}$ per year. The contribution of the ice caps and glaciers on Svalbard to global sea-level change is, therefore, close to 0.01 mm per year as an average value over the last 30 years. This is less negative than earlier estimates based on net mass balance data for individual glaciers.

The net loss of mass through iceberg calving appears to be a very important component of the net mass loss from Svalbard ice masses, even though the ice flow at the 1000 km long calving front is, for the most part, quite low and is certainly rather poorly specified. Our analysis also shows how important the area/altitude distribution within the archipelago is for the sensitivity of the glacier mass balance to climate changes. A diagram of the total net balance and area per altitude increment shows that the present equilibrium line is very close to the bulk of the area (Figure 14.12). Thus, the net surface balance of Svalbard glaciers and ice caps is very sensitive to quite small changes in the equilibrium-line altitude, with a shift of a few tens of metres up or down having a disproportionally large effect on the total mass balance due to the nature of the hypsometric distribution (Figure 14.12).

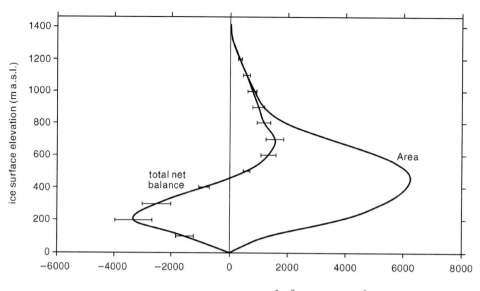

Figure 14.12. The total net balance and area distribution with elevation for Svalbard ice masses. The diagram shows that the mean equilibrium line for the archipelago (i.e. where the curve for total net balance crosses zero) coincides closely with the maximum of the area–altitude curve. This implies that the ice masses of Svalbard are likely to be particularly sensitive to climate-related shifts in equilibrium-line elevation.

14.7 Mass balance of Arctic ice masses: discussion

14.7.1 Problems with field measurements of mass balance

The formation of internal accumulation layers and superimposed ice at the surface of High Arctic glaciers can have a significant impact on the surface mass balance, and makes both direct field investigations of mass balance and remote-sensing analysis complicated. Melt water from the surface snow can percolate into the cold snow pack and refreeze, either as ice lenses in the snow and firn or as a layer of superimposed ice on top of cold impermeable ice. In the firn area, the melt water may also penetrate below the previous year's summer surface and freeze as internal accumulation.

Superimposed ice formation is an important source of accumulation in many Arctic glaciers, and in some it is even the dominant form of accumulation (Koerner, 1970). In the ablation area, the amount of superimposed ice can be measured by stakes drilled into the ice and by artificial reference horizons, but in the accumulation areas empirical modelling based on meteorological input data has often been the only way to estimate the internal accumulation from refreezing of melt water. The difficulties of recognizing and measuring this form of mass gain to glaciers are discussed in some detail in Chapter 2. If such internal accumulation is neglected, the total input of mass to the glacier may be under-estimated, leading to an unrealistically negative net mass balance.

14.7.2 Mass loss from iceberg production and basal ice-shelf melting

The production of icebergs from Arctic ice caps and glaciers with marine margins has rarely been measured routinely, although field surveys of changing ice-cliff height, calving events and surface velocity at its marine margin have allowed the rate of mass loss to be estimated at Hansbreen in Spitsbergen (Jania and Kaczmarska, 1997). A similar approach using field measurements is simply not possible over large areas of the Arctic islands. Other recent work has utilized remote-sensing data sets. Lefauconnier *et al.* (1994) tracked ice-surface features on sequential SPOT imagery to calculate ice velocity, which was combined with field measurements of terminus thickness of Kongsbreen ($625\,km^2$) to yield a calving rate of $0.25\,km^3$ per year. On the Academy of Sciences ice cap in Severnaya Zemlya, a combination of marginal ice-thickness data, derived from airborne 100 MHz radar, and ice velocities around the 200 km length of ice terminal ice cliffs, acquired using synthetic aperture radar (SAR) interferometry (Figure 14.13), has been used to calculate the total mass loss through iceberg production from this $5575\,km^2$ ice cap (Dowdeswell *et al.*, 2002). Total iceberg flux from the ice cap is about $0.65\,km^3$ per year, representing about 35–40% of the overall mass loss, with the remainder coming from surface melting. This is the first time that iceberg calving has been calculated, as opposed to estimated, for the whole of the marine margin of an Arctic ice cap. Where phase coherence is lost at the margins of some fast flowing and heavily crevassed tide-water termini, feature-tracking algorithms can also be used to measure ice motion from time series satellite data (e.g. Rolstad *et al.*, 1997; Scambos *et al.*, 1992).

The proportion of total mass loss that iceberg calving represents on Svalbard, and for the Russian Academy of Sciences ice cap, suggests that this component of total mass balance requires considerable further investigation across the Arctic islands. Most of the outlet glaciers draining the large ice caps on the Canadian Arctic islands end in the sea, and the same is typical of many ice caps in the Russian Arctic (Figure 14.14).

Finally, mass loss or gain at the base of floating glacier ice is significant beneath the extensive floating margins of a number of outlet glaciers of the Greenland ice sheet (see, e.g., Rignot, 1996). However, it is of relatively minor importance in the other parts of the Arctic. Svalbard has no floating glacier margins (Dowdeswell, 1989), and in Franz Josef Land and Severnaya Zemlya only rather limited areas are afloat (Dowdeswell *et al.*, 1994),

---

Figure 14.13. Parameters used to calculate mass loss through iceberg production on the Academy of Sciences ice cap, Severnaya Zemlya, in the Russian Arctic (see Figure 14.1(b)). (a) SAR interferogram containing information on both ice-surface topography and motion in the slant-range geometry of the ERS satellite. Closely spaced interference fringes represent faster flowing ice, and four ice streams are indicated. The pair of SAR images were acquired on September 23 and 24, 1995. The interferometric parallel baseline between the two satellite locations is 1 m. The ramp across the interferogram was removed during subsequent processing, before absolute velocities were calculated. The interferogram of the ice cap is superimposed on a Landsat image of the surrounding terrain and marine waters. (b) The thickness of the ice of the Academy of Sciences ice cap, Severnaya Zemlya, derived from 100 MHz airborne radio-echo sounding data. Contours are at 50 m intervals.

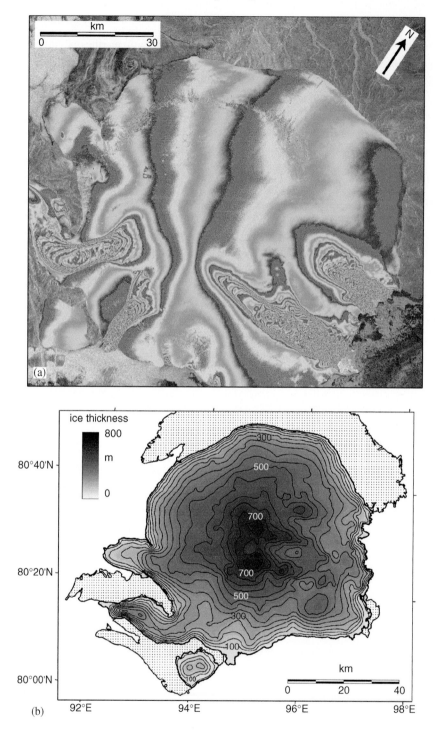

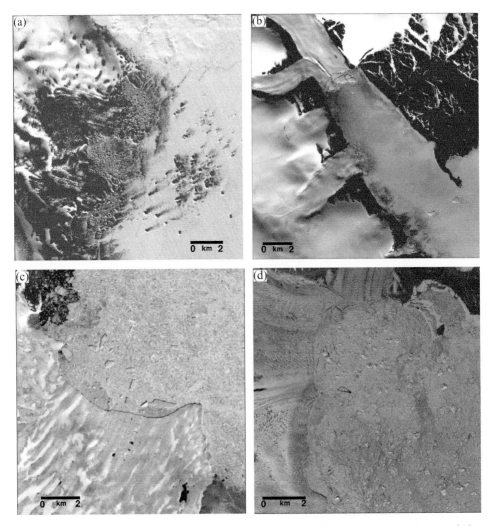

Figure 14.14. Landsat imagery of iceberg calving from four Arctic ice caps. In each part, tabular icebergs can be seen within a cover of sea ice. (a) Icebergs derived from fast flowing outlets on the eastern side of the Academy of Science ice cap, Severnaya Zemlya (Landsat 5 TM 174/001, April 26, 1988). (b) Icebergs calved from the Karpinsky (top left) and University (left) ice caps, Severnaya Zemlya (Landsat 5 TM 164/003, August 26, 1988). (c) Icebergs from the Vilchek Land ice cap in Franz Josef Land (Landsat 5 TM 199/001, July 25, 1986). (d) Icebergs calved from the Prince of Wales ice cap, Ellesmere Island (Figure 14.1(a)) into Talbot Inlet (Landsat 7 TM Panchromatic 42/04, July 9, 1999).

with the $220 \, km^2$ Matusevich ice shelf being by far the largest example (Williams and Dowdeswell, 2001). Many of the small glaciers and ice caps in Greenland beyond the ice sheet terminate on land, and few others are likely to have extensive floating margins. In the Canadian Arctic, recent airborne ice-penetrating radar investigations have shown that some fast flowing outlet glaciers of the Agassiz and Prince of Wales ice caps on Ellesmere Island may have floating termini of a few kilometres in length. Otherwise, extensive floating glacier termini are relatively uncommon. An exception is the ice shelves fringing the Arctic Ocean margin of the Queen Elizabeth Islands, of which the Ward Hunt ice shelf is the largest example at about 440 km^2 (Jeffries, 1992). However, much of the mass of these ice shelves is made up of multi-year shore-fast sea ice, rather than being glacier-nourished.

14.7.3 Airborne and satellite remote sensing of ice-surface characteristics

A continuing problem with field investigations of glacier mass balance is that they are expensive and time-consuming, yet result in data from only a limited number of Arctic ice caps and glaciers that are, for the most part, small. Measurements of snow accumulation over wider areas have, in Svalbard, attempted to set these detailed observations on individual glaciers in a wider context (e.g. Winther *et al.*, 1998). However, it is mainly through satellite remote sensing that our understanding of the geographical patterns and complexities of Arctic glacier mass balance are likely to be enhanced in the coming years.

Landsat and other visible and near-infrared sensors have been used for a number of years to map the general shape of the late-summer snow line on Arctic ice caps (Figure 14.15) (see, e.g., Dowdeswell and Drewry, 1989). Satellite SAR amplitude data, producing measurements of microwave backscatter, have also been employed to examine the snow and ice facies or zones on Arctic ice caps and glaciers (e.g. Engeset and Weydahl, 1998; Rees, Dowdeswell and Diament, 1995). Attempts have also been made, usually using winter SAR data to avoid melting snow from obscuring the underlying patterns, to tie specific backscatter signatures and patterns to the position of the equilibrium line and the firn line (Engeset and Wehdahl, 1998). The firn line is generally more easily identified than the equilibrium line on SAR imagery, and this may be a more effective tool in mass balance studies (König, Winther and Isaksson, 2001). Although these satellite methods have met with some success in outlining the general form of the ice and snow zones, they usually require calibration with field observations in order to refine the relationship between visible surface brightness or microwave backscatter and more quantitative estimates of equilibrium-line altitiude and the changing zonation of snow and ice facies through the melt season.

A different approach to understanding the mass balance of Arctic glaciers and ice caps relies on measurements of changing ice-surface elevation from airborne and satellite-mounted altimeters. By this method, density-corrected volume change is measured directly. Satellite radar altimetry has been used very successfully on the large ice sheets of Antarctica and Greenland (e.g. Shepherd *et al.*, 2001), yielding an accuracy of 20–30 cm. However, the relatively large footprint of the ERS satellite radar altimeters, combined with the relatively steep ice-surface slopes on many of the smaller Arctic glaciers, has meant that measurements of the elevation of even the largest Arctic ice caps, of the order of 10 000 km^2, have not been possible.

High accuracy airborne scanning laser altimetric measurements, capable of repeat measurements of surface elevation to within about ±10 cm, have also been acquired over several

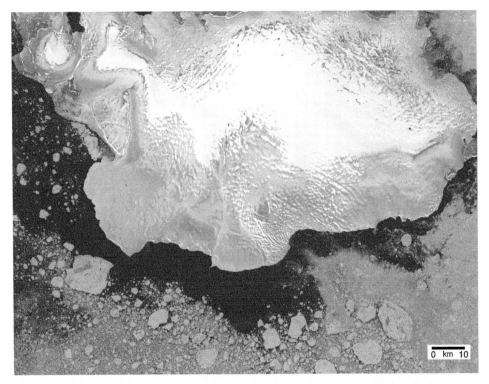

Figure 14.15. Landsat image of the snow line, with exposed ice at lower elevations, on the Austfonna ice cap in eastern Svalbard. The distribution of snow and ice on the ice cap can be compared with the ice-surface contours shown in Figure 14.9. The Landsat 5 multispectral scanner (MSS) image was acquired on August 1, 1993 (Path/Row 211/003).

Canadian Arctic ice caps and Svalbard ice masses, as well as in Greenland and Iceland (Krabill *et al.*, 1995). The results of the Canadian work are being analysed, and a second series of flights to provide repeated profiles on several Svalbard glaciers and ice caps took place in May 2002.

It is likely that satellite altimetry will provide the next major step forward in measuring changes in the surface elevation of Arctic ice caps and glaciers and, hence, their changing volume. This will be possible as a result of the much smaller footprint of the altimeters such as ICESat, launched in January, 2003, which has a footprint of 70 m (Zwally *et al.*, 2002). However, such work will need to be combined with continuing investigations of mass loss through iceberg production, using satellite interferometric and feature-tracking measurements of ice-marginal velocity, in order to specify this component of overall mass balance. In addition, field measurements of mass balance at individual glaciers and ice caps in the Arctic should continue, both in their own right as an important and long time series of detailed measurements, and to provide field evidence important in calibrating and guiding the interpretation of remote-sensing observations.

14.8 Conclusions

Dyurgerov and Meier (1997a) calculated that glaciers and ice caps in the Arctic islands, excluding Greenland, had an average net mass balance of -67 mm per year (-0.067 m per year) during the period 1961 to 1990. This contributed a mean of about 0.05 mm per year to global sea-level rise. They also suggested that this value was almost 20% of the total produced from glaciers and ice caps outside the great ice sheets. About 5% more may have been contributed by the ice caps and glaciers on Greenland, beyond the margins of the ice sheet. The calculations of Dyurgerov and Meier are based on a similar data set of Arctic glacier net mass balance to those observations shown in Figures 14.3–14.8.

We do not go on, from our presentation of ice-cap and glacier net mass balance records from the islands of Arctic Canada, Greenland, Svalbard and the Russian North, to undertake a similar calculation of the overall sea-level contribution of these ice masses. This would simply be to repeat the analysis of Dyurgerov and Meier. Instead, we have highlighted, through the case study of Svalbard glaciers and ice caps, some of the problems associated with calculating the overall net mass balance of even one of these archipelagos. We derived a value of 0.01 mm per year as our best estimate of the contribution of Svalbard ice masses to sea-level rise. This is about 20% of the total for the Arctic islands, as calculated by Dyurgerov and Meier (1997a). However, we have indicated that, although our value is derived not just from net mass balance observations at a set of glaciers, but also from other measurements of accumulation and the way that it varies spatially across the archipelago and with altitude, it is still a very imperfect estimate. The mass loss from iceberg calving is clearly a very important part of the overall mass balance equation for the very many Arctic glaciers and ice-cap outlets that end in the adjacent seas (Figure 14.14). It has, however, been calculated for a very small number for Arctic glaciers and ice caps.

Important future steps in enhancing our understanding of the mass balance of Arctic glaciers and ice caps are, firstly, to assess the contribution of iceberg production to mass loss for a suite of ice caps and major glacier systems in each of the Arctic islands. Secondly, we need to utilize forthcoming data sets on the accurate and repeated altimetric measurement of Arctic ice masses of 10^2 to 10^4 km^2, which a new generation of satellite altimeters will provide. This will, for the first time, allow the production of synoptic measurements of ice-volume change across all the Arctic islands. At the same time, it should be recognized that the long series of conventional field measurements of mass balance are the baseline data set from which progress can be made, not least by continuing these measurements and utilizing field observations to calibrate and understand the new satellite data sets.

References

Bahr, D. B., Meier, M. F. and Peckham, S. 1997. The physical basis of glacier volume area scaling. *J. Geophys. Res.* **102**, 20 355–62.

Baldwin, M. P. and Dunkerton, T. J. 1999. Propagation of the Arctic oscillation from the stratosphere to the troposphere. *J. Geophys. Res.* **104**, 30 937–46.

Black, D. E., Peterson, L. C., Overpeck, J. T., Kaplan, A., Evans, M. N. and Kashgarian, M. 1999. Eight centuries of North Atlantic ocean variability. *Science* **286**, 1709–13.

Bradley, R. S. and England, J. 1978. Recent climatic fluctuations of the Canadian High Arctic and their significance for glaciology. *Arctic & Alpine Res.* **10**, 715–31.

Cattle, H. and Crossley, J. 1995. Modelling Arctic climate change. *Phil. Trans. Roy. Soc. Lond. Series A* **352**, 201–13.

Cogley, J. G. and Adams, W. P. 1998. Mass balance of glaciers other than ice sheets. *J. Glaciol.* **44**, 315–25.

Cogley, J. G., Adams, W. P., Ecclestone, M. A., Jung-Rothenhäusler, F. and Ommanney, C. S. L. 1995. Mass balance of Axel Heiberg Island Glaciers 1960–1991. National Hydrology Research Institute, Science Report 6, Environment Canada.

Dowdeswell, J. A. 1986. Drainage-basin characteristics of Nordaustlandet ice caps, Svalbard. *J. Glaciol.* **32**, 31–8.

  1989. On the nature of Svalbard icebergs. *J. Glaciol.* **35**, 224–34.

  1995. Glaciers in the High Arctic and recent environmental change. *Phil. Trans. Roy. Soc. Lond. Series A* **352**, 321–34.

Dowdeswell, J. A. and Collin, R. L. 1990. Fast-flowing outlet glaciers on Svalbard ice caps. *Geology* **18**, 778–81.

Dowdeswell, J. A. and Drewry, D. J. 1989. The dynamics of Austfonna, Nordaustlandet, Svalbard: surface velocities, mass balance and subglacial melt water. *Ann. Glaciol.* **12**, 37–45.

Dowdeswell, J. A., Drewry, D. J. and Simões, J. C. 1990. Comment on: '6000-year climate records in an ice core from the Høghetta ice dome in northern Spitsbergen'. *J. Glaciol.* **36**, 353–6.

Dowdeswell, J. A., Gorman, M. R., Glazovsky, A. F. and Macheret, Y. Y. 1994. Evidence for floating ice shelves in Franz Josef Land, Russian High Arctic. *Arctic & Alpine Res.* **26**, 86–92.

Dowdeswell, J. A. *et al.* 1997. The mass balance of circum-Arctic glaciers and recent climate change. *Quat. Res.* **48**, 1–14.

Dowdeswell, J. A., Unwin, B., Nuttall, A.-M. and Wingham, D. J. 1999. Velocity structure, flow instability and mass flux on a large Arctic ice cap from satellite radar interferometry. *Earth & Planetary Sci. Lett.* **167**, 131–40.

Dowdeswell, J. A. *et al.* 2002. Form and flow of the Academy of Sciences ice cap, Severnaya Zemlya, Russian High Arctic. *J. Geophys. Res.* **107**, 10.1029/2000/JB000129.

Dyurgerov, M. B. and Meier, M. F. 1997a. Year-to-year fluctuations of global mass balance of small glaciers and their contribution to sea-level change. *Arctic & Alpine Res.* **29**, 392–402.

  1997b. Mass balance of mountain and subpolar glaciers: a new global assessment for 1961–1990. *Arctic & Alpine Res.* **29**, 379–91.

Engeset, R. V. and Weydahl, D. J. 1998. Analysis of glaciers and geomorphology on Svalbard using multitemporal RES-1 SAR images. *IEEE Trans. Geosci. & Remote Sensing* **36**, 1879–87.

Fleming, K. M., Dowdeswell, J. A. and Oerlemans, J. 1997. Modelling the mass balance of north-west Spitsbergen glaciers and responses to climate change. *Ann. Glaciol.* **24**, 203–10.

Førland, E, Hanssen-Bauer, I. and Nordli, Ø. 1997. Climate statistics and long-term series of temperature and precipitation at Svalbard and Jan Mayen. DNMI report **21/97** KLIMA, Norwegian Meteorological Institute, Oslo.

Glazovskiy, A. F. 1996. Russian Arctic. In Jania, J. and Hagen, J. O., eds., *Mass Balance of Arctic Glaciers*. International Arctic Science Committee Report no. 5, pp. 44–53.

Gordiyenko, F. G., Kotlyakov, V. M., Punning, Y-K. M. and Vairmae, R. 1981. Study of a 200 m ice core from the Lomonosov ice plateau on Spitsbergen and the paleoclimatic implications. *Polar Geog. & Geol.* **5**, 242–51.

Grove, J. M. 1988. *The Little Ice Age*. London, Methuen.

Hagen, J. O. 1996a. Svalbard. In Jania, J. and Hagen, J. O., eds., *Mass Balance of Arctic Glaciers*. International Arctic Science Committee Report no. **5**, pp. 30–8.

1996b. Recent trends in mass balance of glaciers in Scandinavia and Svalbard. *Memoirs of the National Institute of Polar Research*, Tokyo, special issue **51**, 343–54.

Hagen, J. O. and Liestøl, O. 1990. Long-term glacier mass balance investigations in Svalbard. *Ann. Glaciol.* **14**, 102–6.

Hagen, J. O., Liestøl, O., Roland, E. and Jørgensen, T. 1993. *Glacier Atlas of Svalbard and Jan Mayen*. Norsk Polarinstitutt Meddelelser **129**.

Hagen, J. O., Melvold, K., Pinglot, F. and Dowdeswell, J. A. 2003. On the net mass balance of the glaciers and ice caps in Svalbard, Norwegian Arctic. *Arctic, Antarctic & Alpine Res.* **35**, 264–70.

Hamilton, G. S. and Dowdeswell, J. A. 1996. Controls on glacier surging in Svalbard. *J. Glaciol.* **42**, 157–68.

Hurrell, J. W., 1995. Decadal trends in the North Atlantic oscillation: regional temperatures and precipitation. *Science* **269**, 676–9.

Hurrell, J. W. and Van Loon, H. 1997. Decadal variations in climate associated with the North Atlantic oscillation. *Climate Change* **36**, 301–26.

Jacobs, J. D., Simms, E. L. and Simms, A. 1997. Recession of the southern part of Barnes ice cap, Baffin Island, Canada, between 1961 and 1993, determined from digital mapping of Landsat TM. *J. Glaciol.* **43**, 98–102.

Jania, J. and Hagen, J. O. eds. 1996. *Mass Balance of Arctic Glaciers*. International Arctic Science Committee Report no. 5.

Jania, J. and Kaczmarska, M. 1997. Hans Glacier – a tidewater glacier in southern Spitsbergen: summary of some results. In Van der Veen, C. J., ed., *Calving Glaciers: Report of a Workshop, February 28–March 2, 1997*. BPRC report no. 15. Byrd Polar Research Center, The Ohio State University, pp. 95–104.

Jeffries, M. O. 1992. Arctic ice shelves and ice islands: origin, growth and disintegration, physical characteristics, structural-stratigraphic variability, and dynamics. *Rev. Geophys.* **30**, 245–67.

Jiskoot, H., Murray, T. and Boyle, P. 2000. Controls on the distribution of surge-type glaciers in Svalbard. *J. Glaciol.* **46**, 412–22.

Johannesson, T., Raymond, C. F. and Waddington, E. 1989. Time-scale for adjustment of glaciers to changes in mass balance. *J. Glaciol.* **35**, 355–69.

Johnsen, S. J., Dansgaard, W., Clausen, H. B. and Langway, C. C. 1970. Climatic oscillations 1200–2000 AD. *Nature* **227**, 482–3.

König, M., Winther, J.-G. and Isaksson, E. 2001. Measuring snow and glacier ice properties from satellite. *Rev. Geophys.* **39**, 1–27.

Koerner, R. M. 1970. Some observations on superimposition of ice on the Devon Island ice cap, N. W. T., Canada. *Geograf. Ann.* **52A**, 57–67.

1977. Devon Island ice cap: core stratigraphy and paleoclimate. *Science* **196**, 15–18.

1996. Canadian Arctic. In Jania, J. and Hagen, J. O., eds., *Mass Balance of Arctic Glaciers*. International Arctic Science Committee Report no. 5, pp. 13–21.

1997. Some comments on climatic reconstructions from ice cores drilled in areas of high melt. *J. Glaciol.* **43**, 90–7.

2002. *Glaciers of Canada: Glaciers of the Arctic Islands.* US Geological Survey, Professional Paper **1386-J**, pp. 111–46.

Koerner, R. M. and Fisher, D. A. 1990. A record of Holocene summer climate from a Canadian high-Arctic ice core. *Nature* **343**, 630–1.

Koerner, R. M. and Paterson, W. S. B. 1974. Analysis of a core through the Meighan ice cap, Arctic Canada, and its paleoclimatic implications. *Quat. Res.* **4**, 253–63.

Koryakin, V. S. 1988. *Ledniki Arktiki [Arctic Glaciers].* Moscow, Nauka.

Kotlyakov, V. M., Korotkov, I. M., Nikolayev, V. I., Petrov, V. N., Barkov, N. I. and Klement'yev, O. L. 1989. Reconstruction of the Holocene climate from the results of ice-core studies on the Vavilov Dome, Severnaya Zemlya. *Materialy Glyatsiologicheskikh Issledovaniy* **67**, 103–8.

Kotlyakov, V. M., Zagorodnov, V. S. and Nikolayev, V. I. 1990. Drilling on ice caps in the Soviet Arctic and on Svalbard and prospects of ice core treatment. In Kotlyakov, V. M. and Sokolov, V. Y., eds., *Arctic Research: Advances and Prospects*, vol. 2 Moscow, Nauka, pp. 5–18.

Krabill, W. B., Thomas, R. H., Martin, C. F., Swift, R. N. and Frederick, E. B. 1995. Accuracy of airborne laser altimetry over the Greenland ice sheet. *Int. J. Remote Sensing* **16**, 1211–22.

Lefauconnier, B. and Hagen, J. O. 1990. Glaciers and climate in Svalbard: statistical analysis and reconstruction of the Brøggerbreen mass balance for the last 77 years. *Ann. Glaciol.* **14**, 148–52.

Lefauconnier, B., Hagen, J. O. and Rudant, J. P. 1994. Flow speed and calving rate of Kongsbreen Glacier, 79°N, Spitsbergen, Svalbard, using SPOT images. *Polar Res.* **13**, 59–65.

Liestøl, O. 1969. Glacier surges in West Spitsbergen. *Can. J. Earth Sci.* **6**, 895–7.

1973. Glaciological work in 1971. *Norsk Polarinstitutt årbok 1971*, Oslo.

1993. *Glaciers of Svalbard, Norway.* US Geological Survey, Professional Paper 1386-E, pp. 127–51.

Meier, M. F. 1984. Contribution of small glaciers to global sea level. *Science* **226**, 1418–21.

Meier, M. F. and Bahr, D. B. 1996. Counting glaciers: use of scaling methods to estimate the number and size distribution of the glaciers of the world. In Colbeck, S. C., ed., *Glaciers, Ice Sheets and Volcanoes: A Tribute to Mark F. Meier.* CRREL Special Report 96–27. Hanover, New Hampshire, US Army.

Melvold, K. and Hagen, J. O. 1998. Evolution of a surge-type glacier in its quiescent phase: Kongsvegen, Spitsbergen, 1964–95. *J. Glaciol.* **44**, 394–404.

Paterson, W. S. B. *et al.* 1977. An oxygen isotope climatic record from the Devon Island ice cap, Arctic Canada. *Nature* **266**, 508–11.

Pinglot, J. F. *et al.* 1999. Accumulation in Svalbard glaciers deduced from ice cores with nuclear tests and Chernobyl reference layers. *Polar Res.* **18**, 315–321.

Pinglot, J. F., Hagen, J. O., Melvold, K., Eiken, T. and Vincent, C. 2001. A mean net accumulation pattern derived from radioactive layers and radar soundings on Austfonna, Nordaustlandet, Svalbard. *J. Glaciol.* **47**, 555–66.

Raper, S. C. B., Brown, O. and Braithwaite, R. J. 2000. A geometric glacier model suitable for sea-level change calculations. *J. Glaciol.* **46**, 357–68.

Reeh, N., Olesen, O. B., Thomsen, H. H., Starzer, W. and Bøggild, C. E. 2001. Mass balance parameterisation for Hans Tausen Iskappe, Peary Land, North Greenland. *Meddelelser om Grønland, Geosci.* **39**, 57–69.

Rees, W. G., Dowdeswell, J. A. and Diament, A. D. 1995. Analysis of ERS-1 synthetic aperture radar data from Nordaustlandet, Svalbard. *Int. J. Remote Sensing* **16**, 905–24.

Rignot, E. 1996. Tidal motion, ice velocity and melt rate of Petermann Gletscher, Greenland, measured from radar interferometry. *J. Glaciol.* **42**, 476–85.

Rolstad, C., Amlien, J., Hagen, J. O. and Lunden, B. 1997. Visible and near-infrared digital images for determination of ice velocities and surface elevation during a surge on Osbornebreen, a tidewater glacier in Svalbard. *Ann. Glaciol.* **24**, 255–61.

Scambos, T. A., Dutkiewicz, M. J., Wilson, J. C. and Bindschadler, R. A. 1992. Application of image cross-correlation to the measurement of glacier velocity using satellite image data. *Remote Sensing Environ.* **42**, 177–86.

Schytt, V. 1969. Some comments on glacier surges in eastern Svalbard. *Can. J. Earth Sci.* **6**, 867–73.

Shepherd, A., Wingham, D. J., Mansley, J. A. D. and Corr, H. F. J. 2001. Inland thinning of Pine Island Glacier, West Antarctica. *Science* **291**, 862–4.

Tarussov, A. 1992. The Arctic from Svalbard to Severnaya Zemlya: climatic reconstructions from ice cores. In Bradley, R. S. and Jones, P. D., eds., *Climate Since A. D. 1500*. London, Routledge, pp. 505–16.

Thompson, D. W. J. and Wallace, J. M. 1998. The Arctic oscillation signature in the wintertime geopotential height and temperature fields. *Geophys. Res. Lett.* **25**, 1297–300.

Warrick, R. A., Le Provost, C., Meier, M. F., Oerlemans, J. and Woodworth, P. L. 1996. Changes in sea level. In Houghton, J. T. *et al.* eds., *Climate Change 1995: The Science of Climate Change*. Cambridge University Press, pp. 359–405.

Weidick, A. 1995. *Greenland*. US Geological Survey, Professional Paper 1386-C.

Weidick, A. and Morris, E. 1998. Local glaciers surrounding the continental ice sheets. In Haeberli, W., Hoelzle, M. and Suter, S., eds., *Introduction to the Second Century of Worldwide Glacier Monitoring: Prospects and Strategies*. Studies and Reports in Hydrology vol. 56, Paris, UNESCO, pp. 197–207.

Werner, A. 1993. Holocene moraine chronology, Spitsbergen, Svalbard: lichenometric evidence for multiple Neoglacial advances in the Arctic. *The Holocene* **3**, 12–137.

Williams, M. and Dowdeswell, J. A., 2001. Historical fluctuations of the Matusevich ice shelf, Severnaya Zemlya, Russian High Arctic. *Arctic, Antarctic & Alpine Res.* **33**, 211–22.

Winther, J.-G., Bruland, O., Sand, K., Killingtveit, Å and Marechal., D. 1998. Snow accumulation distribution on Spitsbergen, Svalbard, in 1997. *Polar Res.* **17**, 155–64.

Woo, M.-K., Heron, R., Marsh, P. and Steer, P. 1983. Comparison of weather station snowfall with winter snowfall accumulation in high arctic basins. *Atmos. & Ocean* **21**, 312–25.

Zwally, H. J. *et al.* 2002. ICESat's laser measurements of polar ice, atmosphere, ocean and land. *J. Geodyn.* **34** (3–4), 405–45.

15

Glaciers and ice caps: historical background and strategies of world-wide monitoring

WILFRIED HAEBERLI

Glaciology and Geomorphodynamics Group, Geography Department,
University of Zurich

15.1 Introduction

Throughout the history of modern science, glaciers and ice caps have not only been a source of fascination but also a key element in discussions about Earth evolution and climate change. The discovery of the Ice Age in the late eighteenth and the nineteenth centuries significantly contributed to the understanding of the evolutionary development of the Earth; it also demonstrated the possibility of important climatic changes involving dramatic environmental effects at a global scale. Today, glaciers and ice caps clearly reflect secular warming at a high rate and at a global scale; they are considered key indicators within global climate-related observing systems for early detection of trends potentially related to the greenhouse effect (Figure 15.1; IPCC, 2001). This chapter discusses the historical background, the observational data basis and related monitoring strategies. It also gives some examples, predominantly from low latitude glaciers. More detailed treatment of the theoretical and methodological background can be found in the Chapters 2, 4 and 6. An example of measurements in the Arctic is given in Chapter 14.

15.2 Historical background of world-wide glacier monitoring

The internationally co-ordinated collection of information about on-going glacier changes was initiated in 1894 with the foundation of the International Glacier Commission at the Sixth International Geological Congress in Zurich, Switzerland. It was hoped that the long-term observation of glaciers would provide answers to the questions about global uniformity and terrestrial or extra-terrestrial forcing of past, on-going and potential future climate and glacier changes (Forel, 1895).

Reports first appeared annually, but became less frequent during the first half of the twentieth century. Results of quantitative measurements were presented for a limited number of cases only; the most detailed data were available for glaciers of the European Alps and

The author expresses his deep appreciation to the great number of friends, colleagues and officials in many national as well as international organizations for helping over many years to build up – and freely exchanging on – a certainly not perfect, but nevertheless unique, source of information. Hans Oerlemans and Tony Payne contributed constructive comments on this chapter.

Mass Balance of the Cryosphere: Observations and Modelling of Contemporary and Future Changes, eds. Jonathan L. Bamber and Antony J. Payne. Published by Cambridge University Press. © Cambridge University Press 2003.

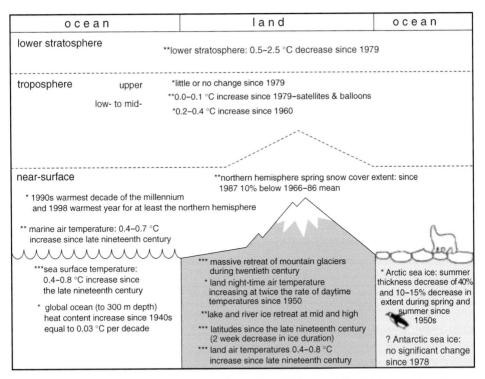

Figure 15.1. Observed variations of the temperature indicators. Likelihoods: ***virtually certain (probability>99%); **very likely (probability ≥ 90% but≤ 99%); *likely (probability >66% but ≤ 90%);? medium likelihood (probability) >33% but ≤ 66%). (Taken from fig. 2.39a from IPCC (2001).)

Scandinavia (Haeberli, Hoelzle and Suter, 1998). In 1967, the Permanent Service on the Fluctuations of Glaciers (PSFG) was established as one of the services of the Federation of Astronomical and Geophysical Services (FAGS) of the International Council of Scientific Unions (ICSU). This resulted in the publication of the *Fluctuations of Glaciers* at five-yearly intervals. With the first volume (IAHS(ICSI)/UNESCO, 1967), mass balance data from various countries, including the USSR, USA and Canada, entered the reports for the first time, thus forming the essential link between climate fluctuations and glacier length changes. In the second volume (IAHS(ICSI)/UNESCO, 1973), length variation data – showing signs of intermittent glacier advance – from the USA and the USSR, as well as from other countries, complemented the corresponding records from the Alps, Scandinavia and Iceland, where most glaciers continued to retreat. The third and fourth volumes (IAHS(ICSI)/UNESCO, 1977; IAHS (ICSI)/UNESCO, 1985) witnessed a major step toward standardization and computer-based processing of data. Length variation data from the southern hemisphere (Peru, Argentina, Kenya, New Zealand and others) were now also regularly included. Glacier re-advances were reported from various parts of the world, especially from the Alps, where mass balances had been predominantly positive since the mid 1960s. For the first time, therefore, empirical information started to become available about glacier reactions to well-documented and strong signals in mass balance history.

Since 1986, world-wide collection of standardized observations on changes in mass, volume, area and length of glaciers with time (glacier fluctuations), as well as statistical information on the distribution of perennial surface ice in space (glacier inventories), has been co-ordinated by the World Glacier Monitoring Service (WGMS). This work is primarily being carried out under the auspices of the International Commission on Snow and Ice (ICSI\IAHS) and the Federation of Astronomical and Geophysical Services (FAGS\ICSU). The *Fluctuations of Glaciers* series now also contains information on special events, such as glacier surges, eruptions of glacier-clad volcanoes, outbursts of ice- and moraine-dammed lakes or major rock-fall events in glacierized areas (IAHS(ICSI)/UNEP/UNESCO, 1988, 1993, 1998). An overview of the World Glacier Inventory was given in 1989 (IAHS(ICSI)/UNEP/UNESCO, 1989). Mass balance data have been reported in a biennial bulletin and on the internet[1] (IAHS(ICSI)/UNEP/UNESCO (2001) and earlier issues) since 1991.

The data collected from such systematic observations clearly reveal the fact that glacier shrinkage at decadal to secular timescales is a fast and world-wide phenomenon (Figure 15.2). It contributes to the observed rise in eustatic sea level, influences the water cycle and affects natural hazards in cold mountain areas. Moreover, it reflects an additional energy flux to the Earth–atmosphere interface which can be quantified in terms of the involved latent heat exchange, and, hence, constitutes a key indicator of on-going climate change (Haeberli *et al.*, 1998). In fact, the anthropogenic influences on the atmosphere may now and for the first time represent a major contributing factor to the observed glacier shrinkage at a global scale (Haeberli *et al.*, 1999). Based on these results of long-term observations, the fluctuation of mountain glaciers is recognized as a high confidence indicator of air temperature trends and as a valuable element of a strategy for early detection of possible man-induced climate changes. The Terrestrial Observation Panel for Climate (TOPC) therefore recommended (WMO, 1997) that glacier mass/volume and area be monitored as part of the global climate observing system (GCOS) established in 1992 by the World Meteorological Organization (WMO), the Intergovernmental Oceanographic Commission (IOC of UNESCO), the United Nations Environment Programme (UNEP) and the International Council of Scientific Unions (ICSU). A global hierarchical observing strategy (GHOST) consisting of several tiers was developed to be used for all GCOS terrestrial variables (see section 15.5). The main goals of long-term glacier observations generally relate to aspects of process understanding, model validation, change detection and impact assessments.

Process understanding provides the basis for physical models. It enables adequate interpretation of the direct signals from mass change and of the much more easily measured, but indirect, indications from cumulative changes in glacier length (Haeberli, 1996). It also helps to discriminate temperate maritime-type glaciers with high mean annual air temperatures at

---

[1] For WGMS, see http://www.geo.unizh.ch/wgms/, and for GCOS/GTOS see http://193.135.216.2/web/gcos/gcoshone.html.

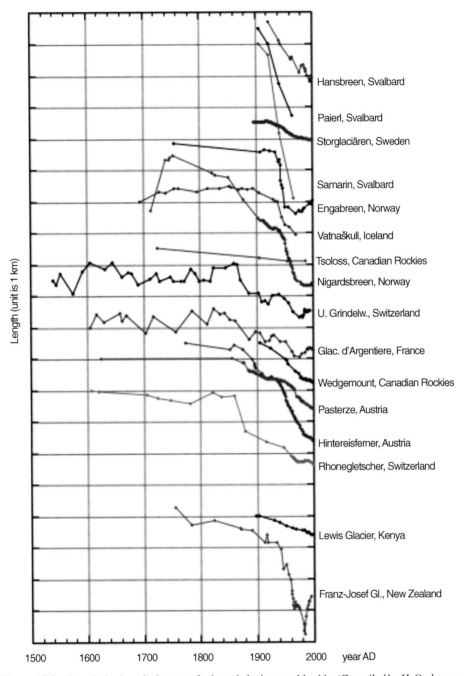

Figure 15.2. Cumulative length changes of selected glaciers world-wide. (Compiled by H. Oerlemans from WGMS data with some additions.)

the equilibrium-line altitude (for instance, the ice caps and valley glaciers of Patagonia and Iceland, the western Cordillera of North America, or the coastal mountains of Norway and New Zealand) from polythermal or cold glaciers with low mean annual air temperatures at the equilibrium-line altitude, existing under dry polar and continental climatic conditions (for instance, in northern Alaska, arctic Canada, subarctic Russia, parts of the Andes near the Atacama Desert, or in many central-Asian mountain chains). The latter have much shorter ablation seasons, a weaker snow–albedo feedback with correspondingly smaller mass balance gradients and weaker mass turnover; they usually terminate far beyond the timber line and are associated with periglacial permafrost; such glaciers may react to atmospheric temperature rise by runoff from the ablation area but warming (not mass loss) of cold firn in the accumulation area (Haeberli, 1983; Oerlemans, 1993).

Model validation and verification today, and in the near future, primarily relate to calibration and tests in remote areas with little or no meteorological information: verification of complex patterns of change in scenarios simulated by coupled atmosphere–ocean general circulation and regional climate models (AOGCMs and RCMs; Beniston *et al.*, 1997) is a focus of interest. An important parameter for the potential application of glacier inventory information to GCM/RCM validation is mean glacier elevation as contained in detailed glacier inventories. This easily determined parameter is a rough approximation to equilibrium-line altitude (ELA). As such, it is connected with continentality and hence with annual precipitation, mass balance gradient (activity index), mass turnover, englacial temperature and glacier/permafrost relations.

Change detection strategies relating to the atmospheric temperature rise caused by anthropogenic greenhouse forcing include secular rates of change in energy fluxes at the Earth–atmosphere interface, natural (pre-industrial) variability in these energy fluxes and the possible accelerating trends of on-going and potential future changes. Glacier changes can be determined quantitatively over various time intervals and expressed as corresponding energy fluxes with their long-term variability. This permits direct comparison with other effects of natural and estimated anthropogenic greenhouse forcing (Haeberli, 1996).

Impact assessments can be made at a global scale (for instance on sea level; Gregory and Oerlemans, 1998; Warrick *et al.*, 1996) and at various regional to local scales, relating to such problems as water resources for irrigation in semi-arid mountains, slope instability in deglaciated terrain, or landscape deterioration in connection with tourism (Fitzharris *et al.*, 1996; Haeberli and Burn, 2002). The main challenge is, in fact, to adapt to high and accelerating rates of environment evolution.

15.3 Observed conditions and trends

Glacier inventory data serve as a statistical basis for extrapolating the results of observations or model calculations concerning individual glaciers (Oerlemans, 1994). Scaling relationships can be used to link the distribution of surface areas as contained in regional glacier inventories to global and regional distributions of other properties such as glacier volumes or characteristic thicknesses, flow velocities or response times (Bahr, 1997). Based on this approach, Meier and Bahr (1996) estimated the total number (160 000), area (680 000 km^2), volume (180 000 km^3) and sea-level equivalent (0.5 m) of glaciers and ice caps world-wide. The relative contribution of polar, sub-polar, temperate-maritime and temperate-continental climatic regions can also be assessed. Dyurgerov and Meier (1997a) analysed the characteristics of the mass balance observation network with respect to global glacier distribution and noted the main size categories and areas under-represented or not represented at all; for instance, the Karakorum, Tibetan Plateau, Kunlun, South-east Pamir and Hindu Kush, and the Patagonian icefields (cf. also the detailed statistical analysis presented by Cogley and Adams, 1998).

Trends in long time series of cumulative glacier length and volume changes represent convincing evidence of fast climatic change at a global scale, since the retreat of mountain glaciers during the twentieth century is striking all over the world. Regional overviews are given by: Casassa *et al.* (1998) for South America; Ommanney *et al.* (1998) for North America; Hagen *et al.* (1998) for Europe; Tsvetkov *et al.* (1998) for Asia; Hastenrath and Chinn (1998) for Africa and New Zealand; and Weidick and Morris (1998) for local glaciers surrounding the continental ice sheets. Total retreat of glacier termini is commonly measured in kilometres for larger glaciers and in hundreds of metres for smaller ones. Characteristic average rates of glacier thinning are a few decimetres per year for temperate glaciers, and centimetres to one decimetre per year for glaciers in continental areas with firn areas below melting temperature. At retreating glacier termini, the total secular surface lowering is up to several hundred metres. The apparent homogeneity of the signal at the secular timescale, however, contrasts with great variability at local\regional scales and over shorter time periods of years to decades (Letréguilly and Reynaud, 1990). Intermittent periods of mass gain and glacier advance during the second half of the twentieth century have been reported from various mountain chains (IAHS(ICSI)/UNEP/UNESCO, 1988, 1993, 1998), especially in areas of abundant precipitation such as southern Alaska, Norway and New Zealand. Glaciers in the European Alps, on the other hand, have lost about 30 to 40% of their surface area and around 50% of their volume since the middle of the nineteenth century (i.e. the end of the Little Ice Age, see Table 15.1; Haeberli and Hoelzle, 1995). The recent emergence of a stone-age man from cold ice on a high altitude ridge of the Oetztal Alps is a striking illustration which confirms that the extent of Alpine ice is probably less today than during the past 5000 years (Figure 15.3; see Haeberli *et al.*, 1999).

The main problem with interpreting world-wide glacier mass balance evolution relates to the methods of averaging and extrapolating the small sample of reported values in a few areas of the world to large unmeasured areas. Several attempts have recently been undertaken to apply (a) all available mass balance measurements, (b) selected glaciers with long-term records, (c) area-weighting using glacier inventory data, (d) averages from reference glaciers in a number of mountain ranges, (e) spatial interpolation based on global ice-extent data and correlations between mass balance time series, (f) laser altimetry flights and GPS surveys on selected flow lines for comparison with integrated geometric

Table 15.1. *Analysis of glacier inventory data for the European Alps.*

Situation 1970\80	
Total glacierized area 1970/80	2909 km^2
Total glacier volume 1970/80	*c.*130 km^3
Sea-level equivalent	*c.*0.35 mm
Number of glaciers >0.2 km^2	1763
Average mass balance 1850–1970/80	−0.25 m/Jahr
Average mass balance 1980–2000	−0.60 m/Jahr
Area reduction 1850–1970/80	35–40%
Mass loss 1850–1970/80	50% of 1850
Mass loss 1980–2000	*c.*25% of 1970/80
Simulation (IPCC, medium scenario of accelerated warming)	
Area reduction 1970/80–2025	*c.*30% of 1970/80
Mass loss 1970/80–2025	*c.*50% of 1970/80
Area reduction 1970/80–2100	80–90% of 1970/80
Mass loss 1970/80–2100	90–95% of 1970/80

Source: Haeberli and Hoelzle (1995), updated.

changes, and (g) cumulative glacier length changes as combined with glacier inventory data (Cogley and Adams, 1998; Dyurgerov and Meier, 1997b; Echelmeyer *et al.*, 1996; Haeberli and Hoelzle, 1995; IAHS(ICSI)/UNEP/UNESCO, 2001; Oerlemans, 1994; Rabus and Echelmeyer, 1998). The results from the various extrapolation schemes only deviate slightly from each other (typically <0.1 m per year where comparable), and all confirm the order of magnitude (a few decimetres per year) characterizing world-wide annual ice-thickness loss during recent decades (see Figure 15.4 and Table 15.2). Dyurgerov and Meier (1997b) found a recent increase in ice loss in close correlation with global air temperature anomalies. Glaciers in continental-type climatic regions appear to have decreased steadily, whereas maritime-type glaciers in humid areas show important temporal and regional variability with recent growth tendencies being observed in areas around the northern Atlantic. Cogley and Adams (1998) confirm the correlation between global averages of air temperature and glacier mass balance. In their compilation, annual thickness losses steadily increased from near zero in 1960–1970 to about 0.3 m in 1980–1990. Glacier mass balances in the European Alps were strongly negative during the extremely warm decades 1980–2000: with an average value of −0.6 m water equivalent. This is considerably more than the mean secular average as derived from comparison with historical high precision maps (Table 15.3) and from cumulative glacier length change (cf. Table 15.1 and the corresponding calculations by Haeberli and Hoelzle (1995)). The Alpine ice cover may, in fact, have lost about 25%

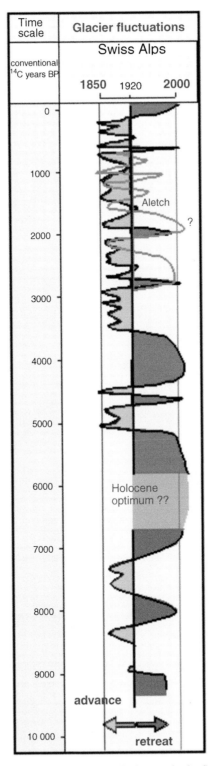

Figure 15.3. Chronology of Holocene glacier length changes in the Swiss Alps. (Compiled by M. Maisch.)

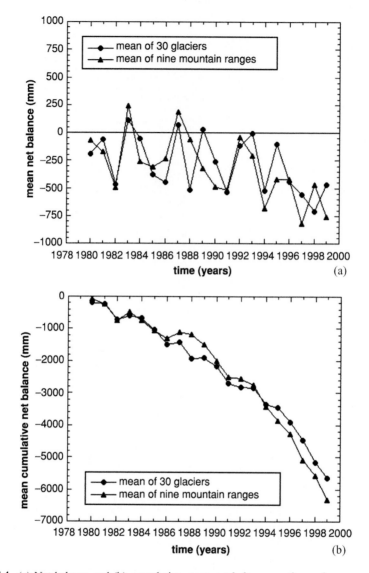

Figure 15.4. (a) Net balance and (b) cumulative mean net balance continuously measured for the period 1980 to 1999 on 30 glaciers in nine mountain ranges. (From IAHS (ICSI)/UNESCO (2000).)

of its volume as estimated for the 1970s (Table 15.2). Recent acceleration of annual mass losses is also reported from the Brooks Range, Alaska, by Rabus and Echelmeyer (1998), for the Tien Shan by Cao (1998) and for the Andes of Argentina by Leiva and Cabrera (1996).

The cumulative mass balances reported for individual glaciers (IAHS(ICSI)/ UNEP/UNESCO, 2001) not only reflect regional climatic variability but also marked

Table 15.2. *Glacier mass balances during the years 1980–1999 from nine mountain ranges.*

Year	Casc.	Alaska	Andes	Svalb.	Scand.	Alps	Altai	Cauc.	Tien S.	Mean
1980	−972	1400	300	−475	−1180	418	−10	380	−482	−69
1981	−967	775	360	−505	194	−16	−213	−910	−271	−172
1982	−337	−245	−2420	−10	−185	−887	−460	420	−337	−495
1983	−606	15	3700	−220	756	−460	197	−970	−220	243
1984	−109	−395	−1240	−705	194	12	307	210	−666	−265
1985	−1541	515	340	−515	−451	−411	200	−380	−581	−313
1986	−1011	−60	1510	−265	−249	−1010	73	−500	−594	−234
1987	−1703	535	950	230	925	−699	183	1540	−258	189
1988	−1305	395	2430	−505	−1215	−610	333	520	−626	−64
1989	−875	−1440	−1260	−345	1911	−893	117	40	−177	−324
1990	−834	−1555	−1530	−585	1196	−1101	107	340	−454	−490
1991	−595	−260	−1050	115	80	−1227	−480	−310	−903	−514
1992	−1400	−210	1740	−120	1162	−1158	−127	−130	−108	−39
1993	−1755	−1170	−290	−955	1174	−459	227	1100	286	−204
1994	−1225	−660	−1860	−140	171	−920	−240	−840	−410	−680
1995	−1588	−785	−950	−785	589	17	60	40	−407	−423
1996	−61	−950	−1180	−75	−639	−411	−140	−150	−207	−423
1997	−129	−2120	−2880	−570	−470	−227	−123	270	−1160	−823
1998	−2155	−135	2890	−725	221	−1611	−1110	−1000	−574	−466
1999	820	−1095	−4260	−350	−123	−636	−113	−560	−511	−759
Mean	−917	−372	−235	−375	203	−614	−60	−44	−433	−316

Glaciers are as follows. Cascade Mountains: Place, South Cascade; Svalbard: Austre Brøggerbreen, Midtre Lovénbreen; Andes: Enchaurren Norte; Alaska: Gulkana, Wolverine; Scandinavia: Engabreen, Ålfotbreen, Nigardsbreen, Gråsubreen, Storbreen, Hellstugubreen, Hardangerjøkulen, Storglaciären; Alps: Saint Sorlin, Sarennes, Silvretta, Gries, Sonnblickkees, Vernagtferner, Kesselwandferner, Hintereisferner, Caresèr; Altai: no. 125, Maliy Aktru, Leviy Aktru; Caucasus: Djankuat; Tien Shan: Ts. Tuyuksuyskiy, Urumqihe S. no. 1.

Table 15.3. *Geodetically/photogrammetrically determined secular mass balances of Alpine glaciers.*

Glacier	Observation period	Co-ordinates	Median elevation (m a.s.l.)	Surface area (km^2)	b (m per year) (water equivalant)
Rhone	1882–1987	4637/0824	2940	17.38	−0.25
Vernagt	1889–1979	4653/1049	3228	09.55	−0.19
Guslar	1889–1979	4651/1048	3143	03.01	−0.26
N. Schnee	1892–1979	4725/1059	2690	00.39	−0.35
S. Schnee	1892–1979	4724/1058	2604	00.18	−0.57
Hintereis	1894–1979	4648/1046	3050	09.70	−0.41

Sources: Chen and Funk (1990); Finsterwalder and Rentsch (1980); IAHS (ICSI)/UNEP/UNESCO (1988).

differences in the sensitivity of the observed glaciers. There is considerable spatio-temporal variability over short time periods: glaciers around the North Atlantic, for instance, exhibited considerable mass increase during the recent past, and the sensitivity of glaciers in maritime climates is generally up to an order of magnitude higher than the sensitivity of glaciers in arid mountains (Oerlemans and Fortuin, 1992). Statistical analysis indicates that spatial correlations typically have a critical range of about 500 km (Cogley and Adams, 1998; Rabus and Echelmeyer, 1998), and tend to increase markedly with increased length of time period under consideration (as it applies to meteorological variables in general): decadal to secular trends are comparable beyond the scale of individual mountain ranges with continentality of the climate being the main classifying factor (Letréguilly and Reynaud, 1990) besides individual hypsometric effects (Furbish and Andrews, 1984; Tangborn, Fountain and Sikonia, 1990).

Rates of change and acceleration trends comparable to the ones observed during the past 100 years must have taken place before, within the framework of Holocene glacier fluctuations, and hence during times of weak anthropogenic forcing. In analogy to the glacier shrinkage documented during the twentieth century, the Holocene record of Alpine glacier advance/retreat (Figure 15.3) mirrors a (regional, global?) pre-industrial variablility of integrated secular to millennial energy flux toward or from the Earth surface. As indicated by the finding of the Oetztal ice man, the 'warm' or 'high energy' limit of this Holocene variability range may now have been reached, and continuation of the observed trend could soon lead to conditions beyond those occurring during the Holocene precedence. Even though detailed understanding of the processes involved is not yet available, the possibility can no longer be excluded that anthropogenic influences on the atmosphere could now, for the first time, represent a major contributing factor to the observed glacier shrinkage. Wallinga and Van de Val (1998) show that Rhonegletscher (Switzerland) could disappear within decades if the presently observed trend continues into the coming centuries. The extensive modelling study by Oerlemans *et al.* (1998) indeed confirms that this could be the case for many if not most other glaciers of the presently current world-wide mass balance network.

15.4 Concepts for data analysis

Both monitoring and modelling the response of glaciers to climatic change are integral and mutually dependent parts of advanced data-analysis concepts and, hence, of modern observational programmes. They require basic knowledge about glacier characteristics and a description of the physical processes involved. A variety of approaches at different levels of sophistication and process understanding can be applied, depending on the questions to be dealt with and on the detail of glacier parameters available.

Analysis of global area data sets from glacier inventories increasingly uses scaling and volume/area relationships for rough estimates of total volumes etc., especially with respect to sea-level effects (IPCC, 2001). Volume/area correlations, however, relate a variable (area) which is contained within the other (volume) and, hence, suppress the information on the true scatter of ice-thickness data. They tend to hide not only the large uncertainties from limited field measurements of ice thickness but also the related physical processes such as the influence of shear stress, slope, mass balance gradient, mass turnover, englacial temperature and continentality of climate. From a statistical point of view, area/thickness scaling would be preferable to area/volume scaling, but the true physical influences include slope/thickness relations as a function of flow-line length, climatic continentality, mass balance gradients and corresponding mass turnover/mass flux. Digital terrain information will hopefully soon enable such more physically based approaches, even in remote regions where only preliminary inventories without any elevation data currently exist.

With the information in detailed glacier inventories (highest and lowest point, area and length), continuity approaches in combination with assumed step changes in mass balance can be applied for backward as well as forward calculations of glacier characteristics and climate-change effects over time periods of a few decades. Such time intervals correspond to the length of the dynamic response time as it is characteristic for smaller mountain glaciers. A corresponding parameterization scheme proposed by Haeberli and Hoelzle (1995) uses an inverted ice-flow law (strain-rate dependent stress) to derive ice thickness and volume from basal shear stress, slope and vertical glacier extent determining mass turnover and, hence, characteristic strain rate. One of the important results is that the response time primarily depends on the surface slope of the glaciers and typically ranges within a decade to a century for medium-sized ice bodies in mountain areas. The rough and simple approach enables realistic quantitative reconstruction of past multi-decadal to secular mass balances from historical and Holocene fluctuations of glacier length, but neglects transient effects and possible feedbacks.

For more *sophisticated model studies* concerning the response of glaciers to climatic change, a distinction must be made between glacier dynamics and the relation between mass balance and meteorological conditions (Oerlemans, 2001). A number of glaciers have been studied with flow-line models. Such models calculate the glacier properties along a flow line with a numerical method and deliver the temporal variation of the geometry. However, models that calculate explicitly the ice flow need a lot of input data, which is only available for a handful of glaciers. When the interest is in the global picture, simpler methods are needed to estimate the characteristic physical properties of glaciers and ice caps of which only a few geometric characteristics are known (e.g. Haeberli and Hoelzle, 1995; Johanesson, Raymond and Waddington, 1989; Oerlemans, 2001).

The complex chain of dynamic processes linking glacier mass balance and length changes is at present numerically simulated for only a few individual glaciers, which have been studied in great detail (for instance, Greuell, 1992; Oerlemans and Fortuin, 1992; Schmeits and Oerlemans, 1997). A new possibility is to fit dynamically mass balance histories to present-day geometries and historical length change measurements of long-observed glaciers using time-dependent flow models (Oerlemans *et al.*, 1998). This approach not only provides important insights concerning mass balances during past periods which are not documented by direct measurements, but also indicates details of potential time-dependent future evolution including feedbacks from effects of flow dynamics. Many of the thereby considered glaciers with extensive observational networks and abundant input data for modelling could, however, disappear within a few decades if warming continues or even accelerates. Some help may be expected from the combination of repeated high resolution geometric image analysis (photogrammetry, high resolution satellite imagery) and GIS technology which allows for detailed analysis of ice flow and glacier mass balance. The kinematic boundary condition at the glacier surface can be used to derive mass balance at individual points as a function of changes in the surface elevation and horizontal/vertical velocities. Vertical velocity can, in turn, be estimated from basal slope, basal ice velocity and surface strain. In a pilot study on the tongue of Gries glacier (Swiss Alps), the applicability of the relation for modelling area-wide ice flow and mass balance distribution was tested (Kääb and Funk, 1999; cf. also Gudmundsson and Bauder, 1999; Hubbard *et al.*, 2000; Kääb, 2000). So far, however, the applicability of the method seems to be somewhat restricted.

Concerning the question of how quickly glaciers and ice caps may vanish and what their contribution to sea-level change may be, the IPCC Third Assessment Report (IPCC, 2001, especially chap. 11) summarizes some of the most important findings. This assessment is mainly based on *coupling glacier models with GCMs* as developed by Gregory and Oerlemans (1998). Depending on whether area reduction is taken into account or not, the sea-level contribution over the twenty-first century is estimated at a few centimetres to one or two decimetres. Such values represent 10 to 40% of the entire mass contained today in glaciers and ice caps. Because of their thickness, large glaciers will certainly continue to exist into the twenty-second century. Small glaciers with limited altitudinal range, however, may disappear very quickly and in places dramatically change the landscape as well as the water cycle and geomorphological processes (slope stability) in many mountain areas. In the European Alps, for instance, half the ice mass still present in the 1970s and 1980s may have melted by the year 2025, and a considerable part of this scenario has already become reality since about 1980. The interpretation of such extreme regional aspects is assisted by the use of statistically downscaled AOGCMs together with seasonal sensitivity characteristics (SSC) on mass balance models of intermediate complexity. A corresponding study by Reichert, Bengtsson and Oerlemans (2001) demonstrates that mass balances in Norway and Switzerland, respectively, are highly correlated with decadal variations in the North Atlantic oscillation (NAO). This mechanism, which is entirely due to internal variations in the climate system, can explain the strong contrast between recent mass gains for some Scandinavian glaciers as compared with the marked ice losses observed in the European Alps.

Regional scaling with advanced AOGCM calculations reflects part but not all of current process understanding. In particular, two fundamental physical aspects still await inclusion into simulations and assessments: the *firn/ice temperature* effect and the *size/dynamics effect*. Firn warming relates to latent heat exchange involved with percolation and re-freezing of surface melt water; this process makes the rate of firn warming considerably higher than corresponding air temperature change (Haeberli and Alean, 1985; Hooke, Gould and Brzozowski, 1983). Once the firn becomes temperate, mass loss starts taking place with continued warming of the air. This means that the mass balance sensitivity of large firn areas in the Canadian Arctic or in Central Asia, etc., could (a) strongly increase during the coming decades and thereby (b) reduce the regional differences in sensitivity. The large and relatively flat glaciers around the Gulf of Alaska or in Patagonia, which produce the most important melt-water contribution to sea-level rise, have dynamic response times beyond the century scale, and cannot dynamically adjust by tongue retreat to rapid forcing but rather waste down with little area loss. This, in turn, causes the mass balance/altitude feedback to become important. A cumulative surface lowering of about 50 to 100 m within a century or so could, indeed, easily increase the mass balance sensitivity by a factor of 2, correspondingly doubling the surface lowering and, hence, lead to a runaway effect. The corresponding growth in size of the ablation area on such glaciers would probably by far over-compensate the effect of shrinking total areas on small glaciers elsewhere. This means that the sensitivity of the main melt-water producers is likely to increase strongly during the coming decades and strengthen regional differences accordingly. The effects on sea level would, however, be reduced to some degree by the fact that important parts of such large maritime melt-water producers are below sea level.

15.5 Strategies of global climate-related glacier observations

Within the framework of the global climate-related observing systems GTOS/GCOS, a global hierarchical observing strategy (GHOST) was developed to be used for all terrestrial variables. According to this system of tiers, the regional to global representativeness in space and time of the records relating to glacier mass and area should be assessed by more numerous observations of glacier length changes as well as by compilations of regional glacier inventories repeated at time intervals of a few decades – the typical dynamic response time of mountain glaciers. The corresponding global network of glacier sites called the global terrestrial network – glaciers (GTN–G) is expected to evolve over time. It will be structured to allow global and regional analyses of glacier changes and to take advantage of different intensities of measurements at various sites. The initial role of the glacier network is primarily to detect long-term climate change through its impact on glaciers, particularly on a regional basis. With reference to the tier system proposed for global terrestrial observations, the following sites and reported observations are envisioned (Haeberli, Barry and Cihlar, 2000).

Tier 1

Tier 1 sites should have large transects; reporting details are to be determined. These major, intensive experimental sites are set up to emphasize detailed measurements and process understanding across environmental gradients. They should be located with a primary emphasis on spatial diversity. Capturing the range of the major glacier types is a critical priority, but the location within the regions will be opportunistic. Although all Tier 1 data and research findings are important to GCOS/GTOS, special attention should be given to long-term measurements. Tier 1 sites encompass large experimental areas, and various adjustments are required before they can become part of a long-term monitoring programme. The long-term measurements will be a subset of those made during the initial experimental period, but the transition from intensive field studies to continuous monitoring requires careful planning. Some of the observed glaciers (for instance, those in the Pyrenees, Alps, Scandinavia and Svalbard) provide large transects and could later form part of Tier 1 observations.

Tier 2

Tier 2 sites will provide extensive and process-oriented glacier mass balance studies within major climatic zones, with annual reporting. The sites make possible glacier mass balance studies within the major climatic zones. Ideally, Tier 2 sites should be located near the centre of the range of environmental conditions (though not necessarily near the geographical centre) of the zone which they are representing. The actual locations will depend more on existing infrastructure and logistical feasibility rather than on strict spatial guidelines, but there is a need to capture a broad range of climatic zones. There are about ten glaciers with intensive research and observation activities that represent Tier 2 sites. Storglaciären in northern Sweden is an example of such a site.

Tier 3

Tier 3 will provide regional glacier mass change within major mountain systems, i.e. reduced stake networks, with annual reporting. Tier 3 sites are intended to sample the range of environmental variation present in the glaciers within climatic zones or regions. There is no requirement for spatial representativeness of the glaciers in this tier. There are numerous potential Tier 3 sites (about 50 glaciers where annual mass balance studies are conducted) to reflect regional patterns of glacier mass change within major mountain systems, but they may not be optimally distributed. As a result, some glacier types may have more potential Tier 3 sites than are needed for GCOS/GTOS. Other types may have too few sites, or none at all, and thus GCOS/GTOS will need to stimulate efforts to enhance and balance the network.

Tier 4

This level will comprise long-term observations of glacier length change data with a minimum of about ten sites within each of the mountain ranges, selected according to size and dynamic response: pluri-annual reporting (frequency to be determined). At this level, spatial representativeness is the highest priority. Approximately 800 glaciers where only length is measured are compatible with Tier 4. Because access is infrequent, they can be located wherever necessary to ensure representativeness. The locations of Tier 4 sites should be based on statistical considerations. It is impractical to prescribe one statistical design for all countries. Hence, individual participating organizations would be responsible for locating the sites, and may choose either a systematic or a stratified random approach (or a combination, depending on the variable or the glacier system). For the glacier network, long-term observations of glacier length change at about ten sites within each of the mountain ranges will be selected according to size and dynamic response from the existing set of sites where glacier length is monitored.

Tier 5

Glacier inventories will be repeated at time intervals of a few decades by using satellite remote sensing: there will be continuous upgrading and analyses of existing and newly available data. For the most part, these fields include glacier inventories repeated at time intervals of a few decades by using satellite remote sensing. Satellite observations are usually for area averages (for areas $<10^2$ to $>10^7\,m^2$, depending on the sensor and the glacier variable), while ground observations are point values. Some variables require surface observations, even for Tier 5. The implementation of Tier 5 requires international collaboration, both in the space and ground components, to produce the required data sets. The preparation of data products from satellite measurements must be based on a long-term programme of data acquisition, archiving, product generation and quality control. Discussions are now underway in the Committee on Earth Observation Satellites (CEOS) to set up such a system. In particular, co-ordination is needed with the GLIMS project which will map changes in the areas of selected glaciers world-wide. Glacier observations made from space-borne instruments are being used in various research investigations. However, routine glacier monitoring (Tier 5) using the advanced space-borne thermal emission and reflection radiometer (ASTER) has been proposed by the US Geological Survey (Flagstaff, AZ), in conjunction with the EROS Data Centre, the National Snow and Ice Data Centre and regional analysis centres.

A network of 60 glaciers representing Tiers 2 and 3 is now established. This step closely corresponds to the data compilation published so far with the biennial *Glacier Mass Balance Bulletin* and also guarantees annual reporting in electronic form. Such a sample of reference glaciers provides information on presently observed rates of change in glacier mass, corresponding acceleration trends and regional distribution patterns. Long-term changes in glacier length must be used to assess the representativity of the small sample of values measured during a few decades with the evolution at a global scale and during previous time periods. As explained above, this can be done by (1) inter-comparison between curves of cumulative glacier length change from geometrically similar glaciers; (2) application of continuity considerations for assumed step changes between steady-state conditions reached after the dynamic response time; and (3) dynamic fitting of time-dependent flow models to present-day geometries and observed long-term length change. New detailed glacier inventories are now being compiled in areas not covered so far or, for comparison, as a repetition of earlier inventories. This task is greatly facilitated by the launching of the ASTER\GLIMS programme (Kieffer *et al.*, 2000). Remote sensing at various scales (satellite imagery, aerophotogrammetry) and GIS technologies must be combined with digital terrain information (Kääb *et al.*, 2002; Paul *et al.*, 2002) in order to overcome the difficulties of earlier satellite-derived preliminary inventories (area determination only) and to reduce the cost and time of compilation. In this way, it should be feasible to reach the goals of global observing systems in the years to come.

References

Bahr, D. B. 1997. Global distribution of glacier properties: a stochastic scaling paradigm. *Water Resources Res.* **33/7**, 1669–79.

Beniston, M., Haeberli, W., Hoelzle, M. and Taylor, A. 1997. On the potential use of glacier and permafrost observations for verification of climate models. *Ann. Glaciol.* **25**, 400–6.

Cao, M. S. 1998. Detection of abrupt changes in glacier mass balance in the Tien Shan Mountains. *J. Glaciol.* **44** (147), 352-8.

Casassa, G., Espizua, L. E., Francou, B., Ribstein, P., Ames, A. and Alean, J. 1998. Glaciers in South America. In Haeberli, W., Hoelzle, M. and Suter, S. eds., *Into the Second Century of Worldwide Glacier Monitoring: Prospects and Strategies*. Studies and Reports in Hydrology 56. Paris, UNESCO, pp. 125–46.

Chen, J. and Funk, M. 1990. Mass balance of Rhonegletscher during 1882/83–1986/87. *J. Glaciol.* **36** (123), 199–209.

Cogley, J. G. and Adams, W. P. 1998. Mass balance of glaciers other than the ice sheets. *J. Glaciol.* **44** (147), 315–25.

Dyurgerov, M. B. and Meier, M. F. 1997a. Mass balance of mountain and subpolar glaciers: a new global assessment for 1961–1990. *Arctic & Alpine Res.* **29** (4), 379–91.

1997b. Year-to-year fluctuations of global mass balance of small glaciers and their contribution to sea level. *Arctic & Alpine Res.* **29**, 392–402.

Echelmeyer, K. A. *et al.* 1996. Airborne surface profiling of glaciers: a case-study in Alaska. *J. Glaciol.* **42** (142), 3–9.

Finsterwalder, R. and Rentsch, H. 1980. Zur Höhenänderung von Ostalpengletschern in Zeitraum 1969–1979. *Zeits. Gletscherkunde Glazialgeol.* **16** (1), 111–15.

Fitzharris, B. B. *et al.* 1996. *The Cryosphere: Changes and their Impacts. Climate Change 1995: Impacts, Adaptations and Mitigation of Climate Change: Scientific-Technical Analyses.* Contribution of Working Group II to the Second Assessment Report of the Intergovernmental Panel on Climate Change. Cambridge University Press, pp. 241–65.

Forel, F.-A. 1895. Les variations périodiques des glaciers. Discours préliminaire. *Arch. Sci. Phys. Naturelles, Genève* **XXXIV**, 209–29.

Furbish, D. J. and Andrews, J. T. 1984. The use of hypsometry to indicate long-term stability and response of valley glaciers to changes in mass transfer. *J. Glaciol.* **30** (105), 199–211.

Gregory, J. M. and Oerlemans, J. 1998. Simulated future sea-level rise due to glacier melt based on regionally and seasonally resolved temperature changes. *Nature* **391**, 474–6.

Greuell, W. 1992. Hintereisferner, Austria: mass balance reconstruction and numerical modelling of historical length variation. *J. Glaciol.* **38** (129), 233–44.

Gudmundsson, G. H. and Bauder, A. 1999. Towards an indirect determination of the mass-balance distribution of glaciers using the kinematic boundary condition. *Geograf. Ann.* **81A** (4), 575–83.

Haeberli, W. 1983. Permafrost – glacier relationships in the Swiss Alps today and in the past. *Proceedings of the Fourth International Conference on Permafrost*, pp. 415–20.
1996. Glacier fluctuations and climate change detection. *Geograf. Fis. Din. Quat.* **18**, 191–9.

Haeberli, W. and Alean, J. 1985. Temperature and accumulation of high altitude firn in the Alps. *Ann. Glaciol.* **6**, 161–3.

Haeberli, W. and Burn, C. 2002. Natural hazards in forests – glacier and permafrost effects as related to climate change. In: Sidle, R.C., ed., *Environmental Change and Geomorphic Hazards in Forests.* IUFRO Research Series **9**, Oxford, CABI Publishing, pp. 167–202.

Haeberli, W. and Hoelzle, M. 1995. Application of inventory data for estimating characteristics of and regional climate-change effects on mountain glaciers: a pilot study with the European Alps. *Ann. Glaciol.* **21**, 206–12. Russian translation in *Data of Glaciol. Studies, Moscow* **82**, 116–24.

Haeberli, W., Hoelzle, M. and Suter, S. (eds.) 1998. Into the second century of worldwide glacier monitoring: prospects and strategies. Studies and Reports in Hydrology 56. Paris, UNESCO.

Haeberli, W., Barry, R. and Cihlar, J. 2000. Glacier monitoring within the global climate observing system. *Ann. Glaciol.* **31**, 241–6.

Haeberli, W., Frauenfelder, R., Hoelzle, M. and Maisch, M. 1999. On rates and acceleration trends of global glacier mass changes. *Geograf. Ann.* **81A**, 585–91.

Hagen, J. O., Zanon, G. and Martínez de Pisón, E. 1998. Glaciers in Europe. In Haeberli, W., Hoelzle, M. and Suter, S., eds., *Into the Second Century of Worldwide Glacier Monitoring: Prospects and Strategies.* Studies and Reports in Hydrology 56, Paris, UNESCO, pp. 147–66.

Hastenrath, S. and Chinn, T. J. H. 1998. Glaciers in Africa and New Zealand. In Haeberli, W., Hoelzle, M. and Suter, S., eds., *Into the Second Century of Worldwide Glacier Monitoring: Prospects and Strategies.* Studies and Reports in Hydrology 56, Paris, UNESCO, pp. 167–75.

Hooke, R., LeB., Gould, J. E. and Brzozowski, J. 1983. Near-surface temperatures near and below the equilibrium line on polar and subpolar glaciers. *Zeits. Gletscherkunde & Glazialgeol.* **19**, 1–25.

Hubbard, A. *et al.* 2000. Glacier mass-balance determination by remote sensing and high-resolution modelling. *J. Glaciol.* **46** (154), 491–8.

IAHS(ICSI)/UNEP/UNESCO. 1988. Haeberli, W. and Müller, P., eds., *Fluctuations of Glaciers 1980–1985.* Paris.

 1989. Haeberli, W., Bösch, H., Scherler, K., Østrem, G. and Wallén, C. C., eds., *World Glacier Inventory – Status 1988.* Nairobi.

 1993. Haeberli, W. and Hoelzle, M., eds., *Fluctuations of Glaciers 1985–1990.* Paris.

 1998. Haeberli, W., Hoelzle, M., Suter, S. and Frauenfelder, R., eds., *Fluctuations of Glaciers 1990–1995.* Zurich.

 2001. Haeberli, W., Frauenfelder, R. and Hoelzle, M., eds., *Glacier Mass Balance Bulletin no. 6.* World Glacier Monitoring Service, University and ETH Zurich.

IAHS(ICSI)/UNESCO. 1967. (Kasser, P., ed., *Fluctuations of Glaciers 1959–1965.* Paris.

 1973. Kasser, P., ed., *Fluctuations of Glaciers 1965–1970.* Paris.

 1977. Müller, F., ed., *Fluctuations of Glaciers 1970–1975.* Paris.

 1985. Haeberli, W., ed., *Fluctuations of Glaciers 1975–1980.* Paris.

IPCC. 2001. *Climate Change 2001. The Scientific Basis.* Contribution of Working Group I to the Third Assessment Report of the Intergovernmental Panel on Climate Change, Cambridge University Press.

Johannesson, T., Raymond, C. F. and Waddington, E. D. 1989. Time-scale for adjustment of glaciers to changes in mass balance. *J. Glaciol.* **35** (121), 355–69.

Kääb, A. 2000. Photogrammetric reconstruction of glacier mass balance using a kinematic ice-flow model: a 20-year time series on Grubengletscher, Swiss Alps. *Ann. Glaciol.* **31**, 45–52.

Kääb, A. and Funk, M. 1999. Modelling mass balance using photogrammetric and geophysical data: a pilot study at Griesgletscher, Swiss Alps. *J. Glaciol.* **45** (151), 575–83.

Kääb, A., Paul, F., Maisch, M., Hoelzle, M. and Haeberli, W. 2002. The new remote-sensing-derived Swiss glacier inventory: II. First results. *Ann. Glaciol.* **34**, 362–6.

Kieffer, H. *et al.* 2000. New eyes in the sky measure glaciers and ice sheets. *EOS, Trans. Am. Geophys. Union* **81/24**, 265, 270–1.

Leiva, J. C. and Cabrera, G. A. 1996. Glacier mass balance analysis and reconstruction in the Cajon del Rubio, Mendoza, Argentina. *Zeits. Gletscherkunde & Glazialgeol.* **32**, 101–7.

Letréguilly, A. and Reynaud, L. 1990. Space and time distribution of glacier mass balance in the northern hemisphere. *Arctic & Alpine Res.* **22** (1), 43–50.

Meier, M. F. and Bahr, D. B. 1996. Counting glaciers: use of scaling methods to estimate the number and size distribution of the glaciers on the world. In Colbeck, S. C., ed., *Glaciers, Ice Sheets and Volcanoes: a Tribute to Mark F. Meier.* CRREL Special Report 96–27, pp. 1–120.

Oerlemans, J. 1993. A model for the surface balance of ice masses: part I. Alpine glaciers. *Zeits. Gletscherkunde & Glazialgeol.* **27/28**, 63–83.

1994. Quantifying global warming from the retreat of glaciers. *Science* **264**, 243–5.

2001. *Glaciers and Climatic Change.* Rotterdam, Balkema.

Oerlemans, J. and Fortuin, J. P. F. 1992. Sensitivity of glaciers and small ice caps to greenhouse warming. *Science* **258**, 115–18.

Oerlemans, J., *et al.* 1998. Modelling the response of glaciers to climate warming. *Climate Dyn.* **14**, 267–74.

Ommanney, C. S. L., Demuth, M. and Meier, M. F. 1998. Glaciers in North America. In Haeberli, W., Hoelzle, M. and Suter, S. eds., *Into the Second Century of Worldwide Glacier Monitoring: Prospects and Strategies.* Studies and Reports in Hydrology, **56**, Paris, UNESCO, pp. 113–23.

Paul, F., Kääb, A., Maisch, M., Kellenberger, T. and Haeberli, W. 2002. The new remote-sensing-derived Swiss glacier inventory: I. Methods. *Ann. Glaciol.* **34**, 355–61.

Rabus, B. T. and Echelmeyer, K. A. 1998. The mass balance of McCall Glacier, Brooks Range, Alaska, U. S. A.; its regional relevance and implications for climate change in the Arctic. *J. Glaciol.* **44** (147), 333–51.

Reichert, B. K., Bengtsson, L. and Oerlemans, J. 2001. Midlatitude forcing mechanisms for glacier mass balance investigated using general circulation models. *J. Climate* **14**, 3767–84.

Schmeits, M. J. and Oerlemans, J. 1997. Simulation of the historical variations in length of Unterer Grindelwaldgletscher, Switzerland. *J. Glaciol.* **43** (143), 152–64.

Tangborn, W. V., Fountain, A. G. and Sikonia, W. G. 1990. Effect of area distribution with altitude on glacier mass balance – a comparison of North and South Klawatti Glaciers, Washington State, U.S.A. *Ann. Glaciol.* **14**, 278–82.

Tsvetkov, D. G. Osipova, G. B., Xie, Z., Wang, Z., Ageta, Y. and Baast, P. 1998. Glaciers in Asia. In Haeberli, W., Hoelzle, M. and Suter, S., eds. *Into the Second Century of Worldwide Glacier Monitoring: Prospects and Strategies.* Studies and Reports in Hydrology 56, Paris, UNESCO, pp. 177–96.

Wallinga, J. and Van de Wal, R. S. W. 1998. Sensitivity of Rhonegletscher, Switzerland, to climate change: experiments with a one-dimensional flowline model. *J. Glaciol.*, **44** (147), 383–93.

Warrick, R.A. *et al.* 1996. *Changes in Sea Level. Climate Change 1995: The Science of Climate Change.* Contribution of Working Group I to the Second Assessment Report of the Intergovernmental Panel on Climate Change. Cambridge University Press, pp. 359–405.

Weidick, A. and Morris, E. 1998. Local glaciers surrounding the ice sheets. In Haeberli, W., Hoelzle, M. and Suter, S. eds., *Into the Second Century of Worldwide Glacier Monitoring: Prospects and Strategies.* Studies and Reports in Hydrology 56, Paris, UNESCO, pp. 197–207.

WMO. 1997. GCOS/GTOS plan for terrestrial climate-related observation. GCOS 32, version 2.0, WMO/TD-796, UNEP/DEIA/TR, 97–7.

16

Glaciers and the study of climate and sea-level change

MARK B. DYURGEROV AND MARK F. MEIER
Institute of Arctic and Alpine Research (INSTAAR),
University of Colorado

16.1 Introduction

Glacier variations have been of interest for hundreds of years because they can be sensitive indicators of changes in climate. More recently, the role of glacier runoff on the hydrology of mountain regions and the impact of glacier wastage on global sea level have become active areas of scientific effort.

We analyse observational data and the state of health of mountain and sub-polar glaciers for the last several decades, and connection of their changes to climate fluctuations and the global water cycle. We deal here with all glaciers on Earth, excluding the Greenland and Antarctic ice sheets. This analysis is mainly based on our most recently updated time series of mass balance components (Dyurgerov (2002); see http://instaar.colorado.edu/other/occ_papers.html). Every effort has been made to include data from all global sources of information, to check data quality and to eliminate errors.

The quantity and quality of data are far better for the northern hemisphere (especially Europe, Canada, USA and the former Soviet Union (FSU), than for the southern hemisphere. About 70% of the measurements have been carried out in Scandinavia, the Alps, the mountains of the USA, Canada, and the FSU, and the other 30% are sparsely distributed in many other mountain and sub-polar regions (Figure 16.1). For many regions, data on glacier regime are almost completely lacking (e.g., Karakorum, Tibetan Plateau, Northern Himalayas, Hindukush and Hindurage, Sayany, south-eastern Siberia, Koryak Range and other ranges in north-eastern Siberia, Aleutian Islands, De Long Islands, Jan Mayen Island, Tierra del Fuego and several regions in the Andes). Very few data are available for individual glaciers and ice caps around the Antarctic and Greenland ice sheets.

The histogram in Figure 16.2. shows that the number of mass balance measurements in the world is currently decreasing, even though some new programmes have been organized in South America and Iceland. A serious problem is that several long-term mass balance programmes were discontinued in the FSU (Caucasus, Polar Ural, Pamir, Tien Shan, Djungariya, Severnaya Zemlya) for economic reasons. In the Alps (Limmern and Plattalva glaciers), East Africa (Lewis glacier) and Canada (Sentinel glacier), long-term programmes

Mass Balance of the Cryosphere: Observations and Modelling of Contemporary and Future Changes, eds. Jonathan L. Bamber and Antony J. Payne. Published by Cambridge University Press. © Cambridge University Press 2003.

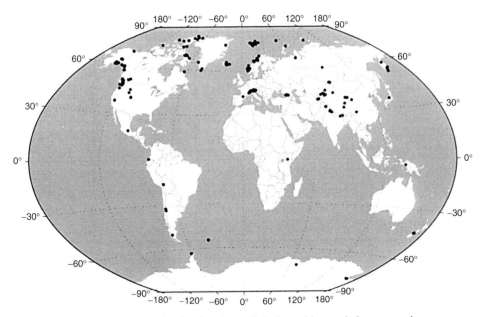

Figure 16.1. Map showing locations of glaciers with mass balance records.

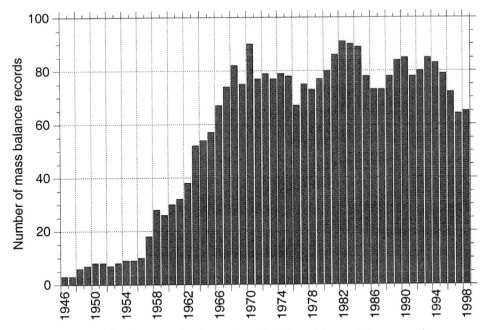

Figure 16.2. Histogram showing number of glaciers with mass balance records.

have also been discontinued. These cancellations have affected global glacier monitoring and related programmes.

Preference in our analysis has been given to data obtained by standard glaciological methods (see, e.g., Mayo, Meier and Tangborn, 1972; Meier, 1962; Østrem and Brugman, 1991). Indirect measurements and estimations, such as those derived by mapping, hydrological methods or calculated by climatic models, are generally avoided here. A few exceptions have been made, e.g. the mass balance of Grosse Aletsch glacier has been measured by a hydrological method and included in the results of mass balance measurements by the World Glacier Monitoring Service, WGMS (IAHS (ICSI)-UNESCO, *Fluctuations of Glaciers*; listed here as *FoG*, 1967, 1973, 1977, 1985, 1988, 1993, 1998). Also, recent volume change information obtained by laser altimeter profiling of glaciers in Alaska and adjacent Canada (Arendt *et al.*, 2002) have been included.

It is important to note that collections of glacier mass balance records have been published and/or used for near-global analysis in previous decades by Braithwaite (2001), Cogley and Adams (1998), Cogley *et al.* (1995), Collins (1984), Dyurgerov and Meier (1997a,b), Ommanney (in Østrem and Brugman, 1991) and Wood (1988).

16.2 Concept and terms

The linkage between climate variation and glacier fluctuation involves the fluxes of ice and water to and from the glacier surface, the resulting net mass balance at the surface, and a dynamic response that alters the glacier geometry and ultimately feeds back to the mass balance (Meier, 1965). The dynamic response time for glaciers typically ranges from decades to a century or more (Jóhannesson, Raymond and Waddington, 1989; Nye, 1960). It is virtually impossible to calculate response times for all of the glaciers in the world, let alone those used in this analysis; the response times are not simple functions of glacier size (Bahr *et al.*, 1998). Harrison *et al.* (2001) also point out that response time is affected by the choice of areas over which mass balance is calculated. For mass balance fluctuations at time scales much shorter than the response time, the correction for dynamics becomes small, but for longer trends the dynamic response becomes appreciable. Elsberg *et al.* (2001) point out that mass balance measurements applied to a known glacier surface that is continuously changing produce correct measures of the water balance (and thus the contribution to sea level), but are less perfect measures of the relation of these changes to the varying climate. A better measure to relate to climate change is the *reference surface mass balance* in which the glacier area (and the area-altitude distribution) is held constant during the period of study (Elsberg *et al.*, 2001).

We discuss glacier mass balance data for three objectives. One is the glacier regime itself in terms of variables listed in Table 16.1. Another application is the climatic interpretation of these data. The third application is to the glacier contribution to global water cycle and sea-level rise. We use 'conventional' annual/net mass balance data, which means that the values include continuous change in glacier area due to climate and dynamic adjustment. In only a few cases do we use 'reference surface' mass balance data, which means that

Table 16.1. *Variables used in this study and their definitions.*

Mass balance data are in water equivalent units.

Variable	Unit	Definition	Data explanation	Selected sources
S	km^2	surface area of an entire glacier determined from a topographic map or digital elevation model	published in *FoG* volumes, or taken from many other sources, published or archived	*FoG* (vols. 1–7); Dyurgerov (2002).
s	km^2	surface area of an elevation band (may vary between 10 and 200 m)	published in *FoG* volumes, or taken from many other sources, published or archived	*FoG* (vols.1–7); Dyurgerov (2002)
b	mm or m per year	specific values; we do not differentiate between net (b_n) and annual mass balances (b_a)	the difference between mass gain and loss measured in the majority of cases by the glaciological method (stakes and pits) at many locations, averaged for each elevation band s, and then calculated for the entire glacier area, S[a]	*FoG* (vols. 1–7); Dyurgerov (2002)
b_w	mm or m per year	specific winter mass balance	the resulting mass gain of a glacier, mainly due to snow accumulation over period from October 1 to May 31 in northern hemisphere[a]	*FoG* (vols.1–7); Dyurgerov and Meier (1999); Dyurgerov (2002)
c_c	mm or m per year	net accumulation of snow, firn and ice over hydrologic year	change in glacier mass determined above the ELA at the end of a hydrologic year[a]	*FoG* (vol. 1–7); Dyurgerov and Meier (1999); Dyurgerov (2002)
c_t	mm or m per year	annual accumulation at time, t, usually the end of hydrologic year	accumulation of snow, firn, ice over an entire glacier determined at the end of a hydrologic year[a]	*FoG* (vols. 1–7); Dyurgerov and Meier (1999); Dyurgerov (2002)
b_s	mm or m per year	specific summer mass balance	the resulting mass losses by a glacier, mainly due to snow-ice melting and evaporation over the period June 1 to September 30 in northern hemisphere	*FoG* (vols. 1–7); Dyurgerov and Meier (1999); Dyurgerov (2002)

Symbol	Units			References
a_a	mm or m per year	net ablation of snow, firn and ice over hydrologic year	change in glacier mass determined below the ELA at the end of a hydrologic year	FoG (vols. 1–7); Dyurgerov and Meier (1999); Dyurgerov (2002)
a_t	mm or m per year	annual ablation at time, t, usually the end of hydrologic year	ablation of snow, firn, ice over an entire glacier determined at the end of a hydrologic year	FoG (vol. 1–7); Dyurgerov and Meier (1999); Dyurgerov (2002)
b_{20}	mm or m per year	annual or net mass balance averaged for time series of 20 years and longer	calculated from the time series of mass balance of all glaciers with time series of 20 years and longer	Dyurgerov (2002).
b_{30}	mm or m per year	the same as above for time series 30 years and longer	the same as above for time series of 30 years and longer	Dyurgerov (2002)
α	mm or m per year in water equivalent	mass turnover, $\alpha = (|b_w| + |b_s|)/2$, or $(b_w - b_s)/2$ in case b_s is considered negative value	calculated from the data of direct measurements of seasonal mass balance components, b_w and b_s	Meier (1984); Dyurgerov and Meier (1999); Dyurgerov (2002).
ELA	m or km above sea level	equilibrium-line altitude, the boundary between two parts of a glacier, accumulation (mass gain area) and ablation (mass loss area) determined at the very end of ablation season	determined from direct measurements or from the graph showing the mass balance patterns distributed vs. elevation	FoG (vols. 1–7); Dyurgerov (2002).
AAR	%	accumulation area ratio $S_{ac}/S \times 100$, where S_{ac} is the surface area of a glacier above ELA, and S is the surface area of entire glacier, both determined from a map	accumulation area at the very end of ablation season determined from ELA position and corresponding topographic map; or directly by remote sensing or any other methods	Meier (1962); Mayo et al. (1972); Østrem and Brugman, (1991); Dyurgerov (2002).
⟨ ⟩	averaged values			

[a]Including, theoretically, refrozen and liquid water trapped in deep firn layers, below the reference layer of previous summer (internal accumulation.)

an initial area of a glacier at the beginning of the programme is used for mass balance calculations (Elsberg *et al.*, 2001).

From more than 280 mass balance time series available in our completed data set, about 230 time series are relatively short, less than 20 years. We found that over these short periods changes in area and surface topography are small, and the differences between *reference* and *conventional* mass balances are relatively small. In 48 time series the duration of continuous measurements is 20 years and longer, with the longest of 40–50 years. Substantial changes in area and surface topography have been found for these glaciers. For all of these time series we have initial and repeated surveys of area change, and, for many, the changing distribution with elevation. For 40 long-term time series with annual mass balance data we introduce surface area annually, thus there is no difference between *conventional* and *reference* mass balances in these time series. For the majority of these glaciers, surface areas were reported in publications as step-wise changes corresponding to the years of mapping or other repeated measurements (e.g., geodetic, GPS). In these cases, the changes happened smoothly. According to this we re-calculated year-by-year glacier areas by linear interpolation between successive surveys. Corresponding to these, annual mass balances (seasonal mass balance components as well) were re-calculated as well as possible (Dyurgerov, 2002).

We were unable to determine annual mass balances on the changing areas for eight glaciers with time series of 20 years and longer. Three of these glaciers are in the Arctic (White Glacier, Meighen and South ice caps), and do not show visible change in area. Silvretta glacier in the Swiss Alps shows no change in area because this glacier has been close to a steady-state with a mass balance of only 40 mm per year averaged over 39 years, 1960–1999. For four other glaciers (Lemon Creek, Juneau ice field; St Sorlin, Alps; Kara Batkak, Tien Shan; and Enchauren Norte, Central Andes), area change may be substantial as their averaged mass balance was negative, but data on area change were not published. For these four glaciers the systematic error in application mass balance data to calculate cumulative water loss may be of the order of 10–20%, according to an estimation made for South Cascade glacier (Elsberg *et al.*, 2001).

Less than 300 glaciers, out of a current total of more than 160 000 glaciers on Earth (Meier and Bahr, 1996), have been measured at one time or another in order to determine annual or multi-annual mass balances. To calculate mass balance, which is closely related to volume change, glacier area and its distribution by elevation has to be determined. Along with the annual (or net) mass balance measurements, seasonal (winter and summer) mass balance components were determined on many glaciers. Seasonal balances show mass gain and loss over cold (period of accumulation) and warm (period of ablation) parts of a hydrological year (usually defined as October 1 to September 30 in the northern hemisphere). Winter and summer mass balance components contain important climatic information on processes causing glacier volume change in different geographical regions (Dyurgerov and Meier, 1999, 2000). Other meaningful parameters were measured in the process of mass balance studies, including equilibrium-line altitude (ELA) and accumulation area ratio (AAR). These also indicate changes in response to climate fluctuations or weather conditions (Kuhn, 1980).

We apply commonly used terms and definitions to describe the regime of glaciers and change in time in relation to climate. Changes in glacier volume (cumulative mass balance) are interpreted in terms of water equivalent as residuals in the water balance equation, and to show glacier contributions to the water cycle in scales of local and regional basins, and to the global scale as the glacier contribution to sea-level change (Meier, 1969, 1984).

16.3 Glacier area and change

The total area of glaciers on Earth is not certain. Many individual glaciers are situated close to but separate from the main ice sheet of Antarctica; these occur in marginal climatic environments and have time constants of response to climate change different from those of the ice sheets. Thus they belong to the global population of small glaciers, not the major ice sheet. The boundaries between them and the ice sheet are sometimes hard to determine. In many cases these ice masses are joined to ice shelves or connected to the major ice sheets by ice shelves, e.g. Spartan glacier in Alexandra Island, or Berkner Island, which is surrounded by the Ronne-Filchner ice shelf. The aggregate area of these local ice caps around Antarctica may comprise from 70×10^3 km^2 to 700×10^3 km^2 (perhaps 5% of the entire area of the Antarctic ice sheet; see Weidick and Morris (1998)).

Meier and Bahr (1996) calculated a figure of 680×10^3 km^2 for the total area of 'small' glaciers, and this is the figure used here. This, however, is probably a minimum estimate because: (1) Meier and Bahr used scaling from the existing world glacier inventory data which in many areas did not include the smallest glaciers, and (2) the area of Antarctic glaciers actively involved in the hydrological cycle could be considerably larger than the 70×10^3 km^2 they included.

Glacier area as recorded in the database changes for two main reasons: the first is new findings, including new maps or updating of previous surveys (e.g. updated inventory of New Zealand glaciers, see Hastentrath and Chinn (1998)); and the second is the observed change of glacier area due to climate change (e.g. glaciers in tropics, see Kaser (1999)). A new inventory of Nepal and Bhutan glaciers was completed in 2001, which gives new and more precise results of 5324 km^2 (Mool, Bajracharya and Joshi, 2001a), and 1317 km^2 (Mool *et al.*, 2001b), respectively, compared with about 6000 km^2 and 1500 km^2 for the first and the second regions published in a previous inventory (IAHS, 1989).

Total surface area of glaciers covered by measurements has been the subject of large fluctuations from year to year, from less than 3000 km^2 to more than 8000 km^2 (Figure 16.3). One of the reasons for these fluctuations has been the infrequent measurement of large glaciers with short-term (one or a few years) of mass balance records, such as Barnes ice cap (area 3090 km^2, in 1962–64), Vavilova ice cap (area 1817 km^2 in 1974–81 and 1986–88), Columbia glacier (area 1090 km^2 in 1978), and others. The aggregate area of glaciers covered by mass balance measurements averaged 5725 km^2 for the 1961 to 1998 period, which is only 0.8–0.9% of the area of all small glaciers on Earth.

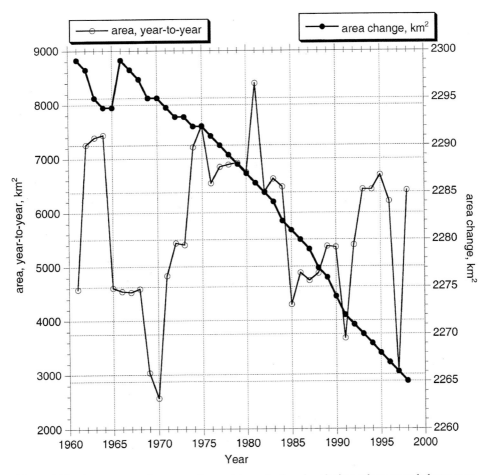

Figure 16.3. Aggregate surface area of mountain and sub-polar glaciers where mass balance measurements were carried out showing all records including those with large annual variability (left hand scale) and area change (right hand scale) calculated for the 36 glaciers having long-term mass balance records.

To estimate the area change due to climate forcing, 36 glaciers with mass balance records of 38 years (1961–1998) have been selected from the database (Dyurgerov, 2002). The aggregate area of these glaciers decreased from 2299 to 2265 km² , or 1.5% (Figure 16.3). Large differences in area reduction have been observed in different regions. In the European Alps it has been estimated as much as 35–40% over 1850–1970\80, or 0.3% per year (Haeberli, Maisch and Paul, 2002). The larger glaciers may have experienced relatively smaller changes in area, even though their thinning rate may have been much larger. The large glaciers bordering the Gulf of Alaska have shown area decrease rates from the mid 1950s to the mid 1990s of 0.02% per year, but the rate has increased to 0.07% per year since the mid 1990s (Arendt *et al.*, 2002). In some regions of central Asia, shrinkage in area is much faster and reached 1% per year (Makarevich and Liu, 1995), and may be even more extreme in the tropics (Hastenrath and Greischnar, 1997; Kaser, 1999). Extrapolating these results to other small glaciers, the total loss of glacier area during the 1961–1998 period is of the order of 10^4 km^2.

16.4 Glacier regime

16.4.1 Seasonal mass balance components

Several components of glacier mass balance are used in glaciological measurements and data presentation (*FoG*, 1967, 1973, 1977, 1985, 1988, 1993, 1998). The seasonal mass balance components, specific winter mass balance (b_w) and specific summer mass balance (b_s), are especially important because they relate to climatic variables of winter precipitation and summer temperature. We have also considered annual snow-ice accumulation (c_t) and annual ablation (a_t). c_t is the true estimation of annual amount of precipitation as snow/ice (b_w minus winter ablation plus summer accumulation) deposited on a glacier surface; a_t is the true estimation of yearly surface ice loss by runoff and evaporation but generally excluding ice calving. All mass balance components are expressed in water equivalent units. Thus, $b_w + b_s = c_t + a_t = b$, which is the annual or net mass balance (b_s and a_t are considered here as negative values). The difference between c_t and b_w and between a_t and b_s may be substantial in some geographical locations (Figures 16.5(a),(b)). It tends to be greater in regions with continental climate conditions, where snow accumulation and ablation may occur simultaneously (e.g., Shumskiy glacier, Figures 16.4 and 16.5). In extremely continental climate conditions with very low total precipitation and/or monsoon climates, as in the mountain ranges of central Asia, more than 75% of the precipitation may occur on a glacier in summer, mostly as snow (Ageta and Higuchi, 1984). This makes the difference between b_w and c_t large compared with maritime climate conditions (Xie Zichu *et al.*, 1999).

The traditional Alpine system of glacier mass balance measurements and data presentation is still in use on the majority of glaciers in the Alps and in the Himalayas, particularly in India and Pakistan (see *FoG*, tables 'C'). That system uses 'net accumulation' (a_c) and 'net ablation' (a_a), which are not seasonal components but net mass balances averaged over the accumulation and ablation areas, respectively. The differences between them and b_w and b_s are large, as is shown for Shumskiy glacier (Figures 16.4 and 16.5) and for Storglaciären (Figure 16.6). These a_c and a_a components permit calculation of the net mass balance, but they cannot be used directly for climate analysis. The advantage of using two or more pairs of components is the fact that a correlation between them may allow one to reconstruct information of climatic significance (Figures 16.4(b), 16.5(b)), such as winter balance (or winter precipitation), summer balance (or summer temperature and precipitation) or annual quantities of these variables. Unfortunately, such comparisons cannot be performed world-wide as simultaneous measurements of these variables are limited to a few glaciers.

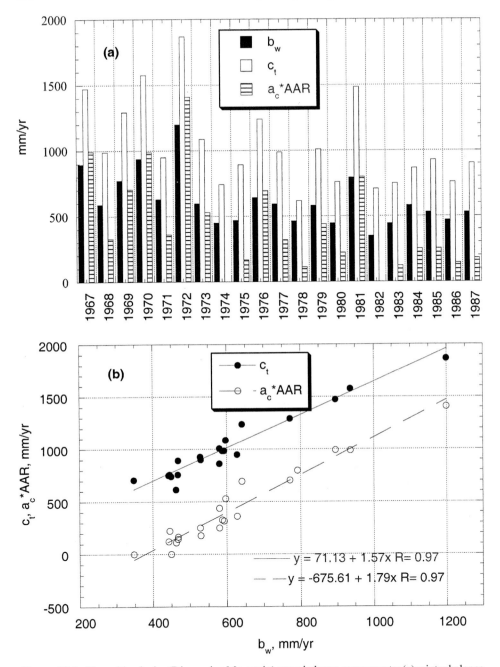

Figure 16.4. Shumskiy glacier (Djungariya Mountain) mass balance components: (a) winter balance (b_w), annual accumulation (c_t) and net accumulation recalculated as averages for the entire glacier area (a_c*AAR); (b) linear regression between b_w and the two other components.

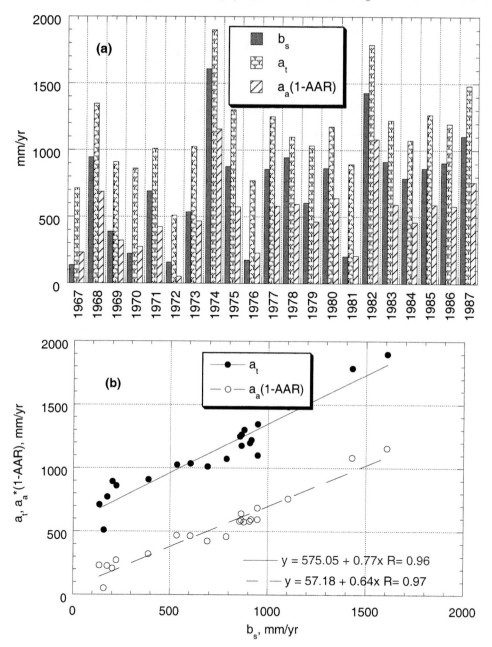

Figure 16.5. Shumskiy glacier mass balance components: (a) summer balance (b_s), annual ablation (a_t) and net ablation recalculated for entire glacier area a_a (1-AAR); (b) linear correlation between b_s and the two other components.

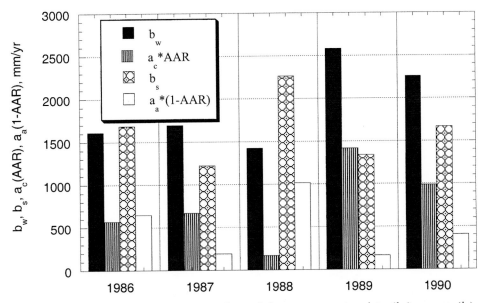

Figure 16.6. Storglaciären (Sweden) seasonal mass balance components: winter (b_w), summer (b_s), net accumulation (a_c^*AAR and net ablation (a_a^* (1−AAR); a_c and a_a recalculated as averages for the entire area of the glacier. The coefficients of regression between b_w and a_c^*(AAR), and b_s vs. a_a^*(1−AAR) are both 0.96.

16.4.2 Change of seasonal components in time

Winter and summer mass balances show strong inter-annual fluctuations (Figures 16.7(a), (b)), and also significant trends, including an increase of b_w from the 1960s, and b_s since the middle of the 1970s. Maximum values of b_w and b_s appear to have increased since the 1960s (Figures 16.7(c), (d)). This is caused by the introduction to the global sample of the results from several very small glaciers, e.g. results from studies in the Alps (Europe), North Cascade and Front Ranges (USA), and Hamagury Yuki (Japan). The mass balance of very small glaciers, particularly of snow patches, fluctuates from year-to-year much more than that of large glaciers (Glazyrin, Kamnyanskiy and Perziger, 1993; Kuhn, 1995). Annual mass balance and seasonal components weighted by area give more realistic estimates of large-scale glacier regimes.

16.4.3 Annual and/or net mass balance

Most published mass balance values include only mass lost by processes on the glacier surface or in the uppermost annual layer of snow/ice deposited during the hydrological year. A portion of melt water may escape this upper layer but may be trapped in deep layers of firn. This forms an internal accumulation (Shumskiy, 1964). According to several studies, internal accumulation may range from perhaps 5% (maritime climate conditions) up

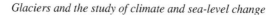

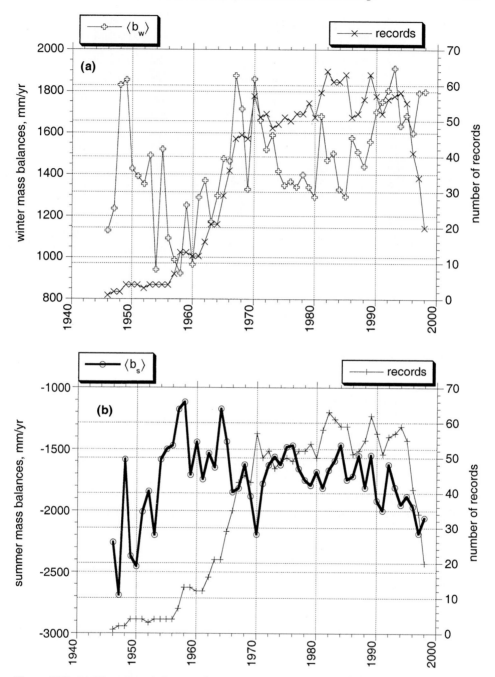

Figure 16.7. (a) Winter mass balance and number of measured glaciers (records). (b) Summer mass balance and number of measured glaciers. (c) Extremes (minimum, maximum) and standard deviation values of winter mass balance. (d) Extremes (minimum, maximum) and standard deviation values of summer mass balance.

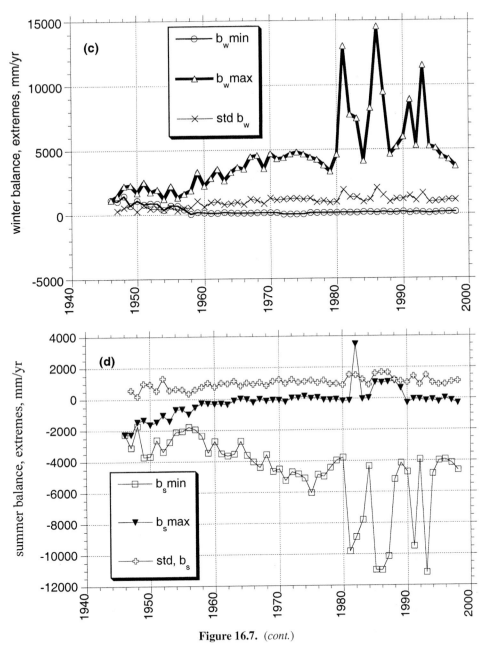

Figure 16.7. (*cont.*)

to 64% (cold and dry continental climates) of annual accumulation (Bazhev, 1997; Golubev, 1976; Trabant and Mayo, 1985). Different methods have been used to calculate the amount of internal accumulation (Bazhev, 1997; Golubev, 1976; Krenke, 1982; Shumskiy, 1964; Trabant and Mayo, 1985). This was included in mass balance results for several glaciers and regions, e.g. Djankuat glacier in the Caucasus; glaciers in Altai, Tien Shan and Pamirs; and McCall glacier in the Brooks Range, Alaska. For the majority of glaciers used for analysis in our data set, the information on internal accumulation has not been published and comprises one of the systematic but poorly known errors in the calculation of annual/net mass balance.

Another important component is the mass loss due to ice calving, either into water or onto land. This has not been included here due to the lack of measurements. This component may be substantial for only few glaciers in our data set. Thus, strictly speaking, mass balance measured by the glaciological method is not necessarily equal to volume change, but may be nearly identical to it for most glaciers, where the calving and other components are small. Annual (b_a) and net (b_n) balances (Mayo *et al.*, 1972) may also differ from year to year, although the differences are not likely to be substantial for longer-term averages.

Globally averaged glacier mass balance

A close relationship exists between glacier mass balance time series combined from all records (b_{all}), records of 20 years and longer (b_{20}), and those with records of 30 years and longer (b_{30}), as shown in Figure 16.8. Thus it is possible to use all time series with study temporal changes over longer periods of time compared with time series of b_{20} (beginning in 1966) and b_{30}, (beginning in 1969).

To obtain accurate knowledge of the glacier contribution to the global water cycle (including sea-level change) it is crucial to calculate characteristics of glacier regimes at large scales, such as river basins or mountain ranges. The surface area and average values of variables must be known to accomplish this task. These are known poorly, and are the main sources of systematic error for the global assessment. The main problems here are as follows.

(1) Glacier inventory data exist for only for 40% or less of all glaciers in the world (Meier and Bahr, 1996). The most recent inventory compilations list 71 558 glaciers (http://www.geo.unizh.ch/wgms).

(2) Glacier area changes over time are not always reported.

(3) The area of individual glaciers around the Antarctic ice sheet has yet to be determined, and may lie between the wide limits of $70 \times 10^3 \, km^2$ to $700 \times 10^3 \, km^2$ (Weidick and Morris, 1998). Shumskiy (1969) estimated that the ice caps around the West Antarctica ice sheet have a total area of about $164 \times 10^3 \, km^2$. In our most recent global mass balance assessment we conservatively used an area of $70 \times 10^3 \, km^2$ to calculate the mass balance of these glaciers (Dyurgerov and Meier, 1997a).

(4) The global estimation may be biassed toward maritime climate conditions as more than 60% of long-term mass balance records are from Alps, Scandinavia and north-western North America.

(5) The lack of data on mass balances of very large glaciers (Alaska, central Asia, Patagonia ice fields), which may have different mass balance characteristics compared with the small and medium-size glaciers that are commonly used for mass balance study.

(6) Very small glaciers, those less than $1 \, km^2$ in area, are poorly represented in the inventory, and their regime may be different from that of larger glaciers (Kuhn, 1995). Their extreme turnover

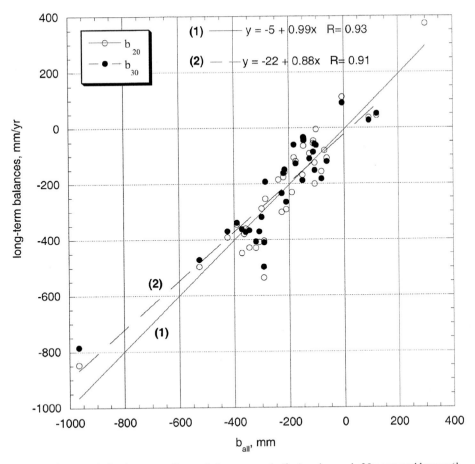

Figure 16.8. Correlation between all mass balance records (b_{all}) and records 20 years and longer (b_{20} and 1), 30 years and longer (b_{30}, 2).

value (up to 20 m per year in water equivalent) and huge inter-annual variability are demonstrated by data from very small glaciers or snow patches (e.g. Hamagury Yuki in the Japanese Alps and the Abramov snow patch, Pamir). We do not know, even roughly, the area of these 'snow patches – nearly glaciers' (translated from Pertziger (1981)). These may be a small fraction of the global coverage, but in some regions they are important water sources.

How precise are mass balance data?

The usual approach to the estimation of data quality is to compare data obtained by one method with one or more independent measurements. Data obtained by the glaciological method can be compared with the results of repeated geodetic/topographic surveys. Discrepancies between the two methods may have different signs (plus and minus) for different glaciers and for different periods. For some glaciers the discrepancy is large (Blue glacier, USA; Storbreen, Norway), but for others it is very small (Careser, Italy; Djankuat, Russia).

There are several reasons for the discrepancies, including ignoring internal accumulation, under-estimation of annual ice-ablation rate (these two, in some cases, may partly eliminate each other), sinking of ablation stakes into ice or snow, errors in maps, extrapolation in data of point measurements by the glaciological method to the entire glacier, and others (Andreassen, 1999; Conway, Rasmussen and Marshall, 1999; Fountain and Vecchia, 1999; Golubev *et al.*, 1978; Jansson, 1999; Krimmel, 1999; Østrem and Haakensen, 1999; Trabant and Mayo, 1985). One important conclusion from these comparisons and discussions is that the topographic method may serve as a control for many cases.

The case of Alfotbreen in south-west Norway is of special interest. The calculated mass balance for this glacier can be applied to about 25% of the entire Alfotbreen ice cap area (Østrem and Haakensen, 1999). The accumulation area of this ice cap is shared by several ice streams (Hansebreen and others) which flow down from the Alfotbreen ice field. This is a case where the ice divide between the different ice streams may migrate in response to changes of snow accumulation, especially in extreme years. Due to the change in time of the position of the ice divide, part of the ice flow may change direction from one basin to the other; stealing one part from the other. This was noted for Blue glacier (Waddington and Marriott, 1986, p. 176): 'Using an average divide position for all balance years is also not correct. Extremes of net balance, occurring in years when the divide was far from its average position, would introduce errors into the budget calculations.' Popovnin (1996) has also observed ice divide migration for two glaciers in Caucasus (Djankuat and Lekzyr) having a common accumulation area along the main mountain divide. Another remarkable example is of two glaciers in the South Patagonian ice field, O'Higgins and Pio XI, which share an ice divide. The first lost surface area of $50 \, \text{km}^2$ and the second gained $60 \, \text{km}^2$ between 1945 and 1995 (Warren and Aniya, 1999). This reveals that mass balance measured by the glaciological method can only be compared with the mapping method with caution, keeping in mind the possible change in an ice divide. This and several other sources of error have recently been analysed and estimated for Blue glacier (Conway *et al.*, 1999).

The topographic (mapping) method is also not free from errors. Haakensen (1986) has shown that three main sources of errors are common for mass change determined from this method, namely uncertainty in map compilation, height determination, paper shrinkage and drafting. Haakensen determined these errors for Hellstugubreen and Grasubreen as 2.3 m in water equivalent over the 1968–80 and 1968–84 periods, correspondingly. Compared with the cumulative mass balance loss for Hellstugubreen, this comprises 34%, and 42% for Storbreen.

A new methodology has been recently introduced to measure change in glacier volume using airborne laser altimetry. The altimetry system, used most successfully to measure glaciers in Alaska, consists of a laser range-finder mounted in a small aircraft, a gyro to measure the orientation of the laser and a kinematic global positioning system (GPS) for continuous measurement of aircraft position (Echelmeyer *et al.*, 1996). Profiles are flown along centre lines of the main stream and tributaries of a glacier at an elevation of 50 to 300 m above the surface. These profiles are compared with contours on large-scale topographic maps or to previous profiles. Differences in elevation are calculated at profile or contour-line intersection points. To determine the area altitude distribution digital elevation models derived from the same maps are used. The volume change is calculated assuming that a measured elevation change can be applied to the entire glacier area within the corresponding elevation band. The change in elevation is then integrated over the original area altitude distribution. Average thickness change for the entire glacier is calculated by dividing the total volume change by the average of the previous and new glacier areas. This technique has most recently been applied to determine the volume change (multi-year mass balances) for several Alaskan glaciers (Arendt *et al.*, 2002). In 1995 the team from the Geophysical Institute, University of Alaska, measured 67 glaciers in seven geographic regions of Alaska, and calculated the volume change from the mid 1950s to the mid 1990s. In 2001 they re-measured 28 glaciers. Thus, glacier volume change was calculated for two periods, early and recent (Arendt *et al.*, 2002).

We have compared these results with mass balances measured in Alaska since 1965 by traditional glaciological methods. We used the annual mass balance for three bench-mark glaciers in Alaska, namely Gulkana (Alaska Range), Wolverine (Kenai Mountains) and Lemon Creek (Coast Mountains North). From these three glaciers we calculated the area-weighted mass balance (specific net balance in water equivalent). We compared these average mass balances with those determined by the laser altimetry method. We used the same periods of time for comparisons: 1965–1995 and 1995–1999 (the last year with available data). For bench-mark glaciers we used annual data. For glaciers measured by laser altimetry we adjusted these to the same two periods by extracting averaged annual mass balances from the cumulative values over the 1955–95 and 1995–2001 periods (Meier and Dyurgerov, 2002). The results of our comparisons (Figure 16.9) show that for these observed glaciers, at least, laser altimeter results and mass balance observations agree within the limits of combined errors of both methods. The laser altimeter error was estimated by Arendt *et al.* (2002) as about 15% over earlier period, and about 30% over the recent period.

Glacier mass balance change in time

Spatial averages of mass balance values show large inter-annual fluctuations, especially in the period 1946 to 1960 when measurements were carried out on fewer than 20–30 glaciers annually (Figure 16.10(a)). Standard root-mean-square (rms) error values were very large also and exceeded annual balance values in some years (Figure 16.10(b)). This was the reason why only the period starting in 1961 was used in the previous global analysis by Dyurgerov and Meier (1997a, b).

Annual mass balances calculated for all glaciers, for the series of records longer than 20 years, and for those with series longer than 30 years, show the same tendencies to increasingly negative values starting in the middle of 1970s, with an accelerating negative rate beginning at the end of 1980s (Figures 16.10(a–c)).

Extreme values calculated for relatively long time series (20 years and longer) show increases of maxima and minima (Figure 16.10(c)), and, due to this, increases of standard deviation values and rms variance in the mass balance averages, especially since the 1980s

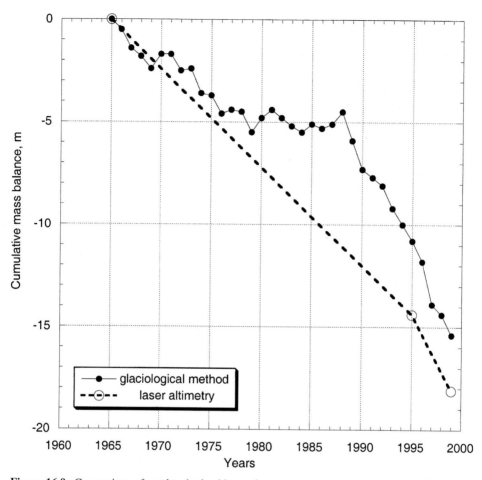

Figure 16.9. Comparison of results obtained by surface mass balance observations and from laser altimeter measurements (Arendt *et al.*, 2002), averaged for three observed Alaskan glaciers. Laser altimeter results were arbitrarily cumulated from 1965 to be comparable with the surface balance observations.

(Figures 16.10(b), (c)). This is possibly connected with data uncertainty, but is more likely due to climatic fluctuations.

16.4.4 Equilibrium-line altitude (ELA) and accumulation area ratio (AAR)

All ELA series show great variability from year to year with differences of several hundreds of metres between maximum and minimum ELA values corresponding to balance, years with highly negative or positive mass balance respectively. The relatively small number of observations is an obvious restriction to the analysis of long-term changes in ELA. As seen

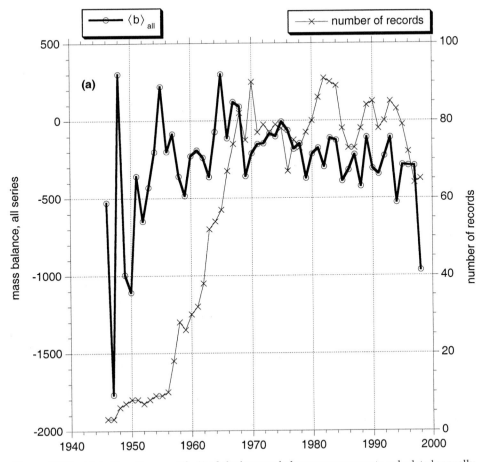

Figure 16.10. (a) Arithmetic mean ($\langle b \rangle_{all}$) of glacier mass balance measurements, calculated annually since the beginning of measurements, together with the number of records. (b) Glacier mass balance calculated for glaciers with mass balance records 20 years and longer (b_{20}) and root-mean-squared (b_{20rms}). (c) Extreme mass balance values for glaciers with mass-balance records 20 years and longer.

in Figure 16.11(a), the number of observations have now reached an acceptable statistical level.

There were more than 20 ELA measurements in 1962, and the inter-annual fluctuation has decreased and appears to be increasingly reliable for use in global analysis. Since the 1960s the standard deviation and rms values show relative stability in time, with rms values stabilizing at about 150–170 m with the relative error of around 6–7% (Figure 16.11(b)). The increase in ELA from 1961 to 1998 is about 200 m and corresponds to the increase in negative values of mass balance.

Only in the mid 1960s did the number of measurements of AAR reach the level of statistical acceptability required to calculate reliable average and standard deviation values. Mean values are shown in Figure 16.12(a) for all glaciers and also for the long-term series.

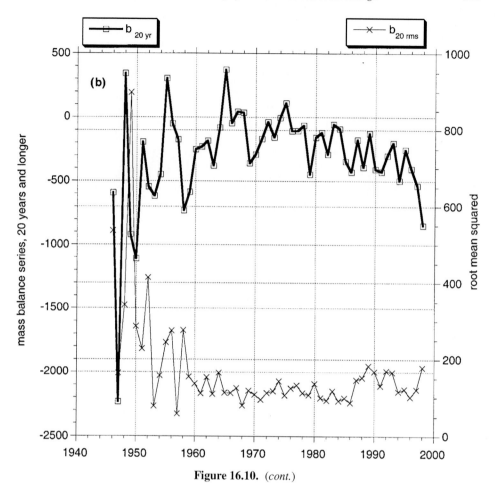

Figure 16.10. (*cont.*)

These demonstrate a steady decrease of AAR from about 60 to 50% over this relatively short period of time. Standard deviation fluctuates greatly from year to year (Figure 16.12(b)), but the rms error shows stability at the level of 6 to 4% (about 10% relative to the mean AAR value).

It is more convenient for many studies to use AAR, rather than ELA. AAR is a non-dimensional parameter, which is easier to use when comparing glaciers in different locations. AAR can be determined more precisely in comparison to ELA, as, for many glaciers, at the end of a summer season snow-cover patterns do not show a simple dependence on elevation, and the spotty distribution causes a problem in determining ELA. Obviously, there has to be a close relationship between changes in AAR and ELA over time. A shift in ELA of 200 m between 1961 and 1998 relates to an AAR decrease of about 10%.

There are more than 160 time series of annual/net mass balance supported by AAR data, measured in more than 30 mountain ranges and sub-polar islands in the world; of

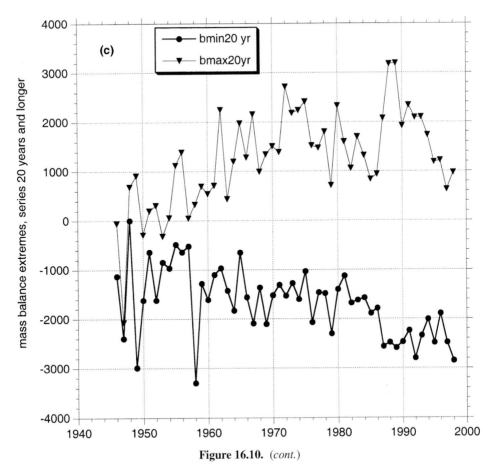

Figure 16.10. (*cont.*)

these, more than 90% are in the northern hemisphere. We have chosen time series of five years and longer of mass balance b and accumulation area ratio AAR for 96 glaciers for further analysis. For 16 of these glaciers, no reliable linkage between AAR and b has been established, for several possible reasons, including short elevation range covered by empirical data of AAR and strong outliers (snow-pattern distribution does not show clear elevation dependences), or a range in inter-annual fluctuation of ELA that has exceeded the range of elevation of a glacier. These glaciers may not be appropriate for mass balance monitoring. The other 80 glaciers demonstrate close, reliable relationships between AAR and b, and we used these to calculate regressions to estimate a steady-state AAR_0 ($b = 0$). We found that the commonly accepted idea of a 'steady-state' mode has to be applied with caution, as analysis of experimental data shows a large spatial–temporal variability in AAR_0. This analysis confirms that, in colder and dryer climate conditions, glaciers require larger accumulation areas to compensate for ice wastage below the ELA. Glaciers in the

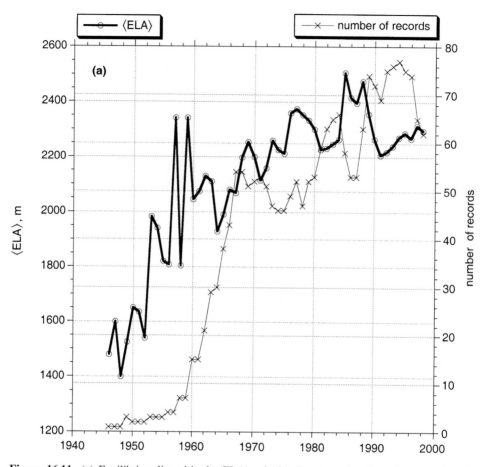

Figure 16.11. (a) Equilibrium-line altitude (ELA) calculated as annual arithmetic means for all measurements together with number of records. (b) Standard deviations, and root-mean-squared error.

Canadian Arctic and in southern Siberia (Altai), show the largest AAR_0 (70% and larger). It is less for glaciers in western Svalbard (50–60%). In the Alps, Alaska and north-west USA the AAR_0 show large variability, between 40 and 60%. AAR_0 has shown similar values for long time series of three glaciers in Tien Shan and Pamir (central Asia). Average AAR_0 for all time series is 58%, which exceeds the average AAR long-term value ($\langle AAR \rangle$) by 7%. This is another indication of the present global wastage (negative mass balances) of glaciers. The standard deviation of AAR_0 is about 9%, and the coefficient of variation, Cv, is 0.15. The mean coefficient of correlation between AAR and b in 80 time series is 0.92. Gaussian probability curves for these AAR_0 fit the empirical data for 80 glaciers well.

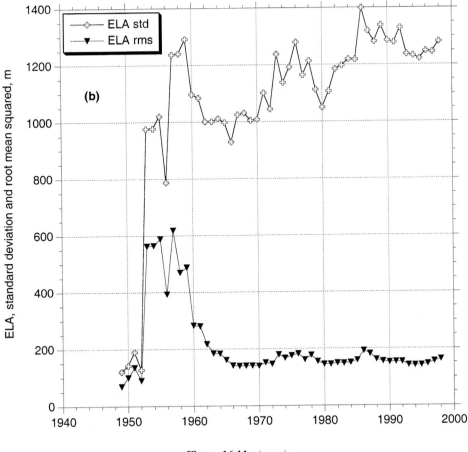

Figure 16.11. (*cont.*)

16.5 Spatial pattern of glacier volume changes

The 36 longest, directly measured time series of mass balance, with accuracies of about ±0.1 to 0.2 m in water equivalent per year, are presented in Figure 16.13. Several remarkable features are recognizable.

- Trends are different within each of the regions and between regions, with some glaciers gaining mass at the time others are shrinking; see, for instance, the Alps and Scandinavia, Figures 16.13 (c) and (d).
- The differences between cumulative values of b for the period 1961 to 1998 between individual glaciers in mainland North America reaches 40 m, and 30 m in the Alps and in Scandinavia (including Svalbard), suggesting differences in climatic conditions.
- Glaciers in cold and dry regions (e.g. the Canadian Arctic) demonstrate trends of shrinkage which are internally rather consistent, but with relatively low changes in b due to the precipitation regime (Figure 16.13(a)). Koerner and Lundgaard (1995), who are responsible for obtaining most of these

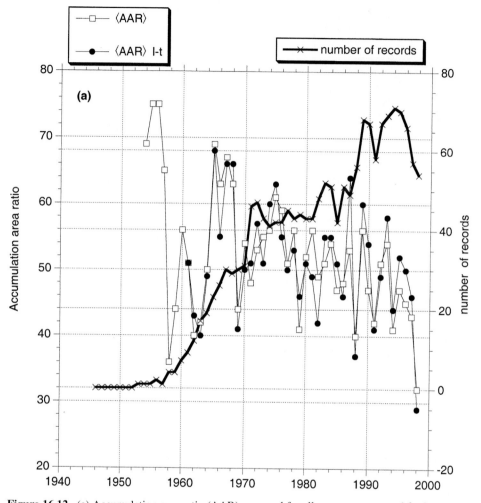

Figure 16.12. (a) Accumulation area ratio (AAR) averaged for all measurements and for long-term (l-t) records. (b) Standard deviation for all (AAR std) and long-term records (std AAR l-t), and root-mean-square errors for all glaciers (AAR rms); rms are the same as for long-term time series.

data, imply that the 'warming trend' indicated by changes in these glaciers in the last 100 years, '... is part of natural variability of climate rather than due to anthropogenic effects.' Without commenting on the cause of this warming, we note that the volume-change trends in this region are consistent with those elsewhere in the world, and that a global cause seems likely. The Canadian Arctic sample represents cold ice caps where thickness change is not as large as in more maritime regions, but tiny ice caps in the Arctic are shrinking rapidly (Bradley and Serreze, 1987; Welker *et al.*, 2002).

Glaciers situated at high altitudes in Asia (low temperature and relatively dry climates) also show a common trend of reducing volume (Figure 16.13(e)), especially glaciers in Pamir and Tien Shan. On the other hand, glaciers in some moist, maritime regions (e.g.

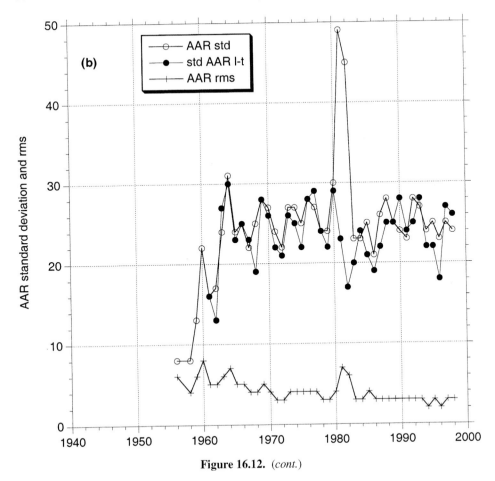

Figure 16.12. (*cont.*)

Djhankuat, Alfotbreen, Nigardsbreen, Hardangerjokulen) are stable or growing (Figures 16.13 (c) and (d)).

In addition to these long and continuous time series of mass balances, we consider all direct measurements of glacier mass balances because a strong correlation exists between long-term time series and shorter series for all 280 glaciers (Figure 16.8). We use b_{all} to estimate a mean global value (Figure 16.13(f)). This was done by averaging the mass balances of all glaciers within five major glacier regions (and in a number of sub regions), then calculating a hemispheric average, weighting each region by the glacier area in that region. The calculated averaged annual decrease in glacier thickness is -133 mm per year in water equivalent. This specific value multiplied by the area of all small glaciers, 680×10^3 km^2, gives a volume change of about -90 km^3 per year or about 3.4×10^3 km^3 of volume loss over 38 years (1961–98).

Obviously, this is only an estimate of the global sum of glacier wastage, because the data are sparse and not homogeneously distributed. However, this estimate includes all

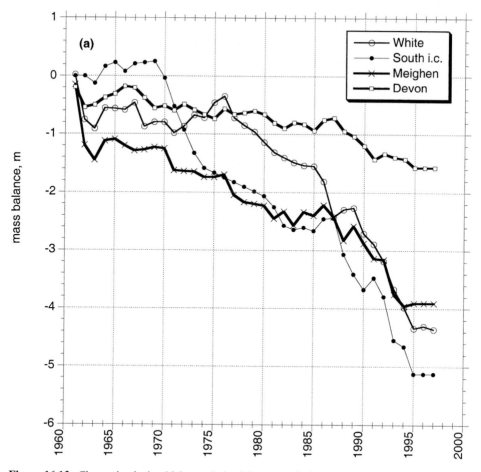

Figure 16.13. Change in glacier thickness derived from mass balance measurements on 36 glaciers in five regions in the northern hemisphere in recent decades. (a) Canadian Arctic archipelago; (b) Alaska, north west USA and Canada; (c) Scandinavia and Svalbard; (d) European Alps; (e) Pamir, Caucasus, Tien Shan, Altai and Kamchatka; (f) archipelagos, continents and global.

available data and recognizes the substantial differences in different regions. It is difficult to estimate the error in the cumulative sum because of the possibility of both random and systematic errors (Cogley and Adams, 1998). The apparent variances in individual years cause a standard error of estimate of the total of only 30 mm or 0.5% (at the 99% probability level), but other errors surely raise this number to at least several per cent. We note that Oerlemans' calculation of globally averaged mass balance (Oerlemans, 1999) by a very different method matches our results closely.

Airborne laser altimetry of 67 Alaskan glaciers in the mid 1990s compared with topographic maps from the 1950s have shown that the average thickness change over this period was -0.52 m per year, or 52 ± 15 km^3 per year extrapolated over the entire Alaska/adjacent Canada glacier area, taken as 90 000 km^2. Repeat measurements of 28 glaciers from the mid

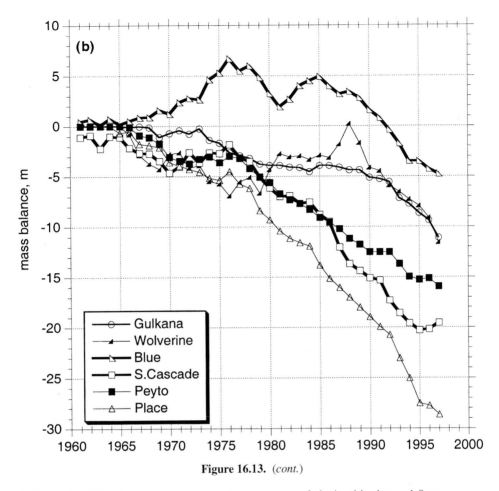

Figure 16.13. (*cont.*)

1990s to 2000–2001 suggest an increased average rate of glacier thinning, -1.8 m per year, or 96 ± 35 km^3 per year (Arendt *et al.*, 2002). About 75% of theses changes over both periods is accounted for by a few large glaciers bordering the Gulf of Alaska in the Chugach, St Elias Mountains and Coast Ranges.

Additional evidence of pervasive glacier wastage is shown by the decrease in the average value of the AAR along with the increase in ELA. This has exposed larger ice areas with low albedo and increased ice melting, with a further tendency to reduce glacier volume, a positive feedback pointed out by Bodvarsson (1955).

16.6 Glacier mass balance and climate variability

On short timescales, e.g. annual, glacier mass balances respond rapidly to changes in climate. Analysis of our time series of mass balance show that within the confidence level of 0.95 there is about a one year lag between volume changes in consecutive years. This is to

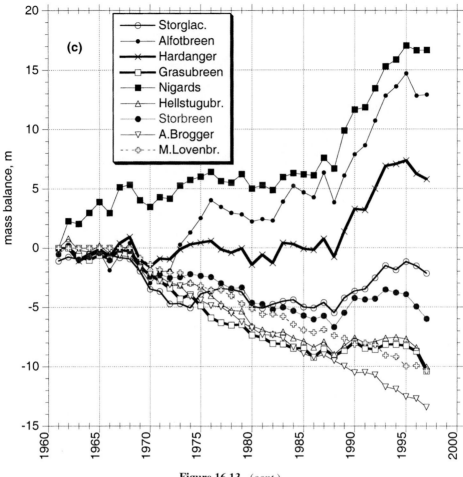

Figure 16.13. (*cont.*)

be expected because the albedo effect of a non-zero balance year may have some carry-over effect to the next year (Meier, 1969). Concentrating on short-term variations minimizes the problem of multi-year or longer dynamic response times (Jóhannesson *et al.*, 1989) because *b* is always related to the instantaneous glacier area. Annual changes in volume can thus be considered to be almost simultaneous with annual changes in weather. Thus, short-term glacier volume changes can be attributed to: (1) changes in atmospheric circulation patterns (atmospheric pressure fields) at regional or global scales; and/or (2) local weather patterns, such as changes in wind regime and local precipitation trajectories, snow avalanches, changes in albedo, moraine cover and others, as influenced by peculiarities of glacier topography and size.

Time series showing both increased melting and accumulation (Figure 16.7 (a), (b)) are especially interesting. The increase in both accumulation and ablation does not seem to

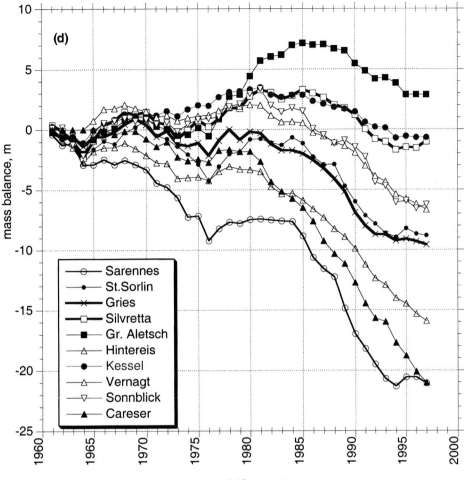

Figure 16.13. (*cont.*)

have been noted for previous periods of observation. Earlier analyses (e.g., Thorarinsson, 1940) did not show such a phenomenon. This may be because previous workers did not have as complete and detailed data sets. But it is also possible that the relationship between glacier regime and climate has changed in recent decades due to global warming. The annual turnover or amplitude $\alpha = (b_\mathrm{w} - b_\mathrm{s})/2$ is one characterization of the glacier regime in relation to climate (Meier, 1984). Our calculation shows that α has not been constant over the period of study but has increased substantially (Figure 16.14). This increase is possibly related to increased heat energy absorbed, accompanied by increased moisture production, because the amount of snow accumulation has increased in high mountain and in sub-polar regions (Dyurgerov and Meier, 2000). The increase in α is also accompanied by increases in the inter-annual variances (standard deviations) of b_w (Figure 16.14).

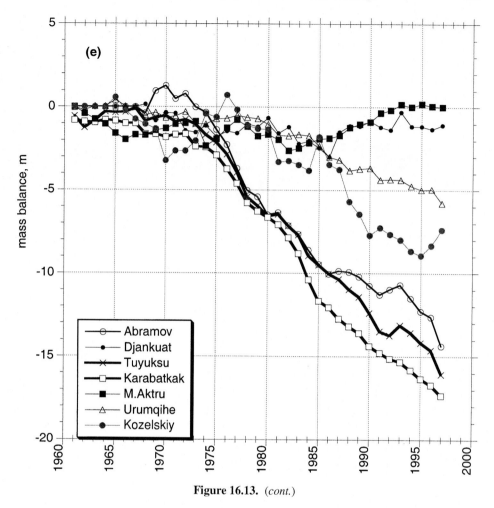

Figure 16.13. (*cont.*)

The α, b_w time series (Figure 16.14) suggest a significant shift during the mid 1970s. The cumulative sum of departures from the mass balance (b_{20}) of -157 mm per year, as averages over the climatologic reference period 1961–90, show a strong change at the end of the 1980s toward decreasing glacier volume (Figure 16.15). This shift corresponds to a shift in climate reported in several recent publications (Hansen *et al.*, 1999; Jones *et al.*, 1999; Trenberth, 1999), which suggest an abrupt shift in the basic state of the atmosphere–ocean climate system over the North Atlantic Ocean.

The changes in glacier mass balances, as well as global temperature, toward the end of 1980s may be unprecedented (Trenberth, 1999). This period bears examination because it may provide insight into the behaviour of the climate/glacier relationship in the future. This period included the warmest decade and several of the warmest years in the history of the instrumental record (Jones *et al.*, 1999).

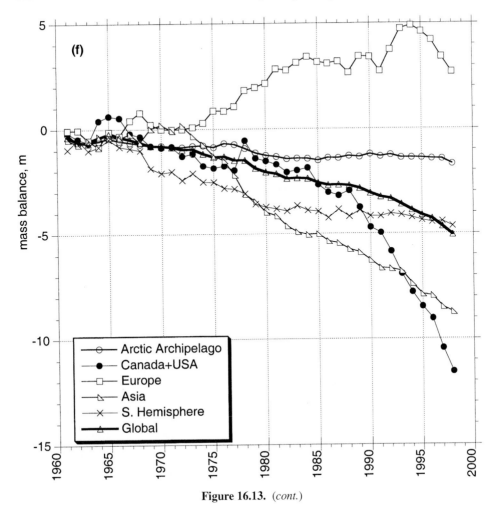

Figure 16.13. (*cont.*)

16.6.1 Glacier mass balance sensitivity to air temperature

Oerlemans and Fortuin (1992) pointed out that the sensitivity of glacier volume changes to temperature ($\partial b/\partial T$) is a function of precipitation (high precipitation regions have higher sensitivity). Our global average data, showing increases in both melting (related to temperature) and accumulation (related to precipitation), are in agreement with this result. The increase in b_w is particularly remarkable because it has occurred in spite of a reduction in the size of the accumulation area (AAR). Thus, significantly increased precipitation at high altitudes is indicated.

The annual balance sensitivity to temperature ($\partial b/\partial T$) is used for most projections of glacier wastage and its contribution to sea-level rise. Typical published values of mass balance sensitivity, unadjusted for precipitation, range from about -0.3 to $1\,\mathrm{m}/^{\circ}\mathrm{C}$ per

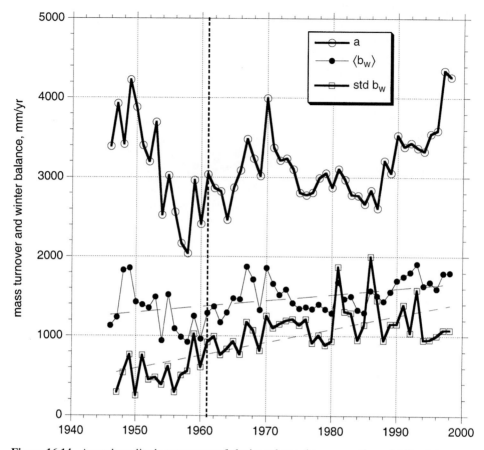

Figure 16.14. Annual amplitude or turnover of glacier volume change $\alpha = (b_w - b_s)/2$, winter mass balance (b_w) and its standard deviations calculated from long-term series (about 30 glaciers). The time period emphasized in our study is marked by the vertical dashed line.

year, with an average of about $-0.7\,\mathrm{m/°C}$ per year (e.g. Kuhn, 1993; Oerlemans *et al.*, 1998). Using the observed change in glacier volumes, this suggests a temperature rise of $0.34\,°\mathrm{C}$, or $0.009\,°\mathrm{C}$ per year over the period of time 1961 to 1988. This is a more rapid warming than the global average surface temperature change for 1901 to 1998 of $0.0065\,°\mathrm{C}$ per year (Oerlemans *et al.*, 1998). Oerlemans (1994) uses glacier dynamics modelling of measured glacier retreats, scaled by region, to estimate an annual temperature rise of $0.62\,°\mathrm{C}$ from 1884 to 1978. Because of the range of volume-change variability among glaciers (Figure 16.12), sensitivity values derived from a limited number of glaciers must be used with caution. The sensitivities suggest that the recent rise in air temperature in glacier regions is somewhat greater than the modelled global average (Oerlemans *et al.*, 1998), which is derived largely from low altitude gauges, and that warming is accelerating.

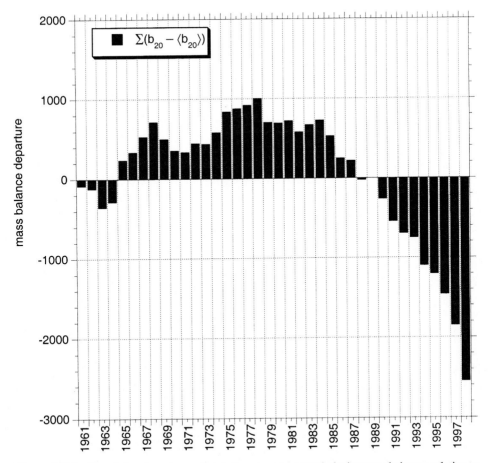

Figure 16.15. Cumulative departures of globally averaged annual glacier mass balances relative to 1961–90 'climatological normal' period, $\langle b_{20} \rangle$. Long-term time series 20 and more years (b_{20}) have been used.

We calculated mass balance sensitivity using all time series for northern hemisphere glaciers in respect to observed northern hemisphere annual air temperature. Our result is -0.2 m/°C per year, which is substantially less than our previous calculation (for global sensitivity) of -0.37 m/°C per year for the period 1977 to 1997 (Dyurgerov and Meier, 2000) and also values published by Oerlemans (1993) of -0.39 m/°C per year and by R. J. Braithwaite (private communication, 2001) of -0.41 m/°C per year. In this calculation we have used updated mass balance results and did not include several short-time series of mass balance from the southern hemisphere. The sensitivity is different for different glaciers. Oerlemans and Reichert (2000) have shown that the glacier sensitivity to climate depends on mass turnover. In regions with dry climate conditions and small mass turnover, summer temperature is the most important factor. In wetter climates, with larger mass turnover, glacier mass balance is very sensitive to the change in amount of precipitation, particularly in spring and autumn. Thus, comparisons have to be made for the same samples.

Note, this a different measure of sensitivity than that used by the IPCC (Warrick *et al.*, 1995), which involves a change between two steady-states.

16.6.2 Atmospheric circulation and mass balance change

The spatial and temporal variability in observed mass balance time series may be studied in connection with atmospheric circulation patterns. These studies focus on the relations between local or regional climate and the mass balance of glaciers (e.g., Hodge *et al.*, 1998; McCabe and Fountain, 1995; McCabe *et al.*, 2000; Meier *et al.*, 1980; Trupin, Meier and Wahr, 1992; Walters and Meier, 1989). These studies are useful for understanding the physical interactions between climate and glaciers on regional to global scales. The processes inter-relating atmospheric circulation, surface meteorological parameters (e.g. air temperature and precipitation) and glacier mass balance are very complex. We found that the spatial covariance of glacier annual mass balances may range from strong to weak, and positive to negative, over the northern hemisphere (Dyurgerov and Meier, 2000). Distant glaciers may correlate more strongly than neighbouring glaciers, showing the existence of teleconnections involving regional atmospheric circulation patterns. The correlation structure of mass balance with atmospheric pressure anomalies is partly explained by changes in the winter balance. A principal components analysis (Jonston, 1980) shows that 46% of the b_w variability is explained by the first two principal components, which are also correlated with the Arctic oscillation index and the southern oscillation index (McCabe *et al.*, 2000). This analysis also explains the recent growth in certain maritime glaciers, such as those in south-western Scandinavia and Iceland (Figure 16.13 (c)). The other components of variability may be explained by summer mass balance b_s and local glacier properties.

16.7 Glacier mass balance and sea-level rise

Ice melt is an important component of relative sea-level rise (RSL). If current rates of sea-level rise were to last for decades or hundreds of years, the socioeconomic effect and environmental consequences would be dramatic (Trenberth, 1999; Warrick *et al.*, 1995). The 2001 Intergovernmental Panel of Climate Change (IPCC) estimate of observed RSL is 1.0–2.0 mm per year (Church *et al.*, 2001). This is caused partly by negative glacier mass balances (e.g., Dyurgerov and Meier, 1997a; Meier, 1984; Warrick *et al.*, 1995). The globally averaged mass balance can be converted into units of sea-level change (362 km^3 of water increases sea level by 1 mm). There are many problems in converting mass balances of individual glaciers to RSL (Warrick *et al.*, 1995). The continuous time series of updated results created here is one of the necessary steps required to perform such calculations more accurately. However, glacier mass balance calculations may under-estimate the actual volume loss because the measurements by standard glaciological methods do not include

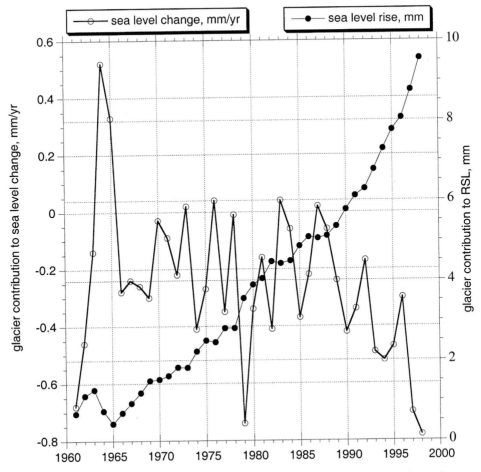

Figure 16.16. Year-by-year fluctuations and cumulative values of change in glacier volume of mountain and sub-polar glaciers expressed in terms of relative sea-level change (RSL).

iceberg calving, which may constitute about 3–5 km³ per year for the entire Arctic in additional mass loss; the iceberg calving component requires further investigation (see Chapter 14).

For the period 1961 to 1998 glaciers lost a volume equivalent to a sea-level rise of about 10 mm (Figure 16.16). This is about 20% of the observed RSL. The data in Figure 16.16 show that the annual rate of glacier wastage has increased. The area-weighted average annual mass balance changed from −82 mm per year during the period 1961 to 1976, to −125 mm per year during the 1977–87 period and to −217 mm per year during 1988–98 (Table 16.2); this is equivalent to a change in the rate of sea-level rise due to glacier wastage from 0.15 to 0.24 to 0.41 mm per year. The repeated laser altimetry of Alaskan glaciers (Arendt *et al.*, 2002) supports these results. Over the period from the mid 1950s to the

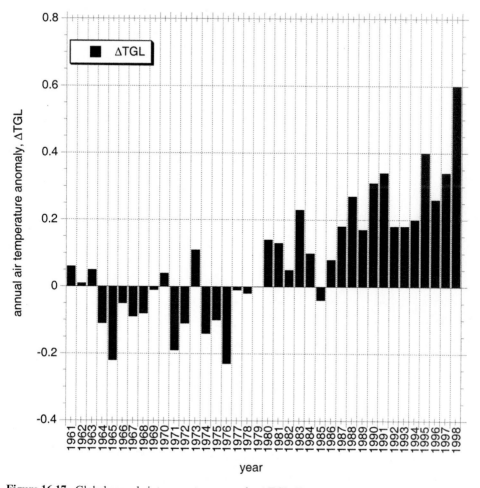

Figure 16.17. Global annual air temperature anomaly, ΔTGL (Jones *et al.*, 1999). The coefficient of correlation between ΔTGL and b_{all} is 0.68 (0.71 between ΔTGL and b_{20}).

mid 1990s, the change in sea level due to changes in Alaskan/adjacent Canada glaciers is estimated at 0.14 ± 0.04 mm per year, and this increased to 0.27 ± 0.10 mm per year from the mid 1990s to 2000–2001. It is interesting that mass balance sensitivity with respect to annual air temperature shows changes in these three periods as well. In particular, the period 1977 to 1987 was exceptional. The mass turnover at this period experienced huge inter-annual variability, and mass balance had positive values for six out of 11 years, which was unusual for the entire period of mass balance study since 1961. The resulting sensitivity was positive due to this. One possible explanation is that this was the most humid period, characterized by a larger amount of precipitation, perhaps only in the high elevations. Global temperature has shown a strong rise since the end of the 1970s, Figure 16.17 (Hansen *et al.*, 1999; Jones *et al.*, 1999). The third period recognized in Table 16.2, 1988 to 1998, again

Table 16.2. *Globally averaged mass balance components, for three periods: 1971–76, 1977–87 and 1988–98.*

Mass balance component	1961−76		1977−87		1988−98	
b_{a-w}, mm per year	−82		−125		−217	
b_{20}, mm per year	−93	(33)	−208	(39)	−400	(36)
b_w, mm per year	1480	(40)	1430	(58)	1700	(51)
b_s, mm per year	1650	(40)	1680	(58)	1890	(51)
α, mm per year	1565	(40)	1555	(58)	1795	(51)
ELA, m	2150	(42)	2330	(56)	2290	(71)
AAR, %	55	(23)	51	(32)	46	(34)
Sensitivity, m/°C per year	−0.96		1.44		−0.73	

Values in parentheses are the number of time series used for averaging.
Mass balance data in water equivalent.

shows large sensitivity with a negative sign, which means that glaciers in our sample have become more sensitive to further increase in annual air temperature.

16.8 Conclusions

Observational data leave no doubt that global glacier wastage has been accelerating since 1970. All components of glacier regime show change, especially since the mid 1970s, and the change has been much stronger since about 1988, with little delay from the corresponding changes in climate. The largest negative mass balance values were observed in north-western North America, particularly in Alaska, and in central Asia. These regions, combined, include about 200×10^3 km^2 of glacier area, which is about 30% of global glacier area outside the two ice sheets. Their fraction of the globally averaged glacier wastage (negative mass exchange) has been more than 70% over 1961 to 1998, and has increased since the end of 1980s up to 80%, implying that the glacier contribution to the water cycle on land substantially increased during global warming.

The increase in ELA and decrease in AAR are strong and also relate to change in climate, but the rate varies spatially. Calculated steady-state values of AAR_0 range from 40 to 70%. Following these climate-related changes, glacier surface area has been diminishing at a rate of about 0.04% per year over the period 1961 to 1998, with a much faster decrease in the tropics and central Asia. The average AAR for all time-series appeared to be less than AAR_0, implying that glaciers need dozens of years of shrinking to adjust their surface area to the present-day climate.

Important consequences of these changes are diminishing permanent water storage on land surface and intensification of the global water cycle with the increase in annual rate of glacier contribution to sea-level rise from 0.15 (1961–76) to 0.24 (1977–87), and 0.41 mm per year in the 1988–98 period.

Our knowledge of glacier regime is incomplete: many regions have no observations, or available time series are very short. Glacier area is under-estimated; very small glaciers, as well as the largest, are not included in the analysis due to the lack of observations.

The experimental data on glacier regime components may not be enough to make accurate calculations of volume change on hemispheric or global scales for many years, but will remain a valuable source of information on the processes related to climate and changes in the water cycle. We need actively to support new approaches to this difficult problem, including innovated technology, new methods and more realistic modelling.

References

Ageta, Y. and Higuchi, K. 1984. Estimation of mass balance components of a summer-accumulation type glacier in Nepal Himalaya. *Geograf. Ann.* **66A** (3), 249–55.

Andreassen, L. M. 1999. Comparing traditional mass balance measurements with long-term volume change extracted from topographical maps: a case study of Storbreen glacier in Jotunheimen, Norway, for the period 1940–1997. *Geograf. Ann.* **81** (4), 467–76.

Arendt, A., Echelmeyer, K., Harrison, W. D., Lingle, G., and Valentine, V. 2002. Rapid wastage of Alaska glaciers and their contribution to rising sea level. *Science* **297** (5580), 382–6.

Bahr, D. B., Pfeffer, W. T., Sassolas, C., and Meier, M. F. 1998. Response time of glaciers as a function of size and mass balance. 1. Theory. *J. Geophys. Res.* **103** (B5), 9777–82.

Bazhev, A. B. 1997. Methods determining the internal infiltration accumulation of glaciers. In *34 Selected Papers on the Main Ideas of the Soviet Glaciology*. Moscow, pp. 371–81.

Bodvarsson, G. 1955. On the flow of ice-sheets and glaciers. *Jökull* **5**, 1–8.

Bradley, R. S. and Serreze, M. 1987. Mass balance of two High Arctic plateau ice caps. *J. Glaciol.* **33** (113), 123–8.

Braithwaite, R. J. 2001. Glacier mass balance: the first 50 years of international monitoring. *Prog. Phys. Geog.* **26** (1), 76–95.

Church, J. A. *et al.* 2001. In Houghton, J. T. *et al.*, eds., *Climate Change 2001, The Scientific Basis. Contribution of Working Group 1 to the Third Assessment Report of the Intergovernmental Panel on Climate Change*. Cambridge University Press, pp. 641–93.

Cogley, J. G. and Adams, W. P. 1998. Mass balance of glaciers other than the ice sheets. *J. Glaciol.* **44** (147), 315–25.

Cogley, J. G., Adams, W. P., Ecclestone, M. A., Jung-Rothenhauser, F. and Ommanney, C. S. L. 1995. *Mass Balance of Axel Heiberg Island Glacier 1960–1991*. National Hydrology Research Institute science report no.6. Saskatoon, NHRI, p. 168.

Collins, D. N. 1984. Water and mass balance measurements in glacierized drainage basins. *Geograf. Ann.* **66A**, 197–214.

Conway, H., Rasmussen, L.A. and Marshall, H.-P. 1999. Annual mass balance of Blue
    Glacier, USA: 1955–97. *Geograf. Ann.* **81A** (4), 509–20.
Dyurgerov, M. B. 2002. Glacier mass balance and regime: data of measurements and
    analysis. Institute of Arctic and Alpine Research occasional paper 55, p. 268. Also
    web site at INSTAAR: http://instaar.colorado.edu/other/occ_papers.html).
Dyurgerov, M. B. and Meier, M. F. 1997a. Mass balance of mountain and subpolar glaciers:
    a new global assessment for 1961–1990. *Arctic & Alpine Res.* **29** (4), 379–91.
    1997b. Year-to-year fluctuation of global mass balance of small glaciers and their
    contribution to sea level changes. *Arctic & Alpine Res.* **29** (4), 392–401.
    1999. Analysis of winter and summer glacier mass balances. *Geograf. Ann.* **81A** (4),
    541–54.
    2000. Twentieth century climate change: evidence from small glaciers. *Proc. Natl Acad.
    Sci. USA* **97** (4), 1406–11.
Echelmeyer, K. 1996. Airborne elevation profiling of glaciers: a case study in Alaska.
    J. Glaciol. **42** (142), 538–47.
Elsberg, D. H., Harrison, W. D., Echelmeyer, K. A. and Krimmel, R. M. 2001.
    Quantifying the effect of climate and surface change on glacier mass balance.
    J. Glaciol. **47** (159), 649–58.
Fountain, A. G. and Vecchia., A. 1999. How many stakes are required to measure the mass
    balance of a glacier? *Geograf. Ann.* **81A** (4), 563–9.
Glazyrin, G. E., Kamnyanskiy, G. M. and Perziger, F. I. 1993. *Reshym lednika Abramova
    (The regime of Abramov glacier)*. Sankt-Petersburg, Hydrometeoizdat, p. 228 (in
    Russian).
Golubev, G. N. 1976. *Glacier Hydrology*. Leningrad, Hydrometeoizdat.
Golubev, G. N. *et al.* 1978. *Djhankuat Glacier. Central Caucasus. Water-Ice and Heat
    Balances of Glacier-Mountain Basins*. Leningrad, Hydrometeoizdat, p. 184 (in
    Russian).
Haakensen, N. 1986. Glacier mapping to confirm results from mass balance
    measurements. *Ann. Glaciol.* **8**, 73–7.
Haeberli, W., Maisch, M. and Paul, F. 2002. Mountain glaciers in global climate-related
    observation networks. *Bull. World Met. Org.* **51** (1), 1–8.
Hansen, J., Ruedy, R., Glascoe, J. and Sato, M. 1999. GISS analysis of surface
    temperature change. *J. Geophys. Res.* **104** (D24), 30 997–1022.
Harrison, W. D., Elsberg, D. H., Echelmeyer, K. A. and Krimmel, R. M. 2001. On the
    characterization of glacier response by a single time-scale. *J. Glaciol.* **47** (159),
    659–64.
Hastenrath, S. and Chinn, T. 1998. Glaciers in Africa and New Zealand. In Haeberli, W.,
    Hoelzle, M. and Suter, S., eds., *Into the Second Century of World Glacier
    Monitoring – Prospects and Strategies. A contribution to the IHP and the GEMS*.
    Prepared by the World Glacier Monitoring Service. UNESCO Publishing, ch. 12,
    pp. 167–75.
Hastenrath, S. and Greischnar, L. 1997. Glacier recession on Kilimanjaro, East Africa,
    1912–89. *J. Glaciol.* **43** (145), 455–9.
Hodge, S. M., Trabant, D., Krimmel, R. M., Heinrichs, T. A., March, R. S. and Josberger,
    E. G. 1998. Climate variations and changes in mass of three glaciers in western
    North America. *J. Climate* **11**, 2161–79.
IAHS (ICSI)-UNESCO. 1967. *Fluctuations of Glaciers (FOG) 1959–1965*, vol. I: Zürich:
    Compiled for the Permanent Service on the Fluctuations of Glaciers of the
    IUGG-FAGS/ICSU by P. Kasser. Paris, p. 52.

1973. *Fluctuations of Glaciers (FOG) 1965–1970*, vol. II. Zürich: Compiled for the Permanent Service on the Fluctuations of Glaciers of the IUGG-FAGS/ICSU by P. Kasser. Paris, p. 357.

1977. *Fluctuations of Glaciers (FOG) 1970–1975*, vol. III. Zürich: Compiled for the Permanent Service on the Fluctuations of Glaciers of the IUGG-FAGS/ICSU by F. Müller. Paris, p. 269.

1985. *Fluctuations of Glaciers (FOG) 1975–1980*, vol. IV. Zürich: Compiled for the Permanent Service on the Fluctuations of Glaciers of the IUGG-FAGS/ICSU by W. Haeberli. Paris, p. 265.

IAHS (ICSI)-UNEP-UNESCO. 1988. *Fluctuations of Glaciers (FOG) 1980–1985,* vol. V. Zürich: World Glacier Monitoring Service. Compiled by W. Haeberli and P. Müller. Paris, p. 290.

1989. *World Glacier Inventory. Status 1988.* Haeberli, W., Bosch, H., Scherler, K., Østrem, G. and Wallén, C. C., eds. Paris, World Glacier Monitoring Service, p. 290.

1993. *Fluctuations of Glaciers (FOG) 1985–1990*, vol. VI. Zürich: World Glacier Monitoring Service. Compiled by W. Haeberli and M. Hoelzle. Paris, p. 322.

1998. *Fluctuations of Glaciers (FOG) 1990–1995*, vol. VII. Zürich: World Glacier Monitoring Service. Compiled by W. Haeberli, M. Hoelzle, S. Suter and R. Frauenfelder Paris, p. 296.

Jansson, P. 1999. Effect of uncertanties in measured variables on the calculated mass balance of Storglaciären. *Geograf. Ann.* **81A** (4), 633–42.

Jóhannessen, T., Raymond, C. F. and Waddington, E. D. 1989. Time-scale for adjustment of glaciers to change in mass balance. *J. Glaciol.* **35** (121), 355–69.

Jones, P. D., New, M., Parker, D. E., Martin, S. and Rigor, I. G. 1999. Surface air temperature and its changes over the past 150 years. *Rev. Geophys.* **37**, 173–99.

Jonston, R. J. 1980. *Multivariate Statistical Analysis in Geography.* New York, Longman, p. 280.

Kaser, G. 1999. A review of the modern fluctuations of tropical glaciers. *Global Planetary Change* **22** (1–4), 93–103.

Koerner, R. M. and Lundgaard, L. 1995. Glaciers and global warming. *Essais Géog. Phys. Quat.* **49** (3), 429–54.

Krenke, A. N. 1982. *Mass Turnover in the Glacier Systems in the Territory of the Soviet Union.* Leningrad, Hydrometeoizdat. (In Russian.)

Krimmel, R. 1999. Analysis of difference between direct and geodetic mass balance measurements at South Cascade Glacier, Washington. *Geograf. Ann.* **81A** (4), 653–8.

Kuhn, M. 1980. *Climate and Glaciers.* IAHS, Publ. **131**, pp. 3–20.

1993. Possible future contributions to sea level change from small glaciers. In Warrick, R. A., Barrow, E. M. and Wigley, T. M. L., eds., *Climate and Sea Level Change Observations.* Cambridge University Press, pp. 134–43.

1995. The mass balance of very small glaciers. *Zeits. Gletscherkunde & Glazialgeol.* **31** (1), 171–9.

McCabe, G. J. and Fountain, A. G. 1995. Relations between atmospheric circulation and mass balance of South Cascade Glacier, Washington, USA. *Arctic & Alpine Res.* **27**, 226–33.

McCabe, G. J., Fountain, A. G. and Dyurgerov, M. B. 2000. Effects of the 1976–77 climate transition on the mass balance of northern hemisphere glaciers. *Arctic, Antarctic & Alpine Res.* **32** (1), 64–72.

Makarevich, K. G. and Liu Chaochai. 1995. Izmeneniya oledeneniya Tyan Shanya v 20 veke (Tien Shan glaciation change in the 20th century). In Dyurgerov, M., Chaohai,

Liu, Zichu, Xie, eds., *Tien-Shan Glaciers (Oledenenie Tyan Shanya)*. Moscow, Publishing House VINITI, p. 233 (in Russian).

Mayo, L. R., Meier, M. F. and Tangborn, W. V. 1972. A system to combine stratigraphic and annual mass-balance systems: a contribution to the international hydrological decade. *J. Glaciol.* **11** (61), 3–14.

Meier, M. F. 1962. Proposed definitions for glacier mass budget terms. *J. Glaciol.* **4** (33), 252–61.

     1965. Glaciers and climate. In Wright, H. E. and Frey, D. G. eds., *The Quaternary of the United States*. Princeton University Press.

     1969. Glaciers and water supply. *J. Am. Water Works Assoc.* **61** (1), 8–12.

     1984. Contribution of small glaciers to global sea level. *Science* **226** (4681), 1418–21.

Meier, M. F. and Bahr, D. B. 1996. Counting glaciers: use of scaling methods to estimate the number and size distribution of the glaciers of the world. In Colbeck, S. C., ed., *Glaciers, Ice Sheets and Volcanoes: A Tribute to Mark F. Meier*. CRREL Special Report, pp. 89–94.

Meier, M. F. and Dyurgerov, M. B. 2002. Sea-level rise: how Alaska affects the world. *Science* **297** (5580), 350–1.

Meier., M. F., Mayo, L., Trabant, D. and Krimmel, R. 1980. Comparison of mass balance and runoff at four glaciers in the United States, 1966 to 1977. *Proceedings of the Academy of Sciences of the USSR*. Soviet Geophysical Committee, Moscow. In *Materialy Glyatsiologicheskikh (Data of Glaciological Studies)*, vol. 38, pp. 214–16.

Mool, P. K., Bajracharya, S. R. Joshi, S. P. 2001a. Inventory of glaciers, glacial lakes and glacial lake outburst floods. In *Monitoring and Early Warning Systems in the Hindu Kush-Himalayan Region*. Nepal, ICIMOD, p. 365.

Mool, P. K., Wangda, D., Bajracharya, S. R., Kunzang, K., Gurung, D. R., Joshi, S. P. 2001b. Inventory of glaciers, glacial lakes and glacial lake outburst floods. In *Monitoring and Early Warning Systems in the Hindu Kush-Himalayan Region*. Bhutan, ICIMOD, p. 227.

Nye, J. F. 1960. The response of glaciers and ice-sheets to seasonal and climatic changes. *Proc. Roy. Soc. London Series A* **256** (1287), 559–84.

Oerlemans, J. 1993. Modelling of glacier mass balance. In Peltier, W. R., ed., *Ice in the Climate System*. Berlin, Springer, pp. 101–16.

     1994. Quantifying global warming from the retreat of glaciers. *Science* **264**, 243–5.

     1999. Comments on 'Mass balance of glaciers other than the ice sheets' by Cogley and Adams. *J. Glaciol.* **45** (150), 397–8.

Oerlemans, J. and Fortuin, J. P. F. 1992. Sensitivity of glaciers and small ice caps to greenhouse warming. *Science* **258**, 115–17.

Oerlemans, J. and Reichert, B. K. 2000. Relating glacier mass balance to meteorological data by using a seasonal sensitivity characteristic. *J. Glaciol.* **46** (152), 1–6.

Oerlemans, J. *et al.*, 1998. Modeling of response of glaciers to climate warming. *Climate Dyn.* **14**, 267–74.

Østrem, G. and Brugman, M. 1991. *Glacier Mass-Balance Measurements. A manual for Field and Office Work*. National Hydrology Research Institute science report no. 4. Saskatoon, NHRI, p. 224.

Østrem, G. and Haakensen, N. 1999. Map comparison or traditional mass-balance measurements: which method is better? *Geograf. Ann.* **81A** (4), 703–11.

Pertziger, F. I. 1981. The internal property and mass balance of snowpatch-nearly-glacier. Conference Proceeding, Alma-Aty. Institute of Geography Kazakh Academy of Sciences, pp. 54–9. (In Russian).

Popovnin, V. V. 1996. Modern evolution of the Djankuat Glacier in the Caucasus. *Zeits. Gletscherkunde & Glazialgeol.* **31** (2), 15–23.

Shumskiy, P. A. 1964. *Principles of Structural Glaciology*. Translated from the Russian by D. Kraus. Dover, New York. (Original publication 1955.)

    1969. Glaciation. In *Atlas of Antarctica*, vol. 2. Leningrad, Hydrometeoizdat) pp. 367–400. (In Russian).

Thorarinsson, S. 1940. Present glacier shrinkage, and eustatic changes of sea-level. *Geograf. Ann.* **22**, 131–59.

Trabant, D. and Mayo, L. 1985. Estimation and effects of internal accumulation of five glaciers in Alaska. *Ann. Glaciol.* **6**, 113–17.

Trenberth, K. T. 1999. The extreme weather events of 1997 and 1998. *Consequences. The Nature & Implications of Environmental Change* **5** (1), 3–15.

Trupin, A. S., Meier, M. F. and Wahr, L. M. 1992. Effect of melting glaciers on the Earth's rotation and gravitational field: 1965–1984. *Geophys. J. Int.* **108**, 1–15.

Waddington, E. D. and Marriott, R. T. 1986. Ice divide migration at Blue Glacier. *Ann. Glaciol.* **8**, 175–6.

Walters, R. A. and Meier, M. F. 1989. Variability of glacier mass balances in western North America. In *Aspects of Climate Variability in the Pacific and the Western Americas*. Geophysical Monographs 55. American Geophysical Union, pp. 365–74.

Warren, C. and Aniya, M. 1999. The calving glaciers of southern South America. *Global & Planetary Change* **22** (1–4), 59–77.

Warrick, R. A., Provost, C. L., Meier, M. F., Oerlemans, J. and Woodworth, P. L. 1995. Changes in sea level. In *Climate Change 1995. The Science of Climate Change*. Contribution of Working Group 1 to the Second Assessment. Report of the Intergovernmental Panel on Climate Change (IPCC). Cambridge, University Press, p. 572.

Weidick, A. and Morris, E. 1998. Local glaciers surrounding continental ice sheets. In Haeberli, W., Hoelzle, M. and Suter, S., eds., *Into the Second Century of World Glacier Monitoring – Prospects and Strategies. A contribution to the IHP and the GEMS*. Prepared by the World Glacier Monitoring Service. UNESCO Publishing, pp. 197–205.

Welker, J. M., Fahnstock, J. T., Henry, G. H. R., O'Dea, K. W. and Piper, R. E. 2002. Microbial activity discovered in previously ice-entombed Arctic ecosystems. *EOS, Trans. Am. Geophys. Union* **83** (26), 281, 284.

Wood, F. B. 1988. Global alpine glacier trends, 1960s to 1980s. *Arctic & Alpine Res.*, **20** (4), 404–13.

Xie Zichu, Han Jiankang, Liu Chaohai and Liu Sciyin. 1999. Measurement and estimative models of glacier mass balance in China. *Geograf. Ann.* **81A** (4), 791–6.

17

Conclusions, summary and outlook

ANTONY J. PAYNE AND JONATHAN L. BAMBER
School of Geographical Sciences, University of Bristol

17.1 Summary of findings

17.1.1 Sea ice

A variety of indicators suggest that during the latter half of the twentieth century the Arctic has undergone substantial climate change (Serreze *et al.*, 2000). One of the key indicators is sea-ice extent and thickness, both of which have shown a measurable and disturbing decrease during the last half of the twentieth century (see Chapter 8). Interestingly, the rate of decrease appears to be at a maximum during summer (see Figure 12.8). The pre-satellite time series (before 1972) has a larger error bar on it, but nonetheless paints a consistent and compelling picture. Extrapolation forward in time suggests that there could, potentially, be no summer sea ice in the Arctic within 50 years. This will have major impacts on energy and moisture exchange, and consequently on the climate of the northern hemisphere. The consequences of such dramatic changes in sea-ice cover are the subject of a number of general circulation model (GCM) studies, and it is currently too early to say what the implications of these changes might be.

The sea-ice record for the Southern Ocean is less temporally extensive and, essentially, limited to the satellite era. Additionally, the seasonal variation in extent is much greater than in the Arctic, increasing the noise on any long-term signal that may be present. In contrast to the Arctic, there is no measurable trend in sea-ice mass balance, although some regional variations have been noted from passive microwave data covering the last 20 years (1979 to 1999; see Parkinson (2002)).

Modelling investigations of future trends based on a 'business as usual' scenario for greenhouse gas emissions are discussed in detail in Chapter 9. There is significant variation between models, but there is general agreement that an acceleration in mass loss will take place over the next century. Some simulations predict that the mean annual areal extent could be as low 2×10^6 km^2 by 2100 for both hemispheres (see Figure 9.9): a reduction of about a factor of 7 compared with the present day.

Mass Balance of the Cryosphere: Observations and Modelling of Contemporary and Future Changes, eds. Jonathan L. Bamber and Antony J. Payne. Published by Cambridge University Press. © Cambridge University Press 2003.

17.1.2 Greenland

Over the last decade a concerted effort by the NASA-funded Program for Regional Climate Assessment (PARCA) has reduced, by about a factor of 2, the uncertainty in the present-day mass balance of the ice sheet. The results indicate that above 2000 m elevation (i.e. above the equilibrium-line altitude), the ice sheet is close to balance, but that below this altitude, in the ablation zone, there is extensive thinning leading to a net negative balance for the ice sheet as a whole, which is conservatively estimated to be 50 km^3 per year, equivalent to 0.13 mm per year of sea-level rise. The results for the upper part of the ice sheet were derived from two independent sources. The first was from measurements of elevation change (dh/dt) from satellite and airborne altimetry (covering a relatively short time period between ∼1978 and 1999). The second source was from ground-based measurements of ice flux along the 2000 m elevation contour, which covers a slightly longer time period, assumed to be from 1970 to 1995 (see Figure 10.12). The high coastal thinning observed has been measured over the shortest time period, between about 1993 and 1999. It is difficult, therefore, to know how long this negative imbalance has existed. Furthermore, the thinning rates cannot be explained by increased ablation alone, and the implication is that there is a dynamic component to the signal. Because the thinning is spatially extensive, occurring in several drainage basins that are, in effect, dynamically uncoupled from each other, it is suggested that the dynamic response may be climatically induced by, for example, increased surface melt reaching the bed, causing enhanced basal sliding (Zwally *et al.*, 2002). If true, this is of profound significance, as it suggests that the dynamic response time of the ice sheet is, potentially, much shorter than previously believed. It should be noted that such a rapid dynamic response to changing climate has not, to date, been satisfactorily incorporated in modelling studies, as the mechanisms are poorly understood and constrained.

There have been several attempts to model the Greenland ice sheet, both through the last glacial–inter-glacial cycle and in terms of its response to anthropogenic climate change over the coming millennium. A number of common themes emerge. The first relates to the state of balance of the ice sheet prior to the industrial age and the onset of any associated anthropogenic climate change. Simulations of the ice sheet through successive glacial–inter-glacial cycles predict that the volume changes associated with deglaciation were complete by this time and that the ice sheet's volume was approximately in equilibrium with the pre-industrial climate (contributing 1 mm per century to global sea level; see Huybrechts and De Wolde (1999)). While the ice sheet may have been in steady-state in terms of overall volume, this does not preclude the existence of regional inheritance effects, which are likely to be most important at the divides because of the very slow flow rates and long response times there (Thomas *et al.*, 2001). Geomorphological evidence from the area can be used to test model predictions through this period of deglaciation. A good example is the retreat history in West Greenland, where comparison is, on the whole, favourable (Van Tatenhove, Van der Meer and Huybrechts, 1995).

The second point to emerge from these modelling efforts is that the dynamics of ice flow must be incorporated in models of the future evolution of the ice sheet. It is insufficient to treat the geometry of the ice sheet as fixed and to determine surface mass balance diagnostically. It is, however, also insufficient only to include the reduction in ice thickness resulting from a negative surface mass balance: horizontal ice flow must also be included. That flow effects are important, even on centennial timescales, is of great importance, and implies that future efforts in improving our predictions of Greenland's response to climate change should not concentrate solely on processes directly affecting the surface mass balance. This conclusion is further supported by the evidence of a possible dynamic response to increased surface melt (Zwally *et al.*, 2002).

Huybrechts and De Wolde (1999) predict, for their middle scenario (with atmospheric carbon dioxide reaching an equilibrium at four times its present concentration), a monotonic reduction in ice-sheet volume (and area) that amounts to about 3% by 2130 but growing to 40% by the year 3000. This implies that the ice-sheet margin will retreat significantly (200–300 km) from its present, coastal location, and that the ice sheet will separate into northern and southern domes. Their high scenario (with atmospheric carbon dioxide reaching an equilibrium at eight times its present concentration) results in the complete loss of the southern dome.

A number of issues still remain, however, before these predictions of the future evolution of the Greenland ice sheet can be accepted. They can be divided into three main groups. The first relates to the parameterization of surface mass balance and, in particular, ablation, for which the degree-day method is most commonly used. Ritz, Fabre and Letreguilly (1996) attempted to find the day-degree parameterization (as well as ice-flow parameterizations) that produced a best fit to the present-day ice sheet's volume and area after allowing the ice sheet to evolve through a glacial–inter-glacial cycle. Their best-fit values for the day-degree coefficients were several times higher than the values obtained from field-based experiments. In fact, simulations that used the field-based estimates resulted in a 36% over-prediction in ice volume. Ritz *et al.* (1996) identify as a potential cause the inadequate spatial resolution of the current generation of ice-sheet models near the margins, where climate and ablation rates vary dramatically. A second point related to the modelling of surface mass balance is that of melt-water refreezing within the snow pack. This process is treated in a very crude manner in the current generation of models. It seems likely that these parameterizations and their use in warming climates may well need to be revisited (Janssens and Huybrechts, 2000). Parameterizations that appear to work under the present-day climate regime may well incur large errors in the warmer climates of the coming millennium.

A second set of issues surrounding the Greenland modelling work is related to the flow of the ice sheet. Huybrechts and De Wolde (1999) indicate that the flow cannot be ignored, even on centennial timescales. Observational evidence implies that the ice sheet is responding most rapidly (in terms of ice-surface lowering) where flow is most rapid, in outlet glaciers near the coast (Abdalati *et al.*, 2001). No attempt has yet been made to determine whether these spatial patterns in the response of the ice sheet are reflected in model output. It may well be that the current spatial resolution of ice-sheet models is too coarse to capture these effects; however, they may well be important in determining the response of the ice sheet to climate warming. A related point is that the most obvious glaciological feature on the ice sheet (the North-East Greenland ice stream) is not captured by any of the current generation of models. This may, again, be a scale-related problem or may reflect a fundamental deficiency in the physics used within these models. In either event, this defect could be said to raise doubts as to the validity of the models.

The third and final set of issues is related to the model forcing employed. The ice sheet is thought to lose roughly half of its accumulated ice through the process of iceberg calving at a number of tide-water outlet glaciers (Church *et al.*, 2001). The models used to simulate this effect (which occurs on spatial scales well below the resolution of the models) are extremely crude and vary from the use of a water-depth dependent law, to the total removal of ice beyond a certain water-depth limit. As mentioned earlier, parameterizations that appear to work for the present day may be erroneous when applied to future configurations of the ice sheet. For example, calving could become less important as less of the shrinking ice margin is in direct contact with the sea, or more important as outlet glaciers retreat into over-deepened troughs.

Most predictions of the future response of Greenland to anthropogenic climate change have been made by way of a two-stage procedure. An atmosphere–ocean model is used to convert carbon-dioxide emission scenarios into predicted regional warming scenarios. The latter are then used to force (via surface mass balance changes) the ice-sheet model. Such a strategy ignores the possibility for coupling between the ice sheet and the atmosphere–ocean system. While several of the feedbacks that have been highlighted in this book may not be triggered by partial deglaciation over centennial timescales (for instance, albedo and planetary wave blocking), others may well be crucial (for instance, the release of melt water into the North Atlantic and its impact on deep-water formation). It is clear, therefore, that simulations of future ice-sheet evolution need to be part of a coupled atmosphere–ocean–ice model.

17.1.3 Antarctica

A complex picture emerges for the Antarctic ice sheet, and it is best to discuss separately the West Antarctic ice sheet (WAIS) and the East Antarctic ice sheet (EAIS). The latter (and by far the larger ice mass) appears to be close to balance. The best available estimates are based on repeat satellite altimetry, for which data are only available from the early 1990s onward. In the case of the EAIS, the signal measured from repeat observations between 1992 and 1996 falls below the limit associated with natural snow-fall variability. Longer records will therefore be required to determine whether there is an underlying trend in these data; however, these results imply that century-scale imbalance is unlikely (Wingham *et al.*, 1998).

A much more dynamic picture is beginning to emerge from the WAIS. The Wingham *et al.* (1998) analysis identifies that the combined Thwaites and Pine Island drainage basin is experiencing ice-surface lowering at a rate in excess of that explicable solely in terms of snow-fall variability. A dynamic forcing is therefore likely. More detailed studies of

Thwaites and Pine Island glaciers (PIG), as well as Smith glacier, indicate that the grounding line of PIG retreated at the rate of 1.2 km per year during the 1990s (Rignot, 1998), and that this retreat was accompanied by ice-surface lowering (at the rate of ∼1 m per year) up to 150 km inland of the grounding line (Shepherd *et al.*, 2001). The latter effect appears to be related to a general acceleration in the vicinity of PIG (Rignot *et al.*, 2002), and a similar drawdown is observed on Thwaites (∼0.5 m per year) and Smith (∼3.2 m per year) glaciers.

It is important to note that the fairly detailed knowledge that we now have about processes occurring in the Thwaites and Pine Island drainage basin is derived almost entirely from satellite observations. The dynamic forcing causing the observed acceleration and ice-surface drawdown could be associated with one (or more) process(es). Firstly, it could be a dynamic response to long-term climate change (such as that associated with the transition from the last Ice Age). A reaction to recent climate change is thought unlikely because of the long response times associated with the Antarctic ice sheet. Secondly, it could be forced by recent changes at the grounding line of the glaciers. The idea that marine-based ice sheets are particularly susceptible to instabilities initiated by ice-shelf thinning or loss has a long history within the glaciological literature (Thomas and Bentley, 1978), although its applicability has recently been questioned (Hindmarsh and Le Meur, 2001). Finally, the changes could be associated with internal instabilities within the ice-flow system. This latter possibility has been the subject of much discussion in the context of the Siple Coast ice streams.

The study of the Siple Coast ice streams has a far longer history, and has been accomplished using a mixture of satellite, airborne, field and modelling techniques. They have clearly had a complex evolution, and there are instances of ice-stream shutdown, for instance ice stream C at roughly 140 years ago (Retzlaff and Bentley, 1993) and the Siple ice stream roughly 450 years ago (Conway, Catania and Raymond, 2002). There is also evidence for the widening, as well as headward and lateral migration, of the ice streams. A recently published estimate of the area's overall mass balance (Joughin and Tulaczyk, 2002) is strongly positive, with average ice thickening equal to approximately 25% of the annual accumulation rate. The majority of this signal is associated with the stagnation of ice stream C.

The cause of the changes in ice-stream flow is currently the topic of much debate within the literature. A consensus has yet to emerge, although it is clear that subglacial sediment and water flow, as well as the ice streams' thermal regime and force balance, all play a role in determining their stability. It is also apparent that (i) the basal thermal regime of most of the Siple Coast ice streams is negative, with a tendency toward the freezing-on of subglacial melt water (Hulbe and MacAyeal, 1999); (ii) the gravitational forcing of ice-steam flow is principally balanced by resistance at their margins (Whillans and Van der Veen, 1997); and (iii) extreme changes in the importance of basal resistance can be affected by changes in the porosity of the underlying soft sediment (Tulazyk, Kamb and Engelhardt, 2000).

Large-scale numerical models of the Antarctic ice sheet tend to ignore much of the detailed dynamics highlighted in the above discussion of the WAIS. Exceptions are Hulbe and MacAyeal (1999) and Payne (1998), who do not, however, attempt to model the growth and decay of the ice sheet. Huybrechts (2002) models the evolution of the Antarctic ice sheet through the last glacial cycle. He finds the characteristic late deglaciation (beginning around 10 000 years BP) and the 'swinging gate' mode of deglaciation in the Ross Sea sector, both of which agree with interpretations based on the area's geomorphology (Conway *et al.*, 1999). This agreement should give us some confidence in predictions from the model of the Antarctic ice sheet's future evolution (Huybrechts and De Wolde, 1999). These predictions imply slight ice-sheet growth during the next few hundred years, which is caused by warmer, moister polar air masses. Ice-sheet dynamics becomes important on the longer, millennial timescale, and contributes to the grounding-line retreat of the WAIS (especially in the Siple Coast area) and a positive contribution to sea-level rise. The timescale and amplitude of this dynamic response is, however, highly dependent on both the prescribed amount of ice melt from the base of floating ice shelves and their temperature.

The present-day WAIS is thought to be still responding to deglaciation from the last Ice Age (and is currently one-third of its size during the last Ice Age); in particular, there is speculation that the retreat of the grounding line in the Siple Coast area has yet to reach equilibrium with inter-glacial climate and sea level. This is supported by numerical modelling (Huybrechts, 2002) as well as interpretation based on field evidence (Bindschadler and Vornberger, 1998; Conway *et al.*, 1999). However, the eventual equilibrium position of the grounding line is unclear, although the complete collapse of the WAIS through entirely non-anthropogenic climate and sea-level change is thought unlikely.

These predictions highlight areas where the physics of the current generation of large-scale models must be improved. Three themes can be identified. The first concern is raised by Hindmarsh and Le Meur (2001), who suggest that the dynamics governing the response of marine ice sheets to sea-level change are those of a neutral equilibrium. This contrasts with land-based ice masses, and has the implication that details of a numerical model's solution technique may affect the delicate balance at the grounding line and lead to artefacts in the response of the modelled ice sheet. They argue that successful simulation of past retreat patterns (for instance, deglaciation after the last Ice Age) does not guarantee accurate prediction of future response. This is because ice-sheet models omit several possible mechanisms that can cause grounding-line retreat, the importance of which may vary. In particular, they highlight issues related to the parameterization of ice slip over the underlying substrate. This leads to the second area of concern associated with the large-scale models, namely their inability to simulate the variability mentioned in connection with the WAIS above.

The unique dynamics of the WAIS are not incorporated into the current generation of large-scale ice-sheet models in a particularly rigorous fashion (an exception is the recent work of Ritz, Rommelaere and Dumas, 2001). Features such as the marked variability of ice flow on relatively short (decadal to centennial) timescales are not captured by these models. It could be argued that these variations are merely noise around the large-scale response to climate and sea-level change; however, this view is presently no more than an assertion and has yet to be backed by any rigorous analysis. On the contrary, Hindmarsh and Le Meur (2001) regard this variability as crucial in determining the large-scale response.

A third aspect of the observational record that has yet to be incorporated into these models is the recent collapse of many of the ice shelves fringing the Antarctic peninsula (Vaughan and Doake, 1996). The mechanism responsible for this collapse is thought to be related to the formation of surface melt-water ponds that structurally weaken the ice shelves (Scambos *et al.*, 2000). The major differences between the climatic regime and thickness of the small, peninsula ice shelves, and the Ross and Filchner-Ronne ice shelves, imply that the collapse mechanisms may not be replicated on the larger ice shelves.

17.1.4 Arctic ice masses

The ice caps and glaciers of the Arctic islands make up around 45% of the total volume of ice outside Antarctica and Greenland, and may, therefore, be a significant contributor to glacial sea-level rise. They comprise, however, a large number of ice masses, ranging in size, typically, between 10^2 and 10^4 km^2. Very few of these ice masses possess any sort of reliable mass balance data. Consequently, it has been necessary to extrapolate what limited data do exist. Most of these data come from heterogeneously distributed field measurements. The best estimate for the mass balance of Arctic ice masses suggests that they have a slight negative balance, which contributes 0.05 mm per year to global sea-level rise. Iceberg calving is an important, yet poorly determined, component of the mass budget of Arctic ice caps and glaciers. Any improvement in mass balance estimates will have to address this issue.

17.1.5 Glaciers and ice caps

Although the amount of water stored in the glaciers and ice caps of the world is minute (0.5 m of sea-level equivalent) in comparison to that contained in the two ice sheets (totalling 68.3 m), their rapid response times (typically below 100 years) mean that the bulk of the cryosphere's contribution to anthropogenic sea-level rise over the coming century will come from these types of ice mass (Church *et al.*, 2001). The short response time of glaciers comes about through the relatively high ratio of their mass throughput (as typically measured by ablation rates) to their storage (volume or typical thickness) in comparison to ice sheets (Jóhanneson, Raymond and Waddington, 1989). This response time can be lengthened significantly by feedbacks initiated by the glaciers' response, most noticeably via changes in their hypsometry (and hence mass balance) and changes in their patterns of flow (Oerlemans, 1997). The amplitude of their response is particularly large, in comparison to other components of the Earth's climate system, for several reasons. These include the dramatic difference between the albedo of snow and ice, and other natural surfaces, as well as the fact that glacier temperature cannot rise above the melting point of ice. This latter effect reduces the potential for energy loss via enhanced long-wave counter radiation, and hence weakens a possible negative feedback typical of other Earth surfaces.

The task of characterizing the contribution of glaciers and ice caps to observed sea-level rise can be addressed by three complementary strategies, all of which are dogged by major problems, perhaps the largest of which is the huge number (over 160 000) of glaciers world-wide. This makes direct measurement of glacier mass balance very difficult. Glacier inventories record (amongst other characteristics, see Chapter 15) the area of a glacier and can be used to track areal changes through time. The number of glaciers included in such inventories has risen rapidly over recent years with the advent of advanced satellite sensors; however, the hurdle of estimating volume change from this information still remains. In addition, the scope for extending this record back in time to help explain historical sea-level rise is obviously limited to the satellite era.

A second strategy uses the glaciers for which mass balance data are available. The number of suitable glaciers now dwindles to a minute fraction of the total. Glaciers with continuous records longer than 20 years total about 40, while those with records over five years are approximately 100 (Dyurgerov and Meier, 1997). Clearly, a statistical extrapolation technique is required to say something about the total population of glaciers from this very small sample. Unfortunately, the sample is not only small but is also very biassed, in particular toward small glaciers in certain areas of the globe (most notably maritime areas such as Patagonia, New Zealand and Alaska are all proportionally under-represented; Dyurgerov and Meier (1997)). A statistical approach based on weighted, regional averages was employed by Dyurgerov and Meier (1997) to produce their estimate of 0.25 ± 0.10 mm per year for the average rate of sea-level rise between 1961 and 1990. In the future, statistical averaging that incorporates information on a glacier's local climate, geometry, topographic setting and flow characteristics may become viable (e.g., Meier, 1984).

The final strategy is more model-orientated. This is, of course, essential if the aim is to predict future contributions to sea level as well as to quantify historical ones. Two examples of this approach are work by Zuo and Oerlemans (1997) and Oerlemans *et al.* (1998). The former uses a more parameterized approach in which empirical relationships between mass balance and mean air temperatures are derived from a sample of 12 glaciers as a function of precipitation (differentiating dry continental and wet maritime locations). These parameterizations are then used to predict glacier volume change in 100 regionally based classes, each class being driven by the observed air temperature time series for that region. It is reassuring that this analysis predicted a rate of sea-level rise between 1961 and 1990 (0.22 ± 0.07 mm per year) that is in close agreement with Dyurgerov and Meier (1997), despite using a very different technique. A modification of the technique that used output from an atmosphere–ocean GCM produced a rate of 0.26 mm per year (Gregory and Oerlemans, 1998).

The above strategies all regard glaciers as essentially responding to climate change in a passive manner. In reality, the future response of a glacier to an initial mass loss will be affected firstly by its altered hypsometry (including its reduced extent) and secondly by the changes in internal flow regime generated by its new geometry. Huybrechts and De Wolde (1999) have used the terms static and dynamic responses for these different effects, while a fixed-geometry response is one that ignores both. On the scale of an individual glacier, this approach clearly ignores some important feedbacks that will control the timescale and eventual magnitude of the response to climate change. However, over the globally aggregated scales at which predictions are made for a sea-level contribution, it is unclear how important these omissions are. There exists a wide range of studies where the response of individual glaciers has been modelled by coupling a surface mass balance model with an ice-flow model. Oerlemans *et al.* (1998) attempt to standardize many of these individual studies using 12 glaciers and a range of future warming scenarios. The results imply that the hypsometry of individual glaciers does indeed play an important role in governing their response, to the extent that no direct relationship between glacier size and volumetric response emerged from the analysis. Although this approach clearly has the soundest physical basis of those discussed, the information required for its successful application prohibits its use except for a very limited number of glaciers.

   The assessment of historical, and predicted future contributions to sea-level rise from glaciers is still in its infancy. The effects have clearly been large, and are likely to continue to be important over the coming century; however, many challenges still remain in their accurate quantification.

17.2 Current uncertainties

17.2.1 *Weaknesses in observational data of the cryosphere*

For many of the smaller glaciers of the world, the methods for monitoring their mass balance have remained, essentially, unchanged since measurements on them began. There is still no satisfactory means for using satellite data to measure volume changes of these ice masses. Nonetheless, the global land ice monitoring from space (GLIMS) project aims to provide a definitive database of the areal extent of land ice on the planet using new visible imaging sensors such as ASTER and the ETM+ on board Landsat 7 (see Chapters 4 and 15). Changes in areal extent can, and have been, used to determine the state of balance of glaciers. With over 160 000 glaciers identified, this is a major challenge, and, to date, only a handful of glaciers have anything more than a 20 year record of *in situ* observations (Chapter 16).

   For the great ice sheets of Antarctica and Greenland, major progress has been made since 1990 in reducing the uncertainty in their mass balance. This progress has come from a number of sources. Firstly, the 1990s saw the development of radar interferometry as a tool for measuring ice velocities, grounding-line locations and surface topography. The launch of ERS-1 in 1991 finally provided extensive coverage of the ice sheets with accurate satellite radar altimeter elevation measurements, and laser altimetry combined with kinematic global positioning systems (GPS) is now a proven technology for measuring centimetric elevation changes. In addition, throughout the 1990s a concerted and co-ordinated programme of research funded by NASA (and known as PARCA, the Program for Arctic Regional Climate Assessment) combined the new satellite data sets with an extensive field campaign (cf. Chapter 10). Most of the effort was focussed on Greenland, and has resulted in a reduction in the uncertainty in the mass balance for the whole ice sheet by about a factor of 2. Above the equilibrium line (and hence the ablation zone) the reduction is much better than this: the uncertainty in elevation change has been reduced from around 10 cm per year to 1–2 cm per year. It is the large errors in the measured and modelled ablation rates that currently limit the overall error budget for the ice sheet. PARCA results have highlighted other serious gaps in our knowledge. Accumulation rates, for example, were shown to be both spatially and temporally highly variable, with most of the elevation change identified by radar altimetry resulting from this high, natural variability (McConnell *et al.*, 2000). High thinning rates were observed near the margins of the ice sheet, but their cause remains unexplained (Abdalati *et al.*, 2001). Since 1996, a wealth of new evidence has been published related to the mass balance of Antarctica. These results are, however, constrained by (i) a relatively poor knowledge of accumulation rates, (ii) a short time record of observations given the long response time of the ice sheet, and (iii) a dearth of observations for a huge expanse of about 2000 km diameter covering central East Antarctica (the area beyond current satellite coverage).

Ice sheets and glaciers are integrators of climate, and reflect, therefore, a long-term response to climate change. Sea ice, on the other hand, responds rapidly. The consequence of this is that for a trend to be meaningful and unambiguous it must cover a sufficiently long time period. What is this time period? Decadal atmospheric oscillations (such as the North Atlantic oscillation and Arctic oscillation) are known to affect Arctic sea-ice extent. To reduce the influence of these oscillations on any underlying trend, a multi-decadal record is clearly essential. This does exist for the Arctic, but is of variable quality and reliability, and there are significant differences between different algorithms used (Vinnikov *et al.*, 1999). For the Southern Ocean, the temporal record is even more limited. Furthermore, the satellite measurements to date have focussed almost exclusively on extent and concentration. Although these are important, they cannot be used alone to estimate changes in mass balance. For this, ice thickness is required, which has, to date, largely been an elusive goal.

17.2.2 Uncertainties in model boundary conditions and physics

The predictive ability of the current generation of ice-sheet and glacier models is hampered by two main factors. The first is uncertainty in model boundary conditions, and here there are two main gaps in our knowledge, the first being a lack of detailed information on the bedrock topography that is crucial in determining ice thickness and, hence, patterns of ice flow. The main means by which bedrock topography is determined is by radio-echo sounding (RES). This is an expensive and time-consuming activity. There are huge gaps in the coverage of the Antarctic ice sheet (in particular in East Antarctica). The recent BEDMAP compilation represents a major step forward (Lythe and Vaughan, 2001); however, approximately one-quarter of the continent still remains unsurveyed. The coverage of the Greenland ice sheet is far more comprehensive (thanks in part to the recent PARCA programme). In addition, only a minute fraction of the ice caps and glaciers of the world have been surveyed. The prospect of having reasonable RES coverage for most of the larger ice caps in the northern hemisphere appears fairly good. It seems likely, however, that the number of glaciers with RES data will always be extremely limited. A means of determining bedrock topography indirectly is clearly a priority, and is discussed further in the following section.

The second boundary condition that requires further attention concerns the meteorological variables of precipitation, near-surface air temperature and incoming short-wave radiation from which surface models (such as those reviewed in Chapter 5) determine the mass balance of an ice mass. It is clearly impractical to obtain the necessary information from field-based meteorological measurements because of the huge areas (ice sheets) and number of sites (glaciers) involved. Two avenues are therefore available. These are either to employ satellite-based observations or the re-analysed results of GCMs of the atmosphere–ocean. There are a number of other data requirements that are less pressing but still of importance, such as knowledge of the geothermal heat flux under ice sheets.

We now discuss deficiencies in the physics employed by the current generation of terrestrial land-ice models. A topic that lies between model physics and the previous boundary condition discussion relates to the statistical methods used to extrapolate from a small sample of glaciers to their total, global population. This is clearly necessary if estimates of present and future sea-level contributions are to be made; however, the extrapolation is difficult because the sample is both extremely small and biassed in a number of ways. It seems likely that the number of glaciers either observed or, more particularly, modelled will remain very small. In the latter case, this is principally because of the data constraints discussed above. A rigorous means of statistical extrapolation is therefore required.

The mass balance input to glaciers and ice sheets appears to be well covered (Chapter 5) by a variety of empirical and physically based approaches, although the quantification of melt water that is refrozen within the snow pack (and hence not lost to the ice mass) still remains an issue. A related issue is the quantification of the amount of ice lost by melting from the underside of floating ice shelves, as well as an understanding of the large-scale controls on this process (i.e. the meso-scale circulation of water masses within the subshelf cavity). Similarly, further improvement is necessary in our ability to quantify and predict the amount of ice lost by the calving of icebergs (e.g., Van der Veen, 2002).

Perhaps the area that is in most need of development is that of quantifying the basal slip experienced by an ice mass (either by sliding over a rigid substrate or by the deformation of an intermediate sediment layer) and understanding the factors that control it. Basal slip is significant for most ice masses that are not frozen to their bed, and is critical in controlling the evacuation of mass from the Antarctic ice sheet via ice streams. However, the physics used in most predictive models of glaciers and ice sheets is no more than a crude parameterization. This is despite the very wide literature on the topic or perhaps because of the wide range of approaches and results reported in the literature. In particular, many models require detailed information on the depth and extent of sediment underlying an ice mass which is not readily available. Nonetheless, progress on the development of a universal basal slip methodology and its incorporation into numerical ice-mass models is a pressing concern. Other areas where progress is required include an effective methodology for incorporating the effects of fabric evolution on the rheology of ice.

Finally, it should be stressed that the steps needed to create a numerical model from its continuum mechanics formulation still require very careful attention. This has been highlighted recently by the suggestion that marine ice masses may experience neutral equilibrium, with the consequence that observed model behaviour may, in part, be an artefact of the discretization techniques used in the model's construction (Hindmarsh, 1993).

17.3 Future trends in research

17.3.1 Observations

For the first time, both the European (ESA) and American (NASA) space agencies have announced satellite missions with a primary focus on observations of the cryosphere and, in particular, mass balance of land and sea ice. The two missions are ICESat (NASA) and CryoSat (ESA), with launch dates of December 2002 and 2004, respectively. ICESat will carry a single instrument: the geosciences laser altimeter system (GLAS). This is a dual-frequency laser system with the design specification aimed to deliver elevations over ice-covered terrain accurate to 10 cm (Chapters 4 and 12). The science objectives of ICESat are to provide a time series of elevation change measurements over land ice and freeboard (and hence thickness) of sea ice. CryoSat will carry a radar altimeter with interferometric capability, improving the ability to determine sea-ice freeboard and allowing useful measurements over the steeper sloping margins of the ice sheets and ice caps down to a size of around $10^4 \, \text{km}^2$ to be obtained.

Both these missions will provide unique data. Nonetheless, airborne laser altimeter data obtained as part of PARCA, for example, could be used to extend the time series back to the early 1990s for the Greenland ice sheet and other ice masses in the Arctic, and it will be possible to do the same for CryoSat using ERS and the ENVISAT radar altimeter for the flatter, central portions of Antarctica and Greenland. Thus, despite being new instruments, for some ice masses at least there will be a short time series to build on. ICESat has a footprint size of around 70 m and, as a consequence, should also provide useful elevation data over smaller ice masses, including valley glaciers. Thus, for the first time, it may be possible to measure, accurately and directly, changes in elevation of sub-polar and alpine glaciers from space. This footprint size may also allow direct measurement of ablation rates near the margins of Greenland and other ice masses. In addition, gravity missions have been launched (such as the gravity recovery and climate experiment: GRACE). These missions provide information on the mass of ice contained in the ice sheets, and, when combined with elevation change data from altimetry, will further reduce the uncertainty in mass balance.

As mentioned earlier, one of the biggest errors in the mass budget of the ice sheets is associated with accumulation rates. Two new approaches may offer major improvements in this area, particularly in elucidating the spatial pattern. The first of these is the development of a high resolution radar for mapping near surface (100–300 m) internal layers within the snow pack. These internal layers are believed to be isochronous, and, if they are dated using ice cores at a small number of sites, can be used to determine mean accumulation rates over spatially extensive areas. Such a system was tested at the north GRIP site as part of PARCA (Kanagaratnam *et al.*, 2001). The second approach also developed within PARCA utilizes passive microwave radiometer data to map accumulation rates within the dry snow zone (Winebrenner, Arthern and Shuman, 2001). The method requires the use of a low frequency channel at around 6.6 GHz. This had, until recently, two drawbacks: (i) the only radiometer to carry this channel was SMMR, which stopped operation in 1987, and (ii) lower frequency channels are also lower resolution, and for SMMR this was 136×89 km (see Chapter 4). In May 2002, however, the satellite AQUA was launched, carrying a radiometer with a channel at 6.9 GHz and resolution of about half that of SMMR. This opens up the possibility of spatially extensive and comprehensive satellite observations of accumulation for most of Antarctica and much of Greenland.

17.3.2 Modelling

Two trends can be identified in the modelling literature. One relates to the type of model being employed, and the second to their use. The sophistication of models of the cryosphere inevitably increases with time; in particular, the level of physical detail incorporated into numerical models is improving, and the amount of coupling between these and models of the other components of the climate system is increasing. Chapters 5, 6 and 7 document this trend toward increased model detail. This progression has, of course, been aided by the ever-increasing availability of computer processing power, as well as by the analysis of inadequacies in the earlier generations of model identified by comparing model output with observations. Examples include the recent development of glacier models that incorporate a fuller representation of the stress balance rather than assuming that gravitational stress is always balanced by basal traction alone (see, e.g., Blatter, Clarke and Colinge, 1998).

The importance of cryospheric processes in the global climate system is increasingly being recognized as models of specific components of the cryosphere are incorporated into coupled climate models at both global and regional scales. Sea-ice modelling is certainly more advanced in this respect because of the short response time of sea ice. All coupled global ocean–atmosphere GCMs contain some sort of sea-ice component. Meso-scale climate models of the polar regions are increasingly looking toward explicit coupling with sea ice (e.g., Vavrus, 1999).

Ice sheets are passive components of the climate system on the typical decadal timescales of these simulations; however, they become active components of the system over longer (millennial) timescales. Models aimed at understanding the long-term evolution of the climate system through an Ice Age have previously been hampered by the excessive computing time required for such experiments. Recent progress in the development and application of Earth system models of intermediate complexity (EMICs; Claussen *et al.*, 2002) has allowed these long timescales to be explored, and has inevitably involved coupling to models of the continental ice sheets.

The second trend in the modelling literature is in the use made of these models. The massive increase in the amount of data on ice-surface topography and velocity from satellite observations has yet to be fully exploited by the modelling community. The previously unimaginable quantity and quality of these data make inverse modelling a strong possibility. Traditional ice-mass modelling has been forward in nature, in which a particular set of model equations are integrated given a set of initial and boundary conditions. The resultant predictions are then compared with the available observational evidence to test the model. The aim of inverse modelling is to determine what sets of boundary (and initial) conditions are consistent with the observations, given a particular set of model equations. A classic example would be to determine the set of boundary conditions operating at the bed of a glacier that is consistent with the observed surface topography and velocities. This type of model requires very detailed observational information, which is beginning to appear. Early examples of this approach applied to ice streams include work by MacAyeal, Bindschadler and Scambos (1995) using a detailed forward model and a control methods inversion technique, and by Gudmundsson, Raymond and Bindschadler (1998) using theoretically derived transfer functions based on a linearized model. In principle, it may be possible to invert surface information for bedrock topography (note the difficulties discussed above in obtaining this information from direct observation) as well as for basal traction and ice rheology. It seems clear that this type of data assimilation will play a major role in the application of modelling to the cryosphere over the next decade. The identification of internal layers (assumed to be isochronous) in RES data promises to add a further dimension to the observational database against which models can be either tested (in the forward sense) or inverted to determine information on ice rheology, basal slip or past accumulation rates.

17.4 Concluding remarks

Remarkable progress has been made in our ability both to measure and model the mass balance of the cryosphere. This progress is due to new satellite sensors/technology, concerted programmes of field measurements such as PARCA and our improved understanding of the physical processes controlling the behaviour of the cryosphere. There remain, however, many challenges: model simulations of future land- and sea-ice evolution differ markedly, and the observational record is, in general, disappointingly short, and, without a suitable proxy, it is, in most cases, difficult to see how to extend it. For most of us the polar regions are remote and distant locales that are out of sight and perhaps out of mind. They are, however, highly sensitive to climate change with strong feedbacks at play. Dramatic and disturbing trends have already been observed in the Arctic and along the Antarctic peninsula. We ignore these changes at our peril.

References

Abdalati, W. *et al.* 2001. Outlet glacier and margin elevation changes: near-coastal thinning of the Greenland ice sheet. *J. Geophys. Res.-Atmos.* **106** (D24), 33 729–41.
Bindschadler, R. and Vornberger, P. 1998. Changes in the West Antarctic ice sheet since 1963 from declassified satellite photography. *Science* **279** (5351), 689–92.

Blatter, H., Clarke, G. K. C. and Colinge, J. 1998. Stress and velocity fields in glaciers: Part II. Sliding and basal stress distribution. *J. Glaciol.* **44** (148), 457–66.

Church, J. A. *et al.* 2001. In Houghton, J. T. *et al.*, eds., *Changes in Sea-level. IPCC Third Scientific Assesment of Climate Change*. Cambridge University Press, pp. 640–93.

Claussen, M. *et al.* 2002. Earth system models of intermediate complexity: closing the gap in the spectrum of climate system models. *Climate Dyn.* **18** (7), 579–86.

Conway, H., Hall, B. L., Denton, G. H., Gades, A. M. and Waddington, E. D. 1999. Past and future grounding-line retreat of the West Antarctic ice sheet. *Science* **286** (5438), 280–3.

Conway, H., Catania, G., and Raymond, C. F. 2002. Switch of flow direction in an Antarctic ice stream. *Nature* **419**, 465–7.

Dyurgerov, M. B. and Meier, M. F. 1997. Mass balance of mountain and subpolar glaciers: a new global assessment for 1961–1990. *Arctic & Alpine Res.* **29** (4), 379–91.

Gregory, J. M. and Oerlemans, J. 1998. Simulated future sea-level rise due to glacier melt based on regionally and seasonally resolved temperature changes. *Nature* **391** (6666), 474–6.

Gudmundsson, G. H., Raymond, C. F. and Bindschadler, R. 1998. The origin and longevity of flow stripes on Antarctic ice streams. *Ann. Glaciol.* **27**, 145–52.

Hindmarsh, R. C. A. 1993. Modeling the dynamics of ice sheets. *Prog. Phys. Geog.* **17** (4), 391–412.

Hindmarsh, R. C. A. and Le Meur, E. 2001. Dynamical processes involved in the retreat of marine ice sheets. *J. Glaciol.* **47** (157), 271–82.

Hulbe, C. L. and MacAyeal, D. R. 1999. A new numerical model of coupled inland ice sheet, ice stream, and ice shelf flow and its application to the West Antarctic ice sheet. *J. Geophys. Res. Solid Earth* **104** (B11), 25 349–66.

Huybrechts, P. 2002. Sea-level changes at the LGM from ice-dynamic reconstructions of the Greenland and Antarctic ice sheets during the glacial cycles. *Quat. Sci. Rev.* **21** (1–3), 203–31.

Huybrechts, P. and De Wolde, J. 1999. The dynamic response of the Greenland and Antarctic ice sheets to multiple-century climatic warming. *J. Climate* **12** (8), 2169–88.

Janssens, I. and Huybrechts, P. 2000. The treatment of meltwater retention in mass-balance parameterizations of the Greenland ice sheet. *Ann. Glaciol.* **31**, 133–40.

Jóannesson, T., Raymond, C. and Waddington, E. 1989. Time-scale for adjustment of glaciers to changes in mass balance. *J. Glaciol.* **35** (121), 355–69.

Joughin, I. and Tulaczyk, S. 2002. Positive mass balance of the Ross ice streams, West Antarctica. *Science* **295** (5554), 476–80.

Kanagaratnam, P., Gogineni, S. P., Gundestrup, N. and Larsen, L. 2001. High-resolution radar mapping of internal layers at the North Greenland ice core project. *J. Geophys. Res. Atmos.* **106** (D24), 33 799–811.

Lythe, M. B. and Vaughan, D. G. 2001. BEDMAP: a new ice thickness and subglacial topographic model of Antarctica. *J. Geophys. Res. Solid Earth* **106** (B6), 11 335–51.

MacAyeal, D. R., Bindschadler, R. A. and Scambos, T. A. 1995. Basal friction of ice-stream-E, West Antarctica. *J. Glaciol.* **41** (138), 247–62.

McConnell, J. R. *et al.* 2000. Changes in Greenland ice sheet elevation attributed primarily to snow accumulation variability. *Nature* **406** (6798), 877–9.

Meier, M. F. 1984. Contribution of small glaciers to global sea-level. *Science* **226** (4681), 1418–21.

Oerlemans, J. 1997. Climate sensitivity of Franz Josef Glacier, New Zealand, as revealed by numerical modeling. *Arctic & Alpine Res.* **29** (2), 233–9.

Oerlemans, J. *et al.* 1998. Modelling the response of glaciers to climate warming. *Climate Dyn.* **14** (4), 267–74.

Parkinson, C. L. 2002. Trends in the length of the Southern Ocean sea-ice season, 1979–99. *Ann. Glaciol.* **34**, 435–40.

Payne, A. J. 1998. Dynamics of the Siple Coast ice streams, West Antarctica: results from a thermomechanical ice sheet model. *Geophys. Res. Lett.* **25** (16), 3173–6.

Retzlaff, R. and Bentley, C. R. 1993. Timing of stagnation of ice stream-C, West Antarctica, from short-pulse radar studies of buried surface crevasses. *J. Glaciol.* **39** (133), 553–61.

Rignot, E. J. 1998. Fast recession of a West Antarctic glacier. *Science* **281** (5376), 549–51.

Rignot, E. J., Vaughan, D. G., Schmeltz, M., Dupont, T. and MacAyeal, D. 2002. Acceleration of Pine Island and Thwaites glaciers, West Antarctica. *Ann. Glaciol.* **34**, 189–94.

Ritz, C., Fabre, A. and Letreguilly, A. 1996. Sensitivity of a Greenland ice sheet model to ice flow and ablation parameters: consequences for the evolution through the last climatic cycle, *Climate Dyn.* **13** (1), 11–24.

Ritz, C., Rommelaere, V. and Dumas, C. 2001. Modeling the evolution of Antarctic ice sheet over the last 420 000 years: implications for altitude changes in the Vostok region. *J. Geophys. Res. Atmos.* **106** (D23), 31 943–64.

Scambos, T. A., Hulbe, C., Fahnestock, M. and Bohlander, J. 2000. The link between climate warming and break-up of ice shelves in the Antarctic peninsula. *J. Glaciol.* **46** (154), 516–30.

Serreze, M. C. *et al.* 2000. Observational evidence of recent change in the northern high-latitude environment. *Climatic Change* **46** (1–2), 159–207.

Shepherd, A., Wingham, D. J., Mansley, J. A. D. and Corr, H. F. J. 2001. Inland thinning of Pine Island glacier, West Antarctica. *Science* **291** (5505), 862–4.

Thomas, R. H. *et al.* 2001. Mass balance of higher-elevation parts of the Greenland ice sheet. *J. Geophys. Res.-Atmos.* **106** (D24), 33 707–16.

Thomas, R. H. and Bentley, C. R. 1978. A model for Holocene retreat of the West Antarctic ice sheet. *Quat. Res.* **10**, 150–70.

Tulaczyk, S., Kamb, W. B. and Engelhardt, H. F. 2000. Basal mechanics of ice stream B, West Antarctica 1. Till mechanics. *J. Geophys. Res. Solid Earth* **105** (B1), 463–81.

Van der Veen, C. J. 2002. Calving glaciers, *Prog. Phys. Geog.* **26** (1), 96–122.

Van Tatenhove, F. G. M., Van der Meer, J. J. M. and Huybrechts, P. 1995. Glacial-geological geomorphological research in West Greenland used to test an ice-sheet. *Quat. Res.* **44** (3), 317–27.

Vaughan, D. G. and Doake, C. S. M. 1996. Recent atmospheric warming and retreat of ice shelves on the Antarctic peninsula. *Nature* **379** (6563), 328–31.

Vavrus, S. J. 1999. The response of the coupled arctic sea ice-atomospheric system to orbital forcing and ice motion at 6 kyr and 115kyr BP. *J. Climate* **12** (3), 873–6.

Vinnikov, K. Y. *et al.* 1999. Global warming and northern hemisphere sea ice extent. *Science* **286** (5446), 1934–7.

Whillans, I. M. and Van der Veen, C. J. 1997. The role of lateral drag in the dynamics of ice stream B, Antarctica. *J. Glaciol* **43** (144), 231–7.

Winebrenner, D. P., Arthern, R. J. and Shuman, C. A. 2001. Mapping Greenland accumulation rates using observations of thermal emission at 4.5-cm wavelength. *J. Geophys. Res. Atmos.* **106** (D24), 33 919–34.

Wingham, D. J., Ridout, A. J., Scharroo, R., Arthern, R. J. and Shum, C. K. 1998. Antarctic elevation change from 1992 to 1996. *Science* **282** (5388), 456–8.

Zuo, Z. and Oerlemans, J. 1997. Contribution of glacier melt to sea-level rise since AD 1865: a regionally differentiated calculation. *Climate Dyn.* **13** (12), 835–45.

Zwally, H. J., Abdalati, W., Herring, T., Larson, K., Saba, J. and Steffen, K. 2002. Surface melt-induced acceleration of Greenland ice-sheet flow. *Science* **297**, 218–20.

Index

Printed in the United States
By Bookmasters